AF619040

Birkhäuser

Progress in Nonlinear Differential Equations and Their Applications

PNLDE Subseries in Control

Volume 98

More information about this series at https://link.springer.com/bookseries/15137

Jérôme Le Rousseau • Gilles Lebeau • Luc Robbiano

Elliptic Carleman Estimates and Applications to Stabilization and Controllability, Volume II

General Boundary Conditions on Riemannian Manifolds

Jérôme Le Rousseau
Laboratoire analyse, géométrie et applications
Université Sorbonne Paris-Nord, CNRS,
Université Paris 8
Villetaneuse, France

Gilles Lebeau
Laboratoire Jean Dieudonné
Université de Nice Sophia-Antipolis
Nice, France

Luc Robbiano
Université Paris-Saclay, UVSQ, CNRS
Laboratoire de Mathématiques de Versailles
Versailles, France

ISSN 1421-1750 ISSN 2374-0280 (electronic)
Progress in Nonlinear Differential Equations and Their Applications
ISSN 2524-4639 ISSN 2524-4647 (electronic)
PNLDE Subseries in Control
ISBN 978-3-030-88672-1 ISBN 978-3-030-88670-7 (eBook)
https://doi.org/10.1007/978-3-030-88670-7

Mathematics Subject Classification: 35Q93, 35B45, 35B60, 35L05, 35L20, 35K05, 35K10, 35K20, 35J15, 35J25, 35S15, 58J05, 58J32, 93B05, 93B07, 93D15

This book is published under the imprint Birkhäuser, www.birkhauser-science.com by the registered company Springer Nature Switzerland AG
The registered company address is: Gewerbestrasse 11, 6330 Cham, Switzerland

Contents

CHAPTER 1

Introduction to Volume 2

Contents

In Volume 1, Carleman estimates for a second-order elliptic operators P were derived in a regular open set Ω of the Euclidean space, away from the boundary and at the boundary. For the latter case, derivation was made with Dirichlet boundary conditions.[1] Such an estimation takes the form

$$\tau^{3/2}\|e^{\tau\varphi}u\|_{L^2(\Omega)} + \tau^{1/2}\|e^{\tau\varphi}\nabla_x u\|_{L^2(\Omega)} + \tau^{1/2}|e^{\tau\varphi}\partial_\nu u_{|\partial\Omega}|_{L^2(\partial\Omega)} \leq C\big(\|e^{\tau\varphi}Pu\|_{L^2(\Omega)} + \tau^{3/2}|e^{\tau\varphi}u_{|\partial\Omega}|_{L^2(\partial\Omega)} + \tau^{1/2}|e^{\tau\varphi}u_{|\partial\Omega}|_{H^1(\partial\Omega)}\big), \tag{1.0.1}$$

where u is smooth or H^2 supported in V a bounded neighborhood of a point of $\overline{\Omega}$. The function φ is smooth and is the so-called Carleman *weight* function. The parameter τ is positive and chosen such that $\tau \geq \tau_* > 0$, for τ_* sufficiently large. The constants $C > 0$ and τ_* do depend on the geometry, that is Ω and V, and the coefficients of P and the weight function. In the derivation of (1.0.1), the choice of the weight function is essential; two properties are important: (1) the sub-ellipticity of the conjugated operator $P_\varphi = e^{\tau\varphi}Pe^{-\tau\varphi}$ and (2) the condition $\partial_\nu\varphi_{|\partial\Omega} < 0$. Necessity and sufficiency of these conditions are discussed in Volume 1.

[1]Estimations involving both the Neumann and Dirichlet traces were also derived, but stronger estimates are obtained if one considers a particular boundary operator.

J. Le Rousseau et al., *Elliptic Carleman Estimates and Applications to Stabilization and Controllability, Volume II*, PNLDE Subseries in Control 98, https://doi.org/10.1007/978-3-030-88670-7_1

Carleman estimates have been intensively used to obtain unique continuation properties. In Volume 1, we focused on this application, in particular on showing how Carleman estimates as in (1.0.1) are used to yield quantified versions of this property. Further applications were also given: stabilization of the damped wave equation and null-controllability of the heat equation.

1.1. Main Content

In Volume 2, one of our goals is to extend Carleman estimates to more general boundary conditions, say Neumann, Robin, etc. A framework is given by the so-called Lopatinskiĭ–Šapiro condition. It encompasses all aforementioned conditions.

The Lopatinskiĭ–Šapiro condition is of geometrical nature. It thus makes sense also to extend the analysis to Riemannian manifolds, for which the Laplace–Beltrami operator $P_0 = -\Delta_g$ appears as a natural second-order elliptic operator. Since Carleman estimates are local in nature and can be patched together, their derivation will be carried out in local coordinates. For an operator like the Laplace–Beltrami operator, the change of coordinates is straightforward. Lopatinskiĭ–Šapiro boundary conditions are also appealing as they are key in the understanding of the resolvability of elliptic boundary problems. Basically, such problems are of Fredholm type if and only if boundary conditions fulfill the Lopatinskiĭ–Šapiro condition.

In the case of Dirichlet boundary conditions, deriving a Carleman estimate on a Riemannian manifold is rather effortless if relying on the results of Chapter 3 in Volume 1. However, to address the derivation of such estimates with general Lopatinskiĭ–Šapiro conditions, techniques from Volume 1 do not work in general. If one carries out proofs as in Chapter 3 in Volume 1, some positivity arguments cannot be made, for instance, in the simple case of Neumann boundary conditions. Yet, an estimate can be proven if one relies on a microlocal partitioning of tangential phase space at the boundary. Each microlocal region then requires a different treatment. It is interesting to note that no boundary condition is needed in some microlocal regimes. In others, the boundary condition is crucial, and one sees that the Lopatinskiĭ–Šapiro condition is precisely the condition that allows one to reach an estimation.

Let us introduce the following norm on $\partial\mathcal{M}$, for $s \in \mathbb{R}_+$:

$$|u|_{\tau,s} \eqsim \tau^s |u|_{L^2(\partial\mathcal{M})} + |u|_{H^s(\partial\mathcal{M})}. \tag{1.1.1}$$

For $\tau > 0$ fixed, this gives a norm of $H^s(\partial\mathcal{M})$ and one can obtain a dual norm $|.|_{\tau,s}$ on $H^{-s}(\partial\mathcal{M})$.

For two functions u and f defined in $\mathcal{M}$, and $s \in \mathbb{R}$, one sets

$$|f \operatorname{tr}(u)|_{\tau,1,s} \eqsim |f u_{|\partial\mathcal{M}}|_{\tau,1+s} + |f \partial_\nu u_{|\partial\mathcal{M}}|_{\tau,s}. \tag{1.1.2}$$

Let V be a bounded open set of $\mathcal{M}$. In some neighborhood of $\partial\mathcal{M} \cap V$, one can use $(m', z) \in \partial\mathcal{M} \times [0, Z]$ for some $Z > 0$. The variable z gives a normal direction to $\partial\mathcal{M}$. One can then introduce the following norms, for $r \in \mathbb{N}$ and $s \in \mathbb{R}_+$:

(1.1.3)

$$\|u\|_{\tau,0,s} = \Big(\int_0^Z |v(.,z)|^2_{\tau,s}\, dz\Big)^{1/2}, \qquad \|u\|_{\tau,r,s} \asymp \sum_{0\leq j\leq r} \|D_z^j u\|_{\tau,0,r+s-j}.$$

If $s = 0$, one writes $\|u\|_{\tau,r}$ in place of $\|u\|_{\tau,r,0}$.

For P with $P_0 = -\Delta_g$ for principal part, B a boundary operator of order k, and φ a weight function, the local Carleman estimates we shall obtain take the form

(1.1.4)

$$\tau^{-1/2}\|e^{\tau\varphi}u\|_{\tau,2} + |e^{\tau\varphi}\operatorname{tr}(u)|_{1,1/2} \lesssim \|e^{\tau\varphi}Pu\|_{L^2(\mathcal{M})} + |e^{\tau\varphi}Bu_{|\partial\mathcal{M}}|_{\tau,3/2-k}.$$

In the case B is the Dirichlet boundary operator, such an estimate is stronger than (1.0.1), by several aspects.

From Volume 1, we shall use the pseudo-differential techniques of Chapter 2, in particular positivity inequalities of Gårding type. Here, in the case of a half-space, we shall provide more material on pseudo-differential operators including such inequalities. We shall also often use the patching techniques for Carleman estimates presented in Sections 3.5 and 3.6 of Volume 1.

On the application side, the subjects we cover are very similar to those treated in Volume 1: quantified unique continuation, logarithmic stabilization of the wave equation, and null-controllability of the heat equation. Yet, all these applications are treated by means of the new estimates derived in this volume. Quantified unique continuation is proven in the case of general Lopatinskiĭ–Šapiro conditions. Stabilization of the wave equation is treated in the case of a boundary damping, first, associated with a Neumann boundary condition and, second, for a general boundary operator that fulfills the Lopatinskiĭ–Šapiro condition and yields a selfadjoint Laplace–Beltrami operator. The case of inner damping as in Chapter 6 of Volume 1 is not treated here; yet it can be adapted from that chapter to the case of Riemannian manifolds and Lopatinskiĭ–Šapiro boundary conditions. Null-controllability of the heat equation is also proven in the case of a Lopatinskiĭ–Šapiro condition that yields a selfadjoint Laplace–Beltrami operator.

In Volume 1, Chapter 4 focused on optimality aspects for Carleman estimates and one aspect was left untouched: the optimality of the norms associated with the boundary terms. Yet, the estimates obtained in Volume 1 and the estimate in (1.1.4) are not optimal with that respect. Here, in Chap. 14, we present estimations with an improvement with respect to the boundary terms. They take the form

$$\tau^{-1/2}\|e^{\tau\varphi}u\|_{\tau,2} + \tau^{-1/4}|e^{\tau\varphi}\operatorname{tr}(u)|_{1,1/2} \lesssim \|e^{\tau\varphi}Pu\|_{L^2(\mathcal{M})} + \tau^{-1/4}|e^{\tau\varphi}Bu_{|\partial\mathcal{M}}|_{\tau,3/2-k}. \tag{1.1.5}$$

If compared to (1.1.4), there is an additional $\tau^{-1/4}$ in front of the boundary terms, in particular the term involving the boundary operator B on the right-hand side. The proof of such an estimate is quite delicate and requires the use of more advanced tools from pseudo-differential calculus. Here, we shall prove that this improvement is final as such estimates are optimal with respect to the boundary terms.

1.2. Outline

Figure 1.1 describes the interrelation of the different parts and chapters in Volume 2.

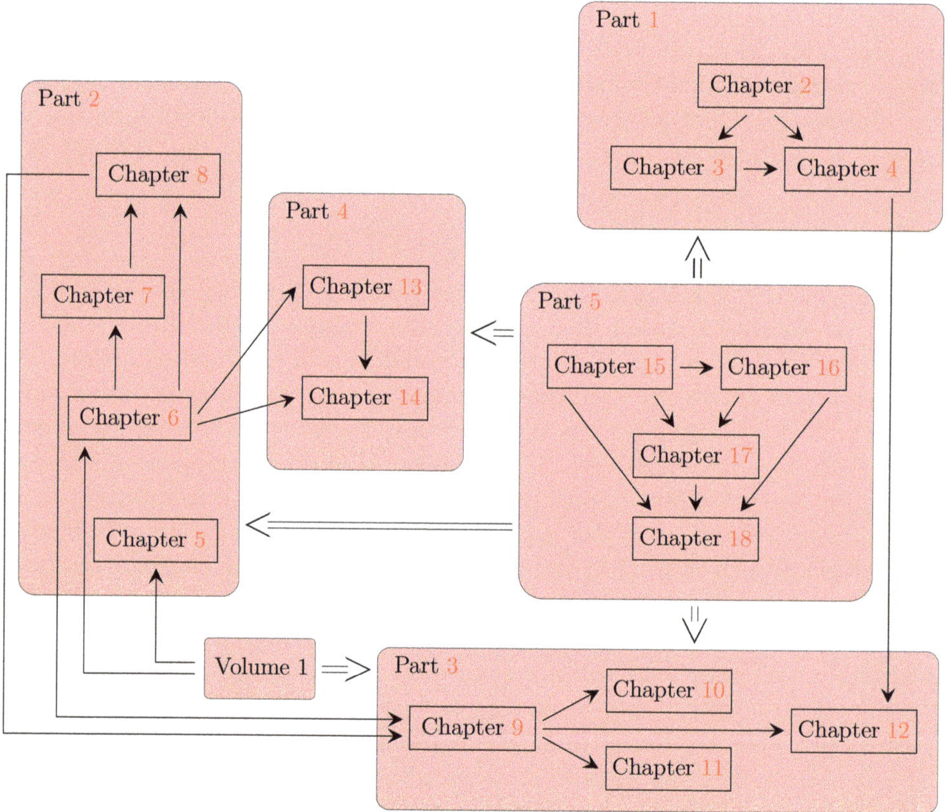

FIGURE 1.1. Interrelation of the parts and chapters in Volume 2

1.2.1. Part 1. Chapters 2–4 form Part 1 that is devoted to the study of Lopatinskiĭ–Šapiro boundary conditions for the Laplace–Beltrami operator. They are presented in Chap. 2. In particular, a characterization is proved in the case of a first-order boundary operator, in all dimensions $d \geq 2$. The case $d = 2$ appears quite different from the cases $d = 3$ and $d \geq 4$.

In Chap. 3, for P an elliptic operator with $P_0 = -\Delta_g$ for principal part and B a boundary operator of order k, we consider the map $L : u \mapsto (Pu, Bu_{|\partial\mathcal{M}})$ from $H^2(\mathcal{M})$ into $L^2(\mathcal{M}) \oplus H^{3/2-k}(\partial\mathcal{M})$. We prove that L is a Fredholm operator if and only if B fulfills the Lopatinskiĭ–Šapiro condition. We also show that the Lopatinskiĭ–Šapiro condition yields *elliptic* property for L, meaning that regularity of Pu and $Bu_{|\partial\mathcal{M}}$ yields maximal regularity for u. We also perform some Fredholm index computations for L in the case of B is of order $k \leq 1$. As mentioned above, the case $d = 2$ yields the most intricate case as topology plays an important rôle.

In Chap. 4, we characterize boundary operators B of order less than or equal to one for an elliptic operator P to be selfadjoint if P has $P_0 = -\Delta_g$ for principal part. Useful tools such as a Green formula for functions with fairly low regularity are also provided. Spectral properties for P are given and the nonhomogeneous elliptic boundary value problem is described.

1.2.2. Part 2. Chapters 5–8 form Part 2 that is devoted to the derivation of Carleman estimates for the Laplace–Beltrami operator $-\Delta_g$ on a Riemannian manifold.

In Chap. 5, we show how the results of Chapter 3 of Volume 1 yield estimates on a Riemannian manifold in the case of Dirichlet boundary conditions. Local and global estimations are given.

Chapters 6 and 7 prepare for the derivation of Carleman estimates for general boundary conditions. The geometry considered in Chap. 6 is that of a half-space, that is, what one obtains when considering local coordinates near a boundary point. In this geometry, we introduce additional classes of pseudo-differential operators; they act like a differential operator in the normal variable and like a pseudo-differential operator in the tangential variables. Adapted Sobolev norms for the half-space are introduced. Quadratic forms built with these pseudo-differential operators are presented. A notion of principal symbol is given as well as a positivity Gårding-type estimation. The remaining of Chaps. 6 is devoted to the derivation of estimations for elliptic first-order pseudo-differential operators on a half-space. These estimations are obtained through Fourier-multiplier techniques. Their quality depends on a root localization in the complex plane. The case of sub-elliptic first-order operators, permitting a real root, is also given. In this latter case, the derivation is quite similar to the derivation of Carleman estimates at the boundary given in Chapter 3 of Volume 1.

In Chap. 7, Sobolev norms with a large parameter on manifolds are introduced. They are necessary for the formulation of Carleman estimates on a manifold, in particular in the case of fractional orders. This aspect was omitted in Volume 1 where Sobolev space on manifold (like a boundary) was rather treated intuitively (see Remark 3.21 in Volume 1). The norms introduced in Chap. 7 are based on the more systematic presentation of Sobolev spaces on Riemannian manifolds given in Chap. 18. They take the forms of the norms given in (1.1.1)–(1.1.3).

Chapter 8 is the cornerstone of Volume 2. First, the Lopatinskiĭ–Šapiro condition is adapted to the conjugated operators $B_\varphi = e^{\tau\varphi} B e^{-\tau\varphi}$ and $P_\varphi = e^{\tau\varphi} P e^{-\tau\varphi}$ for which an estimation with the loss of a half-derivative is sought. The form of the condition after conjugation is very similar to that given in Chap. 2. If B fulfills the Lopatinskiĭ–Šapiro condition for P, we provide sufficient (and sometimes necessary) conditions on the Carleman weight function φ for B_φ to fulfill the Lopatinskiĭ–Šapiro condition for P_φ. Carleman estimates are derived under this Lopatinskiĭ–Šapiro condition and the now usual sub-ellipticity property.

We prove local estimates using normal geodesic coordinates where $x = (x', x_d)$. The principal symbol of P_φ reads $p_\varphi(x, \xi, \tau) = \big(\xi_d - \gamma_1(x, \xi', \tau)\big)\big(\xi_d - \gamma_2(x, \xi', \tau)\big)$. Depending on the sign of the imaginary parts of the roots γ_1 and γ_2, different proof strategies are used. Tangential phase space is thus cut into various pieces. In particular, if $\gamma_1 \neq \gamma_2$, then microlocally both roots are smooth and the principal part of P_φ can be written as the product of two first-order operators. For one factor, the sub-elliptic estimation obtained in Chap. 6 is used. For the second factor, one exploits the Lopatinskiĭ–Šapiro condition. If roots cross, from choices made on the sign of $\partial_\nu \varphi$, this crossing happens in the lower complex half-plane. In such case, we rely on quadratic forms in the half-space $\mathbb{R}^d_+$ and associated microlocal Gårding inequality as introduced in Chap. 6 yielding an elliptic estimation. The microlocal estimates one obtains are patched together to produce a local estimate. These local estimates are in the form given by (1.1.4). In turn, a global estimate can be achieved by the patching of such local estimates.

As in Chapter 3 of Volume 1, we also prove estimates without assuming any boundary condition. There, both the Dirichlet and Neumann traces were assumed known. In Chap. 8 in Volume 1, we consider known the Dirichlet trace and the trace $Bu_{|\partial\mathcal{M}}$, for $B = \partial_\nu + B'$ with B' some tangential differential operator of order one, under the assumption that B fulfills the Lopatinskiĭ–Šapiro condition.[2] In Chapter 3, we obtained a local Carleman

[2] In that case, we do not ask the Lopatinskiĭ–Šapiro condition to hold for the conjugated operators but the operators B and P.

of the form

$$(1.2.1)\quad \tau^{1/2}\|e^{\tau\varphi}u\|_{\tau,1} \lesssim \|e^{\tau\varphi}Pu\|_{L^2(\mathcal{M})} + \tau^{1/2}|e^{\tau\varphi}\partial_\nu u_{|\partial\mathcal{M}}|_{L^2(\partial\mathcal{M})} + \tau^{3/2}|e^{\tau\varphi}u_{|\partial\mathcal{M}}|_{L^2(\partial\mathcal{M})} + \tau^{1/2}|e^{\tau\varphi}u_{|\partial\mathcal{M}}|_{H^1(\partial\mathcal{M})},$$

with the left-hand side formulated with the norm introduced in (1.1.3). Here, a similar estimation is obtained. Yet, through the microlocal arguments we use, we obtain

$$(1.2.2)\quad \tau^{-1/2}\|e^{\tau\varphi}u\|_{\tau,2} \lesssim \|e^{\tau\varphi}Pu\|_{L^2(\mathcal{M})} + |e^{\tau\varphi}Bu_{|\partial\mathcal{M}}|_{\tau,1/2} + \tau^{3/2}|e^{\tau\varphi}u_{|\partial\mathcal{M}}|_{L^2(\partial\mathcal{M})},$$

which in turn can be weakened to the form

$$\tau^{1/2}\|e^{\tau\varphi}u\|_{\tau,1} \lesssim \|e^{\tau\varphi}Pu\|_{L^2(\mathcal{M})} + \tau^{1/2}|e^{\tau\varphi}Bu_{|\partial\mathcal{M}}|_{L^2(\partial\mathcal{M})} + \tau^{3/2}|e^{\tau\varphi}u_{|\partial\mathcal{M}}|_{L^2(\partial\mathcal{M})}.$$

In the case B is the Neumann operator, note the improvement: on the right-hand side, $\tau^{3/2}|e^{\tau\varphi}u_{|\partial\mathcal{M}}|_{L^2(\partial\mathcal{M})} + \tau^{1/2}|e^{\tau\varphi}u_{|\partial\mathcal{M}}|_{H^1(\partial\mathcal{M})}$ is replaced by simply $\tau^{3/2}|e^{\tau\varphi}u_{|\partial\mathcal{M}}|_{L^2(\partial\mathcal{M})}$. Such an improvement is useful in applications, for instance, in Chaps. 10 and 11 that deal with the stabilization of the wave equation through boundary damping.

1.2.3. Part 3. Chapters 9–12 form Part 3 that is devoted to applications of Carleman estimates.

In Chap. 9, we address quantified unique continuation issues. As opposed to the counterpart chapter of Volume 1, Chapter 5, the analysis in Chap. 9 is performed on a Riemannian manifold. Away from boundaries, results from Volume 1 can be adapted. At boundaries, with the estimates of Chap. 8, we are able to derive unique continuation estimates under Lopatinskiĭ–Šapiro boundary conditions. As far as we know, the results of Chap. 9 at the boundaries are not available in the literature in such generality.

In Chap. 10, we prove the logarithmic stabilization of the following boundary-damped wave equation

$$(1.2.3)\quad \begin{cases} \partial_t^2 y - \Delta_g y = 0 & \text{in } (0,+\infty)\times\mathcal{M}, \\ \partial_\nu y + \alpha\partial_t y = 0 & \text{on } (0,+\infty)\times\partial\mathcal{M}, \\ y_{|t=0} = y^0,\ \partial_t y_{|t=0} = y^1 & \text{in } \mathcal{M}. \end{cases}$$

As in Chapter 6 of Volume 1, stabilization is proven through a resolvent estimate for the generator of the associated semigroup that itself relies on the quantified unique continuation results of Chap. 9. Here, the generator is defined on $\mathscr{H} = H^1(\mathcal{M}) \oplus L^2(\mathcal{M})$. Observe that constant functions are trivial solutions of (1.2.3). Associated with these constant functions is the one-dimensional kernel $\mathscr{N}$ of the generator. For the proof of the stabilization result, one has to reduce the analysis to an invariant subspace of $\mathscr{H}$ that is in direct sum with $\mathcal{N}$.

In Chap. 11, the previous result is extended to the wave equation with more general boundary damping. The nature of the wave operator is also extended. One considers

$$\begin{cases} \partial_t^2 y + Py = 0 & \text{in } (0,+\infty)\times\mathcal{M},\\ By + \alpha\partial_t y = 0 & \text{on } (0,+\infty)\times\partial\mathcal{M},\\ y_{|t=0} = y^0,\ \partial_t y_{|t=0} = y^1 & \text{in } \mathcal{M}, \end{cases}$$

with $P = -\Delta_g + R_1$ where R_1 is a first-order differential operator and $B = \partial_\nu + B'$ both chosen such that B and P fulfill the Lopatinskiĭ–Šapiro condition and such that P is selfadjoint (additional assumptions are needed in the case $d = 2$). The strategy follows also that of Chapter 6 in Volume 1. Like in Chap. 10, the analysis is carried out in a reduced space excluding the kernel of the generator of the associated semigroup. Of course, the result of Chap. 10 is a particular case of Chap. 11. We however chose to consider separately the case of a Neumann boundary damping as it is treated in publications. As far as we know, the result of Chap. 11 is not available in the literature in such generality for a first-order boundary damping condition.

Chapter 12 is the counterpart of Chapter 7 of Volume 1. We prove a spectral inequality for the operator $P = -\Delta_g + R_1$ where R_1 is as above along with a boundary operator $B = \partial_\nu + B'$ such that P is selfadjoint. It reads

$$\|w\|_{L^2(\mathcal{M})} \leq K e^{K\sqrt{\mu}} \|w\|_{L^2(\omega)}, \qquad w \in \operatorname{span}\{\phi_j;\ \mu_j \leq \mu\}, \tag{1.2.4}$$

where $(\phi_j)_{j\in\mathbb{N}}$ is a Hilbert basis of $L^2(\mathcal{M})$ formed by eigenfunctions of P, with $0 < \mu_0 \leq \mu_1 \leq \cdots$ the associated *real* and *positive* eigenvalues. Like in Chapter 7 in Volume 1, we deduce from this spectral inequality a null-controllability result for the associated heat equation. The spectral inequality (1.2.4) with such a general boundary condition is an original result.

1.2.4. Part 4. Chapters 13 and 14 form Part 4 that is devoted to more advanced results on Carleman estimates.

It is quite usual to face elliptic problems with a solution u that only lies in $H^1(\mathcal{M})$. Then, $P_0 u \in H^{-1}(\mathcal{M})$, with $P_0 = -\Delta_g$. Up to Chap. 13, Carleman estimates were devoted to the case $Pu \in L^2(\mathcal{M})$. If $Pu \in H^{-1}(\mathcal{M})$ on a manifold, along with boundary conditions that fulfill the Lopatinskiĭ–Šapiro condition, a Carleman estimate can also be derived. If $B = B^{k-1}\partial_\nu u_{|\partial\mathcal{M}} + B^k u_{|\partial\mathcal{M}}$, where B^{k-1} and B^k are smooth differential operators on $\partial\mathcal{M}$ of order $k-1$ and k, respectively, it takes the form

$$\begin{aligned} \tau^{1/2}\|e^{\tau\varphi}u\|_{\tau,1} + \tau|e^{\tau\varphi}u_{|\partial\mathcal{M}}|_{\tau,1/2} &\lesssim \|e^{\tau\varphi}F_0\|_{L^2(\mathcal{M})} + \tau\|e^{\tau\varphi}F\|_{L^2V(\mathcal{M})} \\ &\quad + \tau\big|e^{\tau\varphi_{|\partial\mathcal{M}}}\big(B^{k-1}(\partial_\nu u + g(F,\nu))_{|\partial\mathcal{M}} + B^k u_{|\partial\mathcal{M}}\big)\big|_{\tau,1/2-k}, \end{aligned}$$

where $Pu = F_0 + \operatorname{div}_g F$. The norms $\|.\|_{\tau,n}$ and $|.|_{\tau,\ell/2}$ are as introduced in (1.1.1)–(1.1.3) (see Chap. 7 for details). Their version on a half-space is given in Sect. 6.2. The proof of this estimation has some similarities with those performed in Chap. 8. With the principal symbol of P_φ reading $p_\varphi(x,\xi,\tau) = \big(\xi_d - \gamma_1(x,\xi',\tau)\big)\big(\xi_d - \gamma_2(x,\xi',\tau)\big)$, strategies vary depending on the position of the two roots in the complex plane. Microlocal estimates are derived based in part on the first-order estimations obtained in Chap. 6. Here, as in (1.2.2), we also prove an estimate without assuming any boundary condition, that is, considering known the Dirichlet trace and a trace $Bu_{|\partial\mathcal{M}}$, for $B = \partial_\nu + B'$ with B' some tangential differential operator of order one. One assumes that this boundary operator B fulfills the Lopatinskiĭ–Šapiro condition.[3] We believe that Carleman estimates with source terms in H^{-1} in Chap. 13 were only known in the literature for Dirichlet and Neumann boundary conditions.

In Chap. 14, Carleman inequalities are improved with respect to the estimation of the volume norm. On the one hand, in Chap. 8, one finds an estimate of the form

$$\tau^{-1/2}\|e^{\tau\varphi}u\|_{\tau,2} + |e^{\tau\varphi_{|\partial\mathcal{M}}}\operatorname{tr}(u)|_{\tau,1,1/2} \lesssim \|e^{\tau\varphi}Pu\|_{L^2(\mathcal{M})} + |e^{\tau\varphi}Bu_{|\partial\mathcal{M}}|_{\tau,3/2-k}.$$

On the other hand, in Chap. 14, one obtains

$$\tau^{-1/2}\|e^{\tau\varphi}u\|_{\tau,2} + \tau^{-1/4}|e^{\tau\varphi_{|\partial\mathcal{M}}}\operatorname{tr}(u)|_{\tau,1,1/2} \lesssim \|e^{\tau\varphi}Pu\|_{L^2(\mathcal{M})} + \tau^{-1/4}|e^{\tau\varphi}Bu_{|\partial\mathcal{M}}|_{\tau,3/2-k}. \tag{1.2.5}$$

Here also, the Lopatinskiĭ–Šapiro condition is assumed. The second estimate is better in the sense that the volume norm $\|e^{\tau\varphi}u\|_{\tau,2}$ on the left-hand side is estimated thanks to a boundary term with an additional factor $\tau^{-1/4}$. The proof of this second estimate relies on an improved version of an inequality obtained for a first-order factor under the sub-ellipticity condition in Chap. 6. The improvement one obtains is precisely a factor $\tau^{-1/4}$ for the boundary term. The proof of this first-order operator estimate relies on the use of more advanced pseudo-differential calculus, namely the Hörmander calculus, and a fine Fefferman–Phong inequality proved by J.-M. Bony. The application of that inequality generalizes the use of the Gårding inequality we have made elsewhere for the proof of this type of estimate.

Here, we also prove that the $\tau^{-1/4}$ factor gain in (1.2.5) is optimal, thus completing the optimality results of Chapter 4 of Volume 1. Carleman estimates as in (1.2.5) were only known in the literature for Dirichlet boundary conditions. The proof of their optimality is moreover original.

[3] As for (1.2.2), we do not ask the Lopatinskiĭ–Šapiro condition to hold for the conjugated operators but the operators B and P.

In addition, in Chap. 14, Carleman estimates with source terms in H^{-1} as in Chap. 13 are also considered, and we also prove estimates without assuming any boundary condition.

1.2.5. Part 5. Chapters 15–18 form Part 5 that is devoted to background material on geometry and basic analysis on manifolds.

Chapter 15 provides the necessary background on differential geometry: manifolds, submanifolds, partitions of unity, tangent and cotangent vectors, tensors and associated bundles. We also provide elements of symplectic geometry.

In Chap. 16, we provide theories of integrations on manifolds: integration of d-forms on oriented manifolds, in particular with the Stokes formula, and integration of densities on non-orientable manifolds. We also present some aspects of Radon measures and distribution densities and of differential operators on manifolds.

In Chap. 17, we provide basic aspect of Riemannian geometry, like gradient and divergence associated with the metric and thus the Laplace–Beltrami operator. In connection with Chap. 16, we present how the metric yields a natural density that is very useful to integrate functions. Geodesics can be defined locally as distance minimizing curves. Here, as is also done classically, first, we use the Levi-Civita connection to define geodesics and the geodesic flow, and, second, we prove the distance minimizing property. The Levi-Civita connection is also used to define intrinsic high-order derivatives, including the Hessian of a function and of a tensor field. Normal geodesic coordinates play an important role in proofs of Carleman estimates in Volumes 1 and 2. Yet, in Section 9.4 of Volume 1, they are only defined locally near a point of the boundary. In Chap. 17, we show how such coordinates can be introduced in a neighborhood of a bounded part of the boundary, which can be handy, for instance, for compact manifolds, as it then yields a global parametrization of neighborhood of the boundary.

In Chap. 18, we review Sobolev spaces on Riemannian manifolds. Intrinsic derivatives based on the Levi-Civita connection provide intrinsic norms, without relying on partition of unity and local charts. Traces theorems are presented. We also consider the Laplace problem with, first, Dirichlet boundary conditions, second, Neumann boundary conditions, and third, mixed conditions, that is, Dirichlet boundary conditions on some connected components of the boundary and Neumann boundary conditions on the others. This third case is useful in the Fredholm index computations performed in Chap. 3. Lifting maps are studied in relation with these three boundary value problems as well as some spaces with traces properties. Finally, we show at the end of Chap. 18 how properties obtained for operators on a Riemannian manifold can be used for second-order elliptic operators in an open set in the Euclidean space.

1.3. Acknowledgement

In addition to their institutions, the authors wish to thank the Institut Henri-Poincaré in Paris and the Laboratoire de Mathématiques d'Orsay for their hospitality on numerous occasions during the preparation of this book.

They wish to thank Rémi Buffe, Camille Laurent, Matthieu Léautaud, Kim-Dang Phung, and Emmanuel Zongo for useful suggestions and discussions on some chapters.

Part 1

General Boundary Conditions

CHAPTER 2

Lopatinskiĭ–Šapiro Boundary Conditions

Contents

2.1. Introduction

In Volume 1, on some bounded smooth open set Ω, we considered a second-order elliptic operator $P = P(x, D)$ with a principal part of the form

$$\sum_{1\leq i,j\leq d} D_i(p^{ij}(x)D_j), \qquad \text{with} \quad \sum_{1\leq i,j\leq d} p^{ij}(x)\xi_i\xi_j \geq C|\xi|^2,$$

where $p^{ij} \in \mathscr{C}^\infty(\overline{\Omega};\mathbb{R})$ are such that $p^{ij} = p^{ji}$, $1 \leq i, j \leq d$. Along with such an elliptic operator we considered Dirichlet boundary conditions. In the present chapter, and in this second volume for the most part, we provide a more comprehensive and geometrical point of view by working directly on a Riemannian manifold. The goal of the present short chapter is to introduce the broad class of boundary conditions known under the name Lopatinskiĭ–Šapiro condition.

Here, $(\mathcal{M}, g)$ will thus denote a d-dimensional smooth compact Riemannian manifold. We refer to Chaps. 15–17 for the notions of differential and Riemannian geometry that are needed in what follows. In the case Ω

J. Le Rousseau et al., *Elliptic Carleman Estimates and Applications to Stabilization and Controllability, Volume II*, PNLDE Subseries in Control 98, https://doi.org/10.1007/978-3-030-88670-7_2

is a smooth open set in $\mathbb{R}^d$, setting $\mathcal{M} = \overline{\Omega}$ allows one to cover this case in the presentation below; see Sect. 18.9. The elliptic boundary value problem under consideration is thus the following

$$Pu = f \in L^2(\mathcal{M}), \qquad Bu_{|\partial\mathcal{M}} = f_\partial, \tag{2.1.1}$$

where $P = -\Delta_g + R_1$, with Δ_g the Laplace–Beltrami operator on $\mathcal{M}$, with R_1 a first-order differential operator on $\mathcal{M}$ with bounded coefficients, and with B a smooth differential boundary operator on $\partial\mathcal{M}$. The source term f_∂ lies in a prescribed space depending on the nature of the operator B.

REMARK 2.1. Note that it can be convenient to consider the operator B as a differential operator defined in a neighborhood of the boundary. Both points of view are equivalent as one can perform a smooth extension of the coefficients of B.

Near a point $m^0 \in \partial\mathcal{M}$, if one uses normal geodesic coordinates (x', x_d) where $\partial\mathcal{M}$ is locally given by $\{x_d = 0\}$ (see Sect. 17.6) the Laplace–Beltrami operator takes the form

$$-\Delta_g = D_d^2 + R(x, D_{x'}) + Q_1(x, D), \tag{2.1.2}$$

where $R(x, D_{x'})$ is a second-order differential operator and $Q_1(x, D)$ is a first-order differential operator. Thus, formally one can use the equation $Pu = f$ to reduce the form of the operator B to the case where the order of D_d is at most one. We shall thus assume that B is an operator of order $\beta \in \mathbb{N}$ with yet differentiation transverse to $\partial\mathcal{M}$ of order less than or equal to one. Denoting ν the unique unitary outward pointing vector field along $\partial\mathcal{M}$ orthogonal to $T_m\partial\mathcal{M}$, one then finds that B takes the form

$$B = B^k + B^{k-1}\partial_\nu, \tag{2.1.3}$$

on every connected component $\mathcal{N}$ of $\partial\mathcal{M}$, with B^k and B^{k-1} smooth differential operators on $\mathcal{N}$ of order k and $k-1$, respectively, with $k \le \beta$. Naturally, if B is of order zero on $\mathcal{N}$ one has $B^{k-1} = 0$.

2.1.1. Observations Concerning the Need for Boundary Conditions. On $(0, +\infty)$ consider the first-order operator $L = D_t - \lambda\rho$, with $\rho \in \mathbb{C}$ and $\lambda > 0$. The parameter λ is meant to become large.

Let $v \in \overline{\mathscr{S}}(\mathbb{R}^d_+)$. One wishes to investigate whether one can achieve the estimate

$$\|v\|_{L^2(\mathbb{R}_+)} \lesssim \|Lv\|_{L^2(\mathbb{R}_+)}. \tag{2.1.4}$$

First, assume that $\operatorname{Im}\rho < 0$. One computes

$$\begin{aligned} \operatorname{Re}(Lv, iv)_{L^2(\mathbb{R}_+)} &= \operatorname{Re}(D_t v, iv)_{L^2(\mathbb{R}_+)} - \lambda \operatorname{Im}\rho \|v\|^2_{L^2(\mathbb{R}_+)} \\ &= -\frac{1}{2}\int_{\mathbb{R}_+} \partial_t |v|^2 dt + \lambda |\operatorname{Im}\rho| \|v\|^2_{L^2(\mathbb{R}_+)} \\ &= \frac{1}{2}|v(0)|^2 + \lambda |\operatorname{Im}\rho| \|v\|^2_{L^2(\mathbb{R}_+)}. \end{aligned} \tag{2.1.5}$$

With the Cauchy–Schwarz inequality and the Young inequality we obtain for any $\varepsilon > 0$

$$\lambda \|v\|^2_{L^2(\mathbb{R}_+)} + |v(0)|^2 \lesssim \varepsilon\lambda \|v\|^2_{L^2(\mathbb{R}_+)} + (\varepsilon\lambda)^{-1} \|Lv\|^2_{L^2(\mathbb{R}_+)}.$$

With $\varepsilon > 0$ chosen sufficiently small one concludes that

$$\lambda \|v\|_{L^2(\mathbb{R}_+)} + \lambda^{1/2} |v(0)| \lesssim \|Lv\|_{L^2(\mathbb{R}_+)}. \tag{2.1.6}$$

In this first case, not only do we obtain (2.1.4) but, better yet, we also estimate the trace of v at $t = 0^+$.

Second, assume that $\operatorname{Im} \rho > 0$. Consider $\chi \in \mathscr{C}^\infty_c(\mathbb{R})$ with $\chi(t) = 1$ for $|t| \leq 1$ and $\chi(t) = 0$ for $t \geq 2$. Set $v(t) = \chi(t)e^{i\lambda\rho t}$ and observe that $Lv(t) = -i\chi'(t)e^{i\lambda\rho t}$.

On the one hand one has

$$\begin{aligned} \|v\|^2_{L^2(\mathbb{R}_+)} &= \int_0^2 \chi^2(t) e^{-2\lambda \operatorname{Im} \rho t}\, dt \\ &\geq \int_0^1 e^{-2\lambda \operatorname{Im} \rho t}\, dt = \frac{1}{2\lambda \operatorname{Im} \rho}\big(1 - e^{-2\lambda \operatorname{Im} \rho}\big). \end{aligned}$$

On the other hand, one has

$$\begin{aligned} \|Lv\|^2_{L^2(\mathbb{R}_+)} &= \int_0^2 \big(\chi'(t)\big)^2 e^{-2\lambda \operatorname{Im} \rho\, t}\, dt \\ &\lesssim \int_1^2 e^{-2\lambda \operatorname{Im} \rho\, t}\, dt = \frac{1}{2\lambda \operatorname{Im} \rho}\big(e^{-2\lambda \operatorname{Im} \rho} - e^{-4\lambda \operatorname{Im} \rho}\big). \end{aligned}$$

If $\lambda \to +\infty$, one thus finds

$$\|v\|_{L^2(\mathbb{R}_+)} \gtrsim (2\lambda \operatorname{Im} \rho)^{-1/2} \quad \text{and} \quad \|Lv\|_{L^2(\mathbb{R}_+)} \lesssim (2\lambda \operatorname{Im} \rho)^{-1/2} e^{-\lambda \operatorname{Im} \rho}.$$

This ruins any hope of having an estimate of the form of (2.1.4). Yet, one has

$$\lambda \|v\|_{L^2(\mathbb{R}_+)} \lesssim \|Lv\|_{L^2(\mathbb{R}_+)} + \lambda^{1/2} |v(0)|. \tag{2.1.7}$$

For a proof, compute $\operatorname{Re}(Lv, -iv)_{L^2(\mathbb{R}_+)}$ and argue as above for (2.1.5) and (2.1.6).

Consider now the Laplace–Beltrami operator $-\Delta_g$ in the normal geodesic coordinates provided above, that is, as in (2.1.2), $-\Delta_g = D_d^2 + R(x, D_{x'}) + Q_1(x, D)$. To simplify, assume here that $R(x, D_{x'})$ is a constant coefficient operator. Then, up to a Fourier transformation in the x' variables, one obtains for the principal part the operator

$$\hat{P} = D_d^2 + R(\xi') = D_d^2 + |\xi'|^2 \tilde{R}(\xi'),$$

with $\tilde{R}(\xi') = R(\xi'/|\xi'|)$, which we write

$$\hat{P} = L^+ L^-, \qquad L^\pm = D_d \mp |\xi'| \tilde{R}(\xi')^{1/2}.$$

With $\lambda = |\xi'|$, $x_d = t$, and the two cases considered above, one finds that for the factor L^- an estimation as in (2.1.6) can be obtained. For the factor

L^+ one can only obtain an estimation as in (2.1.7). Combined together one obtains an estimation either of the form

$$\lambda^2 \|v\|_{L^2(\mathbb{R}_+)} + \lambda^{1/2} |D_d v(0)| \lesssim \|\hat{P} v\|_{L^2(\mathbb{R}_+)} + \lambda^{3/2} |v(0)|,$$

or the form

$$\lambda^2 \|v\|_{L^2(\mathbb{R}_+)} + \lambda^{3/2} |v(0)| \lesssim \|\hat{P} v\|_{L^2(\mathbb{R}_+)} + \lambda^{1/2} |D_d v(0)|.$$

This simple example shows that if an elliptic operator can be written as a product of several factors as above, factors of the form $D_d - \lambda\rho$ with $\operatorname{Im}\rho < 0$ yield an estimation without requiring any boundary term; factors of the form $D_d - \lambda\rho$ with $\operatorname{Im}\rho > 0$ yet require a boundary term. The number of factors of the second kind yields the number of required boundary conditions.

In that framework, if given some boundary operators, the Lopatinskiĭ–Šapiro condition states their compatibility with the different factors $D_d - \lambda\rho$ with $\operatorname{Im}\rho > 0$.

If one considers again the Laplace–Beltrami, one sees that only one boundary condition is needed, as is classically known. As we shall see below the Lopatinskiĭ–Šapiro condition gives a relation between the chosen boundary operator and the factor L^+ given above.

2.1.2. Outline. In Sect. 2.2 we introduce the Lopatinskiĭ–Šapiro boundary condition. Then, in Sect. 2.3 we focus our interest to the case where B is a first-order operator with complex coefficients and provide a full description of such operators that fulfill the Lopatinskiĭ–Šapiro conditions.

The actual analysis of the elliptic boundary value problem (2.1.1) is, however, postponed to the next chapter.

2.2. Lopatinskiĭ–Šapiro Condition

For $m \in \partial\mathcal{M}$, the vector space $N_m^*(\partial\mathcal{M})$ denotes the conormal space at m given by (see Sect. 15.5 for details)

$$N_m^*\partial\mathcal{M} = \{\omega \in T_m^*\mathcal{M};\ \omega(u) = \langle \omega, u\rangle = 0, \text{ for all } u \in T_m\partial\mathcal{M}\}.$$

The conormal bundle of $\partial\mathcal{M}$ is given by

$$N^*\partial\mathcal{M} = \bigcup_{m\in\partial\mathcal{M}} \{m\} \times N_m^*\partial\mathcal{M}.$$

Consider ν to be the unique outward pointing vector field on $\mathcal{M}$ along $\partial\mathcal{M}$ such that, for all $m \in \partial\mathcal{M}$, $g_m(\nu_m, \nu_m) = 1$ and $g_m(\nu_m, u) = 0$ for all $u \in T_m\partial\mathcal{M}$, that is, ν is unitary and orthogonal to $T_m\partial\mathcal{M}$ in the sense of g. We then set $n = \nu^\flat$ the associated outward pointing one-form on $\mathcal{M}$ along $\partial\mathcal{M}$. We refer to Sect. 17.1.3 for the musical isomorphisms and notation. It is a nonvanishing section of $N^*\partial\mathcal{M}$. In particular note that $N_m^*\partial\mathcal{M} = \mathbb{R}n_m$ and thus $T_m^*\mathcal{M} = \mathbb{R}n_m \oplus T_m^*\partial\mathcal{M}$ (see Sect. 15.5). Moreover, $N_m^*\partial\mathcal{M}$ is orthogonal to $T_m^*\partial\mathcal{M}$ for the inner product $(.,.)_g$ for cotangent vectors associated with the metric at m, see (17.1.4) in Sect. 17.1.

We recall that $p(m,\omega) = |\omega|_g^2 = g_m(\omega^\sharp, \omega^\sharp)$ is the principal symbol of the (positive) Laplace–Beltrami operator $P = -\Delta_g$ (see Sect. 17.2 with the notation introduced in Sect. 17.1). In a local chart $\mathcal{C} = (O, \kappa)$, its representative reads

$$p^{\mathcal{C}}(x,\xi) = \sum_{1 \leq i,j \leq d} g^{\mathcal{C},ij}(x)\xi_i\xi_j.$$

For $(m,\omega') \in T^*\partial\mathcal{M}$, we set

$$\check{p}(m,\omega',z) = p(m,\omega' - zn_m) = |\omega' - zn_m|_g^2,$$

that is a monic polynomial function of degree two in z by Proposition 16.13 and (16.3.1). As we have $(n_m,\omega')_g = \langle \nu_m, \omega' \rangle = 0$ we find

$$\check{p}(m,\omega',z) = z^2|n_m|_g^2 + |\omega'|_g^2 = z^2 + |\omega'|_g^2 = \ell^-(z)\ell^+(z),$$

with

$$\ell^+ = (z - i|\omega'|_g) \quad \text{and} \quad \ell^- = (z + i|\omega'|_g).$$

We denote by $b(m,\omega)$ the principal symbol of the boundary operator B defined on $\partial\mathcal{M}$. On a connected component $\mathcal{N}$ of $\partial\mathcal{M}$ where B is of order $k \leq \beta$, with B in the form given in (2.1.3), and $b^k(m,\omega')$ and $b^{k-1}(m,\omega')$ the respective principal symbols of B^k and B^{k-1}, we have

$$b(m,\omega) = b^k(m,\omega') + ib^{k-1}(m,\omega')\langle \omega, \nu_m \rangle,$$

for $\omega \in T_m^*(\mathcal{M})$ with $\omega = \omega' + \omega_n n_m$ for some $\omega' \in T_m^*(\partial\mathcal{M})$ and $\omega_n = \langle \omega, \nu_m \rangle$.

For $(m,\omega') \in T^*\partial\mathcal{M}$, we set

$$\check{b}(m,\omega',z) = b(m,\omega' - zn_m) = b^k(m,\omega') - ib^{k-1}(m,\omega')z,$$

that is a polynomial function in z of order less than or equal to one.

We may now state the Lopatinskiĭ–Šapiro condition first microlocally, second at one point, and third locally.

Definition 2.2 (Lopatinskiĭ–Šapiro Condition for the Laplace–Beltrami Operator). *Let $(m,\omega') \in T^*\partial\mathcal{M}$ with $\omega' \neq 0$. One says that the Lopatinskiĭ–Šapiro condition holds for (P,B) at (m,ω') if for any polynomial function $f(z)$ with complex coefficients there exists $c \in \mathbb{C}$ and a polynomial function $g(z)$ with complex coefficients such that, for all $z \in \mathbb{C}$,*

$$f(z) = c\,\check{b}(m,\omega',z) + g(z)\ell^+(z). \tag{2.2.1}$$

One says that the Lopatinskiĭ–Šapiro condition holds for (P,B) at $m \in \partial\mathcal{M}$ if it holds at (m,ω') for all $\omega' \in T_m^*\partial\mathcal{M}$ with $\omega' \neq 0$. If $\Gamma \subset \partial\mathcal{M}$, one says that the Lopatinskiĭ–Šapiro condition holds for (P,B) on Γ if it holds at m for all $m \in \Gamma$.

The choice of the polynomial $\ell^+(z) = z - i|\omega'|_g$ is in connection with the nonnegative sign of the imaginary part of its roots. This can be motivated by the observations made in Sect. 2.1.1. The formulation of the Lopatinskiĭ–Šapiro condition for general operators makes this more explicit (see Remark 2.7 below).

With the Euclidean division of polynomials, we see that it suffices to consider the polynomial function $f(z)$ to be of degree zero in (2.2.1). Observe also that the Lopatinskiĭ–Šapiro condition holds if and only if for any $f(z)$ the complex number $i|\omega'|_g$ is a root of the polynomial function $f(z) - c\check{b}(m, \omega', z)$ for some $c \in \mathbb{C}$. We thus have the following proposition.

PROPOSITION 2.3. *Let $(m, \omega') \in T^*\partial\mathcal{M}$ with $\omega' \neq 0$. The Lopatinskiĭ–Šapiro condition holds for (P, B) at (m, ω') if and only if*

$$\check{b}(m, \omega', i|\omega'|_g) = b^k(m, \omega') + b^{k-1}(m, \omega')|\omega'|_g \neq 0.$$

REMARK 2.4.

(1) With the equivalent formulation of the Lopatinskiĭ–Šapiro condition given by Proposition 2.3 one sees that if it holds at $(m, \omega') \in T^*\partial\mathcal{M}$ with $\omega' \neq 0$, then there exists a conic neighborhood of (m, ω') in $T^*\partial\mathcal{M}$ where it holds also. Similarly, if the Lopatinskiĭ–Šapiro condition holds at $m \in \partial\mathcal{M}$ (meaning it holds for all (m, ω') with $\omega' \in T^*_m\partial\mathcal{M} \setminus \{0\}$), then there exists a neighborhood of $m \in \partial\mathcal{M}$ where it holds also.
(2) Observe that the Lopatinskiĭ–Šapiro condition is written here without any use of local coordinates. It is thus a geometrical condition. As the principal parts of the operators P and B are geometrical objects (see Sect. 16.3.1), if a particular local chart is chosen, then the Lopatinskiĭ–Šapiro condition for (P, B) can be equivalently expressed in this chart.

We list examples of classical boundary conditions that fit the Lopatinskiĭ–Šapiro framework.

EXAMPLES 2.5.

(1) **The Dirichlet boundary condition.**
In this case, B is a zero-order operator given by $Bu = u$, that is, a zero-order operator. Then, $b(m, \omega) = 1$ and $\check{b}(m, \omega', z) = 1$. By Proposition 2.3, we see that the Lopatinskiĭ–Šapiro condition holds at any (m, ω').

If now B is of order 0, that is, $b(m, \omega) = b(m)$, if the Lopatinskiĭ–Šapiro condition holds at some (m, ω'), then $b(m)$ cannot vanish by Proposition 2.3. If the Lopatinskiĭ–Šapiro condition holds on some open set Γ of $\partial\mathcal{M}$ we thus see that up to dividing by a nonvanishing function we recover the Dirichlet boundary condition.

(2) **The Neumann and Robin boundary conditions.**
In the Neumann case, $Bu = \partial_\nu u = \nu(u)$. Then, $b(m,\omega) = i\langle \omega, \nu_m\rangle$ and $\check{b}(m,\omega',z) = -iz$. As $\check{b}(m,\omega',i|\omega'|_g) = |\omega'|_g$, we see that the Lopatinskiĭ–Šapiro condition holds at any (m,ω') if $\omega' \neq 0$. The Robin boundary condition is a natural generalization with $Bu(m) = (\partial_\nu u)(m) + a(m)u(m)$. As the principal symbol of B is the same as that in the Neumann case, the conclusion follows.

(3) **An oblique boundary condition.**
This is a generalization of the Robin boundary condition. In this case $Bu(m) = g_m(\nabla_g u, v) + a(m)u(m) = du(m)(v) + a(m)u(m)$ for v a *real* vector field on $\mathcal{M}$ along $\partial\mathcal{M}$. This precisely covers the case of a first-order boundary operator with a purely imaginary principal symbol. We have $b(m,\omega) = i\langle \omega, v_m\rangle$. Let us assume moreover that $\langle n_m, v_m\rangle \neq 0$ at any $m \in \partial\mathcal{M}$. We have $\check{b}(m,\omega',z) = i\langle \omega', v_m\rangle - iz\langle n_m, v_m\rangle$ and
$$\check{b}(m,\omega',i|\omega'|_g) = i\langle \omega', v_m\rangle + |\omega'|_g\langle n_m, v_m\rangle.$$
Since $\langle n_m, v_m\rangle \neq 0$ and $|\omega'|_g \neq 0$, we see that $\check{b}(m,\omega',i|\omega'|_g) \neq 0$. The Lopatinskiĭ–Šapiro condition thus holds at any (m,ω') if $\omega' \neq 0$.
However, observe that if $\langle n_m, v_m\rangle = 0$ for some $m \in \partial\mathcal{M}$, that is, $v_m \in T_m\partial\mathcal{M}$, we then have $\check{b}(m,\omega',i|\omega'|_g) = i\langle \omega', v_m\rangle$ that vanishes for some $\omega' \in T^*_m\partial\mathcal{M}$ if $d \geq 3$. The condition $\langle n_m, v_m\rangle \neq 0$ is thus necessary for the oblique boundary operator to fulfill the Lopatinskiĭ–Šapiro condition if $d \geq 3$.

(4) **Ventcel boundary condition.**
This is an example of a second-order boundary condition. In this case $Bu_{|\partial\mathcal{M}} = \partial_\nu u_{|\partial\mathcal{M}} - h\Delta_{g_\partial} u_{|\partial\mathcal{M}}$, where Δ_{g_∂} is the Laplace–Beltrami operator on $\partial\mathcal{M}$ associated with g_∂, the induced metric on $\partial\mathcal{M}$, and h is some function defined on $\partial\mathcal{M}$. The principal symbol of B is $b(m,\omega) = h(m)|\omega'|^2_{g_{\partial m}}$ and thus $\check{b}(m,\omega',z) = h(m)|\omega'|^2_{g_{\partial m}}$. Consequently, by Proposition 2.3, the Lopatinskiĭ–Šapiro condition holds for (P,B) at $m \in \partial\mathcal{M}$ if and only if $h(m) \neq 0$.

EXAMPLE 2.6. A setting where the Lopatinskiĭ–Šapiro condition does not hold is given by the following classical example. On the unit disc $\mathbb{D}$ consider $P = -\Delta = -\partial_x^2 - \partial_y^2$ that reads $P = -r^{-1}\partial_r(r\partial_r) - r^{-2}\partial_\theta^2$ in polar coordinates. In the latter coordinates, define the boundary operator $Bu_{|\partial\mathbb{D}} = \partial_r u_{|\partial\mathbb{D}} + i\partial_\theta u_{|\partial\mathbb{D}}$. As one can readily check the Lopatinskiĭ–Šapiro condition holds nowhere on $\partial\mathbb{D}$. In Chap. 3 we prove that the Lopatinskiĭ–Šapiro condition is equivalent to having the operator $L : H^2(\mathbb{D}) \to L^2(\mathbb{D}) \oplus H^{1/2}(\partial\mathbb{D})$ given by $Lu = (Pu, Bu_{|\partial\mathbb{D}})$ of Fredholm type; see Theorem 3.1. Definition and basic properties of Fredholm operators are given in Section 11.5 of Volume 1. One can see that the present operator is not Fredholm in agreement with the Lopatinskiĭ–Šapiro condition not holding. In fact, the operator reads $P = -4\partial\overline{\partial}$ with $\partial = \partial_z = (\partial_x - i\partial_y)/2$ and $\overline{\partial} = \partial_{\overline{z}} = (\partial_x + i\partial_y)/2$. Moreover, we have

$$\overline{\partial} = \frac{1}{2}e^{i\theta}(\partial_r + ir^{-1}\partial_\theta).$$

Consequently any H^2-holomorphic function f on a neighborhood of $\overline{\mathbb{D}}$, solution to $\overline{\partial} f = 0$, is also solution to $Pf = 0$ and $Bf_{|\partial\mathbb{D}} = 0$. As the space of such functions is of infinite dimension, the operator L is not Fredholm.

REMARK 2.7 (Lopatinskiĭ–Šapiro Condition for General Elliptic Differential Operators). For a general elliptic differential operator Q of degree $2k$ on $\mathcal{M}$, with principal symbol $q(m,\omega)$ one defines the following polynomial in z

$$\check{q}(m,\omega',z) = q(m,\omega' - zn_m),$$

and one denotes its complex roots by $\gamma_j(m,\omega')$, $1 \leq j \leq 2k$. One sets

$$\check{q}^+(m,\omega',z) = \prod_{\operatorname{Im}\gamma_j(m,\omega')\geq 0} (z - \gamma_j(m,\omega')).$$

Given boundary operators $B_1,\dots,B_k$ in a neighborhood of $\partial\mathcal{M}$, with principal symbols $b_j(m,\omega)$, $j = 1,\dots,k$, one also sets

$$\check{b}_j(m,\omega',z) = b_j(m,\omega' - zn_m).$$

Let $(m,\omega') \in T^*\partial\mathcal{M}$ with $\omega' \neq 0$. One says that the Lopatinskiĭ–Šapiro condition holds for $(Q, B_1,\dots,B_k)$ at (m,ω') if for any polynomial function $f(z)$ with complex coefficients there exists $c_1,\dots,c_k \in \mathbb{C}$ and a polynomial function $g(z)$ with complex coefficients such that, for all $z \in \mathbb{C}$,

$$f(z) = \sum_{1\leq j\leq k} c_j\,\check{b}_j(m,\omega',z) + g(z)\check{q}^+(m,\omega',z). \tag{2.2.2}$$

Definition 2.2 gives precisely this condition in the case of the Laplace–Beltrami operator.

In Examples 2.5-(2) and 2.5-(3), the boundary operator is of order one and we note that the principal symbol b satisfies $b(m,\omega) \in c\mathbb{R}$ for some fixed $c \in \mathbb{C}$. We now consider the general case of boundary operator of order one.

2.3. First-Order Boundary Operator with Complex Coefficients

The following proposition provides a classification of first-order differential boundary operators B that yield Lopatinskiĭ–Šapiro conditions along with the operator P, depending on the dimension of $\mathcal{M}$.

PROPOSITION 2.8. *Let the boundary operator B be of order $\beta = 1$ with nonvanishing principal symbol*

$$b(m,\omega) = \langle\omega, t_m\rangle + i\langle\omega, v_m\rangle,$$

where t, v are two real vector fields on $\mathcal{M}$ along $\partial\mathcal{M}$. We write $v_m = v_m^\nu \nu_m + v_m'$, $t_m = t_m^\nu \nu_m + t_m'$, with $v_m^\nu, t_m^\nu \in \mathbb{R}$ and $v_m', t_m' \in T_m\partial\mathcal{M}$. Depending on the dimension $d \geq 2$ of $\mathcal{M}$ we have the following results.

***Case* $\mathbf{d = 2}$.** *The Lopatinskiĭ–Šapiro condition holds at m if and only if*

$$t_m' + i v_m' \neq i(t_m^\nu + i v_m^\nu) X', \tag{2.3.1}$$

for any $X' \in T_m\partial\mathcal{M}$ with $|X'|_{g_\partial} = 1$.

***Case* $\mathbf{d = 3}$.** *The Lopatinskiĭ–Šapiro condition holds at m if and only if*

$$|t_m'|_{g_\partial} < |v_m^\nu| \quad or \quad |v_m'|_{g_\partial} < |t_m^\nu|, \tag{2.3.2}$$

or

$$g_m(t_m, v_m)^2 \neq \big(|t_m'|_{g_\partial}^2 - (v_m^\nu)^2\big)\big(|v_m'|_{g_\partial}^2 - (t_m^\nu)^2\big). \tag{2.3.3}$$

***Case* $\mathbf{d \geq 4}$.** *The Lopatinskiĭ–Šapiro condition holds at m if and only if*

$$|t_m'|_{g_\partial} < |v_m^\nu| \quad or \quad |v_m'|_{g_\partial} < |t_m^\nu|, \tag{2.3.4}$$

or

$$g_m(t_m, v_m)^2 > \big(|t_m'|_{g_\partial}^2 - (v_m^\nu)^2\big)\big(|v_m'|_{g_\partial}^2 - (t_m^\nu)^2\big). \tag{2.3.5}$$

Observe that one only assumes that the operator B does not degenerate into a zero-order operator at some point m: one thus excludes that t_m and v_m vanish simultaneously.

PROOF. For $\omega' \in T_m^*\partial\mathcal{M}$ one has

$$\check{b}(m, \omega', z) = \langle \omega', t_m' \rangle + i\langle \omega', v_m' \rangle - z(t_m^\nu + i v_m^\nu).$$

By proposition 2.3, if $\omega' \neq 0$, the Lopatinskiĭ–Šapiro condition does not hold for (P, B) at (m, ω') if and only if $\check{b}(m, \omega', i|\omega'|_{g_\partial}) = 0$. By homogeneity, note that it suffices to consider $\omega' \in S_m^*\partial\mathcal{M}$, that is, $|\omega'|_{g_\partial} = 1$. In such case, we find that the Lopatinskiĭ–Šapiro condition does not hold at (m, ω') if and only if

$$\langle \omega', t_m' \rangle = -v_m^\nu \quad \text{and} \quad \langle \omega', v_m' \rangle = t_m^\nu. \tag{2.3.6}$$

Case 1: $d = 2$. As $\dim T_m\partial\mathcal{M} = 1$, from (2.3.6) we find $t_m' = -v_m^\nu \omega'^\sharp$ and $v_m' = t_m^\nu \omega'^\sharp$. Here $\omega' \in S_m^*\partial\mathcal{M}$. This precisely means that the Lopatinskiĭ–Šapiro holds at m if and only if one has (2.3.1).

Case 2: $d = 3$. If either $|t'_m|_{g_\partial} < |v^\nu_m|$ or $|v'_m|_{g_\partial} < |t^\nu_m|$, then (2.3.6) cannot hold for any $\omega' \in S^*_m\partial\mathcal{M}$. Thus, in this case, the Lopatinskiĭ–Šapiro condition holds at m.

Consider now the case $|t'_m|_{g_\partial} \geq |v^\nu_m|$ and $|v'_m|_{g_\partial} \geq |t^\nu_m|$. Each equation in (2.3.6) has nontrivial solutions. We seek conditions so that such solutions can coincide.

First, if $t'_m = 0$, one has $v^\nu_m = 0$ and then there exists such a common solution; moreover one has

$$0 = g_m(t_m, v_m)^2 = \big(|t'_m|^2_{g_\partial} - (v^\nu_m)^2\big)\big(|v'_m|^2_{g_\partial} - (t^\nu_m)^2\big).$$

Hence, in such case where the Lopatinskiĭ–Šapiro condition does not hold at m conditions (2.3.2) and (2.3.3) are not fulfilled. The same reasoning applies to the case $v'_m = 0$.

Second, we consider the case $t'_m \neq 0$ and $v'_m \neq 0$. Assume that there exists $\omega' \in S^*_m\partial\mathcal{M}$ such that (2.3.6) holds. We pick $\tilde{\omega}' \in S^*_m\partial\mathcal{M}$ so that $(\omega', \tilde{\omega}')$ forms an orthonormal basis of $T^*_m(\partial\mathcal{M})$. Then, $(\omega'^\sharp, \tilde{\omega}'^\sharp)$ forms an orthonormal basis of $T_m(\partial\mathcal{M})$.

We set $T' = t'_m/|t'_m|_{g_\partial}$ and $V' = v'_m/|v'_m|_{g_\partial}$ and we set $\theta_{T'}$ and $\theta_{V'}$ as the angles between $\omega'^\sharp$ and T' and V', respectively, in the frame $(\omega'^\sharp, \tilde{\omega}'^\sharp)$. Then, $\theta = \theta_{V'} - \theta_{T'}$ is the angle between T' and V'. From (2.3.6) we have

$$\cos\theta_{T'} = -\frac{v^\nu_m}{|t'_m|_{g_\partial}} \quad \text{and} \quad \cos\theta_{V'} = \frac{t^\nu_m}{|v'_m|_{g_\partial}}.$$

This is illustrated in Fig. 2.1. We have

$$\sin\theta_{T'} = \pm\Big(1 - \frac{(v^\nu_m)^2}{|t'_m|^2_{g_\partial}}\Big)^{1/2} \quad \text{and} \quad \sin\theta_{V'} = \pm\Big(1 - \frac{(t^\nu_m)^2}{|v'_m|^2_{g_\partial}}\Big)^{1/2},$$

and thus

$$(2.3.7)\quad (g_\partial)_m(T', V') = \cos\theta$$
$$= -\frac{v^\nu_m t^\nu_m}{|t'_m|_{g_\partial}|v'_m|_{g_\partial}} \pm \Big(1 - \frac{(v^\nu_m)^2}{|t'_m|^2_{g_\partial}}\Big)^{1/2}\Big(1 - \frac{(t^\nu_m)^2}{|v'_m|^2_{g_\partial}}\Big)^{1/2},$$

that is,

$$\frac{g_m(t_m, v_m)}{|t'_m|_{g_\partial}|v'_m|_{g_\partial}} = \pm\Big(1 - \frac{(v^\nu_m)^2}{|t'_m|^2_{g_\partial}}\Big)^{1/2}\Big(1 - \frac{(t^\nu_m)^2}{|v'_m|^2_{g_\partial}}\Big)^{1/2}.$$

This is precisely the negation of condition (2.3.3). We have thus found that if $|t'_m|_{g_\partial} \geq |v^\nu_m|$ and $|v'_m|_{g_\partial} \geq |t^\nu_m|$, and if (2.3.3) holds (implying that $t'_m \neq 0$ and $v'_m \neq 0$), then the Lopatinskiĭ–Šapiro condition holds at m.

Conversely, let us assume that $|t'_m|_{g_\partial} \geq |v^\nu_m|$, $|v'_m|_{g_\partial} \geq |t^\nu_m|$, $t'_m \neq 0$, and $v'_m \neq 0$ and moreover

$$(2.3.8)\qquad g_m(t_m, v_m)^2 = \big(|t'_m|^2_{g_\partial} - (v^\nu_m)^2\big)\big(|v'_m|^2_{g_\partial} - (t^\nu_m)^2\big)$$

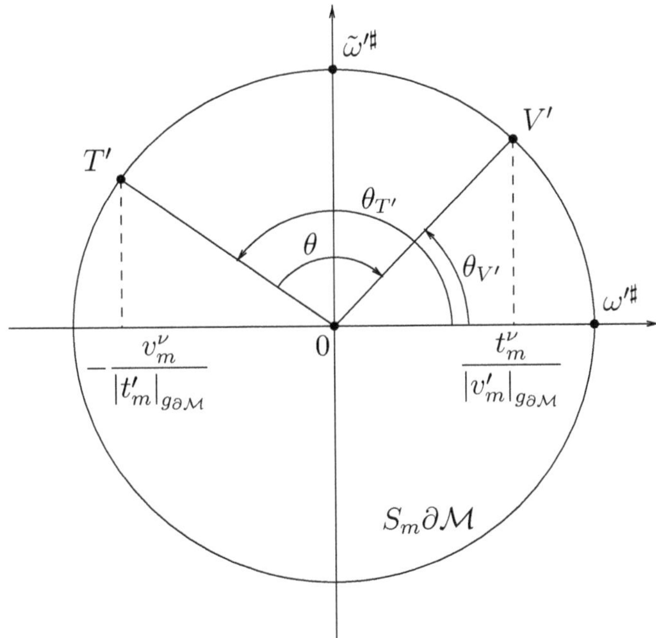

FIGURE 2.1. Geometry associated with a solution of (2.3.6) in the case $t_m' \neq 0$ and $v_m' \neq 0$

holds. As above, we set $T' = t_m'/|t_m'|_{g_\partial}$ and $V' = v_m'/|v_m'|_{g_\partial}$. We also pick $U' \in T_m\partial\mathcal{M}$ such that (T', U') is an orthonormal basis of $T_m\partial\mathcal{M}$. Then, we have $V' = \cos(\theta)T' + \sin(\theta)U'$ for some $\theta \in [0, 2\pi)$. We also set

$$a = \frac{v_m^\nu}{|t_m'|_{g_\partial}} \quad \text{and} \quad b = \frac{t_m^\nu}{|v_m'|_{g_\partial}}.$$

Condition (2.3.8) reads $\big(\cos(\theta) + ab\big)^2 = (1 - a^2)(1 - b^2)$ or equivalently

$$a^2 + b^2 + 2ab\cos(\theta) = \sin^2(\theta). \tag{2.3.9}$$

From (2.3.6) we seek ω' of the form $\omega' = -a(T')^\flat + y(U')^\flat$ for some $y \in \mathbb{R}$ such that $a^2 + y^2 = 1$ and $\langle \omega', V' \rangle = -a\cos(\theta) + y\sin(\theta) = b$.

If $\sin(\theta) \neq 0$. We then set $y = \sin(\theta)^{-1}\big(b + a\cos(\theta)\big)$. With (2.3.9) we find $a^2 + y^2 = 1$. We thus found a solution to (2.3.6).

If $\sin(\theta) = 0$. If $\theta = 0$, then $V' = T'$ and (2.3.9) yields $a = -b$. If $\theta = \pi$, then $V' = -T'$ and (2.3.9) yields $a = b$. In both cases, $\omega' = -a(T')^\flat + \sqrt{1 - a^2}(U')^\flat$ is a solution to (2.3.6) such that $|\omega'|_{g_\partial} = 1$.

In both cases, the Lopatinskiĭ–Šapiro condition does not hold. This concludes the case $d = 3$.

Case 3: $d \geq 4$. As for the previous case, if either $|t_m'|_{g_\partial} < |v_m^\nu|$ or $|v_m'|_{g_\partial} < |t_m^\nu|$, then (2.3.6) cannot hold for any $\omega' \in S_m^*\partial\mathcal{M}$. Thus, in this case, the Lopatinskiĭ–Šapiro condition holds at m.

Consider now the case $|t'_m|_{g_\partial} \geq |v^\nu_m|$ and $|v'_m|_{g_\partial} \geq |t^\nu_m|$.

First, if $t'_m = 0$, we have $v^\nu_m = 0$ and then there exists ω' solution to both equations in (2.3.6); moreover, we have

$$0 = g_m(t_m, v_m)^2 = \big(|t'_m|^2_{g_\partial} - (v^\nu_m)^2\big)\big(|v'_m|^2_{g_\partial} - (t^\nu_m)^2\big).$$

Hence in such case where the Lopatinskiĭ–Šapiro condition does not hold at m conditions (2.3.4)–(2.3.5) are not fulfilled. The same reasoning applies to the case $v'_m = 0$.

Second, we consider the case $t^\nu_m = v^\nu_m = 0$. Then, as $\dim T_m\partial\mathcal{M} \geq 3$, there exists $\omega' \in S^*_m\partial\mathcal{M}$ such that $\langle \omega', t'_m\rangle = \langle \omega', v'_m\rangle = 0$. In such case the Lopatinskiĭ–Šapiro condition does not hold. Observe that both condition in (2.3.4) clearly do not hold. With $t^\nu_m = v^\nu_m = 0$ (2.3.5) reads $(g_\partial)_m(t'_m, v'_m)^2 > |t'_m|^2_{g_\partial}|v'_m|^2_{g_\partial}$, contradicting the Cauchy–Schwarz inequality. We have thus found that in this case where the Lopatinskiĭ–Šapiro condition does not hold at m conditions (2.3.4) and (2.3.5) are not fulfilled.

Third, we consider the case $t'_m \neq 0$ and $v'_m \neq 0$ and $(t^\nu_m, v^\nu_m) \neq (0,0)$. Assume that there exists $\omega' \in S^*_m\partial\mathcal{M}$ such that both equations in (2.3.6) hold. We then define $\alpha' \in T^*_m\partial\mathcal{M}$ as the orthogonal projection of ω' onto $\operatorname{span}((t'_m)^\flat, (v'_m)^\flat)$. We then have $|\alpha'|_{g_\partial} \leq 1$ and

$$\langle \alpha', t'_m\rangle = -v^\nu_m \quad \text{and} \quad \langle \alpha', v'_m\rangle = t^\nu_m.$$

As $(t^\nu_m, v^\nu_m) \neq (0,0)$ we have $\alpha' \neq 0$. We set $T' = t'_m/|t'_m|_{g_\partial}$, $V' = v'_m/|v'_m|_{g_\partial}$, and $A' = \alpha'/|\alpha'|_{g_\partial}$.

We have

$$\langle A', T'\rangle = -\frac{v^\nu_m}{|\alpha'|_{g_\partial}|t'_m|_{g_\partial}} \quad \text{and} \quad \langle A', V'\rangle = \frac{t^\nu_m}{|\alpha'|_{g_\partial}|v'_m|_{g_\partial}}.$$

The analysis that led to (2.3.7) applies here with ω' replaced by A' yielding

$$\Big(\cos\theta + \frac{ab}{|\alpha'|^2_{g_\partial}}\Big)^2 = \Big(1 - \frac{a^2}{|\alpha'|^2_{g_\partial}}\Big)\Big(1 - \frac{b^2}{|\alpha'|^2_{g_\partial}}\Big), \tag{2.3.10}$$

where $\cos\theta = (g_\partial)_m(T', V')$ and

$$a = \frac{v^\nu_m}{|t'_m|_{g_\partial}} \quad \text{and} \quad b = \frac{t^\nu_m}{|v'_m|_{g_\partial}}.$$

With some simple computations we then find

$$a^2 + b^2 + 2ab\cos(\theta) = |\alpha'|^2_{g_\partial}\sin^2(\theta) \leq \sin^2(\theta).$$

Performing the above computations backward (formally with $|\alpha'|_{g_\partial}$ now equal to one) we obtain

$$(\cos\theta + ab)^2 \leq (1 - a^2)(1 - b^2),$$

that in turn reads

$$g_m(t_m, v_m)^2 \leq \big(|t'_m|^2_{g_\partial} - (v^\nu_m)^2\big)\big(|v'_m|^2_{g_\partial} - (t^\nu_m)^2\big).$$

This is precisely the negation of condition (2.3.5). We have thus found that if $|t'_m|_{g_\partial} \geq |v^\nu_m|$ and $|v'_m|_{g_\partial} \geq |t^\nu_m|$, if $t'_m \neq 0$, $v'_m \neq 0$ and $(t^\nu_m, v^\nu_m) \neq (0,0)$, and if (2.3.5) holds, then the Lopatinskiĭ–Šapiro condition holds at m.

Conversely, let us assume that $|t'_m|_{g_\partial} \geq |v^\nu_m|$, $|v'_m|_{g_\partial} \geq |t^\nu_m|$, $t'_m \neq 0$, $v'_m \neq 0$, $(t^\nu_m, v^\nu_m) \neq (0,0)$, and moreover

$$g_m(t_m, v_m)^2 \leq \big(|t'_m|^2_{g_\partial} - (v^\nu_m)^2\big)\big(|v'_m|^2_{g_\partial} - (t^\nu_m)^2\big). \tag{2.3.11}$$

As above, we set $T' = t'_m/|t'_m|_{g_\partial}$ and $V' = v'_m/|v'_m|_{g_\partial}$. We also set

$$a = \frac{v^\nu_m}{|t'_m|_{g_\partial}} \quad \text{and} \quad b = \frac{t^\nu_m}{|v'_m|_{g_\partial}}.$$

Condition (2.3.11) reads $\big(\cos(\theta) + ab\big)^2 \leq (1-a^2)(1-b^2)$ or equivalently

$$a^2 + b^2 + 2ab\cos(\theta) \leq \sin^2(\theta). \tag{2.3.12}$$

The argument is different depending on the dimension of $\operatorname{span}(T', V')$.

Case $\dim \operatorname{span}(T', V') = 1$. If $V' = T'$, then $\cos(\theta) = 1$ and thus (2.3.12) yields $(a+b)^2 \leq 0$, that is, $a = -b$. If $V' = -T'$, then $\cos(\theta) = -1$ and thus (2.3.12) yields $(a-b)^2 \leq 0$, that is $a = b$. In both cases, $\omega' = -a(T')^\flat + \tilde{\omega}'$ is a solution to (2.3.6) such that $|\omega'|_{g_\partial} = 1$ if $\tilde{\omega}'$ is chosen orthogonal to T' (and thus V') and $|\tilde{\omega}'|^2_{g_\partial} = 1 - a^2$.

Case $\dim \operatorname{span}(T', V') = 2$. We pick $U' \in \operatorname{span}(T', V')$ such that (T', U') is an orthonormal basis of $\operatorname{span}(T', V')$. Then, we have $V' = \cos(\theta)T' + \sin(\theta)U'$ for some $\theta \in [0, 2\pi)$. From (2.3.6) we seek $\omega' = \alpha' + \tilde{\omega}'$ with α' of the form

$$\alpha' = -a(T')^\flat + y(U')^\flat,$$

for some $y \in \mathbb{R}$ such that $a^2 + y^2 \leq 1$ and $\langle \omega', V' \rangle = -a\cos(\theta) + y\sin(\theta) = b$, and $\tilde{\omega}'$ orthogonal to $\operatorname{span}(T', V')$ with $|\tilde{\omega}'|^2_{g_\partial} = 1 - a^2 - y^2$. We then set $y = \sin(\theta)^{-1}\big(b + a\cos(\theta)\big)$. With (2.3.12) we find $a^2 + y^2 \leq 1$. We thus found a solution to (2.3.6).

In both cases, the Lopatinskiĭ–Šapiro condition does not hold. This concludes the case $d = 4$. ∎

The Lopatinskiĭ–Šapiro condition can be very restrictive in the choices of the operator B. For instance, cases of purely tangential operators B are very limited if $d \geq 3$ as presented in the following remark and proposition.

REMARK 2.9 (Cases of Boundary Operators with Purely Tangential Action).

(1) In the case $d = 2$, if $(t^\nu_m, v^\nu_m) = (0,0)$, then condition (2.3.1) always holds since $t'_m + i v'_m$ does not vanish. A typical example is $\mathcal{M}$ given by the closed unit disc $\mathbb{D}$, equipped with the flat metric, and with polar coordinates (r, θ) the operator $B = v$ with v given by $v = \partial_\theta$ we see that $v^\nu_m = 0$ and v'_m does not vanish. The proposition shows that in such a case the Lopatinskiĭ–Šapiro condition holds.

(2) In the case $d = 3$, if $(t^{\nu}_m, v^{\nu}_m) = (0,0)$, then one can prove that necessarily the connected component of m in $\partial\mathcal{M}$ is a torus. This result is in fact contained in Proposition 2.10 below.
(3) In the case $d \geq 4$, the form of conditions (2.3.4) and (2.3.5) shows that the occurrence of a boundary operators with purely tangential action is prohibited under the Lopatinskiĭ–Šapiro condition. In fact, in such case we find that $|g_\partial(t'_m, v'_m)| > |t'_m|_{g_\partial}|v'_m|_{g_\partial}$ in contradiction with the Cauchy–Schwarz inequality.

PROPOSITION 2.10. *Let $\mathcal{M}$ be of dimension $d = 3$. Let the boundary operator B be of order $\beta = 1$ with nonvanishing principal symbol*

$$b(m, \omega) = \langle \omega, t_m \rangle + i\langle \omega, v_m \rangle,$$

where t, v are two real vector fields on $\mathcal{M}$ along $\partial\mathcal{M}$. We write $v_m = v^{\nu}_m \nu_m + v'_m$, $t_m = t^{\nu}_m \nu_m + t'_m$, with $v^{\nu}_m, t^{\nu}_m \in \mathbb{R}$ and $v'_m, t'_m \in T_m\partial\mathcal{M}$.

Let $\mathcal{N}$ be a connected component of $\partial\mathcal{M}$. Assume that the Lopatinskiĭ–Šapiro condition holds for (P, B) at every point of the closed surface[1] *$\mathcal{N}$ and moreover there exists $m_0 \in \mathcal{N}$ such that*

$$|t'_{m_0}|_{g_\partial} \geq |v^{\nu}_{m_0}| \quad \text{and} \quad |v'_{m_0}|_{g_\partial} \geq |t^{\nu}_{m_0}| \tag{2.3.13}$$

and

$$g_{m_0}(t_{m_0}, v_{m_0})^2 < \big(|t'_{m_0}|^2_{g_\partial} - (v^{\nu}_{m_0})^2\big)\big(|v'_{m_0}|^2_{g_\partial} - (t^{\nu}_{m_0})^2\big). \tag{2.3.14}$$

Then, $\mathcal{N}$ is a torus and conditions (2.3.13) *and* (2.3.14) *hold at every point of $\mathcal{N}$. Moreover, t' and v' form a frame for $\mathcal{N}$, that is,* $\operatorname{span}(t'_m, v'_m) = T_m\mathcal{M}$ *for all $m \in \mathcal{N}$.*

EXAMPLE 2.11. An example of manifold that fits the framework of Proposition 2.10 is given by, for $0 < R < r_0$,

$$\mathcal{M} = \{(r, \theta, z);\ (r - r_0)^2 + z^2 \leq R^2\} \subset \mathbb{R}^3,$$

in cylindrical coordinates, equipped with the Euclidean metric inherited from $\mathbb{R}^3$. Then $\partial\mathcal{M}$ is a torus, and for $B = v + it$ with $v_m = \partial_\theta$ and $t_m = -z\partial_r + (r - r_0)\partial_z$ we see that $v^{\nu}_m = t^{\nu}_m = 0$ and $\dim \operatorname{span}(t'_m, v'_m) = 2$.

PROOF OF PROPOSITION 2.10. We proceed by contradiction. Assume that $m_1 \in \mathcal{N}$ is such that

$$|t'_{m_1}|_{g_\partial} < |v^{\nu}_{m_1}| \quad \text{or} \quad |v'_{m_1}|_{g_\partial} < |t^{\nu}_{m_1}|$$

or

$$g_{m_1}(t_{m_1}, v_{m_1})^2 \geq \big(|t'_{m_1}|^2_{g_\partial} - (v^{\nu}_{m_1})^2\big)\big(|v'_{m_1}|^2_{g_\partial} - (t^{\nu}_{m_1})^2\big).$$

[1]By closed surface one means a compact manifold without boundary of dimension two.

Let now $\gamma : [0,1] \to \mathcal{N}$ be a continuous path such that $\gamma(0) = m_0$ and $\gamma(1) = m_1$. Set

$$S = \{s \in [0,1]; (2.3.13)\text{and}(2.3.14) \text{ holds at } \gamma(s') \text{ for all } s' \in [0,s]\}.$$

Set $\overline{s} = \sup S$. We have $\overline{s} < 1$ and, by a continuity argument, (2.3.13) holds at $\overline{m} = \gamma(\overline{s})$ and

$$g_{\overline{m}}(t_{\overline{m}}, v_{\overline{m}})^2 \leq \big(|t'_{\overline{m}}|^2_{g_\partial} - (v^\nu_{\overline{m}})^2\big)\big(|v'_{\overline{m}}|^2_{g_\partial} - (t^\nu_{\overline{m}})^2\big).$$

Since the Lopatinskiĭ–Šapiro condition holds everywhere on $\mathcal{N}$, from Proposition 2.8 we deduce that

$$g_{\overline{m}}(t_{\overline{m}}, v_{\overline{m}})^2 < \big(|t'_{\overline{m}}|^2_{g_\partial} - (v^\nu_{\overline{m}})^2\big)\big(|v'_{\overline{m}}|^2_{g_\partial} - (t^\nu_{\overline{m}})^2\big). \tag{2.3.15}$$

Thus condition (2.3.14) holds for $\overline{s} \leq s \leq \overline{s} + \varepsilon$ for some $\varepsilon > 0$. For any $\overline{s} < s \leq \overline{s} + \varepsilon$ since (2.3.13) and (2.3.14) does not hold this means that

$$\text{either } \ |t'_m|_{g_\partial} < |v^\nu_m| \ \text{ or } \ |v'_m|_{g_\partial} < |t^\nu_m| \qquad \text{for } m = \gamma(s).$$

Thus, either there exists $(s_n)_n$ with $\overline{s} < s_n \leq \overline{s} + \varepsilon$ and $s_n \to \overline{s}$ such that $|t'_{m_n}|_{g_\partial} < |v^\nu_{m_n}|$ for $m_n = \gamma(s_n)$ or there exists $(s'_n)_n$ with $\overline{s} < s'_n \leq \overline{s} + \varepsilon$ and $s'_n \to \overline{s}$ such that $|v'_{m'_n}|_{g_\partial} < |t^\nu_{m'_n}|$ for $m'_n = \gamma(s'_n)$. Since we have

$$|t'_{\overline{m}}|_{g_\partial} \geq |v^\nu_{\overline{m}}| \ \text{ and } \ |v'_{\overline{m}}|_{g_\partial} \geq |t^\nu_{\overline{m}}|,$$

this implies that

$$\text{either } \ |t'_{\overline{m}}|_{g_\partial} = |v^\nu_{\overline{m}}| \ \text{ or } \ |v'_{\overline{m}}|_{g_\partial} = |t^\nu_{\overline{m}}|.$$

Then the r.h.s. of (2.3.15) vanishes at $\overline{m}$ yielding a contradiction. Hence, conditions (2.3.13) and (2.3.14) hold at every point of $\mathcal{N}$.

By Lemma 2.12 given below, used in the case $s = 0$, one finds for every $m \in \mathcal{N}$

$$|(g_\partial)_m(t'_m, v'_m)| < |t'_m|_{g_\partial} |v'_m|_{g_\partial}.$$

This strict estimate in the Cauchy–Schwarz inequality yields $\dim \operatorname{span}(t'_m, v'_m) = 2$.

With t' and v', the closed surface $\mathcal{N}$ exhibits nonvanishing real vector fields. By a classical result of differential topology the Euler characteristic of $\mathcal{N}$ is zero [310, Corollary 39.8]. By the classification of closed surfaces [245, page 30], this leaves two possibilities: $\mathcal{N}$ is either a torus or a Klein bottle.

As $\operatorname{span}(t'_m, v'_m) = T_m\mathcal{M}$ for all $m \in \mathcal{N}$, then the vector fields t' and v' form a frame. In particular, this yields an orientation on the surface $\mathcal{N}$; see Sect. 16.1.1. This rules out the possibility of $\mathcal{N}$ being a Klein bottle as this latter surface is not orientable. ∎

LEMMA 2.12. *Under the assumptions of Proposition 2.10, for $s \in [0,1]$ define the real vector fields $t_{(s)}$ and $v_{(s)}$ by*

$$(t_{(s)})_m = s t^\nu_m \nu_m + t'_m, \qquad (v_{(s)})_m = s v^\nu_m \nu_m + v'_m.$$

Then, for any $m \in \mathcal{N}$ and any $s \in [0,1]$

$$g_m\big((t_{(s)})_m, (v_{(s)})_m\big)^2 < \big(|t'_m|^2_{g_\partial} - (s v^\nu_m)^2\big)\big(|v'_m|^2_{g_\partial} - (s t^\nu_m)^2\big). \tag{2.3.16}$$

PROOF. After simple computations, condition (2.3.14) reads

$$\begin{aligned}(v^\nu_m)^2 |v'_m|^2_{g_\partial} + (t^\nu_m)^2 |t'_m|^2_{g_\partial} + 2 t^\nu_m v^\nu_m (g_\partial)_m(t'_m, v'_m)\\ < |t'_m|^2_{g_\partial} |v'_m|^2_{g_\partial} - (g_\partial)_m(t'_m, v'_m)^2.\end{aligned}$$

With the Cauchy–Schwarz inequality one can estimate the l.h.s. from below as follows

$$\begin{aligned}&(v^\nu_m)^2 |v'_m|^2_{g_\partial} + (t^\nu_m)^2 |t'_m|^2_{g_\partial} + 2 t^\nu_m v^\nu_m (g_\partial)_m(t'_m, v'_m)\\ &\geq (v^\nu_m)^2 |v'_m|^2_{g_\partial} + (t^\nu_m)^2 |t'_m|^2_{g_\partial} - 2|t^\nu_m v^\nu_m| |v'_m|_{g_\partial} |t'_m|_{g_\partial}\\ &\geq \big(|t^\nu_m| |t'_m|_{g_\partial} - |v^\nu_m| |v'_m|_{g_\partial}\big)^2\\ &\geq 0.\end{aligned}$$

Thus, for $s \in [0,1]$

$$\begin{aligned}s^2\big((v^\nu_m)^2 |v'_m|^2_{g_\partial} + (t^\nu_m)^2 |t'_m|^2_{g_\partial} + 2 t^\nu_m v^\nu_m (g_\partial)_m(t'_m, v'_m)\big)\\ < |t'_m|^2_{g_\partial} |v'_m|^2_{g_\partial} - (g_\partial)_m(t'_m, v'_m)^2,\end{aligned}$$

which is precisely (2.3.16). ∎

2.4. Notes

The now called Lopatinskiĭ–Šapiro conditions (with various spellings and author orders) originate from the 1953 articles of Y. B. Lopatinskiĭ [238] and Z. Y. Šapiro [323]. The conditions where further studied in the works of S. Agmon, A. Douglis and L. Nirenberg [3, 4] and M. S. Agranovič and A. S. Dynin [5]. In fact, elliptic problems on a bounded open set or a manifold are well-posed or at least of Fredholm type (finite dimension kernel and co-kernel) if proper boundary conditions are imposed. These conditions are precisely the Lopatinskiĭ–Šapiro boundary conditions. This is recalled in Chap. 3. Dirichlet, Neumann, Robin-type boundary conditions are all particular cases of Lopatinskiĭ–Šapiro conditions in the elliptic case. A good reference for the study of general elliptic operators of arbitrary order along with Lopatinskiĭ–Šapiro conditions is the work of L. Hörmander [175, Section 20.1].

Note that for nonelliptic problems, Lopatinskiĭ–Šapiro conditions do not provide all conditions under which a problem is well-posed. For example, the mixed boundary value problem for the wave equation with Cauchy data at $t = 0$ and a homogeneous Neumann boundary condition for $(t,m) \in [0,+\infty) \times \partial\mathcal{M}$ is well-posed. However, the Neumann boundary condition $\partial_\nu u_{|[0,+\infty)\times\partial\mathcal{M}} = 0$ fails to satisfy the Lopatinskiĭ–Šapiro criterium in the case of the wave operator $\partial_t^2 - \Delta_g$.

Here, we only treat the case of second-order elliptic operators as in the whole book. Lopatinskiĭ–Šapiro conditions are stated in Definition 2.2 in the form of an algebraic criterium and given in the simple form $\check{b}(m, \omega', i|\omega'|_g) \neq 0$ in Proposition 2.3. In Sect. 2.3 we fully describe first-order boundary operators that fulfill the Lopatinskiĭ–Šapiro conditions. For higher-order elliptic operators the Lopatinskiĭ–Šapiro conditions are stated in Remark 2.7. The conditions given therein can in fact be written as a rank condition for some matrix built from the principal symbols of the elliptic operator and the boundary operator.

CHAPTER 3

Fredholm Properties of Second-Order Elliptic Operators

Contents

J. Le Rousseau et al., *Elliptic Carleman Estimates and Applications to Stabilization and Controllability, Volume II*, PNLDE Subseries in Control 98, https://doi.org/10.1007/978-3-030-88670-7_3

3.1. Setting and Main Result

On a smooth compact Riemannian manifold $(\mathcal{M}, g)$ with boundary, we set $P = -\Delta_g + R_1$, where Δ_g denotes the Laplace–Beltrami operator whose form is recalled in Sect. 17.2 and R_1 is a first-order differential operator on $\mathcal{M}$.

We consider a differential boundary operator B of order $\beta \in \mathbb{N}$ defined on $\partial\mathcal{M}$ (or equivalently, in a neighborhood of $\partial\mathcal{M}$, see Remark 2.1) with differentiation transverse to $\partial\mathcal{M}$ of order less than or equal to one. If $\partial\mathcal{M}$ is not connected, the order of the operator may vary from one connected component to the other. Recall that the number of connected components of $\partial\mathcal{M}$ is finite since $\mathcal{M}$ is compact. Consider $\mathcal{N}$ a connected component of $\partial\mathcal{M}$ where B is of order k; then on $\mathcal{N}$ the boundary operator takes the form $B = B^k + B^{k-1}\partial_\nu$, where both B^k and B^{k-1} are differential operators on $\mathcal{N}$ of order k and $k-1$, respectively. If $u \in H^2(\partial\mathcal{M})$, then $u_{|\mathcal{N}} \in H^{3/2}(\mathcal{N})$ and $\partial_\nu u_{|\mathcal{N}} \in H^{1/2}(\mathcal{N})$ by the trace formulae of Theorem 18.25. With the usual continuity result for differential operators recalled in Proposition 18.23, one then finds

$$Bu_{|\mathcal{N}} = B^k u_{|\mathcal{N}} + B^{k-1}\partial_\nu u_{|\mathcal{N}} \in H^{3/2-k}(\mathcal{N}). \tag{3.1.1}$$

We denote by $b(m, \omega)$, $m \in \partial\mathcal{M}$ and $\omega \in T^*_m\mathcal{M}$, its principal symbol. We shall be interested in cases where the Lopatinskiĭ–Šapiro condition of Definition 2.2 holds on the whole $\partial\mathcal{M}$. If B is of order zero on a connected component of $\partial\mathcal{M}$, then the condition is of Dirichlet type by Example 2.5-(1). The elliptic problem with this type of condition on the whole $\partial\mathcal{M}$ is treated in Sect. 18.6, partly with the analysis based on that of Section 10.1 of Volume 1 in the case $P = \sum_{1\leq i,j\leq d} D_i(p^{ij}(x)D_j)$ in a regular open set of $\mathbb{R}^d$. There, we invoke results proven by means of a variational formulation of the elliptic problem. The analysis proposed in the present chapter applies to this case and to all boundary conditions that fit the Lopatinskiĭ–Šapiro framework. However, we do not prove well-posedness here (as it may not hold) but rather consider Fredholm properties.

On each connected component of $\partial\mathcal{M}$, the differential boundary operator B is of order $k \in \{0, \dots, \beta\}$. For each $k \in \{0, \dots, \beta\}$, we denote by ${}^k\partial\mathcal{M}$ the union of all the connected components of $\partial\mathcal{M}$ where the order of B is exactly k. For $r \in \mathbb{R}$, we set

$$H^{(r)}_B(\partial\mathcal{M}) = \bigoplus_{0\leq k\leq\beta} H^{r-k}({}^k\partial\mathcal{M}). \tag{3.1.2}$$

Observe that ${}^k\partial\mathcal{M}$ may possibly be empty for some $k \in \{0, \dots, \beta\}$.

For $m \in \mathbb{N}$, we study the Fredholm property of the operator

$$\begin{aligned} (3.1.3) \qquad L : H^{m+2}(\mathcal{M}) &\to H^m(\mathcal{M}) \oplus H_B^{(m+3/2)}(\partial\mathcal{M}) \\ u &\mapsto (Pu, Bu_{|^0\partial\mathcal{M}}, \dots, Bu_{|^\beta\partial\mathcal{M}}), \end{aligned}$$

which is bounded by (3.1.1). Fredholm operators are recalled in Section 11.5 of Volume 1. Having the case of the Neumann boundary condition in mind, this is quite natural since in such case the map L is neither injective nor surjective, but rather Fredholm of index zero, and well-posedness only makes sense after applying quotients with respect to finite dimensional subspaces; see Sect. 18.7.

If ${}^1\partial\mathcal{M} \neq \emptyset$, basic examples of boundary operators of order one for which the Lopatinskiĭ–Šapiro condition holds are given in Examples 2.5-(2) to 2.5-(3) and the general case is analyzed in Sect. 2.3. If ${}^2\partial\mathcal{M} \neq \emptyset$, an example of a boundary operator of order two is given in Example 2.5-(4).

The main result we prove in this chapter is the following.

THEOREM 3.1. *The operator L is Fredholm if and only if (P, B) fulfills the Lopatinskiĭ–Šapiro condition on $\partial\mathcal{M}$.*

To prove Theorem 3.1, we shall first establish the following result. By Theorem 11.7 of Volume 1, this implies that the Lopatinskiĭ–Šapiro condition is sufficient for the Fredholm property of L to hold.

PROPOSITION 3.2. *Let $m \in \mathbb{N}$. Assume that (P, B) fulfills the Lopatinskiĭ–Šapiro condition on $\partial\mathcal{M}$. There exists a bounded linear operator*

$$M : H^m(\mathcal{M}) \oplus H_B^{(m+3/2)}(\partial\mathcal{M}) \to H^{m+2}(\mathcal{M})$$

such that

$$ML = \mathrm{Id}_{H^{m+2}(\mathcal{M})} + K^\ell \quad \text{and} \quad LM = \mathrm{Id}_{H^m(\mathcal{M}) \oplus H_B^{(m+3/2)}(\partial\mathcal{M})} + K^r,$$

where both operators

$$\begin{aligned} &K^\ell : H^{m+2}(\mathcal{M}) \to H^{m+3}(\mathcal{M}) \\ &K^r : H^m(\mathcal{M}) \oplus H_B^{(m+3/2)}(\partial\mathcal{M}) \to H^{m+1}(\mathcal{M}) \oplus H_B^{(m+5/2)}(\partial\mathcal{M}) \end{aligned}$$

are bounded.

By the Rellich–Kondrachov theorem (see Theorem 18.7), K^ℓ is compact from $H^{m+2}(\mathcal{M})$ into itself and K^r is compact from $H^m(\mathcal{M}) \oplus H_B^{(m+3/2)}(\partial\mathcal{M})$ into itself.

The proof of Proposition 3.2 is performed in Sect. 3.3 based on the analysis in a half-space of Sect. 3.2. The second part of Theorem 3.1, that is, the necessity of Lopatinskiĭ–Šapiro conditions, is proven in Sect. 3.4.

Theorem 3.1 states in particular that L is not Fredholm if the Lopatinskiĭ–Šapiro condition does not hold. This is illustrated in Example 2.6

A corollary of Theorem 3.1 is the following one.

COROLLARY 3.3. *Assume that (P, B) fulfills the Lopatinskiĭ–Šapiro condition on $\partial\mathcal{M}$, and let $m \in \mathbb{N}$. Then, there exists $C > 0$ such that*

$$\|u\|_{H^{m+2}(\mathcal{M})} \leq C\big(\|Pu\|_{H^m(\mathcal{M})} + |Bu_{|\partial\mathcal{M}}|_{H_B^{(m+3/2)}(\partial\mathcal{M})} + \|u\|_{L^2(\mathcal{M})}\big),$$

for all $u \in H^{m+2}(\mathcal{M})$.

A proof is given in Appendix 3.A.1. Note that this result gives a precise form for operator K_1 and the space Z_1 in the first inequality in Proposition 11.10 of Volume 1 if applied to the present case.

The following proposition states that the Fredholm index of L is independent of the chosen value of m. This explains why we chose to write L in place of $L^{(m)}$.

PROPOSITION 3.4. *The values of* $\operatorname{nul} L$, $\operatorname{def} L$, *and* $\operatorname{ind}(L)$ *are independent of the chosen value $m \in \mathbb{N}$. Note that in fact* $\ker(L)$ *is independent of m.*

We refer to Appendix 3.A.2 for a proof.

In Sect. 3.5 in the case $\beta \leq 1$, we compute the index of L in some cases: in dimension $d \geq 3$ the index is zero, and in dimension $d = 2$ this is not always the case.

3.2. Analysis in a Half-Space

3.2.1. Local Setting Near a Point of the Boundary. At a point $m^0 \in \partial\mathcal{M}$, we consider a local chart $\mathcal{C} = (O, \kappa)$ characterized by normal geodesic coordinates; see Sect. 17.6. We recall that $\kappa(O)$ is an open set of $\overline{\mathbb{R}^d_+}$ and

$$\kappa(\partial\mathcal{M} \cap O) = \{x_d = 0\} \cap \kappa(O),$$

with $x = (x', x_d)$, $x' \in \mathbb{R}^{d-1}$ and $x_d \in \mathbb{R}$. The local representative of P takes the form

$$P^{\mathcal{C}}(x, D) = D_d^2 + R(x, D') + r_0(x) D_d,$$

where $R(x, D') \in \mathscr{D}^2_{\mathsf{T}}$ is a tangential differential operator with x_d as a parameter and r_0 is a smooth function. Note that O meets only one component of $\partial\mathcal{M}$. Thus, $O \cap \partial\mathcal{M} \subset {}^k\partial\mathcal{M}$ for some $k \in \{0, \ldots, \beta\}$.

In what follows, we drop the superscript $\mathcal{C}$ since there is no possible confusion. We thus write $P(x, D)$ and $B(x, D)$ in place of $P^{\mathcal{C}}(x, D)$ and $B^{\mathcal{C}}(x, D)$, respectively.

If we denote by $r(x, \xi')$ the principal symbol of $R(x, D')$, in the sense of standard operators (see Section 2.12 of Volume 1), then $\xi' \mapsto r(x, \xi')$ is a positive definite quadratic form and, for some $C > 0$,

$$r(x, \xi') \geq C|\xi'|^2, \qquad x \in \kappa(O), \xi' \in \mathbb{R}^{d-1}.$$

For convenience, the coefficients of the operator $R(x, D')$ and the function $r_0(x)$ are extended to $\mathbb{R}^d$, yet preserving the elliptic property[1] of $R(x, D')$. We pick $\rho(x, \xi') \in S^1_{\mathsf{T}}$ such that $\rho(x, \xi') > 0$ and

$$\rho(x, \xi') = r(x, \xi')^{1/2} \qquad \text{for } |\xi'| \geq 1.$$

REMARK 3.5. As a reminder, on the manifold level, for $|\xi'| \geq 1$, we have $\rho(x, \xi') = |\omega'|_g$, if $x = (x', 0) = \kappa(m)$ with $m \in \partial\mathcal{M}$ and if $(\xi', 0)$ is the local representative of $\omega' \in T^*_m(\partial\mathcal{M})$. This is useful below to write the Lopatinskiĭ–Šapiro condition in the local coordinates.

The principal symbol of $P(x, D)$ is given by

$$p(x, \xi) = \xi_d^2 + r(x, \xi'),$$

and one has $p(x, \xi) \gtrsim |\xi|^2$ for $x \in \mathbb{R}^d$, $\xi \in \mathbb{R}^d$. For $|\xi'| \geq 1$, we have

$$p(x, \xi) = \big(\xi_d - i\rho(x, \xi')\big)\big(\xi_d + i\rho(x, \xi')\big). \tag{3.2.1}$$

As mentioned above, $O \cap \partial\mathcal{M} \subset {}^k\partial\mathcal{M}$ for some $k \in \{0, \dots, \beta\}$, meaning that the boundary operator B is exactly of order k in $O \cap \partial\mathcal{M}$. In the chosen local coordinates, we may write, in a neighborhood of (or at) the boundary,

$$B(x, D) = B^k(x, D') - iB^{k-1}(x, D')D_d, \tag{3.2.2}$$

with $B^k(x, D')$ and $B^{k-1}(x, D')$ differential operators in the x' variables of order k and $k - 1$, respectively, with principal symbols $b^k(x, \xi') \in S^k_{\mathsf{T}}$ and $b^{k-1}(x, \xi') \in S^{k-1}_{\mathsf{T}}$. Note that if $k = 0$, we take $b^1(x, \xi') = 1$ and $b^{-1}(x, \xi') = 0$ as it may be obtained upon division by a suitable nonvanishing function since the Lopatinskiĭ–Šapiro condition is assumed to hold; see Example 2.5-(1).

According to Proposition 2.3 and Remark 3.5, the Lopatinskiĭ–Šapiro condition reads

$$\big(b^k + b^{k-1}\rho\big)_{|x_d=0^+}(x', \xi') \neq 0, \tag{3.2.3}$$

for all $(x', 0) \in \kappa(O \cap \partial\mathcal{M})$, and all $\xi' \in \mathbb{R}^{d-1}$, with $|\xi'| \geq 1$.

3.2.2. Action of a Parametrix on a Half-Space. Let $u \in H^{m+2}(\mathbb{R}^d_+)$ and set $f = P(x, D)u$. Denoting by $Y = Y(x_d)$ the Heaviside function in the x_d variable, we set $\underline{u} = Yu$ and $\underline{f} = Yf$; we have $\underline{u}, \underline{f} \in L^2(\mathbb{R}^d)$. We also set $\gamma^{\mathsf{D}}(u) = u_{|x_d=0}$ and $\gamma^{\mathsf{N}}(u) = \partial_\nu u_{|x_d=0} = -\partial_d u_{|x_d=0} = -iD_d u_{|x_d=0}$. Observing that

$$D_d\underline{u} = -i\delta_{|x_d=0} \otimes \gamma^{\mathsf{D}}(u) + YD_d u$$
$$D_d^2\underline{u} = -\delta'_{x_d=0} \otimes \gamma^{\mathsf{D}}(u) + \delta_{x_d=0} \otimes \gamma^{\mathsf{N}}(u) + YD_d^2 u,$$

[1] Such an extension is carried out explicitly at the beginning of the proof of Theorem 2.28 in Section 2.A.6.1 of Volume 1.

we obtain

$$P(x,D)\underline{u} = \underline{f} - \delta'_{x_d=0} \otimes \gamma^{\mathsf{D}}(u) + \delta_{x_d=0} \otimes \big(\gamma^{\mathsf{N}}(u) - ir_0\gamma^{\mathsf{D}}(u)\big),$$

in the sense of distributions in $\mathbb{R}^d$. This leads us to define

$$\tilde{\gamma}^{\mathsf{N}} = \gamma^{\mathsf{N}} - ir_0\gamma^{\mathsf{D}},$$

which is bounded from $H^{m+2}(\mathbb{R}^d_+)$ into $H^{m+1/2}(\mathbb{R}^{d-1})$. We thus have

$$P(x,D)\underline{u} = \underline{f} - \delta'_{x_d=0} \otimes \gamma^{\mathsf{D}}(u) + \delta_{x_d=0} \otimes \tilde{\gamma}^{\mathsf{N}}(u).$$

We now use a parametrix of $P(x,D)$ in $\mathbb{R}^d$. By Proposition 2.33 of Volume 1 (adapted to standard operators), as $p(x,\xi) \gtrsim |\xi|^2$ for $(x,\xi) \in \mathbb{R}^d \times \mathbb{R}^d$, there exists $q(x,\xi) \in S^{-2}(\mathbb{R}^d \times \mathbb{R}^d)$, with $p(x,\xi)^{-1}$ as principal symbol for $|\xi| \geq 1$, such that $Q = \operatorname{Op}(q) \in \Psi^{-2}(\mathbb{R}^d)$ and

$$QP = \operatorname{Id} - R \quad \text{and} \quad PQ = \operatorname{Id} - R', \tag{3.2.4}$$

with $R, R' \in \Psi^{-\infty}(\mathbb{R}^d) = \cap_{N\in\mathbb{N}}\Psi^{-N}(\mathbb{R}^d)$.

This yields

$$\underline{u} = Q\underline{f} - Q(\delta'_{x_d=0} \otimes \gamma^{\mathsf{D}}(u)) + Q\big(\delta_{x_d=0} \otimes \tilde{\gamma}^{\mathsf{N}}(u)\big) + R\underline{u}. \tag{3.2.5}$$

We define

$$\begin{aligned} \mathscr{Q} : L^2(\mathbb{R}^d_+) &\to H^2(\mathbb{R}^d_+) \\ f &\mapsto \big(Q\underline{f}\big)_{|\mathbb{R}^d_+}. \end{aligned} \tag{3.2.6}$$

It is well defined and continuous since $Q\underline{f} \in H^2(\mathbb{R}^d)$ and the restriction of a H^2-function in $\mathbb{R}^d$ to $\mathbb{R}^d_+$ yields (continuously) a function in $H^2(\mathbb{R}^d_+)$. Thus,

$$\|\mathscr{Q}(f)\|_{H^2(\mathbb{R}^d_+)} \lesssim \|f\|_{L^2(\mathbb{R}^d_+)}.$$

For $h \in \mathscr{S}'(\mathbb{R}^{d-1})$ and $j \in \{-1,0\}$, we also define

$$\mathscr{Q}^{(j)}(h) = \big(Q(\delta^{(j+1)}_{x_d=0} \otimes h)\big)_{|\mathbb{R}^d_+} \in \mathscr{D}'(\mathbb{R}^d_+), \tag{3.2.7}$$

where $\delta^{(0)}_{x_d=0} = \delta_{x_d=0}$ and $\delta^{(1)}_{x_d=0} = \delta'_{x_d=0}$. As Q maps $\mathscr{S}'(\mathbb{R}^d)$ into itself, the operators $\mathscr{Q}^{(j)}$ are well defined. From (3.2.5), we have

$$u = \mathscr{Q}(f) - \mathscr{Q}^{(0)}(\gamma^{\mathsf{D}}(u)) + \mathscr{Q}^{(-1)}(\tilde{\gamma}^{\mathsf{N}}(u)) + \big(R\underline{u}\big)_{|\mathbb{R}^d_+}. \tag{3.2.8}$$

The following lemma improves the bound of $\mathscr{Q}$ and provides bounds for the operators $\mathscr{Q}^{(j)}$, for $j = -1, 0$.

LEMMA 3.6. *Let $m \in \mathbb{N}$.*

(1) *The map $\mathscr{Q}$ is bounded from $H^m(\mathbb{R}^d_+)$ into $H^{m+2}(\mathbb{R}^d_+)$ and*

$$P(x,D)\mathscr{Q} = \operatorname{Id} + K,$$

with K bounded from $H^m(\mathbb{R}^d_+)$ into $H^N(\mathbb{R}^d_+)$ for any $N \in \mathbb{N}$.

(2) *If* $j \in \{-1, 0\}$, *the map* $\mathscr{Q}^{(j)}$ *is bounded from* $H^{m+j-1/2}(\mathbb{R}^{d-1})$ *into* $H^m(\mathbb{R}^d_+)$ *and*

$$P(x, D)\mathscr{Q}^{(j)} = K^{(j)},$$

with $K^{(j)}$ *bounded from* $H^{m+j-1/2}(\mathbb{R}^{d-1})$ *into* $H^N(\mathbb{R}^d_+)$ *for any* $N \in \mathbb{N}$.

We refer to Appendix 3.A.3 for a proof.

REMARK 3.7. Note that we shall first use that $\mathscr{Q}^{(j)}$ is bounded from $H^{m+j+3/2}(\mathbb{R}^{d-1})$ into $H^{m+2}(\mathbb{R}^d_+)$. However, the finer result of Lemma 3.6 is of use in Sect. 3.6.

3.2.3. The Calderón Projector. Let $\phi \in H^m(\mathbb{R}^d_+)$ and $h \in H^{m+j+3/2}(\mathbb{R}^{d-1})$ for $j \in \{-1, 0\}$. With the regularity obtained in Lemma 3.6, that is, $\mathscr{Q}(\phi), \mathscr{Q}^{(j)}(h) \in H^{m+2}(\mathbb{R}^{d-1})$, the traces of $\mathscr{Q}(\phi)$, $\mathscr{Q}^{(j)}(h)$ can be considered at $x_d = 0^+$. With the trace formula of Theorem 18.25, we may thus introduce the following bounded maps:

(3.2.9)

$$\begin{aligned} \mathscr{Q}_{\mathsf{D}} : H^m(\mathbb{R}^d_+) &\to H^{m+3/2}(\mathbb{R}^{d-1}) \\ \phi &\mapsto \gamma^{\mathsf{D}}\big(\mathscr{Q}(\phi)\big), \end{aligned} \qquad \begin{aligned} \mathscr{Q}_{\mathsf{N}} : H^m(\mathbb{R}^d_+) &\to H^{m+1/2}(\mathbb{R}^{d-1}) \\ \phi &\mapsto \tilde{\gamma}^{\mathsf{N}}\big(\mathscr{Q}(\phi)\big), \end{aligned}$$

and, for $j \in \{-1, 0\}$,

$$\begin{aligned} \mathscr{Q}^{(j)}_{\mathsf{D}} : H^{m+j+3/2}(\mathbb{R}^{d-1}) &\to H^{m+3/2}(\mathbb{R}^{d-1}) \\ h &\mapsto \gamma^{\mathsf{D}}\big(\mathscr{Q}^{(j)}(h)\big), \end{aligned} \tag{3.2.10}$$

and

$$\begin{aligned} \mathscr{Q}^{(j+1)}_{\mathsf{N}} : H^{m+j+3/2}(\mathbb{R}^{d-1}) &\to H^{m+1/2}(\mathbb{R}^{d-1}) \\ h &\mapsto \tilde{\gamma}^{\mathsf{N}}\big(\mathscr{Q}^{(j)}(h)\big). \end{aligned} \tag{3.2.11}$$

The shift $j \to j+1$ in the definition of $\mathscr{Q}^{(j+1)}_{\mathsf{N}}$ is to be justified in Sect. 3.2.4. In fact, there one proves that $\mathscr{Q}^{(k)}_{\mathsf{N}} \in \Psi^k(\mathbb{R}^{d-1})$, $k = 0, 1$. We also have $\mathscr{Q}^{(j)}_{\mathsf{D}} \in \Psi^j(\mathbb{R}^{d-1})$, $j = -1, 0$.

We also introduce the following operators:

$$\mathsf{C}^{\mathsf{D},\mathsf{D}} = -\mathscr{Q}^{(0)}_{\mathsf{D}}, \quad \mathsf{C}^{\mathsf{D},\mathsf{N}} = \mathscr{Q}^{(-1)}_{\mathsf{D}}, \quad \mathsf{C}^{\mathsf{N},\mathsf{D}} = -\mathscr{Q}^{(1)}_{\mathsf{N}}, \quad \mathsf{C}^{\mathsf{N},\mathsf{N}} = \mathscr{Q}^{(0)}_{\mathsf{N}}, \tag{3.2.12}$$

and the matrix operator

$$\mathsf{C} = \begin{pmatrix} \mathsf{C}^{\mathsf{D},\mathsf{D}} & \mathsf{C}^{\mathsf{D},\mathsf{N}} \\ \mathsf{C}^{\mathsf{N},\mathsf{D}} & \mathsf{C}^{\mathsf{N},\mathsf{N}} \end{pmatrix}, \tag{3.2.13}$$

which is bounded from $H^{m+3/2}(\mathbb{R}^{d-1}) \oplus H^{m+1/2}(\mathbb{R}^{d-1})$ into itself. From (3.2.8), computing the Dirichlet and Neumann traces at $x_d = 0^+$, we obtain the following identity:

$$(\operatorname{Id} - \mathsf{C}) \begin{pmatrix} \gamma^{\mathsf{D}}(u) \\ \tilde{\gamma}^{\mathsf{N}}(u) \end{pmatrix} = \begin{pmatrix} \mathscr{Q}_{\mathsf{D}}(f) \\ \mathscr{Q}_{\mathsf{N}}(f) \end{pmatrix} + R_{\mathsf{C}}(u), \tag{3.2.14}$$

for $u \in H^{m+2}(\mathbb{R}^d_+)$ and $f = P(x,D)u$, and where R_{C} is bounded from $H^{m+2}(\mathbb{R}^d_+)$ into $\big(H^N(\mathbb{R}^{d-1})\big)^2$, for any $N \in \mathbb{N}$.

The following properties hold.

PROPOSITION 3.8. *Let $m \in \mathbb{N}$. The matrix operator $\mathsf{C}^2 - \mathsf{C}$ maps $H^{m+3/2}(\mathbb{R}^{d-1}) \oplus H^{m+1/2}(\mathbb{R}^{d-1})$ into $\big(H^N(\mathbb{R}^{d-1})\big)^2$, for any $N \in \mathbb{N}$. The map*

$$\phi \mapsto \mathsf{C}\begin{pmatrix} \mathscr{Q}_{\mathsf{D}}(\phi) \\ \mathscr{Q}_{\mathsf{N}}(\phi) \end{pmatrix} \tag{3.2.15}$$

is bounded from $H^m(\mathbb{R}^d_+)$ into $\big(H^N(\mathbb{R}^{d-1})\big)^2$, for any $N \in \mathbb{N}$.

Since $\mathsf{C}^2 = \mathsf{C}$ up to a regularizing operator, one says that C is a projector. It is often referred to as the Calderón projector associated with the elliptic operator $P(x,D)$.

Observe that the regularizing aspect of the map (3.2.15) can easily be perceived upon applying the operator C to identity (3.2.14) and using the projector property of C.

We refer to Appendix 3.A.4 for a proof of Proposition 3.8.

From Proposition 3.8, computing $\mathsf{C}^2 - \mathsf{C}$, we find that the operators

$$\mathsf{C}^{\mathsf{D},\mathsf{N}}\mathsf{C}^{\mathsf{N},\mathsf{D}} + \mathsf{C}^{\mathsf{D},\mathsf{D}}(\mathsf{C}^{\mathsf{D},\mathsf{D}} - \mathrm{Id}) : H^{m+3/2}(\mathbb{R}^{d-1}) \to H^N(\mathbb{R}^{d-1}) \tag{3.2.16}$$

and

$$\mathsf{C}^{\mathsf{D},\mathsf{D}}\mathsf{C}^{\mathsf{D},\mathsf{N}} + \mathsf{C}^{\mathsf{D},\mathsf{N}}(\mathsf{C}^{\mathsf{N},\mathsf{N}} - \mathrm{Id}) : H^{m+1/2}(\mathbb{R}^{d-1}) \to H^N(\mathbb{R}^{d-1}) \tag{3.2.17}$$

are bounded for any $N \in \mathbb{N}$.

3.2.4. Pseudo-Differential Symbol Computations. With the following lemma, we further detail the structures of the operators $\mathscr{Q}^{(j)}$, for $j = -1, 0$.

LEMMA 3.9. *Let $h \in \mathscr{S}(\mathbb{R}^{d-1})$. For $x_d \geq 0$, we have*

$$\mathscr{Q}^{(-1)}(h)(x) = Q(\delta_{x_d=0} \otimes h)(x_d, x') = \mathrm{Op}\big(q^{(-1)}(x_d)\big)h(x'),$$

and

$$\mathscr{Q}^{(0)}(h)(x) = Q(\delta'_{x_d=0} \otimes h)(x_d, x') = \mathrm{Op}\big(q^{(0)}(x_d)\big)h(x'),$$

with $q^{(-1)}(x_d) = t^{(-1)}(x_d)e(x_d)$ and $q^{(0)}(x_d) = t^{(0)}(x_d)e(x_d)$ such that

(1) *$e(x_d)(x',\xi') = e^{-x_d\rho(x,\xi')}$.*

(2) *$t^{(-1)}(x_d)$ and $t^{(0)}(x_d)$ are smooth with values in $S^{-1}(\mathbb{R}^{d-1} \times \mathbb{R}^{d-1})$ and $S^0(\mathbb{R}^{d-1} \times \mathbb{R}^{d-1})$, respectively.*

(3) *Their principal symbols are*

$$t^{(-1)}_{-1}(x_d)(x',\xi') = \frac{1}{2}\rho^{-1}(x,\xi'), \qquad t^{(0)}_0(x_d)(x',\xi') = -\frac{1}{2},$$

respectively.

A proof of Lemma 3.9 is given in Appendix 3.A.5.

Observe that $e(x_d) \in S^0(\mathbb{R}^{d-1} \times \mathbb{R}^{d-1})$. However, differentiation with respect to x_d affects the symbol order; $e(x_d)(x', \xi')$ is thus not a tangential symbol in the sense given in Section 2.12 of Volume 1. The same applies to $q^{(-1)}(x_d)$ and $q^{(0)}(x_d)$. There is however no obstruction to consider the pseudo-differential operators $\operatorname{Op}\big(q^{(-1)}(x_d)\big)$ and $\operatorname{Op}\big(q^{(0)}(x_d)\big)$ as in Lemma 3.9.

LEMMA 3.10. *We have the following properties:*

(1) *For $x_d \geq 0$, $x_d \mapsto e(x_d)$ is bounded with values in $S^0(\mathbb{R}^{d-1} \times \mathbb{R}^{d-1})$ and continuous with values in $S^1(\mathbb{R}^{d-1} \times \mathbb{R}^{d-1})$.*

(2) *Moreover, there exists $C_0 > 0$ such that the map $x_d \mapsto e^{C_0 x_d} e(x_d)$ with values in $S^0(\mathbb{R}^{d-1} \times \mathbb{R}^{d-1})$ is bounded.*

(3) *For $n \in \mathbb{N}$ and $x_d \geq 0$, $\partial_d^n e(x_d) = a(x_d) e(x_d)$ and $x_d \mapsto a(x_d)$ is smooth with values in $S^n(\mathbb{R}^{d-1} \times \mathbb{R}^{d-1})$ and $x_d \mapsto \partial_d^n e(x_d)$ is bounded with values in $S^n(\mathbb{R}^{d-1} \times \mathbb{R}^{d-1})$.*

In Sect. 3.2.3, we saw that with the regularity obtained in Lemma 3.6, the traces of $\mathscr{Q}^{(-1)}(h)$ and $\mathscr{Q}^{(0)}(h)$ can be considered at $x_d = 0^+$. With oscillatory integrals as defined in Section 2.4 of Volume 1 and with Lemma 3.9, the action of the operator $\mathscr{Q}^{(j)}$, $j = -1, 0$, reads

$$
\begin{aligned}
\mathscr{Q}^{(j)}(h)(x) &= \operatorname{Op}\big(q^{(j)}(x_d)\big) h(x') \\
&= (2\pi)^{1-d} \iint_{\mathbb{R}^{d-1} \times \mathbb{R}^{d-1}} e^{i(x'-y')\cdot\xi'} q^{(j)}(x_d)(x', \xi') h(y') dy' d\xi'.
\end{aligned}
$$

With Remark 2.13 in Volume 1, oscillatory integral regularization allows one to perform limits or differentiation under the sum sign. With Lemma 3.10, we thus obtain

(3.2.18)
$$
\mathscr{Q}_{\mathsf{D}}^{(j)}(h) = \big(\operatorname{Op}\big(q^{(j)}(x_d)\big) h\big)_{|x_d=0^+}(x') = \operatorname{Op}\big(q_{\mathsf{D}}^{(j)}\big) h(x'), \quad j = -1, 0,
$$

with $q_{\mathsf{D}}^{(j)} = t^{(j)}(0) = q^{(j)}(0)$, with principal symbols given by

$$
\begin{aligned}
q_{\mathsf{D},-1}^{(-1)}(0)(x', \xi') &= \frac{1}{2} \rho_{|x_d=0^+}^{-1}(x', \xi') \in S^{-1}(\mathbb{R}^{d-1} \times \mathbb{R}^{d-1}), \\
q_{\mathsf{D},-0}^{(0)}(0)(x', \xi') &= -\frac{1}{2} \in S^0(\mathbb{R}^{d-1} \times \mathbb{R}^{d-1}).
\end{aligned}
$$

We also obtain

(3.2.19)
$$
\begin{aligned}
\mathscr{Q}_{\mathsf{N}}^{(k)}(h) &= \tilde{\gamma}^{\mathsf{N}} \operatorname{Op}\big(q^{(k-1)}(x_d) h \\
&= -i\big(D_d \operatorname{Op}\big(q^{(k-1)}(x_d)\big) h\big)_{|x_d=0^+}(x') \\
&\quad - i r_0 \big(\operatorname{Op}\big(q^{(k-1)}(x_d)\big) h\big)_{|x_d=0^+}(x') \\
&= \operatorname{Op}\big(q_{\mathsf{N}}^{(k)}\big) h(x'),
\end{aligned}
$$

for $k = 0, 1$, where $q_{\mathsf{N}}^{(k)} \in S^k(\mathbb{R}^{d-1} \times \mathbb{R}^{d-1})$ with principal symbols given by

$$q_{\mathsf{N},0}^{(0)}(x', \xi') = \frac{1}{2} \in S^0(\mathbb{R}^{d-1} \times \mathbb{R}^{d-1}),$$
$$q_{\mathsf{N},1}^{(1)}(x', \xi') = -\frac{1}{2}\rho_{|x_d=0^+}(x', \xi') \in S^1(\mathbb{R}^{d-1} \times \mathbb{R}^{d-1}).$$

With the above characterizations, we find that the Calderón projector C is a matrix pseudo-differential operator. Its matrix symbol is given by

$$\begin{pmatrix} -q_{\mathsf{D}}^{(0)}(0) & q_{\mathsf{D}}^{(-1)}(0) \\ -q_{\mathsf{N}}^{(1)}(0) & q_{\mathsf{N}}^{(0)}(0) \end{pmatrix} (x', \xi'),$$

yielding the principal symbol

$$\mathsf{c}(x', \xi') = \frac{1}{2} \begin{pmatrix} 1 & \rho_{|x_d=0^+}^{-1} \\ \rho_{|x_d=0^+} & 1 \end{pmatrix} (x', \xi'),$$

and one can check that $\mathsf{c}(x', \xi')^2 = \mathsf{c}(x', \xi')$, in agreement with Proposition 3.8.

3.2.5. Expressing All Traces from the Boundary Condition. In the considered chart $\mathcal{C} = (O, \kappa)$, the boundary operator B of order k and P fulfill the Lopatinskiĭ–Šapiro condition in $\kappa(\partial\mathcal{M} \cap O)$. We consider $B^k(x, D')$ and $B^{k-1}(x, D')$ as in (3.2.2). We denote $B^k(x', D') = B^k(x', x_d = 0, D')$ and $B^{k-1}(x', D') = B^{k-1}(x', x_d = 0, D')$, with coefficients smoothly extended in the whole $\mathbb{R}^{d-1}$, preserving the operator orders. Similarly, we set $r_0(x') = r_0(x', x_d = 0)$. We then set

$$L_\partial = \begin{pmatrix} \mathrm{Id} - \mathsf{C}^{\mathsf{D},\mathsf{D}} & -\mathsf{C}^{\mathsf{D},\mathsf{N}} \\ B^k(x', D') + iB^{k-1}(x', D')r_0(x') & B^{k-1}(x', D') \end{pmatrix}. \tag{3.2.20}$$

We set $H^{r,s}(\mathbb{R}^{d-1}) = H^r(\mathbb{R}^{d-1}) \oplus H^s(\mathbb{R}^{d-1})$. We see that L_∂ is a bounded operator from $H^{s,s-1}(\mathbb{R}^{d-1})$ into $H^{s,s-k}(\mathbb{R}^{d-1})$, for any $s \in \mathbb{R}$.

We choose $\chi_\partial \in \mathscr{C}_c^\infty(\mathbb{R}^{d-1})$ such that $\operatorname{supp}(\chi_\partial) \subset \kappa(\partial\mathcal{M} \cap O)$.

LEMMA 3.11. *There exist operators M_∂, K_∂^ℓ, and K_∂^r*

$$M_\partial : H^{s,s-k}(\mathbb{R}^{d-1}) \to H^{s,s-1}(\mathbb{R}^{d-1}),$$
$$K_\partial^\ell : H^{s,s-1}(\mathbb{R}^{d-1}) \to H^{s+1,s}(\mathbb{R}^{d-1}),$$
$$K_\partial^r : H^{s,s-k}(\mathbb{R}^{d-1}) \to H^{s+1,s+1-k}(\mathbb{R}^{d-1}),$$

which are bounded for any $s \in \mathbb{R}$, and such that the following hold:

(1) *For $U_\partial \in H^{s,s-1}(\mathbb{R}^{d-1})$ and $F_\partial \in H^{s,s-k}(\mathbb{R}^{d-1})$, we have*

$$M_\partial L_\partial U_\partial = \chi_\partial U_\partial + K_\partial^\ell U_\partial, \quad \text{and} \quad L_\partial M_\partial F_\partial = \chi_\partial F_\partial + K_\partial^r F_\partial.$$

(2) *The operators M_∂, K_∂^ℓ, and K_∂^r are 2×2 matrix operators*

$$M_\partial = \begin{pmatrix} M_\partial^{11} & M_\partial^{12} \\ M_\partial^{21} & M_\partial^{22} \end{pmatrix}, \qquad K_\partial^\bullet = \begin{pmatrix} K_\partial^{\bullet,11} & K_\partial^{\bullet,12} \\ K_\partial^{\bullet,21} & K_\partial^{\bullet,22} \end{pmatrix}, \quad \bullet = \ell, r,$$

where

(a) $M_\partial^{11} \in \Psi^0(\mathbb{R}^{d-1})$, $M_\partial^{12} \in \Psi^{-k}(\mathbb{R}^{d-1})$, $M_\partial^{21} \in \Psi^1(\mathbb{R}^{d-1})$, *and* $M_\partial^{22} \in \Psi^{1-k}(\mathbb{R}^{d-1})$;

(b) $K_\partial^{\ell,11}, K_\partial^{\ell,22} \in \Psi^{-1}(\mathbb{R}^{d-1})$, $K_\partial^{\ell,12} \in \Psi^{-2}(\mathbb{R}^{d-1})$, *and* $K_\partial^{\ell,21} \in \Psi^0(\mathbb{R}^{d-1})$;

(c) $K_\partial^{r,11}, K_\partial^{r,22} \in \Psi^{-1}(\mathbb{R}^{d-1})$, $K_\partial^{r,12} \in \Psi^{-k-1}(\mathbb{R}^{d-1})$, *and* $K_\partial^{r,21} \in \Psi^{k-1}(\mathbb{R}^{d-1})$.

We refer to Appendix 3.A.6 for a proof of Lemma 3.11.

For $u \in H^{m+2}(\mathbb{R}^d_+)$, if we consider the first row in (3.2.14) and set g to be the first component of ${}^t(\mathscr{Q}_\mathsf{D} f, \mathscr{Q}_\mathsf{N} f) + R_\mathsf{C}(u)$, we obtain $F_\partial = L_\partial U_\partial$, where

$$U_\partial = \begin{pmatrix} \gamma^\mathsf{D}(u) \\ \tilde{\gamma}^\mathsf{N}(u) \end{pmatrix} \in H^{m+3/2, m+1/2}(\mathbb{R}^{d-1})$$

and

$$F_\partial = \begin{pmatrix} g \\ B(x,D)u_{|x_d=0^+} \end{pmatrix} \in H^{m+3/2, m+3/2-k}(\mathbb{R}^{d-1}).$$

If we apply Lemma 3.11, we thus see that we can recover all the traces of u expressed in U_∂, in the region where $\chi_\partial \equiv 1$, from the Lopatinskiĭ–Šapiro boundary data and g up to a regularizing operator.

REMARK 3.12. In the case of Dirichlet boundary conditions, with $B^1 = 1$ and $B^0 = 0$, observe that the second row of the operator M_∂ yields the principal part of the so-called Dirichlet-to-Neumann map, here in the case of a interior source term g.

3.3. Proof of the Fredholm Property

Here, we consider the case ${}^k\partial\mathcal{M} \neq \emptyset$ for all $k \in \{0, \dots, \beta\}$. Other cases for which ${}^k\partial\mathcal{M} = \emptyset$ for k in a subset of $\{0, \dots, \beta\}$ can be treated similarly and yield simpler matrix operators in the proof. Recall that the notation ${}^k\partial\mathcal{M}$, $k = 0, \dots, \beta$, is introduced in Sect. 3.1.

For each $m \in \mathcal{M} \setminus \partial\mathcal{M}$, there exists a chart $\mathcal{C}^m = (\kappa^m, O^m)$ such that $m \in O^m$ and $O^m \cap \partial\mathcal{M} = \emptyset$. For each $m \in \partial\mathcal{M}$, there exists a chart $\mathcal{C}^m = (\kappa^m, O^m)$ such that $m \in O^m$ and O^m meets only one of the connected components of $\partial\mathcal{M}$, and κ^m yields normal geodesic coordinates as introduced in Sect. 17.6. Because of the compactness of $\mathcal{M}$, there exists $(m_j)_{j\in J}$ with J finite such that

(1) $\mathcal{M} = \cup_{j\in J} O^{m_j}$.
(2) The set J is partitioned into $J = J_{\text{int}} \cup J_\partial$.
(3) One has $j \in J_{\text{int}}$ if and only if $O^{m_j} \cap \partial\mathcal{M} = \emptyset$.
(4) the set J_∂ reads $J_\partial = {}^0J_\partial \cup \dots \cup {}^\beta J_\partial$, and $j \in {}^kJ_\partial$ if and only if $O^{m_j} \cap {}^k\partial\mathcal{M} \neq \emptyset$.

To ease notation, we write $\mathcal{C}^j = (\kappa^j, O^j)$ in place of $\mathcal{C}^{m_j} = (\kappa^{m_j}, O^{m_j})$.

With Theorem 15.14, we introduce a $\mathscr{C}^\infty$-partition of unity $(\psi^j)_{j\in J}$ subordinated to the open covering $(O^j)_{j\in J}$, that is,

(1) For all $j \in J$, we have $\psi^j \in \mathscr{C}^\infty(\mathcal{M})$, $0 \le \psi^j \le 1$, and $\operatorname{supp}(\psi^j) \subset O^j$.
(2) We have $\sum_{j\in J} \psi^j = 1$ on $\mathcal{M}$.

For each $j \in J$, we also consider $\varphi^j \in \mathscr{C}^\infty(\mathcal{M})$ such that $\operatorname{supp}(\varphi^j) \subset O^j$ and $\varphi^j \equiv 1$ on a neighborhood of $\operatorname{supp}(\psi^j)$. In each chart $\mathcal{C}^j$, the local representatives of ψ^j and φ^j are denoted by $\check{\psi}^j$ and $\check{\varphi}^j$, respectively, that is,

$$\check{\psi}^j = \big((\kappa^j)^{-1}\big)^*\psi^j, \qquad \check{\varphi}^j = \big((\kappa^j)^{-1}\big)^*\varphi^j.$$

Because of the supports of ψ^j and φ^j, we have $\operatorname{supp}(\check{\psi}^j) \subset \operatorname{supp}(\check{\varphi}^j) \subset \kappa^j(O^j)$ and $\check{\varphi}^j \equiv 1$ on a neighborhood of $\operatorname{supp}(\check{\psi}^j)$.

We now construct the operator $M : H^m(\mathcal{M}) \oplus H_B^{(m+3/2)}(\partial\mathcal{M}) \to H^{m+2}(\mathcal{M})$ of Proposition 3.2, in the form

$$M = \sum_{j\in J} \psi^j M^j, \tag{3.3.1}$$

with M^j constructed in Sects. 3.3.1 and 3.3.2 below, first in the local chart $\mathcal{C}^j$ using the analysis performed in a half-space in Sect. 3.2 and, second, lifted back to the manifold.

We consider $\mathsf{f} \in H^m(\mathcal{M})$ and $\mathsf{h} = ({}^0\mathsf{h}, \dots, {}^\beta\mathsf{h}) \in H_B^{(m+3/2)}(\partial\mathcal{M})$. For each $j \in J$, we let f^j be the local representative of f in $\mathcal{C}^j$, that is,

$$\mathsf{f}^j = \big((\kappa^j)^{-1}\big)^*\mathsf{f} \in H^m\big(\kappa^j(O^j)\big). \tag{3.3.2}$$

If moreover $j \in J_\partial$ and $j \in {}^kJ_\partial$, that is, O^j meets ${}^k\partial\mathcal{M}$, we also let ${}^k\mathsf{h}^j$ be the local representative of ${}^k\mathsf{h}$ in $\mathcal{C}^j$, that is,

$${}^k\mathsf{h}^j = \big((\kappa^j)^{-1}\big)^*\, {}^k\mathsf{h} \in H^{m+3/2-k}\big(\kappa^j(\partial\mathcal{M} \cap O^j)\big). \tag{3.3.3}$$

3.3.1. Charts Away from the Boundary. Here, $j \in J_{\text{int}}$. The chart $\mathcal{C}^j$ does not meet the boundary.

If $P^{\mathcal{C}^j}(x, D)$ is the representative of P in $\kappa^j(O^j)$, we extend its coefficients to the whole $\mathbb{R}^d$ to form a second-order elliptic operator. We denote by $Q \in \Psi^{-2}(\mathbb{R}^d)$ a parametrix as given by Proposition 2.33 of Volume 1, that is,

$$QP^{\mathcal{C}^j}(x, D) = \operatorname{Id} + R, \quad P^{\mathcal{C}^j}(x, D)Q = \operatorname{Id} + R', \tag{3.3.4}$$

with $R, R' \in \Psi^{-\infty}(\mathbb{R}^d) = \cap_{N\in\mathbb{N}} \Psi^{-N}(\mathbb{R}^d)$. We define the operator $M^{\mathcal{C}^j}$ as

$$M^{\mathcal{C}^j} = \check{\varphi}^j Q \check{\varphi}^j, \tag{3.3.5}$$

which is bounded from $H^m(\mathbb{R}^d)$ into $H^{m+2}(\mathbb{R}^d)$. We then define the $1 \times (\beta + 2)$ matrix operator

$$M^j = \Big((\kappa^j)^* M^{\mathcal{C}^j} \big((\kappa^j)^{-1}\big)^* \;\; 0 \;\; \cdots \;\; 0\Big), \tag{3.3.6}$$

which is bounded from $H^m(\mathcal{M}) \oplus H_B^{(m+3/2)}(\partial\mathcal{M})$ into $H^{m+2}(\mathcal{M})$.

With the map L defined in (3.1.3), we have the following lemma.

LEMMA 3.13. *Let $j \in J_{int}$. We have*

$$\psi^j L M^j = \psi^j \operatorname{Id} + K^{r,j} \quad \text{and} \quad \psi^j M^j L = \psi^j + K^{\ell,j}, \tag{3.3.7}$$

where $K^{r,j}$ is bounded from $H^m(\mathcal{M}) \oplus H_B^{(m+3/2)}(\partial\mathcal{M})$ into $H^{m+1}(\mathcal{M}) \oplus H_B^{(m+5/2)}(\partial\mathcal{M})$ and $K^{\ell,j}$ is bounded from $H^{m+2}(\mathcal{M})$ into $H^{m+3}(\mathcal{M})$.

Note that because of the support of ψ^j, one has $\psi^j h = 0$ implying that the first equality in (3.3.7) actually reads

$$\psi^j L M^j \begin{pmatrix} \mathsf{f} \\ \mathsf{h} \end{pmatrix} = \psi^j \begin{pmatrix} \mathsf{f} \\ 0 \end{pmatrix} + K^{r,j} \begin{pmatrix} \mathsf{f} \\ \mathsf{h} \end{pmatrix}.$$

PROOF. We compute

$$\psi^j P M^j = (\kappa^j)^* \check{\psi}^j P^{\mathcal{C}^j}(x, D) \begin{pmatrix} M^{\mathcal{C}^j} & 0 & \cdots & 0 \end{pmatrix} ((\kappa^j)^{-1})^*.$$

Using that $P^{\mathcal{C}^j}(x, D)$ is a local operator and the support properties of $\check{\psi}^j$ and $\check{\varphi}^j$ and (3.3.4), we find that

$$\begin{aligned} \check{\psi}^j P^{\mathcal{C}^j}(x, D) M^{\mathcal{C}^j} &= \check{\psi}^j P^{\mathcal{C}^j}(x, D) \check{\varphi}^j Q \check{\varphi}^j = \check{\psi}^j P^{\mathcal{C}^j}(x, D) Q \check{\varphi}^j. \\ &= \check{\psi}^j \check{\varphi}^j + \check{\psi}^j K_1^{r,\mathcal{C}^j} \check{\varphi}^j = \check{\psi}^j + \check{\psi}^j K_1^{r,\mathcal{C}^j} \check{\varphi}^j, \end{aligned}$$

where $K_1^{r,\mathcal{C}^j}$ is bounded from $H^m(\mathbb{R}^d)$ into $H^N(\mathbb{R}^d)$ for any $N \in \mathbb{N}$. We then define the operator

$$K_{1,1}^{r,j} = (\kappa^j)^* \check{\psi}^j K_1^{r,\mathcal{C}^j} \check{\varphi}^j ((\kappa^j)^{-1})^*$$

that is bounded from $H^m(\mathcal{M})$ into $H^{m+1}(\mathcal{M})$.

Because of the support of φ^j, we have

$$B M^j \begin{pmatrix} \mathsf{f} \\ \mathsf{h} \end{pmatrix} \Big|_{\partial\mathcal{M}} = 0.$$

Setting the $(\beta + 2) \times (\beta + 2)$ matrix operator

$$K^{r,j} = \begin{pmatrix} K_{1,1}^{r,j} & 0 & \cdots & 0 \\ 0 & 0 & \cdots & 0 \\ \vdots & \vdots & \ddots & \vdots \\ 0 & 0 & \cdots & 0 \end{pmatrix},$$

we obtain the first result.

For the second result, we write

$$\begin{aligned} \psi^j M^j L &= \psi^j (\kappa^j)^* \check{\varphi}^j Q \check{\varphi}^j P^{\mathcal{C}^j}(x, D) ((\kappa^j)^{-1})^* \\ &= (\kappa^j)^* \check{\psi}^j Q \check{\varphi}^j P^{\mathcal{C}^j}(x, D) ((\kappa^j)^{-1})^*. \end{aligned}$$

Let $\hat{\varphi}^j \in \mathscr{C}_c^\infty(\kappa(O^j))$ be such that $\hat{\varphi}^j \equiv 1$ on a neighborhood of $\operatorname{supp}(\check{\psi}^j)$ and $\check{\varphi}^j \equiv 1$ on a neighborhood of $\operatorname{supp}(\hat{\varphi}^j)$. We then have

$$\check{\psi}^j Q \check{\varphi}^j P^{C^j}(x, D) = \check{\psi}^j Q \check{\varphi}^j P^{C^j}(x, D)\hat{\varphi}^j + K_1^{\ell,j},$$

with $K_1^{\ell,j} = \check{\psi}^j Q \check{\varphi}^j P^{C^j}(x, D)(1 - \hat{\varphi}^j) \in \Psi^{-N}(\mathbb{R}^d)$ for any $N \in \mathbb{N}$, by symbol calculus using the support properties of $\check{\psi}^j$ and $\hat{\varphi}^j$. Using that $P^{C^j}(x, D)$ is a local operator and (3.3.4), we then write

$$\begin{aligned}\check{\psi}^j Q \check{\varphi}^j P^{C^j}(x, D)\hat{\varphi}^j &= \check{\psi}^j Q P^{C^j}(x, D)\hat{\varphi}^j = \check{\psi}^j \hat{\varphi}^j + \check{\psi}^j K_2^{\ell,j}\hat{\varphi}^j \\ &= \check{\psi}^j + \check{\psi}^j K_2^{\ell,j}\hat{\varphi}^j,\end{aligned}$$

where $K_2^{\ell,j} \in \Psi^{-N}(\mathbb{R}^d)$ for any $N \in \mathbb{N}$. We then conclude as above. ∎

3.3.2. Charts at the Boundary. Here, $j \in {}^k J_\partial$ for some $k \in \{0, \dots, \beta\}$ that gives the order of the boundary operator equal to k.

We choose $\chi_\partial \in \mathscr{C}_c^\infty(\mathbb{R}^{d-1})$ as introduced above Lemma 3.11 such that $\chi_\partial \equiv 1$ on a neighborhood of $\operatorname{supp}(\check{\varphi}^j)$.

With the operator $\mathscr{Q}_\mathsf{D}$ introduced in (3.2.9), we define the following 3×2 matrix operator:

$$M_a = \begin{pmatrix} 1 & 0 \\ \mathscr{Q}_\mathsf{D} & 0 \\ 0 & 1 \end{pmatrix},$$

which is bounded from $H^m(\mathbb{R}^d_+) \oplus H^{m+3/2-k}(\mathbb{R}^{d-1})$ into $H^m(\mathbb{R}^d_+) \oplus H^{m+3/2}(\mathbb{R}^{d-1}) \oplus H^{m+3/2-k}(\mathbb{R}^{d-1})$. With the 2×2 matrix operator M_∂ given by Lemma 3.11, we define the following 3×3 matrix operator:

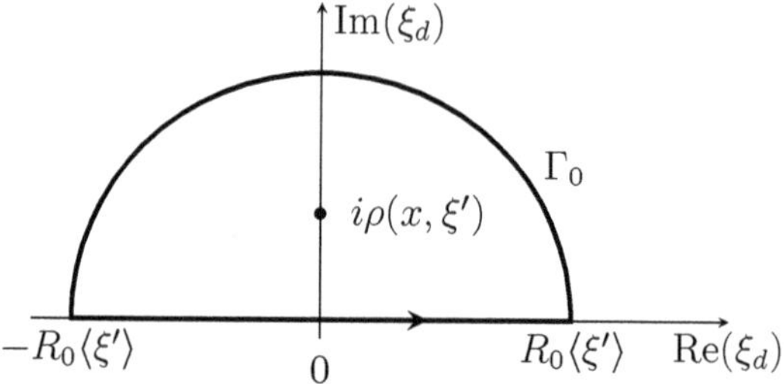

which is bounded from $H^m(\mathbb{R}^d_+) \oplus H^{m+3/2}(\mathbb{R}^{d-1}) \oplus H^{m+3/2-k}(\mathbb{R}^{d-1})$ into $H^m(\mathbb{R}^d_+) \oplus H^{m+3/2}(\mathbb{R}^{d-1}) \oplus H^{m+1/2}(\mathbb{R}^{d-1})$. With the operator $\mathscr{Q}$ introduced in (3.2.6) and the operators $\mathscr{Q}^{(-1)}$, $\mathscr{Q}^{(0)}$ introduced in (3.2.7), we define the 1×3 matrix operator

$$M_c = \begin{pmatrix} \mathscr{Q} & -\mathscr{Q}^{(0)} & \mathscr{Q}^{(-1)} \end{pmatrix} \tag{3.3.8}$$

that is bounded from $H^m(\mathbb{R}^d_+) \oplus H^{m+3/2}(\mathbb{R}^{d-1}) \oplus H^{m+1/2}(\mathbb{R}^{d-1})$ into $H^{m+2}(\mathbb{R}^d_+)$. We then define the operator M^{C^j} as

$$M^{C^j} = \check{\varphi}^j M_c M_b M_a \check{\varphi}^j, \tag{3.3.9}$$

which is bounded from $H^m(\mathbb{R}^d_+) \oplus H^{m+3/2-k}(\mathbb{R}^{d-1})$ into $H^{m+2}(\mathbb{R}^d_+)$. Set the $2 \times (\beta+2)$ matrices

$$W^k = \begin{pmatrix} 1 & 0 & \cdots & 0 & 0 & 0 & \cdots & 0 \\ 0 & 0 & \cdots & 0 & 1 & 0 & \cdots & 0 \end{pmatrix},$$

where the '1' in the second row lies in the $(k+2)$th column. We then set the $1 \times (\beta+2)$ matrix operator

$$M^j = (\kappa^j)^* M^{C^j} \left((\kappa^j)^{-1}\right)^* W^k \tag{3.3.10}$$

that is bounded from $H^m(\mathcal{M}) \oplus H_B^{(m+3/2)}(\partial\mathcal{M})$ into $H^{m+2}(\mathcal{M})$. Recall that the value of k is given by the chart and corresponds to the order of the boundary operator in that chart. Observe that M^j is well defined as an operator on functions defined in $\mathcal{M}$ because of the two occurrences of the cutoff function $\check{\varphi}^j$ in the definition of M^{C^j}.

With the map L defined in (3.1.3), we have the following lemma.

LEMMA 3.14. *Let $j \in {}^kJ_\partial$. We have*

$$\psi^j L M^j = \psi^j \operatorname{Id} + K^{r,j} \quad \textit{and} \quad \psi^j M^j L = \psi^j + K^{\ell,j},$$

where $K^{r,j}$ is bounded from $H^m(\mathcal{M}) \oplus H_B^{(m+3/2)}(\partial\mathcal{M})$ into $H^{m+1}(\mathcal{M}) \oplus H_B^{(m+5/2)}(\partial\mathcal{M})$ and $K^{\ell,j}$ is bounded from $H^{m+2}(\mathcal{M})$ into $H^{m+3}(\mathcal{M})$.

We refer to Appendix 3.A.7 for a proof of Lemma 3.14.

3.3.3. Conclusion of the Proof of the Fredholm Property. Here, we prove Proposition 3.2. With Lemmata 3.13 and 3.14, we write

$$LM = \sum_{j\in J} \left(\psi^j L M^j + [L, \psi^j] M^j\right) = \operatorname{Id} + K^r,$$

recalling that $\sum_{j\in J} \psi^j \equiv 1$, with $K^r = \sum_{j \in J} \left(K^{r,j} + [L,\psi^j]M^j\right)$, where the operators $K^{r,j}$ are bounded maps from $H^m(\mathcal{M}) \oplus H_B^{(m+3/2)}(\partial\mathcal{M})$ into $H^{m+1}(\mathcal{M}) \oplus H_B^{(m+5/2)}(\partial\mathcal{M})$. If $j \in J_{\text{int}}$, we have

$$[L, \psi^j] v = \begin{pmatrix} [P, \psi^j] v \\ 0 \end{pmatrix}.$$

If $j \in {}^kJ_\partial$, we have

$$[L, \psi^j] v = \begin{pmatrix} [P, \psi^j] v \\ [B, \psi^j] v_{|\partial\mathcal{M}} \end{pmatrix}.$$

As $[P, \psi^j]$ is a differential operator of order 1 and as $[B, \psi^j]$ is of order $k-1$ if $k \geq 1$ and vanishes if $k = 0$, we find that $[L, \psi^j] M^j$ also maps $H^m(\mathcal{M}) \oplus H_B^{(m+3/2)}(\partial\mathcal{M})$ into $H^{m+1}(\mathcal{M}) \oplus H_B^{(m+5/2)}(\partial\mathcal{M})$.

Similarly, with Lemmata 3.13 and 3.14, we write

$$ML = \sum_{j \in J} \psi^j M^j L = \operatorname{Id} + K^\ell,$$

with $K^\ell = \sum_{j\in J} K^{\ell,j}$ that is bounded from $H^{m+2}(\mathcal{M})$ into $H^{m+3}(\mathcal{M})$. ■

3.4. Necessity of the Lopatinskiĭ–Šapiro Conditions

Here, we prove that if the map L defined in (3.1.3) is Fredholm, then (P, B) fulfills the Lopatinskiĭ–Šapiro condition.

We consider the bounded map

$$
\begin{aligned}
{}^{\mathsf{D}}L_P : H^{m+2}(\mathcal{M}) &\to H^m(\mathcal{M}) \oplus H^{m+3/2}(\partial\mathcal{M}) \\
u &\mapsto (Pu, u_{|\partial\mathcal{M}}),
\end{aligned}
$$

that is, the case of Dirichlet boundary conditions. As the Lopatinskiĭ–Šapiro condition holds on $\partial\mathcal{M}$ for such boundary conditions, by Proposition 3.2, there exists[2] a bounded linear operator ${}^{\mathsf{D}}M_P : H^m(\mathcal{M}) \oplus H^{m+3/2}(\partial\mathcal{M}) \to H^{m+2}(\mathcal{M})$ such that

$$
\begin{aligned}
{}^{\mathsf{D}}M_P {}^{\mathsf{D}}L_P &= \mathrm{Id}_{H^{m+2}(\mathcal{M})} + {}^{\mathsf{D}}K_P^\ell \quad \text{and} \\
{}^{\mathsf{D}}L_P {}^{\mathsf{D}}M_P &= \mathrm{Id}_{H^m(\mathcal{M})\oplus H^{m+3/2}(\partial\mathcal{M})} + {}^{\mathsf{D}}K_P^r,
\end{aligned}
$$

where both operators

$$
\begin{aligned}
&{}^{\mathsf{D}}K_P^\ell : H^{m+2}(\mathcal{M}) \to H^{m+3}(\mathcal{M}), \\
&{}^{\mathsf{D}}K_P^r : H^m(\mathcal{M}) \oplus H^{m+3/2}(\partial\mathcal{M}) \to H^{m+1}(\mathcal{M}) \oplus H^{m+5/2}(\partial\mathcal{M})
\end{aligned}
$$

are bounded.

Let $k \in \{0, \dots, \beta\}$ and $m^0 \in {}^k\partial\mathcal{M}$, the connected component of $\partial\mathcal{M}$ where B is of order k. Let $\mathcal{C} = (O, \kappa)$ be a local chart such that $m^0 \in O$. We also consider $\overline{U} \subset O$ an open set of $\mathcal{M}$ with $m^0 \in U$, and $\varphi, \tilde{\varphi} \in \mathscr{C}^\infty(\mathcal{M})$ with $\operatorname{supp}(\tilde{\varphi}) \subset O$, $\tilde{\varphi} \equiv 1$ on a neighborhood of φ, $\varphi \equiv 1$ on a neighborhood of $\overline{U}$.

We consider $h \in H^{m+3/2}(\partial\mathcal{M})$ supported in $\partial\mathcal{M} \cap U$, and we define

$$
u = {}^{\mathsf{D}}m_P(h) = {}^{\mathsf{D}}M_P \begin{pmatrix} 0 \\ h \end{pmatrix} \in H^{m+2}(\mathcal{M}),
$$

and we also consider the operators

$$
{}^{\mathsf{D}}k_P^{r,1} = \begin{pmatrix} 1 & 0 \end{pmatrix} {}^{\mathsf{D}}K_P^r \begin{pmatrix} 0 \\ 1 \end{pmatrix}, \qquad {}^{\mathsf{D}}k_P^{r,2} = \begin{pmatrix} 0 & 1 \end{pmatrix} {}^{\mathsf{D}}K_P^r \begin{pmatrix} 0 \\ 1 \end{pmatrix},
$$

which are bounded from $H^{m+3/2}(\partial\mathcal{M})$ into $H^{m+1}(\mathcal{M})$ and from $H^{m+3/2}$ $(\partial\mathcal{M})$ into $H^{m+5/2}(\partial\mathcal{M})$, respectively. We have $Pu = {}^{\mathsf{D}}k_P^{r,1}(h)$. We set

$$
v = \varphi u = \varphi\, {}^{\mathsf{D}}m_P(h),
$$

and we compute

$$
Pv = [P, \varphi]\, {}^{\mathsf{D}}m_P(h) + \varphi {}^{\mathsf{D}}k_P^{r,1} h = \tilde{\varphi} K_1 \tilde{\varphi}_{|\partial\mathcal{M}} h,
$$

---

[2]Note that this follows also from Theorem 18.40 and Theorem 11.15 of Volume 1 as $P + \Delta_g$ is a first-order differential operator.

where $K_1 : H^{m+3/2}(\partial\mathcal{M}) \to H^{m+1}(\mathcal{M})$ is bounded. We also have

$$v_{|\partial\mathcal{M}} = h + \tilde{\varphi}_{|\partial\mathcal{M}}{}^{\mathsf{D}}k_P^{r,2}\tilde{\varphi}_{|\partial\mathcal{M}}h.$$

To ease notation, we also denote by h and v their local representatives in the chart $\mathcal{C}$. We then write $v = \mathsf{m}h$ with $\mathsf{m} = (\kappa^{-1})^*\varphi\,{}^{\mathsf{D}}m_P\varphi\kappa^* : H^{m+3/2}(\mathbb{R}^{d-1}) \to H^{m+2}(\mathbb{R}^d_+)$ bounded. We set

$$K = (\kappa^{-1})^*\tilde{\varphi}K_1\tilde{\varphi}_{|\partial\mathcal{M}}\kappa^*, \qquad k = (\kappa^{-1})^*\tilde{\varphi}_{|\partial\mathcal{M}}{}^{\mathsf{D}}k_P^{r,2}\tilde{\varphi}_{|\partial\mathcal{M}}\kappa^*,$$

and we have

$$P(x,D)v = Kh, \quad \text{and} \quad v_{|x_d=0^+} = h + kh, \tag{3.4.1}$$

with $K : H^{m+3/2}(\mathbb{R}^{d-1}) \to H^{m+1}(\mathbb{R}^d_+)$ and $k : H^{m+3/2}(\mathbb{R}^{d-1}) \to H^{m+5/2}(\mathbb{R}^{d-1})$ bounded, and where $P(x,D)$ denotes $P^{\mathcal{C}}(x,D)$, the local representative of P in $\mathcal{C}$. Note that K is compact from $H^{m+3/2}(\mathbb{R}^{d-1})$ into $H^m(\mathbb{R}^d_+)$ and k is compact from $H^{m+3/2}(\mathbb{R}^{d-1})$ into itself because of the cutoff functions in its definition and the Rellich–Kondrachov theorem (see Theorem 18.7).

Similarly, we denote by $B(x,D)$ the local representative of B in $\mathcal{C}$, which takes the form

$$B(x,D) = B^k(x,D') - iB^{k-1}(x,D')D_d,$$

with $B^k(x,D')$ and $B^{k-1}(x,D')$ as described in (3.2.2).

From the analysis of Sects. 3.2.2 and 3.2.3, we find with identity (3.2.14)

$$(\operatorname{Id} - \mathsf{C})\begin{pmatrix}\gamma^{\mathsf{D}}(v)\\ \tilde{\gamma}^{\mathsf{N}}(v)\end{pmatrix} = \mathsf{K}h, \tag{3.4.2}$$

where

$$\mathsf{K} = \begin{pmatrix}\mathscr{Q}_{\mathsf{D}}\\ \mathscr{Q}_{\mathsf{N}}\end{pmatrix} \circ K + R_{\mathsf{C}} \circ \mathsf{m} : H^{m+3/2}(\mathbb{R}^{d-1}) \to H^{m+5/2}(\mathbb{R}^{d-1}) \oplus H^{m+3/2}(\mathbb{R}^{d-1})$$

is bounded. With K_1 denoting the first component of the 2×1 matrix operator K, from (3.4.2), we have

$$(\operatorname{Id} - \mathsf{C}^{\mathsf{D},\mathsf{D}})\gamma^{\mathsf{D}}(v) - \mathsf{C}^{\mathsf{D},\mathsf{N}}\tilde{\gamma}^{\mathsf{N}}(v) = \mathsf{K}_1 h.$$

We recall that the principal symbol of $\mathsf{C}^{\mathsf{D},\mathsf{N}}$ is $\rho_{|x_d=0^+}^{-1}/2$, which is an elliptic symbol of order -1. With a parametrix as given by Proposition 2.33 of Volume 1 adapted to standard operators, we find

$$\tilde{\gamma}^{\mathsf{N}}(v) = G\gamma^{\mathsf{D}}(v) + \mathsf{K}_1' h,$$

where $G \in \Psi^1(\mathbb{R}^{d-1})$ with principal symbol equal to $\rho_{|x_d=0^+}(x',\xi')$ and where $\mathsf{K}_1' : H^{m+3/2}(\mathbb{R}^{d-1}) \to H^{m+3/2}(\mathbb{R}^{d-1})$ is bounded.

We now compute

$$\begin{aligned} B(x,D)v_{|x_d=0^+} &= B^k(x',0,D')\gamma^{\mathsf{D}}(v) - iB^{k-1}(x',0,D')D_d v_{|x_d=0^+} \\ &= \big(B^k(x',0,D') + iB^{k-1}(x',0,D')r_0(x',0)\big)\gamma^{\mathsf{D}}(v) \\ &\quad + B^{k-1}(x',0,D')\tilde{\gamma}^{\mathsf{N}}(v) \\ &= E\gamma^{\mathsf{D}}(v) + B^{k-1}(x',0,D')\mathsf{K}_1' h, \end{aligned}$$

with $E = \big(B^k(x',0,D') + iB^{k-1}(x',0,D')r_0(x',0)\big) + B^{k-1}(x',0,D')G \in \Psi^k(\mathbb{R}^{d-1})$ with symbol $e = b^k(x',0,\xi') + b^{k-1}(x',0,\xi')\rho_{|x_d=0^+}(x',\xi')$. With (3.4.1), we write

$$B(x,D)v_{|x_d=0^+} = Eh + \mathsf{K}_1'' h,$$

where $\mathsf{K}_1'' : H^{m+3/2}(\mathbb{R}^{d-1}) \to H^{m+5/2-k}(\mathbb{R}^{d-1})$ is bounded. Because of the support of v, for $\check{\varphi} = (\kappa^{-1})^* \tilde{\varphi}$, we have

$$B(x,D)v_{|x_d=0^+} = \check{\varphi} B(x,D)v_{|x_d=0^+} = \check{\varphi} E h + \check{\varphi}\mathsf{K}_1'' h.$$

Note that $\check{\varphi}\mathsf{K}_1''$ is compact from $H^{m+3/2}(\mathbb{R}^{d-1})$ into $H^{m+3/2-k}(\mathbb{R}^{d-1})$ by the Rellich–Kondrachov theorem.

Since $L = (P,B)$ is Fredholm, by Proposition 11.10 and Remark 11.11 of Volume 1, there exists $K^\ell : H^{m+2}(\mathcal{M}) \to H^{m+2}(\mathcal{M})$ compact such that

$$\begin{aligned} \|w\|_{H^{m+2}(\mathcal{M})} &\lesssim \|P(x,D)w\|_{H^m(\mathcal{M})} + |B(x,D)w_{|\partial\mathcal{M}}|_{H_B^{(m+3/2)}(\partial\mathcal{M})} \\ &\quad + \|K^\ell w\|_{H^{m+2}(\mathcal{M})}, \end{aligned}$$

for $w \in H^{m+2}(\mathcal{M})$. Applied to $w = \kappa^* v$, with the notation abuses mentioned above, this yields

$$\begin{aligned} \|v\|_{H^{m+2}(\mathbb{R}^d_+)} &\lesssim \|P(x,D)v\|_{H^m(\mathbb{R}^d_+)} + |B(x,D)v_{|x_d=0^+}|_{H^{m+3/2-k}(\mathbb{R}^{d-1})} \\ &\quad + \|K^\ell \kappa^* v\|_{H^{m+2}(\mathcal{M})}. \end{aligned}$$

As we have

$$|h|_{H^{m+3/2}(\mathbb{R}^{d-1})} \lesssim \|v\|_{H^{m+2}(\mathbb{R}^d_+)} + |kh|_{H^{m+3/2}(\mathbb{R}^{d-1})},$$

this gives, with (3.4.1),

$$\begin{aligned} |h|_{H^{m+3/2}(\mathbb{R}^{d-1})} &\lesssim |\check{\varphi} E h|_{H^{m+3/2-k}(\mathbb{R}^{d-1})} + \|Kh\|_{H^m(\mathbb{R}^d_+)} + |kh|_{H^{m+3/2}(\mathbb{R}^{d-1})} \\ &\quad + \|K^\ell \kappa^* \mathsf{m} h\|_{H^{m+2}(\mathbb{R}^d_+)} + \|\check{\varphi}\mathsf{K}_1'' h\|_{H^{m+3/2-k}(\mathbb{R}^{d-1})}. \end{aligned}$$

Here, h is any smooth function supported in $\kappa(\partial\mathcal{M} \cap U)$. Note that $\check{\varphi}\mathsf{K}_1''$ is compact from $H^{m+3/2}(\mathbb{R}^{d-1})$ into $H^{m+3/2-k}(\mathbb{R}^{d-1})$ by the Rellich–Kondrachov theorem. As $m^0 \in U$, for $x^0 = \kappa(m^0)$, by Theorem 2.57 of Volume 1, this implies that $e(x^0,\xi') \gtrsim |\xi'|^k$, for $|\xi'| \geq 1$ as $\check{\varphi} \equiv 1$ in a neighborhood of x^0. This precisely means that the Lopatinskiĭ–Šapiro condition holds for (P,B) at m^0 as recalled in (3.2.3). ∎

3.5. Some Index Computations

Let $P = -\Delta_g + R_1$ and B be as in Sect. 3.1 and L be the map from $H^{m+2}(\mathcal{M})$ into $H^m(\mathcal{M}) \oplus H_B^{(m+3/2)}(\partial\mathcal{M})$ defined in (3.1.3), with $m \in \mathbb{N}$. By Proposition 3.4, for index computation, it suffices to only consider the case $m = 0$.

Since R_1 is a first-order differential operator, it is compact from $H^2(\mathcal{M})$ into $L^2(\mathcal{M})$; thus, by Theorem 11.15 of Volume 1 for the computation of the index, we simply take $R_1 = 0$.

In the present section, we restrict the analysis to the case $\beta \leq 1$. Recalling the notation of Sect. 3.1, we denote by ${}^k\partial\mathcal{M}$ the union of the connected components of $\partial\mathcal{M}$ where B is of order k, $k = 0, 1$. For better reading, in this section, we shall write

$$H^{3/2,1/2}(\partial\mathcal{M}) = H^{3/2}({}^0\partial\mathcal{M}) \oplus H^{1/2}({}^1\partial\mathcal{M})$$

in place of $H_B^{(3/2)}(\partial\mathcal{M})$, compare with (3.1.2).

For the computation of the index, dimension is of importance in connection with the result of Proposition 2.8.

3.5.1. Dimension $d \geq 4$.

This is the simplest case to analyze.

THEOREM 3.15. *Let $d \geq 4$ and $\beta \leq 1$. If (P, B) is such that the Lopatinskiĭ–Šapiro condition holds at $\partial\mathcal{M}$, then* $\operatorname{ind}(L) = 0$.

Case 1: Dirichlet-Like Boundary Condition. If ${}^0\partial\mathcal{M} \neq \emptyset$, as the Lopatinskiĭ–Šapiro condition holds there, then it is equivalent to consider Dirichlet boundary conditions upon division by a nonvanishing function on ${}^0\partial\mathcal{M}$; see Example 2.5-(1).

Case 2: Neumann-Like Boundary Condition. If ${}^1\partial\mathcal{M} \neq \emptyset$, there, we write the principal symbol of B as

$$b(m, \omega) = \langle \omega, t_m \rangle + i\langle \omega, v_m \rangle, \tag{3.5.1}$$

where t and v are two real vector fields on $\mathcal{M}$ along $\partial\mathcal{M}$. We write $v_m = v_m^\nu \nu_m + v_m'$, $t_m = t_m^\nu \nu_m + t_m'$, with $v_m^\nu, t_m^\nu \in \mathbb{R}$ and $v_m', t_m' \in T_m\partial\mathcal{M}$. Recall that v_m^ν and t_m^ν cannot vanish simultaneously as pointed out in Remark 2.9-(3). With $t^\nu + iv^\nu$ nowhere vanishing, upon division by this term, we may reduce the analysis to the case where $t^\nu = 1$ and $v^\nu = 0$ everywhere. We then have $B = \partial_\nu + B'$ with B' with $t_m' + iv_m'$ for principal symbol. By Proposition 2.8, the Lopatinskiĭ–Šapiro condition holding for (P, B) on ${}^1\partial\mathcal{M}$ reads

$$|v_m'|_{g_\partial} < 1 \quad \text{or} \quad g_\partial(t_m', v_m')^2 > |t_m'|_{g_\partial}^2 \left(|v_m'|_{g_\partial}^2 - 1\right), \tag{3.5.2}$$

for all $m \in {}^1\partial\mathcal{M}$.

PROOF OF THEOREM 3.15. On ${}^0\partial\mathcal{M}$, if it is nonempty, we have a Dirichlet-like case, and we define the operator $B_{(s)}$ to be constant equal to $B_{(s)} u_{|{}^0\partial\mathcal{M}} = u_{|{}^0\partial\mathcal{M}}$.

On ${}^1\partial\mathcal{M}$, if it is nonempty, we have a Neumann-like case, we have $B = \partial_\nu + B'$, and we define the operator $B_{(s)}$ to be

$$B_{(s)} = \partial_\nu + sB', \qquad s \in [0,1],$$

which has $b_{(s)}(m,\omega) = \langle \nu_m, \omega\rangle + \langle \omega, (t'_{(s)})_m + i(v'_{(s)})_m\rangle$ for principal symbol, with $t'_{(s)} = st'$ and $v'_{(s)} = sv'$.

We define

$$\begin{aligned} L_{(s)} : H^2(\mathcal{M}) &\to L^2(\mathcal{M}) \oplus H^{3/2,1/2}(\partial\mathcal{M}) \\ u &\mapsto (Pu, B_{(s)}u_{|{}^0\partial\mathcal{M}}, B_{(s)}u_{|{}^1\partial\mathcal{M}}). \end{aligned}$$

We claim that the Lopatinskiĭ–Šapiro condition holds for $(P, B_{(s)})$ for all $s \in [0,1]$. Thus, $L_{(s)}$ is Fredholm by Theorem 3.1. As $s \mapsto L_{(s)}$ is continuous on $[0,1]$ with values in $\mathscr{B}\big(H^2(\mathcal{M}), L^2(\mathcal{M}) \oplus H^{3/2,1/2}(\partial\mathcal{M})\big)$, then every $L_{(s)}$ lies in one connected component of the open set $F\mathscr{B}\big(H^2(\mathcal{M}), L^2(\mathcal{M}) \oplus H^{3/2,1/2}(\partial\mathcal{M})\big)$, the set of Fredholm operators in $\mathscr{B}\big(H^2(\mathcal{M}), L^2(\mathcal{M}) \oplus H^{3/2,1/2}(\partial\mathcal{M})\big)$; see Theorem 11.13 of Volume 1. By Theorem 11.14 (also in Volume 1), the index of $L_{(s)}$ is independent of $s \in [0,1]$. If $s = 0$, $L_{(s)}$ coincides with

(1) either a purely Dirichlet–Laplace boundary problem if $\partial\mathcal{M} = {}^0\partial\mathcal{M}$; then, the index is zero by Theorem 18.40;
(2) or a purely Neumann–Laplace boundary problem if $\partial\mathcal{M} = {}^1\partial\mathcal{M}$; then, the index is zero by Theorem 18.56;
(3) or a mixed Dirichlet–Neumann boundary problem otherwise; then, the index is zero by Theorem 18.63.

We now prove the claim formulated above. For the Dirichlet-like case, it is obvious. For the Neumann-like case on ${}^1\partial\mathcal{M}$, we argue as follows.

First, consider $m \in {}^1\partial\mathcal{M}$ where $|v'_m|_{g_\partial} < 1$. Then, for all $s \in [0,1]$, we have $|(v'_{(s)})_m|_{g_\partial} < 1$. Thus, the Lopatinskiĭ–Šapiro condition holds for $(P, B_{(s)})$ at m for all $s \in [0,1]$ as (3.5.2) holds at such a point.

Second, consider the case where $g_\partial(t'_m, v'_m)^2 > |t'_m|^2_{g_\partial}\big(|v'_m|^2_{g_\partial} - 1\big)$. Then, for $s \in (0,1]$, one has

$$g_\partial\big((t'_{(s)})_m, (v'_{(s)})_m\big)^2 = s^4 g_\partial(t'_m, v'_m)^2 > |(t'_{(s)})_m|^2_{g_\partial}\big(|(v'_{(s)})_m|^2_{g_\partial} - s^2\big),$$

implying

$$g_\partial\big((t'_{(s)})_m, (v'_{(s)})_m\big)^2 > |(t'_{(s)})_m|^2_{g_\partial}\big(|(v'_{(s)})_m|^2_{g_\partial} - 1\big).$$

Thus, the Lopatinskiĭ–Šapiro condition holds for $(P, B_{(s)})$ at m for all $s \in (0,1]$ as (3.5.2) holds at such a point. Observe now that for $s_0 > 0$ close to 0, we have $|(v'_{(s_0)})_m|_{g_\partial} < 1$, meaning that we can conclude using the first case for $s \in [0, s_0]$. ■

3.5.2. Dimension $d = 3$. This case has some similarity with the case $d \geq 4$. Its analysis is however more involved.

THEOREM 3.16. *Let $d = 3$ and $\beta \leq 1$. If (P, B) is such that the Lopatinskiĭ–Šapiro condition holds at $\partial\mathcal{M}$, then* $\operatorname{ind}(L) = 0$.

Below, we face three disjoint cases. Connected components of ${}^0\partial\mathcal{M}$ are treated as in the case $d \geq 4$. For the connected components of ${}^1\partial\mathcal{M}$, the analysis needs to be split into two subcases.

Case 1: Dirichlet-Like Boundary Condition. If ${}^0\partial\mathcal{M} \neq \emptyset$, there the boundary operator B can be replaced by the Dirichlet boundary operator; see Sect. 3.5.1 in the case $d \geq 4$.

We now consider ${}^1\partial\mathcal{M}$ if it is nonempty. Let $\mathcal{N}$ be one of the connected components of ${}^1\partial\mathcal{M}$. There, we write the principal symbol of B as in (3.5.1). As the Lopatinskiĭ–Šapiro condition holds for (P, B) on $\mathcal{N}$, according to Propositions 2.8 and 2.10, we now face two disjoint cases.

Case 2: Neumann-Like Boundary Condition. For all $m \in \mathcal{N}$, one has either

$$|t'_m|_{g_\partial} < |v^\nu_m| \quad \text{or} \quad |v'_m|_{g_\partial} < |t^\nu_m| \tag{3.5.3}$$

or

$$g_m(t_m, v_m)^2 > \big(|t'_m|^2_{g_\partial} - (v^\nu_m)^2\big)\big(|v'_m|^2_{g_\partial} - (t^\nu_m)^2\big). \tag{3.5.4}$$

Observe that (3.5.3)–(3.5.4) coincide exactly with the characterization of the Lopatinskiĭ–Šapiro condition in the case $d \geq 4$ in Proposition 2.8. Thus, the procedure of the proof of Theorem 3.15 applies. The principal symbol of $B_{|\mathcal{N}}$ given in (3.5.1) reads

$$b(m, \omega) = (t^\nu_m + i v^\nu_m)\omega^n_m + \langle \omega', t'_m \rangle + i\langle \omega', v'_m \rangle.$$

Then, $t^\nu_m + i v^\nu_m$ cannot vanish, and upon a division by this term, we take $B_{|\mathcal{N}}$ of the form

$$B_{|\mathcal{N}} = \partial_\nu + B'_{|\mathcal{N}}. \tag{3.5.5}$$

If we set

$$B_{(s)} = \partial_\nu + s B'_{|\mathcal{N}}, \qquad s \in [0, 1], \tag{3.5.6}$$

we reach the conclusion of the following lemma.

LEMMA 3.17. *The Lopatinskiĭ–Šapiro condition holds for $(P, B_{(s)})$ on $\mathcal{N}$ and all $s \in [0, 1]$.*

Observe that the Neumann boundary conditions fall into this case and are given by $B_{(0)}$. Hence, the name "Neumann-like" we use to describe this case.

Case 3: Tangential-Like Boundary Condition on a Torus. For all $m \in \mathcal{N}$, one has

$$|t'_m|_{g_\partial} \geq |v^\nu_m| \text{ and } |v'_m|_{g_\partial} \geq |t^\nu_m| \tag{3.5.7}$$

and

$$g_m(t_m, v_m)^2 < \big(|t'_m|^2_{g_\partial} - (v^\nu_m)^2\big)\big(|v'_m|^2_{g_\partial} - (t^\nu_m)^2\big), \tag{3.5.8}$$

and in this case $\mathcal{N}$ is a torus.

On $\mathcal{N}$, the operator B reads

$$B_{|\mathcal{N}} = (t^\nu_m + iv^\nu_m)\partial_\nu + B'_{|\mathcal{N}}, \tag{3.5.9}$$

meaning that $B'_{|\mathcal{N}}$ is a first-order differential operator on $\mathcal{N}$ with principal symbol

$$b'(m, \omega') = \langle \omega', t'_m \rangle + i\langle \omega', v'_m \rangle, \quad (m, \omega') \in T^*\mathcal{N}.$$

Consider now the boundary operator, for $s \in [0, 1]$

$$B_{(s)} = s(t^\nu_m + iv^\nu_m)\partial_\nu + B'_{|\mathcal{N}}. \tag{3.5.10}$$

LEMMA 3.18. *The Lopatinskiĭ–Šapiro condition holds for* $(P, B_{(s)})$ *on* $\mathcal{N}$ *for all* $s \in [0, 1]$.

PROOF. For $s \in [0, 1]$, set $(t_{(s)})_m = st^\nu_m \nu_m + t'_m$ and $(v_{(s)})_m = sv^\nu_m \nu_m + v'_m$. One then finds $|t'_m|_{g_\partial} > |v^\nu_m| \geq |sv^\nu_m|$, $|v'_m|_{g_\partial} > |t^\nu_m| \geq |st^\nu_m|$, and

$$g_m\big((t_{(s)})_m, (t_{(s)})_m\big)^2 < \big(|t'_m|^2_{g_\partial} - (sv^\nu_m)^2\big)\big(|v'_m|^2_{g_\partial} - (st^\nu_m)^2\big),$$

by Lemma 2.12. Together, these three conditions yield the conclusion by Proposition 2.8. ∎

The path $s \mapsto B_{(s)}$ initiated from B has the purely tangential operator B' for end point. Hence, the name 'Tangential-like' we use to describe this case.

By Proposition 2.10, one has $\operatorname{span}(t'_m, v'_m) = T_m\mathcal{N}$, meaning that the first-order operator B' is elliptic on $\mathcal{N}$.

LEMMA 3.19. *Let* $\mathcal{N}$ *be a torus. Consider the first-order differential operator* $W = X + iY + V$, *where* X *and* Y *are smooth real vector fields and* $V \in \mathscr{C}^\infty(\mathcal{N}; \mathbb{C})$. *Assume that* $\operatorname{span}(X_m, Y_m) = T_m\mathcal{N}$ *for every* $m \in \mathcal{N}$. *Consider* W *as a bounded operator from* $H^{s+1}(\mathcal{N})$ *into* $H^s(\mathcal{N})$, $s \in \mathbb{R}$. *Then,* W *is Fredholm and* $\operatorname{ind}(W) = 0$.

PROOF. As W is elliptic and $\mathcal{N}$ does not have a boundary, by using a parametrix in every chart of an atlas, as is done in Sect. 3.3.1, one can adapt the proof of the Fredholm property of Sect. 3.3.3 without having to deal with boundaries. The order of the operator plays essentially no *role.*

A (smooth) torus is by definition diffeomorphic to $\mathbb{S}^1 \times \mathbb{S}^1$. We may thus use global coordinates and use $\mathcal{N} = (\mathbb{R}/2\pi\mathbb{Z})^2$. We denote the coordinates on $\mathcal{N}$ by (x, y) and $W = X + iY = (a\partial_x + b\partial_y) + i(c\partial_x + d\partial_y)$, where

a, b, c, and d are smooth periodic functions. As $V : H^{s+1}(\mathcal{N}) \to H^s(\mathcal{N})$ is a compact operator, by Theorem 11.15 of Volume 1, we may simply take $V = 0$ to compute the index of W.

As $\dim \operatorname{span}(X_m, Y_m) = 2$, one has $a+ic \neq 0$ everywhere. Upon dividing by this nonvanishing function, we may reduce the computation of the index to that of $W' = \partial_x + (\alpha + i\beta)\partial_y$, where α and β are smooth real valued periodic functions and $\beta \neq 0$ everywhere.

As $\beta \neq 0$, we have $\beta > 0$ everywhere or $\beta < 0$ everywhere. If $\beta < 0$, the change of variables $(x, y) \mapsto (x, -y)$ yields $\beta > 0$. We thus simply assume that $\beta > 0$. Let $W_t = \partial_x + (t\alpha + i(t\beta + (1-t)))\partial_y$ for $t \in [0, 1]$, then we have $W_1 = W'$, $W_0 = \partial_x + i\partial_y$, and $t\beta + (1-t) \geq \min(\beta, 1) > 0$. Then, $W_t : H^{s+1}(\mathcal{N}) \to H^s(\mathcal{N})$ is an elliptic operator for every $t \in [0, 1]$. It is Fredholm and, by Theorem 11.14 in Volume 1, W' and $\partial_x + i\partial_y$ have the same index.

To compute the index of $\partial_x + i\partial_y$, it is more convenient to consider $W_\mu = \partial_x + i\partial_y + \mu$ with $\mu \in \mathbb{R} \setminus \mathbb{Z}$. The two operators share the same index by Theorem 11.15 in Volume 1. Working with Fourier series, we have

$$W_\mu u = \sum_{(n,k)\in\mathbb{Z}^2} (in - k + \mu) u_{n,k} e^{i(nx+ky)}, \qquad u = \sum_{(n,k)\in\mathbb{Z}^2} u_{n,k} e^{i(nx+ky)}.$$

As $(in - k + \mu) \neq 0$ for all $(n, k) \in \mathbb{Z}^2$, we deduce that W_μ is invertible from $H^{s+1}(\mathcal{N})$ to $H^s(\mathcal{N})$, which concludes the proof. ■

An example of a geometry where the three cases, "Dirichlet-like," "Neumann-like," and "Tangential-like" can occur is given in Fig. 3.1.

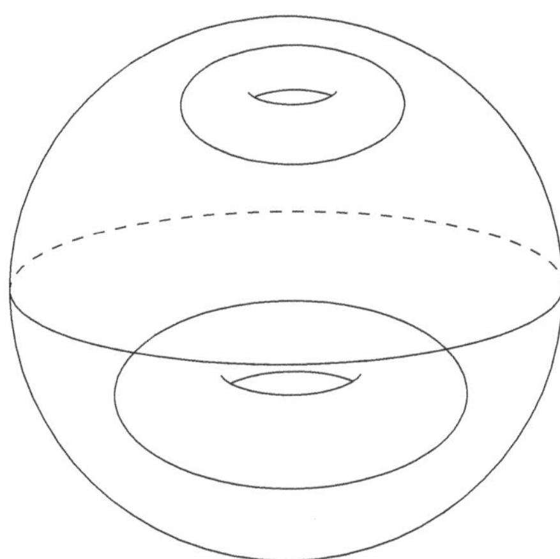

FIGURE 3.1. Example of a manifold of dimension 3 with a boundary with several connected components. Here, two solid tori are removed from a ball. One component of $\partial\mathcal{M}$ is a sphere and two other components are tori

PROOF OF THEOREM 3.16. We write ${}^1\partial\mathcal{M} = {}^{\mathsf{N}}\partial\mathcal{M} \cup {}^{\mathsf{T}}\partial\mathcal{M}$, with ${}^{\mathsf{N}}\partial\mathcal{M}$ and ${}^{\mathsf{T}}\partial\mathcal{M}$ as follows:

(1) The manifold ${}^{\mathsf{N}}\partial\mathcal{M}$ is the union of the connected components of $\partial\mathcal{M}$ characterized by "Neumann-like" boundary conditions, that is, where (3.5.3)–(3.5.4) hold.
(2) The manifold ${}^{\mathsf{T}}\partial\mathcal{M}$ is the union of the connected components of $\partial\mathcal{M}$ that are tori characterized by "tangential-like" boundary conditions, that is, where (3.5.7)–(3.5.8) hold.

On ${}^{0}\partial\mathcal{M}$, if it is nonempty, we define the operator $B_{(s)}$ to be constant equal to $B_{(s)}u_{|{}^0\partial\mathcal{M}} = u_{|{}^0\partial\mathcal{M}}$.

On ${}^{\mathsf{N}}\partial\mathcal{M}$, if it is nonempty, on each of its connected components, we define $B_{(s)}$, $s \in [0,1]$, by first reducing $B_{|{}^{\mathsf{N}}\partial\mathcal{M}}$ to the form given in (3.5.5) and then setting $B_{(s)}$ according to (3.5.6).

On ${}^{\mathsf{T}}\partial\mathcal{M}$, if it is nonempty, on each of its connected components, we define $B_{(s)}$, $s \in [0,1]$, according to (3.5.10).

The boundary operator $B_{(s)}$, defined in that manner on each connected component of $\partial\mathcal{M}$, fulfills the following properties:

(1) The map $s \mapsto B_{(s)}$ is continuous from $[0,1]$ into $\mathscr{B}(H^2(\mathcal{M}), H^{3/2,1/2}(\partial\mathcal{M}))$.
(2) The Lopatinskiĭ–Šapiro condition holds for $(P, B_{(s)})$ on $\partial\mathcal{M}$ for all $s \in [0,1]$;
(3) The end point $s = 1$ is such that $B_{(1)} = B$.
(4) The end point $s = 0$ is such that
   (a) On ${}^{0}\partial\mathcal{M}$, we have the Dirichlet boundary operator $B_{(0)}u = u_{|{}^0\partial\mathcal{M}}$;
   (b) On ${}^{\mathsf{N}}\partial\mathcal{M}$, we have the Neumann boundary operator $B_{(0)}u = \partial_\nu u_{|{}^{\mathsf{N}}\partial\mathcal{M}}$;
   (c) On each connected component of ${}^{\mathsf{T}}\partial\mathcal{M}$, we have a first-order tangential boundary operator $B_{(0)}u = B' u_{|{}^{\mathsf{T}}\partial\mathcal{M}}$ with B' as defined in (3.5.9).

We set

$$L_{(s)} : H^2(\mathcal{M}) \to L^2(\mathcal{M}) \oplus H^{3/2,1/2}(\partial\mathcal{M})$$
$$u \mapsto (Pu, B_{(s)}u)$$

and find that $L_{(s)}$ is Fredholm by Theorem 3.1. As $s \mapsto L_{(s)}$ is continuous on $[0,1]$ with values in $\mathscr{B}(H^2(\mathcal{M}), L^2(\mathcal{M}) \oplus H^{3/2,1/2}(\partial\mathcal{M}))$, then every $L_{(s)}$ lies in one connected component of the open set $F\mathscr{B}(H^2(\mathcal{M}), L^2(\mathcal{M}) \oplus H^{3/2,1/2}(\partial\mathcal{M}))$, the set of Fredholm operators in $\mathscr{B}(H^2(\mathcal{M}), L^2(\mathcal{M}) \oplus H^{3/2,1/2}(\partial\mathcal{M}))$; see Theorem 11.13 of Volume 1. By Theorem 11.14 also in Volume 1, the index of $L_{(s)}$ is independent of $s \in [0,1]$. The conclusion thus follows from Lemma 3.20 below. ■

LEMMA 3.20. *The index of the limit operator*

$$L_{(0)} : H^2(\mathcal{M}) \to L^2(\mathcal{M}) \oplus H^{3/2,1/2}(\partial\mathcal{M})$$
$$u \mapsto (Pu, B_{(0)}u)$$

is zero.

PROOF. If ${}^0\partial\mathcal{M} \cup {}^\mathsf{T}\partial\mathcal{M} = \emptyset$, then the limit operator $L_{(0)}$ corresponds to a pure Neumann problem. By Theorem 18.56, its index is zero.

We now assume that ${}^0\partial\mathcal{M} \cup {}^\mathsf{T}\partial\mathcal{M} \neq \emptyset$. We set ${}^\mathsf{D}\partial\mathcal{M} = {}^0\partial\mathcal{M} \cup {}^\mathsf{T}\partial\mathcal{M}$. We consider an auxiliary boundary value elliptic problem where Neumann boundary conditions are imposed on ${}^\mathsf{N}\partial\mathcal{M}$, if nonempty, and Dirichlet boundary conditions on ${}^\mathsf{D}\partial\mathcal{M}$. We denote by M the mixed Dirichlet–Neumann lifting map of Definition 18.60 associated with this auxiliary problem. This map is well defined precisely because ${}^\mathsf{D}\partial\mathcal{M} \neq \emptyset$; see Sect. 18.8. It has the regularity given by Proposition 18.62.

If ${}^\mathsf{T}\partial\mathcal{M}$ is nonempty, we denote by ${}^\mathsf{T}\partial\mathcal{M}^1, \ldots, {}^\mathsf{T}\partial\mathcal{M}^n$ its connected components, which are all tori. On ${}^\mathsf{T}\partial\mathcal{M}^j$, denote $B_j' = B_{(0)|{}^\mathsf{T}\partial\mathcal{M}^j}$. This is a tangential boundary operator. By Lemma 3.19, the bounded operator $B_j' : H^{3/2}({}^\mathsf{T}\partial\mathcal{M}^j) \to H^{1/2}({}^\mathsf{T}\partial\mathcal{M}^j)$ is Fredholm and $\operatorname{ind}(B_j') = 0$. We then set

$$K_\partial = \{({}^\mathsf{D}h, {}^\mathsf{N}h) \in H^{3/2,1/2}_{\mathsf{DN}}(\partial\mathcal{M});\ {}^\mathsf{D}h_{|{}^0\partial\mathcal{M}} = 0,$$
$${}^\mathsf{D}h_{|{}^\mathsf{T}\partial\mathcal{M}^j} \in \ker(B_j'),\ j = 1, \ldots, n, \text{ and } {}^\mathsf{N}h = 0\}$$

and $K = \mathsf{M}(K_\partial)$. The space $H^{3/2,1/2}_{\mathsf{DN}}(\partial\mathcal{M})$ is defined in (18.8.2). We have $\dim K = \sum_{1 \le j \le n} \dim \ker(B_j')$. We claim that $K = \ker(L_{(0)})$. Let $h \in K_\partial$ and $u = \mathsf{M}(h)$, then we find that $L_{(0)}u = 0$; in fact, using that B_j' is a tangential differential operator, note that $\left(B_{(0)}u\right)_{|{}^\mathsf{T}\partial\mathcal{M}^j} = B_j' u_{|{}^\mathsf{T}\partial\mathcal{M}^j} = 0$. Thus, $K \subset \ker(L_{(0)})$. Conversely, let $u \in \ker(L_{(0)})$. Then, $Pu = 0$, $u_{|{}^0\partial\mathcal{M}} = 0$, and $\partial_\nu u_{|{}^\mathsf{N}\partial\mathcal{M}} = 0$. We also have $\left(B_{(0)}u\right)_{|{}^\mathsf{T}\partial\mathcal{M}^j} = B_j' u_{|{}^\mathsf{T}\partial\mathcal{M}^j} = 0$, $j = 1, \ldots, n$. Thus, $u_{|{}^\mathsf{T}\partial\mathcal{M}^j} \in \ker(B_j')$, meaning that $u \in K$.

We now consider the range of $L_{(0)}$. We define the following subspace of $L^2(\mathcal{M}) \oplus H^{3/2,1/2}(\partial\mathcal{M})$:

$$R_\partial = \{(\mathsf{f}, \mathsf{h});\ \mathsf{f} \in L^2(\mathcal{M}),\ \mathsf{h} = ({}^0\mathsf{h}, {}^1\mathsf{h}) \in H^{3/2,1/2}(\partial\mathcal{M})$$
$$\text{with } {}^0\mathsf{h} \in H^{3/2}({}^0\partial\mathcal{M}), {}^1\mathsf{h}_{|{}^\mathsf{N}\partial\mathcal{M}} \in H^{1/2}({}^\mathsf{N}\partial\mathcal{M}),$$
$$\text{and}\, {}^1\mathsf{h}_{|{}^\mathsf{T}\partial\mathcal{M}^j} \in \operatorname{Ran}(B_j'),\ j = 1, \ldots, n\}.$$

We have $\operatorname{Ran}(L_{(0)}) \subset R_\partial$. Conversely, let $(\mathsf{f}, \mathsf{h}) \in R_\partial$. We set ${}^\mathsf{N}h = {}^1\mathsf{h}_{|{}^\mathsf{N}\partial\mathcal{M}} \in H^{1/2}({}^\mathsf{N}\partial\mathcal{M})$ and ${}^\mathsf{T}h^j = {}^1\mathsf{h}_{|{}^\mathsf{T}\partial\mathcal{M}^j} \in \operatorname{Ran}(B_j')$, that is, ${}^\mathsf{T}h^j = B_j'(u^j)$ with $u^j \in H^{3/2}({}^\mathsf{T}\partial\mathcal{M}^j)$. We now define $h = ({}^\mathsf{D}h, {}^\mathsf{N}h) \in H^{3/2,1/2}_{\mathsf{DN}}(\partial\mathcal{M})$ given by

$$
\begin{aligned}
{}^{\mathsf{D}}h_{|{}^0\partial\mathcal{M}} &= {}^0\mathsf{h} \in H^{3/2}({}^0\partial\mathcal{M}),\\
{}^{\mathsf{D}}h_{|{}^{\mathsf{T}}\partial\mathcal{M}} &= u_j \in H^{3/2}({}^{\mathsf{T}}\partial\mathcal{M}^j),\\
{}^{\mathsf{N}}h_{|{}^{\mathsf{N}}\partial\mathcal{M}} &= {}^{\mathsf{N}}h \in H^{1/2}({}^{\mathsf{N}}\partial\mathcal{M}).
\end{aligned}
$$

We denote by ${}^{\mathsf{M}}L$ the map

$$
\begin{aligned}
{}^{\mathsf{M}}L : H^2(\mathcal{M}) &\to L^2(\mathcal{M}) \oplus H^{3/2,1/2}_{\mathsf{DN}}(\partial\mathcal{M})\\
u &\mapsto \big(Pu, u_{|{}^{\mathsf{D}}\partial\mathcal{M}}, \partial_\nu u_{|{}^{\mathsf{N}}\partial\mathcal{M}}\big).
\end{aligned}
$$

It is a homeomorphism by Theorem 18.63. Thus, there exists a unique $u \in H^2(\mathcal{M})$ such that ${}^{\mathsf{M}}Lu = h$. We then find the $L_{(0)}u = (\mathsf{f}, \mathsf{h})$. Hence, $R_\partial = \operatorname{Ran}(L_{(0)})$ and

$$
\begin{aligned}
\operatorname{codim}\operatorname{Ran}(L_{(0)}) &= \sum_{1\le j\le n} \operatorname{codim}\operatorname{Ran}(B_j') = \sum_{1\le j\le n} \dim\ker(B_j')\\
&= \dim\ker(L_{(0)}),
\end{aligned}
$$

since $\operatorname{ind}(B_j') = 0$ by Lemma 3.19. This concludes the proof. ∎

3.5.3. Dimension $d = 2$. This is the most intricate case as topology plays an important *role*. We exhibit cases where the index is zero and cases where this index is nonzero. As above, we restrict our analysis to the case $\beta \le 1$.

Recalling the notation of Sect. 3.1, we denote by ${}^0\partial\mathcal{M}$ the union of the connected components of $\partial\mathcal{M}$ where the order of the boundary operator B is $k = 0$ and by ${}^1\partial\mathcal{M}$ the union of the connected components of $\partial\mathcal{M}$ where $k = 1$. As $d = 2$, the manifold ${}^1\partial\mathcal{M}$ is a finite union of one-dimensional manifolds without boundaries all diffeomorphic to $\mathbb{S}^1$. Let $\mathcal{N}$ be one of these connected components, and let $Y' \in \mathscr{C}^\infty V(\mathcal{N})$ be a unitary vector field (in the sense of the metric g_∂). We write the principal symbol of B on $\mathcal{N}$ as

$$
b(m,\omega) = \langle \omega, t_m\rangle + i\langle \omega, v_m\rangle, \tag{3.5.11}
$$

where t and v are two real vector fields on $\mathcal{M}$ along $\mathcal{N}$. We write $v_m = v_m^\nu \nu_m + v_m'$, $t_m = t_m^\nu \nu_m + t_m'$, with $v_m^\nu, t_m^\nu \in \mathbb{R}$ and $v_m', t_m' \in T_m\partial\mathcal{M}$. By Proposition 2.8, the Lopatinskiĭ–Šapiro condition holds on $\mathcal{N}$ if and only if

$$
t_m' + iv_m' \neq \pm i(t_m^\nu + iv_m^\nu)Y_m', \quad m \in \mathcal{N}.
$$

We shall present some geometrical reduction that may allow one to compute the index of L.

3.5.3.1. *Some Admissible Deformations.* A function $f \in \mathscr{C}^\infty(\mathcal{N};\mathbb{C})$ is called a Morse function if it has no degenerate critical point. Morse functions form a dense subset of $\mathscr{C}^\infty(\mathcal{N})$ in the $\mathscr{C}^k$-topology [260, Corollary 6.8], $k \in \mathbb{N}$. Thus, there exists a Morse function ρ close to $m \mapsto t^\nu_m + iv^\nu_m$ in the $\mathscr{C}^0$-topology such that

$$t'_m + iv'_m \neq \pm i\big(s(t^\nu_m + iv^\nu_m) + (1-s)\rho(m)\big)Y'_m, \quad s \in [0,1],\ m \in \mathcal{N}.$$

Thus, if we replace t and v by $t' + (\operatorname{Re}\rho)\nu$ and $v' + (\operatorname{Im}\rho)\nu$, the operator (P,B) is also Fredholm as the Lopatinskiĭ–Šapiro condition holds and its index is unchanged by Theorem 11.14 of Volume 1.

Now, if $m^0 \in \mathcal{N}$ is such that $\rho(m^0) = 0$, then such a point is isolated[3]. Since ρ is complex valued, there exists $\tilde\rho \in \mathscr{C}^\infty(\mathcal{N};\mathbb{C})$ close to ρ in the $\mathscr{C}^0$-topology such that

$$t'_m + iv'_m \neq \pm i\big(s\rho(m) + (1-s)\tilde\rho(m)\big)Y'_m, \quad s \in [0,1],\ m \in \mathcal{N},$$

and moreover $\tilde\rho$ does not vanish on $\mathcal{N}$. Arguing as above, we may thus replace t and v by $t' + (\operatorname{Re}\tilde\rho)\nu$ and $v' + (\operatorname{Im}\tilde\rho)\nu$. Since $\tilde\rho \neq 0$ upon dividing by $\tilde\rho$, we may reduce the analysis to only considering boundary operators that take the form

$$B = \partial_\nu + B', \tag{3.5.12}$$

with B' a first-order tangential differential operator with principal symbol given by $b'(m,\omega') = \langle \omega', t'_m\rangle + i\langle \omega', v'_m\rangle$. The Lopatinskiĭ–Šapiro condition then reads

$$t'_m + iv'_m \neq \pm iY'_m, \qquad m \in \mathcal{N}.$$

Since Y' does not vanish, we may write $t' = \alpha Y'$ and $v' = \beta Y'$. The Lopatinskiĭ–Šapiro condition is then

$$\alpha(m) + i\beta(m) \neq \pm i \qquad m \in \mathcal{N}. \tag{3.5.13}$$

We define the closed curve $\gamma_\mathcal{N} : \mathcal{N} \to \mathbb{C}$ given by $\gamma_\mathcal{N}(m) = \alpha(m) + i\beta(m)$. This reduction can be performed on all connected components of $\partial\mathcal{M}$.

On the one hand, the following lemma allows one to deform the boundary condition toward Neumann boundary conditions on $\mathcal{N}$ under favorable topological conditions.

LEMMA 3.21 (Deformation Toward Neumann Boundary Conditions). *Assume that the close curve $\gamma_\mathcal{N}$ lies in $\mathbb{C}^{**} = \mathbb{C} \setminus \{-i, i\}$ and is homotopic to 0 in $\mathbb{C}^{**}$. Then, there exists a continuous family of first-order differential operators $B_{(s)}$ on $\mathcal{N}$, $s \in [0,1]$ such that*

(1) $B_{(1)} = B$ *and* $B_{(0)} = \partial_\nu$.

[3] In fact, one faces two cases. Either m^0 is not a critical point; then since $\dim\mathcal{N} = 1$, the function ρ cannot vanish near m^0. Or m^0 is a critical point; then, as ρ is a Morse function, it is nondegenerate; then, either $\operatorname{Re}\rho$ or $\operatorname{Im}\rho$ is nondegenerate, and since $\dim\mathcal{N} = 1$ again, this implies that the either $\operatorname{Re}\rho$ or $\operatorname{Im}\rho$ cannot vanish near m^0.

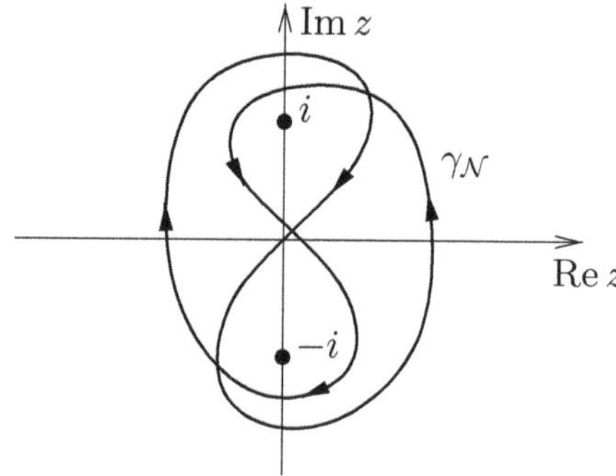

FIGURE 3.2. Example of a curve $\gamma_{\mathcal{N}}$ that does not fit the assumption of Lemma 3.21. However, both the winding numbers of i and $-i$ are zero

(2) *The Lopatinskiĭ–Šapiro condition holds on $\mathcal{N}$ for $(P, B_{(s)})$ for all $s \in [0,1]$.*

PROOF. There exists $\gamma_{(s),\mathcal{N}} : [0,1] \times \mathcal{N} \to \mathbb{C}^{**}$ such that $\gamma_{(1),\mathcal{N}} = \gamma_{\mathcal{N}}$ and $\gamma_{(0),\mathcal{N}}(m) = 0$ for all $m \in \mathcal{N}$. One sees that the boundary operator $B_{(s)} = \partial_\nu + \gamma_{(s),\mathcal{N}} Y'$ is such that the Lopatinskiĭ–Šapiro condition holds on $\mathcal{N}$ for $(P, B_{(s)})$ for all $s \in [0,1]$. With the definition of $B_{(s)}$ for $s \in [0,1]$, all the listed properties are fulfilled. ■

REMARK 3.22. Note that for the assumption of this lemma to hold, it is not sufficient that both the winding numbers of i and $-i$ with respect to the closed curve $\gamma_{\mathcal{N}}$ be zero. Figure 3.2 provides an example.

On the other hand, the following lemma allows one to deform the boundary condition toward the case of tangential boundary conditions.

LEMMA 3.23 (Deformation Toward Tangential Boundary Conditions). *Assume that the close curve $\gamma_{\mathcal{N}}$ is homotopic to $\mu : \mathbb{S}^1 \ni z \mapsto 2z^k$, for some $k \in \mathbb{Z}$, then there exists a continuous family of first-order differential operators $B_{(s)}$ on $\mathcal{N}$, $s \in [0,1]$ such that*

(1) *$B_{(1)} = B$ and $B_{(0)}$ is a nonvanishing tangential operator on $\mathcal{N}$.*
(2) *The Lopatinskiĭ–Šapiro condition holds on $\mathcal{N}$ for $(P, B_{(s)})$ for all $s \in [0,1]$.*

PROOF. By assumption, there exists $\gamma_{(s),\mathcal{N}} : [1/2, 1] \times \mathcal{N} \to \mathbb{C} \setminus \{-i, i\}$ such that $\gamma_{(1),\mathcal{N}} = \gamma_{\mathcal{N}}$ and $\gamma_{(1/2),\mathcal{N}} = \mu$; one sees that the boundary operator $B_{(s)} = \partial_\nu + \gamma_{(s),\mathcal{N}} Y'$ is such that the Lopatinskiĭ–Šapiro condition holds on $\mathcal{N}$ for $(P, B_{(s)})$ for all $s \in [1/2, 1]$.

Next, for $s \in [0, 1/2]$, we set $B_{(s)} = 2s\partial_\nu + \mu Y'$. One sees that the Lopatinskiĭ–Šapiro condition holds on $\mathcal{N}$ for $(P, B_{(s)})$ for all $s \in [0, 1/2]$.

With the definition of $B_{(s)}$ for $s \in [0,1]$, all the listed properties are fulfilled. ■

LEMMA 3.24. *Consider the first-order differential operator* $W = X + iY + V$, *where* X *and* Y *are smooth real vector fields and* $V \in \mathscr{C}^\infty(\mathcal{N}; \mathbb{C})$. *Assume that* $X + iY$ *is nonvanishing. Consider* W *as a bounded operator from* $H^{s+1}(\mathcal{N})$ *into* $H^s(\mathcal{N})$, $s \in \mathbb{R}$. *Then,* W *is Fredholm and* $\operatorname{ind}(W) = 0$.

PROOF. As $\mathcal{N}$ is diffeomorphic to $\mathbb{S}^1$ parameterized by $\theta \in \mathbb{R}/2\pi\mathbb{Z}$, upon division by a smooth nonvanishing function, we may consider $W = \partial_\theta + q$. Upon perturbation by a compact operator, by Theorem 11.5 of Volume 1, it is equivalent to prove that $\partial_\theta + 1/2$ is Fredholm with index equal to zero. By Fourier series, one finds that this operator is in fact invertible. ■

3.5.3.2. *Computation of the Index in Some Topological Situations.*

THEOREM 3.25. *We use the notation of the previous section. We assume that for every connected component* $\mathcal{N}$ *of* ${}^1\partial\mathcal{M}$, *we either have*

(1) *The close curve* $\gamma_\mathcal{N}$ *lies in* $\mathbb{C}^{**} = \mathbb{C} \setminus \{-i, i\}$ *and is homotopic to* 0 *in* $\mathbb{C}^{**}$.
(2) *The close curve* $\gamma_\mathcal{N}$ *is homotopic to* $\mu : \mathbb{S}^1 \ni z \mapsto 2z^k$.

Then, the index of $L = (P, B) : H^2(\mathcal{M}) \to L^2(\mathcal{M}) \oplus H^{3/2,1/2}(\partial\mathcal{M})$ *is zero.*

PROOF. On each connected component of ${}^1\partial\mathcal{M}$, we use the continuous deformation given by either Lemma 3.21 or Lemma 3.23. As the Lopatinskiĭ–Šapiro condition holds along that deformation, the continuously deformed operator L remains Fredholm and its index remains unchanged by Theorem 11.4 of Volume 1. We thus end up with either Neumann boundary conditions or tangential boundary conditions on the different connected components of ${}^1\partial\mathcal{M}$. On ${}^0\partial\mathcal{M}$, as in the case $d = 3$, we may simply replace the boundary operator by the Dirichlet operator. One then sees that the setting is exactly the same setting as we have obtained in the case $d = 3$. The proof of Lemma 3.20 can be adapted *mutatis mutandis*, for instance, by replacing the use of Lemma 3.19 by that of Lemma 3.24. ■

The topological assumption in the previous theorem is far from being trivial. The following result shows that in the case $d = 2$, any value in $\mathbb{Z}$ can be obtained as the index of the operator L for a properly chosen boundary operator B. This example shows also the importance of the winding number of $+i$ and $-i$ with respect to $\gamma_\mathcal{N}$ for the value of the index.

PROPOSITION 3.26. *Let* $\mathbb{D}$ *be the unit disc in* $\mathbb{C}$, *and let* $k \in \mathbb{Z}$. *We denote by* r *the complex modulus and by* θ *the complex argument. Let* Δ *be the Laplace operator with the flat metric on* $\mathbb{D}$, *that is,* $\Delta = \partial_x^2 + \partial_y^2 = 4\partial\bar{\partial}$ *and set* $Bu_{|\partial\mathbb{D}} = (\partial_r u + \alpha\partial_\theta u)_{|r=1}$ *with* $\alpha(\theta) = i + e^{ik\theta}$. *Then,* $L = (P, B)$ *is Fredholm and* $\operatorname{ind}(L) = -k$.

In this example, the closed path $\gamma_\mathcal{N}$ for $\mathcal{N} = \partial\mathbb{D}$ is given by $\gamma_\mathcal{N}(\theta) = \alpha(\theta)$. The closed path thus goes k times around $+i$ counter clockwise. The Lopatinskiĭ–Šapiro condition holds by (3.5.13), and thus L is Fredholm by Theorem 3.1.

For the proof of Proposition 3.26, it is convenient to use the following lemma.

LEMMA 3.27. *Define the operator* $\tilde{B} : w \mapsto B\,\mathsf{D}(w_{|\partial\mathbb{D}})$ *that is bounded from* $H^{3/2}(\partial\mathbb{D})$ *into* $H^{1/2}(\partial\mathbb{D})$. *The operator* $\tilde{B}$ *is Fredholm. Moreover,* $\operatorname{nul} L = \operatorname{nul} \tilde{B}$ *and* $\operatorname{def} L = \operatorname{def} \tilde{B}$.

Recall that D is the Dirichlet lifting map. On $\mathbb{D}$, if $h \in H^{3/2}(\partial\mathbb{D})$, then $\mathsf{D}(h)$ is the unique H^2 solution of

$$\Delta u = 0, \qquad u_{|\partial\mathbb{D}} = h;$$

see Definition 18.37 and Proposition 18.39.

PROOF. Denote by R_0 the resolvent map associated with the homogeneous Dirichlet problem, that is, if $f \in L^2(\mathbb{D})$, then $v = \mathsf{R}_0 f$ is the unique solution in $H^2(\mathbb{D})$ of

$$\Delta v = f, \qquad v_{|\partial\mathbb{D}} = 0.$$

First, we consider the kernels of L and $\tilde{B}$. We see that $\mathsf{D}(\ker \tilde{B}) = \ker L$. As D is injective and $\ker L$ is finite dimensional, we obtain $\dim\ker \tilde{B} = \dim\ker L$.

Second, let us exhibit the connection between the ranges of the two maps. Let $(f,h) \in L^2(\mathbb{D}) \times H^{1/2}(\partial\mathbb{D})$ be in $\operatorname{Ran}(L)$. There exists $u \in H^2(\mathbb{D})$ such that $Lu = (f,h)$. We have $u = \mathsf{R}_0 f + \mathsf{D}u_{|\partial\mathbb{D}}$, leading to $h = Bu_{|\partial\mathbb{D}} = B\mathsf{R}_0 f_{|\partial\mathbb{D}} + \tilde{B}(u_{|\partial\mathbb{D}})$. We thus find $h - B\mathsf{R}_0 f_{|\partial\mathbb{D}} \in \operatorname{Ran}(\tilde{B})$.

Conversely, let $(f,h) \in L^2(\mathbb{D}) \times H^{1/2}(\partial\mathbb{D})$ be such that $h - B\mathsf{R}_0 f_{|\partial\mathbb{D}} \in \operatorname{Ran}(\tilde{B})$. Let then $w \in H^{3/2}(\partial\mathbb{D})$ be such that $\tilde{B}w = h - B\mathsf{R}_0 f_{|\partial\mathbb{D}}$. If we set $u = \mathsf{R}_0 f + \mathsf{D}w$, then $Pu = f$ and

$$Bu_{|\partial\mathbb{D}} = B\mathsf{R}_0 f_{|\partial\mathbb{D}} + B\mathsf{D}w_{|\partial\mathbb{D}} = B\mathsf{R}_0 f_{|\partial\mathbb{D}} + \tilde{B}w = h.$$

Thus, $(f,h) \in \operatorname{Ran}(L)$. We have thus found that

$$(f,h) \in \operatorname{Ran}(L) \quad \Leftrightarrow \quad h - B\mathsf{R}_0 f_{|\partial\mathbb{D}} \in \operatorname{Ran}(\tilde{B}),$$

or equivalently $k \in \operatorname{Ran}(\tilde{B}) \quad \Leftrightarrow \quad (f, k + B\mathsf{R}_0 f_{|\partial\mathbb{D}}) \in \operatorname{Ran}(L)$, for all $f \in L^2(\mathbb{D})$. In particular, this gives $\operatorname{Ran}(\tilde{B})$ closed in $H^{1/2}(\mathbb{D})$.

Assume that $(g,\ell) \in L^2(\mathbb{D}) \oplus H^{-1/2}(\partial\mathbb{D})$ is orthogonal to $\operatorname{Ran}(L)$ in the sense of duality. Then,

$$\langle f, g\rangle_{L^2(\mathbb{D}),L^2(\mathbb{D})} + \langle k + B\mathsf{R}_0 f_{|\partial\mathbb{D}}, \ell\rangle_{H^{1/2}(\mathbb{D}),H^{-1/2}(\mathbb{D})} = 0,$$

for all $f \in L^2(\mathbb{D})$ and all $k \in \operatorname{Ran}(\tilde{B})$. This is equivalent to having

$$g = -\mathsf{R}_0^* B^* \ell \quad \text{and} \quad \ell \perp \operatorname{Ran}(\tilde{B}).$$

This yields $\operatorname{codim}(\operatorname{Ran}(L)) = \operatorname{codim}(\operatorname{Ran}(\tilde{B}))$. ∎

PROOF OF PROPOSITION 3.26. For the computation of $\operatorname{nul} L$, we characterize $\ker L$. For the computation of $\operatorname{def} L$, we compute in fact $\operatorname{def} \tilde{B}$ and we use Lemma 3.27.

Let $u \in \ker L$, that is, $u \in H^2(\mathbb{D})$ with $\Delta u = 0$ and $Bu_{|\partial\mathbb{D}} = 0$. As $u \in H^2(\mathbb{D})$, then $u_{|\partial\mathbb{D}} \in H^{3/2}(\partial\mathbb{D})$. We may thus write $u_{|\partial\mathbb{D}}$ in the form

$$u_{|\partial\mathbb{D}} = \sum_{n\in\mathbb{Z}} u_{(n)} e^{in\theta}, \quad \text{with } \big(\langle n\rangle^{3/2} u_{(n)}\big) \in \ell^2(\mathbb{Z}).$$

Then, u is of the form

$$u = \sum_{n\geq 0} u_{(n)} z^n + \sum_{n\geq 1} u_{(-n)} \overline{z}^n = \sum_{n\in\mathbb{Z}} u_{(n)} r^{|n|} e^{in\theta},$$

as this is solution of $\Delta u = 0$ with the Dirichlet trace given above and this solution is unique. To see this, use that $\Delta = 4\partial\overline{\partial}$, with $\partial = \partial_z = (\partial_x - i\partial_y)/2$ and $\overline{\partial} = \partial_{\overline{z}} = (\partial_x + i\partial_y)/2$. We now compute

$$\begin{aligned} Bu_{|\partial\mathbb{D}} &= \sum_{n\geq 1} n u_{(n)} e^{in\theta}(1 + i\alpha) + \sum_{n\geq 1} n u_{(-n)} e^{-in\theta}(1 - i\alpha) \\ &= i \sum_{n\geq 1} n u_{(n)} e^{i(n+k)\theta} + 2 \sum_{n\geq 1} n u_{(-n)} e^{-in\theta} - i \sum_{n\geq 1} n u_{(-n)} e^{i(k-n)\theta} \\ &= i \sum_{n\in\mathbb{Z}} n u_{(n)} e^{i(n+k)\theta} - 2 \sum_{n\leq -k-1} (n+k) u_{(n+k)} e^{i(n+k)\theta}. \end{aligned}$$

Having $Bu_{|\partial\mathbb{D}} = 0$ reads

$$n u_{(n)} = 0 \ \text{ for } n \geq -k, \qquad i n u_{(n)} - 2(n+k) u_{(n+k)} = 0 \ \text{ for } n \leq -k-1.$$

Consider the case $k \geq 0$. With the first condition, we have $u_{(n)} = 0$ if $n \geq -k$ and if $n \neq 0$. With the second equation, we conclude that $u_{(n)} = 0$ for all $n \neq 0$. Hence, $\ker(L) = \{u = u_{(0)}\}$ and $\operatorname{nul} L = 1$.

Consider now the case $k \leq -1$. With the first condition, we have $u_{(n)} = 0$ if $n \geq -k = |k|$. With the second equation, for $n = 0$, we find that $u_{(k)} = 0$ and then $u_{(jk)} = 0$, $j \in \mathbb{N}$, by induction. If $|k| \geq 2$, we also find

$$u_{(n+jk)} = \big(\frac{i}{2}\big)^j \frac{n}{n+2k} u_{(n)}$$

for $1 \leq n \leq |k|-1$ and $j \in \mathbb{N}$. An element of $\ker L$ is thus fully characterized by the values of $u_{(0)}, \dots, u_{(|k|-1)}$, and the resulting values of $(u_{(n)})$ yield a converging series and $u \in H^2(\mathbb{D})$. Hence, $\operatorname{nul} L = |k|$.

Second, we prove that $\operatorname{def}(\tilde{B}) = k+1$ if $k \geq 0$ and $\operatorname{def}(\tilde{B}) = 0$ if $k \leq -1$. We compute the adjoint of $\tilde{B}$ using $L^2(\partial\mathbb{D})$ as a pivot space. For $w = \sum_{n\in\mathbb{Z}} w_{(n)} e^{in\theta}$ with $\big(\langle n\rangle^{3/2} w_{(n)}\big) \in \ell^2(\mathbb{Z})$ using the computations made above, we have

$$(\tilde{B}w, \ell) = \sum_{n\in\mathbb{Z}} (\tilde{B}w)_{(n)} \overline{\ell_{(n)}} = \sum_{n\in\mathbb{Z}} w_{(n)} \big(\overline{-in\ell_{(n+k)}}\big) + \sum_{n\leq -1} w_{(n)} \big(\overline{-2n\ell_{(n)}}\big)$$

for $\ell = \sum_{n\in\mathbb{Z}} \ell_{(n)} e^{in\theta} \in H^{-1/2}(\partial\mathbb{D})$, that is, $(\langle n\rangle^{-1/2}\ell_{(n)}) \in \ell^2(\mathbb{Z})$. This gives $\tilde{B}^*\ell = v = \sum_{n\in\mathbb{Z}} v_{(n)} e^{in\theta}$ with

$$v_{(n)} = -in\ell_{(n+k)} \text{ if } n \geq 0, \quad \text{and} \quad v_{(n)} = -in\ell_{(n+k)} - 2n\ell_{(n)} \text{ if } n \leq -1.$$

Observe that these explicit formulae give the expected boundedness of $\tilde{B}^*$ from $H^{-1/2}(\partial\mathbb{D})$ into $H^{-3/2}(\partial\mathbb{D})$.

Consider the case $k = 0$. If $\tilde{B}^*\ell = 0$, then ℓ is fully determined by $\ell_{(0)}$. Consider the case $k > 0$. If $\tilde{B}^*\ell = 0$, then ℓ is fully determined by $\ell_{(0)}, \dots, \ell_{(k)}$. In fact, $\ell_{(n)} = 0$ for $n \geq k+1$ and $\ell_{(-kr+j)} = (-i/2)^r \ell_{(j)}$, $j = 0, \dots, k-1$, $r \in \mathbb{N}$, leading to a convergent series.

For $k \geq 0$, we thus have $\dim\ker(\tilde{B}^*) = k+1$. As $\overline{\mathrm{Ran}(\tilde{B})} = (\ker \tilde{B}^*)^\perp$, and as here $\overline{\mathrm{Ran}(\tilde{B})} = \mathrm{Ran}(\tilde{B})$ since $\tilde{B}$ is Fredholm, we find that $\mathrm{def}(\tilde{B}) = k+1$.

Consider now the case $k \leq -1$. If $\tilde{B}^*\ell = 0$, then $\ell_{(n)} = 0$ for $n \geq -|k|+1$. We also find that

$$\ell_{(-j|k|)} = (2i)^{(j-1)} \ell_{(-|k|)}, \quad j \in \mathbb{N}^*, \tag{3.5.14}$$

and if $|k| \geq 2$, $\ell_{n-j|k|} = 0$ for $1 \leq n \leq |k|-1$ and $j \in \mathbb{N}$. As we have $\langle n\rangle^{-1/2}\ell_{(n)} \in \ell^2(\mathbb{Z})$, condition (3.5.14) implies $\ell_{(-|k|)} = 0$ and thus $\ell = 0$. We thus have $\dim\ker(\tilde{B}^*) = 0$. Arguing as above, this gives $\mathrm{def}(\tilde{B}) = 0$. ∎

3.6. Additional Regularity Results

Here, we consider a function $u \in L^2(\mathcal{M})$ such that $Pu \in H^{m-1}(\mathcal{M})$ with $m \geq 1$. Then, $u \in \mathcal{W}_P(\mathcal{M})$, and from Lemma 18.32, we know that both the Dirichlet and Neumann traces can be defined and $\gamma^{\mathsf{D}}(u) \in H^{-1/2}(\partial\mathcal{M})$ and $\gamma^{\mathsf{N}}(u) \in H^{-3/2}(\partial\mathcal{M})$.

Here, we wish to prove that additional regularity can be reached in the case of smoother boundary data by means of a boundary operator B that fulfills the Lopatinskiĭ–Šapiro conditions.

THEOREM 3.28. *Let $\ell, \ell' \in \mathbb{N}$ with $\ell \geq 1$, and let $u \in L^2(\mathcal{M})$ be such that $Pu \in H^{\ell-1}(\mathcal{M})$. Let B be a boundary operator as in Sect. 3.1 such that Lopatinskiĭ–Šapiro condition holds for (P, B) on $\partial\mathcal{M}$. If $Bu_{|\partial\mathcal{M}} \in H_B^{(\ell'+1/2)}(\partial\mathcal{M})$, then $u \in H^{\min(\ell,\ell')+1}(\mathcal{M})$.*

This result is consistent with Corollary 18.41 obtained in the case $P = -\Delta_g$ along with Dirichlet boundary conditions.

PROOF. From standard elliptic theory, the function u is $H^{\ell+1}$ away from the boundary. We thus only consider the issue of its regularity in some neighborhood of $\partial\mathcal{M}$.

First, we use normal geodesic coordinates as given by Theorem 17.22: there exist V an open set of $\mathcal{M}$ neighborhood of $\partial\mathcal{M}$, $z_0 > 0$, and a diffeomorphism Φ, such that

$$\begin{aligned}\Phi : \partial\mathcal{M} \times [0, z_0) &\to V\\ (m', z) &\mapsto \Phi(m', z).\end{aligned}$$

The pullback of the metric takes the form

$$\Phi^* g_{(m',z)} = g'_{m'}(z) \otimes 1_z + d_z \otimes d_z,$$

where $g'(z)$ is a Riemannian metric on $\partial\mathcal{M}$ that smoothly depends on z. In these coordinates, by Lemma 18.68, the function u has the following regularity:

(3.6.1)

$$u \in H^2\big([0, z_0); H^{-2}(\partial\mathcal{M})\big) \cap H^1\big([0, z_0); H^{-1}(\partial\mathcal{M})\big) \cap L^2\big((0, z_0) \times \partial\mathcal{M}\big).$$

Second, let $m^0 \in \partial\mathcal{M}$, and let $\mathcal{C}_\partial = (O_\partial, \kappa_\partial)$ be a local chart of $\partial\mathcal{M}$ such that $m^0 \in O_\partial$. This allows us to form a local chart $\mathcal{C} = (O, \kappa)$ with $O = \Phi\big(O_\partial \times [0, z_0)\big)$ and $\kappa(m) = (\kappa(m'), z)$ if $m = \Phi(m', z)$. With $\psi \in \mathscr{C}^\infty(\mathcal{M})$ with $\operatorname{supp}(\psi) \subset O$, we set $v = \psi u$. By (3.6.1), we have

$$v \in H^2\big([0, z_0); H^{-2}(\partial\mathcal{M})\big) \cap H^1\big([0, z_0); H^{-1}(\partial\mathcal{M})\big) \tag{3.6.2}$$

and $\gamma^{\mathsf{D}}(v) = \psi_{|z=0}\gamma^{\mathsf{D}}(u) \in H^{-1/2}(\partial\mathcal{M})$ and $\gamma^{\mathsf{N}}(v) \in H^{-3/2}(\partial\mathcal{M})$ by the Leibniz rule given in Lemma 18.32. Since $[P, \psi]$ is a first-order differential operator on $\mathcal{M}$, with (3.6.1), we find

$$f = Pv = \psi Pu + [P, \Psi]v \in L^2\big([0, z_0); H^{-1}(\partial\mathcal{M})\big).$$

We still denote by v and f the representative of v and f in the chart $\mathcal{C}$. As we did in Sect. 3.2, we write $P(x, D)$ and $B(x, D)$ in place of $P^{\mathcal{C}}(x, D)$ and $B^{\mathcal{C}}(x, D)$, respectively.

We observe that the regularity properties of v given in (3.6.2) allow one to carry out the analysis of Sect. 3.2.2. We may thus write

$$\underline{v} = Q\underline{f} - Q\big(\delta'_{x_d=0} \otimes \gamma^{\mathsf{D}}(v)\big) + Q\big(\delta_{x_d=0} \otimes \tilde{\gamma}^{\mathsf{N}}(v)\big) + R\underline{v}.$$

Note that $\underline{f} \in L^2\big(\mathbb{R}; H^{-1}(\mathbb{R}^{d-1})\big) \subset H^{-1}(\mathbb{R}^d)$. As $Q \in \Psi^{-2}(\mathbb{R}^d)$, we have $Q\underline{f} \in H^1(\mathbb{R}^d)$. Note also that

$$\begin{aligned}Q\big(\delta'_{x_d=0} \otimes \gamma^{\mathsf{D}}(v)\big)_{|\mathbb{R}^d_+} &= \mathscr{Q}^{(0)}\big(\gamma^{\mathsf{D}}(v)\big) \quad \text{and}\\ Q\big(\delta_{x_d=0} \otimes \tilde{\gamma}^{\mathsf{N}}(v)\big)_{|\mathbb{R}^d_+} &= \mathscr{Q}^{(-1)}\big(\tilde{\gamma}^{\mathsf{N}}(v)\big),\end{aligned}$$

since $\mathscr{Q}^{(0)}$ and $\mathscr{Q}^{(-1)}$ are both well defined on $\mathscr{S}'(\mathbb{R}^{d-1})$. We thus have

$$v = \big(Q\underline{f}\big)_{|\mathbb{R}^d_+} - \mathscr{Q}^{(0)}\big(\gamma^{\mathsf{D}}(v)\big) + \mathscr{Q}^{(-1)}\big(\tilde{\gamma}^{\mathsf{N}}(v)\big) + \big(R\underline{v}\big)_{|\mathbb{R}^d_+}. \tag{3.6.3}$$

By Lemma 3.6, we have $\mathscr{Q}^{(0)}\big(\gamma^{\mathsf{D}}(v)\big)$ and $\mathscr{Q}^{(-1)}\big(\tilde{\gamma}^{\mathsf{N}}(v)\big)$ in $L^2(\mathbb{R}^d_+)$ and $P\mathscr{Q}^{(0)}\big(\gamma^{\mathsf{D}}(v)\big) \in H^N(\mathbb{R}^d_+)$ and $P\mathscr{Q}^{(-1)}\big(\tilde{\gamma}^{\mathsf{N}}(v)\big) \in H^N(\mathbb{R}^d_+)$ for any $N \in \mathbb{N}$. Observe that the proof of Lemma 18.32 in Sect. 18.A.1 can be adapted *mutatis mutandis* in the case $\partial\mathcal{M}$ is replaced[4] by $\mathbb{R}^{d-1}$. The Dirichlet traces of $\mathscr{Q}^{(0)}(\gamma^{\mathsf{D}}(v))$ and $\mathscr{Q}^{(-1)}(\tilde{\gamma}^{\mathsf{N}}(v))$ are thus well defined, and they lie in $H^{-1/2}(\mathbb{R}^{d-1})$. From the regularity of $Q\underline{f}$, we thus have

$$\gamma^{\mathsf{D}}(v) + \gamma^{\mathsf{D}}\big(\mathscr{Q}^{(0)}(\gamma^{\mathsf{D}}(v))\big) - \gamma^{\mathsf{D}}\big(\mathscr{Q}^{(-1)}(\tilde{\gamma}^{\mathsf{N}}(v))\big) \in H^{1/2}(\mathbb{R}^{d-1}).$$

As $\mathscr{Q}^{(0)}_{\mathsf{D}} \in \Psi^0(\mathbb{R}^{d-1})$ and $\mathscr{Q}^{(-1)}_{\mathsf{D}} \in \Psi^{-1}(\mathbb{R}^{d-1})$ as shown in Lemma 3.9, their action on function in $H^{-1/2}(\mathbb{R}^{d-1})$ and $H^{-3/2}(\mathbb{R}^{d-1})$ makes sense, and we find

$$\big(\mathrm{Id} + \mathscr{Q}^{(0)}_{\mathsf{D}}\big)(\gamma^{\mathsf{D}}(v)) - \mathscr{Q}^{(-1)}_{\mathsf{D}}(\tilde{\gamma}^{\mathsf{N}}(v)) \in H^{1/2}(\mathbb{R}^{d-1}),$$

which reads

$$\big(\mathrm{Id} - \mathsf{C}^{\mathsf{D},\mathsf{D}}\big)(\gamma^{\mathsf{D}}(v)) - \mathsf{C}^{\mathsf{D},\mathsf{N}}(\tilde{\gamma}^{\mathsf{N}}(v)) \in H^{1/2}(\mathbb{R}^{d-1}). \tag{3.6.4}$$

As we have $Bu_{|\partial\mathcal{M}} \in H^{(\ell'+1/2)}_B(\partial\mathcal{M})$, if $m^0 \in {}^k\partial\mathcal{M}$, that is, the operator B is of order k on the connected component of $\partial\mathcal{M}$ where m^0 lies, we have $Bv_{|{}^k\partial\mathcal{M}} \in H^{\ell'+1/2-k}({}^k\partial\mathcal{M})$. In local coordinates, this reads

(3.6.5)
$$\begin{aligned} H^{\ell'+1/2-k}(\mathbb{R}^{d-1}) \ni B^k(x',0,D')v_{|x_d=0^+} &- iB^{k-1}(x',0,D')D_d v_{|x_d=0^+} \\ &= \big(B^k(x',0,D') + iB^{k-1}(x',0,D')r_0(x',x_d=0)\big)\gamma^{\mathsf{D}}(v) \\ &\quad + B^{k-1}(x',0,D')\tilde{\gamma}^{\mathsf{N}}(v). \end{aligned}$$

Together, (3.6.4) and (3.6.5) read

$$L_\partial \begin{pmatrix} \gamma^{\mathsf{D}}(v) \\ \tilde{\gamma}^{\mathsf{N}}(v) \end{pmatrix} \in H^{1/2}(\mathbb{R}^{d-1}) \oplus H^{\ell'+1/2-k}(\mathbb{R}^{d-1}) = H^{1/2,\ell'+1/2-k}(\mathbb{R}^{d-1}),$$

with the notation introduced in the beginning of Sect. 3.2.5. By Lemma 3.11, we then obtain that

$$\gamma^{\mathsf{D}}(v) \in H^{1/2}(\mathbb{R}^{d-1}) \quad \text{and} \quad \tilde{\gamma}^{\mathsf{N}}(v) \in H^{-1/2}(\mathbb{R}^{d-1}).$$

Thus, by Lemma 3.6, we have $\mathscr{Q}^{(0)}(\gamma^{\mathsf{D}}(v))$ and $\mathscr{Q}^{(-1)}(\tilde{\gamma}^{\mathsf{N}}(u))$ in $H^1(\mathbb{R}^d_+)$. Thus, by (3.6.3), we find that $v \in H^1(\mathbb{R}^d_+)$. If $\ell' = 0$, this result is precisely the result we sought.

We may now assume $\ell' \geq 1$. Starting from the beginning, we now find that $f = Pv \in L^2(\mathbb{R})$. Then, $\mathscr{Q}(f) = \big(Q\underline{f}\big)_{|\mathbb{R}^d_+} \in H^2(\mathbb{R}^d_+)$ by Lemma 3.6 and

$$v = \mathscr{Q}(f) - \mathscr{Q}^{(0)}(\gamma^{\mathsf{D}}(v)) + \mathscr{Q}^{(-1)}(\tilde{\gamma}^{\mathsf{N}}(v)) + \big(R\underline{v}\big)_{|\mathbb{R}^d_+}. \tag{3.6.6}$$

Arguing as above, we find that

$$L_\partial \begin{pmatrix} \gamma^{\mathsf{D}}(v) \\ \tilde{\gamma}^{\mathsf{N}}(v) \end{pmatrix} \in H^{3/2}(\mathbb{R}^{d-1}) \oplus H^{\ell'+1/2-k}(\mathbb{R}^{d-1}) = H^{3/2,\ell'+1/2-k}(\mathbb{R}^{d-1}),$$

[4]Observe that we did mention this replacement in one of the key results: Lemma 18.69.

implying by Lemma 3.11 that, as $\ell' + 1/2 \geq 3/2$,

$$\gamma^{\mathsf{D}}(v) \in H^{3/2}(\mathbb{R}^{d-1}) \quad \text{and} \quad \tilde{\gamma}^{\mathsf{N}}(v) \in H^{1/2}(\mathbb{R}^{d-1}).$$

This yields $v \in H^2(\mathbb{R}^d_+)$ by (3.6.6). If $\min(\ell', \ell) = 1$, this result is precisely the result we sought.

If now $\min(\ell', \ell) \geq 2$, we iterate the argument we used above. This leads to $v \in H^{\min(\ell',\ell)+1}(\mathbb{R}^d_+)$, and thus we find $u \in H^{\min(\ell',\ell)+1}(\mathcal{M})$. ■

The result of Theorem 3.28 may very well fail to hold if the Lopatinskiĭ–Šapiro condition is not fulfilled. To see this, let us use Example 2.6 with $\mathcal{M} = \mathbb{D}$, the unit disc, $P = -\Delta = -\partial_x^2 - \partial_y^2$ and $Bu_{|\partial\mathbb{D}} = \partial_r u_{|\partial\mathbb{D}} + i\partial_\theta u_{|\partial\mathbb{D}}$. There we saw that the Lopatinskiĭ–Šapiro condition does not hold for (P, B). Let us use polar coordinates on $\mathbb{D}$ and consider the function given by $u(r, \theta) = \log(1 - re^{i\theta}) = \log(1 - z)$. This function is in $L^2(\mathbb{D})$ (see below). The function u is also holomorphic in $\mathbb{D}$. As a result, it satisfies $Pu = 0$ in $\mathscr{D}'(\mathbb{D})$ since $P = -4\partial\overline{\partial}$ with $\overline{\partial} = \partial_{\overline{z}} = (\partial_x + i\partial_y)/2$. Consequently, $u \in \mathcal{W}_P(\mathbb{D})$ with this space defined in (18.6.4). By Lemma 18.32, the Dirichlet and Neumann traces of u are well defined in $H^{-1/2}(\partial\mathbb{D})$ and $H^{-3/2}(\partial\mathbb{D})$, respectively. Thus, $Bu_{|\partial\mathbb{D}}$ is well defined in $H^{-3/2}(\partial\mathbb{D})$. The following proposition shows that a regularity result as in Theorem 3.28 does not hold.

Proposition 3.29. *The function u satisfies $Pu = 0$ and $Bu_{|\partial\mathbb{D}} = 0$, meaning that $Bu_{|\partial\mathbb{D}} \in \cap_{s\in\mathbb{R}} H^s(\partial\mathbb{D})$. Yet, $u \in L^2(\mathbb{D}) \setminus H^1(\mathbb{D})$.*

Proof. The L^2-norm is given by

$$\|u\|^2_{L^2(\mathbb{D})} = \int_0^{2\pi}\int_0^1 r|\log(1 - re^{i\theta})|^2 dr d\theta.$$

The only singularity in the integrand occurs at $r = 1$ and $\theta = 0$. There, we have $r|\log(1 - re^{i\theta})|^2 \sim \log(1 - r)^2$ that is integrable. Thus, $\|u\|_{L^2(\mathbb{D})} < \infty$.

We have

$$\|u\|^2_{H^1(\mathbb{D})} \eqsim \|u\|^2_{L^2(\mathbb{D})} + \int_0^{2\pi}\int_0^1 \big(|\partial_r u|^2 + r^{-2}|\partial_\theta u|^2\big) r dr d\theta.$$

This can be seen by observing that the Riemannian metric associated with Δ is $g = dr^2 + r^2 d\theta^2$ in the (r, θ) coordinates. Then, $\nabla_g u = (\partial_r u)\partial_r + r^{-2}(\partial_\theta u)\partial_\theta$ by (17.2.2). We compute $\partial_r u(r, \theta) = -e^{i\theta}/(1 - re^{i\theta})$. Observe then that

$$r|\partial_r u(r, \theta)|^2 \sim \frac{1}{(1 - r)^2} \quad \text{as}(r, \theta) \to (1, 0),$$

implying that $\int_0^{2\pi}\int_0^1 |\partial_r u|^2 r dr d\theta = +\infty$.

We now prove that $\partial_r u_{|r=1} + i\partial_\theta u_{|r=1} = 0$. First, observe that these traces make sense by Theorem 18.32 as $u \in \mathcal{W}_P(\mathbb{D})$.

As we have $\partial_r + ir^{-1}\partial_\theta = 2e^{-i\theta}\overline{\partial}$ and as u is holomorphic in $\mathbb{D}$, we find $(\partial_r + ir^{-1}\partial_\theta)u = 0$ for $r < 1$. From (18.A.3) and (18.A.4) in the proof of Lemma 18.32 given in Sect. 18.A.1, we obtain $\partial_r u_{|r=1} + i\partial_\theta u_{|r=1} = 0$ by letting $r \to 1^-$. ■

3.7. Notes

The Lopatinskiĭ–Šapiro boundary conditions presented in Chap. 2 are here studied from the point of view of the associated boundary value problem as is done in the seminal works of Y. B. Lopatinskiĭ [238] and Z. Y. Šapiro [323]. Here, we only treat the case of second-order elliptic operators as in the whole book. A good reference for the study of general elliptic operators of arbitrary order along with Lopatinskiĭ–Šapiro conditions is the work of L. Hörmander [175, Section 20.1]. Here, we have chosen to consider this analysis only for a second order which makes it simpler to expose. For further details, we refer the reader to Chapter 20 in [175] and to the bibliographical notes therein.

Lopatinskiĭ–Šapiro conditions are necessary and sufficient for the boundary value problem to be Fredholm. Both the necessary and the sufficient parts of the equivalence are treated with care here. Moreover, as presented in Sect. 3.6, the Lopatinskiĭ–Šapiro conditions give the proper framework to study how, for the boundary value problem, the regularity of the source terms, in the interior or at the boundary, gives the regularity of the solution.

Concerning the value of the index of a boundary value problem, we do not attempt to make any connection with the celebrated Atiyah–Singer theorem [43, 44]. For this connection, we again refer to Section 20.3 in [175]. However, in the case a boundary operator of order at most one, we show how a continuous deformation of the boundary operator can lead to a boundary problem with the same index and for which the index is easily computed, e.g., in the case of Dirichlet or Neumann boundary conditions. In dimension $d \geq 3$, we then find a zero index. In dimension $d = 2$, examples of boundary value problems with a nonzero index are provided.

Appendix

3.A. Proof of Some Technical Results

3.A.1. An *A Priori* Estimate Consequence of the Fredholm Property. Here, we prove Corollary 3.3. By Theorem 3.1, if the Lopatinskiĭ–Šapiro condition holds, then the operator L given in (3.1.3) is Fredholm. First, $\ker(L)$ is finite dimensional, say $\dim\ker(L) = \operatorname{nul} L = J$, and is given by $\ker(L) = \operatorname{span}\{u_j; 1 \leq j \leq J\}$ for some $u_j \in H^{m+2}(\mathcal{M})$. Second, $\operatorname{Ran}(L)$ has a finite codimension, say $\operatorname{codim}\operatorname{Ran}(L) = \operatorname{def} L = N$, and is given by $\operatorname{Ran}(L) = \cap_{1\leq n\leq N}\phi_n^\perp$ for some $\phi_n \in H^m(\mathcal{M}) \oplus H_B^{(m+3/2)}(\partial\mathcal{M})$. We then define

$$\begin{aligned}\tilde{L} : H^{m+2}(\mathcal{M}) \oplus \mathbb{C}^N &\to H^m(\mathcal{M}) \oplus H_B^{(m+3/2)}(\partial\mathcal{M}) \oplus \mathbb{C}^J\\ (u,\alpha) &\mapsto \Big(Lu + \sum_{1\leq n\leq N} \alpha_n\phi_n, (u,u_1)_{L^2(\mathcal{M})}, \dots, (u,u_J)_{L^2(\mathcal{M})}\Big).\end{aligned}$$

We claim that $\tilde{L}$ is bijective. In fact, if $\tilde{L}(u,\alpha) = 0$, then $u \in \big(\ker(L)\big)^{\perp}$ since $(u, u_j)_{L^2(\mathcal{M})} = 0$, $j = 1, \dots, J$. As Lu and $\sum_{1\le n\le N} \alpha_n \phi_n$ lie in orthogonal subspaces of $H^m(\mathcal{M}) \oplus H_B^{(m+3/2)}(\partial\mathcal{M})$, we thus have $Lu = 0$ and $\sum_{1\le n\le N} \alpha_n\phi_n = 0$ implying that $u = 0$ and $\alpha_1 = \cdots = \alpha_N = 0$. We thus see that $\tilde{L}$ is injective.

If $u = u_j$ and $\alpha = 0$, then $\tilde{L}(u,\alpha) = (0, e^j)$ with $e^1, \dots, e^J$ forming the canonical basis of $\mathbb{C}^J$. If $F \in H^m(\mathcal{M}) \oplus H_B^{(m+3/2)}(\partial\mathcal{M})$, it takes the form $F = F_1 + F_2$ with $F_1 \in \operatorname{Ran}(L)$ and $F_2 \in \operatorname{span}\{\phi_n;\ 1 \le n \le N\}$. Thus, there exist $u \in H^{m+2}(\mathcal{M})$ and $\alpha \in \mathbb{C}^N$ such that $Lu = F_1$ and $\sum_{1\le n\le N} \alpha_n\phi_n = F_2$; that is, $\tilde{L}(u,\alpha) = (F, z)$ with some $z \in \mathbb{C}^J$. By linearity, we see that $\tilde{L}$ is surjective.

By the closed graph theorem, the inverse map $\tilde{L}^{-1}$ is bounded yielding

$$\begin{aligned}\|u\|_{H^{2+m}(\mathcal{M})} + |\alpha|_{\mathbb{C}^N} &\lesssim \|\tilde{L}(u,\alpha)\|_{H^m(\mathcal{M})\oplus H_B^{(m+3/2)}(\partial\mathcal{M})\oplus\mathbb{C}^J}\\ &\lesssim \Big\|Lu + \sum_{1\le n\le N} \alpha_n\phi_n\Big\|_{H^m(\mathcal{M})\oplus H_B^{(m+3/2)}(\partial\mathcal{M})}\\ &\quad + \big|\big((u,u_1)_{L^2(\mathcal{M})}, \dots, (u,u_J)_{L^2(\mathcal{M})}\big)\big|_{\mathbb{C}^J}.\end{aligned}$$

If we now choose $\alpha = 0$, the conclusion follows by the Cauchy–Schwarz inequality. ■

3.A.2. Fredholm Index Independent of the Regularity Level. Here, we prove Proposition 3.4.

Here, we denote by $L^{(m)}$ the operator from $H^{m+2}(\mathcal{M})$ into $H^m(\mathcal{M}) \oplus H_B^{(m+3/2)}(\partial\mathcal{M})$ defined in (3.1.3). We also denote by $M^{(m)}$, $K^{(m),\ell}$, $K^{(m),r}$ the operators given by Proposition 3.2. The operator $K^{(m),\ell}$ is bounded from $H^{m+2}(\mathcal{M})$ into $H^{m+3}(\mathcal{M})$. Note also that since P and B are differential operators, we have

$$L^{(m)}{}_{|H^{m+3}(\mathcal{M})} = L^{(m+1)}. \tag{3.A.1}$$

Let $m \in \mathbb{N}$. With (3.A.1), one has $\ker(L^{(m+1)}) \subset \ker(L^{(m)})$. Let now $u \in \ker(L^{(m)})$, meaning $u \in H^{m+2}(\mathcal{M})$, $Pu = 0$, and $Bu = 0$. By Proposition 3.2, we have $u = -K^{(m),\ell}u$. With the above observation, we find that $u \in H^{m+3}(\mathcal{M})$. This means that $u \in \ker(L^{(m+1)})$. We have thus obtained that $\ker(L^{(m+1)}) = \ker(L^{(m)})$. Consequently, $\operatorname{nul} L^{(m+1)} = \operatorname{nul} L^{(m)}$.

With (3.A.1), one has $\operatorname{Ran}(L^{(m+1)}) \subset \operatorname{Ran}(L^{(m)})$. As $n = \operatorname{def} L^{(m+1)} < \infty$, there exist $\phi_1, \dots, \phi_n$ linearly independent continuous forms on $H^{m+1}(\mathcal{M}) \oplus H_B^{(m+5/2)}(\partial\mathcal{M})$ such that $\operatorname{Ran}(L^{(m+1)}) = \cap_{1\le j\le n} \ker(\phi_j)$. As $H^{m+1}(\mathcal{M}) \oplus H_B^{(m+5/2)}(\partial\mathcal{M})$ is dense in $H^m(\mathcal{M}) \oplus H_B^{(m+3/2)}(\partial\mathcal{M})$, the maps $\phi_1, \dots, \phi_n$ can be uniquely extended as continuous forms on this latter space; we denote by $\tilde{\phi}_1, \dots, \tilde{\phi}_n$ these extensions. Let now $u \in H^{m+2}(\mathcal{M})$ and $(u_n)_n \subset H^{m+3}(\mathcal{M})$ be a sequence that converges to u in $H^{m+2}(\mathcal{M})$. Then, for $j = 1, \dots, n$, we have $0 = \phi_j(Pu_n, Bu_n) = \tilde{\phi}_j(Pu_n, Bu_n) \to \tilde{\phi}_j(Pu, Bu)$. We thus find that $\operatorname{Ran}(L^{(m)}) \subset \cap_{1\le j\le n} \ker(\tilde{\phi}_j)$. Thus, $\operatorname{def} L^{(m)} \ge \operatorname{def} L^{(m+1)}$.

As $n' = \operatorname{def} L^{(m)} < \infty$, there exist $\psi_1, \dots, \psi_{n'}$ linearly independent continuous forms on $H^m(\mathcal{M}) \oplus H_B^{(m+3/2)}(\partial\mathcal{M})$ such that $\operatorname{Ran}(L^{(m)}) = \cap_{1\leq j\leq n'} \ker(\psi_j)$. Since $H^{m+1}(\mathcal{M}) \oplus H_B^{(m+5/2)}(\partial\mathcal{M})$ is dense in $H^m(\mathcal{M}) \oplus H_B^{(m+3/2)}(\partial\mathcal{M})$, the restrictions $\hat{\psi}_1, \dots, \hat{\psi}_{n'}$ of these forms to $H^{m+1}(\mathcal{M}) \oplus H_B^{(m+5/2)}(\partial\mathcal{M})$ are also linearly independent. We thus find that $\operatorname{Ran}(L^{(m+1)}) \subset \cap_{1\leq j\leq n'} \ker(\hat{\psi}_j)$. Thus, $\operatorname{def} L^{(m+1)} \geq \operatorname{def} L^{(m)}$.

3.A.3. Sobolev Regularity of the Parametrix Action. Here, we prove Lemma 3.6. First, we consider the operator $\mathscr{Q}$. Let $\phi \in H^m(\mathbb{R}^d_+)$ and set $v = \mathscr{Q}\phi$. Let $\psi, \tilde{\psi} \in \mathscr{C}^\infty(\mathbb{R}^d_+)$ both vanishing for $0 \leq x_d \leq 1$ be such that $\tilde{\psi} \equiv 1$ on a neighborhood of $\operatorname{supp}(\psi)$. We can consider ψv as a function on the whole $\mathbb{R}^d$ that vanishes on $\mathbb{R}^d_-$; we have

$$\psi v = \psi Q \underline{\phi} = \psi Q(\tilde{\psi}\phi) + \psi Q(1 - \tilde{\psi})\underline{\phi}.$$

As $\tilde{\psi}\phi \in H^m(\mathbb{R}^d)$ and $Q \in \Psi^{-2}(\mathbb{R}^d)$, we have $\psi Q(\tilde{\psi}\phi) \in H^{m+2}(\mathbb{R}^d)$. As $\psi Q(1 - \tilde{\psi}) \in \Psi^{-N}(\mathbb{R}^d)$ for any $N \in \mathbb{N}$ by symbol calculus, we find that $\psi Q(1 - \tilde{\psi})\underline{\phi} \in H^{m+2}(\mathbb{R}^d)$. Moreover, the map $\phi \mapsto \psi\mathscr{Q}\phi$ is bounded from $H^m(\mathbb{R}^d_+)$ into $H^{m+2}(\mathbb{R}^d_+)$. We thus see that the result holds away from $\{x_d = 0\}$. In what follows, we show how this Sobolev regularity holds up to the boundary by means of an iterative procedure that relies on the structure of the operator $P(x, D)$.

As $Q \in \Psi^{-2}(\mathbb{R}^d)$, by Proposition 2.52 of Volume 1 (adapted to standard operators), we have

$$\|\Lambda^m_{\mathsf{T}} Q\underline{\phi}\|_{H^2(\mathbb{R}^d)} \lesssim \|\Lambda^m_{\mathsf{T}} \underline{\phi}\|_{L^2(\mathbb{R}^d)} \lesssim \|\phi\|_{H^m(\mathbb{R}^d_+)}.$$

Consequently, we have $\Lambda^m_{\mathsf{T}} v \in H^2(\mathbb{R}^d_+)$, that is,

$$v \in \bigcap_{0\leq k\leq 2} H^k\big(\mathbb{R}_+; H^{m+2-k}(\mathbb{R}^{d-1})\big). \tag{3.A.2}$$

Moreover, we have $\sum_{0\leq k\leq 2} \|v\|_{H^k(\mathbb{R}_+;H^{m+2-k}(\mathbb{R}^{d-1}))} \lesssim \|\phi\|_{H^m(\mathbb{R}^d_+)}$.

With (3.2.4), we have

$$P(x, D)Q\underline{\phi} = \underline{\phi} + R'\underline{\phi},$$

where $R' \in \Psi^{-\infty}(\mathbb{R}^d \times \mathbb{R}^d)$. For $w \in \mathscr{S}'(\mathbb{R}^d)$, we set $\mathscr{R}'w = \big(R'w\big)_{|\mathbb{R}^d_+}$. For any $N \in \mathbb{N}$, we have $\|\mathscr{R}'\underline{\phi}\|_{H^N(\mathbb{R}^d_+)} \lesssim \|\underline{\phi}\|_{L^2(\mathbb{R}^d)} = \|\phi\|_{L^2(\mathbb{R}^d_+)}$. Since the operator $P(x, D)$ is local, we obtain

$$P(x, D)v = \big(P(x, D)Q\underline{\phi}\big)_{|\mathbb{R}^d_+} = \phi + \mathscr{R}'\underline{\phi}.$$

We set $K\phi = \mathscr{R}'\underline{\phi}$.

We set $w = (1 - \psi)v$, that is, we localize v near the boundary. From (3.A.2), we have

$$w \in \bigcap_{0 \leq k \leq 2} H^k\big(\mathbb{R}_+; H^{m+2-k}(\mathbb{R}^{d-1})\big). \tag{3.A.3}$$

Moreover, we have $\sum_{0 \leq k \leq 2} \|w\|_{H^k(\mathbb{R}_+; H^{m+2-k}(\mathbb{R}^{d-1}))} \lesssim \|\phi\|_{H^m(\mathbb{R}^d_+)}$. We set $\alpha(x) = \int_0^{x_d} r_0(x', \sigma) d\sigma$, and we have $P(x, D) = D_d^2 + r_0 D_d + R(x, D') = e^{-\alpha} D_d\big(e^{\alpha} D_d\big) + R(x, D')$. We have

$$P(x, D)w = (1 - \psi)P(x, D)v - [P(x, D), \psi]v.$$

Above we saw that v is H^{m+2} in the support of ψ, we thus find that $P(x, D)w \in H^m(\mathbb{R}^d_+)$ and $\|P(x, D)w\|_{H^m(\mathbb{R}^d_+)} \lesssim \|\phi\|_{H^m(\mathbb{R}^d_+)}$.

From (3.A.3), we obtain

$$D_d\big(e^{\alpha} D_d w\big) \in \bigcap_{0 \leq k \leq 2} H^k\big(\mathbb{R}_+; H^{m-k}(\mathbb{R}^{d-1})\big),$$

as e^{α} is bounded in the support of w. Moreover, we have

$$\sum_{0 \leq k \leq 2} \|D_d\big(e^{\alpha} D_d\big)\|_{H^k(\mathbb{R}_+; H^{m-k}(\mathbb{R}^{d-1}))} \lesssim \|\phi\|_{H^m(\mathbb{R}^d_+)}.$$

With Lemma 18.69 applied twice, we conclude that

$$v \in \bigcap_{0 \leq k \leq 4} H^k\big(\mathbb{R}_+; H^{m+2-k}(\mathbb{R}^{d-1})\big)$$

and $\sum_{0 \leq k \leq 4} \|v\|_{H^k(\mathbb{R}_+; H^{m+2-k}(\mathbb{R}^{d-1}))} \lesssim \|\phi\|_{H^m(\mathbb{R}^d_+)}$.

If we iterate this argument j times then we find

$$v \in \bigcap_{0 \leq k \leq 2+2j} H^k\big(\mathbb{R}_+; H^{m+2-k}(\mathbb{R}^{d-1})\big)$$

and $\sum_{0 \leq k \leq 2+2j} \|v\|_{H^k(\mathbb{R}_+; H^{m+2-k}(\mathbb{R}^{d-1}))} \lesssim \|\phi\|_{H^m(\mathbb{R}^d_+)}$. If $2 + 2j \geq m + 2$, then $v \in H^{m+2}(\mathbb{R}^d_+)$ and $\|v\|_{H^{m+2}(\mathbb{R}^d_+)} \lesssim \|\phi\|_{H^m(\mathbb{R}^d_+)}$.

Second, we consider the operator $\mathscr{Q}^{(j)}$ for some $j \in \{-1, 0\}$. Let $h \in H^{m+j-1/2}(\mathbb{R}^{d-1})$ and set $v^{(j)} = \mathscr{Q}^{(j)}h$. With $\psi, \tilde{\psi} \in \mathscr{C}^\infty(\mathbb{R}^d_+)$ chosen as at the beginning of the proof, we write

$$\psi v^{(j)} = \psi Q(\delta^{(j+1)}_{x_d=0} \otimes h) = \psi Q\big((1 - \tilde{\psi})\delta^{(j+1)}_{x_d=0} \otimes h\big).$$

As $\psi Q(1 - \tilde{\psi}) \in \Psi^{-N}(\mathbb{R}^d)$ for any $N \in \mathbb{N}$ by symbol calculus, we find that $\psi v^{(j)} \in H^m(\mathbb{R}^d)$, and moreover the map $h \mapsto \psi \mathscr{Q}^{(j)} h$ is bounded from $H^{m+j-1/2}(\mathbb{R}^{d-1})$ into $H^m(\mathbb{R}^d_+)$.

Set $k^{(j)} = \delta^{(j+1)}_{x_d=0} \otimes h$. Its Fourier transform is given by $(i\xi_d)^{j+1}\hat{h}(\xi')$. For $r > j - 1/2$ and $s \in \mathbb{R}$, we compute

$$\|\Lambda_{\mathsf{T}}^{s+r} k^{(j)}\|^2_{H^{-r-2}(\mathbb{R}^d)} = \int_{\mathbb{R}^d} \langle \xi' \rangle^{2(s+r)} \langle \xi \rangle^{-2r-4} \xi_d^{2(j+1)} |\hat{h}(\xi')|^2 d\xi$$
$$\lesssim \int_{\mathbb{R}^{d-1}} \langle \xi' \rangle^{2(s+j)-1} |\hat{h}(\xi')|^2 d\xi' = |h|^2_{H^{s+j-1/2}(\mathbb{R}^{d-1})},$$

using that, for $a > 0$,

$$\int_{\mathbb{R}} (a^2 + \xi_d^2)^{-r-2} \xi_d^{2(j+1)} d\xi_d = a^{2(j-r)-1} \int_{\mathbb{R}} (1 + \eta^2)^{-r-2} \eta^{2(j+1)} \partial\eta \lesssim a^{2(j-r)-1},$$

if $r > j - 1/2$. In particular, with $0 < \varepsilon < 1/2$, we have $r = j - 1/2 + \varepsilon < 0$, and with $s = 0$, we find that $\Lambda_{\mathsf{T}}^{j-1/2+\varepsilon} k^{(j)} \in H^{-j-3/2-\varepsilon}(\mathbb{R}^d)$ and

$$\text{(3.A.4)} \qquad \|k^{(j)}\|_{H^{-2}(\mathbb{R}^d)} \leq \|\Lambda_{\mathsf{T}}^{j-1/2+\varepsilon} k^{(j)}\|_{H^{-j-3/2-\varepsilon}(\mathbb{R}^d)} \lesssim |h|_{H^{j-1/2}(\mathbb{R}^{d-1})}.$$

As $Q \in \Psi^{-2}(\mathbb{R}^d)$, by Proposition 2.52 of Volume 1 (adapted to standard operators), we obtain

$$\|\Lambda_{\mathsf{T}}^{m+r} Q k^{(j)}\|_{H^{-r}(\mathbb{R}^d)} \lesssim \|\Lambda_{\mathsf{T}}^{m+r} k^{(j)}\|_{H^{-r-2}(\mathbb{R}^d)} \lesssim |h|_{H^{m+j-1/2}(\mathbb{R}^{d-1})}.$$

Since $r > j - 1/2$ with $j = -1, 0$, we may pick $r = 0$, and we obtain

$$\|v^{(j)}\|_{L^2(\mathbb{R}_+; H^m(\mathbb{R}^{d-1}))} \lesssim |h|_{H^{m+j-1/2}(\mathbb{R}^{d-1})}.$$

Arguing as above, we have

$$P(x,D)v^{(j)} = \big(P(x,D)Qk^{(j)}\big)_{|\mathbb{R}^d_+} = \mathscr{R}'(k^{(j)}),$$

using that $\big(k^{(j)}\big)_{|\mathbb{R}^d_+} = 0$. With (3.A.4), for any $N \in \mathbb{N}$, we obtain

$$\|\mathscr{R}' k^{(j)}\|_{H^N(\mathbb{R}^d_+)} \lesssim |h|_{H^{m+j-1/2}(\mathbb{R}^{d-1})}.$$

We set $K^{(j)} h = \mathscr{R}' k^{(j)}$.

Choosing $N \geq m$, we obtain $P(x,D)v^{(j)} \in H^m(\mathbb{R}^d)$. We set $w^{(j)} = (1-\psi)v^{(j)}$. Arguing as we did for $w = (1-\psi)v$, and using the structure of $P(x,D)$, we obtain

$$D_d\big(e^\alpha D_d w^{(j)}\big) \in L^2\big(\mathbb{R}_+; H^{m-2}(\mathbb{R}^{d-1})\big),$$

with, moreover, $\|D_d\big(e^\alpha D_d w^{(j)}\big)\|_{L^2(\mathbb{R}_+; H^{m-2}(\mathbb{R}^{d-1}))} \lesssim |h|_{H^{m+j-1/2}(\mathbb{R}^{d-1})}$. With Lemma 18.69 applied twice, we conclude that $w^{(j)} \in H^2\big(\mathbb{R}_+; H^{m-2}(\mathbb{R}^{d-1})\big)$ and $\|w^{(j)}\|_{H^2(\mathbb{R}_+; H^{m-2}(\mathbb{R}^{d-1}))} \lesssim |h|_{H^{m+j+3/2}(\mathbb{R}^{d-1})}$. With an interpolation argument [74, 236], we obtain that

$$w^{(j)} \in \bigcap_{0 \leq k \leq 2} H^k\big(\mathbb{R}_+; H^{m-k}(\mathbb{R}^{d-1})\big)$$

and $\sum_{0 \leq k \leq 2} \|w^{(j)}\|_{H^k(\mathbb{R}_+; H^{m-k}(\mathbb{R}^{d-1}))} \lesssim |h|_{H^{m+j-1/2}(\mathbb{R}^{d-1})}$. From that point, we proceed as we did for v from (3.A.2). ∎

3.A.4. Properties of the Calderón Projector. Here, we prove Proposition 3.8.

Let $h^{(0)} \in H^{m+3/2}(\mathbb{R}^{d-1})$ and $h^{(-1)} \in H^{m+1/2}(\mathbb{R}^{d-1})$. We set $v^{(j)} = \mathscr{Q}^{(j)} h^{(j)}$ for $j = -1, 0$. We have $v^{(j)} \in H^{m+2}(\mathbb{R}^d_+)$ by Lemma 3.6 and $P(x,D)v^{(j)} = K^{(j)} h^{(j)}$, with $K^{(j)}$ bounded from $H^{m+3/2+j}(\mathbb{R}^{d-1})$ into $H^N(\mathbb{R}^d_+)$, for any $N \in \mathbb{N}$. By (3.2.8), we then have

$$v^{(j)} = -\mathscr{Q}^{(0)}(\gamma^{\mathsf{D}}(v^{(j)})) + \mathscr{Q}^{(-1)}(\tilde{\gamma}^{\mathsf{N}}(v^{(j)})) + \mathscr{Q}(K^{(j)}h^{(j)}) + \big(R\underline{v}^{(j)}\big)_{|\mathbb{R}^d_+},$$

where $h^{(j)} \mapsto \mathscr{Q}(K^{(j)}h^{(j)}) + \big(R\underline{v}^{(j)}\big)_{|\mathbb{R}^d_+}$ is bounded from $H^{m+3/2+j}(\mathbb{R}^{d-1})$ into $H^N(\mathbb{R}^d_+)$, for any $N \in \mathbb{N}$. Computing both the Dirichlet γ^{D} and the modified Neumann traces $\tilde{\gamma}^{\mathsf{N}}$ at $x_d = 0^+$, we obtain

$$\begin{aligned}
\big(\mathrm{Id} + \mathscr{Q}^{(0)}_{\mathsf{D}}\big)(\gamma^{\mathsf{D}}(v^{(j)})) - \mathscr{Q}^{(-1)}_{\mathsf{D}}(\tilde{\gamma}^{\mathsf{N}}(v^{(j)})) &= Y^{(j)}_{\mathsf{D}}(h^{(j)}),\\
\big(\mathscr{Q}^{(1)}_{\mathsf{N}}\big)(\gamma^{\mathsf{D}}(v^{(j)})) + \big(\mathrm{Id} - \mathscr{Q}^{(0)}_{\mathsf{N}}\big)(\tilde{\gamma}^{\mathsf{N}}(v^{(j)})) &= Y^{(j)}_{\mathsf{N}}(h^{(j)}),
\end{aligned}$$

with $Y^{(j)}_{\mathsf{D}}$ and $Y^{(j)}_{\mathsf{N}}$ both bounded from $H^{m+3/2+j}(\mathbb{R}^{d-1})$ into $H^N(\mathbb{R}^{d-1})$, for any $N \in \mathbb{N}$. As $\gamma^{\mathsf{D}}(v^{(j)}) = \mathscr{Q}^{(j)}_{\mathsf{D}}(h^{(j)})$ and $\tilde{\gamma}^{\mathsf{N}}(v^{(j)}) = \mathscr{Q}^{(j+1)}_{\mathsf{N}}(h^{(j)})$ and using the definition of the entries of the matrix operator C, we have, in the case $j = -1$,

$$\begin{aligned}
\big(\mathrm{Id} - \mathsf{C}^{\mathsf{D},\mathsf{D}}\big)\mathsf{C}^{\mathsf{D},\mathsf{N}} - \mathsf{C}^{\mathsf{D},\mathsf{N}}\mathsf{C}^{\mathsf{N},\mathsf{N}} &= Y^{(-1)}_{\mathsf{D}},\\
-\,\mathsf{C}^{\mathsf{N},\mathsf{D}}\mathsf{C}^{\mathsf{D},\mathsf{N}} + \big(\mathrm{Id} - \mathsf{C}^{\mathsf{N},\mathsf{N}}\big)\mathsf{C}^{\mathsf{N},\mathsf{N}} &= Y^{(-1)}_{\mathsf{N}},
\end{aligned}$$

and, in the case $j = 0$,

$$\begin{aligned}
\big(\mathrm{Id} - \mathsf{C}^{\mathsf{D},\mathsf{D}}\big)\mathsf{C}^{\mathsf{D},\mathsf{D}} - \mathsf{C}^{\mathsf{D},\mathsf{N}}\mathsf{C}^{\mathsf{N},\mathsf{D}} &= -Y^{(0)}_{\mathsf{D}},\\
-\,\mathsf{C}^{\mathsf{N},\mathsf{D}}\mathsf{C}^{\mathsf{D},\mathsf{D}} + \big(\mathrm{Id} - \mathsf{C}^{\mathsf{N},\mathsf{N}}\big)\mathsf{C}^{\mathsf{N},\mathsf{D}} &= -Y^{(0)}_{\mathsf{N}}.
\end{aligned}$$

We then observe that this reads

$$\begin{pmatrix} \mathrm{Id} - \mathsf{C}^{\mathsf{D},\mathsf{D}} & -\mathsf{C}^{\mathsf{D},\mathsf{N}} \\ -\mathsf{C}^{\mathsf{N},\mathsf{D}} & \mathrm{Id} - \mathsf{C}^{\mathsf{N},\mathsf{N}} \end{pmatrix} \begin{pmatrix} \mathsf{C}^{\mathsf{D},\mathsf{D}} & \mathsf{C}^{\mathsf{D},\mathsf{N}} \\ \mathsf{C}^{\mathsf{N},\mathsf{D}} & \mathsf{C}^{\mathsf{N},\mathsf{N}} \end{pmatrix} = (\mathrm{Id} - \mathsf{C})\mathsf{C} = \begin{pmatrix} -Y^{(0)}_{\mathsf{D}} & Y^{(-1)}_{\mathsf{D}} \\ -Y^{(0)}_{\mathsf{N}} & Y^{(-1)}_{\mathsf{N}} \end{pmatrix}.$$

As the operator on the r.h.s. maps $H^{m+3/2}(\mathbb{R}^{d-1}) \oplus H^{m+1/2}(\mathbb{R}^{d-1})$ into $\big(H^N(\mathbb{R}^{d-1})\big)^2$, we have reached the first result.

We now consider $\phi \in H^m(\mathbb{R}^d_+)$. Set $v = \mathscr{Q}(\phi)$. We have $v \in H^{m+2}(\mathbb{R}^d_+)$ by Lemma 3.6 and $P(x,D)v = \phi + K\phi$ with $\phi \mapsto K\phi$ bounded from $H^m(\mathbb{R}^d_+)$ into $H^N(\mathbb{R}^d_+)$, for any $N \in \mathbb{N}$. Applying identity (3.2.14), we obtain

$$\big(\mathrm{Id} - \mathsf{C}\big)\begin{pmatrix} \gamma^{\mathsf{D}}(v) \\ \gamma^{\mathsf{N}}(v) \end{pmatrix} = \begin{pmatrix} \mathscr{Q}_{\mathsf{D}}(\phi) \\ \mathscr{Q}_{\mathsf{N}}(\phi) \end{pmatrix} + \begin{pmatrix} \mathscr{Q}_{\mathsf{D}}(K\phi) \\ \mathscr{Q}_{\mathsf{N}}(K\phi) \end{pmatrix} + R_{\mathsf{C}}(v).$$

Observe that

$$\phi \mapsto \begin{pmatrix} \mathscr{Q}_{\mathsf{D}}(K\phi) \\ \mathscr{Q}_{\mathsf{N}}(K\phi) \end{pmatrix} + R_{\mathsf{C}}(v)$$

maps $H^m(\mathbb{R}^d_+)$ into $\big(H^N(\mathbb{R}^{d-1})\big)^2$ for any $N \in \mathbb{N}$. Since $\gamma^{\mathsf{D}}(v) = \mathscr{Q}_{\mathsf{D}}(\phi)$ and $\tilde{\gamma}^{\mathsf{N}}(v) = \mathscr{Q}_{\mathsf{N}}(\phi)$, the result follows. ■

3.A.5. Pseudo-Differential Form of the Action on the Traces. Here, we prove Lemma 3.9. We set

$$w = \delta_{x_d=0} \otimes h \quad \text{and} \quad w_\varepsilon(x) = \psi_\varepsilon(x_d) h(x') \in \mathscr{S}(\mathbb{R}^d),$$

with $\psi_\varepsilon(s) = \varepsilon^{-1}\psi(\varepsilon^{-1}s)$ with $\varepsilon > 0$ and $\psi \in \mathscr{C}_c^\infty(\mathbb{R})$ with $\operatorname{supp}(\psi) \subset \mathbb{R}_-$ and $\int \psi = 1$. As w_ε converges to w in $\mathscr{S}'(\mathbb{R}^d)$ and as Q maps $\mathscr{S}'(\mathbb{R}^d)$ into itself continuously by Proposition 2.19 of Volume 1 (see Remark 2.55 also in Volume 1), we have

$$Q(\delta_{x_d=0} \otimes h) = \lim_{\varepsilon \to 0} Q(w_\varepsilon) \quad \text{in } \mathscr{S}'(\mathbb{R}^d).$$

As Q maps $\mathscr{S}(\mathbb{R}^d)$ into itself, we have $Qw_\varepsilon \in \mathscr{S}(\mathbb{R}^d)$, and, in the sense of oscillatory integrals (see Theorem 2.11 of Volume 1), we have

(3.A.5)
$$\begin{aligned} Q(w_\varepsilon)(x) &= (2\pi)^{-d} \iint_{\mathbb{R}^d \times \mathbb{R}^d} e^{i(x-y)\cdot\xi} q(x,\xi) w_\varepsilon(y)\, dy d\xi \\ &= (2\pi)^{-(d-1)} \iint_{\mathbb{R}^{d-1} \times \mathbb{R}^{d-1}} e^{i(x'-y')\cdot\xi'} q_\varepsilon^{(-1)}(x_d)(x',\xi') h(y')\, dy' d\xi', \end{aligned}$$

where $q_\varepsilon^{(-1)}(x_d)(x',\xi')$ is a function given by

$$q_\varepsilon^{(-1)}(x_d)(x',\xi') = (2\pi)^{-1} \iint_{\mathbb{R} \times \mathbb{R}} e^{i(x_d-y_d)\xi_d} q(x,\xi)\psi_\varepsilon(y_d)\, dy_d d\xi_d,$$

using the Fubini theorem that applies to oscillatory integrals. Observe that the integrand has a compact support in y_d and behaves like $\langle \xi_d \rangle^{-2}$ as $|\xi_d| \to \infty$ and thus this formula is well defined in the sense of the Lebesgue integral. In particular, we have continuity with respect to (x_d, x', ξ') by the Lebesgue dominated-convergence theorem. Moreover, differentiation with respect to (x', ξ') under the integral sign can be performed yielding a smooth function with respect to these tangential variables. We can regularize the oscillatory integral defining $q_\varepsilon^{(-1)}(x_d)(x',\xi')$. Using that

$$i(1 + \partial_{y_d}) e^{i(x_d-y_d)\xi_d} = (\xi_d + i) e^{i(x_d-y_d)\xi_d},$$

we obtain, with integration by parts,

$$\begin{aligned} q_\varepsilon^{(-1)}(x_d)(x',\xi') &= i(2\pi)^{-1} \iint_{\mathbb{R} \times \mathbb{R}} e^{i(x_d-y_d)\xi_d} q(x,\xi) \frac{{}^t(1+\partial_{y_d})\big(\psi_\varepsilon(y_d)\big)}{(\xi_d + i)}\, dy_d d\xi_d \\ &= i^k (2\pi)^{-1} \iint_{\mathbb{R} \times \mathbb{R}} e^{i(x_d-y_d)\xi_d} q(x,\xi) \frac{{}^t(1+\partial_{y_d})^k\big(\psi_\varepsilon(y_d)\big)}{(\xi_d + i)^k}\, dy_d d\xi_d, \end{aligned}$$

for $k \in \mathbb{N}$. With this last form, we see that the integrand now behaves like $\langle \xi_d \rangle^{-2-k}$ as $|\xi_d| \to \infty$. This allows for differentiations with respect to x_d. We conclude that $q_\varepsilon^{(-1)}(x_d)(x',\xi')$ is smooth with respect to (x_d, x', ξ').

Knowing now that $q_\varepsilon^{(-1)}(x_d)(x',\xi')$ is smooth with respect to all its variables, we now study its symbol properties. For that, it suffices to consider $|\xi'| \geq 1$. In such case, by (3.2.1), we have

$$q(x,\xi) = p(x,\xi)^{-1}\tilde{q}(x,\xi) = \frac{\tilde{q}(x,\xi)}{\big(\xi_d - i\rho(x,\xi')\big)\big(\xi_d + i\rho(x,\xi')\big)}, \tag{3.A.6}$$

where $\tilde{q}(x,\xi)$ has 1 for principal symbol. The iterative construction of $\tilde{q}$ in the proof of Proposition 2.34 in Appendix 2.A.7 of Volume 1 shows that here $\tilde{q}(x,\xi) - 1$ is a rational function in the variable ξ.

We then observe that the integrand is meromorphic in the upper complex half-plane with a pole at $\xi_d = i\rho(x,\xi')$. We consider $x_d \geq 0$. Because of the support of ψ, we have $x_d - y_d > 0$ and thus $\operatorname{Re}(i(x_d - y_d)\xi_d) \leq 0$ for $\operatorname{Im}\xi_d \geq 0$. A classical change of contour of integration then gives

$$q_\varepsilon^{(-1)}(x_d)(x',\xi') = (2\pi)^{-1} \int_{\mathbb{R}} \int_{\Gamma_0} e^{i(x_d - y_d)\xi_d} q(x,\xi)\psi_\varepsilon(y_d)\, d\xi_d dy_d,$$

where Γ_0 is the closed path in $\mathbb{C}$, oriented counter-clockwise, composed with

(1) the real interval $[-R_0\langle\xi'\rangle, R_0\langle\xi'\rangle]$;
(2) the upper half-circle centered at 0 of radius $R_0\langle\xi'\rangle$,

with $R_0 \geq 1 + |\rho(x,\xi')|/\langle\xi'\rangle$, for $x \in \mathbb{R}^d_+$ and $\xi' \in \mathbb{R}^{d-1}$.

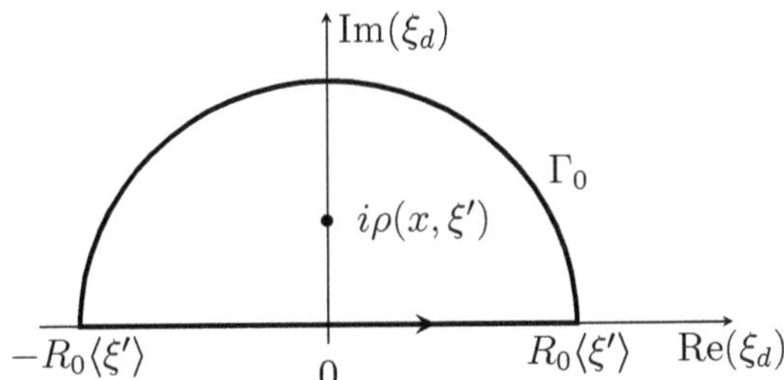

With (3.A.6), the residue theorem yields

$$q_\varepsilon^{(-1)}(x_d)(x',\xi') = t^{(-1)}(x_d)(x',\xi')e_\varepsilon(x_d)(x',\xi'),$$

where $t^{(-1)}(x_d)$ is smooth with values in $S^{-1}(\mathbb{R}^{d-1}\times\mathbb{R}^{d-1})$, with $\frac{1}{2}\rho^{-1}(x',x_d,\xi')$ as principal symbol, and

$$e_\varepsilon(x_d)(x',\xi') = \int_{\mathbb{R}} e^{-(x_d - y_d)\rho(x,\xi')}\psi_\varepsilon(y_d)dy_d = \int_{\mathbb{R}} e^{-(x_d - \varepsilon s)\rho(x,\xi')}\psi(s)ds.$$

In $\operatorname{supp}(\psi)$, we have $s < 0$ and thus $\varepsilon s\rho(x,\xi') \leq 0$ since $\rho(x,\xi') \geq 0$, yielding $|e^{\varepsilon s\rho(x,\xi')}| \leq 1$. The Lebesgue dominated-convergence theorem then gives

$$e_\varepsilon(x_d)(x',\xi') \to e(x_d)(x',\xi') = e^{-x_d\rho(x,\xi')}, \qquad \text{as } \varepsilon \to 0,$$

and we obtain

$$q^{(-1)}(x_d)(x',\xi') = \lim_{\varepsilon\to 0} q_\varepsilon^{(-1)}(x_d)(x',\xi') = t^{(-1)}(x_d)(x',\xi')e^{-x_d\rho(x,\xi')}.$$

We see that $q^{(-1)}$ is a smooth function.

Observe that $\partial_{x'}^{\alpha}\partial_{\xi'}^{\beta}\big(e^{-(x_d-\varepsilon s)\rho(x,\xi')}\big)$ is a linear combination of terms of the form

$$(x_d-\varepsilon s)^k e^{-(x_d-\varepsilon s)\rho(x,\xi')}\partial_{x'}^{\alpha_1}\partial_{\xi'}^{\beta_1}\rho(x,\xi')\cdots\partial_{x'}^{\alpha_k}\partial_{\xi'}^{\beta_k}\rho(x,\xi'),$$

with $\alpha_1+\cdots+\alpha_k=\alpha$ and $\beta_1+\cdots+\beta_k=\beta$. Let $x_d\geq 0$. As $s<0$ and $\rho(x,\xi')\gtrsim\langle\xi'\rangle$, we find that

$$\big|\partial_{x'}^{\alpha}\partial_{\xi'}^{\beta}\big(e^{-(x_d-\varepsilon s)\rho(x,\xi')}\big)\big|\lesssim (x_d-\varepsilon s)^k\langle\xi'\rangle^{k-|\beta|}e^{-(x_d-\varepsilon s)\langle\xi'\rangle}\lesssim\langle\xi'\rangle^{-|\beta|},$$

where the estimation is independent of $\varepsilon>0$. We thus find that $x_d\mapsto e_\varepsilon(x_d)$ is a bounded map with values in $S^0(\mathbb{R}^{d-1}\times\mathbb{R}^{d-1})$. Similarly, we have

$$|\partial_{x'}^{\alpha}\partial_{\xi'}^{\beta}e(x_d)(x',\xi')|\lesssim\langle\xi'\rangle^{-|\beta|}.$$

We compute $(e-e_\varepsilon)(x_d)(x',\xi')=e(x_d)(x',\xi')\sigma_\varepsilon(x_d)(x',\xi')$ with

$$\sigma_\varepsilon(x_d)(x',\xi')=\int_{\mathbb{R}}(1-e^{\varepsilon s\rho(x,\xi')})\psi(s)\,ds.$$

We have $|1-e^{\varepsilon s\rho(x,\xi')}|\lesssim\varepsilon\rho(x,\xi')\lesssim\varepsilon\langle\xi'\rangle$, since $|1-e^{\theta}|\leq|\theta|$ if $\theta\leq 0$. We thus have

$$|\sigma_\varepsilon(x_d)(x',\xi')|\lesssim\varepsilon\langle\xi'\rangle.$$

With a computation similar to that performed above,

$$|\partial_{x'}^{\alpha}\partial_{\xi'}^{\beta}\sigma_\varepsilon(x_d)(x',\xi')|\lesssim\varepsilon\langle\xi'\rangle^{1-|\beta|}.$$

For $x_d\geq 0$, we thus find that the sequence $\big(\sigma_\varepsilon(x_d)\big)_\varepsilon$ with values in $S^0(\mathbb{R}^{d-1}\times\mathbb{R}^{d-1})$ converges to 1 for the topology of the symbol class S^1. This implies that the sequence $\big(q_\varepsilon^{(-1)}(x_d)\big)_\varepsilon$ with values in $S^{-1}(\mathbb{R}^{d-1}\times\mathbb{R}^{d-1})$ converges to $q^{(-1)}(x_d)$, itself in the same symbol space for the topology of the symbol class S^0.

Since the oscillatory integral (3.A.5) is continuous with respect to $q_\varepsilon^{(-1)}(x_d)$ lying in $S^0(\mathbb{R}^{d-1}\times\mathbb{R}^{d-1})$ by Theorem 2.11 of Volume 1, we find that we may let $\varepsilon\to 0$ in (3.A.5), and we obtain

(3.A.7)
$$\begin{aligned}Q(w)(x)&=(2\pi)^{-(d-1)}\iint_{\mathbb{R}^{d-1}\times\mathbb{R}^{d-1}}e^{i(x'-y')\cdot\xi'}q^{(-1)}(x_d)(x',\xi')h(y')\,dy'd\xi'\\&=\operatorname{Op}\big(q^{(-1)}(x_d)\big)h(x').\end{aligned}$$

Similarly, we set

$$w'=\delta'_{x_d=0}\otimes h\quad\text{and}\quad w'_\varepsilon(x)=\psi'_\varepsilon(x_d)h(x')\in\mathscr{S}(\mathbb{R}^d),$$

with $\psi'_\varepsilon(s)=\varepsilon^{-2}\psi'(\varepsilon^{-1}s)$ with $\varepsilon>0$. As w'_ε converges to w' in $\mathscr{S}'(\mathbb{R}^d)$, we have

$$Q(\delta'_{x_d=0}\otimes h)=\lim_{\varepsilon\to 0}Q(w'_\varepsilon)\quad\text{in }\mathscr{S}'(\mathbb{R}^d),$$

and we obtain
(3.A.8)

$$Q(w'_\varepsilon)(x) = (2\pi)^{-d} \iint\limits_{\mathbb{R}^d\times\mathbb{R}^d} e^{i(x-y)\cdot\xi} q(x,\xi) w'_\varepsilon(y)\, dy d\xi$$
$$= (2\pi)^{-(d-1)} \iint\limits_{\mathbb{R}^{d-1}\times\mathbb{R}^{d-1}} e^{i(x'-y')\cdot\xi'} q_\varepsilon^{(0)}(x_d)(x',\xi') h(y')\, dy' d\xi',$$

where $q_\varepsilon^{(0)}(x', x_d, \xi')$ is a function given by

$$q_\varepsilon^{(0)}(x_d)(x',\xi') = (2\pi)^{-1} \iint\limits_{\mathbb{R}\times\mathbb{R}} e^{i(x_d-y_d)\xi_d} q(x,\xi)\psi'_\varepsilon(y_d)\, dy_d d\xi_d.$$

Regularizing the oscillatory integral as is done above for $q_\varepsilon^{(-1)}(x_d)(x',\xi')$ in the analysis of the funtion $Q(w_\varepsilon)$ defined in (3.A.5), we obtain $q_\varepsilon^{(0)}(x_d)(x',\xi')$ is smooth with respect to all variables. We then consider $|\xi'| \geq 1$, allowing one to use (3.A.6) and to perform the same change of contour of integration as above and obtain

$$q_\varepsilon^{(0)}(x_d)(x',\xi') = (2\pi)^{-1} \int\limits_{\mathbb{R}} \int\limits_{\Gamma_0} e^{i(x_d-y_d)\xi_d} q(x,\xi)\psi'_\varepsilon(y_d)\, d\xi_d dy_d.$$

An integration by parts gives

$$q_\varepsilon^{(0)}(x_d)(x',\xi') = i(2\pi)^{-1} \int\limits_{\mathbb{R}} \int\limits_{\Gamma_0} e^{i(x_d-y_d)\xi_d} q(x,\xi)\xi_d\psi_\varepsilon(y_d)\, d\xi_d dy_d.$$

Proceeding as above, first with the residue theorem, and second letting ε go to zero, we obtain that

$$Q(\delta'_{x_d=0}\otimes h) = \operatorname{Op}\big(q^{(0)}(x_d)\big) h(x'),$$

with

$$q^{(0)}(x_d)(x',\xi') = \lim_{\varepsilon\to 0} q_\varepsilon^{(0)}(x', x_d, \xi') = t^{(0)}(x_d)(x',\xi') e^{-x_d\rho(x,\xi')},$$

where $t^{(0)}(x_d)$ is smooth with values in $S^0(\mathbb{R}^{d-1}\times\mathbb{R}^{d-1})$, with $-\frac{1}{2}$ as principal symbol. ∎

3.A.6. Recovery of the Traces up to Regularizing Operators. Here, we prove Lemma 3.11. With $\Lambda_{\mathsf{T}}^s = \operatorname{Op}_{\mathsf{T}}(\langle\xi'\rangle^s)$, $s\in\mathbb{R}$, we define

$$\tilde{L} = \begin{pmatrix} 1 & 0 \\ 0 & \Lambda_{\mathsf{T}}^{-k} \end{pmatrix} L_\partial \begin{pmatrix} 1 & 0 \\ 0 & \Lambda_{\mathsf{T}}^{1} \end{pmatrix},$$

which is a 2×2 matrix operator with each entry in $\Psi^0(\mathbb{R}^{d-1})$. The associated principal symbol is

$$\tilde{\ell}(x',\xi') = \begin{pmatrix} 1/2 & -\rho^{-1}_{|x_d=0^+}(x',\xi')\langle\xi'\rangle/2 \\ \langle\xi'\rangle^{-k} b^k_{|x_d=0^+}(x',\xi') & \langle\xi'\rangle^{1-k} b^{k-1}_{|x_d=0^+}(x',\xi') \end{pmatrix},$$

with each entry made of a symbol in $S^0(\mathbb{R}^{d-1}\times\mathbb{R}^{d-1})$. Observe that

$$\det\big(\tilde{\ell}(x',\xi')\big) = \frac{1}{2}\langle\xi'\rangle^{1-k}\rho^{-1}_{|x_d=0^+}(x',\xi')\big(b^k + b^{k-1}\rho\big)_{|x_d=0^+}(x',\xi') \neq 0,$$

for $x \in \kappa(\partial\mathcal{M} \cap O)$ and $|\xi'| \geq 1$, by the Lopatinskiĭ–Šapiro property for (P, B) written in (3.2.3) in the local coordinates. As a Result, $\tilde{L}$ (respectively, $\tilde{\ell}(x', \xi')$) is elliptic in $\kappa(\partial\mathcal{M} \cap O)$.

We set $\tilde{m}(x', \xi') = \chi_\partial(x')\big(\tilde{\ell}(x', \xi')\big)^{-1}$ and $\tilde{M} = \operatorname{Op}(\tilde{m})$ that is a 2×2 matrix operator with entries in $\Psi^0(\mathbb{R}^{d-1})$. By symbol calculus, we have

$$\tilde{M}\tilde{L} = \chi_\partial + R^\ell_{-1} \quad \text{and} \quad \tilde{L}\tilde{M} = \chi_\partial + R^r_{-1}, \tag{3.A.9}$$

where R^ℓ_{-1} and R^r_{-1} are two 2×2 matrix operators with entries in $\Psi^{-1}(\mathbb{R}^{d-1})$.

We set

$$M_\partial = \begin{pmatrix} 1 & 0 \\ 0 & \Lambda_{\mathsf{T}}^1 \end{pmatrix} \tilde{M} \begin{pmatrix} 1 & 0 \\ 0 & \Lambda_{\mathsf{T}}^{-k} \end{pmatrix}.$$

As we have

$$L_\partial = \begin{pmatrix} 1 & 0 \\ 0 & \Lambda_{\mathsf{T}}^k \end{pmatrix} \tilde{L} \begin{pmatrix} 1 & 0 \\ 0 & \Lambda_{\mathsf{T}}^{-1} \end{pmatrix},$$

we obtain

$$M_\partial L_\partial U_\partial = \begin{pmatrix} 1 & 0 \\ 0 & \Lambda_{\mathsf{T}}^1 \end{pmatrix} \tilde{M}\tilde{L} \begin{pmatrix} 1 & 0 \\ 0 & \Lambda_{\mathsf{T}}^{-1} \end{pmatrix} = \begin{pmatrix} 1 & 0 \\ 0 & \Lambda_{\mathsf{T}}^1 \end{pmatrix} (\chi_\partial + R^\ell_{-1}) \begin{pmatrix} 1 & 0 \\ 0 & \Lambda_{\mathsf{T}}^{-1} \end{pmatrix}.$$

By symbol calculus, we have

$$\begin{pmatrix} 1 & 0 \\ 0 & \Lambda_{\mathsf{T}}^1 \end{pmatrix} \chi_\partial \begin{pmatrix} 1 & 0 \\ 0 & \Lambda_{\mathsf{T}}^{-1} \end{pmatrix} = \begin{pmatrix} \chi_\partial & 0 \\ 0 & \Lambda_{\mathsf{T}}^1 \chi_\partial \Lambda_{\mathsf{T}}^{-1} \end{pmatrix} = \chi_\partial + K^\ell_{1,\partial},$$

with

$$K^\ell_{1,\partial} = \begin{pmatrix} 0 & 0 \\ 0 & K^{\ell,22}_{1,\partial} \end{pmatrix},$$

with $K^{\ell,22}_{1,\partial} \in \Psi^{-1}(\mathbb{R}^{d-1})$. We also have

$$\begin{pmatrix} 1 & 0 \\ 0 & \Lambda_{\mathsf{T}}^1 \end{pmatrix} R^\ell_{-1} \begin{pmatrix} 1 & 0 \\ 0 & \Lambda_{\mathsf{T}}^{-1} \end{pmatrix} = K^\ell_{2,\partial} = \begin{pmatrix} K^{\ell,11}_{2,\partial} & K^{\ell,12}_{2,\partial} \\ K^{\ell,21}_{2,\partial} & K^{\ell,22}_{2,\partial} \end{pmatrix},$$

where $K^{\ell,11}_{2,\partial}, K^{\ell,22}_{2,\partial} \in \Psi^{-1}(\mathbb{R}^{d-1})$, $K^{\ell,12}_{2,\partial} \in \Psi^{-2}(\mathbb{R}^{d-1})$, and $K^{\ell,21}_{2,\partial} \in \Psi^0(\mathbb{R}^{d-1})$. Setting $K^\ell_\partial = K^\ell_{1,\partial} + K^\ell_{2,\partial}$, we see that K^ℓ_∂ has the sought operator matrix structure, and

$$M_\partial L_\partial U_\partial = \chi_\partial U_\partial + K^\ell_\partial U_\partial,$$

for $U_\partial \in H^{s,s-1}(\mathbb{R}^{d-1})$. We note that the operator K^ℓ_∂ maps $H^{s,s-1}(\mathbb{R}^{d-1})$ in $H^{s+1,s}(\mathbb{R}^{d-1})$.

We also have

$$L_\partial M_\partial = \begin{pmatrix} 1 & 0 \\ 0 & \Lambda_{\mathsf{T}}^k \end{pmatrix} \tilde{L}\tilde{M} \begin{pmatrix} 1 & 0 \\ 0 & \Lambda_{\mathsf{T}}^{-k} \end{pmatrix} = \begin{pmatrix} 1 & 0 \\ 0 & \Lambda_{\mathsf{T}}^k \end{pmatrix} (\chi_\partial + R^r_{-1}) \begin{pmatrix} 1 & 0 \\ 0 & \Lambda_{\mathsf{T}}^{-k} \end{pmatrix}.$$

Proceeding as above, we find $L_\partial M_\partial = \chi_\partial + K^r_\partial$, with K^r_∂ given by

$$\begin{pmatrix} 0 & 0 \\ 0 & \Lambda^k_\mathsf{T}\chi_\partial\Lambda^{-k}_\mathsf{T} - \chi_\partial \end{pmatrix} + \begin{pmatrix} 1 & 0 \\ 0 & \Lambda^k_\mathsf{T} \end{pmatrix} R^r_{-1} \begin{pmatrix} 1 & 0 \\ 0 & \Lambda^{-k}_\mathsf{T} \end{pmatrix}.$$

We thus find that K^r_∂ has the matrix structure described in the lemma and maps $H^{s,s-k}(\mathbb{R}^{d-1})$ in $H^{s+1,s+1-k}(\mathbb{R}^{d-1})$. ■

3.A.7. Local Right and Left Inverses up to a Compact Operator. Here, we prove Lemma 3.14. We compute

$$\psi^j P M^j = (\kappa^j)^* \check{\psi}^j P^{\mathcal{C}^j}(x,D) M^{\mathcal{C}^j} ((\kappa^j)^{-1})^* W^k.$$

Using that $P^{\mathcal{C}^j}(x,D)$ is a local operator and the support properties of $\check{\psi}^j$ and $\check{\varphi}^j$, we find that

$$\check{\psi}^j P^{\mathcal{C}^j}(x,D) M^{\mathcal{C}^j} = \check{\psi}^j P^{\mathcal{C}^j}(x,D) \check{\varphi}^j M_c M_b M_a \check{\varphi}^j = \check{\psi}^j P^{\mathcal{C}^j}(x,D) M_c M_b M_a \check{\varphi}^j.$$

By Lemma 3.6, we have

$$P^{\mathcal{C}^j}(x,D) M_c = \begin{pmatrix} \mathrm{Id} + K & -K^{(0)} & K^{(-1)} \end{pmatrix},$$

where the operators

$$K : H^m(\mathbb{R}^d_+) \to H^{m+1}(\mathbb{R}^d_+),$$
$$K^{(j)} : H^{m+j+3/2}(\mathbb{R}^{d-1}) \to H^{m+1}(\mathbb{R}^d_+), \ j = -1, 0,$$

are all bounded. As $M_b M_a \check{\varphi}^j$ is a bounded map from $H^m(\mathbb{R}^d_+) \oplus H^{m+3/2-k}(\mathbb{R}^{d-1})$ into $H^m(\mathbb{R}^d_+) \oplus H^{m+3/2}(\mathbb{R}^{d-1}) \oplus H^{m+1/2}(\mathbb{R}^{d-1})$, we conclude that, if $(f,h) \in H^m(\mathbb{R}^d_+) \oplus H^{m+3/2-k}(\mathbb{R}^{d-1})$, we have

$$\check{\psi}^j P^{\mathcal{C}^j}(x,D) M^{\mathcal{C}^j} \begin{pmatrix} f \\ h \end{pmatrix} = \check{\psi}^j f + \check{\psi}^j K_1^{r,\mathcal{C}^j} \check{\varphi}^j \begin{pmatrix} f \\ h \end{pmatrix}, \tag{3.A.10}$$

where

$$K_1^{r,\mathcal{C}^j} : H^m(\mathbb{R}^d_+) \oplus H^{m+3/2-k}(\mathbb{R}^{d-1}) \to H^{m+1}(\mathbb{R}^d_+).$$

With $(\mathsf{f}^j, {}^k\mathsf{h}^j)$ as introduced in (3.3.2)–(3.3.3), we have obtained

$$\begin{aligned} \psi^j P M^j \begin{pmatrix} \mathsf{f} \\ \mathsf{h} \end{pmatrix} &= (\kappa^j)^* \check{\psi}^j P^{\mathcal{C}^j}(x,D) M^{\mathcal{C}^j} \begin{pmatrix} \mathsf{f}^j \\ {}^k\mathsf{h}^j \end{pmatrix} \\ &= (\kappa^j)^* \check{\psi}^j \mathsf{f}^j + (\kappa^j)^* \check{\psi}^j K_1^{r,\mathcal{C}^j} \check{\varphi}^j \begin{pmatrix} \mathsf{f}^j \\ {}^k\mathsf{h}^j \end{pmatrix} \\ &= \psi^j \mathsf{f} + K_1^{r,j} \begin{pmatrix} \mathsf{f} \\ \mathsf{h} \end{pmatrix}, \end{aligned} \tag{3.A.11}$$

where $K_1^{r,j} : H^m(\mathcal{M}) \oplus H_B^{(m+3/2)}(\partial\mathcal{M}) \to H^{m+1}(\mathcal{M})$ is bounded.

We now compute

$$\left(\psi^j B M^j \begin{pmatrix} \mathsf{f} \\ \mathsf{h} \end{pmatrix}\right)\Big|_{{}^k\partial\mathcal{M}} = (\kappa^j)^* \left(\check{\psi}^j B^{\mathcal{C}^j}(x,D) M^{\mathcal{C}^j} \begin{pmatrix} \mathsf{f}^j \\ {}^k\mathsf{h}^j \end{pmatrix}\right)\Big|_{x_d=0^+}. \tag{3.A.12}$$

Note that this function vanishes on the other connected components of $\partial\mathcal{M}$. Arguing as above, we write

$$\check{\psi}^j B^{C^j}(x', 0, D) M^{C^j} = \check{\psi}^j B^{C^j}(x, D) M_c M_b M_a \check{\varphi}^j,$$

which we compute with the following lemma.

LEMMA 3.30. *Let $(f, h) \in H^m(\mathbb{R}^d_+) \oplus H^{m+3/2-k}(\mathbb{R}^{d-1})$ and set*

$$v = M_c M_b M_a \check{\varphi}^j \begin{pmatrix} f \\ h \end{pmatrix} \in H^{m+2}(\mathbb{R}^d_+).$$

We have

$$\begin{pmatrix} \gamma^{\mathsf{D}}(v) \\ \tilde{\gamma}^{\mathsf{N}}(v) \end{pmatrix} = M_\partial \begin{pmatrix} \mathscr{Q}_{\mathsf{D}}(\check{\varphi}^j f) \\ \check{\varphi}^j h \end{pmatrix} + \hat{K}^r \check{\varphi}^j \begin{pmatrix} f \\ h \end{pmatrix}, \tag{3.A.13}$$

where

$$\hat{K}^r : H^m(\mathbb{R}^d_+) \oplus H^{m+3/2-k}(\mathbb{R}^{d-1}) \to H^{m+5/2}(\mathbb{R}^{d-1}) \oplus H^{m+3/2}(\mathbb{R}^{d-1})$$

is a bounded operator.

A proof of Lemma 3.30 is given below.

With the notation $B^j(x', D') = B^j(x', x_d = 0, D')$, for $j = k$ and $j = k-1$, we have

$$\begin{aligned} \check{\psi}^j B^{C^j}(x, D) M^{C^j} \begin{pmatrix} f \\ h \end{pmatrix} &= \check{\psi}^j \begin{pmatrix} B^k(x', D') & -iB^{k-1}(x', D') \end{pmatrix} \begin{pmatrix} \gamma^{\mathsf{D}}(v) \\ D_d v_{|x_d=0^+} \end{pmatrix} \\ &= \check{\psi}^j L_{\partial,2} \begin{pmatrix} \gamma^{\mathsf{D}}(v) \\ \tilde{\gamma}^{\mathsf{N}}(v) \end{pmatrix}, \end{aligned}$$

with v as in Lemma 3.30 and

$$L_{\partial,2} = \begin{pmatrix} B^k(x', D') + iB^{k-1}(x', D') r_0(x', x_d = 0) & B^{k-1}(x', D') \end{pmatrix}.$$

The matrix $L_{\partial,2}$ is the second row of the operator L_∂ given in (3.2.20), with (3.A.13) and Lemma 3.11, we obtain

$$\begin{aligned} \check{\psi}^j B^{C^j}(x, D) M^{C^j} \begin{pmatrix} f \\ h \end{pmatrix} &= \check{\psi}^j \chi_\partial \check{\varphi}^j h + \check{\psi}^j K_2^{r,C^j} \check{\varphi}^j \begin{pmatrix} f \\ h \end{pmatrix} \\ &= \check{\psi}^j h + \check{\psi}^j K_2^{r,C^j} \check{\varphi}^j \begin{pmatrix} f \\ h \end{pmatrix}, \end{aligned}$$

using the support properties of χ_∂ and $\check{\varphi}^j$ and where

$$K_2^{r,C^j} : H^m(\mathbb{R}^d_+) \oplus H^{m+3/2-k}(\mathbb{R}^{d-1}) \to H^{m+5/2-k}(\mathbb{R}^{d-1})$$

is a bounded operator.

With $(\mathsf{f}^j, {}^k\mathsf{h}^j)$ as introduced in (3.3.2)–(3.3.3) and with (3.A.12), we have obtained

$$\text{(3.A.14)} \quad \left(\psi^j B M^j \begin{pmatrix}\mathsf{f}\\ \mathsf{h}\end{pmatrix}\right)\Big|_{{}^k\partial\mathcal{M}} = (\kappa^j)^*\left(\check{\psi}^j B^{\mathcal{C}^j}(x,D) M^{\mathcal{C}^j}\begin{pmatrix}\mathsf{f}^j\\ {}^k\mathsf{h}^j\end{pmatrix}\right)\Big|_{x_d=0^+}$$
$$= (\kappa^j)^*\check{\psi}^j\ {}^k\mathsf{h}^j + (\kappa^j)^*\check{\psi}^j K_2^{r,\mathcal{C}^j}\check{\varphi}^j\begin{pmatrix}\mathsf{f}^j\\ {}^k\mathsf{h}^j\end{pmatrix}$$
$$= \psi^j\ {}^k\mathsf{h} + K_2^{r,j}\begin{pmatrix}\mathsf{f}\\ \mathsf{h}\end{pmatrix},$$

where $K_2^{r,j} : H^m(\mathcal{M}) \oplus H_B^{(m+3/2)}(\partial\mathcal{M}) \to H^{m+5/2-k}({}^k\partial\mathcal{M})$ is bounded.

With $K_1^{r,j}$ defined in (3.A.11) and $K_2^{r,j}$ defined in (3.A.14), we define the $(\beta+2)\times(\beta+2)$ matrix operator

$$K^{r,j} = \begin{pmatrix} K_1^{r,j} \\ 0 \\ \vdots \\ 0 \\ K_2^{r,j} \\ 0 \\ \vdots \\ 0 \end{pmatrix},$$

where the $1\times(\beta+2)$ matrix operator $K_2^{r,j}$ is in the $k+2$ line. Note that the zeros in the matrices stand for the zero 1×3 matrix operator. We recall that k is the order of the boundary operator on the only connected component of $\partial\mathcal{M}$ that meets the local chart $\mathcal{C}^j$. The operator $K^{r,j}$ is bounded from $H^m(\mathcal{M}) \oplus H_B^{(m+3/2)}(\partial\mathcal{M})$ into $H^{m+1}(\mathcal{M}) \oplus H_B^{(m+5/2)}(\partial\mathcal{M})$. With (3.A.11) and (3.A.14), we have thus reached the conclusion of the first result of Lemma 3.14.

For the second result, we write

$$\psi^j M^j L = \psi^j (\kappa^j)^*\check{\varphi}^j M_c M_b M_a \check{\varphi}^j \begin{pmatrix} P^{\mathcal{C}^j}(x,D) \\ \gamma^{\mathsf{D}} B^{\mathcal{C}^j}(x,D)\end{pmatrix} ((\kappa^j)^{-1})^*$$
$$= (\kappa^j)^*\check{\psi}^j M_c M_b M_a \check{\varphi}^j \begin{pmatrix} P^{\mathcal{C}^j}(x,D) \\ \gamma^{\mathsf{D}} B^{\mathcal{C}^j}(x,D)\end{pmatrix} ((\kappa^j)^{-1})^*.$$

For $\mathsf{u} \in H^{m+2}(\mathcal{M})$, we set ${}^t(\mathsf{f},\mathsf{h}) = L\mathsf{u}$, that is,

$$\mathsf{f} = P\mathsf{u} \in H^m(\mathcal{M}), \quad \mathsf{h} = ({}^0\mathsf{h}, \ldots, {}^\beta\mathsf{h})$$
$$= (B\mathsf{u}_{|{}^0\partial\mathcal{M}}, \ldots, B\mathsf{u}_{|{}^\beta\partial\mathcal{M}}) \in H_B^{(m+3/2)}(\partial\mathcal{M}).$$

We then set

$$\mathsf{u}^j = \big((\kappa^j)^{-1}\big)^*\mathsf{u}, \quad \mathsf{f}^j = \big((\kappa^j)^{-1}\big)^*\mathsf{f}, \quad {}^k\mathsf{h}^j = \big((\kappa^j)^{-1}\big)^{*k}\mathsf{h}, \quad k = 0, \dots, \beta,$$
$$u = \breve{\varphi}^j\mathsf{u}^j, \quad f = \breve{\varphi}^j\mathsf{f}^j, \quad h = \breve{\varphi}^j_{|x_d=0^+}{}^k\mathsf{h}^j.$$

We introduce $\tilde{\varphi}^j \in \mathscr{C}^\infty(\overline{\mathbb{R}^d_+})$ be such that $\operatorname{supp}(\tilde{\varphi}^j) \subset \kappa^j(O^j)$, $\tilde{\varphi}^j \equiv 1$ on $\operatorname{supp}(\breve{\varphi}^j)$, and $\chi_\partial \equiv 1$ on $\operatorname{supp}(\tilde{\varphi}^j_{|x_d=0^+})$. We then have

(3.A.15)
$$\begin{aligned} P^{\mathcal{C}^j}(x,D)u &= \breve{\varphi}^j P^{\mathcal{C}^j}(x,D)\mathsf{u}^j + [P^{\mathcal{C}^j}(x,D), \breve{\varphi}^j]\tilde{\varphi}^j\mathsf{u}^j = \breve{\varphi}^j\mathsf{f}^j + K_1^\ell\tilde{\varphi}^j\mathsf{u}^j \\ &= f + K_1^\ell\tilde{\varphi}^j\mathsf{u}^j, \end{aligned}$$

and

(3.A.16)
$$\begin{aligned} B^{\mathcal{C}^j}(x,D)u_{|x_d=0^+} &= \breve{\varphi}^j B^{\mathcal{C}^j}(x,D)\mathsf{u}^j_{|x_d=0^+} + [B^{\mathcal{C}^j}(x,D), \breve{\varphi}^j]\tilde{\varphi}^j\mathsf{u}^j_{|x_d=0^+} \\ &= \breve{\varphi}^j_{|x_d=0^+}{}^k\mathsf{h}^j + K_2^\ell\tilde{\varphi}^j\mathsf{u}^j \\ &= h + K_2^\ell\tilde{\varphi}^j\mathsf{u}^j, \end{aligned}$$

where K_1^ℓ is bounded from $H^{m+2}(\mathbb{R}^d_+)$ into $H^{m+1}(\mathbb{R}^d_+)$ and K_2^ℓ is bounded from $H^{m+2}(\mathbb{R}^d_+)$ into $H^{m+5/2-k}(\mathbb{R}^{d-1})$. With (3.2.8) and (3.A.15), we have

$$u = \mathscr{Q}(f) - \mathscr{Q}^{(0)}(\gamma^{\mathsf{D}}(u)) + \mathscr{Q}^{(-1)}(\tilde{\gamma}^{\mathsf{N}}(u)) + K_3^\ell\tilde{\varphi}^j\mathsf{u}^j, \tag{3.A.17}$$

with $K_3^\ell w = \mathscr{Q}K_1^\ell w + (R\underline{w})_{|\mathbb{R}^d_+}$ that is bounded from $H^{m+2}(\mathbb{R}^d_+)$ into $H^{m+3}(\mathbb{R}^d_+)$. Computing the Dirichlet trace of (3.A.17) (or equivalently, considering the first row in identity (3.2.14)), we have

$$\begin{pmatrix}\mathrm{Id} - \mathsf{C}^{\mathsf{D},\mathsf{D}} & -\mathsf{C}^{\mathsf{D},\mathsf{N}}\end{pmatrix}\begin{pmatrix}\gamma^{\mathsf{D}}(u) \\ \tilde{\gamma}^{\mathsf{N}}(u)\end{pmatrix} = \mathscr{Q}_{\mathsf{D}}(f) + K_4^\ell\tilde{\varphi}^j\mathsf{u}^j,$$

with $K_4^\ell = \gamma^{\mathsf{D}}K_3^\ell$. We see that K_4^ℓ is bounded from $H^{m+2}(\mathbb{R}^d_+)$ into $H^{m+5/2}(\mathbb{R}^{d-1})$. We write (3.A.16) in the following form:

$$L_{\partial,2}\begin{pmatrix}\gamma^{\mathsf{D}}(u) \\ \tilde{\gamma}^{\mathsf{N}}(u)\end{pmatrix} = h + K_2^\ell\tilde{\varphi}^j\mathsf{u}^j,$$

where $L_{\partial,2}$ is the second row of the matrix L_∂ defined in (3.2.20). We thus have

$$L_\partial\begin{pmatrix}\gamma^{\mathsf{D}}(u) \\ \tilde{\gamma}^{\mathsf{N}}(u)\end{pmatrix} = \begin{pmatrix}\mathscr{Q}_{\mathsf{D}}(f) \\ h\end{pmatrix} + \begin{pmatrix}K_4^\ell \\ K_2^\ell\end{pmatrix}\tilde{\varphi}^j\mathsf{u}^j.$$

With Lemma 3.11, we then have, using the support of u,

$$\begin{pmatrix}\gamma^{\mathsf{D}}(u) \\ \tilde{\gamma}^{\mathsf{N}}(u)\end{pmatrix} = \chi_\partial\begin{pmatrix}\gamma^{\mathsf{D}}(u) \\ \tilde{\gamma}^{\mathsf{N}}(u)\end{pmatrix} = M_\partial\begin{pmatrix}\mathscr{Q}_{\mathsf{D}}(f) \\ h\end{pmatrix} + K_5^\ell\tilde{\varphi}^j\mathsf{u}^j,$$

where K_5^ℓ is bounded from $H^{m+2}(\mathbb{R}_+^d)$ into $H^{m+5/2}(\mathbb{R}^{d-1}) \oplus H^{m+3/2}(\mathbb{R}^{d-1})$. With the operator M_c defined in (3.3.8) and (3.A.17), we write

$$u = M_c \begin{pmatrix} f \\ \gamma^{\mathsf{D}}(u) \\ \tilde{\gamma}^{\mathsf{N}}(u) \end{pmatrix} + K_3^\ell \tilde{\varphi}^j \mathsf{u}^j = M_c M_b \begin{pmatrix} f \\ \mathscr{Q}_{\mathsf{D}}(f) \\ h \end{pmatrix} + K_6^\ell \tilde{\varphi}^j \mathsf{u}^j$$

where K_6^ℓ is bounded from $H^{m+2}(\mathbb{R}_+^d)$ into $H^{m+3}(\mathbb{R}_+^d)$. We then obtain

$$\check{\psi}^j u = \check{\psi}^j M^{\mathcal{C}^j} \begin{pmatrix} \mathsf{f}^j \\ {}^k\mathsf{h}^j \end{pmatrix} + \check{\psi}^j K_6^\ell \tilde{\varphi}^j \mathsf{u}^j,$$

that is

$$\psi^j \mathsf{u} = \psi^j M^j \begin{pmatrix} \mathsf{f} \\ \mathsf{h} \end{pmatrix} + K^{\ell,j} \mathsf{u},$$

where $K^{\ell,j}$ is bounded from $H^{m+2}(\mathcal{M})$ into $H^{m+3}(\mathcal{M})$. ∎

PROOF OF LEMMA 3.30. With the forms of the operators M_a, M_b, and M_c, we have

$$v = \mathscr{Q}(\check{\varphi}^j f) + \begin{pmatrix} -\mathscr{Q}^{(0)} & \mathscr{Q}^{(-1)} \end{pmatrix} M_\partial \begin{pmatrix} \mathscr{Q}_{\mathsf{D}}(\check{\varphi}^j f) \\ \check{\varphi}^j h \end{pmatrix}.$$

With (3.2.9) and (3.2.10) and the definition of the Calderón projector in (3.2.13), computing $\gamma^{\mathsf{D}}(v)$ and $\tilde{\gamma}^{\mathsf{N}}(v)$, we find

$$\begin{pmatrix} \gamma^{\mathsf{D}}(v) \\ \tilde{\gamma}^{\mathsf{N}}(v) \end{pmatrix} = \begin{pmatrix} \mathscr{Q}_{\mathsf{D}}(\check{\varphi}^j f) \\ \mathscr{Q}_{\mathsf{N}}(\check{\varphi}^j f) \end{pmatrix} + \mathsf{C} M_\partial \begin{pmatrix} \mathscr{Q}_{\mathsf{D}}(\check{\varphi}^j f) \\ \check{\varphi}^j h \end{pmatrix}.$$

We set

$$\begin{pmatrix} r^{\mathsf{D}} \\ r^{\mathsf{N}} \end{pmatrix} = \begin{pmatrix} \gamma^{\mathsf{D}}(v) \\ \tilde{\gamma}^{\mathsf{N}}(v) \end{pmatrix} - M_\partial \begin{pmatrix} \mathscr{Q}_{\mathsf{D}}(\check{\varphi}^j f) \\ \check{\varphi}^j h \end{pmatrix},$$

and we have

$$\begin{pmatrix} r^{\mathsf{D}} \\ r^{\mathsf{N}} \end{pmatrix} = \begin{pmatrix} \mathscr{Q}_{\mathsf{D}}(\check{\varphi}^j f) \\ \mathscr{Q}_{\mathsf{N}}(\check{\varphi}^j f) \end{pmatrix} + (\mathsf{C} - \mathrm{Id}) M_\partial \begin{pmatrix} \mathscr{Q}_{\mathsf{D}}(\check{\varphi}^j f) \\ \check{\varphi}^j h \end{pmatrix}.$$

We first compute ${}^t(r^{\mathsf{D}}, \mathsf{C}^{\mathsf{D,N}} r^{\mathsf{N}})$. We have

$$\begin{aligned} \begin{pmatrix} \mathrm{Id} & 0 \\ 0 & \mathsf{C}^{\mathsf{D,N}} \end{pmatrix} (\mathsf{C} - \mathrm{Id}) &= \begin{pmatrix} \mathsf{C}^{\mathsf{D,D}} - \mathrm{Id} & \mathsf{C}^{\mathsf{D,N}} \\ \mathsf{C}^{\mathsf{D,N}} \mathsf{C}^{\mathsf{N,D}} & \mathsf{C}^{\mathsf{D,N}} (\mathsf{C}^{\mathsf{N,N}} - \mathrm{Id}) \end{pmatrix} \\ &= \begin{pmatrix} \mathsf{C}^{\mathsf{D,D}} - \mathrm{Id} & \mathsf{C}^{\mathsf{D,N}} \\ -\mathsf{C}^{\mathsf{D,D}} (\mathsf{C}^{\mathsf{D,D}} - \mathrm{Id}) & -\mathsf{C}^{\mathsf{D,D}} \mathsf{C}^{\mathsf{D,N}} \end{pmatrix} + R \\ &= \begin{pmatrix} \mathrm{Id} & 0 \\ 0 & -\mathsf{C}^{\mathsf{D,D}} \end{pmatrix} \begin{pmatrix} \mathsf{C}^{\mathsf{D,D}} - \mathrm{Id} & \mathsf{C}^{\mathsf{D,N}} \\ \mathsf{C}^{\mathsf{D,D}} - \mathrm{Id} & \mathsf{C}^{\mathsf{D,N}} \end{pmatrix} + R, \end{aligned}$$

with (3.2.16) and (3.2.17), where

$$R : H^{m+3/2}(\mathbb{R}^{d-1}) \oplus H^{m+1/2}(\mathbb{R}^{d-1}) \to \big(H^{-N}(\mathbb{R}^{d-1})\big)^2$$

is bounded for any $N \in \mathbb{N}$. We thus find

$$\begin{pmatrix} r^{\mathsf{D}} \\ \mathsf{C}^{\mathsf{D},\mathsf{N}} r^{\mathsf{N}} \end{pmatrix} = \begin{pmatrix} \mathscr{Q}_{\mathsf{D}}(\breve{\varphi}^j f) \\ \mathsf{C}^{\mathsf{D},\mathsf{N}} \mathscr{Q}_{\mathsf{N}}(\breve{\varphi}^j f) \end{pmatrix} + \begin{pmatrix} \mathrm{Id} & 0 \\ 0 & -\mathsf{C}^{\mathsf{D},\mathsf{D}} \end{pmatrix} \begin{pmatrix} \mathsf{C}^{\mathsf{D},\mathsf{D}} - \mathrm{Id} & \mathsf{C}^{\mathsf{D},\mathsf{N}} \\ \mathsf{C}^{\mathsf{D},\mathsf{D}} - \mathrm{Id} & \mathsf{C}^{\mathsf{D},\mathsf{N}} \end{pmatrix} M_{\partial} \begin{pmatrix} \mathscr{Q}_{\mathsf{D}}(\breve{\varphi}^j f) \\ \breve{\varphi}^j h \end{pmatrix} + R' \begin{pmatrix} \breve{\varphi}^j f \\ \breve{\varphi}^j h \end{pmatrix}$$

with

$$R' : H^m(\mathbb{R}^d_+) \oplus H^{m+3/2-k}(\mathbb{R}^{d-1}) \to \big(H^{-N}(\mathbb{R}^{d-1})\big)^2$$

bounded for any $N \in \mathbb{N}$. The matrix $\big(\mathrm{Id} - \mathsf{C}^{\mathsf{D},\mathsf{D}} \quad -\mathsf{C}^{\mathsf{D},\mathsf{N}}\big)$ is the first row of the matrix L_{∂} in (3.2.20). With Lemma 3.11, we obtain

$$\begin{pmatrix} r^{\mathsf{D}} \\ \mathsf{C}^{\mathsf{D},\mathsf{N}} r^{\mathsf{N}} \end{pmatrix} = \begin{pmatrix} (1-\chi_{\partial}) \mathscr{Q}_{\mathsf{D}}(\breve{\varphi}^j f) \\ \mathsf{C}^{\mathsf{D},\mathsf{N}} \mathscr{Q}_{\mathsf{N}}(\breve{\varphi}^j f) + \mathsf{C}^{\mathsf{D},\mathsf{D}} \chi_{\partial} \mathscr{Q}_{\mathsf{D}}(\breve{\varphi}^j f) \end{pmatrix} + K^r_a \begin{pmatrix} \breve{\varphi}^j f \\ \breve{\varphi}^j h \end{pmatrix},$$

where K^r_a is bounded from $H^m(\mathbb{R}^d_+) \oplus H^{m+3/2-k}(\mathbb{R}^{d-1})$ into $\big(H^{m+5/2}(\mathbb{R}^{d-1})\big)^2$.

We introduce $\tilde{\varphi}^j, \tilde{\varphi}^{j\prime} \in \mathscr{C}^\infty(\overline{\mathbb{R}^d_+})$ be such that

1. $\mathrm{supp}(\tilde{\varphi}^j) \cup \mathrm{supp}(\tilde{\varphi}^{j\prime}) \subset \kappa^j(O^j)$;
2. $\tilde{\varphi}^j \equiv 1$ on a neighborhood of $\mathrm{supp}(\breve{\varphi}^j)$;
3. $\tilde{\varphi}^{j\prime} \equiv 1$ on a neighborhood of $\mathrm{supp}(\tilde{\varphi}^j)$; and
4. $\chi_{\partial} \equiv 1$ on a neighborhood of $\mathrm{supp}(\tilde{\varphi}^{j\prime}_{|x_d=0^+})$.

We write $(1-\chi_{\partial}) \mathscr{Q}_{\mathsf{D}}(\breve{\varphi}^j f) = K^r_b(\breve{\varphi}^j f)$, with $K^r_b : w \mapsto (1-\chi_{\partial}) \mathscr{Q}_{\mathsf{D}}(\tilde{\varphi}^j w)$. We now prove that K^r_b is regularizing. We write

$$(1-\chi_{\partial}) \mathscr{Q}_{\mathsf{D}}(\tilde{\varphi}^j w) = (1-\chi_{\partial}) \Big((1-\tilde{\varphi}^{j\prime}) Q(\tilde{\varphi}^j \underline{w}) \Big)_{|x_d=0^+},$$

and we observe that the operator $(1-\tilde{\varphi}^{j\prime}) Q \tilde{\varphi}^j \in \Psi^{-N}(\mathbb{R}^d)$ for any $N \in \mathbb{N}$. We thus find that K^r_b is bounded from $H^m(\mathbb{R}^d_+)$ into $H^{m+5/2}(\mathbb{R}^{d-1})$.

We now write

$$\mathsf{C}^{\mathsf{D},\mathsf{D}} \chi_{\partial} \mathscr{Q}_{\mathsf{D}}(\breve{\varphi}^j f) = \mathsf{C}^{\mathsf{D},\mathsf{D}} \mathscr{Q}_{\mathsf{D}}(\breve{\varphi}^j f) - \mathsf{C}^{\mathsf{D},\mathsf{D}} K^r_b(\breve{\varphi}^j f).$$

Since the first row of the map in (3.2.15) is precisely

$$\mathsf{C}^{\mathsf{D},\mathsf{N}} \mathscr{Q}_{\mathsf{N}} + \mathsf{C}^{\mathsf{D},\mathsf{D}} \mathscr{Q}_{\mathsf{D}},$$

we find that

$$\mathsf{C}^{\mathsf{D},\mathsf{N}} \mathscr{Q}_{\mathsf{N}}(\breve{\varphi}^j f) + \mathsf{C}^{\mathsf{D},\mathsf{D}} \chi_{\partial} \mathscr{Q}_{\mathsf{D}}(\breve{\varphi}^j f) = K^r_c(\breve{\varphi}^j f),$$

where K^r_c is bounded from $H^m(\mathbb{R}^d_+)$ into $H^{m+5/2}(\mathbb{R}^{d-1})$. Setting

$$K_2^{r\prime} \begin{pmatrix} f \\ h \end{pmatrix} = \begin{pmatrix} K^r_b \\ K^r_c \end{pmatrix} f + K^r_a \begin{pmatrix} f \\ h \end{pmatrix},$$

we obtain that $K_2^{r\prime}$ is bounded from $H^m(\mathbb{R}^d_+) \oplus H^{m+3/2-k}(\mathbb{R}^{d-1})$ into $\big(H^{m+5/2}(\mathbb{R}^{d-1})\big)^2$ and

$$\begin{pmatrix} r\mathsf{D} \\ \mathsf{C}^{\mathsf{D},\mathsf{N}} r\mathsf{N} \end{pmatrix} = K_2^{r\prime} \check{\varphi}^j \begin{pmatrix} f \\ h \end{pmatrix}.$$

Since $\mathsf{C}^{\mathsf{D},\mathsf{N}} \in \Psi^{-1}(\mathbb{R}^{d-1})$ with nonvanishing principal symbol $\rho^{-1}_{|x_d=0^+}(x',\xi')/2$, we may apply a parametrix as given by Proposition 2.33 of Volume 1 yielding

$$\begin{pmatrix} r\mathsf{D} \\ r\mathsf{N} \end{pmatrix} = \hat{K}^r \check{\varphi}^j \begin{pmatrix} f \\ h \end{pmatrix},$$

with $\hat{K}^r$ as in the statement of the lemma. ∎

CHAPTER 4

Selfadjoint Operators Under General Boundary Conditions

Contents

4.1. Introduction and Setting

Let $(\mathcal{M}, g)$ be a connected smooth compact Riemannian manifold with boundary. We consider a second-order differential operator P on $\mathcal{M}$ given by $P = -\Delta_g + R_1$ where R_1 is a first-order differential operator. We also consider a smooth differential boundary operator B that takes the form $B = B^k + B^{k-1}\partial_\nu$ on each connected component $\mathcal{N}$ of $\partial\mathcal{M}$ with B^k and B^{k-1} smooth differential operators on (a neighborhood of) $\mathcal{N}$ of order k and $k-1$, respectively, with $k \leq \beta \in \mathbb{N}$. Thus, the operator B is of order β with differentiation transverse to $\partial\mathcal{M}$ of order less than or equal to one as in Chap. 2.

J. Le Rousseau et al., *Elliptic Carleman Estimates and Applications to Stabilization and Controllability, Volume II*, PNLDE Subseries in Control 98, https://doi.org/10.1007/978-3-030-88670-7_4

We introduce the unbounded operator P on $L^2(\mathcal{M})$ with domain

$$D(\mathsf{P}) = \{u \in L^2(\mathcal{M});\ Pu \in L^2(\mathcal{M}) \text{ and } Bu_{|\partial\mathcal{M}} = 0\}, \tag{4.1.1}$$

and given by $\mathsf{P}u = Pu$ for $u \in D(\mathsf{P})$. This definition is sensible: if $u \in L^2(\mathcal{M})$ is such that $Pu \in L^2(\mathcal{M})$, then $u \in \mathcal{W}_P(\mathcal{M})$, with this space defined in Sect. 18.6.2, and from Lemma 18.32 we know that both the Dirichlet and Neumann traces can be defined and $\gamma^{\mathsf{D}}(u) = u_{|\partial\mathcal{M}} \in H^{-1/2}(\partial\mathcal{M})$ and $\gamma^{\mathsf{N}}(u) = \partial_\nu u_{|\partial\mathcal{M}} \in H^{-3/2}(\partial\mathcal{M})$; then $Bu_{|\partial\mathcal{M}} \in H_B^{(-1/2-k)}(\partial\mathcal{M})$; see (3.1.2) for the definition of this space. Without any further requirement one sees that the operator P is closed.

LEMMA 4.1. *The unbounded operator* $(\mathsf{P}, D(\mathsf{P}))$ *is closed.*

PROOF. Let (u_n) be a sequence in $D(\mathsf{P})$ that converges to u in $L^2(\mathcal{M})$ and such that $f_n = \mathsf{P}u_n = Pu_n$ converges to some f in $L^2(\mathcal{M})$. Observe that $Pu = f$ in ${}^0\mathscr{D}'(\mathcal{M})$. Thus, u lies in the space $\mathcal{W}_P(\mathcal{M})$. Moreover, $(u_n) \subset \mathcal{W}_P(\mathcal{M})$ and it converges to u in the topology of that space. The Dirichlet and Neumann trace operators γ^{D} and γ^{N} are defined on $\mathcal{W}_P(\mathcal{M})$, and moreover continuous with values in $H^{-1/2}(\partial\mathcal{M})$ and $H^{-3/2}(\partial\mathcal{M})$, respectively. We thus have the convergence of $\gamma^{\mathsf{D}}(u_n)$ to $\gamma^{\mathsf{D}}(u)$ in $H^{-1/2}(\partial\mathcal{M})$ and that of $\gamma^{\mathsf{N}}(u_n)$ to $\gamma^{\mathsf{N}}(u)$ in $H^{-3/2}(\partial\mathcal{M})$. In particular Bu_n converges to Bu in $H_B^{(-1/2-k)}(\partial\mathcal{M})$. As $Bu_n = 0$ we conclude that $Bu = 0$. ∎

In this chapter, we shall furthermore assume that the Lopatinskiĭ–Šapiro holds for (P, B) on the whole $\partial\mathcal{M}$; see Definition 2.2. By Theorem 3.28 we then find

$$D(\mathsf{P}) = \{u \in H^2(\mathcal{M});\ Bu_{|\partial\mathcal{M}} = 0\}. \tag{4.1.2}$$

Our goal in this short chapter is to characterize the joint properties of the operators P and B for the unbounded operator $(\mathsf{P}, D(\mathsf{P}))$ to be selfadjoint. We choose to concentrate on the case $\beta \leq 1$, that is, on boundary operators of order less than one.

4.2. Selfadjointness

We note that ${}^0\mathscr{D}_c^\infty(\mathcal{M}) \subset D(\mathsf{P})$ implying that $D(\mathsf{P})$ is dense in $L^2(\mathcal{M})$. We recall that ${}^0\mathscr{D}_c^\infty(\mathcal{M})$ is the space of smooth functions on $\mathcal{M}$ with support away from $\partial\mathcal{M}$ (see Section 8.1.3 of Volume 1). As a result the adjoint of P is well defined; see Section 11.4 (also in Volume 1).

For $(\mathsf{P}, D(\mathsf{P}))$ to be selfadjoint it is first required to be a closed operator. We saw above that this holds without requiring the Lopatinskiĭ–Šapiro condition. A second requirement for $(\mathsf{P}, D(\mathsf{P}))$ to be selfadjoint is symmetry, that is,

$$(Pu, w)_{L^2(\mathcal{M})} = (u, Pw)_{L^2(\mathcal{M})}, \qquad u, w \in D(\mathsf{P}).$$

This implies constraints on both the structures of P and B.

Above all, if $(\mathsf{P}, D(\mathsf{P}))$ is symmetric, then P is itself symmetric on ${}^0\mathscr{D}_c^\infty(\mathcal{M})$. As $P = -\Delta_g + R_1$ and Δ_g are symmetric on ${}^0\mathscr{D}_c^\infty(\mathcal{M})$ this implies that R_1 has to be symmetric on ${}^0\mathscr{D}_c^\infty(\mathcal{M})$.

LEMMA 4.2. *Let R_1 be a first-order differential operator on $\mathcal{M}$ and be symmetric on ${}^0\mathscr{D}_c^\infty(\mathcal{M})$. Then $R_1 = iV + f$, where V is a smooth real vector field and f is a smooth function such that* $\operatorname{Im} f = \operatorname{div}_g V/2$.

PROOF. The operator R_1 is of the form $R_1 = X + f$ with X a complex smooth vector field. If R_1 is symmetric on ${}^0\mathscr{D}_c^\infty(\mathcal{M})$, then $R_1 = {}^t\overline{R_1} = {}^t\overline{X} + \overline{f}$. Yet, by Proposition 17.7 we have ${}^t\overline{X} = -\overline{X} - \operatorname{div}_g \overline{X}$. This yields $X = -\overline{X}$ and $f - \overline{f} = -\operatorname{div}_g \overline{X}$. The result follows. ■

With the structure of R_1 given by Lemma 4.2 we now consider $u, w \in H^2(\mathcal{M})$. With Proposition 18.29 and the Green formula given in Proposition 18.30 we then obtain

(4.2.1)
$$(Pu, w)_{L^2(\mathcal{M})} = (u, Pw)_{L^2(\mathcal{M})} - (\gamma^{\mathsf{N}}(u), \gamma^{\mathsf{D}}(w))_{L^2(\partial\mathcal{M})} + (\gamma^{\mathsf{D}}(u), \gamma^{\mathsf{N}}(w))_{L^2(\partial\mathcal{M})} + i(\gamma^{\mathsf{D}}(g(V,\nu)u), \gamma^{\mathsf{D}}(w))_{L^2(\partial\mathcal{M})}.$$

Symmetry of P is thus equivalent to having

$$0 = -(\gamma^{\mathsf{N}}(u), \gamma^{\mathsf{D}}(w))_{L^2(\partial\mathcal{M})} + (\gamma^{\mathsf{D}}(u), \gamma^{\mathsf{N}}(w))_{L^2(\partial\mathcal{M})} + i(\gamma^{\mathsf{D}}(g(V,\nu)u), \gamma^{\mathsf{D}}(w))_{L^2(\partial\mathcal{M})}, \tag{4.2.2}$$

if $u, w \in D(\mathsf{P})$.

On each connected component of $\partial\mathcal{M}$ the differential boundary operator B is either of order $k = 0$ or 1. For $k = 0$ or 1, we denote by ${}^k\partial\mathcal{M}$, the union of all the connected components of $\partial\mathcal{M}$ where the order of B is equal to k. On a connected component $\mathcal{N}$ of ${}^0\partial\mathcal{M}$ under the Lopatinskiĭ–Šapiro condition it is equivalent to consider $Bu_{|\mathcal{N}} = u_{|\mathcal{N}}$, that is, a Dirichlet boundary operator. See Example 2.5-(1). In such case, that connected component yields a vanishing contribution to (4.2.2).

On a connected component $\mathcal{N}$ of ${}^1\partial\mathcal{M}$, conditions on B are required for both the Lopatinskiĭ–Šapiro condition and the symmetry condition (4.2.2) to hold. In fact, as we shall see below, the boundary condition can then be reduced to the form $B = \partial_\nu + B'$, with B' a first-order tangential differential operator.

Our first main result is the following one.

THEOREM 4.3. *Let $d \geq 2$ and let $P = -\Delta_g + R_1$, with R_1 a smooth first-order differential operator on $\mathcal{M}$, and B a boundary operator of order less than or equal to one such that (P, B) satisfies the Lopatinskiĭ–Šapiro condition on $\partial\mathcal{M}$. The unbounded operator $(\mathsf{P}, D(\mathsf{P}))$ is selfadjoint on $L^2(\mathcal{M})$ if and only if*

(1) *$R_1 = iV + f$ with V is a smooth real vector field and f a smooth complex valued function on $\mathcal{M}$ such that* $\operatorname{Im} f = \operatorname{div}_g V/2$;

(2) *On every connected component* $\mathcal{N}$ *of* ${}^1\partial\mathcal{M}$ *then the condition* $Bu_{|\mathcal{N}} = 0$ *is equivalent*
   (a) *either to a homogeneous Dirichlet boundary condition* $u_{|\mathcal{N}} = 0$;
   (b) *or to the form* $\partial_\nu u_{|\mathcal{N}} + (iX'u + hu)_{|\mathcal{N}} = 0$ *with* X' *a real valued vector field and* h *a complex valued function on* $\mathcal{N}$ *with the condition*

$$2\,\mathrm{Im}\,h - \mathrm{div}_g\,X' + \gamma^{\mathsf{D}}_{\mathcal{N}}(g(V,\nu)) = 0.$$

   In the case $d = 2$, *we moreover have* $|X'|_{g_\partial} \neq 1$. *In the case* $d \geq 3$, *we moreover have* $|X'|_{g_\partial} < 1$.

Here and in what follows, we denote the Dirichlet and Neumann trace operators on a connected component $\mathcal{N}$ of $\partial\mathcal{M}$ by $\gamma^{\mathsf{D}}_{\mathcal{N}}$ and $\gamma^{\mathsf{N}}_{\mathcal{N}}$.

Before proving this result we first show that the reduction of B into the form $\partial_\nu + B'$ stated in item (2) can be carried out. This is based on both having P symmetric and the Lopatinskiĭ–Šapiro condition holding.

4.3. Preliminary Result on Symmetry

Here, we first prove that if the Lopatinskiĭ–Šapiro condition holds and if the operator $(\mathsf{P}, D(\mathsf{P}))$ is symmetric, then on any connected component $\mathcal{N}$ of $\partial\mathcal{M}$ the boundary condition $Bu_{|\mathcal{N}} = 0$ is equivalent

(1) either to a homogeneous Dirichlet boundary condition $u_{|\mathcal{N}} = 0$;
(2) or to the form $(\partial_\nu u + B'u)_{|\mathcal{N}} = 0$ with B' a first-order tangential differential operator.

The case $d \geq 4$ is the simplest to treat. The cases $d = 3$ and $d = 2$ require more analysis. After this reduction to simple forms for B on each connected component of $\partial\mathcal{M}$ we show that in the first-order case $\partial_\nu + B'$, the symmetry for the operator $(\mathsf{P}, D(\mathsf{P}))$ gives some constraints on the tangential operator B'.

4.3.1. Dimension $d \geq 4$. In dimension $d \geq 4$, if the Lopatinskiĭ–Šapiro condition holds, then by Proposition 2.8 a boundary operator of order one on a connected component $\mathcal{N}$ of $\partial\mathcal{M}$ necessarily takes the form $B = \alpha\partial_\nu + B'$, where α does not vanish and B' is a first-order tangential operator (the argument based on (2.3.4) and (2.3.5) is given in Remark 2.9-(3)). Upon dividing by α we may thus assume that $B = \partial_\nu + B'$. Note that for the case $d \geq 4$ we obtain the expected form without using the symmetry of P. The next two sections treat the cases $d = 3$ and $d = 2$. In these two cases we shall exploit the symmetry of P.

4.3.2. Dimension $d = 3$. We consider a connected component $\mathcal{N}$ of ${}^1\partial\mathcal{M}$. There the boundary operator B takes the form

$$B = a\partial_\nu + B', \tag{4.3.1}$$

where B' is a first-order tangential differential operator, that is, $B' = X' + iY' + h$ where X', Y' are two real valued vector fields on $\mathcal{N}$ and h is a complex valued function. The principal symbol of B is given by

$$b(m, \omega) = ia\omega_n + i\langle X', \omega'\rangle - \langle Y', \omega'\rangle,$$

for $\omega = \omega' + \omega_n n_m$, with $\omega' \in T_m^*\mathcal{N}$, $\omega_n \in \mathbb{R}$, and n_m with $n = \nu^\flat$, that is, the unitary outgoing cotangent vector. In the statement of Proposition 2.8 we have $t = t' + t_\nu \nu$ and $v = v' + v_\nu \nu$ here given by $t_\nu = -\operatorname{Im} a$, $t' = -Y'$, $v_\nu = \operatorname{Re} a$, $v' = X'$. By that proposition and by Proposition 2.10 two cases arise if the Lopatinskiĭ–Šapiro condition holds on $\mathcal{N}$.

(1) For all $m \in \mathcal{N}$ we have either

$$|t'_m|_{g_\partial} < |v_m^\nu| \ \text{ or } \ |v'_m|_{g_\partial} < |t_m^\nu| \tag{4.3.2}$$

or

$$g_m(t_m, v_m)^2 > \big(|t'_m|^2_{g_\partial} - (v_m^\nu)^2\big)\big(|v'_m|^2_{g_\partial} - (t_m^\nu)^2\big). \tag{4.3.3}$$

In particular, the complex valued function a does not vanish on $\mathcal{N}$. Note that this condition coincides with the condition one finds in the case $d \geq 4$ (see Proposition 2.8).

(2) For all $m \in \mathcal{N}$ we have

$$|t'_m|_{g_\partial} > |v_m^\nu| \ \text{ and } \ |v'_m|_{g_\partial} > |t_m^\nu| \tag{4.3.4}$$

and

$$g_m(t_m, v_m)^2 < \big(|t'_m|^2_{g_\partial} - (v_m^\nu)^2\big)\big(|v'_m|^2_{g_\partial} - (t_m^\nu)^2\big). \tag{4.3.5}$$

In this second case, the complex valued function a may vanish. Moreover, $\operatorname{span}(t'_m, v'_m) = \operatorname{span}(X'_m, Y'_m) = T_m\mathcal{N}$.

The following lemma shows that in the case of a symmetric operator the second case does not occur.

LEMMA 4.4. *Let $d = 3$. Assume that the Lopatinskiĭ–Šapiro condition holds for (P, B) and that the unbounded operator $(\mathsf{P}, D(\mathsf{P}))$ is symmetric. Let $\mathcal{N}$ be a connected component of $\partial\mathcal{M}$ where B is given by (4.3.1). Then either the boundary condition on $\mathcal{N}$ is equivalent to having a homogeneous Dirichlet condition or conditions (4.3.2)–(4.3.3) hold. In the latter case, the function a does not vanish.*

If the boundary condition on $\mathcal{N}$ is equivalent to a homogeneous Dirichlet condition we shall rather consider it as a zero-order boundary condition.

In the case of a true first-order boundary operator, with this lemma we see that if $(\mathsf{P}, D(\mathsf{P}))$ is symmetric the first case, that is (4.3.2) and (4.3.3), only occurs. Then the function a does not vanish on $\mathcal{N}$. Upon dividing by a we may thus assume that $B = \partial_\nu + B'$. This reduction corresponds to item (2) of Theorem 4.3.

In this reduced form, since the principal symbol is now $b(m,\omega) = i\omega_n + i\langle X', \omega'\rangle - \langle Y', \omega'\rangle$, that is, $t_\nu = 0$, $v_\nu = 1$, $t' = -Y'$, and $v' = X'$. Conditions (4.3.2) and (4.3.3) now read

$$|t'_m|_{g_\partial} < 1 \text{ or } (g_\partial)_m(t'_m, v'_m)^2 > \big(|t'_m|^2_{g_\partial} - 1\big)|v'_m|^2_{g_\partial}. \tag{4.3.6}$$

PROOF. Let us assume that (4.3.4) and (4.3.5) hold. If so, we first prove that $a = 0$ meaning that $B = B'$ and second we prove that the purely tangential boundary condition is equivalent to a homogeneous Dirichlet condition.

We proceed by contradiction and we assume that $a \neq 0$. We may then choose an open set W in $\mathcal{N}$ where $|a| \geq C > 0$ for some $C > 0$. Let $u, w \in D(\mathsf{P})$ be such that both their Dirichlet and Neumann traces vanish on all the connected components of $\partial\mathcal{M}$ but $\mathcal{N}$. Such function can be constructed by means of Theorem 18.25 once the two traces are also prescribed on $\mathcal{N}$. If we prescribe $\gamma^{\mathsf{D}}_{\mathcal{N}}(u)$ and $\gamma^{\mathsf{D}}_{\mathcal{N}}(w)$ in $H^{3/2}(\mathcal{N})$ with support in W, then we have $\gamma^{\mathsf{N}}_{\mathcal{N}}(u) = -a^{-1}B'\gamma^{\mathsf{D}}_{\mathcal{N}}(u)$ and $\gamma^{\mathsf{N}}_{\mathcal{N}}(w) = -a^{-1}B'\gamma^{\mathsf{D}}_{\mathcal{N}}(w)$ and the functions u and w are in $D(\mathsf{P})$ as $Bu_{|\mathcal{N}} = Bw_{|\mathcal{N}} = 0$. By (4.2.2) we have

$$\begin{aligned} 0 = (a^{-1}B'\gamma^{\mathsf{D}}_{\mathcal{N}}(u), \gamma^{\mathsf{D}}_{\mathcal{N}}(w))_{L^2(\mathcal{N})} &- (\gamma^{\mathsf{D}}_{\mathcal{N}}(u), a^{-1}B'\gamma^{\mathsf{D}}_{\mathcal{N}}(w))_{L^2(\mathcal{N})} \\ &+ i(\gamma^{\mathsf{D}}_{\mathcal{N}}(g(V,\nu)u), \gamma^{\mathsf{D}}_{\mathcal{N}}(w))_{L^2(\mathcal{N})}, \end{aligned}$$

which reads

$$0 = \big(Q\gamma^{\mathsf{D}}_{\mathcal{N}}(u), \gamma^{\mathsf{D}}_{\mathcal{N}}(w)\big)_{L^2(\mathcal{N})}, \tag{4.3.7}$$

with $Q = a^{-1}B' - {}^t\overline{B'}\,\overline{a^{-1}} + i\gamma^{\mathsf{D}}_{\mathcal{N}}(g(V,\nu))$. It is a first-order differential operator on $\mathcal{N}$. Thus $Q = Z + k$ with Z a complex valued vector field and k a complex valued function. With $B' = X' + iY'$, by Proposition 17.7 we have

$$\begin{aligned} a^{-1}B' - {}^t\overline{B'}\,\overline{a^{-1}} &= a^{-1}(X' + iY' + h) + (X' - iY' - \overline{h})\overline{a^{-1}} \\ &\quad + (\operatorname{div}_g X' - i\operatorname{div}_g Y')\overline{a^{-1}} \\ &= 2\operatorname{Re}(a^{-1})X' - 2\operatorname{Im}(a^{-1})Y' + [X' - iY', \overline{a^{-1}}] \\ &\quad + a^{-1}h - \overline{ha^{-1}} \\ &\quad + (\operatorname{div}_g X' - i\operatorname{div} Y')\overline{a^{-1}}. \end{aligned}$$

Thus $Z = 2\operatorname{Re} a^{-1}X' - 2\operatorname{Im} a^{-1}Y'$ and

$$k = i\gamma^{\mathsf{D}}_{\mathcal{N}}(g(V,\nu)) + [X' - iY', \overline{a^{-1}}] + (\operatorname{div}_g X' - i\operatorname{div} Y')\overline{a^{-1}} + 2i\operatorname{Im}(a^{-1}h).$$

As $\gamma^{\mathsf{D}}_{\mathcal{N}}(u)$ and $\gamma^{\mathsf{D}}_{\mathcal{N}}(w)$ can be chosen arbitrary with support in W, from (4.3.7) we find that Q vanishes on any function supported in W. This gives $k = 0$ and $Z = 0$. Yet, $\operatorname{rank}(X', Y') = 2$ this implies that $\operatorname{Re} a^{-1} = \operatorname{Im} a^{-1} = 0$ in W; a contradiction.

We now know that $a = 0$. The operator $B = B'$. It thus only acts tangentially on $\mathcal{N}$. We have $B' = X' + iY' + h$. It is a bounded operator from $H^{3/2}(\mathcal{N})$ into $H^{1/2}(\mathcal{N})$.

If B' is injective,[1] then, the boundary condition $Bu_{|\mathcal{N}} = B'u_{|\mathcal{N}} = 0$ is equivalent to having $u_{|\mathcal{N}} = 0$, the homogeneous Dirichlet boundary condition.

Let us assume now that B' is not injective and let us reach a contradiction. This will conclude the proof. In such case, there exists $\varphi \in H^{3/2}(\mathcal{N})$ such that $\varphi \neq 0$ and $B'\varphi = 0$. We choose $u, w \in H^2(\mathcal{M})$ as above with vanishing traces on the other connected components of the boundary and on $\mathcal{N}$ we choose

$$\gamma_{\mathcal{N}}^{\mathsf{D}}(u) = \varphi, \quad \gamma_{\mathcal{N}}^{\mathsf{N}}(u) = 0, \quad \gamma_{\mathcal{N}}^{\mathsf{D}}(w) = 0, \quad \gamma_{\mathcal{N}}^{\mathsf{N}}(w) = \varphi. \tag{4.3.8}$$

We have $Bu_{|\mathcal{N}} = Bw_{|\mathcal{N}} = 0$ and thus $u, w \in D(\mathsf{P})$. However, with (4.2.2) that expresses the symmetry of the operator P we obtain $|\varphi|^2_{L^2(\mathcal{N})} = 0$; a contradiction as $\varphi \neq 0$. ∎

4.3.3. Dimension $d = 2$. Let $\mathcal{N}$ be a connected component of ${}^1\partial\mathcal{M}$. It is diffeomorphic to the unit circle $\mathbb{S}^1$. There, the operator B takes the form

$$B = a\partial_\nu + bT' + c, \tag{4.3.9}$$

where T' is a real vector field on $\mathcal{N}$ such that $|T'|_{g_\partial} = 1$ and a, b, c are smooth complex valued functions.

LEMMA 4.5. *Let $d = 2$. Assume that the Lopatinskiĭ–Šapiro condition holds for (P, B) and that the unbounded operator $(\mathsf{P}, D(\mathsf{P}))$ is symmetric. Let $\mathcal{N}$ be a connected component of ${}^1\partial\mathcal{M}$ where B is as given in (4.3.9). Then the boundary condition on $\mathcal{N}$ is equivalent*

1. *either to the homogeneous Dirichlet condition $u_{|\mathcal{N}} = 0$;*
2. *or to $\partial_\nu u_{|\mathcal{N}} + (i\beta T' + h)u_{|\mathcal{N}} = 0$ with β, h smooth functions, β real valued, and T' a real vector field on $\mathcal{N}$, such that*
   - (a) *$|T'|_{g_\partial} = 1$;*
   - (b) *$|\beta| < 1$ on $\mathcal{N}$ or $|\beta| > 1$ on $\mathcal{N}$;*
   - (c) *$2\operatorname{Im} h - \operatorname{div}_g(\beta T') + \gamma_{\mathcal{N}}^{\mathsf{D}}(g(V, \nu)) = 0$ on $\mathcal{N}$.*

The proof shows in fact that either $a \equiv 0$ or a does not vanish on $\mathcal{N}$. Case 1 corresponds to $a \equiv 0$; Case 2 to $a \neq 0$.

PROOF. We set $B' = bT' + c$, with b, c and T' as given in (4.3.9). As B has a nonvanishing principal symbol on $\mathcal{N}$ the open sets $\{a \neq 0\}$ and $\{b \neq 0\}$ form a cover of $\mathcal{N}$. Set W to be an open set of $\mathcal{N}$ such that $W \Subset \{a \neq 0\}$. For $u, w \in D(\mathsf{P})$ such that their traces are supported in W, from (4.2.2), we have

$$0 = (a^{-1}B'\gamma_{\mathcal{N}}^{\mathsf{D}}(u), \gamma_{\mathcal{N}}^{\mathsf{D}}(w))_{L^2(\mathcal{N})} - (\gamma_{\mathcal{N}}^{\mathsf{D}}(u), a^{-1}B'\gamma_{\mathcal{N}}^{\mathsf{D}}(w))_{L^2(\mathcal{N})} + i(\gamma_{\mathcal{N}}^{\mathsf{D}}(g(V, \nu)u), \gamma_{\mathcal{N}}^{\mathsf{D}}(w))_{L^2(\mathcal{N})}.$$

[1] In fact, by Lemma 3.19 it is Fredholm and its index is zero. If injective it is bijective with a bounded inverse by the closed graph theorem.

Such functions can be constructed by means of Theorem 18.25 by having both their Dirichlet and Neumann traces vanish on all the connected components of $\partial\mathcal{M}$ but $\mathcal{N}$. On $\mathcal{N}$ their Dirichlet trace can be chosen arbitrarily with support on W and the Neumann traces given by $\gamma_{\mathcal{N}}^{\mathsf{N}}(u) = -a^{-1}B'\gamma_{\mathcal{N}}^{\mathsf{D}}(u)$. For simplicity we choose $\gamma_{\mathcal{N}}^{\mathsf{D}}(u), \gamma_{\mathcal{N}}^{\mathsf{D}}(w) \in \mathscr{C}_c^\infty(W)$. We then obtain

$$a^{-1}B' - {}^t\overline{B'}a^{-1} + i\gamma_{\mathcal{N}}^{\mathsf{D}}(g(V,\nu)) = 0 \ \text{ in } W,$$

which reads, by Proposition 17.7,

$$\begin{aligned}2\operatorname{Re}(a^{-1}b)T' + [T', \overline{a^{-1}b}] + 2i\operatorname{Im}(a^{-1}c) + \overline{a^{-1}b}\operatorname{div}_g T' \\ + i\gamma_{\mathcal{N}}^{\mathsf{D}}(g(V,\nu)) = 0 \ \text{ in } W.\end{aligned} \tag{4.3.10}$$

This implies $\operatorname{Re}(a^{-1}b) = 0$ in W. With the Lopatinskiĭ–Šapiro condition holding, by Proposition 2.8, we also have $a^{-1}b \neq \pm i$.

Assume that there is a point $m^0 \in W$ such that $a^{-1}b(m^0) \in i(-1,1)$. Then, this remains true for all m in the same connected component in W. In fact, this excludes that a vanishes on $\overline{W}$ as $|b| \geq C > 0$ in a neighborhood of $\{a=0\}$. Thus $\overline{W} = W = \mathcal{N}$. We thus find that a does not vanish on $\mathcal{N}$ in this case and $a^{-1}b(m) \in i(-1,1)$ for all $m \in \mathcal{N}$. Upon dividing B by a it reduces to the form

$$\tilde{B}u_{|\mathcal{N}} = \partial_\nu u_{|\mathcal{N}} + (i\beta T' + h)u_{|\mathcal{N}},$$

with β a real function such that $|\beta| < 1$. From (4.3.10) we also have

$$-[T',\beta] + 2\operatorname{Im} h - \beta\operatorname{div}_g T' + \gamma_{\mathcal{N}}^{\mathsf{D}}(g(V,\nu)) = 0 \ \text{ in } \mathcal{N},$$

which reads by (17.2.6)

$$2\operatorname{Im} h - \operatorname{div}_g(\beta T') + \gamma_{\mathcal{N}}^{\mathsf{D}}(g(V,\nu)) = 0 \ \text{ in } \mathcal{N}. \tag{4.3.11}$$

To the contrary, assume that there is a point $m^0 \in W$ such that $a^{-1}b(m^0) \in i(-\infty,-1)\cup i(1,+\infty)$, the same argument as above shows that b does not vanish on $\mathcal{N}$ and $a^{-1}b(m) \in i(-\infty,-1) \cup i(1,+\infty)$ for all $m \in \mathcal{N}$. Upon dividing B by the b it reduces to the form

$$\hat{B}u_{|\mathcal{N}} = \alpha\partial_\nu u_{|\mathcal{N}} + (iT' + \rho)u_{|\mathcal{N}},$$

with α a real function such that $|\alpha| < 1$. From (4.3.10) we also have

$$-[T',\alpha^{-1}] + 2\alpha^{-1}\operatorname{Im}(\rho) - \alpha^{-1}\operatorname{div}_g T' + \gamma_{\mathcal{N}}^{\mathsf{D}}(g(V,\nu)) = 0$$

in $\{\alpha \neq 0\} = \{a \neq 0\}$. This reads

$$T'(\alpha) + (2\operatorname{Im}(\rho) - \operatorname{div}_g T')\alpha + \gamma_{\mathcal{N}}^{\mathsf{D}}(g(V,\nu))\alpha^2 = 0, \tag{4.3.12}$$

in $\{\alpha \neq 0\}$. By continuity observe in fact this first-order equation holds in $\overline{\{\alpha \neq 0\}}$. Observe also that (4.3.12) holds in $\operatorname{int}\{\alpha = 0\}$. Thus (4.3.12) everywhere on $\mathcal{N}$. By the Cauchy–Lipschitz theorem then either $\alpha \equiv 0$ or α does not vanish on $\mathcal{N}$.

If α does not vanish, meaning that a does not vanish in the first place, upon dividing B by the a it reduces to the form

$$\tilde{B}u_{|\mathcal{N}} = \partial_\nu u_{|\mathcal{N}} + (i\beta T' + h)u_{|\mathcal{N}},$$

with β a real function such that $|\beta| > 1$. From (4.3.10) we also find (4.3.11).

Finally, if $\alpha \equiv 0$, that is $a \equiv 0$, then $\hat{B} = iT' + \rho$. Using the same argument as in the proof of Lemma 4.4 at (4.3.8) we find that this first-order operator is injective meaning that the boundary condition $\hat{B}u_{|\mathcal{N}} = 0$ is equivalent to a homogeneous Dirichlet boundary condition $u_{|\mathcal{N}} = 0$. ∎

4.3.4. Necessary Conditions for Symmetry to Hold. With Lemmata 4.4 and 4.5, for any dimension $d \geq 2$, we now know that if the Lopatinskiĭ–Šapiro condition holds and if the operator $(\mathsf{P}, D(\mathsf{P}))$ is symmetric, then on any connected component $\mathcal{N}$ of $\partial\mathcal{M}$ the boundary condition $Bu_{|\mathcal{N}} = 0$ is equivalent

(1) either to a homogeneous Dirichlet boundary condition $u_{|\mathcal{N}} = 0$;
(2) or to the form $(\partial_\nu u + B'u)_{|\mathcal{N}} = 0$ with B' a first-order tangential differential operator.

In the latter case, the following proposition gives some constraints on the tangential operator B'. We recall that P takes the form $P = -\Delta_g + iV + f$ with V a smooth real vector field and f a smooth complex function on $\mathcal{M}$ with the additional constraint $\operatorname{Im} f = \operatorname{div}_g V/2$ by Lemma 4.2.

PROPOSITION 4.6. *Let $d \geq 2$ and let $\mathcal{N}$ be a connected component of* [1]*$\partial\mathcal{M}$ where $B = \partial_\nu + B'$. If the Lopatinskiĭ–Šapiro condition holds for (P, B) on $\partial\mathcal{M}$ and $(\mathsf{P}, D(\mathsf{P}))$ is symmetric, then*

$$B' - {}^t\overline{B'} + i\gamma_{\mathcal{N}}^{\mathsf{D}}(g(V, \nu)) = 0. \tag{4.3.13}$$

In particular, this means that $B' = iX' + h$ with X' a real valued vector field and h a complex valued function on $\mathcal{N}$ with the condition

$$2\operatorname{Im} h - \operatorname{div}_g X' + \gamma_{\mathcal{N}}^{\mathsf{D}}(g(V, \nu)) = 0. \tag{4.3.14}$$

In the case $d = 2$, we moreover have $|X'|_{g_\partial} \neq 1$. In the case $d \geq 3$, we moreover have $|X'|_{g_\partial} < 1$.

PROOF. Let $u, w \in D(\mathsf{P})$ be such that both their Dirichlet and Neumann traces vanish on all the connected components of $\partial\mathcal{M}$ but $\mathcal{N}$. Such function can be constructed by means of Theorem 18.25 once the two traces are also prescribed on $\mathcal{N}$. We denote the Dirichlet and Neumann trace operators on $\mathcal{N}$ by $\gamma_{\mathcal{N}}^{\mathsf{D}}$ and $\gamma_{\mathcal{N}}^{\mathsf{N}}$.

By (4.2.2) we have

$$\begin{aligned} 0 = &-(\gamma_{\mathcal{N}}^{\mathsf{N}}(u), \gamma_{\mathcal{N}}^{\mathsf{D}}(w))_{L^2(\mathcal{N})} + (\gamma_{\mathcal{N}}^{\mathsf{D}}(u), \gamma_{\mathcal{N}}^{\mathsf{N}}(w))_{L^2(\mathcal{N})} \\ &+ i(\gamma_{\mathcal{N}}^{\mathsf{D}}(g(V, \nu)u), \gamma_{\mathcal{N}}^{\mathsf{D}}(w))_{L^2(\mathcal{N})}. \end{aligned}$$

Because of the imposed boundary condition we have $\gamma_{\mathcal{N}}^{\mathsf{N}}(u) = -B'\gamma_{\mathcal{N}}^{\mathsf{D}}(u)$ and $\gamma_{\mathcal{N}}^{\mathsf{N}}(w) = -B'\gamma_{\mathcal{N}}^{\mathsf{D}}(w)$. Thus, upon prescribing $\gamma_{\mathcal{N}}^{\mathsf{D}}(u)$ and $\gamma_{\mathcal{N}}^{\mathsf{D}}(w)$ in $H^{3/2}(\mathcal{N})$ the functions u and w are uniquely defined. Moreover, we have

$$\begin{aligned}0 = (B'\gamma_{\mathcal{N}}^{\mathsf{D}}(u), \gamma_{\mathcal{N}}^{\mathsf{D}}(w))_{L^2(\mathcal{N})} &- (\gamma_{\mathcal{N}}^{\mathsf{D}}(u), B'\gamma_{\mathcal{N}}^{\mathsf{D}}(w))_{L^2(\mathcal{N})} \\ &+ i(\gamma_{\mathcal{N}}^{\mathsf{D}}(g(V,\nu)u), \gamma_{\mathcal{N}}^{\mathsf{D}}(w))_{L^2(\mathcal{N})},\end{aligned}$$

which reads

$$0 = \big((B' - {}^t\overline{B'} + i\gamma_{\mathcal{N}}^{\mathsf{D}}(g(V,\nu)))\gamma_{\mathcal{N}}^{\mathsf{D}}(u), \gamma_{\mathcal{N}}^{\mathsf{D}}(w)\big)_{L^2(\mathcal{N})}.$$

As $\gamma_{\mathcal{N}}^{\mathsf{D}}(u)$ and $\gamma_{\mathcal{N}}^{\mathsf{D}}(w)$ can be arbitrarily chosen in $H^{3/2}(\mathcal{N})$ we obtain (4.3.13).

Let us write $B' = Y' + h$ where Y' is a complex vector field and h is a smooth complex function on $\mathcal{N}$. Then, by Proposition 17.7 we have

$$B' - {}^t\overline{B'} = Y' + \overline{Y'} + \operatorname{div}_g \overline{Y'} + h - \overline{h}.$$

From (4.3.13) we conclude that $Y' = iX'$ with X' a real vector field on $\mathcal{N}$. Moreover, we find

$$B' - {}^t\overline{B'} = -i \operatorname{div}_g X' + 2i \operatorname{Im} h,$$

which with (4.3.13) yields (4.3.14).

From the characterization of the Lopatinskiĭ–Šapiro condition in Proposition 2.8, we find that $|X'|_{g_\partial} \neq 1$ in the case $d = 2$ and $|X'|_{g_\partial} < 1$ in the case $d \geq 4$.

Finally, in the case $d = 3$ by Lemma 4.4 we have (4.3.2) and (4.3.3) that precisely coincides with the condition one finds in the case $d \geq 4$ in Proposition 2.8. Thus, for the same reason we conclude that $|X'|_{g_\partial} < 1$. ∎

The following lemma will be of use in what follows.

LEMMA 4.7. *If the Lopatinskiĭ–Šapiro condition holds for (P, B) on $\partial\mathcal{M}$ and $(\mathsf{P}, D(\mathsf{P}))$ is symmetric, then*

$$\begin{aligned}(R_1 u, v)_{L^2(\mathcal{M})} &+ \langle B' u_{|\partial\mathcal{M}}, \bar{v}_{|\partial\mathcal{M}}\rangle_{H^{-1/2}({}^1\partial\mathcal{M}), H^{1/2}({}^1\partial\mathcal{M})} \\ &= (u, R_1 v)_{L^2(\mathcal{M})} + \langle u_{|\partial\mathcal{M}}, \overline{B' v}_{|\partial\mathcal{M}}\rangle_{H^{1/2}({}^1\partial\mathcal{M}), H^{-1/2}({}^1\partial\mathcal{M})},\end{aligned}$$

for $u, v \in H^1(\mathcal{M})$ such that $u_{|{}^0\partial\mathcal{M}} = v_{|{}^0\partial\mathcal{M}} = 0$.

The proof follows the arguments of the proofs of Lemma 4.2 and Proposition 4.6, noting that H^1-regularity suffices for the performed computations.

4.4. Sufficient Conditions for Selfadjointness

Here, we finish the proof of Theorem 4.3.

Above in Lemma 4.2 and Proposition 4.6 we saw that under the Lopatinskiĭ–Šapiro condition if the unbounded operator $(\mathsf{P}, D(\mathsf{P}))$ is selfadjoint, then the two properties in Theorem 4.3 hold. We now prove that they are in fact sufficient for selfadjointness to hold.

As explained above, in dimension $d = 2$ or $d = 3$, on a connected component $\mathcal{N}$ of ${}^1\partial\mathcal{M}$, if the boundary condition $Bu_{|\mathcal{N}} = 0$ is equivalent to a homogeneous Dirichlet condition $u_{|\mathcal{N}} = 0$ we choose to use this latter and simpler form of the boundary condition. We thus consider $\mathcal{N}$ to be part of ${}^{0}\partial\mathcal{M}$. We refer to Lemmata 4.4 and 4.5 and their proofs for details.

PROOF OF THE SUFFICIENCY OF THE CONDITIONS IN THEOREM 4.3. The adjoint operator of P is well defined as $D(\mathsf{P})$ is dense in $L^2(\mathcal{M})$ as mentioned at the beginning of Sect. 4.2. Its domain is given by

$$D(\mathsf{P}^*) = \{v \in L^2(\mathcal{M});\ \exists C > 0,\ \forall u \in D(\mathsf{P}),\ |(v, Pu)_{L^2(\mathcal{M})}| \leq C\|u\|_{L^2(\mathcal{M})}\}.$$

Under the properties (1) and (2) in the statement of the theorem one finds that (4.2.2) holds meaning that P is symmetric. Hence, it only remains to prove that $D(\mathsf{P}^*) = D(\mathsf{P})$.

If $v \in D(\mathsf{P}^*)$, one has $\mathsf{P}^* v \in L^2(\mathcal{M})$ and

$$(v, Pu)_{L^2(\mathcal{M})} = (P^* v, u)_{L^2(\mathcal{M})},$$

for all $u \in D(\mathsf{P})$ by the very definition of P^*. We choose $\varphi \in {}^{0}\mathscr{D}_c^\infty(\mathcal{M}) \subset D(\mathsf{P})$. We then have

$$\begin{aligned}(P^* v, \varphi)_{L^2(\mathcal{M})} &= (v, P\varphi)_{L^2(\mathcal{M})} = \langle v, \overline{P\varphi}\mu_g\rangle_{{}^0\mathscr{D}'(\mathcal{M}), {}^1\mathscr{D}_c^\infty(\mathcal{M})} \\ &= \langle {}^t\overline{P} v, \overline{\varphi}\mu_g\rangle_{{}^0\mathscr{D}'(\mathcal{M}), {}^1\mathscr{D}_c^\infty(\mathcal{M})},\end{aligned}$$

where μ_g is the canonical positive density associated with the metric g (see Sect. 17.3).

Now with the assumption made on X and f we have ${}^t\overline{P} = P$. We thus find

$$\langle Pv, \overline{\varphi}\mu_g\rangle_{{}^0\mathscr{D}'(\mathcal{M}), {}^1\mathscr{D}_c^\infty(\mathcal{M})} = (P^* v, \varphi)_{L^2(\mathcal{M})} = \langle P^* v, \overline{\varphi}\mu_g\rangle_{{}^0\mathscr{D}'(\mathcal{M}), {}^1\mathscr{D}_c^\infty(\mathcal{M})}$$

for all $\varphi \in {}^{0}\mathscr{D}_c^\infty(\mathcal{M})$ implying that $Pv = P^* v$ in ${}^0\mathscr{D}'(\mathcal{M})$. As $P^* v \in L^2(\mathcal{M})$ we thus have $Pv \in L^2(\mathcal{M})$. Consequently v lies in the space $\mathcal{W}_P(\mathcal{M})$ defined in Sect. 18.6.2. By Lemma 18.32 the Dirichlet and Neumann traces $\gamma^{\mathsf{D}}(v)$ and $\gamma^{\mathsf{N}}(v)$ are well defined in $H^{-1/2}(\partial\mathcal{M})$ and $H^{-3/2}(\partial\mathcal{M})$, respectively. Moreover, by Lemma 18.33 we have the following Green-like formula

$$\begin{aligned}(Pv, w)_{L^2(\mathcal{M})} &+ \langle \gamma^{\mathsf{N}}(v), \gamma^{\mathsf{D}}(\overline{w})\rangle_{H^{-3/2}(\partial\mathcal{M}), H^{3/2}(\partial\mathcal{M})} \\ &= (v, Pw)_{L^2(\mathcal{M})} + \langle \gamma^{\mathsf{D}}(v), \gamma^{\mathsf{N}}(\overline{w})\rangle_{H^{-1/2}(\partial\mathcal{M}), H^{1/2}(\partial\mathcal{M})} \\ &\quad + \langle \gamma^{\mathsf{D}}(g(X,\nu)v), \gamma^{\mathsf{D}}(\overline{w})\rangle_{H^{-1/2}(\partial\mathcal{M}), H^{1/2}(\partial\mathcal{M})},\end{aligned}$$

for any $w \in H^2(\mathcal{M})$, as here ${}^t\overline{P} = P$. We now take $w \in D(\mathsf{P})$. Since $Pv = P^* v$ and $(Pv, w)_{L^2(\mathcal{M})} = (v, Pw)_{L^2(\mathcal{M})}$ we thus find

$$\begin{aligned}0 = &-\langle \gamma^{\mathsf{N}}(v), \gamma^{\mathsf{D}}(\overline{w})\rangle_{H^{-3/2}(\partial\mathcal{M}), H^{3/2}(\partial\mathcal{M})} + \langle \gamma^{\mathsf{D}}(v), \gamma^{\mathsf{N}}(\overline{w})\rangle_{H^{-1/2}(\partial\mathcal{M}), H^{1/2}(\partial\mathcal{M})} \\ &+ i\langle \gamma^{\mathsf{D}}(g(V,\nu)v), \gamma^{\mathsf{D}}(\overline{w})\rangle_{H^{-1/2}(\partial\mathcal{M}), H^{1/2}(\partial\mathcal{M})},\end{aligned}$$

using that $X = iV$ here.

First, if any, consider a connected component $\mathcal{N}$ of $\partial\mathcal{M}$ where B coincides with the Dirichlet boundary operator, that is, B is of order zero. By Theorem 18.25 pick $w \in H^2(\mathcal{M})$ such that $\gamma^{\mathsf{D}}(w) = 0$, that is, its

Dirichlet trace vanishes on the whole $\partial\mathcal{M}$ and $\partial_\nu w_{|\partial\mathcal{M}\setminus\mathcal{N}} = 0$, that is, its Neumann trace vanishes also except on $\mathcal{N}$. We then have $Bw_{|\partial\mathcal{M}} = 0$ and thus $w \in D(\mathsf{P})$ and we obtain

$$\langle \gamma^{\mathsf{D}}_{\mathcal{N}}(v), \gamma^{\mathsf{N}}_{\mathcal{N}}(\overline{w})\rangle_{H^{-1/2}(\mathcal{N}),H^{1/2}(\mathcal{N})} = 0.$$

As the value of $\gamma^{\mathsf{N}}_{\mathcal{N}}(\overline{w})$ can be chosen arbitrarily in $H^{1/2}(\partial\mathcal{M})$, we obtain that $Bv_{|\mathcal{N}} = \gamma^{\mathsf{D}}_{\mathcal{N}}(v) = 0$.

Second, if any, consider a connected component $\mathcal{N}$ of $\partial\mathcal{M}$ where B is of order one, of the form $B = \partial_\nu + B'$. By Theorem 18.25 pick w such that both its Dirichlet and Neumann traces vanish on all connected components of $\partial\mathcal{M}$ but $\mathcal{N}$. On $\mathcal{N}$ on let $\gamma^{\mathsf{D}}_{\mathcal{N}}(w) = \varphi \in H^{3/2}(\mathcal{N})$ be kept arbitrary and on set $\gamma^{\mathsf{N}}_{\mathcal{N}}(w)$ to be equal to $-B'\varphi$. Thus $Bw_{|\mathcal{N}} = 0$. With such choice we have

$$\begin{aligned} 0 = &-\langle \gamma^{\mathsf{N}}_{\mathcal{N}}(v), \varphi\rangle_{H^{-3/2}(\mathcal{N}),H^{3/2}(\mathcal{N})} - \langle \gamma^{\mathsf{D}}_{\mathcal{N}}(v), B'\varphi\rangle_{H^{-1/2}(\mathcal{N}),H^{1/2}(\mathcal{N})} \\ &+ i\langle \gamma^{\mathsf{D}}_{\mathcal{N}}(g(V,\nu)v), \varphi\rangle_{H^{-1/2}(\mathcal{N}),H^{1/2}(\mathcal{N})}. \end{aligned} \tag{4.4.1}$$

We write, using that $B' - {}^t\overline{B'} + i\gamma^{\mathsf{D}}_{\mathcal{N}}(g(V,\nu)) = 0$

$$\begin{aligned} \langle \gamma^{\mathsf{D}}_{\mathcal{N}}(v), B'\varphi\rangle_{H^{-1/2}(\mathcal{N}),H^{1/2}(\mathcal{N})} &= \langle {}^t\overline{B'}\gamma^{\mathsf{D}}_{\mathcal{N}}(v), \varphi\rangle_{H^{-3/2}(\mathcal{N}),H^{3/2}(\mathcal{N})} \\ &= \langle B'\gamma^{\mathsf{D}}_{\mathcal{N}}(v) \\ &\quad + i\gamma^{\mathsf{D}}_{\mathcal{N}}(g(V,\nu)v), \varphi\rangle_{H^{-3/2}(\mathcal{N}),H^{3/2}(\mathcal{N})}, \end{aligned}$$

which yields with (4.4.1)

$$\langle Bv_{|\mathcal{N}}, \varphi\rangle_{H^{-3/2}(\mathcal{N}),H^{3/2}(\mathcal{N})} = 0.$$

As φ is arbitrary in $H^{3/2}(\mathcal{N})$ we find that $Bv_{|\mathcal{N}} = 0$.

We thus conclude that $Bv_{|\partial\mathcal{M}} = 0$. As we saw above that $Pv \in L^2(\mathcal{M})$, we conclude that $v \in D(\mathsf{P})$ from the definition of $D(\mathsf{P})$ in (4.1.1). ■

4.5. A Green Formula

The following subspace of $L^2(\mathcal{M})$ associated with P is introduced in Sect. 18.6.2:

$$\mathcal{W}_P(\mathcal{M}) = \{u \in L^2(\mathcal{M});\ Pu \in L^2(\mathcal{M})\}. \tag{4.5.1}$$

Note that the action of P on u is to be understood in the sense of distributions, as expressed in (18.3.1), and not as that of the unbounded operator $(\mathsf{P}, D(\mathsf{P}))$ on $L^2(\mathcal{M})$.

Properties of the space $\mathcal{W}_P(\mathcal{M})$ are analyzed in Sect. 18.6.2. In particular, by Lemma 18.32, if $u \in \mathcal{W}_P(\mathcal{M})$, then both its Dirichlet and Neumann traces, $\gamma^{\mathsf{D}}(u)$ and $\gamma^{\mathsf{N}}(u)$, are well defined and

$$\gamma^{\mathsf{D}}(u) = u_{|\partial\mathcal{M}} \in H^{-1/2}(\partial\mathcal{M}) \quad \text{and} \quad \gamma^{\mathsf{N}}(u) = \partial_\nu u_{|\partial\mathcal{M}} \in H^{-3/2}(\partial\mathcal{M}).$$

If the unbounded operator $(\mathsf{P}, D(\mathsf{P}))$ is selfadjoint, meaning that the necessary of sufficient conditions given in Theorem 4.3 hold, one also has the following Green-like formula.

PROPOSITION 4.8. *Let P and B be such that $(\mathsf{P}, D(\mathsf{P}))$ is selfadjoint. If $u \in \mathcal{W}_P(\mathcal{M})$ and $w \in H^2(\mathcal{M})$ we have*

$$\begin{aligned}
&(Pu, w)_{L^2(\mathcal{M})} - (u, Pw)_{L^2(\mathcal{M})} + \langle \gamma^{\mathsf{N}}_{{}^0\partial\mathcal{M}}(u), \gamma^{\mathsf{D}}_{{}^0\partial\mathcal{M}}(\overline{w})\rangle_{H^{-3/2}({}^0\partial\mathcal{M}),H^{3/2}({}^0\partial\mathcal{M})} \\
&\quad + \langle Bu_{|{}^1\partial\mathcal{M}}, \gamma^{\mathsf{D}}_{{}^1\partial\mathcal{M}}(\overline{w})\rangle_{H^{-3/2}({}^1\partial\mathcal{M}),H^{3/2}({}^1\partial\mathcal{M})} \\
&= \langle \gamma^{\mathsf{D}}_{{}^0\partial\mathcal{M}}(u), \gamma^{\mathsf{N}}_{{}^0\partial\mathcal{M}}(\overline{w})\rangle_{H^{-1/2}({}^0\partial\mathcal{M}),H^{1/2}({}^0\partial\mathcal{M})} \\
&\quad + \langle \gamma^{\mathsf{D}}_{{}^1\partial\mathcal{M}}(u), \overline{Bw_{|{}^1\partial\mathcal{M}}}\rangle_{H^{-1/2}({}^1\partial\mathcal{M}),H^{1/2}({}^1\partial\mathcal{M})} \\
&\quad + i\langle \gamma^{\mathsf{D}}_{{}^0\partial\mathcal{M}}(g(V,\nu)u), \gamma^{\mathsf{D}}_{{}^0\partial\mathcal{M}}(\overline{w})\rangle_{H^{-1/2}({}^0\partial\mathcal{M}),H^{1/2}({}^0\partial\mathcal{M})}.
\end{aligned}$$

Recall that ${}^0\partial\mathcal{M}$ is the union of all the connected components of $\partial\mathcal{M}$ where B is the Dirichlet boundary operator and ${}^1\partial\mathcal{M}$ is the union of all the connected components of $\partial\mathcal{M}$ where B is of order one with the properties given in Theorem 4.3.

For both $u, w \in H^2(\mathcal{M})$, with Proposition 4.8 seen in Proposition 4.6 and its proof. we find

$$\begin{aligned}
&(Pu, w)_{L^2(\mathcal{M})} + (\gamma^{\mathsf{N}}_{{}^0\partial\mathcal{M}}(u), \gamma^{\mathsf{D}}_{{}^0\partial\mathcal{M}}(w))_{L^2({}^0\partial\mathcal{M})} + (Bu_{|{}^1\partial\mathcal{M}}, \gamma^{\mathsf{D}}_{{}^1\partial\mathcal{M}}(w))_{L^2({}^1\partial\mathcal{M})} \\
&\quad = (u, Pw)_{L^2(\mathcal{M})} + (\gamma^{\mathsf{D}}_{{}^0\partial\mathcal{M}}(u), \gamma^{\mathsf{N}}_{{}^0\partial\mathcal{M}}(w))_{L^2({}^0\partial\mathcal{M})} \\
&\qquad + (\gamma^{\mathsf{D}}_{{}^1\partial\mathcal{M}}(u), Bw_{|{}^1\partial\mathcal{M}})_{L^2({}^1\partial\mathcal{M})} + i(\gamma^{\mathsf{D}}_{{}^0\partial\mathcal{M}}(g(V,\nu)u), \gamma^{\mathsf{D}}_{{}^0\partial\mathcal{M}}(w))_{L^2({}^0\partial\mathcal{M})}.
\end{aligned}$$

Note also that in the case where both u and w have vanishing Dirichlet traces on ${}^0\partial\mathcal{M}$, we have the simple form, for $u \in \mathcal{W}_P(\mathcal{M})$ and $w \in H^2(\mathcal{M})$,

$$\begin{aligned}
(Pu, w)_{L^2(\mathcal{M})} &+ \langle Bu_{|{}^1\partial\mathcal{M}}, \gamma^{\mathsf{D}}_{{}^1\partial\mathcal{M}}(\overline{w})\rangle_{H^{-3/2}({}^1\partial\mathcal{M}),H^{3/2}({}^1\partial\mathcal{M})} \\
&= (u, Pw)_{L^2(\mathcal{M})} + \langle \gamma^{\mathsf{D}}_{{}^1\partial\mathcal{M}}(u), \overline{Bw_{|{}^1\partial\mathcal{M}}}\rangle_{H^{-1/2}({}^1\partial\mathcal{M}),H^{1/2}({}^1\partial\mathcal{M})},
\end{aligned} \tag{4.5.2}$$

which in the case where u and w are moreover both in $H^2(\mathcal{M})$ gives

$$\begin{aligned}
(Pu, w)_{L^2(\mathcal{M})} &+ (Bu_{|{}^1\partial\mathcal{M}}, \gamma^{\mathsf{D}}_{{}^1\partial\mathcal{M}}(w))_{L^2({}^1\partial\mathcal{M})} \\
&= (u, Pw)_{L^2(\mathcal{M})} + (\gamma^{\mathsf{D}}_{{}^1\partial\mathcal{M}}(u), Bw_{|{}^1\partial\mathcal{M}})_{L^2({}^1\partial\mathcal{M})}.
\end{aligned} \tag{4.5.3}$$

Note that this latter formula is also a consequence of Lemma 4.7.

PROOF. We start from the result of Lemma 18.33 that reads here

$$\begin{aligned}
(Pu, w)_{L^2(\mathcal{M})} &+ \langle \gamma^{\mathsf{N}}(u), \gamma^{\mathsf{D}}(\overline{w})\rangle_{H^{-3/2}(\partial\mathcal{M}),H^{3/2}(\partial\mathcal{M})} \\
&= (u, Pw)_{L^2(\mathcal{M})} + \langle \gamma^{\mathsf{D}}(u), \gamma^{\mathsf{N}}(\overline{w})\rangle_{H^{-1/2}(\partial\mathcal{M}),H^{1/2}(\partial\mathcal{M})} \\
&\quad + i\langle \gamma^{\mathsf{D}}(g(V,\nu)u), \gamma^{\mathsf{D}}(\overline{w})\rangle_{H^{-1/2}(\partial\mathcal{M}),H^{1/2}(\partial\mathcal{M})}.
\end{aligned}$$

On ${}^1\partial\mathcal{M}$ with $B = \partial_\nu + B'$, with B' a first-order tangential differential operator we have

$$Bu_{|{}^1\partial\mathcal{M}} = (\partial_\nu u + B'u)_{|{}^1\partial\mathcal{M}} \in H^{-3/2}({}^1\partial\mathcal{M}),$$
$$Bw_{|{}^1\partial\mathcal{M}} = (\partial_\nu w + B'w)_{|{}^1\partial\mathcal{M}} \in H^{1/2}({}^1\partial\mathcal{M}).$$

The condition in Theorem 4.3 precisely mean ${}^t\overline{B'} = B' + i\gamma^{\mathsf{D}}(g(V,\nu)$ (as seen in Proposition 4.6 and its proof). It follows that

$$\begin{aligned}&\langle \gamma^{\mathsf{N}}_{{}^1\partial\mathcal{M}}(u), \gamma^{\mathsf{D}}_{{}^1\partial\mathcal{M}}(\overline{w})\rangle_{H^{-3/2}({}^1\partial\mathcal{M}),H^{3/2}({}^1\partial\mathcal{M})}\\ &\quad= \langle Bu_{|{}^1\partial\mathcal{M}}, \gamma^{\mathsf{D}}_{{}^1\partial\mathcal{M}}(\overline{w})\rangle_{H^{-3/2}({}^1\partial\mathcal{M}),H^{3/2}({}^1\partial\mathcal{M})}\\ &\qquad- \langle B'u_{|{}^1\partial\mathcal{M}}, \gamma^{\mathsf{D}}_{{}^1\partial\mathcal{M}}(\overline{w})\rangle_{H^{-3/2}({}^1\partial\mathcal{M}),H^{3/2}({}^1\partial\mathcal{M})},\end{aligned}$$

and

$$\begin{aligned}&\langle B'u_{|{}^1\partial\mathcal{M}}, \gamma^{\mathsf{D}}_{{}^1\partial\mathcal{M}}(\overline{w})\rangle_{H^{-3/2}({}^1\partial\mathcal{M}),H^{3/2}({}^1\partial\mathcal{M})}\\ &\quad= \langle \gamma^{\mathsf{D}}_{{}^1\partial\mathcal{M}}(u), {}^tB'\overline{w}_{|{}^1\partial\mathcal{M}}\rangle_{H^{-1/2}({}^1\partial\mathcal{M}),H^{1/2}({}^1\partial\mathcal{M})}\\ &\quad= \langle \gamma^{\mathsf{D}}_{{}^1\partial\mathcal{M}}(u), \overline{B'w}_{|{}^1\partial\mathcal{M}}\rangle_{H^{-1/2}({}^1\partial\mathcal{M}),H^{1/2}({}^1\partial\mathcal{M})}\\ &\qquad- i\langle \gamma^{\mathsf{D}}_{{}^1\partial\mathcal{M}}(u), \gamma^{\mathsf{D}}_{{}^1\partial\mathcal{M}}(g(V,\nu)\overline{w})\rangle_{H^{-1/2}({}^1\partial\mathcal{M}),H^{1/2}({}^1\partial\mathcal{M})}.\end{aligned}$$

We thus find

$$\begin{aligned}&\langle \gamma^{\mathsf{N}}_{{}^1\partial\mathcal{M}}(u), \gamma^{\mathsf{D}}_{{}^1\partial\mathcal{M}}(\overline{w})\rangle_{H^{-3/2}({}^1\partial\mathcal{M}),H^{3/2}({}^1\partial\mathcal{M})}\\ &\qquad- \langle \gamma^{\mathsf{D}}_{{}^1\partial\mathcal{M}}(u), \gamma^{\mathsf{N}}_{{}^1\partial\mathcal{M}}(\overline{w})\rangle_{H^{-1/2}(\partial\mathcal{M}),H^{1/2}(\partial\mathcal{M})}\\ &\quad= \langle Bu_{|{}^1\partial\mathcal{M}}, \gamma^{\mathsf{D}}_{{}^1\partial\mathcal{M}}(\overline{w})\rangle_{H^{-3/2}({}^1\partial\mathcal{M}),H^{3/2}({}^1\partial\mathcal{M})}\\ &\qquad- \langle \gamma^{\mathsf{D}}_{{}^1\partial\mathcal{M}}(u), \overline{Bw}_{|{}^1\partial\mathcal{M}}\rangle_{H^{-1/2}({}^1\partial\mathcal{M}),H^{1/2}({}^1\partial\mathcal{M})}\\ &\qquad+ i\langle \gamma^{\mathsf{D}}_{{}^1\partial\mathcal{M}}(u), \gamma^{\mathsf{D}}_{{}^1\partial\mathcal{M}}(g(V,\nu)\overline{w})\rangle_{H^{-1/2}({}^1\partial\mathcal{M}),H^{1/2}({}^1\partial\mathcal{M})},\end{aligned}$$

which yields the result. ∎

4.6. Spectral Properties

In the case where $(\mathsf{P}, D(\mathsf{P}))$ is selfadjoint, meaning that the assumptions of Theorem 4.3 are fulfilled, we have following spectral decomposition.

THEOREM 4.9. *Let P and B be such that the Lopatinskiĭ–Šapiro condition holds on $\partial\mathcal{M}$ and $(\mathsf{P}, D(\mathsf{P}))$ is selfadjoint. Then, the spectrum of $(\mathsf{P}, D(\mathsf{P}))$ is a sequence of real eigenvalues $(\mu_j)_j$ whose modulus goes to $+\infty$. Each eigenvalue is of finite multiplity. Moreover, there exists a Hilbert basis $(\phi_j)_{j\in\mathbb{N}}$ of $L^2(\mathcal{M})$ made of eigenfunctions of $(\mathsf{P}, D(\mathsf{P}))$, each associated with the eigenvalues μ_j.*

PROOF. For $c \in \mathbb{C}$, we define the bounded operator

$$\begin{aligned}L_c : H^2(\mathcal{M}) &\to L^2(\mathcal{M}) \oplus H^{3/2,1/2}(\partial\mathcal{M})\\ u &\mapsto (Pu + cu, Bu_{|{}^0\partial\mathcal{M}}, Bu_{|{}^1\partial\mathcal{M}}),\end{aligned} \tag{4.6.1}$$

with the notation of (3.1.3) in Chap. 3 with

$$H^{3/2,1/2}(\partial\mathcal{M}) = H_B^{(3/2)}(\partial\mathcal{M}) = H^{3/2}({}^0\partial\mathcal{M}) \oplus H^{1/2}({}^1\partial\mathcal{M}).$$

By Theorem 3.1 the Fredholm properties of this operator L_c is equivalent to having the Lopatinskiĭ–Šapiro condition. Moreover, the index of L_c is zero: in the case $d \geq 4$, this follows from Theorem 3.15; in the case $d = 3$, this follows from Theorem 3.16; in the case $d = 2$, we find that the form of B given in Theorem 4.3 allows one to apply the first point of Theorem 3.25.

First, we consider the case $c = i$. Since $(\mathsf{P}, D(\mathsf{P}))$ is selfadjoint, we have $\ker(i\,\mathrm{Id} + \mathsf{P}) = \{0\}$ by Theorem 11.18 of Volume 1. Hence, the operator L_i is injective and consequently surjective. The inverse $L_i^{-1} : L^2(\mathcal{M}) \oplus H^{3/2,1/2}(\partial\mathcal{M}) \to H^2(\mathcal{M})$ is thus well defined and bounded by the open mapping theorem. Observe then that L_i^{-1} restricted to $L^2(\mathcal{M}) \oplus \{0\}$ provides the inverse of $i\,\mathrm{Id} + \mathsf{P}$. We obtain that $(i\,\mathrm{Id} + \mathsf{P})^{-1} : L^2(\mathcal{M}) \to H^2(\mathcal{M})$ is a bounded operator. As the injection $\iota : H^2(\mathcal{M}) \to L^2(\mathcal{M})$ is compact by the Rellich–Kondrachov theorem (Theorem 18.7), the resolvent map $(i\,\mathrm{Id} + \mathsf{P})^{-1}$ maps $L^2(\mathcal{M})$ into itself in a compact way. In particular, from the spectral properties of compact operators, it has a bounded and at most countable spectrum.

Second, we claim that L_c is injective for some $c \in \mathbb{R}$. In fact, if $0 \neq u \in H^2(\mathcal{M})$ is such that $L_c u = 0$, then $Bu_{|\partial\mathcal{M}} = 0$ and $Pu + iu = (i - c)u$. As $i - c \neq 0$, this means that $1/(i - c)$ is an eigenvalue of $(i\,\mathrm{Id} + \mathsf{P})^{-1}$. The set $\{1/(i - c);\ c \in \mathbb{R}\}$ is the image of the horizontal line $i + \mathbb{R}$ by the inversion map $z \mapsto 1/z$. It is the circle centerer at $-i/2$ with radius $1/2$. Above, we saw that the spectrum of $(i\,\mathrm{Id} + \mathsf{P})^{-1}$ is at most countable. We may thus find $c \in \mathbb{R}$ such that $1/(i - c)$ is not one of its eigenvalues. For such a well-chosen value of c, the map L_c is injective.

With the analysis used above for L_i transposed to L_c, we find that $(c\,\mathrm{Id} + \mathsf{P})^{-1}$ is a well defined compact map on $L^2(\mathcal{M})$. Since $c\,\mathrm{Id} + \mathsf{P}$ is selfadjoint, then so is $(c\,\mathrm{Id} + \mathsf{P})^{-1}$. From the spectral properties of selfadjoint compact operators, we find that its spectrum is formed by a sequence of real eigenvalues of finite multiplicity that accumulate to zero and that zero is not an eigenvalue. Moreover, there exists a Hilbert basis of $L^2(\mathcal{M})$ made of eigenfunctions $(\phi_j)_{j \in \mathbb{N}}$ associated with the sequence of eigenvalues.

We conclude that the spectrum of $(\mathsf{P}, D(\mathsf{P}))$ is formed of a sequence of real eigenvalues whose modulus go to ∞ and any eigenvalue μ is of finite multiplicity, as $(\mu + c)^{-1}$ is then an eigenvalue of $(c\,\mathrm{Id} + \mathsf{P})^{-1}$. ■

As stated in Theorem 4.3 on any connected component $\mathcal{N}$ of $\partial\mathcal{M}$ the boundary operator B is

(1) either (equivalent to) the Dirichlet operator and the part of the boundary where this occurs is denoted by ${}^0\partial\mathcal{M}$;
(2) or of the form $Bu_{|\mathcal{N}} = \partial_\nu u_{|\mathcal{N}} + (iX' + h)u_{|\mathcal{N}}$ where with X' a real valued vector field on $\mathcal{N}$ and h a complex valued function on

$\mathcal{N}$ with some joint condition on P, X', and h. The part of the boundary where this occurs is denoted by ${}^1\partial\mathcal{M}$.

To understand the basic behavior of the sequence of eigenvalues we shall need the following lemma.

LEMMA 4.10. *Let X' be such that $|X'|_{g_\partial} < 1$ on ${}^1\partial\mathcal{M}$. Set $B' = iX' + h$. There exists $0 < C_0 < 1$ and $C > 0$ such that*

$$|(B'\gamma^{\mathsf{D}}(w), \gamma^{\mathsf{D}}(w))_{L^2({}^1\partial\mathcal{M})}| \leq C_0\|\nabla_g w\|^2_{L^2V(\mathcal{M})} + C\|w\|^2_{L^2(\mathcal{M})},$$

for all w in $H^1(\mathcal{M})$.

PROOF. Let us first focus on one of the connected components $\mathcal{N}$ of ${}^1\partial\mathcal{M}$. In a neighborhood of $\mathcal{N}$ we use normal geodesic coordinates $(m', z) \in O = \mathcal{N} \times [0, Z_0)$ as given by Theorem 17.22. In such coordinates the metric g takes the form

$$g_{(m',z)} = g'_{m'}(z) \otimes 1_z + d_z \otimes d_z.$$

Since $\mathcal{N}$ is compact we have

$$|X'_{m'}|_{g_\partial} = |X'_{m'}|_{g'(0)} \leq C_1 < 1,$$

for some $C_1 > 0$ and for all $m' \in \mathcal{N}$.

Let $Y'(z)$ be a smooth real vector field on $\mathcal{N}$ that depends smoothly on the parameter $z \in [0, Z_0)$ and such that $Y'(0) = X'$ and satisfies $|Y'(z)_{m'}|_{g'(z)_{m'}} \leq C_2$ for some $C_1 \leq C_2 < 1$ and for all $(m', z) \in O$. Using (17.2.1), this implies that

$$\begin{aligned} |Y'w(m', z)| &= |g_{(m',z)}(Y'(z)_{m'}, \nabla_g w_{(m',z)})| \\ &= |g'_{m'}(z)(Y'(z)_{m'}, \nabla_{g'(z)} w_{(m',z)})| \\ &\leq C_2|\nabla_{g'(z)} w(m', z)|_{g'(z)_{m'}}. \end{aligned} \tag{4.6.2}$$

We first choose $w \in H^2(O)$ supported near $z = 0$, say in $\mathcal{N} \times [0, Z_0/2)$, and we compute

$$\begin{aligned} &i(X'w_{|z=0^+}, w_{|z=0^+}))_{L^2(\mathcal{N})} \\ &\quad = (D_z Y'(z)w, w)_{L^2(O)} - (Y'(z)w, D_z w)_{L^2(O)} \\ &\qquad + \frac{1}{2}\big(Y'(z)w, wD_z \log\det g\big)_{L^2(O)} \\ &\quad = (Y'(z)D_z w, w)_{L^2(O)} - (Y'(z)w, D_z w)_{L^2(O)} \\ &\qquad + \Big(\big([D_z, Y'(z)] - \frac{1}{2}(D_z \log\det g)Y'(z)\big)w, w\Big)_{L^2(O)}. \end{aligned}$$

By Proposition 17.7 giving the transpose of a vector field and using that $\mathcal{N}$ has no boundary we find

$$(Y'(z)D_z w, w)_{L^2(O)} = -(D_z w, Y'(z)w)_{L^2(O)} - (D_z w, \operatorname{div}_g Y'(z)w)_{L^2(O)}.$$

We thus obtain

$$i(X'w_{|z=0^+}, w_{|z=0^+}))_{L^2(\mathcal{N})} = -2\operatorname{Re}(D_z w, Y'(z)w)_{L^2(O)} + (Rw, w)_{L^2(O)},$$

where R is a first-order differential operator. We now have, for $\varepsilon > 0$ to be fixed below,

$$\begin{aligned} 2|(D_z w, Y'(z)w)_{L^2(O)}| &\leq (1+\varepsilon)^{-1}\|D_z w\|^2_{L^2(O)} + (1+\varepsilon)\|Y'(z)w\|^2_{L^2(O)} \\ &\leq (1+\varepsilon)^{-1}\|D_z w\|^2_{L^2(O)} \\ &\quad + C_2^2(1+\varepsilon)\int_{[0,Z_0)} \|\nabla_{g'(z)} w\|^2_{L^2V(\mathcal{N})} dz, \end{aligned}$$

by (4.6.2). We choose $\varepsilon > 0$ so that $C_3 = \max\big((1+\varepsilon)^{-1}, C_2^2(1+\varepsilon)\big) < 1$. We then have

$$\begin{aligned} 2|(D_z w, Y'(z)w)_{L^2(O)}| &\leq C_3\big(\|D_z w\|^2_{L^2(O)} + \int_{[0,Z_0)} \|\nabla_{g'(z)} w\|^2_{L^2V(\mathcal{N},g'(z))} dz\big) \\ &= C_3\|\nabla_g w\|^2_{L^2V(O)}. \end{aligned}$$

We also write, for any $\eta > 0$,

$$|(Rw, w)_{L^2(O)}| \leq \eta\|\nabla_g w\|^2_{L^2V(O)} + C_\eta\|w\|^2_{L^2(O)}.$$

If we choose $\eta > 0$ such that $C_{\mathcal{N}} = C_3 + \eta < 1$ we then obtain

$$|(X'w_{|z=0^+}, w_{|z=0^+}))_{L^2(\mathcal{N})}| \leq C_{\mathcal{N}}\|\nabla_g w\|^2_{L^2V(O)} + C\|w\|^2_{L^2(O)}. \tag{4.6.3}$$

This estimate extends to any function $w \in H^1(O)$ by density.

Denote by $\mathcal{N}^1, \ldots, \mathcal{N}^n$ the connected components of ${}^1\partial\mathcal{M}$ and by $O^1, \ldots, O^n$ the associated open neighborhoods where we use normal geodesic coordinates as above. These neighborhoods are chosen disjoint. Let $\psi^1, \ldots, \psi^n \in \mathscr{C}^\infty(\mathcal{M})$ be such that $\psi^j \equiv 1$ in a neighborhood of $\mathcal{N}^j$, $0 \leq \psi^j \leq 1$, and $\operatorname{supp}(\psi^j) \subset O^j$, $j = 1, \ldots n$.

Let now $w \in H^1(\mathcal{M})$. Estimate (4.6.3) applies to $\psi^j w$, $j = 1, \ldots, n$. We then write

$$\begin{aligned} (B'\gamma^{\mathsf{D}}(w), \gamma^{\mathsf{D}}(w))_{L^2({}^1\partial\mathcal{M})} &= \sum_{1\leq j\leq n} (B'\gamma^{\mathsf{D}}(\psi^j w), \gamma^{\mathsf{D}}(\psi^j w))_{L^2(\mathcal{N}^j)} \\ &\leq \sum_{1\leq j\leq n} C_{\mathcal{N}^j}\|\nabla_g(\psi^j w)\|^2_{L^2V(O)} + C\|w\|^2_{L^2(O)}, \end{aligned}$$

with $C_{\mathcal{N}^j} < 1$. We set $C_0 = \max_{1\leq j\leq n} C_{\mathcal{N}^j}$. Since the supports of ψ^j, $j = 1, \ldots, n$, are disjoint if we set $\psi = \sum_{1\leq j\leq n} \psi^j$ we obtain

$$\begin{aligned} (B'\gamma^{\mathsf{D}}(w), \gamma^{\mathsf{D}}(w))_{L^2({}^1\partial\mathcal{M})} &\leq C_0\|\nabla_g(\psi w)\|^2_{L^2V(O)} + C\|w\|^2_{L^2(O)} \\ &\leq C_0\|\psi\nabla_g(w)\|^2_{L^2V(O)} + C'\|w\|^2_{L^2(O)}, \end{aligned}$$

using that $[\nabla_g, \psi]$ is a bounded function. As $\psi \leq 1$ the result follows. ∎

From Lemma 4.10 we deduce the following norm equivalence.

COROLLARY 4.11. *Let* $(\mathsf{P}, D(\mathsf{P}))$ *be selfadjoint. In the case* $d = 2$ *assume furthermore that* X' *is such that* $|X'|_{g_\partial} < 1$ *on* ${}^1\partial\mathcal{M}$. *Then, there exists* $C_* > 0$ *and* $C > 0$ *such that*

$$C^{-1}\|u\|^2_{H^1(\mathcal{M})} \leq (\mathsf{P}u, u)_{L^2(\mathcal{M})} + C_*\|u\|^2_{L^2(\mathcal{M})} \leq C\|u\|^2_{H^1(\mathcal{M})}, \quad u \in D(\mathsf{P}).$$

PROOF. Let $u \in D(\mathsf{P})$. With (18.5.5) consequence of the divergence formula of Proposition 18.28, we write

$$(Pu, u)_{L^2(\mathcal{M})} = \|\nabla_g u\|^2_{L^2V(\mathcal{M})} - (\gamma^{\mathsf{N}}(u), \gamma^{\mathsf{D}}(u))_{L^2(\partial\mathcal{M})} + (R_1 u, u)_{L^2(\mathcal{M})}.$$

On ${}^0\partial\mathcal{M}$, the part of $\partial\mathcal{M}$ where B is of order zero, we have $\gamma^{\mathsf{D}}(u) = 0$. On ${}^1\partial\mathcal{M}$, the part of $\partial\mathcal{M}$ where B is of order one, we have $\gamma^{\mathsf{N}}(u) = -B'\gamma^{\mathsf{D}}(u)$ leading to

$$(Pu, u)_{L^2(\mathcal{M})} = \|\nabla_g u\|^2_{L^2V(\mathcal{M})} + (B'\gamma^{\mathsf{D}}(u), \gamma^{\mathsf{D}}(u))_{L^2({}^1\partial\mathcal{M})} + (R_1 u, u)_{L^2(\mathcal{M})}.$$

In the case $d \geq 3$ we have $|X'|_{g_\partial} < 1$ on ${}^1\partial\mathcal{M}$ by Proposition 4.6. In the case $d = 2$ this property is assumed here. By Lemma 4.10 we have for some $0 < C_0 < 1$ and $C_1 > 0$

$$(Pu, u)_{L^2(\mathcal{M})} \geq (1 - C_0)\|\nabla_g u\|^2_{L^2V(\mathcal{M})} - C_1\|u\|^2_{L^2(\mathcal{M})} - |(R_1 u, u)_{L^2(\mathcal{M})}|.$$

For any $0 < \eta < (1 - C_0)$, we have

$$|(R_1 u, u)_{L^2(\mathcal{M})}| \leq \eta\|\nabla_g u\|^2_{L^2V(\mathcal{M})} + C_\eta\|u\|^2_{L^2(\mathcal{M})},$$

yielding $(Pu, u)_{L^2(\mathcal{M})} \geq (1 - C_0 - \eta)\|\nabla_g u\|^2_{L^2V(\mathcal{M})} - (C_1 + C_\eta)\|u\|^2_{L^2(\mathcal{M})}$. The result follows. ∎

From Corollary 4.11 we obtain the following result similar to the classical case of Dirichlet boundary condition as presented in Sect. 10.1 of Volume 1.

THEOREM 4.12. *Let* $(\mathsf{P}, D(\mathsf{P}))$ *be selfadjoint and the* (P, B) *fulfills the Lopatinskiĭ–Šapiro condition. In the case* $d = 2$ *assume furthermore that* X' *is such that* $|X'|_{g_\partial} < 1$ *on* ${}^1\partial\mathcal{M}$. *The eigenvalues counted with multiplicities can be sorted in a nondecreasing sequence*

$$\mu_0 \leq \mu_1 \leq \cdots \leq \mu_n \leq \cdots$$

that goes to $+\infty$.

PROOF. It suffices to prove that there exists $M \in \mathbb{R}$ such that $\mu \geq M$ for any eigenvalue as we know that the sequence of the modulus of the eigenvalues goes to $+\infty$. Let thus μ be an eigenvalue and let ϕ be a unitary eigenfunction. We have $\mu = (P\phi, \phi)_{L^2(\mathcal{M})}$. By Corollary 4.11 we have $\mu \geq C\|\phi\|^2_{H^1(\mathcal{M})} - C_* \geq -C_*$. ∎

We show through an example that having $B = \partial_\nu + iX' + h$ with $|X'|_{g_\partial} > 1$ in a connected component of ${}^1\partial\mathcal{M}$ may have a dramatic effect on the spectrum of $(\mathsf{P}, D(\mathsf{P}))$.

PROPOSITION 4.13. *On the unit disc $\mathbb{D}$ consider $P = -\Delta = -\partial_x^2 - \partial_y^2$ and the boundary operator $B = \partial_\nu + 2i\partial_\theta$. The associated operator $(\mathsf{P}, D(\mathsf{P}))$ is selfadjoint and its spectrum is given the union of two real sequences, one going to $+\infty$ and the other one going to $-\infty$.*

Observe that with such a spectral behavior the operator $(\mathsf{P}, D(\mathsf{P}))$ cannot be the generator of a C_0-semigroup.

PROOF. As one can readily check, from Theorem 4.3 and Proposition 2.8 the unbounded operator $(\mathsf{P}, D(\mathsf{P}))$ is selfadjoint. We use polar coordinates (r, θ) for which the Laplace operator reads $-\Delta = -r^{-1}\partial_r(r\partial_r) - r^{-2}\partial_\theta^2$. Note that B reads $Bu_{|\partial\mathbb{D}} = \partial_r u_{|r=1} + 2i\partial_\theta u_{|r=1}$.

First, set $w_n = r^{2n}e^{in\theta}$. We have $Bw_{n|\partial\mathbb{D}} = 0$ yielding $w_n \in D(\mathsf{P})$. We have $\|w_n\|^2_{L^2(\mathbb{D})} = 2\pi/(4n+1)$. We compute

$$Pw_n = -3n^2 r^{2n-2}e^{in\theta},$$

which leads to

$$(Pw_n, w_n)_{L^2(\mathbb{D})} = \int_0^{2\pi}\int_0^1 (Pw_n)\overline{w_n}\, r dr d\theta = -\frac{3}{2}\pi n,$$

implying the existence of a sequence of eigenvalues that goes to $-\infty$.

Second, set $w_n = \psi(r)e^{in\theta}$, with $\psi \in \mathscr{C}_c^\infty(1/4, 3/4; \mathbb{R})$. We have $Bw_{n|\partial\mathbb{D}} = 0$ yielding $w_n \in D(\mathsf{P})$. We have $\|w_n\|_{L^2(\mathbb{D})} = \text{Cst}$. We compute

$$Pw_n = (-\psi'' - r^{-1}\psi' + r^{-2}n^2\psi)e^{in\theta},$$

which leads to

$$(Pw_n, w_n)_{L^2(\mathbb{D})} = 2\pi \int_0^1 (-\psi'' - r^{-1}\psi')\psi\, r dr + 2\pi n^2 \|r^{-1/2}\psi\|^2_{L^2(0,1)},$$

implying the existence of a sequence of eigenvalues that goes to $+\infty$. ∎

In the case $(\mathsf{P}, D(\mathsf{P}))$ is selfadjoint as above, we now provide some norm equivalence on $H^1(\mathcal{M})$ and on its linear subspace formed by functions with vanishing Dirichlet trace on ${}^0\partial\mathcal{M}$ and some density result of $D(\mathsf{P})$ in that space,

Consider $u \in H^2(\mathcal{M})$ and $v \in H^1(\mathcal{M})$ with $v_{|{}^0\partial\mathcal{M}} = 0$. One has

$$(Pu, v)_{L^2(\mathcal{M})} = (\nabla_g u, \nabla_g v)_{L^2V(\mathcal{M})} + (R_1 u, v)_{L^2(\mathcal{M})} - (\gamma^{\mathsf{N}}(u), \gamma^{\mathsf{D}}(v))_{L^2({}^1\partial\mathcal{M})}. \tag{4.6.4}$$

If moreover $Bu_{|{}^1\partial\mathcal{M}} = 0$ one finds

$$(\mathsf{P}u, v)_{L^2(\mathcal{M})} = (\nabla_g u, \nabla_g v)_{L^2V(\mathcal{M})} + (R_1 u, v)_{L^2(\mathcal{M})} + (B'\gamma^{\mathsf{D}}(u), \gamma^{\mathsf{D}}(v))_{L^2({}^1\partial\mathcal{M})}.$$

Motivated by this computation, for $u, v \in H^1(\mathcal{M})$, we define

$$\tilde{N}(u,v) = (\nabla_g u, \nabla_g v)_{L^2V(\mathcal{M})} + (R_1 u, v)_{L^2(\mathcal{M})} + \langle B'\gamma^{\mathsf{D}}(u), \gamma^{\mathsf{D}}(\bar{v})\rangle_{H^{-1/2}({}^1\partial\mathcal{M}), H^{1/2}({}^1\partial\mathcal{M})}. \tag{4.6.5}$$

By Lemma 4.7 the bilinear form $\tilde{N}(.,.)$ is Hermitian symmetric. One sets

$$N(u) = \tilde{N}(u,u) = \|\nabla_g u\|^2_{L^2V(\mathcal{M})} + (R_1 u, u)_{L^2(\mathcal{M})} + \langle B'\gamma^{\mathsf{D}}(u), \gamma^{\mathsf{D}}(\bar{u})\rangle_{H^{-1/2}({}^1\partial\mathcal{M}), H^{1/2}({}^1\partial\mathcal{M})}, \tag{4.6.6}$$

and $N_\lambda(u) = N(u) + \lambda\|u\|^2_{L^2(\mathcal{M})}$.

We introduce the following space

$$H^1_{\mathsf{D}}(\mathcal{M}) = \{u \in H^1(\mathcal{M});\ u_{|{}^0\partial\mathcal{M}} = 0\},$$

equipped with the usual H^1-norm, yielding a Hilbert space structure. Above we saw that

$$(\mathsf{P}u, v)_{L^2(\mathcal{M})} = \tilde{N}(u,v), \tag{4.6.7}$$

for $u \in D(\mathsf{P})$ and $v \in H^1_{\mathsf{D}}(\mathcal{M})$.

Finally, for $\lambda \in \mathbb{R}$, we set $\mathsf{P}_\lambda = \mathsf{P} + \lambda\,\mathrm{Id}_{L^2(\mathcal{M})}$ with $D(\mathsf{P}_\lambda) = D(\mathsf{P})$.

PROPOSITION 4.14. *Let $(\mathsf{P}, D(\mathsf{P}))$ be selfadjoint and the (P,B) fulfills the Lopatinskiĭ–Šapiro condition. In the case $d = 2$ assume furthermore that X' is such that $|X'|_{g_\partial} < 1$ on ${}^1\partial\mathcal{M}$.*

Let μ_0 be the lowest eigenvalue of $(\mathsf{P}, D(\mathsf{P}))$. Then, $N_\lambda(.)^{1/2}$ is a norm on $H^1_{\mathsf{D}}(\mathcal{M})$ for $\lambda > -\mu_0$ that is equivalent to the usual H^1-norm. Moreover, $D(\mathsf{P})$ is dense in $H^1_{\mathsf{D}}(\mathcal{M})$.

Consequently, the inner product

$$(u,v) \mapsto \tilde{N}(u,v) + \lambda(u,v)_{L^2(\mathcal{M})},$$

is an inner product that yields the Hilbert space structure on $H^1_{\mathsf{D}}(\mathcal{M})$.

PROOF. By Theorem 4.12, with $\lambda > -\mu_0$, the operator $(\mathsf{P}_\lambda, D(\mathsf{P}))$ is selfadjoint and positive. For $u \in D(\mathsf{P})$, with the computation carried out above, one has

$$N_\lambda(u) = (\mathsf{P}_\lambda u, u)_{L^2(\mathcal{M})} = \sum_{j\in\mathbb{N}} (\mu_j + \lambda)|u_j|^2 \geq 0,$$

with $u_j = (u, \phi_j)_{L^2(\mathcal{M})}$, $j \in \mathbb{N}$, with $(\phi_j)_{j\in\mathbb{N}}$ a Hilbert basis of eigenfunctions of $(\mathsf{P}_\lambda, D(\mathsf{P}))$, each associated with the eigenvalues $\mu_j + \lambda$; see Theorem 4.9. This implies that $N_\lambda(.)^{1/2}$ is a norm on $D(\mathsf{P})$ for $\lambda > -\mu_0$. Observe that two values of $\lambda > -\mu_0$ yield two associated norms $N_\lambda(.)^{1/2}$ that are equivalent.

One observes that $N_\lambda(u) \lesssim \|u\|^2_{H^1(\mathcal{M})}$ for $u \in H^1(\mathcal{M})$. By Lemma 4.10 and the Young inequality, there exists $0 < C_1 < 1$ and $C_2 > 0$ such that

$$N_\lambda(u) \geq (1 - C_1)\|\nabla_g u\|^2_{L^2V(\mathcal{M})} + (\lambda - C_2)\|u\|^2_{L^2(\mathcal{M})}.$$

Thus, for $\lambda > \max(C_2, -\mu_0)$ one has $N_\lambda(.)^{1/2}$ equivalent to $\|.\|_{H^1(\mathcal{M})}$ on $H^1(\mathcal{M})$.

The above two observations, put together, show that $N_\lambda(.)^{1/2}$ and $\|.\|_{H^1(\mathcal{M})}$ are equivalent on $D(\mathsf{P})$ if $\lambda > -\mu_0$. Note that this equivalence may not hold on $H^1(\mathcal{M})$ if $\lambda \leq \max(C_2, -\mu_0)$. The norm $N_\lambda(.)^{1/2}$ is associated with the inner product $\tilde{N}_\lambda(u, v) = \tilde{N}(u, v) + \lambda(u, v)_{L^2(\mathcal{M})}$.

The space $H^1_{\mathsf{D}}(\mathcal{M})$ is closed in $H^1(\mathcal{M})$ and one has $D(\mathsf{P}) \subset H^1_{\mathsf{D}}(\mathcal{M})$. Below we show that the orthogonal of $D(\mathsf{P})$ in $H^1_{\mathsf{D}}(\mathcal{M})$ with respect to the inner product $\tilde{N}_\lambda(.,.)$ for $\lambda > \max(C_2, -\mu_0)$ is the trivial set. Consequently, $D(\mathsf{P})$ is dense in $H^1_{\mathsf{D}}(\mathcal{M})$ for the norm $N_\lambda(.)^{1/2}$ (with $\lambda > \max(C_2, -\mu_0)$) and thus also for the norm $\|.\|_{H^1(\mathcal{M})}$.

Let now $\lambda > -\mu_0$. As explained above, there exists $C > c > 0$ such that

$$c\|u\|_{H^1(\mathcal{M})} \leq N_\lambda(u)^{1/2} \leq C\|u\|_{H^1(\mathcal{M})}, \qquad u \in D(\mathsf{P}).$$

If $v \in H^1_{\mathsf{D}}(\mathcal{M})$ and $(v_n)_n \subset D(\mathsf{P})$ converges to v for the norm $\|.\|_{H^1(\mathcal{M})}$ one has $c\|v_n\|_{H^1(\mathcal{M})} \leq N_\lambda(v_n)^{1/2} \leq C\|v_n\|_{H^1(\mathcal{M})}$. Observing that $u \mapsto N_\lambda(u)^{1/2}$ is continuous on $H^1(\mathcal{M})$ one finds, passing to the limit,

$$c\|v\|_{H^1(\mathcal{M})} \leq N_\lambda(v)^{1/2} \leq C\|v\|_{H^1(\mathcal{M})},$$

which proves that $N_\lambda(.)^{1/2}$ is a norm of $H^1_{\mathsf{D}}(\mathcal{M})$, that is moreover equivalent to the norm $\|.\|_{H^1(\mathcal{M})}$.

Consider $v \in H^1_{\mathsf{D}}(\mathcal{M})$ that is in the orthogonal of $D(\mathsf{P})$ with respect to the inner product $\tilde{N}_\lambda(.,.)$ in the case $\lambda > \max(C_2, -\mu_0)$. For any $u \in D(\mathsf{P})$ one has, with Lemma 4.7,

$$\begin{aligned}
0 &= \tilde{N}_\lambda(u, v) \\
&= \lambda(u, v)_{L^2(\mathcal{M})} + (\nabla_g u, \nabla_g v)_{L^2V(\mathcal{M})} + (R_1 u, v)_{L^2(\mathcal{M})} \\
&\quad + (B'\gamma^{\mathsf{D}}(u), \gamma^{\mathsf{D}}(v))_{L^2({}^1\partial\mathcal{M})} \\
&= \lambda(u, v)_{L^2(\mathcal{M})} + (-\Delta_g u, v)_{L^2(\mathcal{M})} + (R_1 u, v)_{L^2(\mathcal{M})} \\
&\quad + (\gamma^{\mathsf{N}}(u) + B'\gamma^{\mathsf{D}}(u), \gamma^{\mathsf{D}}(v))_{L^2({}^1\partial\mathcal{M})} \\
&= (\mathsf{P}_\lambda u, v)_{L^2(\mathcal{M})}.
\end{aligned}$$

Consider the operator

$$\begin{aligned}
L_\lambda : H^2(\mathcal{M}) &\to L^2(\mathcal{M}) \oplus H^{3/2,1/2}(\partial\mathcal{M}) \\
u &\mapsto (Pu + \lambda u, u_{|{}^0\partial\mathcal{M}}, Bu_{|{}^1\partial\mathcal{M}}),
\end{aligned}$$

with the notation of (3.1.3) in Chap. 3 with

$$H^{3/2,1/2}(\partial\mathcal{M}) = H^{(3/2)}_B(\partial\mathcal{M}) = H^{3/2}({}^0\partial\mathcal{M}) \oplus H^{1/2}({}^1\partial\mathcal{M}).$$

This operator is Fredholm with a zero index as explained in the beginning of the proof of Theorem 4.9. With $\lambda > -\mu_0$, this operator is injective and thus surjective. This implies in particular that $\operatorname{Ran}(\mathsf{P}_\lambda) = L^2(\mathcal{M})$. Hence, having $(\mathsf{P}_\lambda u, v)_{L^2(\mathcal{M})} = 0$ for all $u \in D(\mathsf{P})$ gives $v = 0$. ∎

For any $\lambda \in \mathbb{R}$, on $D(\mathsf{P}_\lambda) = D(\mathsf{P})$ the graph norm

$$\|u\|^2_{D(\mathsf{P}_\lambda)} = \|u\|^2_{L^2(\mathcal{M})} + \|\mathsf{P}_\lambda u\|^2_{L^2(\mathcal{M})}$$

is a norm equivalent to the graph norm

$$\|u\|^2_{D(\mathsf{P})} = \|u\|^2_{L^2(\mathcal{M})} + \|\mathsf{P} u\|^2_{L^2(\mathcal{M})}.$$

Since $(\mathsf{P}, D(\mathsf{P}))$ is a closed operator then the inner product

$$(u, v) \mapsto (u, v)_{L^2(\mathcal{M})} + (\mathsf{P}u, \mathsf{P}v)_{L^2(\mathcal{M})}$$

yields a Hilbert space structure on $D(\mathsf{P})$.

LEMMA 4.15. *The norms* $u \mapsto \|u\|_{D(\mathsf{P})}$, $u \mapsto \|u\|_{H^2(\mathcal{M})}$ *and* $u \mapsto \|\mathsf{P}_\lambda u\|_{L^2(\mathcal{M})}$ *for* $\lambda > -\mu_0$ *are equivalent on* $D(\mathsf{P})$.

PROOF. Naturally, one has

$$\|\mathsf{P}_\lambda u\|_{L^2(\mathcal{M})} \lesssim \|u\|_{D(\mathsf{P})} \lesssim \|u\|_{H^2(\mathcal{M})}.$$

For $\lambda > -\mu_0$ one has $\ker(\mathsf{P}_\lambda) = \{0\}$. Thus the Fredholm map (4.6.1) with $c = \lambda$ is injective, and thus bijective as its index is zero and the inverse map $L_\lambda^{-1} : L^2(\mathcal{M}) \oplus H^{3/2,1/2}(\partial\mathcal{M}) \to H^2(\mathcal{M})$ is bounded as recalled at the beginning of the proof of Theorem 4.9. Hence, for $u \in D(\mathsf{P})$ one has $Bu_{|\partial\mathcal{M}} = 0$ implying that $\|u\|_{H^2(\mathcal{M})} \lesssim \|\mathsf{P}_\lambda u\|_{L^2(\mathcal{M})}$. ∎

The Hilbert basis $(\phi_j)_{j\in\mathbb{N}}$ of eigenfunctions of the unbounded selfadjoint operator $(\mathsf{P}, D(\mathsf{P}))$ on $L^2(\mathcal{M})$ allows one to give a description of the spaces $H^1_{\mathsf{D}}(\mathcal{M})$ and $D(\mathsf{P})$, as is done in Section 10.1.2 of Volume 1 in the case of Dirichlet boundary conditions. Naturally, if $u \in L^2(\mathcal{M})$ one has $u = \sum_{j\in\mathbb{N}} u_j \phi_j$, with $u_j = (u, \phi_j)_{L^2(\mathcal{M})}$, and where convergence takes place in $L^2(\mathcal{M})$, meaning that $(u_j)_{j\in\mathbb{N}} \in \ell^2(\mathbb{C})$.

Let $\lambda > -\mu_0$ and set $\lambda_j = \mu_j + \lambda > 0$.

PROPOSITION 4.16. *Let* $u \in L^2(\mathcal{M})$ *and* $u_j = (u, \phi_j)_{L^2(\mathcal{M})}$, $j \in \mathbb{N}$. *First, we have the following equivalences:*

$$u \in D(\mathsf{P}) \quad \Leftrightarrow \quad (\mu_j u_j)_{j\in\mathbb{N}} \in \ell^2(\mathbb{C}) \quad \Leftrightarrow \quad (\lambda_j u_j)_{j\in\mathbb{N}} \in \ell^2(\mathbb{C}),$$

$$u \in H^1_{\mathsf{D}}(\mathcal{M}) \quad \Leftrightarrow \quad (|\mu_j|^{1/2} u_j)_{j\in\mathbb{N}} \in \ell^2(\mathbb{C}) \quad \Leftrightarrow \quad (\lambda_j^{1/2} u_j)_{j\in\mathbb{N}} \in \ell^2(\mathbb{C}).$$

Second, the bilinear map

$$(u, v) \mapsto \sum_{j\in\mathbb{N}} \lambda_j^2 u_j \bar{v}_j$$

is an inner product on $D(\mathsf{P})$ *that yields the same Hilbert space structure on* $D(\mathsf{P})$ *as that given by the* H^2*-inner product and*

$$\|u\|^2_{H^2(\mathcal{M})} \eqsim \|u\|^2_{D(\mathsf{P})} \eqsim \|\mathsf{P}_\lambda u\|^2_{L^2(\mathcal{M})} = \sum_{j\in\mathbb{N}} \lambda_j^2 |u_j|^2,$$

for $u \in D(\mathsf{P})$.

Third, the bilinear map

$$(u, v) \mapsto \sum_{j\in\mathbb{N}} \lambda_j u_j \bar{v}_j$$

is an inner product on H^1_{D} *that yields the same Hilbert space structure on* H^1_{D} *as that given by the* H^1*-inner product and*

$$\|u\|^2_{H^1(\mathcal{M})} \eqsim N_\lambda(u)^2 = \sum_{j\in\mathbb{N}} \lambda_j |u_j|^2,$$

for $u \in H^1_{\mathsf{D}}$.

PROOF. Arguing as in the proof of Theorem 4.9 one sees that P_λ^{-1} is a compact operator on $L^2(\mathcal{M})$. One has $\mathsf{P}_\lambda^{-1}(\phi_j) = \lambda_j^{-1}\phi_j$. Since $\operatorname{Ran}(\mathsf{P}_\lambda^{-1}) = D(\mathsf{P})$ this gives the characterization of $D(\mathsf{P})$, as $u \in D(\mathsf{P})$ reads $u = \sum_{j\in\mathbb{N}} \lambda_j^{-1} v_j \phi_j$ for some $(v_j)_{j\in\mathbb{N}} \in \ell^2(\mathbb{C})$. Moreover one finds that if $u = \sum_{j\in\mathbb{N}} u_j\phi_j \in D(\mathsf{P})$, then $(\lambda_j u_j)_{j\in\mathbb{N}} \in \ell^2(\mathbb{C})$ and $\mathsf{P}_\lambda u = \sum_{j\in\mathbb{N}} \lambda_j u_j \phi_j$. Thus, we have $\|\mathsf{P}_\lambda u\|_{L^2(\mathcal{M})} = \sum_{j\in\mathbb{N}} \lambda_j^2 |u_j|^2$. By Lemma 4.15, $u \mapsto \|\mathsf{P}_\lambda u\|_{L^2(\mathcal{M})}$ is a norm equivalent to the H^2-norm on $D(\mathsf{P})$ and

$$(u, v) \mapsto \sum_{j\in\mathbb{N}} \lambda_j^2 u_j \bar{v}_j$$

is the inner product on $D(\mathsf{P})$ associated with this norm. This concludes the second point of the proposition.

From Proposition 4.14 $u \mapsto N_\lambda(u)$ is a norm on $H^1_{\mathsf{D}}(\mathcal{M})$ and $D(\mathsf{P})$ is dense in $H^1_{\mathsf{D}}(\mathcal{M})$and $(u, v) \mapsto \tilde{N}(u, v) + \lambda(u, v)_{L^2(\mathcal{M})}$, is an inner product that yields the Hilbert space structure on $H^1_{\mathsf{D}}(\mathcal{M})$. If $u, v \in D(\mathsf{P})$, by (4.6.7) one has

$$\tilde{N}(u, v) + \lambda(u, v)_{L^2(\mathcal{M})} = (\mathsf{P}u, v)_{L^2(\mathcal{M})} = \sum_{j\in\mathbb{N}} \lambda_j u_j \bar{v}_j.$$

Then, arguing by density as in the proof of Proposition 10.5 of Volume 1 one obtains the characterization of $H^1_{\mathsf{D}}(\mathcal{M})$ and one proves the third point of the proposition. ∎

With Proposition 4.16 one finds

$$N(u) = \sum_{j\in\mathbb{N}} \mu_j |u_j|^2, \quad \text{and} \quad \tilde{N}(u, v) = \sum_{j\in\mathbb{N}} \mu_j u_j \bar{v}_j,$$

for $u, v \in H^1_{\mathsf{D}}(\mathcal{M})$. If $\mu_0 \geq 0$ one has $N(u) \geq 0$ but $N(.)^{1/2}$ may not be a norm on $H^1_{\mathsf{D}}(\mathcal{M})$ unless $\mu_0 > 0$, that is, if $\ker(\mathsf{P}) = \{0\}$.

However, in the case $\mu_0 \geq 0$ one has the following Cauchy–Schwarz like inequality

$$|\tilde{N}(u, v)| \leq N(u)^{1/2} N(v)^{1/2}, \qquad u, v \in H^1_{\mathsf{D}}(\mathcal{M}). \tag{4.6.8}$$

With Proposition 4.16 if $u \in D(\mathsf{P})$ and $u = \sum_{j\in\mathbb{N}} u_j\phi_j$ one has $\mathsf{P}u = \sum_{j\in\mathbb{N}} \mu_j u_j \phi_j$. A simple consequence is the following result.

PROPOSITION 4.17. *The range of* P *is closed in* $L^2(\mathcal{M})$ *and* $\operatorname{Ran}(\mathsf{P}) = \ker(\mathsf{P})^\perp$.

4.7. Lopatinskiĭ–Šapiro Elliptic Problem

In this section we carry on with the analysis of the operator P in the case (P, B) fulfills the Lopatinskiĭ–Šapiro condition and $(\mathsf{P}, D(\mathsf{P}))$ is selfadjoint. As above if $d = 2$ we further assume that $|X'|_{g_\partial} < 1$ on ${}^1\partial\mathcal{M}$ to have the conclusion of Theorem 4.12.

4.7.1. The Nonhomogeneous Elliptic Problem. Let $k \in \mathbb{N}$. Denote by L the map

$$
\begin{aligned}
(4.7.1) \qquad L : H^{k+2}(\mathcal{M}) &\to H^k(\mathcal{M}) \oplus H^{k+3/2,k+1/2}(\partial\mathcal{M}) \\
u &\mapsto (Pu, u_{|{}^0\partial\mathcal{M}}, Bu_{|{}^1\partial\mathcal{M}}).
\end{aligned}
$$

We set $n = \dim\ker(\mathsf{P})$, meaning that $\{\phi_0, \dots, \phi_{n-1}\}$ is a basis of $E_0 = \ker(\mathsf{P})$. Recall that $\operatorname{nul} L$, $\operatorname{def} L$, and $\operatorname{ind}(L)$ are independent of the chosen value $k \in \mathbb{N}$; see Proposition 3.4. With the property of L recalled after (4.7.10), the range of L, $\operatorname{Ran}(L)$, is closed and of codimension equal to n (see Definition 11.6 of Volume 1) since its index is zero.

For $u \in L^2(\mathcal{M})$ we set $\underline{u} = (\operatorname{Id} - \Pi_{E_0})u$, that is, the projection of u onto $E_0^\perp$, yielding

$$
(4.7.2) \qquad \underline{u} = u - \sum_{j=0}^{n-1} u_j \phi_j = \sum_{j \geq n} u_j \phi_j, \qquad u_j = (u, \phi_j)_{L^2(\mathcal{M})}.
$$

Let $k \in \mathbb{N}$. Naturally, if $u \in H^k(\mathcal{M})$ one has $\underline{u} \in H^k(\mathcal{M})$ because of the H^k-regularity of the eigenfunctions $\phi_0, \dots, \phi_{n-1}$. Setting

$$
(4.7.3) \qquad \underline{H}^k(\mathcal{M}) = \{\underline{u};\ u \in H^k(\mathcal{M})\} = (\operatorname{Id}_{L^2(\mathcal{M})} - \Pi_{E_0})(H^k(\mathcal{M})),
$$

one sees that $\underline{H}^k(\mathcal{M})$ is a closed linear subspace of $H^k(\mathcal{M})$, thus a Hilbert space.

Consider the nonhomogeneous elliptic problem

$$
(4.7.4) \quad Pu = f, \text{ in } \mathcal{M}, \qquad u_{|{}^0\partial\mathcal{M}} = g_0 \text{ in } {}^0\partial\mathcal{M}, \qquad Bu_{|{}^1\partial\mathcal{M}} = g_1 \text{ in } {}^1\partial\mathcal{M}.
$$

If $u \in H^{k+2}(\mathcal{M})$, then $f \in H^k(\mathcal{M})$, $g_0 \in H^{k+3/2}({}^0\partial\mathcal{M})$, and $g_1 \in H^{k+1/2}({}^1\partial\mathcal{M})$. For $j = 0, \dots, n-1$, by Proposition 4.8 one computes

$$
(4.7.5) \quad (f, \phi_j)_{L^2(\mathcal{M})} + (g_1, \gamma^{\mathsf{D}}_{{}^1\partial\mathcal{M}}(\phi_j))_{L^2({}^1\partial\mathcal{M})} - (g_0, \gamma^{\mathsf{N}}_{{}^0\partial\mathcal{M}}(\phi_j))_{L^2({}^0\partial\mathcal{M})} = 0.
$$

Since $\operatorname{codim}\operatorname{Ran}(L) = n$ then one sees that conditions (4.7.5), for $j = 0, \dots, n-1$, characterize $\operatorname{Ran}(L)$ as a subset of $H^k(\mathcal{M}) \oplus H^{k+3/2,k+1/2}(\partial\mathcal{M})$: they are necessary and sufficient conditions for the resolution of (4.7.4).

THEOREM 4.18 (Elliptic Problem—Strong Solutions). *Let* $k \in \mathbb{N}$. *Let* $f \in H^k(\mathcal{M})$, $g_0 \in H^{k+3/2}({}^0\partial\mathcal{M})$, *and* $g_1 \in H^{k+1/2}({}^1\partial\mathcal{M})$ *be such that* (4.7.5) *hold, for* $j = 0, \dots, n-1$. *Then, there exists a unique* $v \in \underline{H}^{k+2}(\mathcal{M})$ *such that* (4.7.4) *holds. Moreover, any other solution* $u \in \mathcal{W}_P(\mathcal{M})$ *of* (4.7.4) *is such that* $u - v \in E_0 = \ker(\mathsf{P})$, *meaning in particular that* $u \in H^{k+2}(\mathcal{M})$. *Finally, there exists* $C > 0$ *such that*

$$\|v\|_{H^{k+2}(\mathcal{M})} \leq C\big(\|f\|_{H^k(\mathcal{M})} + |g_0|_{H^{k+3/2}({}^0\partial\mathcal{M})} + |g_1|_{H^{k+1/2}({}^1\partial\mathcal{M})}\big). \tag{4.7.6}$$

PROOF. Consider $\underline{L} : \underline{H}^{2+k}(\mathcal{M}) \to \operatorname{Ran}(L)$ given by $\underline{L}u = Lu$. Since $\operatorname{Ran}(L)$ is closed it is complete with the norm of $H^k(\mathcal{M}) \oplus H^{k+3/2,k+1/2}(\partial\mathcal{M})$. Hence, $\underline{L}$ is bijective and bounded. By the open mapping theorem $\underline{L}^{-1}$ is bounded and $v = \underline{L}^{-1}(f, g_0, g_1) \in \underline{H}^{2+k}(\mathcal{M})$ is a solution of (4.7.4) and estimate (4.7.6) holds.

Consider now $u \in \mathcal{W}_P(\mathcal{M})$. Then, by Lemma 18.32 one has $Pu \in L^2(\mathcal{M})$ and the traces $u_{|{}^0\partial\mathcal{M}}$ and $Bu_{|{}^1\partial\mathcal{M}}$ make sense in $H^{-1/2}({}^0\partial\mathcal{M})$ and $H^{-3/2}({}^0\partial\mathcal{M})$, respectively. Assume that $Pu = f$, $u_{|{}^0\partial\mathcal{M}} = g_0$, and $Bu_{|{}^1\partial\mathcal{M}} = g_1$. By Theorem 3.28, one has $u \in H^{k+2}(\mathcal{M})$. Observe that $w = u - v \in H^{k+2}(\mathcal{M})$ is such that $Pw = 0$, $w_{|{}^0\partial\mathcal{M}} = 0$, $Bw_{|{}^1\partial\mathcal{M}} = 0$ meaning that $w \in \ker(\mathsf{P})$. ■

Consider now the space

$$\underline{H}^1_{\mathsf{D}}(\mathcal{M}) = \{\underline{u};\ u \in H^1_{\mathsf{D}}(\mathcal{M})\} = (\mathrm{Id}_{L^2(\mathcal{M})} - \Pi_{E_0})(H^1_{\mathsf{D}}(\mathcal{M})).$$

It is a closed linear subspace of $H^1_{\mathsf{D}}(\mathcal{M})$. On the space $\underline{H}^1_{\mathsf{D}}(\mathcal{M})$, one can introduce a variational form of the elliptic problem (4.7.4) in the case $u_{|{}^0\partial\mathcal{M}} = g_0 = 0$.

On $H^1_{\mathsf{D}}(\mathcal{M})$, for $\lambda > -\mu_0$, the bilinear map $(u, v) \mapsto \tilde{N}(u, v) + \lambda(u, v)_{L^2(\mathcal{M})}$ is an inner product that yields the Hilbert structure of the space by Proposition 4.14. Moreover, one has

$$\|u\|^2_{H^1(\mathcal{M})} \eqsim N_\lambda(u) = \sum_{j\in\mathbb{N}} \lambda_j |u_j|^2,$$

by Proposition 4.16. Observe that if $u \in \underline{H}^1_{\mathsf{D}}(\mathcal{M})$, then

$$N_\lambda(u) = \sum_{j\geq n} \lambda_j |u_j|^2 \eqsim \sum_{j\geq n} \mu_j |u_j|^2 = N(u).$$

Thus the bilinear map $(u, v) \mapsto \tilde{N}(u, v)$ yields the Hilbert structure of the space $\underline{H}^1_{\mathsf{D}}(\mathcal{M})$.

THEOREM 4.19 (Elliptic Problem—Variational Form). *Let* $f \in L^2(\mathcal{M})$ *and* $g_1 \in H^{-1/2}({}^1\partial\mathcal{M})$ *be such that*

$$(f, \phi_j)_{L^2(\mathcal{M})} + \langle g_1, \gamma^{\mathsf{D}}_{{}^1\partial\mathcal{M}}(\overline{\phi_j})\rangle_{H^{-1/2}({}^1\partial\mathcal{M}), H^{1/2}({}^1\partial\mathcal{M})} = 0, \quad j = 0, \ldots, n-1. \tag{4.7.7}$$

Then, there exists a unique $v \in \underline{H}^1_{\mathsf{D}}(\mathcal{M})$ *such that*

$$\tilde{N}(v, u) = (f, u)_{L^2(\mathcal{M})} + \langle g_1, \bar{u}_{|{}^1\partial\mathcal{M}}\rangle_{H^{-1/2}({}^1\partial\mathcal{M}), H^{1/2}({}^1\partial\mathcal{M})}, \qquad u \in H^1_{\mathsf{D}}(\mathcal{M}). \tag{4.7.8}$$

Moreover, if $g_1 \in H^{1/2}({}^1\partial\mathcal{M})$, *then* $v \in \underline{H}^2(\mathcal{M})$ *and* v *is the unique solution to the elliptic problem* (4.7.4) *provided by Theorem 4.18 in the case* $g_0 = 0$.

REMARK 4.20. Let $v \in \underline{H}^1_{\mathsf{D}}(\mathcal{M})$ and $j = 0, \dots, n-1$. Then by (4.6.7) we have

$$\tilde{N}(v, \phi_j) = (v, \mathsf{P}\phi_j)_{L^2(\mathcal{M})} = 0, \tag{4.7.9}$$

as $\phi_j \in \ker(\mathsf{P})$. One thus finds that the conditions (4.7.7) are necessary for (4.7.8) to hold.

PROOF. As $u \mapsto (u, f)_{L^2(\mathcal{M})} + \langle u_{|^1\partial\mathcal{M}}, \overline{g_1}\rangle_{H^{1/2}(^1\partial\mathcal{M}), H^{-1/2}(^1\partial\mathcal{M})}$ is bounded on $\underline{H}^1_{\mathsf{D}}(\mathcal{M})$, and as the Hilbert structure of $\underline{H}^1_{\mathsf{D}}(\mathcal{M})$ is given by the inner product $\tilde{N}(.,.)$, there exists $v \in \underline{H}^1_{\mathsf{D}}(\mathcal{M})$ such that

$$\tilde{N}(v, u) = (f, u)_{L^2(\mathcal{M})} + \langle g_1, \bar{u}_{|^1\partial\mathcal{M}}\rangle_{H^{-1/2}(^1\partial\mathcal{M}), H^{1/2}(^1\partial\mathcal{M})}, \quad u \in \underline{H}^1_{\mathsf{D}}(\mathcal{M}).$$

by the Riesz theorem. With conditions (4.7.7) and (4.7.9) one sees that this identity also holds for $u = \phi_j$, $j = 0, \dots, n-1$. Thus (4.7.8) holds for all $u \in H^1_{\mathsf{D}}(\mathcal{M})$.

Consider now $g_1 \in H^{1/2}(^1\partial\mathcal{M})$ and the unique solution $w \in \underline{H}^2(\mathcal{M})$ to (4.7.5) provided by Theorem 4.18 in the case $g_0 = 0$. Let $u \in H^1_{\mathsf{D}}(\mathcal{M})$. Since $w \in H^2(\mathcal{M})$ and $w_{|^0\partial\mathcal{M}} = 0$, with (4.6.5) one finds

$$\begin{aligned}
\tilde{N}(w, u) &= (\nabla_g w, \nabla_g u)_{L^2V(\mathcal{M})} + (R_1 w, u)_{L^2(\mathcal{M})} + (B' w_{|^1\partial\mathcal{M}}, u_{|^1\partial\mathcal{M}})_{L^2(^1\partial\mathcal{M})}\\
&= (\nabla_g w, \nabla_g u)_{L^2V(\mathcal{M})} + (R_1 w, u)_{L^2(\mathcal{M})} + (\partial_\nu w_{|^1\partial\mathcal{M}}, u_{|^1\partial\mathcal{M}})_{L^2(^1\partial\mathcal{M})}\\
&\quad + (g_1, u_{|^1\partial\mathcal{M}})_{L^2(^1\partial\mathcal{M})}\\
&= (Pw, u)_{L^2(\mathcal{M})} + (g_1, u_{|^1\partial\mathcal{M}})_{L^2(^1\partial\mathcal{M})}\\
&= (f, u)_{L^2(\mathcal{M})} + (g_1, u_{|^1\partial\mathcal{M}})_{L^2(^1\partial\mathcal{M})}.
\end{aligned}$$

Consequently, since $w \in \underline{H}^2(\mathcal{M}) \subset \underline{H}^1(\mathcal{M})$, with the uniqueness of v proven above we have $w = v$. ∎

4.7.2. A Boundary Lifting Map. For $\lambda \in \mathbb{R}$, we set $P_\lambda = P + \lambda \operatorname{Id}_{L^2(\mathcal{M})}$ and

$$\begin{aligned}
L_\lambda : H^2(\mathcal{M}) &\to L^2(\mathcal{M}) \oplus H^{3/2,1/2}(\partial\mathcal{M})\\
u &\mapsto (P_\lambda u, u_{|^0\partial\mathcal{M}}, B u_{|^1\partial\mathcal{M}}),
\end{aligned} \tag{4.7.10}$$

as in (3.1.3) in Chap. 3. This map is Fredholm of index zero as seen in the beginning of the proof of Theorem 4.9.

Below we shall consider $\lambda > -\mu_0$. Then, the unbounded operator $\mathsf{P}_\lambda = \mathsf{P} + \lambda \operatorname{Id}_{L^2(\mathcal{M})}$ with domain $D(\mathsf{P})$ is such that $\ker(\mathsf{P}_\lambda) = \{0\}$.

We introduce here a map that is the counterpart of the Dirichlet lifting introduced in Section 10.5 in Volume 1 and in Sect. 18.6.3 in the present volume or the Neumann lifting map introduced in Sect. 18.7.1 or the mixed Dirichlet-Neumann lifting map introduced in Sect. 18.8.2. The setting of the latter map is quite close to that we face here. We shall thus closely follow this section.

For $r, s \in \mathbb{R}$ we set

$$H^{r,s}(\partial\mathcal{M}) = H^r({}^0\partial\mathcal{M}) \oplus H^s({}^1\partial\mathcal{M}). \tag{4.7.11}$$

For a function $w \in H^2(\mathcal{M})$, we set

$$\gamma(w) = (w_{|{}^0\partial\mathcal{M}}, Bw_{|{}^1\partial\mathcal{M}}), \tag{4.7.12}$$

and we have $\gamma(w) \in H^{3/2,1/2}(\partial\mathcal{M})$. By Lemma 18.32, for a function $w \in \mathcal{W}_P(\mathcal{M})$ we have $\gamma(w) \in H^{-1/2,-3/2}(\partial\mathcal{M})$.

Proposition 4.21. *Let $\mathsf{h} \in H^{-1/2,-3/2}(\partial\mathcal{M})$. There exists a unique $u \in \mathcal{W}_P(\mathcal{M})$ such that $P_\lambda u = 0$ and $\gamma(u) = \mathsf{h}$. Moreover, the map*

$$\begin{aligned} H^{-1/2,-3/2}(\partial\mathcal{M}) &\to \mathcal{W}_P(\mathcal{M}) \\ \mathsf{h} &\mapsto u \end{aligned}$$

is bounded.

Definition 4.22 (Boundary Lifting Map). We call the bounded map $\mathsf{M}_{P_\lambda,B} : H^{-1/2,-3/2}(\partial\mathcal{M}) \to \mathcal{W}_P(\mathcal{M})$ given by Proposition 4.21 the boundary lifting map adapted to (P_λ, B).

Proof of Proposition 4.21. Recalling that $\ker(\mathsf{P}_\lambda) = \{0\}$, we denote by $\mathsf{R}_\lambda : L^2(\mathcal{M}) \to H^2(\mathcal{M})$ the resolvent map associated with the homogeneous boundary problem

$$P_\lambda u = f \text{ in } \mathcal{M}, \qquad Bu_{|\partial\mathcal{M}} = 0 \text{ on } \partial\mathcal{M}. \tag{4.7.13}$$

First, we address uniqueness. By linearity, we consider $u \in \mathcal{W}_P(\mathcal{M})$ such that $P_\lambda u = 0$ and $\gamma(u) = 0$. Let $f \in L^2(\mathcal{M})$ and choose $w = \mathsf{R}_\lambda f$. We have $w \in H^2(\mathcal{M})$ and $\gamma(w) = 0$. By Proposition 4.8, or rather (4.5.3), we obtain $(u, f)_{L^2(\mathcal{M})} = (u, P_\lambda w)_{L^2(\mathcal{M})} = 0$. As $f \in L^2(\mathcal{M})$ is arbitrary we conclude that $u = 0$.

Second, we address existence. To ease notation we set $H = H^{-1/2,-3/2}(\partial\mathcal{M})$. For a function $w \in H^2(\mathcal{M})$ set

$$\gamma'(w) = (\partial_\nu w_{|{}^0\partial\mathcal{M}}, w_{|{}^1\partial\mathcal{M}}) \in H' = H^{1/2,3/2}(\partial\mathcal{M}).$$

For $\mathsf{h} = ({}^0h, {}^1h) \in H$ and $\mathsf{k} = ({}^0k, {}^1k) \in H'$, we set

$$\langle k, h\rangle_{H',H} = -\langle {}^0k, {}^0h\rangle_{H^{1/2}({}^0\partial\mathcal{M}),H^{-1/2}({}^0\partial\mathcal{M})} + \langle {}^1k, {}^1h\rangle_{H^{3/2}({}^1\partial\mathcal{M}),H^{-3/2}({}^1\partial\mathcal{M})}.$$

For $\mathsf{h} \in H$, consider then the map

$$\begin{aligned} U : L^2(\mathcal{M}) &\to \mathbb{C} \\ f &\mapsto \langle \gamma' \mathsf{R}_\lambda f, \overline{\mathsf{h}}\rangle_{H',H}. \end{aligned}$$

By the trace formula of Theorem 18.25, this form is bounded and we have

$$|U(f)| \leq |\mathsf{h}|_H \|\mathsf{R}_\lambda f\|_{H^2(\mathcal{M})} \lesssim |\mathsf{h}|_H \|f\|_{L^2(\mathcal{M})}.$$

By the Riesz theorem, there exists $u \in L^2(\mathcal{M})$ such that $U(f) = (f, u)_{L^2(\mathcal{M})}$ for all $f \in L^2(\mathcal{M})$. Moreover, $\|u\|_{L^2(\mathcal{M})} \lesssim |\mathsf{h}|_H$. If $\varphi \in {}^0\mathscr{D}_c^\infty(\mathcal{M})$ and $f = P_\lambda \varphi$, then $\varphi = \mathsf{R}_\lambda f$ and we find

$$U(f) = \langle \gamma'(\varphi), \overline{\mathsf{h}} \rangle_{H',H} = 0.$$

Consequently

$$\begin{aligned}\langle \overline{P_\lambda u}, \varphi \mu_g \rangle_{{}^0\mathscr{D}'(\mathcal{M}), {}^1\mathscr{D}_c^\infty(\mathcal{M})} &= \langle \overline{u}, P_\lambda \varphi \mu_g \rangle_{{}^0\mathscr{D}'(\mathcal{M}), {}^1\mathscr{D}_c^\infty(\mathcal{M})} \\ &= (P_\lambda \varphi, u)_{L^2(\mathcal{M})} = U(f) = 0.\end{aligned}$$

Thus, in the sense of distributions we have $P_\lambda u = 0$. Thus $u \in \mathcal{W}_P(\mathcal{M})$ and $\|u\|_{\mathcal{W}_P(\mathcal{M})} = \|u\|_{L^2(\mathcal{M})} + \|Pu\|_{L^2(\mathcal{M})} \lesssim |\mathsf{h}|_H$.

LEMMA 4.23. *There exists a bounded map* $\mathsf{M}_0 : H' \to H^2(\mathcal{M})$ *such that* $\gamma' \circ \mathsf{M}_0 = \mathrm{Id}_{H'}$ *and* $\gamma \circ \mathsf{M}_0 = 0$.

This is a particular case of Theorem 18.25. A proof follows from Lemmata 18.26 and 18.27 by working locally at the boundary.

Let $k \in H'$. For $v = \mathsf{M}_0 k \in H^2(\mathcal{M})$ as given by Lemma 4.23, we have

$$(P_\lambda v, u)_{L^2(\mathcal{M})} = \langle k, \gamma(\overline{u}) \rangle_{H',H},$$

by the Green formula of Proposition 4.8. We also have

$$\begin{aligned}(P_\lambda v, u)_{L^2(\mathcal{M})} &= U(P_\lambda v) = \langle \gamma' \mathsf{R}_\lambda (P_\lambda v), \overline{h} \rangle_{H',H} \\ &= \langle \gamma'(v), \overline{h} \rangle_{H',H} = \langle k, \overline{h} \rangle_{H',H}.\end{aligned}$$

For all $k \in H'$ we thus find $\langle k, \overline{h} - \gamma(\overline{u}) \rangle_{H',H} = 0$, implying $\gamma(u) = h$, which concludes the existence part of the proof. ∎

PROPOSITION 4.24. *Let* $k \in \mathbb{N}$. *If* $\mathsf{h} \in H^{k-1/2, k-3/2}(\partial\mathcal{M})$, *then* $\mathsf{M}_{P,B}(\mathsf{h}) \in H^k(\mathcal{M})$. *Moreover, for some* $C > 0$, *we have*

$$\|\mathsf{M}_{P,B}(\mathsf{h})\|_{H^k(\mathcal{M})} \le C |\mathsf{h}|_{H^{k-1/2,k-3/2}(\partial\mathcal{M})}.$$

PROOF. The case $k = 0$ is treated above. The proof can be adapted from that of Proposition 18.39 for $k \geq 2$. The case $k = 1$ follows by an interpolation argument [74, 236]. ∎

Set

$$\underline{\mathcal{W}}_P(\mathcal{M}) = \{\underline{u};\ u \in \mathcal{W}_P(\mathcal{M})\} = (\mathrm{Id}_{L^2(\mathcal{M})} - \Pi_{E_0})(\mathcal{W}_P(\mathcal{M})).$$

The following theorem extends Theorem 4.18 to lower regularity for the boundary data.

THEOREM 4.25. *Let* $f \in L^2(\mathcal{M})$, $g_0 \in H^{-1/2}({}^0\partial\mathcal{M})$, *and* $g_1 \in H^{-3/2}({}^1\partial\mathcal{M})$ *be such that*

$$\begin{aligned}0 = (f, \phi_j)_{L^2(\mathcal{M})} &+ \langle g_1, \gamma^{\mathsf{D}}_{{}^1\partial\mathcal{M}}(\phi_j) \rangle_{H^{-3/2}({}^1\partial\mathcal{M}), H^{3/2}({}^1\partial\mathcal{M})} \\ &- \langle g_0, \gamma^{\mathsf{D}}_{{}^0\partial\mathcal{M}}(\phi_j) \rangle_{H^{-1/2}({}^0\partial\mathcal{M}), H^{1/2}({}^0\partial\mathcal{M})},\end{aligned} \tag{4.7.14}$$

for $j = 0, \dots, n-1$. *Then, there exists a unique* $v \in \underline{\mathcal{W}}_P(\mathcal{M})$ *such that* (4.7.4) *holds. Any other solution* $u \in \mathcal{W}_P(\mathcal{M})$ *of* (4.7.4) *is such that* $u - v \in E_0 = \ker(\mathsf{P})$. *Moreover, there exists* $C > 0$ *such that*

$$\|v\|_{L^2(\mathcal{M})} \leq C\big(\|f\|_{L^2(\mathcal{M})} + |g_0|_{H^{-1/2}(^0\partial\mathcal{M})} + |g_1|_{H^{-3/2}(^1\partial\mathcal{M})}\big). \tag{4.7.15}$$

If one has $g_0 \in H^{1/2}(^0\partial\mathcal{M})$ *and* $g_1 \in H^{-1/2}(^1\partial\mathcal{M})$, *then* $v \in \underline{\mathcal{W}}_P(\mathcal{M}) \cap H^1(\mathcal{M})$ *and there exists* $C > 0$ *such that*

$$\|v\|_{H^1(\mathcal{M})} \leq C\big(\|f\|_{L^2(\mathcal{M})} + |g_0|_{H^{1/2}(^0\partial\mathcal{M})} + |g_1|_{H^{-1/2}(^1\partial\mathcal{M})}\big). \tag{4.7.16}$$

This theorem generalizes what is obtained in Sect. 18.6 for the Dirichlet-Laplace problem, in Sect. 18.7 for the Neumann-Laplace problem, and in Sect. 18.8 for the mixed Dirichlet-Neumann-Laplace problem. Note that the cases $^0\partial\mathcal{M} = \emptyset$ or $^1\partial\mathcal{M} = \emptyset$ are contained in the present result. Note, however, that in the case $^1\partial\mathcal{M} = \emptyset$, that is, the Dirichlet-Laplace problem, weaker regularity for f, namely $H^{-1}(\mathcal{M})$, can be considered; see Theorem 18.40.

The last result in the theorem makes sense since $u_{|^0\partial\mathcal{M}} \in H^{1/2}(^0\partial\mathcal{M})$ and $Bu_{|^1\partial\mathcal{M}} \in H^{-1/2}(^1\partial\mathcal{M})$ by Lemma 18.42 since $\mathcal{W}_P(\mathcal{M}) \cap H^1(\mathcal{M}) = \mathcal{W}_g(\mathcal{M}) \cap H^1(\mathcal{M})$.

PROOF. Let $u, v \in \mathcal{W}_P(\mathcal{M})$ be two solutions of (4.7.4). Then $P(u-v) = 0$, $(u-v)_{|^0\partial\mathcal{M}} = 0$ and $B(u-v)_{|^1\partial\mathcal{M}} = 0$. Thus, by Theorem 3.28 one has $u - v \in H^2(\mathcal{M})$ and thus $u - v \in \ker(\mathsf{P})$. This implies the uniqueness of a solution in $\underline{\mathcal{W}}_P(\mathcal{M})$.

Let $\lambda > -\mu_0$ and $w = \mathsf{M}_{P_\lambda, B}(g_0, g_1)$ meaning that $w \in \mathcal{W}_P(\mathcal{M})$ with $Pw = -\lambda w$, $w_{|^0\partial\mathcal{M}} = g_0$ and $Bw_{|^1\partial\mathcal{M}} = g_1$ by Proposition 4.21 and Definition 4.22.

For $j = 0, \dots, n-1$, with the Green formula of Proposition 4.8 we compute

$$\begin{aligned}(Pw, \phi_j)_{L^2(\mathcal{M})} &+ \langle g_1, \gamma^{\mathsf{D}}_{^1\partial\mathcal{M}}(\overline{\phi_j})\rangle_{H^{-1/2}(^1\partial\mathcal{M}), H^{1/2}(^1\partial\mathcal{M})}\\ &= \langle g_0, \gamma^{\mathsf{N}}_{^0\partial\mathcal{M}}(\overline{\phi_j})\rangle_{(L^2(^0\partial\mathcal{M})},\end{aligned}$$

using that $P\phi_j = 0$, $\gamma^{\mathsf{D}}_{^0\partial\mathcal{M}}(\phi_j) = 0$, and $B\phi_{j|^1\partial\mathcal{M}} = 0$. We thus have $(f - Pw, \phi_j)_{L^2(\mathcal{M})} = 0$. Thus by Theorem 4.18 there exists $u \in \underline{H}^2(\mathcal{M})$ such that $Pu = f - Pw$, $u_{|^0\partial\mathcal{M}} = 0$, and $Bu_{|^1\partial\mathcal{M}} = 0$. We set $v = u + w \in \mathcal{W}_P(\mathcal{M})$. We have $Pv = f$, $v_{|^0\partial\mathcal{M}} = g_0$ and $Bv_{|^1\partial\mathcal{M}} = g_1$. We then find that $\underline{v} \in \underline{\mathcal{W}}_P(\mathcal{M})$ and $P\underline{v} = f$, $\underline{v}_{|^0\partial\mathcal{M}} = g_0$ and $B\underline{v}_{|^1\partial\mathcal{M}} = g_1$.

The case $g_0 \in H^{1/2}(^0\partial\mathcal{M})$ and $g_1 \in H^{-1/2}(^1\partial\mathcal{M})$ follows similarly using Proposition 4.24. ∎

Finally we have the following result.

PROPOSITION 4.26. *Let $f \in L^2(\mathcal{M})$ and $g_1 \in H^{-1/2}({}^1\partial\mathcal{M})$ be such that*
(4.7.17)
$$(f, \phi_j)_{L^2(\mathcal{M})} + \langle g_1, \gamma^{\mathsf{D}}_{{}^1\partial\mathcal{M}}(\overline{\phi_j})\rangle_{H^{-1/2}({}^1\partial\mathcal{M}),H^{1/2}({}^1\partial\mathcal{M})} = 0, \qquad j = 0, \dots, n-1.$$
The unique solution in $\underline{H}^1_{\mathsf{D}}(\mathcal{M})$ to (4.7.8) coincides with the unique solution to (4.7.4) provided by Theorem 4.25 in the case $g_0 = 0$.

PROOF. Let $v \in \underline{\mathcal{W}}_P(\mathcal{M}) \cap H^1_{\mathsf{D}}(\mathcal{M})$ be the unique solution to (4.7.5) provided by Theorem 4.25, that is $Pv = f$ and $Bv_{|{}^1\partial\mathcal{M}} = g_1$. Note that $Bv_{|{}^1\partial\mathcal{M}}$ makes sense by Lemma 18.42 since $\mathcal{W}_P(\mathcal{M}) \cap H^1(\mathcal{M}) = \mathcal{W}_g(\mathcal{M}) \cap H^1(\mathcal{M})$. Let $u \in D(\mathsf{P})$. With Lemma 4.7 we have
$$\begin{aligned}
\tilde{N}(v, u) &= (\nabla_g v, \nabla_g u)_{L^2V(\mathcal{M})} + (v, R_1 u)_{L^2(\mathcal{M})} \\
&\quad + \langle v_{|{}^1\partial\mathcal{M}}, \bar{B'} u_{|{}^1\partial\mathcal{M}}\rangle_{H^{-1/2}({}^1\partial\mathcal{M}),H^{1/2}({}^1\partial\mathcal{M})} \\
&= (v, Pu)_{L^2(\mathcal{M})} + \langle v_{|{}^1\partial\mathcal{M}}, \partial_\nu \bar{u}_{|{}^1\partial\mathcal{M}}\rangle_{H^{-1/2}({}^1\partial\mathcal{M}),H^{1/2}({}^1\partial\mathcal{M})} \\
&\quad + \langle v_{|{}^1\partial\mathcal{M}}, \overline{B'u}_{|{}^1\partial\mathcal{M}}\rangle_{H^{-1/2}({}^1\partial\mathcal{M}),H^{1/2}({}^1\partial\mathcal{M})} \\
&= (v, Pu)_{L^2(\mathcal{M})} + \langle v_{|{}^1\partial\mathcal{M}}, \overline{Bu}_{|{}^1\partial\mathcal{M}}\rangle_{H^{-1/2}({}^1\partial\mathcal{M}),H^{1/2}({}^1\partial\mathcal{M})}
\end{aligned}$$
which with (4.5.2) gives
$$\begin{aligned}
\tilde{N}(v, u) &= (Pv, u)_{L^2(\mathcal{M})} + \langle Bv_{|{}^1\partial\mathcal{M}}, \bar{u}_{|{}^1\partial\mathcal{M}}\rangle_{H^{-3/2}({}^1\partial\mathcal{M}),H^{3/2}({}^1\partial\mathcal{M})} \\
&= (f, u)_{L^2(\mathcal{M})} + \langle g_1, \bar{u}_{|{}^1\partial\mathcal{M}}\rangle_{H^{-1/2}({}^1\partial\mathcal{M}),H^{1/2}({}^1\partial\mathcal{M})}.
\end{aligned}$$
Since $D(\mathsf{P})$ is dense in $H^1_{\mathsf{D}}(\mathcal{M})$ one obtains (4.7.8). ∎

Part 2

Carleman Estimates on Riemannian Manifolds

CHAPTER 5

Estimates on Riemannian Manifolds for Dirichlet Boundary Conditions

Contents

This chapter aims to transpose the results obtained in Chapter 3 of Volume 1 to the framework of Riemannian manifolds. Using local coordinates one can use directly the local results obtained in that chapter. Passing from regular open sets to Riemannian manifolds is thus fairly straightforward, up to reproducing some of the patching arguments of Chapter 3.

5.1. Setting

We consider a smooth Riemannian manifold $(\mathcal{M}, g)$. Elementary facts on manifolds and Riemannian manifolds are presented in Chap. 15.

We assume that $\mathcal{M}$ is σ-compact, that is, $\mathcal{M}$ admits an exhaustion by compact sets $(K^n)_n$. A subset L of $\mathcal{M}$ is then said to be bounded if there exists $n \in \mathbb{N}$ such that $L \subset K^n$.

In local coordinates the Laplace–Beltrami operator reads

$$\Delta_g f = (\det g)^{-1/2} \sum_{1 \leq i,j \leq d} \partial_i \big((\det g)^{1/2} g^{ij} \partial_j f \big),$$

J. Le Rousseau et al., *Elliptic Carleman Estimates and Applications to Stabilization and Controllability, Volume II*, PNLDE Subseries in Control 98, https://doi.org/10.1007/978-3-030-88670-7_5

where, at each point of $\mathcal{M}$, (g^{ij}) is the inverse matrix of the metric $g = (g_{ij})$ and $\det g = \det(g_{ij})$.

We set $P_0 = -\Delta_g$ and we shall consider more general operators P of the form $P = P_0 + R_1$ where R_1 is a first-order differential operator with bounded coefficients on $\mathcal{M}$. Those operators are the sum of a bounded function and a locally finite sum of vectors fields with bounded coefficients (See Remark 16.9).

For $\varphi \in \mathscr{C}^\infty(\mathcal{M})$ we introduce the conjugated operator, for $\tau > 0$,

$$P_\varphi = e^{\tau\varphi} P_0 e^{-\tau\varphi}.$$

In each local chart, its representative has the form

$$(\det g)^{-1/2} \sum_{1 \leq i,j \leq d} (D_i + i\tau(d\varphi)_i)\big((\det g)^{1/2} g^{ij} (D_j + i\tau(d\varphi)_j)\big),$$

where $d\varphi$ has $\sum_{1\leq i\leq d}(d\varphi)_i dx^i$ for representative. With such a local representative we may see P_φ as a second-order differential operator with a large parameter on $\mathcal{M}$ (see Sect. 16.3.4). Its principal symbol is the function on the cotangent bundle $T^*\mathcal{M}$ with τ as a parameter, given by

$$p_\varphi(m, \omega, \tau) = (\omega + i\tau d\varphi(m), \omega + i\tau d\varphi(m))_{g_m},$$

with $(.,.)_{g_m}$ defined in (17.1.4). In local coordinates, this reads

$$p_\varphi(x, \xi, \tau) = \sum_{1\leq i,j\leq d} (\xi_i + i\tau(d\varphi(x))_i) g^{ij}(x)(\xi_j + i\tau(d\varphi(x))_j).$$

Following Section 3.2.1 we set

$$P_2 = \frac{1}{2}(P_\varphi + P_\varphi^*), \quad P_1 = \frac{1}{2i}(P_\varphi - P_\varphi^*),$$

that are formally selfadjoint, and we have $P_\varphi = P_2 + iP_1$. Adjoint operators are defined similarly to transpose operators with an additional complex conjugation (see Sect. 16.3.3). They are differential operators of order two with a large parameter. Their principal symbols are given by

$$p_2(m, \omega, \tau) = |\omega|^2_{g_m} - \tau^2 |d\varphi(m)|^2_{g_m},$$
$$p_1(m, \omega, \tau) = 2\tau(\omega, d\varphi(m))_{g_m},$$

with $|.|_{g_m}$ defined in (17.1.4). In local coordinates, they read

$$p_2(x, \xi, \tau) = \sum_{1\leq i,j\leq d} g^{ij}(x)\big(\xi_i\xi_j - \tau^2(d\varphi(x))_i(d\varphi(x))_j\big),$$
$$p_1(x, \xi, \tau) = 2\tau \sum_{1\leq i,j\leq d} g^{ij}(x)\, \xi_i\, (d\varphi(x))_j.$$

The sub-ellipticity property of Definition 3.2 of Volume 1 can be extended to the case of a manifold as follows. The Poisson bracket of two functions on a manifold is recalled in Sect. 15.7.2.

DEFINITION 5.1 (Sub-ellipticity). Let V be a bounded open set in $\mathcal{M}$. We say that the weight function $\varphi \in \mathscr{C}^\infty(\mathcal{M};\mathbb{R})$ and P have the *sub-ellipticity* condition in $\overline{V}$ if $d\varphi \neq 0$ in $\overline{V}$ and if
(5.1.1)
$$\forall (m,\omega) \in T^*\overline{V},\ \forall \tau > 0, \quad p_\varphi(m,\omega,\tau) = 0 \quad \Rightarrow \quad \{p_2, p_1\}(m,\omega,\tau) > 0.$$

Having φ and P satisfying the sub-ellipticity condition of Definition 5.1 in $\overline{V}$ precisely means that in each local chart (O,κ) the representatives of φ and P satisfy the sub-ellipticity condition of Definition 3.2 in $\overline{\kappa(V \cap O)}$.

5.2. Estimates Away from the Boundary

The counterpart of Theorem 3.11 of Volume 1 is the following result.

THEOREM 5.2. *Let $P = P_0 + R_1$ with $P_0 = -\Delta_g$ where R_1 is a first-order differential operator with bounded coefficients. Let V be a bounded open set in $\mathcal{M}$ such that $V \cap \partial\mathcal{M} = \emptyset$, and let φ and P have the sub-ellipticity property of Definition 5.1 in $\overline{V}$; then, there exist $\tau_* > 0$ and $C > 0$ such that*
$$\begin{aligned}\tau^3 \|e^{\tau\varphi} u\|^2_{L^2(\mathcal{M})} + \tau \|e^{\tau\varphi} \mathsf{D}\, u\|^2_{L^2\Lambda^1(\mathcal{M})} + \tau^{-1} \|e^{\tau\varphi} \mathsf{H}\, u\|^2_{L^2\Lambda^2(\mathcal{M})}\\ \leq C \|e^{\tau\varphi} P u\|^2_{L^2(\mathcal{M})},\end{aligned} \tag{5.2.1}$$
for $u \in \mathscr{C}^\infty_c(V)$ and $\tau \geq \tau_*$.

Here, D stands for the covariant derivative on $\mathcal{M}$ and $\mathsf{H} = \mathsf{D}^2$ is the Hessian defined by means of the Levi-Civita connection. For a function such as u we simply have $\mathsf{D}\, u = du$. These notions are recalled in Sects. 17.4 and 17.7. In particular, the gradient of u on $(\mathcal{M}, g)$, given in local coordinates by,
$$(\nabla_g u)^i = \sum_{1 \leq j \leq d} g^{ij} \partial_j u, \quad i = 1, \ldots, d,$$
is related to $\mathsf{D}\, u$ by the relation $\nabla_g u = (\mathsf{D}\, u)^\sharp$ (see Sect. 17.1.3 for the musical isomorphisms on a Riemannian manifold).

The space $L^2\Lambda(\mathcal{M})$ (resp. $L^2\Lambda^2(\mathcal{M})$) is the Hilbert space of L^2 one-forms (resp. 2-covariant tensors) on $\mathcal{M}$. These spaces and the associated norms $\|.\|_{L^2\Lambda(\mathcal{M})}$ and $\|.\|_{L^2\Lambda^2(\mathcal{M})}$ are given in Sect. 18.2. Note that
$$\|e^{\tau\varphi} \mathsf{D}\, u\|_{L^2\Lambda^1(\mathcal{M})} = \|e^{\tau\varphi} \nabla_g u\|_{L^2V(\mathcal{M})},$$
with the space $L^2V(\mathcal{M})$ of L^2-vector fields and its associated norm given in Sect. 18.1.

PROOF. We first observe that it suffices to prove the estimate for P_0 in place of P by adapting Remark 3.13 to the manifold setting.

Let the local charts $\mathcal{C}^i = (O^i, \kappa^i)$, $i \in \mathcal{I}$, form an atlas of $\mathcal{M}$. As V is bounded, that is, contained in a compact set of $\mathcal{M}$ as recalled above, there exists $\mathcal{J} \subset \mathcal{I}$ with $\#\mathcal{J} < \infty$ such that $\overline{V} \subset \cup_{i \in \mathcal{J}} O^i$. Let $(\chi^i)_{i \in \mathcal{J}}$

be a $\mathscr{C}^\infty$-partition of unity of $\overline{V}$ subordinated to this open covering (see Definition 15.4 and Theorem 15.14).

Let $f = P_0 u$. For each $i \in \mathscr{J}$, we let $u^{\mathcal{C}^i}$ (resp. $P_0^{\mathcal{C}^i}$, $f^{\mathcal{C}^i}$, φ, $\chi^{\mathcal{C}^i}$) be the representative of u (resp. P_0, f, $\varphi^{\mathcal{C}^i}$, χ^i) in the chart $\mathcal{C}^i$. We have $P_0^{\mathcal{C}^i} u^{\mathcal{C}^i} = f^{\mathcal{C}^i}$. The operator $P_0^{\mathcal{C}^i}$ is an elliptic second-order differential operator on $\tilde{O}^i = \kappa^i(O^i)$ with smooth real principal part. Setting $v^i = \chi^{\mathcal{C}^i} u^{\mathcal{C}^i} \in \mathscr{C}_c^\infty(\tilde{O}^i)$ we have $P_0^{\mathcal{C}^i} v^i = \chi^{\mathcal{C}^i} f^{\mathcal{C}^i} + [P_0, \chi^{\mathcal{C}^i}] u^{\mathcal{C}^i}$. As $\varphi^{\mathcal{C}^i}$ and $P_0^{\mathcal{C}^i}$ satisfy the sub-ellipticity condition in $\kappa^i(\overline{V \cap O^i})$ that is a neighborhood of $\operatorname{supp}(v^i)$, the Carleman estimate of Theorem 3.11 applies. We can then obtain

$$\begin{aligned}
&\tau^3 \| e^{\tau \varphi^{\mathcal{C}^i}} v^i \|^2_{L^2(\tilde{O}^i)} + \tau \sum_{1 \leq j \leq d} \| e^{\tau \varphi^{\mathcal{C}^i}} D_j v^i \|^2_{L^2(\tilde{O}^i)} \\
&\qquad + \tau^{-1} \sum_{1 \leq j,k \leq d} \| e^{\tau \varphi^{\mathcal{C}^i}} (\partial_j \partial_k - \sum_{1 \leq \ell \leq d} \Gamma^\ell_{jk} \partial_\ell) v^i \|^2_{L^2(\tilde{O}^i)} \\
&\lesssim \| e^{\tau \varphi^{\mathcal{C}^i}} P_0^{\mathcal{C}^i} v^i \|^2_{L^2(\tilde{O}^i)} + \| e^{\tau \varphi^{\mathcal{C}^i}} u^{\mathcal{C}^i} \|^2_{L^2(\tilde{O}^i)} + \sum_{1 \leq j \leq d} \| e^{\tau \varphi^{\mathcal{C}^i}} D_j u^{\mathcal{C}^i} \|^2_{L^2(\tilde{O}^i)},
\end{aligned}$$

for $\tau \geq \tau^i$ for some $\tau^i > 0$ chosen sufficiently large, using that $[P_0, \chi^{\mathcal{C}^i}]$ is a first-order differential operator. The symbol Γ^ℓ_{jk} stands for the Christoffel symbol associated with the metric g and the Levi-Civita connection in the local chart $\mathcal{C}^i$ (see (17.4.11) in Sect. 17.4). The L^2-norms on $\tilde{O}^i$ in the above inequality are for the Lebesgue measure dx in $\mathbb{R}^d$. However, the same inequality holds for dx replaced by $\mu_g^{\mathcal{C}^i}$, the local representative of the Riemannian canonical density μ_g (see Sect. 17.3), that is, $\mu_g^{\mathcal{C}^i} = T_{(\det g^{\mathcal{C}^i})^{1/2}}$, using the notation of Section 8.1.2 of Volume 1, precisely meaning $\mu_g^{\mathcal{C}^i} = (\det g^{\mathcal{C}^i})^{1/2} dx$.

We refer the reader to Sect. 17.1.3 for the musical isomorphisms on a Riemannian manifold and to Sects. 17.4 and 17.7 for the definition of the covariant derivative D. Then, using that, for a function w on $\mathcal{M}$,

- The covariant derivative $\mathsf{D}\, w = dw$ is a one-form, with moreover $(\mathsf{D}\, w)^\sharp = \nabla_g w$, with $\nabla_g w$ the Riemannian gradient, and that both their representatives in $\mathcal{C}^i$ have norms equivalent to that of $D w^{\mathcal{C}^i}$;
- the Hessian $\mathsf{H}\, w = \mathsf{D}^2\, w$ is a 2-covariant tensor and its representative in $\mathcal{C}^i$ is given by $(\mathsf{H}\, w)^{\mathcal{C}}_{jk} = (\partial_j \partial_k - \sum_{1 \leq \ell \leq d} \Gamma^\ell_{jk} \partial_\ell) w^{\mathcal{C}^i}$, see (17.7.2);

on the manifold, we thus obtain

$$\begin{aligned}
\tau^3 \| e^{\tau\varphi} \chi^i u \|^2_{L^2(\mathcal{M})} + \tau \| e^{\tau\varphi}\, \mathsf{D}(\chi^i u) \|^2_{L^2\Lambda^1(\mathcal{M})} + \tau^{-1} \| e^{\tau\varphi}\, \mathsf{H}(\chi^i u) \|^2_{L^2\Lambda^2(\mathcal{M})} \\
\lesssim \| e^{\tau\varphi} P_0 \chi^i u \|^2_{L^2(\mathcal{M})} + \| e^{\tau\varphi} u \|^2_{L^2(\mathcal{M})} + \| e^{\tau\varphi}\, \mathsf{D}\, u \|^2_{L^2\Lambda^1(\mathcal{M})}.
\end{aligned}$$

Commuting χ^i with D, $\mathsf{H} = \mathsf{D}^2$ and P_0 yields the estimation

$$\tau^3 \|e^{\tau\varphi}\chi^i u\|^2_{L^2(\mathcal{M})} + \tau\|e^{\tau\varphi}\chi^i \mathsf{D}\, u\|^2_{L^2\Lambda^1(\mathcal{M})} + \tau^{-1}\|e^{\tau\varphi}\chi^i \mathsf{H}\, u\|^2_{L^2\Lambda^2(\mathcal{M})} \lesssim \|e^{\tau\varphi} P_0 u\|^2_{L^2(\mathcal{M})} + \tau\|e^{\tau\varphi}u\|^2_{L^2(\mathcal{M})} + \|e^{\tau\varphi}\mathsf{D}\, u\|^2_{L^2\Lambda^1(\mathcal{M})},$$

for $\tau \geq \tau^i$. Summing over $i \in \mathcal{J}$, for $\tau \geq \max \tau^i$, recalling that $\#\mathcal{J} < \infty$ and using the properties of the partition of unity, we obtain with the triangular inequality,

$$\begin{aligned}
&\tau^3 \|e^{\tau\varphi}u\|^2_{L^2(\mathcal{M})} + \tau\|e^{\tau\varphi}\mathsf{D}\, u\|^2_{L^2\Lambda^1(\mathcal{M})} + \tau^{-1}\|e^{\tau\varphi}\mathsf{H}\, u\|^2_{L^2\Lambda^2(\mathcal{M})}\\
&\quad\lesssim \tau^3 \sum_{i\in\mathcal{J}} \|e^{\tau\varphi}\chi^i u\|^2_{L^2(\mathcal{M})} + \tau \sum_{i\in\mathcal{J}} \|e^{\tau\varphi}\chi^i \mathsf{D}\, u\|^2_{L^2\Lambda^1(\mathcal{M})}\\
&\qquad + \tau^{-1} \sum_{i\in\mathcal{J}} \|e^{\tau\varphi}\chi^i \mathsf{H}\, u\|^2_{L^2\Lambda^2(\mathcal{M})}\\
&\quad\lesssim \|e^{\tau\varphi} P_0 u\|^2_{L^2(\mathcal{M})} + \tau\|e^{\tau\varphi}u\|^2_{L^2(\mathcal{M})} + \|e^{\tau\varphi}\mathsf{D}\, u\|^2_{L^2\Lambda^1(\mathcal{M})}.
\end{aligned}$$

The result follows by choosing $\tau > 0$ sufficiently large. ■

5.3. Estimates at the Boundary

Let ν be the unique outward pointing vector field along $\partial\mathcal{M}$ such that, for all $m \in \partial\mathcal{M}$, $g_m(\nu_m, \nu_m) = 1$ and $g_m(\nu_m, u) = 0$ for all $u \in T_m\partial\mathcal{M}$, that is, ν is unitary and orthogonal to $T_m\partial\mathcal{M}$ in the sense of g. For a smooth function f on $\mathcal{M}$, we define its normal derivative at $m \in \partial\mathcal{M}$ by

$$\partial_\nu f(m) = \nu_m(f) = df(m)(\nu_m). \tag{5.3.1}$$

The counterparts of the local estimations of Lemmata 3.15 and 3.16 are given by the following results.

LEMMA 5.3. *Let $P = P_0 + R_1$ with $P_0 = -\Delta_g$ where R_1 is a first-order differential operator with bounded coefficients. Let $m^0 \in \partial\mathcal{M}$. Let V^0 be a bounded open set in $\mathcal{M}$ such that $m^0 \in V^0$, and let $\varphi \in \mathscr{C}^\infty(\overline{V^0})$ and P have the sub-ellipticity property of Definition 5.1 in $\overline{V^0}$. Then, there exits an open neighborhood V^1 of m^0 in $\mathcal{M}$ such that $V^1 \subset V^0$ and there exist $\tau_* > 0$ and $C > 0$ such that*

$$\begin{aligned}
\tau^3 \|e^{\tau\varphi}u\|^2_{L^2(\mathcal{M})} + \tau\|e^{\tau\varphi}\mathsf{D}\, u\|^2_{L^2\Lambda^1(\mathcal{M})} &\leq C\big(\|e^{\tau\varphi}Pu\|^2_{L^2(\mathcal{M})}\\
+ \tau^3|e^{\tau\varphi}u_{|\partial\mathcal{M}}|^2_{L^2(\partial\mathcal{M})} + \tau|e^{\tau\varphi}\mathsf{D}'\, u_{|\partial\mathcal{M}}|^2_{L^2\Lambda^1(\partial\mathcal{M})} &+ \tau|e^{\tau\varphi}\partial_\nu u_{|\partial\mathcal{M}}|^2_{L^2(\partial\mathcal{M})}\big),
\end{aligned} \tag{5.3.2}$$

for $u \in \mathscr{C}^\infty(\mathcal{M})$, with $\operatorname{supp}(u) \subset V^1$ *and $\tau \geq \tau_*$.*

Here, D' denotes the covariant derivative on $\partial\mathcal{M}$ and the norms $|.|_{L^2(\partial\mathcal{M})}$ and $|.|_{L^2\Lambda(\partial\mathcal{M})}$ are defined by means of the metric g_∂ inherited on $\partial\mathcal{M}$ from that on $\mathcal{M}$.

On the boundary $\partial\mathcal{M}$, the H^1-norm is given by

$$|w|^2_{H^1(\partial\mathcal{M})} = |w|^2_{L^2(\partial\mathcal{M})} + |\,\mathsf{D}'\, w|^2_{L^2\Lambda^1(\partial\mathcal{M})},$$

(see Sects. 18.1 and 18.2). We then have

$$|e^{\tau\varphi}\,\mathsf{D}'\,u|^2_{L^2\Lambda^1(\partial\mathcal{M})} \lesssim \tau^2|e^{\tau\varphi}u|^2_{L^2(\partial\mathcal{M})} + |e^{\tau\varphi}u_{|\partial\mathcal{M}}|^2_{H^1(\partial\mathcal{M})},$$

implying that the Carleman estimate of the previous theorem can thus be written in the same form as that of Lemma 3.15 of Volume 1:

$$(5.3.3)\quad \tau^3\|e^{\tau\varphi}u\|^2_{L^2(\mathcal{M})} + \tau\|e^{\tau\varphi}\,\mathsf{D}\,u\|^2_{L^2\Lambda^1(\mathcal{M})} \leq C\big(\|e^{\tau\varphi}Pu\|^2_{L^2(\mathcal{M})} + \tau^3|e^{\tau\varphi}u_{|\partial\mathcal{M}}|^2_{L^2(\partial\mathcal{M})} + \tau|e^{\tau\varphi}u_{|\partial\mathcal{M}}|^2_{H^1(\partial\mathcal{M})} + \tau|e^{\tau\varphi}\partial_\nu u_{|\partial\mathcal{M}}|^2_{L^2(\partial\mathcal{M})}\big).$$

LEMMA 5.4. *Let $P = P_0 + R_1$ with $P_0 = -\Delta_g$ where R_1 is a first-order differential operator with bounded coefficients. Let $m^0 \in \partial\mathcal{M}$. Let V^0 be a bounded open set in $\mathcal{M}$ such that $m^0 \in V^0$, and let $\varphi \in \mathscr{C}^\infty(\overline{V^0})$ and P have the sub-ellipticity property of Definition 5.1 in $\overline{V^0}$ and $\partial_\nu\varphi(m^0) < 0$. Then, there exits an open neighborhood V^1 of m^0 in $\mathcal{M}$ such that $V^1 \subset V^0$ and there exist $\tau_* > 0$ and $C > 0$ such that*

$$(5.3.4)\quad \tau^3\|e^{\tau\varphi}u\|^2_{L^2(\mathcal{M})} + \tau\|e^{\tau\varphi}\,\mathsf{D}\,u\|^2_{L^2\Lambda^1(\mathcal{M})} + \tau|e^{\tau\varphi}\partial_\nu u_{|\partial\mathcal{M}}|^2_{L^2(\partial\mathcal{M})} \leq C\big(\|e^{\tau\varphi}Pu\|^2_{L^2(\mathcal{M})} + \tau^3|e^{\tau\varphi}u_{|\partial\mathcal{M}}|^2_{L^2(\partial\mathcal{M})} + \tau|e^{\tau\varphi}\,\mathsf{D}'\,u_{|\partial\mathcal{M}}|^2_{L^2\Lambda^1(\partial\mathcal{M})}\big),$$

for $u \in \mathscr{C}^\infty(\mathcal{M})$, with $\operatorname{supp}(u) \subset V^1$ *and $\tau \geq \tau_*$.*

PROOF OF LEMMA 5.3 (RESP. 5.4). We first observe that it suffices to prove the estimate for P_0 in place of P by adapting Remark 3.13 to the manifold setting.

Choose an open neighborhood $V^1 \subset V^0$ of m^0 in $\mathcal{M}$ that is contained in a local chart $\mathcal{C} = (O, \kappa)$ at the boundary. In this local chart $\mathcal{M} \cap V^1$ is given by $\{x_d \geq 0\} \cap \kappa(O)$ and $\partial\mathcal{M} \cap V^1 = \{x_d = 0\} \cap \kappa(O)$.

In $\kappa(V^1)$, the elliptic problem, through the representatives of the operator and the involved functions, takes precisely the form given in Sections 3.4 and 3.5. The result of Theorem 3.28 of Volume 1 (resp. 3.29) applies to the representative $P_0^{\tilde{\mathcal{C}}}$ of the operator P_0. Under the assumptions of Lemma 5.3 we then have, for $\tau_* > 0$,

$$\tau^3\|e^{\tau\varphi^{\tilde{\mathcal{C}}}}w\|^2_{L^2(\mathbb{R}^d_+)} + \tau\|e^{\tau\varphi^{\tilde{\mathcal{C}}}}Dw\|^2_{L^2(\mathbb{R}^d_+)} \lesssim \|e^{\tau\varphi^{\tilde{\mathcal{C}}}}P_0^{\tilde{\mathcal{C}}}w\|^2_{L^2(\mathbb{R}^d_+)} + \tau^3|e^{\tau\varphi^{\tilde{\mathcal{C}}}}w_{|x_d=0^+}|^2_{L^2(\mathbb{R}^{d-1})} + \tau|e^{\tau\varphi^{\tilde{\mathcal{C}}}}D'w_{|x_d=0^+}|^2_{L^2(\mathbb{R}^{d-1})} + \tau|e^{\tau\varphi^{\tilde{\mathcal{C}}}}D_dw_{|x_d=0^+}|^2_{L^2(\mathbb{R}^{d-1})},$$

for $\tau \geq \tau_*$ and $w \in \overline{\mathscr{C}}_c^\infty(U^1_+)$, with $U^1_+ = \tilde{\kappa}(V^1) \cap \mathbb{R}^d_+$.

For some open set U_+ of $\mathbb{R}^d_+$ the space $\overline{\mathscr{C}}_c^\infty(U_+)$ is defined in (3.4.11) as

$$(5.3.5)\qquad \overline{\mathscr{C}}_c^\infty(U_+) = \{u = v_{|\mathbb{R}^d_+};\, v \in \mathscr{C}_c^\infty(\mathbb{R}^d) \text{ and } \operatorname{supp} v \subset U\},$$

where U is some open set of $\mathbb{R}^d$ such that $U_+ = U \cap \mathbb{R}^d_+$. Note that the space $\overline{\mathscr{C}}_c^\infty(U_+)$ is independent of the choice of U.

As in the proof of Theorem 5.2, we change the Lebesgue measure for the local representative of the Riemannian canonical density μ_g in $\tilde{\kappa}(V^1)$ and similarly we change the Lebesgue measure on $\kappa(\partial\mathcal{M}\cap V^1)$ by the canonical density μ_{g_∂}, which only affects constants in the estimation.

We refer the reader to Sect. 17.1.3 for the musical isomorphisms on a Riemannian manifold and to Sects. 17.4 and 17.7 for the definition of the covariant derivative D and its connection with the Riemannian gradient in the case of functions. Using that $(\mathsf{D}\,u)^\sharp = \nabla_g u$, $(\mathsf{D}'\,u_{|\partial\mathcal{M}})^\sharp = \nabla_{g_\partial} u_{|\partial\mathcal{M}}$ and that their representatives have norms equivalent to that of $Du^{\tilde{C}}$ and $D'u^{\tilde{C}}_{|x_d=0^+}$, respectively, and using that the representative $\partial_\nu u_{|\partial\mathcal{M}}$ is $-\partial_d u^{\tilde{C}}_{|x_d=0^+} = -iD_d u^{\tilde{C}}_{|x_d=0^+}$, we then have,

$$\tau^3\|e^{\tau\varphi}w\|^2_{L^2(\mathcal{M})}+\tau\|e^{\tau\varphi}\,\mathsf{D}\,u\|^2_{L^2\Lambda^1(\mathcal{M})} \lesssim \|e^{\tau\varphi}P_0u\|^2_{L^2(\mathcal{M})}+\tau^3|e^{\tau\varphi}u_{|\partial\mathcal{M}}|^2_{L^2(\partial\mathcal{M})} + \tau|e^{\tau\varphi}\,\mathsf{D}'\,u_{|\partial\mathcal{M}}|^2_{L^2\Lambda^1(\partial\mathcal{M})} + \tau|e^{\tau\varphi}\partial_\nu u_{|\partial\mathcal{M}}|^2_{L^2(\partial\mathcal{M})},$$

for $\tau \geq \tau_*$ and $u \in \mathscr{C}^\infty(\mathcal{M})$ with $\operatorname{supp}(u) \subset V^1$. Here, because of the chosen neighborhood V^1, there is, however, no need for the introduction of partition of unity as opposed to the proof of Theorem 5.2.

We leave to the reader the adaptation of the estimate of Theorem 3.29 under the assumptions of Lemma 5.4. ∎

Patching estimates together, as is done in Section 3.5, we need not reduce the size of the considered open set where the sub-ellipticity condition hold. The following theorems are the counterparts of Theorems 3.28 and 3.29.

THEOREM 5.5. *Let $P = P_0 + R_1$ with $P_0 = -\Delta_g$ where R_1 is a first-order differential operator with bounded coefficients. Let V be a bounded open set in $\mathcal{M}$ and let $\varphi \in \mathscr{C}^\infty(\overline{V})$ and P have the sub-ellipticity property of Definition 5.1 in $\overline{V}$. Then, there exist $\tau_* > 0$ and $C > 0$ such that*

$$\tau^3\|e^{\tau\varphi}u\|^2_{L^2(\mathcal{M})} + \tau\|e^{\tau\varphi}\,\mathsf{D}\,u\|^2_{L^2\Lambda^1(\mathcal{M})} \leq C\big(\|e^{\tau\varphi}Pu\|^2_{L^2(\mathcal{M})} + \tau^3|e^{\tau\varphi}u_{|\partial\mathcal{M}}|^2_{L^2(\partial\mathcal{M})} + \tau|e^{\tau\varphi}\,\mathsf{D}'\,u_{|\partial\mathcal{M}}|^2_{L^2\Lambda^1(\partial\mathcal{M})} + \tau|e^{\tau\varphi}\partial_\nu u_{|\partial\mathcal{M}}|^2_{L^2(\partial\mathcal{M})}\big), \tag{5.3.6}$$

for $u \in \mathscr{C}^\infty(\mathcal{M})$, with $\operatorname{supp}(u) \subset V$ and $\tau \geq \tau_*$.

Note that $\overline{V}\cap\partial\mathcal{M}$ may be empty in the previous statement.

THEOREM 5.6. *Let $P = P_0 + R_1$ with $P_0 = -\Delta_g$ where R_1 is a first-order differential operator with bounded coefficients. Let V be a bounded open set in $\mathcal{M}$ and set $V_\partial = V\cap\partial\mathcal{M}$, and let $\varphi \in \mathscr{C}^\infty(\overline{V})$ and P have the sub-ellipticity property of Definition 5.1 in $\overline{V}$ and $\partial_\nu\varphi < 0$ in $\overline{V_\partial}$ in the case*

$V_\partial \neq \emptyset$. Then, there exist $\tau_* > 0$ and $C > 0$ such that

(5.3.7)

$$\tau^3 \|e^{\tau\varphi} u\|^2_{L^2(\mathcal{M})} + \tau \|e^{\tau\varphi} \operatorname{D} u\|^2_{L^2\Lambda^1(\mathcal{M})} + \tau |e^{\tau\varphi} \partial_\nu u_{|\partial\mathcal{M}}|^2_{L^2(\partial\mathcal{M})} \\ \leq C\big(\|e^{\tau\varphi} P u\|^2_{L^2(\mathcal{M})} + \tau^3 |e^{\tau\varphi} u_{|\partial\mathcal{M}}|^2_{L^2(\partial\mathcal{M})} + \tau |e^{\tau\varphi} \operatorname{D}' u_{|\partial\mathcal{M}}|^2_{L^2\Lambda^1(\partial\mathcal{M})}\big),$$

for $u \in \mathscr{C}^\infty(\mathcal{M})$, with $\operatorname{supp}(u) \subset V$ and $\tau \geq \tau_*$.

Using Lemmata 5.3 and 5.4, the proofs of Theorems 5.5 and 5.6 can be readily adapted from that of Theorem 5.2 and Theorems 3.28 and 3.29, using a partition of unity.

5.4. Global Estimations

We can adapt the results of Sections 3.6.1 and 3.6.2 to the case of a Riemannian manifold.

5.4.1. A Global Estimate with an Inner Observation. Let $(\tilde{\mathcal{M}}, g)$ be a smooth σ-compact Riemannian manifold (possibly not compact) with or without boundary and let $\mathcal{M}$ be a bounded connected open set of $\tilde{\mathcal{M}}$. Let Γ_0 be an open set of $\partial\mathcal{M}$ such that $\partial\mathcal{M}$ is smooth in a neighborhood of $\overline{\Gamma_0}$. Note that, whereas the boundary of $\tilde{\mathcal{M}}$ is smooth, we need not assume $\partial\mathcal{M}$ to be smooth everywhere. Let also ω_0 be an open subset of $\mathcal{M}$. As above P_0 denotes the Laplace–Beltrami operator on $\tilde{\mathcal{M}}$ and $P = P_0 + R_1$ where R_1 is a first-order differential operator with bounded coefficients.

DEFINITION 5.7. A real valued function $\varphi \in \mathscr{C}^\infty(\overline{\mathcal{M}})$ is said to be a global Carleman weight function on $\mathcal{M}$ adapted to Γ_0 and ω_0 if it satisfies

$$\partial_\nu \varphi_{|\partial\mathcal{M}}(m) < 0, \quad \text{for } m \in \overline{\Gamma_0},$$

and if the sub-ellipticity property of Definition 5.1 for (P, φ) is fulfilled in $\overline{\mathcal{M}} \setminus \omega_0$.

The construction of a global weight function φ with these properties can be done by adapting Proposition 3.31 of volume 1 and its proof to the manifold case.

THEOREM 5.8 (Global Carleman Estimate—Inner Observation). *Let $P = P_0 + R_1$ with $P_0 = -\Delta_g$ where R_1 is a first-order differential operator with bounded coefficients. Let also W_0 be a neighborhood of $\partial\mathcal{M} \setminus \Gamma_0$ in $\tilde{\mathcal{M}}$. Let ω be an open set of $\mathcal{M}$ such that $\omega_0 \Subset \omega$. Let $\varphi \in \mathscr{C}^\infty(\overline{\mathcal{M}})$ be a global weight function adapted to Γ_0 and ω_0 in the sense of Definition 5.7. Then, there exist $\tau_* > 0$ and $C \geq 0$ such that*

$$\tau^3 \|e^{\tau\varphi} u\|^2_{L^2(\mathcal{M})} + \tau \|e^{\tau\varphi} \operatorname{D} u\|^2_{L^2\Lambda^1(\mathcal{M})} + \tau |e^{\tau\varphi} \partial_\nu u_{|\Gamma_0}|^2_{L^2(\Gamma_0)} \\ \leq C\big(\|e^{\tau\varphi} P u\|^2_{L^2(\mathcal{M})} + \tau^3 \|e^{\tau\varphi} u\|^2_{L^2(\omega)} \\ + \tau^3 |e^{\tau\varphi} u_{|\Gamma_0}|^2_{L^2(\Gamma_0)} + \tau |e^{\tau\varphi} \operatorname{D}' u_{|\Gamma_0}|^2_{L^2\Lambda^1(\Gamma_0)}\big),$$

for $\tau \geq \tau_*$ and $u \in \mathscr{C}^\infty(\overline{\mathcal{M}})$ vanishing in $W_0 \cap \mathcal{M}$.

The case where $\mathcal{M}$ is itself a smooth connected compact Riemannian manifold and $\Gamma_0 = \partial\mathcal{M}$ yields the following corollary.

COROLLARY 5.9. *Let $\mathcal{M}$ be a smooth bounded connected Riemannian manifold. Let $P = P_0 + R_1$ with $P_0 = -\Delta_g$ where R_1 is a first-order differential operator with bounded coefficients. Let ω, ω_0 be two nonempty open subsets of $\mathcal{M}$ such that $\omega_0 \Subset \omega$.*

Let $\varphi \in \mathscr{C}^\infty(\mathcal{M})$ be a global weight function adapted to $\Gamma_0 = \partial\mathcal{M}$ and ω_0 in the sense of Definition 5.7. Then, there exist $\tau_* > 0$ and $C \geq 0$ such that

$$\begin{aligned}\tau^3\|e^{\tau\varphi}u\|^2_{L^2(\mathcal{M})} + \tau\|e^{\tau\varphi}\,\mathsf{D}\,u\|^2_{L^2\Lambda^1(\mathcal{M})} + \tau|e^{\tau\varphi}\partial_\nu u_{|\partial\mathcal{M}}|^2_{L^2(\partial\mathcal{M})} \\ \leq C\big(\|e^{\tau\varphi}Pu\|^2_{L^2(\mathcal{M})} + \tau^3\|e^{\tau\varphi}u\|^2_{L^2(\omega)} \\ + \tau^3|e^{\tau\varphi}u_{|\partial\mathcal{M}}|^2_{L^2(\partial\mathcal{M})} + \tau|e^{\tau\varphi}\,\mathsf{D}'\,u_{|\partial\mathcal{M}}|^2_{L^2\Lambda^1(\partial\mathcal{M})}\big),\end{aligned}$$

for $u \in \mathscr{C}^\infty(\overline{\mathcal{M}})$ and $\tau \geq \tau_*$.

5.4.2. A Global Estimate with a Boundary Observation. We consider the same setting as in Sect. 5.4.1. Here, We consider an open set Γ_0 of $\partial\mathcal{M}$ such that $\Gamma_0 \Subset \partial\mathcal{M}$.

DEFINITION 5.10. A real valued function $\varphi \in \mathscr{C}^\infty(\overline{\mathcal{M}})$ is said to be a global Carleman weight function on $\mathcal{M}$ adapted to Γ_0 if it satisfies

$$\partial_\nu\varphi_{|\Gamma_0}(m) < 0, \quad \text{for } m \in \overline{\Gamma_0},$$

and if the sub-ellipticity property of Definition 5.1 for (P, φ) is fulfilled in $\overline{\mathcal{M}}$.

THEOREM 5.11 (Global Carleman Estimate—Boundary Observation). *Let $P = P_0 + R_1$ with $P_0 = -\Delta_g$ where R_1 is a first-order differential operator with bounded coefficients. Let Γ_0 and Γ_{obs} be two nonempty open sets of $\partial\mathcal{M}$ such that $\Gamma_{\text{obs}} \setminus \overline{\Gamma_0} \neq \emptyset$, and such that $\partial\mathcal{M}$ is smooth in a neighborhood of $\overline{\Gamma_0} \cup \overline{\Gamma_{\text{obs}}}$. Let also W_0 be a neighborhood of $\partial\mathcal{M} \setminus (\Gamma_0 \cup \Gamma_{\text{obs}})$ in $\tilde{\mathcal{M}}$.*

Let $\varphi \in \mathscr{C}^\infty(\overline{\mathcal{M}})$ be a global weight function adapted to Γ_0 in the sense of Definition 5.10. Then, there exist $\tau_* > 0$ and $C \geq 0$ such that

$$\begin{aligned}\tau^3\|e^{\tau\varphi}u\|^2_{L^2(\mathcal{M})} + \tau\|e^{\tau\varphi}\,\mathsf{D}\,u\|^2_{L^2\Lambda^1(\mathcal{M})} + \tau|e^{\tau\varphi}\partial_\nu u_{|\Gamma_0}|^2_{L^2(\Gamma_0)} \\ \leq C\big(\|e^{\tau\varphi}Pu\|^2_{L^2(\mathcal{M})} + \tau|e^{\tau\varphi}\partial_\nu u_{|\Gamma_{\text{obs}}}|^2_{L^2(\Gamma_{\text{obs}})} \\ + \tau^3|e^{\tau\varphi}u_{|\partial\mathcal{M}}|^2_{L^2(\partial\mathcal{M})} + \tau|e^{\tau\varphi}\,\mathsf{D}'\,u_{|\partial\mathcal{M}}|^2_{L^2\Lambda^1(\partial\mathcal{M})}\big),\end{aligned}$$

for $\tau \geq \tau_*$ and $u \in \mathscr{C}^\infty(\overline{\mathcal{M}})$ vanishing in $W_0 \cap \mathcal{M}$.

For the construction of the global weight function φ one can pick an open set Γ_1 of $\partial\mathcal{M}$ such that $\Gamma_1 \Subset \Gamma_{\text{obs}} \setminus \overline{\Gamma_0}$ and then adapt Proposition 3.39 of volume 1 and its proof to the manifold case.

The case where $\mathcal{M}$ is itself a smooth connected compact Riemannian manifold yields the following corollary.

COROLLARY 5.12. *Let $P = P_0 + R_1$ with $P_0 = -\Delta_g$ where R_1 is a first-order differential operator with bounded coefficients. Let Γ_0, Γ_{obs} be two nonempty open subsets of $\partial\mathcal{M}$ such that $\Gamma_0 \cup \Gamma_{\text{obs}} = \partial\mathcal{M}$ and $\Gamma_{\text{obs}} \setminus \overline{\Gamma_0} \neq \emptyset$. Let $\varphi \in \mathscr{C}^\infty(\mathcal{M})$ be a global weight function adapted to Γ_0 in the sense of Definition 5.10. Then, there exist $\tau_* > 0$ and $C \geq 0$ such that*

$$\begin{aligned}\tau^3 \|e^{\tau\varphi} u\|^2_{L^2(\mathcal{M})} + \tau \|e^{\tau\varphi} \mathsf{D}\, u\|^2_{L^2\Lambda^1(\mathcal{M})} + \tau |e^{\tau\varphi} \partial_\nu u_{|\partial\mathcal{M}}|^2_{L^2(\partial\mathcal{M})} \\ \leq C\big(\|e^{\tau\varphi} P u\|^2_{L^2(\Omega)} + \tau |e^{\tau\varphi} \partial_\nu u_{|\Gamma_{\text{obs}}}|^2_{L^2(\Gamma_{\text{obs}})} \\ + \tau^3 |e^{\tau\varphi} u_{|\partial\mathcal{M}}|^2_{L^2(\partial\mathcal{M})} + \tau |e^{\tau\varphi} \mathsf{D}'\, u_{|\partial\mathcal{M}}|^2_{L^2\Lambda^1(\partial\mathcal{M})}\big),\end{aligned}$$

for $u \in \mathscr{C}^\infty(\overline{\mathcal{M}})$ and $\tau \geq \tau_*$.

CHAPTER 6

Pseudo-Differential Operators on a Half-Space

Contents

In Chapter 2 of Volume 1 we presented the basic results that allowed us to derive estimates away from boundaries and at boundaries in Chapter 3 also in Volume 1, in particular in the case of Dirichlet boundary conditions. In the present chapter, we wish to present additional material on pseudo-differential operators with a large parameter, focusing on the case of operators defined on the half-space $\mathbb{R}^d_+$. This will allow us to introduce the basic tools for the derivation of Carleman estimates at a boundary in the case of general boundary conditions, namely the Lopatinskiĭ–Šapiro condition, that we shall present in Chap. 8.

J. Le Rousseau et al., *Elliptic Carleman Estimates and Applications to Stabilization and Controllability, Volume II*, PNLDE Subseries in Control 98, https://doi.org/10.1007/978-3-030-88670-7_6

6.1. More on Tangential Symbols and Operators

As in Chapter 2 of Volume 1, we consider symbols that depend on a parameter $\tau \geq 1$ meant to be large. For $x \in \mathbb{R}^d$ or $\overline{\mathbb{R}^d_+}$ and $\xi \in \mathbb{R}^d$ we shall often write $\varrho = (x, \xi, \tau)$. Similarly, for $\xi' \in \mathbb{R}^{d-1}$ we shall write $\varrho' = (x, \xi', \tau)$.

As introduced in Section 2.10, for $X = \mathbb{R}^d$ or $X = \overline{\mathbb{R}^d_+}$, we recall that $a(\varrho') \in S^m_{\mathsf{T},\tau}(X \times \mathbb{R}^{d-1})$ if $a(\varrho') \in \mathscr{C}^\infty(X \times \mathbb{R}^{d-1})$ and

$$|\partial_x^\alpha \partial_{\xi'}^\beta a(x, \xi', \tau)| \leq C_{\alpha,\beta} \lambda_{\mathsf{T},\tau}^{m-|\beta|}, \quad x \in X, \ \xi' \in \mathbb{R}^{d-1}, \ \tau \in [1, +\infty), \tag{6.1.1}$$

where $\lambda_{\mathsf{T},\tau} = |(\xi', \tau)| = \big(|\xi'|^2 + \tau^2\big)^{\frac{1}{2}}$. Note that we make clear that τ is a parameter rather than a variable in the notation of the symbol space. Indeed smoothness with respect to τ is not an issue here.

More generally, if $\mathscr{U}$ is a conic open set of $X \times \mathbb{R}^{d-1} \times [1, +\infty)$ we say that $a \in S^m_{\mathsf{T},\tau}$ microlocally in $\mathscr{U}$ if (6.1.1) holds in for $\varrho' = (x, \xi', \tau) \in \mathscr{U}$.

6.1.1. Additional Classes of Symbols.

We consider symbols that behave polynomially in the ξ_d variable, as already presented in Remark 2.46-(2) in Chapter 2.

Definition 6.1. Let $a(\varrho) \in \mathscr{C}^\infty(X \times \mathbb{R}^d)$, with τ as a parameter in $[1, +\infty)$, and $m \in \mathbb{N}$ and $r \in \mathbb{R}$. We say that $a \in S^{m,r}_\tau(X \times \mathbb{R}^d)$ if

$$a(\varrho) = \sum_{j=0}^{m} a_j(\varrho') \xi_d^j, \quad a_j \in S^{m-j+r}_{\mathsf{T},\tau}(X \times \mathbb{R}^{d-1}),$$

for $x \in X$, $\xi \in \mathbb{R}^d$, $\tau \in [1, +\infty)$, and $\xi_d \in \mathbb{R}$.

We also simply write $a \in S^{m,r}_\tau$. If $\mathscr{U}$ is conic open set of $X \times \mathbb{R}^{d-1} \times [1, +\infty)$ we say that $a \in S^{m,r}_\tau$ microlocally for $\varrho' \in \mathscr{U}$ if each a_j is in $S^{m-j+r}_{\mathsf{T},\tau}$ microlocally in $\mathscr{U}$, $j = 0, \ldots, m$.

Note that we have

$$S^{m,r}_\tau \subset S^{m+m',r-m'}_\tau, \qquad m, m' \in \mathbb{N}, \ r \in \mathbb{R}. \tag{6.1.2}$$

We call the principal symbol of a the symbol

$$\sigma(a)(\varrho) = \sum_{j=0}^{m} \sigma(a_j)(\varrho') \xi_d^j,$$

which is a representative of the class of a in $S^{m,r}_\tau / S^{m,r-1}_\tau$.

Note that $S^{m,r}_\tau \not\subset S^{m+r}_\tau$. For example, consider $a(x, \xi, \tau) = \lambda_{\mathsf{T},\tau} \xi_d$ for $\lambda_{\mathsf{T},\tau} \geq 1$. We have $a \in S^{1,1}_\tau \subset S^{2,0}_\tau$ and yet $a \notin S^2_\tau$. In fact observe that differentiating with respect to ξ' yields

$$|\partial_{\xi'}^\alpha a(x, \xi, \tau)| \leq C_\alpha \lambda_{\mathsf{T},\tau}^{1-|\alpha|} |\xi_d|.$$

An estimate of the form of (2.2.1) is, however, not achieved for $|\alpha| \geq 2$.

In Definition 2.6, polyhomogeneous symbols were introduced: we recall that $a \sim \sum_{j \in \mathbb{N}} a_{m-j} \in S^m_{\tau,\mathrm{ph}}$ if $a_{m-j} \in S^{m-j}_\tau$ is homogeneous of degree $m - j$

with respect to (ξ,τ) for each $j\in\mathbb{N}$. Additionally, we define tangential poly-homogeneous symbols. They are characterized by an asymptotic expansion where each term is positively homogeneous (ξ',τ).

Definition 6.2. We shall say that $a\in S^m_{\mathsf{T},\tau,\mathrm{ph}}(X\times\mathbb{R}^{d-1})$ or simply $S^m_{\mathsf{T},\tau,\mathrm{ph}}$ if there exists $a^{(j)}\in S^{m-j}_{\mathsf{T},\tau}$, homogeneous of degree $m-j$ in (ξ',τ) for $|(\xi',\tau)|\geq r_0$, with $r_0\geq 0$, such that

$$a\sim\sum_{j\geq 0}a^{(j)},\quad\text{in the sense that}\quad a-\sum_{j=0}^{N}a^{(j)}\in S^{m-N-1}_{\mathsf{T},\tau}. \tag{6.1.3}$$

A representative of the principal symbol is then given by the first term in the expansion. We denote it by $\sigma(a)$. We have

$$S^m_{\tau,\mathrm{ph}}\subset S^m_{\tau},\quad S^m_{\mathsf{T},\tau,\mathrm{ph}}\subset S^m_{\mathsf{T},\tau}.$$

Then, for $m\in\mathbb{N}$ and $r\in\mathbb{R}$, we shall say that $a(\varrho)\in S^{m,r}_{\tau,\mathrm{ph}}(X\times\mathbb{R}^d)$ or simply $S^{m,r}_{\tau,\mathrm{ph}}$, if

$$a(\varrho)=\sum_{j=0}^{m}a_j(\varrho')\xi_d^j,\quad\text{with } a_j\in S^{m-j+r}_{\mathsf{T},\tau,\mathrm{ph}},\quad \varrho=(\varrho',\xi_d).$$

A representative of the principal symbol is given by $\sum_{j=0}^m\sigma(a_j)(\varrho')\xi_d^j$ and is homogeneous of degree m in (ξ,τ). We have

$$S^{m,r}_{\tau,\mathrm{ph}}\subset S^{m,r}_{\tau}.$$

6.1.2. Corresponding Classes of Operators. For $m\in\mathbb{R}$, as $S^m_{\mathsf{T},\tau,\mathrm{ph}}\subset S^m_{\mathsf{T},\tau}$, the operator $\mathrm{Op}_{\mathsf{T}}(a)=a(x,D',\tau)$ is given in Definition 2.42 of Volume 1 and we write $\mathrm{Op}_{\mathsf{T}}(a)\in\Psi^m_{\mathsf{T},\tau,\mathrm{ph}}(X)$ or simply $\Psi^m_{\mathsf{T},\tau,\mathrm{ph}}$.

The principal symbol of A is $\sigma(A)=\sigma(a)$ in $S^m_{\mathsf{T},\tau,\mathrm{ph}}/S^{m-1}_{\mathsf{T},\tau,\mathrm{ph}}$.

Definition 6.3. For $m\in\mathbb{N}$, $r\in\mathbb{R}$, and $a\in S^{m,r}_{\tau}$ with

$$a(\varrho)=\sum_{j=0}^{m}a_j(\varrho')\xi_d^j,\quad a_j\in S^{m-j+r}_{\mathsf{T},\tau}(X\times\mathbb{R}^{d-1}),$$

we set

$$a(x,D,\tau)=\mathrm{Op}(a)=\sum_{j=0}^{m}a_j(x,D',\tau)D_d^j,$$

and we write $A=\mathrm{Op}(a)\in\Psi^{m,r}_{\tau}(X)$ or simply $\Psi^{m,r}_{\tau}$. The principal symbol of A is $\sigma(A)=\sigma(a)$ in $S^{m,r}_{\tau}/S^{m,r-1}_{\tau}$.

We denote by $\Psi^{m,r}_{\tau,\mathrm{ph}}(X)$ the subclass of these operators associated with symbols in $S^{m,r}_{\tau,\mathrm{ph}}$. The principal symbol of A is then $\sigma(A)=\sigma(a)$ in $S^{m,r}_{\tau,\mathrm{ph}}/S^{m,r-1}_{\tau,\mathrm{ph}}$.

We provide a notion of formal adjoint.

DEFINITION 6.4 (Formal Adjoint). Let $b \in S_{\mathcal{T}}^{m,r}$, with

$$\mathrm{Op}(b) = \sum_{j=0}^{m} \mathrm{Op}_{\mathsf{T}}(b_j) D_d^j, \qquad b_j \in S_{\mathsf{T},\tau}^{m+r-j}.$$

We set

$$\mathrm{Op}(b)^* = \sum_{j=0}^{m} D_d^j \mathrm{Op}_{\mathsf{T}}(b_j)^*.$$

In other words, in this definition we ignore the possible occurrence of boundary terms when performing the operator transposition. Note that here $\mathrm{Op}_{\mathsf{T}}(b_j)^* \in \Psi_{\mathsf{T},\tau}^{m+r-j}$. Observe also that $\mathrm{Op}_{\mathsf{T}}(b_j)^* \in \Psi_{\mathsf{T},\tau,\mathrm{ph}}^{m+r-j}$ if $b_j \in S_{\mathsf{T},\tau,\mathrm{ph}}^{m+r-j}$.

For $a \in S_{\mathsf{T},\tau}^m$ (resp. $S_{\mathsf{T},\tau,\mathrm{ph}}^m$) we have $[D_d, \mathrm{Op}_{\mathsf{T}}(a)] = \mathrm{Op}_{\mathsf{T}}(D_d a) \in \Psi_{\mathsf{T},\tau}^m$ (resp. $\Psi_{\mathsf{T},\tau,\mathrm{ph}}^m$) and more generally, for $j \geq 1$, we have

$$[D_d^j, \mathrm{Op}(a)] = \sum_{k=0}^{j-1} \mathrm{Op}_{\mathsf{T}}(\alpha_k) D_d^k, \qquad \alpha_k \in S_{\mathsf{T},\tau}^m \quad (\text{resp. } S_{\mathsf{T},\tau,\mathrm{ph}}^m),$$

where the symbols α_k involve various derivatives of a in the x_d-direction. As an application we see that if we consider $a_j \in S_{\mathsf{T},\tau}^{m-j+r}$ (resp. $S_{\mathsf{T},\tau,\mathrm{ph}}^{m-j+r}$), then we have

$$\sum_{j=0}^{m} D_d^j \mathrm{Op}_{\mathsf{T}}(a_j) = \sum_{j=0}^{m} \mathrm{Op}_{\mathsf{T}}(\tilde{a}_j) D_d^j,$$

where $\tilde{a}_j \in S_{\mathsf{T},\tau}^{m-j+r}$ (resp. $S_{\mathsf{T},\tau,\mathrm{ph}}^{m-j+r}$) and its principal part satisfies $\sigma(\tilde{a}_j) \equiv a_j$ in $S_{\mathsf{T},\tau}^{m-j+r}/S_{\mathsf{T},\tau}^{m-j+r-1}$ (resp. $S_{\mathsf{T},\tau,\mathrm{ph}}^{m-j+r}/S_{\mathsf{T},\tau,\mathrm{ph}}^{m-j+r-1}$). Hence

$$\sigma\Big(\sum_{j=0}^{m} D_d^j \mathrm{Op}_{\mathsf{T}}(a_j)\Big) = \sum_{j=0}^{m} a_j(\varrho')\xi_d^j \mod S_{\mathcal{T}}^{m,r-1} \quad (\text{resp. } S_{\tau,\mathrm{ph}}^{m,r-1}).$$

From the calculus rules given in Section 2.10 of Volume 1 for tangential operators and the above observations we have the following results on principal symbols.

PROPOSITION 6.5. *Let* $a \in S_{\mathcal{T}}^{m,r}$ *(resp.* $S_{\tau,\mathrm{ph}}^{m,r}$*) and* $b \in S_{\mathcal{T}}^{m',r'}$ *(resp.* $S_{\tau,\mathrm{ph}}^{m',r'}$*) with*

$$a(\varrho) = \sum_{j=0}^{m} a_j(\varrho')\xi_d^j, \quad b(\varrho) = \sum_{j=0}^{m'} b_j(\varrho')\xi_d^j.$$

(1) *We have* $\mathrm{Op}(a)^* \in \Psi_{\mathcal{T}}^{m,r}$ *(resp.* $\Psi_{\tau,\mathrm{ph}}^{m,r}$*) and*

$$\sigma\big(\mathrm{Op}(a)^*\big)(\varrho) \equiv \sum_{j=0}^{m} \overline{a}_j(\varrho')\xi_d^j \in S_{\mathcal{T}}^{m,r}/S_{\mathcal{T}}^{m,r-1} \quad (\textit{resp. } S_{\tau,\mathrm{ph}}^{m,r}/S_{\tau,\mathrm{ph}}^{m,r-1}).$$

Moreover, we have $\mathrm{Op}(a)^* - \mathrm{Op}(\overline{a}) \in \Psi_{\mathcal{T}}^{m,r-1}$ *(resp.* $\Psi_{\tau,\mathrm{ph}}^{m,r-1}$*).*

(2) *We have* $\mathrm{Op}(a)\mathrm{Op}(b) \in \Psi_\tau^{m+m',r+r'}$ *(resp.* $\Psi_{\tau,\mathrm{ph}}^{m+m',r+r'}$*) and*

$$\sigma\big(\mathrm{Op}(a)\mathrm{Op}(b)\big) \equiv \sum_{\substack{0\le j\le m\\ 0\le k\le m'}} (a_j b_k)(\varrho')\xi_d^{j+k} \in S_\tau^{m+m',r+r'}/S_\tau^{m+m',r+r'-1}$$

$$\text{(resp. } S_{\tau,\mathrm{ph}}^{m+m',r+r'}/S_{\tau,\mathrm{ph}}^{m+m',r+r'-1}).$$

We have

$$\mathrm{Op}(a)\mathrm{Op}(b)u - \mathrm{Op}(ab)u \in \Psi_\tau^{m+m',r+r'-1}(\text{resp.}\Psi_{\tau,\mathrm{ph}}^{m+m',r+r'-1}).$$

6.2. Adapted Sobolev Norms and Continuity Results

As in Chapters 2 and 3 of Volume 1, for u and v defined in $\mathbb{R}_+^d = \{x_d > 0\}$, we often use the notation

$$(u,v)_+ = \int_{\mathbb{R}_+^d} u(x)\overline{v(x)}dx, \qquad \|u\|_+ = \|u\|_{L^2(\mathbb{R}_+^d)}.$$

For u and v defined on $\{x_d = 0\}$ we use

$$(u,v)_\partial = \int_{\mathbb{R}^{d-1}} u(x')\bar{v}(x')dx', \qquad |u|_\partial = |u|_{L^2(\mathbb{R}^{d-1})},$$

We recall that $\Lambda_{\mathsf{T},\tau}^s := \mathrm{Op}(\lambda_{\mathsf{T},\tau}^s)$ and we define the following norm for functions in $\mathbb{R}_+^d$, for $m \in \mathbb{N}$ and $s \in \mathbb{R}$,

$$\|u\|_{\tau,m,s}^2 = \|\Lambda_{\mathsf{T},\tau}^s u\|_{\tau,m}^2, \quad u \in \overline{\mathscr{S}}(\mathbb{R}_+^d),$$

with

$$\|v\|_{\tau,m}^2 = \|v\|_{\tau,m,0}^2 = \tau^{2m}\|v\|_+^2 + \|v\|_{H^m(\mathbb{R}_+^d)}^2$$
$$\asymp \sum_{j=0}^m \tau^{2j}\|v\|_{H^{m-j}(\mathbb{R}_+^d)}^2, \quad v \in \overline{\mathscr{S}}(\mathbb{R}_+^d).$$

From pseudo-differential symbol calculus we obtain the following inequality.

PROPOSITION 6.6. *If* $a \in S_\tau^{m,r}$, *with* $m \in \mathbb{N}$ *and* $r \in \mathbb{R}$, *then for* $m' \in \mathbb{N}$ *and* $r' \in \mathbb{R}$ *there exists* $C > 0$ *such that*

$$\|\mathrm{Op}(a)u\|_{\tau,m',r'} \le C\|u\|_{\tau,m+m',r+r'}, \qquad u \in \overline{\mathscr{S}}(\mathbb{R}_+^d).$$

A consequence of this results is the following property.

COROLLARY 6.7. *Let* $m, m' \in \mathbb{N}$ *and* $r \in \mathbb{R}$. *There exists* $C > 0$ *such that*

$$\|u\|_{\tau,m,r} \le C\|u\|_{\tau,m+m',r-m'}, \qquad u \in \overline{\mathscr{S}}(\mathbb{R}_+^d).$$

PROOF. We write $u = \Lambda_{\mathsf{T},\tau}^{m'} v$ with $v = \Lambda_{\mathsf{T},\tau}^{-m'} u$. As we have by (6.1.2)

$$\lambda_{\mathsf{T},\tau}^{m'} \in S_{\mathsf{T},\tau}^{m'} = S_{\tau}^{0,m'} \subset S_{\tau}^{m',0},$$

we obtain

$$\begin{aligned}\|u\|_{\tau,m,r} &= \|\Lambda_{\mathsf{T},\tau}^{m'} v\|_{\tau,m,r} \lesssim \|v\|_{\tau,m+m',r} = \|\Lambda_{\mathsf{T},\tau}^{-m'} u\|_{\tau,m+m',r} \\ &= \|u\|_{\tau,m+m',r-m'},\end{aligned}$$

by Proposition 6.6. ■

The counterpart of Remark 2.27 in Chapter 2.

REMARK 6.8. We observe that we have, for some $C > 0$,

$$\|u\|_{\tau,m,s} \leq C\tau^{-\ell} \|u\|_{\tau,m,s+\ell}, \qquad u \in \overline{\mathscr{S}}(\mathbb{R}^d_+),$$

for $m \in \mathbb{N}$ and $s \in \mathbb{R}$ and $\ell \geq 0$. This implies that $\|u\|_{\tau,m,s} \ll \|u\|_{\tau,m,s+\ell}$ for τ sufficiently large.

For a sufficiently smooth function u defined in $\overline{\mathbb{R}^d_+}$ we set, for $m \in \mathbb{N}$,

$$\operatorname{tr}^m(u) = (u_{|x_d=0^+}, D_d u_{|x_d=0^+}, \dots, D_d^m u_{|x_d=0^+})$$

on $\{x_d = 0\}$ and we define the following norm

$$|\operatorname{tr}^m(u)|^2_{\tau,m,s} = \sum_{j=0}^{m} |\Lambda_{\mathsf{T},\tau}^{m+s-j} D_d^j u_{|x_d=0^+}|^2_{\partial}. \tag{6.2.1}$$

We see that without any ambiguity we may write $|\operatorname{tr}(u)|^2_{\tau,m,s}$ in place of $|\operatorname{tr}^m(u)|^2_{\tau,m,s}$.

Let $u \in \overline{\mathscr{S}}(\mathbb{R}^d_+)$ and set

$$\mathscr{S}_u = \{v \in \mathscr{S}(\mathbb{R}^d);\ u = v_{|\mathbb{R}^d_+}\}.$$

We define, for $r \in \mathbb{R}$,

$$\|u\|_{\tau,r} = \inf_{v \in \mathscr{S}_u} \|v\|_{\tau,r}, \quad \text{with } \|v\|_{\tau,r} = \|\Lambda_\tau^r v\|_{L^2(\mathbb{R}^d)}. \tag{6.2.2}$$

For $r \in \mathbb{N}$, this norm can be proven equivalent to the norm $\|u\|_{\tau,r} = \|u\|_{\tau,r,0}$ defined above, by using a properly designed extension operator [2, Theorem 5.21].

We have the following trace inequality.

PROPOSITION 6.9 (Trace Inequality). *Let $s > 0$. There exists $C > 0$ such that*

$$|u_{|x_d=0^+}|_{\tau,s} \leq C \|u\|_{\tau,s+1/2}, \qquad u \in \overline{\mathscr{S}}(\mathbb{R}^d_+).$$

PROOF. For $v \in \mathscr{S}_u$, denote by $\mathscr{F}v(\xi)$ the Fourier transform of v with respect to x and by $\mathscr{F}'v(\xi', x_d)$ its Fourier transform with respect to x'. We have

$$
\begin{aligned}
|u_{|x_d=0^+}|^2_{\tau,s} &= |v_{|x_d=0}|^2_{\tau,s} \\
&= (2\pi)^{1-d}|\lambda^s_{\mathsf{T},\tau}\mathscr{F}'v(\xi',0)|^2_{\mathbb{R}^{d-1}} = (2\pi)^{-d} \int_{\mathbb{R}^{d-1}} \lambda^{2s}_{\mathsf{T},\tau}|\int_{\mathbb{R}} \mathscr{F}v(\xi)d\xi_d|^2 d\xi' \\
&\leq (2\pi)^{-d} \int_{\mathbb{R}^{d-1}} \lambda^{2s}_{\mathsf{T},\tau}(\int_{\mathbb{R}} \lambda_\tau^{-2s-1}d\xi_d)(\int_{\mathbb{R}} \lambda_\tau^{2s+1}|\mathscr{F}v(\xi)|^2 d\xi_d)d\xi',
\end{aligned}
$$

by the Cauchy–Schwarz inequality. We compute

$$
\lambda^{2s}_{\mathsf{T},\tau}\int_{\mathbb{R}} \lambda_\tau^{-2s-1}d\xi_d = \lambda^{2s}_{\mathsf{T},\tau}\int_{\mathbb{R}}(\lambda^2_{\mathsf{T},\tau}+\xi_d^2)^{-s-1/2}d\xi_d = \int_{\mathbb{R}}(1+\sigma^2)^{-s-1/2}d\sigma < \infty,
$$

which gives $|u_{|x_d=0^+}|_{\tau,s} \lesssim \|v\|_{\tau,s+1/2}$. We conclude with (6.2.2). ∎

COROLLARY 6.10 (Trace Inequality). *Let $m \in \mathbb{N}$ and $s \in \mathbb{R}$. For some $C > 0$, we have*

$$
|\operatorname{tr}(u)|_{\tau,m,s} \leq C\|u\|_{\tau,m+1,s-1/2}, \quad u \in \overline{\mathscr{S}}(\mathbb{R}^d_+).
$$

PROOF. With (6.2.1) we write

$$
|\operatorname{tr}(u)|^2_{\tau,m,s} = \sum_{j=0}^m |\Lambda^{m-j+s}_{\mathsf{T},\tau} D^j_d u_{|x_d=0^+}|^2_\partial = \sum_{j=0}^m |\Lambda^{m-j+s-1/2}_{\mathsf{T},\tau} D^j_d u_{|x_d=0^+}|^2_{\tau,1/2}.
$$

From the trace inequality of Proposition 6.9 we obtain

$$
\begin{aligned}
|\operatorname{tr}(u)|_{\tau,m,s} &\lesssim \sum_{j=0}^m \|\Lambda^{m-j+s-1/2}_{\mathsf{T},\tau} D^j_d u\|_{\tau,1} \lesssim \sum_{j=0}^{m+1} \|\Lambda^{m+1-j+s-1/2}_{\mathsf{T},\tau} D^j_d u\|_+ \\
&\lesssim \|u\|_{\tau,m+1,s-1/2}.
\end{aligned}
$$

∎

6.2.1. A Microlocal Norm Equivalence. One has $\tau \in S^1_{\mathsf{T},\tau}$ yielding, for $s \in \mathbb{R}$, $\|\tau u\|_{0,s} \lesssim \|u\|_{0,s+1}$. However, in a microlocal region where $\tau \approx \lambda_{\mathsf{T},\tau}$, that is, if $|\xi'| \lesssim \tau$ one can expect a norm equivalence. The following proposition states that it is indeed the case, yet up to a remainder term.

PROPOSITION 6.11. *Let U be a bounded open set of $\mathbb{R}^d_+$ and $\mathscr{U}$ be conic open sets of $U \times \mathbb{R}^{d-1} \times \mathbb{R}_+$ with $C_0 > 0$ such that $\lambda_{\mathsf{T},\tau} \leq C_0\tau$ in $\mathscr{U}$. Let $s, s', s'' \in \mathbb{R}$ with $s = s' + s''$. There exist $C, C' > 0$ such that for $\chi \in S^0_{\mathsf{T},\tau}$, homogeneous of degree zero with* $\operatorname{supp}(\chi) \subset \mathscr{U}$, *and $N \in \mathbb{N}$, there exists C_N such that*

$$
\begin{aligned}
&C\tau^{s'}\|\operatorname{Op}_{\mathsf{T}}(\chi)u\|_{\tau,0,s''} - C_N\|u\|_{\tau,0,-N} \\
&\qquad \leq \|\operatorname{Op}_{\mathsf{T}}(\chi)u\|_{\tau,0,s} \leq C'\tau^{s'}\|\operatorname{Op}_{\mathsf{T}}(\chi)u\|_{\tau,0,s''} + C_N\|u\|_{\tau,0,-N},
\end{aligned}
$$

for all $u \in \overline{\mathscr{S}}(\mathbb{R}^d_+)$.

The result of Proposition 6.11 can be generalized.

PROPOSITION 6.12. *Let U be a bounded open set of $\mathbb{R}^d_+$ and $\mathscr{U}$ be conic open sets of $U \times \mathbb{R}^{d-1} \times \mathbb{R}_+$ with $C_0 > 0$ such that $\lambda_{\mathsf{T},\tau} \leq C_0 \tau$ in $\mathscr{U}$. Let $L \in \Psi^{m,r}$, with $m \in \mathbb{N}$ and $r \in \mathbb{R}$, and let $s, s', s'' \in \mathbb{R}$ with $s = s' + s''$. There exist $C, C' > 0$ such that for $\chi \in S^0_{\mathsf{T},\tau}$, homogeneous of degree zero with* $\operatorname{supp}(\chi) \subset \mathscr{U}$, *and $N \in \mathbb{N}$, there exists C_N such that*

$$\begin{aligned} C\tau^{s'} \|L\operatorname{Op}_{\mathsf{T}}(\chi)v\|_{\tau,0,s''} - C_N \|v\|_{\tau,m,-N} \\ \leq \|L\operatorname{Op}_{\mathsf{T}}(\chi)v\|_{\tau,0,s} \leq C'\tau^{s'} \|L\operatorname{Op}_{\mathsf{T}}(\chi)v\|_{\tau,0,s''} + C_N \|v\|_{\tau,m,-N}, \end{aligned}$$

for all $v \in \overline{\mathscr{S}}(\mathbb{R}^d_+)$.

PROOF OF PROPOSITION 6.12. Note that $\mathbb{S}_{\overline{\mathscr{U}}}$ is compact. Thus, there exists $\mathscr{V}$ a conic open set of $U \times \mathbb{R}^{d-1} \times \mathbb{R}_+$ such that $\overline{\mathscr{U}} \subset \mathscr{V}$ and $\lambda_{\mathsf{T},\tau} \leq C_1 \tau$, for some $C_1 > 0$ by homogeneity.

Let $\hat{\chi} \in S^0_{\mathsf{T},\tau}$, homogeneous of degree zero, be such that $\operatorname{supp}(\hat{\chi}) \subset \mathscr{V}$ and $\hat{\chi} = 1$ on $\mathscr{U}$. One has

$$\Lambda^s_{\mathsf{T},\tau} L\operatorname{Op}_{\mathsf{T}}(\chi) = \operatorname{Op}_{\mathsf{T}}(\hat{\chi}) \Lambda^s_{\mathsf{T},\tau} L\operatorname{Op}_{\mathsf{T}}(\chi) \mod \Psi^{m,-\infty}_{\tau},$$

yielding

$$\|L\operatorname{Op}_{\mathsf{T}}(\chi)v\|_{\tau,0,s} \leq \|\operatorname{Op}_{\mathsf{T}}(\hat{\chi}) \Lambda^s_{\mathsf{T},\tau} L\operatorname{Op}_{\mathsf{T}}(\chi)v\|_+ + C_N \|v\|_{\tau,m,-N},$$

for some $C_N > 0$.

We then observe that $\tau^{-s'} \operatorname{Op}_{\mathsf{T}}(\hat{\chi}) \Lambda^s_{\mathsf{T},\tau} \in \Psi^{s''}_{\mathsf{T},\tau}$ giving

$$\|\operatorname{Op}_{\mathsf{T}}(\hat{\chi}) \Lambda^s_{\mathsf{T},\tau} L\operatorname{Op}_{\mathsf{T}}(\chi)v\|_+ \lesssim \tau^{s'} \|L\operatorname{Op}_{\mathsf{T}}(\chi)v\|_{\tau,0,s''}.$$

We have thus obtained

$$\|L\operatorname{Op}_{\mathsf{T}}(\chi)v\|_{\tau,0,s} \leq C\tau^{s'} \|L\operatorname{Op}_{\mathsf{T}}(\chi)v\|_{\tau,0,s''} + C_N \|v\|_{\tau,m,-N},$$

for some $C > 0$ independent of N and χ. With this inequality, exchanging the rôles of s'' and s and changing s' into $-s'$ one obtains the second sought inequality. ■

6.3. Quadratic Forms in a Half-Space and Gårding Inequality

DEFINITION 6.13. Let $u \in \overline{\mathscr{S}}(\mathbb{R}^d_+)$. We say that

$$Q(u) = \sum_{s=1}^{N} (A^s u, B^s u)_+, \qquad A^s = \operatorname{Op}(a^s),\ B^s = \operatorname{Op}(b^s), \tag{6.3.1}$$

is an interior quadratic form of type (m, σ) with smooth coefficients, if for each $s = 1, \dots N$, we have $a^s(\varrho) \in S^{m,\sigma'}_{\tau,\mathrm{ph}}(\mathbb{R}^d_+ \times \mathbb{R}^d)$ and $b^s(\varrho) \in S^{m,\sigma''}_{\tau,\mathrm{ph}}(\mathbb{R}^d_+ \times \mathbb{R}^d)$, with $\sigma' + \sigma'' = 2\sigma$, $\varrho = (x, \xi, \tau)$.

The principal symbol of the quadratic form Q is defined as the class of

$$q(\varrho) = \sum_{s=1}^{N} a^s(\varrho) \overline{b^s}(\varrho) \tag{6.3.2}$$

in $S^{2m,2\sigma}_{\tau,\mathrm{ph}}(\mathbb{R}^d_+ \times \mathbb{R}^d) / S^{2m,2\sigma-1}_{\tau,\mathrm{ph}}(\mathbb{R}^d_+ \times \mathbb{R}^d)$.

REMARK 6.14. Note that σ' and σ'' can vary with $s \in \{1, \dots, N\}$. Their sum yet remains constant equal to 2σ. In what follows we shall not write this dependency explicitly for concision.

Observe that several quadratic forms may share the same principal symbol. As an example, in one dimension, for $N = 1$ we can choose $A = D_d^2 \in \Psi_\tau^{2,0}$ and $B = \Lambda_{\mathsf{T},\tau}^2 \in \Psi_\tau^{0,2} \subset \Psi_\tau^{2,0}$ yielding to $\lambda_{\mathsf{T},\tau}^2 \xi_d^2$ for principal symbol. The choice $A = B = \Lambda_{\mathsf{T},\tau} D_d \in \Psi_\tau^{1,1} \subset \Psi_\tau^{2,0}$ leads to the same principal symbol.

In fact, if $u \in \mathscr{C}_c^\infty(\mathbb{R}_+^d)$, then

$$Q(u) = \sum_{s=1}^{N} ((B^s)^* \circ A^s u, u)_+.$$

The principal symbol of Q thus coincides with the principal symbol of $\sum_{s=1}^N (B^s)^* \circ A^s$. Note that considering test functions with nonvanishing traces at the boundary $x_d = 0^+$ will naturally generate boundary terms when performing such operator transpositions. Such questions will be dealt with below.

For $s = 1, \dots, N$, we have

$$a^s(\varrho) = \sum_{j=0}^{m} a_j^s(\varrho')\xi_d^j, \quad b^s(\varrho) = \sum_{j=0}^{m} b_j^s(\varrho')\xi_d^j, \quad \varrho = (\varrho', \xi_d),\ \varrho' = (x, \xi', \tau),$$

with $a_j^s \in S_{\tau,\text{ph}}^{m-j+\sigma'}$ and $b_j^s \in S_{\tau,\text{ph}}^{m-j+\sigma''}$, and

$$A^s = \text{Op}(a^s) = \sum_{j=0}^{m} A_j^s D_d^j, \qquad A_j^s = \text{Op}_\mathsf{T}(a_j^s),$$
$$B^s = \text{Op}(b^s) = \sum_{j=0}^{m} B_j^s D_d^j, \qquad B_j^s = \text{Op}_\mathsf{T}(b_j^s).$$

Then, for $u \in \overline{\mathscr{S}}(\mathbb{R}_+^d)$, the quadratic form given by (6.3.1) reads

$$Q(u) = \sum_{j=0}^{m} \sum_{k=0}^{m} \left(C_{j,k} D_d^j u, D_d^k u\right)_+,$$

where $C_{j,k}$ are tangential operators given by

$$C_{j,k} = \sum_{s=1}^{N} (B_k^s)^* A_j^s,$$

with symbols

$$c_{jk}(\varrho') = \sum_{s=1}^{N} (b_k^s)^* \circ a_j^s(\varrho') \in S_{\mathsf{T},\tau,\text{ph}}^{2(m+\sigma)-(j+k)}.$$

We have the following lemma that follows directly from Proposition 6.6.

LEMMA 6.15. *We consider the interior quadratic form of type (m,σ), as above,*

$$Q(u) = \sum_{j=0}^{m}\sum_{k=0}^{m} \big(C_{j,k} D_d^j u, D_d^k u\big)_+, \quad C_{j,k} = \mathrm{Op}_{\mathsf{T}}(c_{j,k}),$$
$$c_{j,k} \in S_{\mathsf{T},\tau,\mathrm{ph}}^{2(m+\sigma)-(j+k)}.$$

We have, for some $C>0$,

$$|Q(u)| \leq C \|u\|_{\tau,m,\sigma}^2, \quad u \in \overline{\mathscr{S}}(\mathbb{R}_+^d).$$

Next, we consider the case of a quadratic form with a vanishing principal symbol. Such a result is useful when comparing quadratic forms associated with the same principal symbol.

LEMMA 6.16. *We consider the interior quadratic form of type (m,σ), as above,*

$$Q(u) = \sum_{j=0}^{m}\sum_{k=0}^{m} \big(C_{j,k} D_d^j u, D_d^k u\big)_+, \quad C_{j,k} = \mathrm{Op}_{\mathsf{T}}(c_{j,k}),$$
$$c_{j,k} \in S_{\mathsf{T},\tau,\mathrm{ph}}^{2(m+\sigma)-(j+k)},$$

and we further assume that its principal symbol vanishes, that is,

$$\sum_{\substack{0\leq j,k\leq m\\ j+k=\ell}} c_{j,k} \equiv 0 \mod S_{\mathsf{T},\tau,\mathrm{ph}}^{2(m+\sigma)-\ell-1}, \quad \forall \ell \in \{0,\dots,2m\}.$$

Then the following estimate holds

$$|Q(u)| \leq C\big(\|u\|_{\tau,m,\sigma-1/2}^2 + |\operatorname{tr}(u)|_{\tau,m-1,\sigma+1/2}^2\big), \qquad u \in \overline{\mathscr{S}}(\mathbb{R}_+^d).$$

PROOF. Let $\ell \in \{0,\dots,2m\}$. We introduce $\alpha_\ell = \max(0,\ell-m)$ and $\beta_\ell = \min(m,\ell)$. Note that $\alpha_\ell + \beta_\ell = \ell$. We set

$$I_\ell = \sum_{\substack{0\leq j,k\leq m\\ j+k=\ell}} \big(C_{j,k} D_d^j u, D_d^k u\big)_+ = \sum_{k=\alpha_\ell}^{\beta_\ell} \big(C_{\ell-k,k} D_d^{\ell-k} u, D_d^k u\big)_+.$$

First, we consider $0<\ell<2m$. For $k>\alpha_\ell$ we write

$$\begin{aligned}\big(C_{\ell-k,k} D_d^{\ell-k} u, D_d^k u\big)_+ &= \big(C_{\ell-k,k} D_d^{\ell-k+1} u, D_d^{k-1} u\big)_+ \\ &\quad + \big(\mathrm{Op}(D_d c_{\ell-k,k}) D_d^{\ell-k} u, D_d^{k-1} u\big)_+ \\ &\quad - i\big(C_{\ell-k,k} D_d^{\ell-k} u_{|x_d=0^+}, D_d^{k-1} u_{|x_d=0^+}\big)_\partial,\end{aligned}$$

which by induction yields,

$$\begin{aligned}\big(C_{\ell-k,k} D_d^{\ell-k} u, D_d^k u\big)_+ &= \big(C_{\ell-k,k} D_d^{\beta_\ell} u, D_d^{\alpha_\ell} u\big)_+ \\ &\quad + \sum_{s=1}^{k-\alpha_\ell} \big(\mathrm{Op}(D_d c_{\ell-k,k}) D_d^{\ell-k+s-1} u, D_d^{k-s} u\big)_+ \\ &\quad - i \sum_{s=1}^{k-\alpha_\ell} \big(C_{\ell-k,k} D_d^{\ell-k+s-1} u_{|x_d=0^+}, D_d^{k-s} u_{|x_d=0^+}\big)_\partial.\end{aligned}$$

As $D_d c_{\ell-k,k} \in S^{2(m+\sigma)-\ell}_{\mathsf{T},\tau,\text{ph}}$ we note that

$$
\begin{aligned}
(6.3.3)\qquad \big|\big(\operatorname{Op}(D_d c_{\ell-k,k}) D_d^{\ell-k+s-1} u, D_d^{k-s} u\big)_+\big|
&\lesssim \|\Lambda_{\mathsf{T},\tau}^{m+\sigma+k-\ell-s+\frac12} D_d^{\ell-k+s-1} u\|_+ \|\Lambda_{\mathsf{T},\tau}^{m+\sigma-k+s-\frac12} D_d^{k-s} u\|_+\\
&\lesssim \|\Lambda_{\mathsf{T},\tau}^{m+\sigma+k-\ell-s+\frac12} u\|_{\tau,\ell-k+s-1} \|\Lambda_{\mathsf{T},\tau}^{m+\sigma-k+s-\frac12} u\|_{\tau,k-s}\\
&\lesssim \|u\|_{\tau,\ell-k+s-1,m+\sigma+k-\ell-s+\frac12} \|u\|_{\tau,k-s,m+\sigma-k+s-\frac12}\\
&\lesssim \|u\|^2_{\tau,m-1,\sigma+\frac12},
\end{aligned}
$$

by Corollary 6.7 as $m+k-\ell-s \geq 0$ and $m-1-k+s \geq 0$.

Similarly we write

$$
\big|\big(C_{\ell-k,k} D_d^{\ell-k+s-1} u_{|x_d=0^+}, D_d^{k-s} u_{|x_d=0^+}\big)_\partial\big| \lesssim |\operatorname{tr}(u)|^2_{\tau,m-1,\sigma+\frac12}.
$$

We thus obtain

$$
|I_\ell| - \Big|\sum_{k=\alpha_\ell}^{\beta_\ell} (C_{\ell-k,k} D_d^{\beta_\ell} u, D_d^{\alpha_\ell} u)_+\Big| \lesssim \|u\|^2_{\tau,m-1,\sigma+\frac12} + |\operatorname{tr}(u)|^2_{\tau,m-1,\sigma+\frac12}.
$$

As by assumption we have

$$
\sum_{k=\alpha_\ell}^{\beta_\ell} C_{\ell-k,k} = \sum_{\substack{0\leq j,k\leq m\\ j+k=\ell}} C_{j,k} \in \Psi^{2(m+\sigma)-\ell-1}_{\mathsf{T},\tau,\text{ph}},
$$

we find, as $\alpha_\ell + \beta_\ell = \ell$,

$$
\begin{aligned}
\Big|\sum_{k=\alpha_\ell}^{\beta_\ell} \big(C_{\ell-k,k} D_d^{\beta_\ell} u, D_d^{\alpha_\ell} u\big)_+\Big| &\lesssim \|\Lambda_{\mathsf{T},\tau}^{m+\sigma-\beta_\ell-\frac12} D_d^{\beta_\ell} u\|_+ \|\Lambda_{\mathsf{T},\tau}^{m+\sigma-\alpha_\ell-\frac12} D_d^{\alpha_\ell} u\|_+\\
&\lesssim \|u\|_{\tau,\beta_\ell,m+\sigma-\beta_\ell-\frac12} \|u\|_{\tau,\alpha_\ell,m+\sigma-\alpha_\ell-\frac12}\\
&\lesssim \|u\|^2_{\tau,m,\sigma-\frac12},
\end{aligned}
$$

by Corollary 6.7 since $m-\alpha_\ell \geq 0$ and $m-\beta_\ell \geq 0$. In the case $0<\ell<2m$ we have thus obtained

$$
(6.3.4)\qquad |I_\ell| \lesssim \|u\|^2_{\tau,m,\sigma-\frac12} + |\operatorname{tr}(u)|^2_{\tau,m-1,\sigma+\frac12}.
$$

Second, we consider $\ell = 0$. Then $I_0 = (C_{0,0}u, u)_+$ and as $C_{0,0} \in \Psi^{2(m+\sigma)-1}_{\mathsf{T},\tau,\text{ph}}$ we find $|I_0| \lesssim \|u\|^2_{\tau,m,\sigma-\frac12}$.

Third, we consider $\ell = 2m$. Then we have $I_{2m} = (C_{m,m} D_d^m u, D_d^m u)_+$ with $C_{m,m} \in \Psi^{2\sigma-1}_{\mathsf{T},\tau,\text{ph}}$ yielding $|I_{2m}| \lesssim \|u\|^2_{\tau,m,\sigma-\frac12}$. This concludes the proof. ∎

We have the following microlocal Gårding inequality for the quadratic forms we have introduced.

THEOREM 6.17 (Microlocal Gårding Inequality). *Let K be a compact set of $\overline{\mathbb{R}^d_+}$ and let $\mathscr{U}$ be a conic open set of $\overline{\mathbb{R}^d_+} \times \mathbb{R}^{d-1} \times \mathbb{R}_+$ contained in $K \times \mathbb{R}^{d-1} \times \mathbb{R}_+$. Let also $\chi \in S^0_{\mathsf{T},\tau}$ be homogeneous of degree 0, be such* *that* $\operatorname{supp}(\chi) \subset \mathscr{U}$. *Let* Q *be an interior quadratic form of type* (m, σ) *with (one of the representatives of) its principal symbol* $q \in S^{2m,2\sigma}_{\tau,\mathrm{ph}}$ *satisfying, for some* $C_0 > 0$ *and* $r_0 > 0$,

$$\operatorname{Re} q(\varrho) \geq C_0 \lambda_\tau^{2m} \lambda_{\mathsf{T},\tau}^{2\sigma}, \quad \text{for } \tau \geq r_0, \;\; \varrho = (\varrho', \xi_d), \; \varrho' = (x, \xi', \tau) \in \mathscr{U}, \; \xi_d \in \mathbb{R}.$$

For $0 < C_1 < C_0$ *and* $N \in \mathbb{N}$ *there exist* τ_*, $C > 0$, *and* $C_N > 0$ *such that*

$$\operatorname{Re} Q(\operatorname{Op}_{\mathsf{T}}(\chi) u) \geq C_1 \|\operatorname{Op}_{\mathsf{T}}(\chi) u\|^2_{\tau,m,\sigma} - C |\operatorname{tr}(\operatorname{Op}_{\mathsf{T}}(\chi) u)|^2_{\tau,m-1,\sigma+1/2} - C_N \|u\|^2_{\tau,m,-N},$$

for $u \in \overline{\mathscr{S}}(\mathbb{R}^d_+)$ *and* $\tau \geq \tau_*$.

The important feature of this version of the Gårding inequality is that it concerns functions defined on a half-space. Compare with Theorem 2.29 of Volume 1.

REMARK 6.18. In the case $\mathscr{U} = U_0 \times \mathbb{R}^{d-1} \times \mathbb{R}^+$, with U_0 a bounded open subset of $\overline{\mathbb{R}^d_+}$, by continuity, there exists a U_1 bounded open subset of $\overline{\mathbb{R}^d_+}$ such that U_1 is a neighborhood of $\overline{U_0}$, with $\operatorname{dist}(U_0, \partial U^1) \geq \delta > 0$ and

$$\operatorname{Re} q(\varrho) \geq C_0' \lambda_\tau^{2m} \lambda_{\mathsf{T},\tau}^{2\sigma}, \quad \text{for } \tau \geq r_0,$$

for $C_1 < C_0' < C_0$, where $\varrho = (\varrho', \xi_d)$ with $\varrho' = (x, \xi', \tau) \in U_1 \times \mathbb{R}^{d-1} \times \mathbb{R}^+$ and $\xi_d \in \mathbb{R}$. (See the beginning of the proof of Theorem 2.28 in Section 2.A.6.1 of Volume 1 for a detailed argumentation.) Then, there exist C and $\tau_* > 0$ such that

$$\operatorname{Re} Q(u) \geq C_1 \|u\|^2_{\tau,m,\sigma} - C |\operatorname{tr}(u)|^2_{\tau,m-1,\sigma+1/2}, \tag{6.3.5}$$

for $u \in \overline{\mathscr{S}}(\mathbb{R}^d_+)$ with $\operatorname{supp}(u) \subset U_0$. This is obtained from Theorem 6.17 by choosing $\chi = \chi(x) \in \mathscr{C}^\infty(\mathbb{R}^d)$ with $\operatorname{supp}(\chi_{|x_d>0}) \subset U_1$ and $\chi \equiv 1$ on U_0 and by taking τ sufficiently large. We thus obtain the counterpart of Theorem 2.28 for quadratic forms in a half-space.

Observe that the assumption that U_0 is bounded is not necessary. One could combine the arguments of the proofs of Theorems 2.28 and 6.17 to obtain a direct proof of (6.3.5). Yet, we shall not need such a result here and we leave this proof to the reader.

PROOF OF THEOREM 6.17. We introduce the interior quadratic form

$$\tilde{Q}(u) = \operatorname{Re} Q(u) = \frac{1}{2} \sum_{s=1}^{N} \big((A^s u, B^s u)_+ + (B^s u, A^s u)_+ \big),$$

that we may write in the form of (6.3.1) with $2N$ terms in the sum. Its principal symbol as given by (6.3.2) is then in the class

$$\frac{1}{2} \sum_{s=1}^{N} \big(\overline{b}^s a^s + \overline{a}_s b^s \big) = \operatorname{Re} \sum_{s=1}^{N} \overline{a}_s b^s \in S^{2m,2\sigma}_{\tau,\mathrm{ph}} / S^{2m,2\sigma-1}_{\tau,\mathrm{ph}}.$$

Without any loss of generality we may thus assume that the interior quadratic form Q has a real principal symbol $q(\varrho)$.

We first pick a representative of the principal symbol $q(\varrho)$ in $S^{2m,2\sigma}_{\tau,\mathrm{ph}}/S^{2m,2\sigma-1}_{\tau,\mathrm{ph}}$ that reads

$$q(\varrho) = \sum_{j=0}^{2m} q_j(\varrho')\xi_d^j, \quad q_j \in S^{2m+2\sigma-j}_{\mathsf{T},\tau,\mathrm{ph}}/S^{2m+2\sigma-j-1}_{\mathsf{T},\tau,\mathrm{ph}},$$

for $\varrho = (\varrho', \xi_d)$, and $\varrho' = (x, \xi', \tau)$. We choose q_j^0 a representative of q_j that is homogeneous of degree $2m + 2\sigma - j$ in (ξ', τ) for $\tau \geq r_0$ with $r_0 > 0$ (see Definition 6.2). Then

$$q^0(\varrho) = \sum_{j=0}^{2m} q_j^0(\varrho')\xi_d^j \tag{6.3.6}$$

is a real and homogeneous representative of the principal symbol q, and we have for $C_1 < C_0' < C_0$,

$$q^0(\varrho) \geq C_0' \lambda_\tau^{2m} \lambda_{\mathsf{T},\tau}^{2\sigma}, \quad \varrho' \in \overline{\mathscr{U}}, \quad \xi_d \in \mathbb{R}, \quad \text{for } \tau \geq r_0,$$

with $r_0 > 0$ chosen sufficiently large.

Arguing as in the proof of Theorem 2.29 in Section 2.A.6.2 of Volume 1, there exists $\tilde{\chi} \in S^0_{\mathsf{T},\tau}$, homogeneous of degree 0, such that $\operatorname{supp}(\tilde{\chi}) \subset \mathscr{U}$ and $\tilde{\chi} \equiv 1$ on $\operatorname{supp}(\chi)$.

With $C_1 < L < C_0'$, we see that $q^0(\varrho) - L\lambda_\tau^{2m}\lambda_{\mathsf{T},\tau}^{2\sigma}$ is a polynomial function with real coefficients in the variable ξ_d of order $2m$, that takes positive values for $\xi_d \in \mathbb{R}$, $\varrho' \in \mathscr{U}$ and $\tau \geq r_0$. The leading coefficient $a_0(\varrho') \in S_\tau^{2\sigma}$ is homogeneous of degree 2σ in (ξ', τ), for $\tau \geq r_0$, and $a_0(\varrho') \geq C\lambda_{\mathsf{T},\tau}^{2\sigma}$ for $C > 0$ and $\varrho' \in \overline{\mathscr{U}}$. This gives $\sqrt{a_0} \in S_\tau^\sigma$ microlocally in $\mathscr{U}$. The roots of the polynomial come into conjugated pairs and are functions of the other variables $\varrho' = (x, \xi', \tau) \in \mathscr{U}$. We may thus write

$$q^0(\varrho) - L\lambda_\tau^{2m}\lambda_{\mathsf{T},\tau}^{2\sigma} = a_0(\varrho') f(\varrho)\overline{f}(\varrho), \qquad \varrho = (\varrho', \xi_d),\ \varrho' \in \overline{\mathscr{U}},\ \xi_d \in \mathbb{R},\ \tau \geq r_0,$$

with

$$f(\varrho) = \prod_{i=1}^{m} \big(\xi_d - \rho_i^+(\varrho')\big),$$

where ρ_i^+, $i = 1, \dots, m$, denote the roots with positive imaginary parts.

LEMMA 6.19. *We have $f(\varrho)$ is in $S^{m,0}_{\tau,\mathrm{ph}}$ microlocally for $\varrho' \in \mathscr{U}$ and $\tau \geq r_0 > 0$.*

A proof is given below. Note in particular that this uses the homogeneity of the functions $q_{j,0}$ in (6.3.6).

We have

$$\tilde{\chi}^2(\varrho')\big(q^0(\varrho) - L\lambda_\tau^{2m}\lambda_{\mathsf{T},\tau}^{2\sigma}\big) = \tilde{\chi}^2(\varrho') a_0(\varrho') |f|^2(\varrho) \in S_\tau^{2m,2\sigma}, \tag{6.3.7}$$

for $\varrho = (\varrho', \xi_d)$ with $\varrho' \in \overline{\mathbb{R}^d_+} \times \mathbb{R}^{d-1} \times \mathbb{R}_+$, $\xi_d \in \mathbb{R}$ for $\tau \geq r_0$. Observe that $v \to \|\mathrm{Op}_\mathsf{T}(\tilde{\chi})v\|^2_{\tau,m,\sigma}$ is an interior quadratic form of type (m, σ) with principal symbol $\tilde{\chi}^2 \lambda_\tau^{2m} \lambda_{\mathsf{T},\tau}^{2\sigma}$. We thus see that $Q(\mathrm{Op}_\mathsf{T}(\tilde{\chi})v) - L\|\mathrm{Op}_\mathsf{T}(\tilde{\chi})v\|^2_{\tau,m} - \|\mathrm{Op}(\tilde{\chi}\sqrt{a_0}f)v\|^2_+$ is an interior quadratic form of type (m, σ) with a vanishing principal symbol in $S^{2m,2\sigma}_{\tau,\mathrm{ph}} / S^{2m,2\sigma-1}_{\tau,\mathrm{ph}}$. Lemma 6.16 then yields

$$\text{(6.3.8)} \quad \left| \mathrm{Re}\, Q(\mathrm{Op}_\mathsf{T}(\tilde{\chi})v) - L\|\mathrm{Op}_\mathsf{T}(\tilde{\chi})v\|^2_{\tau,m,\sigma} - \|\mathrm{Op}(\tilde{\chi}\sqrt{a_0}f)v\|^2_+ \right| \lesssim \|v\|^2_{\tau,m,\sigma-1/2} + |\operatorname{tr}(v)|^2_{\tau,m-1,\sigma+1/2},$$

for $v \in \overline{\mathscr{S}}(\mathbb{R}^d_+)$. The triangular inequality then yields

$$\mathrm{Re}\, Q(\mathrm{Op}_\mathsf{T}(\tilde{\chi})v) \geq L\|\mathrm{Op}_\mathsf{T}(\tilde{\chi})v\|^2_{\tau,m,\sigma} - C\big(\|v\|^2_{\tau,m,\sigma-1/2} + |\operatorname{tr}(v)|^2_{\tau,m-1,\sigma+1/2}\big).$$

We now set $v = \mathrm{Op}_\mathsf{T}(\chi)u$. We have $\mathrm{Op}_\mathsf{T}(\tilde{\chi})v = \mathrm{Op}_\mathsf{T}(\chi)u + Ru$ with $R \in \cap_{N\in\mathbb{N}} \Psi^{-N}_{\mathsf{T},\tau}$ by pseudo-differential calculus. We then obtain the sought estimate by taking $\tau > 0$ sufficiently large. ■

PROOF OF LEMMA 6.19. We write

$$p(\varrho', \xi_d) = q^0(\rho) - L\lambda_\tau^{2m}\lambda_{\mathsf{T},\tau}^{2\sigma} = a_0(\varrho') \sum_{0\leq j\leq 2m} b_{2m-j}(\varrho')\xi_d^j,$$

the coefficient $b_{2m-j}(\varrho')$ are, respectively, homogeneous of degree $2m - j$ with respect to (ξ', τ). We have $b_0 \equiv 1$ and we recall that $a_0(\varrho') \geq C\lambda_{\mathsf{T},\tau}^{2\sigma}$ with $C > 0$ for $\varrho' \in \overline{\mathscr{U}}$. If ϱ' lie in the compact set[1] $K \times \mathbb{S}^{d-1}_+$, then by Lemma 6.29, the roots of $\xi_d \mapsto p(\varrho', \xi_d)$ are contained in a bounded open ball $B(0, R_0)$ with $R_0 > 0$.

Let γ^0 be the closed path in $\mathbb{C}$, oriented counter-clockwise, composed with

(1) the real interval $[-R_0, R_0]$;
(2) the upper half circle centered at 0 of radius R_0.

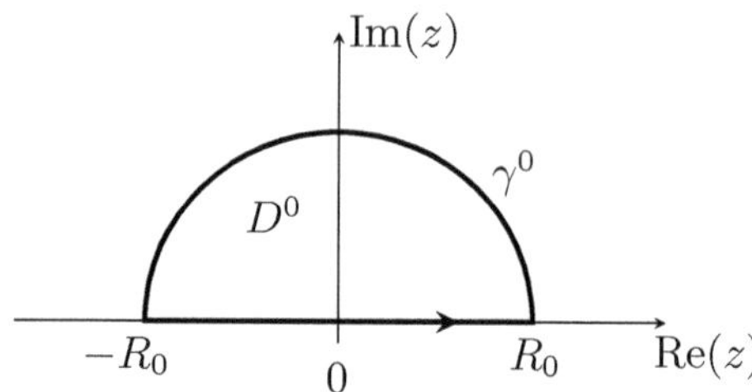

As the polynomial $z \mapsto p(\varrho', z)$ does not have any real root if $\varrho' \in \overline{\mathscr{U}}$, with what precedes, for $\varrho' \in \mathbb{S}_{\overline{\mathscr{U}}}$, with the notation introduced in (1.7.2), the roots of the polynomial that have a positive imaginary part all lie in the

[1]Observe that if ϱ' were to lie in $\mathbb{R}^d \times \mathbb{S}^{d-1}_+$ instead, then one could simply argue that the coefficients $b_{2m-j}(\varrho')$ remain in a compact set since $q^0(\rho) \in S^{2m,2\sigma}_{\tau,\mathrm{ph}}$. One can then still apply Lemma 6.29 and prove that the roots are contained in a bounded set.

interior D^0 of γ^0. In particular, $p(\varrho', z)$ does not vanish if $z \in \gamma^0$. As ϱ' lie in the compact set $\mathbb{S}_{\overline{\mathscr{U}}}$ we find that

$$|p(\varrho', z)| \geq C > 0, \qquad \varrho' \in \mathbb{S}_{\overline{\mathscr{U}}},\ z \in \gamma^0, \tag{6.3.9}$$

for some $C > 0$.

By Proposition 6.24 we then obtain that, for $\ell \in \mathbb{N}$,

$$s_\ell(\varrho') := \sum_{1 \leq i \leq m} \rho_i^+(\varrho')^\ell = \frac{1}{2\pi i} \int_{\gamma^0} \frac{d}{dz} p(\varrho', z) \frac{z^\ell}{p(\varrho', z)} dz$$

is a $\mathscr{C}^\infty$-function of (x, ξ') for $\varrho' \in \mathbb{S}_{\mathscr{U}}$, from the regularity of the integrand on the right-hand-side. By (6.3.9) and using again that ϱ' lie in the compact set $\mathbb{S}_{\overline{\mathscr{U}}}$ we find that $s_\ell(\varrho')$, and all its derivatives with respect to (x, ξ'), are bounded functions of $\varrho' \in \mathbb{S}_{\mathscr{U}}$.

We recall that $f(\varrho', z) = \prod_{1 \leq i \leq m}(z - \rho_i^+(\varrho'))$, that we write

$$f(\varrho', z) = \sum_{0 \leq j \leq m} f_{m-j}(\varrho') z^j,$$

with $f_0 \equiv 1$. By Lemma 6.26 in the appendix below, we have

$$\begin{aligned}
f_1(\varrho') &= -s_1(\varrho'),\\
2f_2(\varrho') &= -\big(s_2(\varrho') + f_1(\varrho')\, s_1(\varrho')\big),\\
3f_3(\varrho') &= -\big(s_3(\varrho') + f_1(\varrho')\, s_2(\varrho') + f_2(\varrho')\, s_1(\varrho')\big),\\
&\ \vdots\\
m\, f_m(\varrho') &= -\big(s_m(\varrho') + f_1(\varrho')\, s_{m-1}(\varrho') + \cdots + f_{m-1}(\varrho')\, s_1(\varrho')\big).
\end{aligned}$$

By a finite induction we find that the coefficients $f_j(\varrho')$ of $f(\varrho', z)$ and all their derivatives with respect to (x, ξ') are bounded functions of $\varrho' \in \mathbb{S}_{\mathscr{U}}$.

According to Lemma 6.30 in the appendix below, the set of the roots $\rho_i^+(\varrho')$ is homogeneous of degree one in the variables (ξ', τ). This implies that $s_\ell(\varrho')$ is homogeneous of degree ℓ in (ξ', τ) for $\varrho' \in \mathscr{U}$. Consequently, the formulae above yield the function $f_j(\varrho')$ to be homogeneous of degree j in (ξ', τ). Hence, we find that $f_j \in S_\tau^j$ microlocally in $\mathscr{U}$, $j = 1, \ldots, m$. This concludes the proof. ■

6.4. Estimates for First-Order Operators

In this section we derive estimates for first-order operators in a half-space of the form $L = D_d - \mathrm{Op}_{\mathsf{T}}(b)$ with $b(\varrho') \in S^1_{\mathsf{T},\tau}$. Depending on the sign of the imaginary part of b, inequalities of different natures are obtained. They are to be used in other chapters below. There, first-order operators will appear through the factorization of a conjugated operator; however, there, the sign of the imaginary part of b will vary. In such occurrence, microlocal estimations can be obtained with proper cutoffs in phase space. We have thus chosen to derive such microlocal estimates here.

6.4.1. A Perfectly Elliptic Estimate. Let $\mathscr{U}$ be a conic open set of $\overline{\mathbb{R}^d_+} \times \mathbb{R}^{d-1} \times \mathbb{R}_+$ and $b(\varrho') \in S^1_{\mathsf{T},\tau}$, with principal symbol $b_1(\varrho')$ homogeneous of degree one, such that, for $C > 0$,

$$\operatorname{Im} b_1(\varrho') \le -C\lambda_{\mathsf{T},\tau}, \qquad \varrho' \in \mathscr{U}.$$

We set $L = D_d - \operatorname{Op}_{\mathsf{T}}(b)$ and we have the following microlocal elliptic estimate.

LEMMA 6.20. *Let $\chi \in S^0_{\mathsf{T},\tau}$, homogeneous of degree 0, be such that $\operatorname{supp}(\chi) \subset \mathscr{U}$. Let $s \in \mathbb{R}$. There exist $C > 0$ and $\tau_* > 0$ such that for any $N \in \mathbb{N}$, there exists $C_N > 0$ such that*

$$\|\operatorname{Op}_{\mathsf{T}}(\chi)u\|_{\tau,1,s} + |\operatorname{Op}_{\mathsf{T}}(\chi)u_{|x_d=0^+}|_{\tau,s+1/2} \le C\|L\operatorname{Op}_{\mathsf{T}}(\chi)u\|_{\tau,0,s} + C_N\|u\|_{\tau,0,-N}, \tag{6.4.1}$$

for $\tau \ge \tau_*$ and $u \in \overline{\mathscr{S}}(\mathbb{R}^d_+)$.

The proof is based on a multiplier method. We set $w = \operatorname{Op}_{\mathsf{T}}(\chi)u$.

PROOF. With an integration by parts, with $r = 2s+1$, we compute

$$\begin{aligned} 2\operatorname{Re}(Lw, i\Lambda^r_{\mathsf{T},\tau}w)_+ &= 2\operatorname{Re}\big((D_d - \operatorname{Op}_{\mathsf{T}}(b))w, i\Lambda^r_{\mathsf{T},\tau}w\big)_+ \\ &= \operatorname{Re}(i(\Lambda^r_{\mathsf{T},\tau}\operatorname{Op}_{\mathsf{T}}(b) - \operatorname{Op}_{\mathsf{T}}(b)^*\Lambda^r_{\mathsf{T},\tau})w, w)_+ \\ &\quad + \operatorname{Re}(\Lambda^r_{\mathsf{T},\tau}w_{|x_d=0^+}, w_{|x_d=0^+})_\partial. \end{aligned}$$

We have $\operatorname{Re}(\Lambda^r_{\mathsf{T},\tau}w_{|x_d=0^+}, w_{|x_d=0^+})_\partial = |w_{|x_d=0^+}|^2_{\tau,s+1/2}$ and the principal symbol σ of the operator $i(\Lambda^r_{\mathsf{T},\tau}\operatorname{Op}_{\mathsf{T}}(b) - \operatorname{Op}_{\mathsf{T}}(b)^*\Lambda^r_{\mathsf{T},\tau})$ is real and satisfies

$$\sigma(\varrho') = -2\lambda^r_{\mathsf{T},\tau}\operatorname{Im} b_1(\varrho') \gtrsim \lambda^{2s+2}_{\mathsf{T},\tau} \quad \text{in a neighborhood of } \operatorname{supp}(\chi).$$

Then, the microlocal Gårding inequality of Theorem 2.50 of Volume 1 yields

$$\operatorname{Re}(Lw, i\Lambda^r_{\mathsf{T},\tau}w)_+ \ge |w_{|x_d=0^+}|^2_{\tau,s+1/2} + C\|\Lambda^{s+1}_{\mathsf{T},\tau}w\|^2_+ - C_N\|\Lambda^{-N}_{\mathsf{T},\tau}u\|^2_+.$$

The Young inequality gives

$$|(Lw, i\Lambda^r_{\mathsf{T},\tau}w)_+| \lesssim \varepsilon^{-1}\|\Lambda^s_{\mathsf{T},\tau}Lw\|^2_{\tau,+} + \varepsilon\|\Lambda^{s+1}_{\mathsf{T},\tau}w\|^2_+,$$

which yields, for ε chosen sufficiently small,

$$|w_{|x_d=0^+}|_{\tau,s+1/2} + \|\Lambda_{\mathsf{T},\tau}w\|_{\tau,0,s+1} \lesssim \|Lw\|_{\tau,0,s} + \|u\|_{\tau,0,-N}.$$

Finally, we write

$$\begin{aligned} \|D_d w\|_{\tau,0,s} &\lesssim \|\big(D_d - \operatorname{Op}_{\mathsf{T}}(b)\big)w\|_{\tau,0,s} + \|w\|_{\tau,0,s+1} \\ &= \|Lw\|_{\tau,0,s} + \|w\|_{\tau,0,s+1}. \end{aligned}$$

Together, these last two inequalities yield the result. ∎

6.4.2. An Elliptic Estimate with a Boundary Observation. Let $\mathscr{U}$ be a conic open set of $\overline{\mathbb{R}^d_+} \times \mathbb{R}^{d-1} \times \mathbb{R}_+$ and $b(\varrho') \in S^1_{\mathsf{T},\tau}$, with principal symbol $b_1(\varrho')$ homogeneous of degree one, such that, for $C > 0$,

$$\operatorname{Im} b_1(\varrho') \geq C\lambda_{\mathsf{T},\tau}, \qquad \varrho' \in \mathscr{U}.$$

We set $L = D_d - \operatorname{Op}_{\mathsf{T}}(b)$ and we have the following microlocal elliptic estimate.

LEMMA 6.21. *Let $\chi \in S^0_{\mathsf{T},\tau}$, homogeneous of degree 0, be such that $\operatorname{supp}(\chi) \subset \mathscr{U}$. Let $s \in \mathbb{R}$. There exist $C > 0$ and $\tau_* > 0$ such that for any $N \in \mathbb{N}$, there exists $C_N > 0$ such that*

$$\|\operatorname{Op}_{\mathsf{T}}(\chi)u\|_{\tau,1,s} \leq C\Big(\|L\operatorname{Op}_{\mathsf{T}}(\chi)u\|_{\tau,0,s} + |\operatorname{Op}_{\mathsf{T}}(\chi)u_{|x_d=0^+}|_{\tau,s+1/2}\Big) + C_N\|u\|_{\tau,0,-N}, \tag{6.4.2}$$

for $\tau \geq \tau_*$, $u \in \overline{\mathscr{S}}(\mathbb{R}^d_+)$.

The proof is very similar to that of Lemma 6.20.

PROOF. With an integration by parts, with $r = 2s+1$, we compute

$$\begin{aligned} 2\operatorname{Re}(Lw, -i\Lambda^r_{\mathsf{T},\tau}w)_+ &= 2\operatorname{Re}\big((D_d - \operatorname{Op}_{\mathsf{T}}(b))w, -i\Lambda^r_{\mathsf{T},\tau}w\big)_+ \\ &= \operatorname{Re}(i(\operatorname{Op}_{\mathsf{T}}(b)^*\Lambda^r_{\mathsf{T},\tau} - \Lambda^r_{\mathsf{T},\tau}\operatorname{Op}_{\mathsf{T}}(b))w, w)_+ \\ &\quad - \operatorname{Re}(\Lambda^r_{\mathsf{T},\tau}w_{|x_d=0^+}, w_{|x_d=0^+})_\partial. \end{aligned}$$

We have $\operatorname{Re}(\Lambda^r_{\mathsf{T},\tau}w_{|x_d=0^+}, w_{|x_d=0^+})_\partial = |w_{|x_d=0^+}|^2_{\tau,s+1/2}$ and the principal symbol σ of the operator $i(\operatorname{Op}_{\mathsf{T}}(b)^*\Lambda^r_{\mathsf{T},\tau} - \Lambda^r_{\mathsf{T},\tau}\operatorname{Op}_{\mathsf{T}}(b))$ is real and satisfies

$$\sigma(\varrho') = 2\lambda^r_{\mathsf{T},\tau}\operatorname{Im} b_1(\varrho') \gtrsim \lambda^{2s+2}_{\mathsf{T},\tau} \quad \text{in a neighborhood of } \operatorname{supp}(\chi).$$

Then, the microlocal Gårding inequality of Theorem 2.50 of Volume 1 yields

$$\operatorname{Re}(Lw, -i\Lambda^r_{\mathsf{T},\tau}w)_+ + |w_{|x_d=0^+}|^2_{\tau,s+1/2} \geq C\|\Lambda^{s+1}_{\mathsf{T},\tau}w\|^2_+ - C_N\|\Lambda^{-N}_{\mathsf{T},\tau}u\|^2_+.$$

The Young inequality gives

$$|(Lw, -i\Lambda^r_{\mathsf{T},\tau}w)_+| \lesssim \varepsilon^{-1}\|\Lambda^s_{\mathsf{T},\tau}Lw\|^2_{\tau,+} + \varepsilon\|\Lambda^{s+1}_{\mathsf{T},\tau}w\|^2_+,$$

which yields, for ε chosen sufficiently small,

$$\|\Lambda_{\mathsf{T},\tau}w\|_{\tau,0,s+1} \lesssim \|Lw\|_{\tau,0,s} + |w_{|x_d=0^+}|_{\tau,s+1/2} + \|u\|_{\tau,0,-N}.$$

Finally, we write

$$\begin{aligned} \|D_dw\|_{\tau,0,s} &\lesssim \|(D_d - \operatorname{Op}_{\mathsf{T}}(b))w\|_{\tau,0,s} + \|w\|_{\tau,0,s+1} \\ &= \|Lw\|_{\tau,0,s} + \|w\|_{\tau,0,s+1}. \end{aligned}$$

Together, these last two inequalities yield the result. ∎

6.4.3. Estimate Under Sub-ellipticity. Let $b(\varrho') \in S^1_{\mathsf{T},\tau}$, with principal symbol $b_1(\varrho')$ homogeneous of degree one, $L = D_d - \mathrm{Op}_{\mathsf{T}}(b)$ with principal symbol $\ell(\varrho) = \xi_d - b_1(\varrho')$. Introduce

$$L_2 = \frac{1}{2}(L + L^*) = D_d - \frac{1}{2}(\mathrm{Op}_{\mathsf{T}}(b) + (\mathrm{Op}_{\mathsf{T}}(b)^*),$$
$$L_1 = \frac{1}{2i}(L - L^*) = -\frac{1}{2i}(\mathrm{Op}_{\mathsf{T}}(b) - \mathrm{Op}_{\mathsf{T}}(b)^*).$$

Observe that $L_2 \in \Psi^{1,0}_{\tau}$ and $L_1 \in \Psi^1_{\mathsf{T},\tau}$ are both formally selfadjoint. Their respective principal symbols are

$$\ell_2(\varrho) = \mathrm{Re}\,\ell(\varrho) = \xi_d - \mathrm{Re}\,b_1(\varrho') \in S^{1,0}_{\tau},$$
$$\ell_1(\varrho') = \mathrm{Im}\,\ell(\varrho) = -\,\mathrm{Im}\,b_1(\varrho') \in S^1_{\mathsf{T},\tau}.$$

We observe that $\frac{1}{2i}\{\overline{\ell}, \ell\} = \{\ell_2, \ell_1\}$ is a tangential symbol.

Let $\mathscr{U}$ be a conic open set of $\overline{\mathbb{R}^d_+} \times \mathbb{R}^{d-1} \times \mathbb{R}_+$ be such that the following sub-ellipticity property holds

(6.4.3)
$$\ell(\varrho) = 0 \;\Rightarrow\; \frac{1}{2i}\{\overline{\ell}, \ell\}(\varrho') = \{\ell_2, \ell_1\}(\varrho') > 0, \qquad \varrho' \in \overline{\mathscr{U}},\ \varrho = (\varrho', \xi_d).$$

Here, the sub-ellipticity property is written in the same form as for a second-order operator (see Definition 3.2 of Volume 1). For the present first-order operator, there is a more natural form. In fact, consider $\varrho' \in \overline{\mathscr{U}}$ such that $\ell_1(\varrho') = 0$. Then, setting $\varrho = (\varrho', \xi_d)$ with $\xi_d = \mathrm{Re}\,b_1(\varrho')$ we have $\ell(\varrho) = 0$ thus implying $\{\ell_2, \ell_1\}(\varrho') > 0$ by (6.4.3). The sub-ellipticity property thus reads

$$\ell_1(\varrho') = 0 \;\Rightarrow\; \{\ell_2, \ell_1\}(\varrho') > 0, \qquad \varrho' \in \overline{\mathscr{U}}. \tag{6.4.4}$$

LEMMA 6.22. *Assume that* (6.4.3) *holds and* $\mathbb{S}_{\overline{\mathscr{U}}}$ *is compact. Let* $\chi \in S^0_{\mathsf{T},\tau}$*, homogeneous of degree* 0*, be such that* $\mathrm{supp}(\chi) \subset \mathscr{U}$*. There exist* $C > 0$ *and* $\tau_* > 0$ *such that for any* $N \in \mathbb{N}$*, there exists* $C_N > 0$ *such that*

$$\tau^{-1/2}\|\mathrm{Op}_{\mathsf{T}}(\chi)v\|_{\tau,1} \leq C\big(\|L\mathrm{Op}_{\mathsf{T}}(\chi)v\|_+ + |\mathrm{Op}_{\mathsf{T}}(\chi)v_{|x_d=0^+}|_{\tau,1/2}\big)$$
$$+ C_N\|v\|_{\tau,0,-N},$$

for $\tau \geq \tau_*$*, and* $v \in \overline{\mathscr{S}}(\mathbb{R}^d_+)$.

PROOF. We set $w = \mathrm{Op}_{\mathsf{T}}(\chi)v$. We compute, with an integration by parts,

$$\|Lw\|_+^2 = \|L_2w\|_+^2 + \|L_1w\|_+^2 + 2\,\mathrm{Re}(L_2w, iL_1w)_+$$
$$= \|L_2w\|_+^2 + \|L_1w\|_+^2 + i([L_2, L_1]w, w)_+ + (L_1w_{|x_d=0^+}, w_{|x_d=0^+})_\partial.$$

We then obtain,

$$\|Lw\|_+^2 + |w_{|x_d=0^+}|^2_{\tau,1/2} \gtrsim \|L_1w\|_+^2 + i([L_2, L_1]w, w)_+.$$

Then, for $\mu > 0$ to be chosen below and for τ chosen sufficiently large so that $\tau^{-1}\mu \leq 1$, we obtain

$$\begin{aligned} \|Lw\|_+^2 + |w_{|x_d=0^+}|^2_{\tau,1/2} &\gtrsim \tau^{-1}\big(\mu\|L_1 w\|_+^2 + i\tau([L_2, L_1]w, w)_+\big) \\ &\gtrsim \tau^{-1}\big((\mu L_1^2 + i\tau[L_2, L_1])w, w\big)_+. \end{aligned} \tag{6.4.5}$$

Next, we observe that the principal symbol of the operator $\mu L_1^2 + i\tau[L_2, L_1] \in \Psi^2_{\mathsf{T},\tau}$ is $\mu|\ell_1|^2 + \tau\{\ell_2, \ell_1\} \in S^2_{\mathsf{T},\tau}$. As we have assumed that $\mathbb{S}_{\overline{\mathscr{U}}}$ is compact, from the sub-ellipticity property (6.4.4), Lemma 3.56 of Volume 1 and the homogeneity of the symbol $\mu|\ell_1|^2 + \tau\{\ell_2, \ell_1\}$, we deduce the following lemma.

LEMMA 6.23. *There exist $\mu_0 > 0$, $C > 0$ such that*

$$\mu|\ell_1|^2(\varrho') + \tau\{\ell_2, \ell_1\}(\varrho') \geq C\lambda^2_{\mathsf{T},\tau}, \qquad \varrho' = (x, \xi', \tau) \in \overline{\mathscr{U}},$$

for $\mu \geq \mu_0$.

Let $\mu = \mu_0$ and $N \in \mathbb{N}$. As $\operatorname{supp}(\chi) \subset \mathscr{U}$, the microlocal Gårding inequality of Theorem 2.50 of Volume 1 yields for some $C > 0$ and $C' > 0$,

$$\operatorname{Re}\big((\mu L_1^2 + i\tau[L_2, L_1])w, w\big)_+ \geq C'\|w\|^2_{\tau,0,1} - C_N\|v\|^2_{\tau,0,-N},$$

for $\tau > 0$ chosen sufficiently large. From (6.4.5) we thus obtain

$$\|Lw\|_+ + |w_{|x_d=0^+}|_{\tau,1/2} + \|v\|_{\tau,0,-N} \gtrsim \tau^{-1/2}\|w\|_{\tau,0,1}.$$

From the form of L we observe that

$$\|D_d w\|_+ \lesssim \|Lw\|_+ + \|w\|_{\tau,0,1},$$

yielding the result. ∎

6.5. Notes

Here, in a half-space, the operators and symbols we treat are differential in the direction normal to the boundary and pseudo-differential in the tangential direction. Symbols of this form already appear in the proof of Carleman estimates at the boundary in Section 3.4.2. With these operators we consider quadratic forms. Away from boundaries, such differential quadratic forms are studied in the work of L. Hörmander [172, Chapter 8]. Quadratic forms in a half-space associated with the symbols we define here can be found in the work of D. Tataru [314] and some joint work with M. Bellassoued [70]. One of the main goals of this chapter is the derivation of the Gårding inequality of Theorem 6.17 that applied to quadratic forms in a half space. It can be found in the works of D. Tataru [314], M. Eller [131] and in [70]. Here we give a microlocal version of the inequality. Note that for pseudo-differential operators obtaining a Gårding inequality on a half-space is not straight-forward. Some sufficient conditions are given in the work of N. Lerner and X. Saint-Raymond [231], with some improvement given by F. Herau [168].

In Sect. 6.4, in preparation for later chapters, we also derive microlocal estimates for first-order operators at a boundary. Decomposition in first-order factors is an approach that goes back to the work of A.P. Calderón [100]. The estimates we write and their proofs can be found in a joint work with N. Lerner [211].

Appendix

6.A. Smooth Factorisation of Polynomials

6.A.1. Some Results from Complex Analysis. Let Ω be an open set of $\mathbb{C}$ and γ be a closed continuous path in Ω such that

(1) $\operatorname{ind}_\gamma(z) = 0$ if $z \in \mathbb{C} \setminus \Omega$.
(2) $\operatorname{ind}_\gamma(z) = 0$ or 1 for $z \in \Omega \setminus \gamma$.

Set $\mathcal{O}_\gamma = \{z \in \Omega \setminus \gamma;\ \operatorname{ind}_\gamma(z) = 1\}$. The following proposition is a consequence of the residue formula. For a holomorphic function in Ω that does not vanish on γ, we denote by $(\zeta_i)_{i \in \mathcal{I}_\gamma}$ the zeros of f in $\mathcal{O}_\gamma$ (counted with multiplicities).

PROPOSITION 6.24. *Let f be a holomorphic function in Ω that does not vanish on γ. Let $\ell \in \mathbb{N}$. We have*

$$\frac{1}{2\pi i} \int_\gamma \frac{f'(z) z^\ell}{f(z)} dz = \sum_{i \in \mathcal{I}_\gamma} \zeta_i^\ell.$$

In particular, the case $\ell = 0$ yields

$$N_\gamma(f) = \#\mathcal{I}_\gamma = \frac{1}{2\pi i} \int_\gamma \frac{f'(z)}{f(z)} dz, \tag{6.A.1}$$

as the number of zeros of f in $\mathcal{O}_\gamma$ (counted with multiplicities).

We also recall the Rouché theorem (see for instance [294]).

THEOREM 6.25 (Rouché Theorem). *Under the same setting as above for γ, if f and g are two holomorphic functions in Ω that satisfy*

$$|f(z) - g(z)| < |f(z)|, \quad \textit{for all } z \in \gamma,$$

then $N_\gamma(f) = N_\gamma(g)$.

6.A.2. An Algebraic Identity. The following result is classical and known as the Newton identities.

LEMMA 6.26 (Newton Identities). *Let $q(z) = \sum_{0 \le k \le n} q_{n-k}\, z^k$ be a monic polynomial, that is, $q_0 = 1$, and let $(r_i)_{1 \le i \le n}$ be its roots (counted with multiplicities). If we set $s_\ell = \sum_{1 \le i \le n} r_i^\ell$, $\ell \in \mathbb{N}$, we have*

$$k q_k + \sum_{1 \le \ell \le k} s_\ell\, q_{k-\ell} = 0, \qquad k \in \{1, \dots, n\}.$$

PROOF. We have $q(z) = \prod_{1\leq k\leq n}(z - r_k)$ and we thus find

$$\sum_{0\leq k\leq n} q_k\, z^k = z^n q(1/z) = \prod_{1\leq k\leq n} (1 - r_k z).$$

Differentiating both sides with respect to z we obtain, after multiplication by z,

$$\sum_{1\leq k\leq n} k q_k\, z^k = - \sum_{1\leq j\leq n} \Big(r_j z \prod_{\substack{1\leq k\leq n\\ k\neq j}} (1 - r_k z)\Big).$$

Choosing $|z|$ sufficiently small so as to have $|r_j z| < 1$ for all j we write, with a Neumann series,

$$\begin{aligned}\sum_{1\leq k\leq n} k q_k\, z^k &= -\Big(\sum_{1\leq j\leq n} \frac{r_j z}{1 - r_j z}\Big) \prod_{1\leq k\leq n} (1 - r_k z)\\ &= -\Big(\sum_{1\leq j\leq n} \sum_{\ell\geq 1} r_j^\ell z^\ell\Big) \sum_{0\leq m\leq n} q_m\, z^m\\ &= -\Big(\sum_{\ell\geq 1} s_\ell z^\ell\Big) \sum_{0\leq m\leq n} q_m\, z^m.\end{aligned}$$

The result then follows by identifying the coefficients of z^k on the right-hand-side. ∎

Setting $q_k = 0$ for $k \geq n+1$, observe that the proof yields moreover that

$$\sum_{1\leq \ell\leq k} s_\ell\, q_{k-\ell} = 0 \quad \text{if } k \geq n+1.$$

6.A.3. Roots Regularity. For $n \in \mathbb{N}$ we shall use the notation $J_n = \{1, \dots, n\}$.

Let $N \in \mathbb{N}$ and $p(t, z)$ be a monic polynomial function in $z \in \mathbb{C}$ of degree N with coefficients that continuously depend on $t \in U$, an open set of a d-dimensional smooth manifold $\mathcal{M}$,

$$p(t, z) = z^N + a_1(t) z^{N-1} + \cdots + a_{N-1}(t) z + a_N(t).$$

Let $t^0 \in U$ and denote by $\alpha^j(t^0)$, $j \in J_n$, with $n \leq N$, the different roots of $z \mapsto p(t^0, z)$ with respective multiplicities m_j, meaning that $m_1 + \cdots + m_n = N$, that is,

$$p(t^0, z) = \prod_{j\in J_n} (z - \alpha^j(t^0))^{m_j}.$$

An important consequence of the Rouché theorem is the following regularity result for the roots of a polynomial function.

LEMMA 6.27. *There exist $U^0 \subset U$ an open neighborhood of t^0 and N functions $t \mapsto \beta^{j,k}(t)$, $j \in J_n$, $k = 1, \dots, m_j$, that enumerate the roots of $p(t, z)$ for $t \in U^0$, that is,*

$$p(t, z) = \prod_{j\in J_n} \prod_{1\leq k\leq m_j} (z - \beta^{j,k}(t)), \qquad t \in U^0,\ z \in \mathbb{C}.$$

and such that

$$\lim_{t \to t^0} \beta^{j,k}(t) = \alpha^j(t^0), \quad j \in J_n,\ k = 1, \ldots, m_j.$$

Observe that the smoothness of the coefficients a_k, $k = 1, \ldots, N$ does not necessarily imply the smoothness of the roots $t \mapsto \beta^{j,k}(t)$, $j \in J_n$, $k = 1, \ldots, m_j$. This is well illustrated in the following example. Let $p(t, z) = z^2 - t$. Then, a pair of continuous roots is given by, for $j = 1, 2$,

$$\alpha^j(t) = (-1)^j \times \begin{cases} \sqrt{t} & \text{if } t \geq 0, \\ i\sqrt{-t} & \text{if } t < 0. \end{cases}$$

One sees that these two roots are not smooth at $t = 0$. Any other choice of functions, say

$$\alpha^j(t) = (-1)^j \times \begin{cases} \sqrt{t} & \text{if } t \geq 0, \\ -i\sqrt{-t} & \text{if } t < 0. \end{cases}$$

does not yield smooth functions at $t = 0$ either. The crossing of the roots is responsible for the lack of smoothness of the roots. Observe that the argument of the proof of Lemma 6.19 can be adapted to prove the following result.

PROPOSITION 6.28. *Let γ be a closed continuous path in $\mathbb{C}$ such that $\operatorname{ind}_\gamma(z) = 0$ or 1 for $z \in \mathbb{C} \setminus \gamma$. Assume that $z \mapsto p(t^0, z)$ does not vanish on γ. Set $J(\gamma) = \{j \in J_n;\ \operatorname{ind}_\gamma(\alpha^j(t^0)) = 1\}$. For U^0 as given by Lemma 6.27 there exists $\delta > 0$ such that $B(t^0, \delta) \subset U^0$ and*

$$\operatorname{ind}_\gamma(\beta^{j,k}(t)) = \operatorname{ind}_\gamma(\alpha^j(t^0)), \qquad j \in J_n,\ k = 1, \ldots, m_j,\ t \in B(t^0, \delta).$$

Moreover, assume that the coefficients of $z \mapsto p(t, z)$ are of class $\mathscr{C}^\ell$ (resp. analytic) in $t \in U$. Then, the coefficients of the polynomial function

$$z \mapsto \prod_{j \in J(\gamma)} \prod_{1 \leq k \leq m_j} (z - \beta^{j,k}(t))$$

are of class $\mathscr{C}^\ell$ (resp. analytic) in $t \in B(t^0, \delta)$.

In particular, isolated roots (of constant multiplicity) are of class $\mathscr{C}^\ell$ (resp. analytic) in $t \in B(t^0, \delta)$.

This result shows that the lack of smoothness of the roots is precisely due to roots crossing. Away from change of multiplicities, roots are smooth functions of the parameter t. If the coefficients of p are moreover analytic in t, then, away from change of multiplicities, the roots are analytic in t.

Note that the existence of $\delta > 0$ in the statement of this proposition is a simple consequence of the limit of the roots $\beta^{j,k}(t)$ as $t \to t^0$, as stated in Lemma 6.27.

The following lemmata are used in the proof of Lemma 6.19.

LEMMA 6.29. *Let $K \subset U$ be a compact set of $\mathcal{M}$. There exists a bounded set B_K where all the roots of the polynomial $z \mapsto p(t, z)$ lie if $t \in K$.*

LEMMA 6.30. *Assume that $\mathcal{M} = \mathcal{M}' \times \mathcal{M}''$ where $\mathcal{M}'$ is a manifold and $\mathcal{M}''$ is a conic submanifold (with or without boundary) of $\mathbb{R}^{d''}$ for some $d'' \in \mathbb{N}$. Accordingly for $t \in \mathcal{M}$ we write $t = (t', t'')$ with $t' \in \mathcal{M}'$ and $t'' \in \mathcal{M}''$. Let then U be a conic open set of $\mathcal{M}$ (with respect to the variable t''). In addition to the hypotheses of Lemma 6.27, assume that each coefficient $a_k(t)$ is homogeneous of degree k in t'' for $|t''| \geq r_0 > 0$. Let $t^0 = (t^{0\prime}, t^{0\prime\prime}) \in U$ with $t^{0\prime\prime} \neq 0$. Then the neighborhood U^0 of Lemma 6.27 can be chosen conic and the set*

$$\mathcal{R}(t) = \{\beta^{j,k}(t);\ j \in J_n,\ k = 1, \ldots, m_j\},$$

is homogeneous of degree one, meaning $\mathcal{R}(t', \lambda t'') = \lambda \mathcal{R}(t', t'')$ for $\lambda > 0$ and $(t', t'') \in U^0$.

PROOF OF LEMMA 6.27. Let $j \in J_n$. We consider a closed circular curve $\gamma^j : [0, 1] \to \mathbb{C}$ with center $\alpha^j(t_0)$, with radius chosen sufficiently small for $\alpha^j(t_0)$ to be the only root of $z \mapsto p(t^0, z)$ in D^j, the interior disk of $\mathcal{C}^j = \gamma^j([0, 1])$. We set $\varepsilon = \min_{z \in \mathcal{C}^j} |p(t^0, z)|/2 > 0$ (we shall omit some cumbersome superscript j in what follows). Let $z \in \mathcal{C}^j$. By continuity of p, there exists a neighborhood $V_z \subset \mathbb{C}$ of z and a neighborhood $U_z \subset U$ of t^0 such that

$$|p(t, \zeta) - p(t^0, z)| < \varepsilon, \qquad t \in U_z,\ \zeta \in V_z.$$

Since $\mathcal{C}^j \subset \cup_{z \in \mathcal{C}} V_z$, and $\mathcal{C}^j$ is compact, we can extract a finite covering with such neighborhoods, viz., there exists $z^1, \ldots, z^\ell \in \mathcal{C}^j$ such that $\mathcal{C}^j \subset \cup_{k=1}^{\ell} V_{z^k}$. Then, $U^j = \cap_{k=1}^{\ell} U_{z^k} \subset U$ defines a neighborhood of t^0 such that for all $z \in \mathcal{C}^j$ and all $t \in U^j$ we have

$$|p(t, z) - p(t^0, z)| < 2\varepsilon \leq |p(t^0, z)|.$$

By the Rouché theorem, for each $t \in U^j$ the polynomial $z \mapsto p(t, z)$ has m_j roots in the disc D^j, that we denote by $\beta^{j,k}(t)$, $k = 1, \ldots, m_j$.

Arguing the same for each $j \in J_n$, the neighborhood U^0 as in the statement of the lemma is simply obtained by setting $U^0 = \cap_{j \in J_n} U^j$.

For each $j \in J_n$, since we can let the radius of the circle $\mathcal{C}^j$ go to zero, meaning that the disk D^j shrinks to the point $\alpha^j(t^0)$, we obtain

$$\lim_{t \to t^0} \beta^{j,k}(t) = \alpha^j(t^0), \quad k = 1, \ldots, m_j,$$

which concludes the proof. ■

PROOF OF LEMMA 6.29. Because of the continuity of the coefficients with respect to the parameter t, we have

$$|p(t, z) - z^N| \leq C_K(1 + |z|)^{N-1},$$

for some $C_K > 0$ if $t \in K$. Then, as there exists $R_K > 0$ such that $|z|^N > C_K(1 + |z|)^{N-1}$ for $|z| \geq R_K$, we see that the roots of $z \mapsto p(t, z)$ are confined in the ball $B(0, R_K)$ uniformly with respect to $t \in K$. ■

PROOF OF LEMMA 6.30. Set $\mathbb{M}'' = \mathcal{M}'' \cap \mathbb{S}^{d''-1}$ and $\mathbb{U} = U \cap (\mathcal{M}' \times \mathbb{M}'')$. We then consider $t^1 = (t^{0\prime}, t^{1\prime\prime}) \in \mathbb{U}$ such that $t^{1\prime\prime} = \nu t^{0\prime\prime}$ for $\nu > 0$. By the homogeneity of the coefficients $a_k(t)$ we have

$$p(t^1, z) = \nu^N p(t^0, \nu^{-1} z) = \nu^N \prod_{j \in J_n} (\nu^{-1} z - \alpha^j(t^{0\prime}, t^{0\prime\prime}))^{m_j} = \prod_{j \in J_n} (z - \nu \alpha^j(t^{0\prime}, t^{0\prime\prime}))^{m_j}. \tag{6.A.2}$$

We thus set $\tilde{\alpha}^j = \nu \alpha^j(t^{0\prime}, t^{0\prime\prime})$, $j \in J_n$, that enumerate the different roots of $z \mapsto p(t^1, z)$, with multiplicity m_j. Lemma 6.27 yields an open set $\mathbb{U}^0 \subset \mathbb{U}$ and N functions $(t', t'') \mapsto \tilde{\beta}^{j,k}(t', t'')$ that enumerate the roots for $(t', t'') \in \mathbb{U}^0$ with continuity at t^1:

$$p(t', t'', z) = \prod_{j \in J_n} \prod_{1 \le k \le m_j} (z - \tilde{\beta}^{j,k}(t', t'')),$$

and

$$\lim_{(t', t'') \to t^1} \tilde{\beta}^{j,k}(t', t'') = \tilde{\alpha}^j, \quad j \in J_n.$$

We then introduce the open set $U^0 = \{(t', \nu t'');\ (t', t'') \in \mathbb{U}^0 \text{ and } \nu > 0\}$ in $\mathcal{M}' \times \mathcal{M}''$ and we define the functions $\beta^{j,k}(t', t'')$ to coincide with $\tilde{\beta}^{j,k}(t', t'')$ on $\mathbb{U}^0$ and to be homogeneous of degree one in t''. Arguing as in (6.A.2) we see that we obtain the sought result. ■

CHAPTER 7

Sobolev Norms with a Large Parameter on a Manifold

Contents

In Sect. 18.2, Sobolev spaces of order k, $k \in \mathbb{N}$, are defined on a Riemannian manifold. In Sect. 18.5, Sobolev spaces of order s, $s \in \mathbb{R}$, are defined on a compact Riemannian manifold without boundary. Here, we define their associated norms with a large parameter τ. We also provide an estimation of the bound of the map $u \mapsto e^{\tau\varphi}u$ with respect to these norms.

7.1. Nonnegative Sobolev Orders on a Manifold

First, we consider a smooth Riemannian manifold $\mathcal{M}$, possibly with a boundary. For $k \in \mathbb{N}$, we define the following norm:

$$\|u\|_{\tau,k}^2 = \tau^{2k}\|u\|_{L^2(\mathcal{M})}^2 + \|u\|_{H^k(\mathcal{M})}^2 \asymp \sum_{j=0}^{k} \tau^{2j}\|u\|_{H^{k-j}(\mathcal{M})}^2.$$

It is the counterpart of the norm defined in (2.6.1) in $\mathbb{R}^d$.

J. Le Rousseau et al., *Elliptic Carleman Estimates and Applications to Stabilization and Controllability, Volume II*, PNLDE Subseries in Control 98, https://doi.org/10.1007/978-3-030-88670-7_7

Second, we consider a smooth Riemannian manifold $\mathcal{N}$ that is compact and without boundary. Then, for $s \in \mathbb{R}_+$, we define

$$\|u\|_{\tau,s}^2 = \tau^{2s}\|u\|_{L^2(\mathcal{N})}^2 + \|u\|_{H^s(\mathcal{N})}^2, \tag{7.1.1}$$

with the norm $\|.\|_{H^s(\mathcal{N})}$ as in Sect. 18.5. In application we have in mind, $\mathcal{N}$ will be equal to $\partial\mathcal{M}$ or a connected component of $\partial\mathcal{M}$, where $\mathcal{M}$ is a smooth compact Riemannian manifold. As in other sections, we denote norms on boundaries by $|.|$ rather than $\|.\|$ for an easier reading of the statements of the results we prove. In place of (7.1.1), we write

$$|u|_{\tau,s}^2 = \tau^{2s}|u|_{L^2(\partial\mathcal{M})}^2 + |u|_{H^s(\partial\mathcal{M})}^2. \tag{7.1.2}$$

Note that for $0 \leq s < r$, we have

$$|u|_{\tau,s} \lesssim \tau^{s-r}|u|_{\tau,r}, \tag{7.1.3}$$

implying $|u|_{\tau,s} \ll |u|_{\tau,r}$ for $\tau > 0$ chosen sufficiently large.

7.2. Manifold Without Boundary

On a smooth Riemannian manifold without boundary $\mathcal{N}$, the previous section applies. We are now interested in the extension of the above Sobolev norms to the case $s < 0$. We proceed similarly to the definition of $H^s(\mathcal{N})$ in Sect. 18.5.

On $\mathcal{N}$, with g its Riemannian metric, we define P_0 has the unbounded operator on $L^2(\mathcal{N})$ with domain

$$D(\mathsf{P}_0) = \{u \in H^1(\mathcal{N});\ \Delta_g u \in L^2(\mathcal{N})\},$$

given by $\mathsf{P}_0 u = -\Delta_g u$. In Sect. 18.4, we find that $D(\mathsf{P}_0) = H^2(\mathcal{N})$ and that P_0 is selfadjoint with spectrum made of eigenvalues with finite multiplicities, given by

$$0 = \mu_0 < \mu_1 \leq \mu_2 \leq \cdots \leq \mu_n \leq \cdots$$

We denote an associated Hilbert basis of eigenfunctions by $(\phi_j)_j$. If $s \in \mathbb{R}$, then $u \in H^s(\mathcal{N})$ if $u = \sum_j u_j\phi_j$, with $(u_j)_j \subset \mathbb{C}$ such that $\big(\mu_j^{-s/2}u_j\big)_j \in \ell^2(\mathbb{C})$; see the beginning of Sect. 18.5 and Proposition 18.16 for the case $s \geq 0$ and Proposition 18.19 for the case $s < 0$. Then, one has

$$\|u\|_{H^s(\mathcal{N})}^2 \eqsim \sum_j (1+\mu_j)^s |u_j|^2.$$

For any $s \in \mathbb{R}$, we now introduce the norm

$$\|u\|_{\tau,H^s(\mathcal{N})}^2 \eqsim \sum_j (1+\tau^2+\mu_j)^s |u_j|^2,$$

where $\tau \geq 0$. Note that this corresponds to considering $\mathsf{P}_\tau = 1 + \tau^2 + \mathsf{P}_0$ in place of P_0 above. For $\tau \geq 0$ kept fixed, the norms $\|.\|_{\tau,H^s(\mathcal{N})}$ and $\|.\|_{H^s(\mathcal{N})}$ are equivalent. For $s \geq 0$, we have

$$(1+\tau^2+\mu_j)^s \eqsim \tau^{2s} + (1+\mu_j)^s,$$

and we recover

$$\|u\|_{\tau,H^s(\mathcal{N})} \eqsim \tau^s \|u\|_{L^2(\mathcal{N})} + \|u\|_{H^s(\mathcal{N})}, \tag{7.2.1}$$

as in (7.1.1). This equivalence in the case $s \geq 0$ allows us to write $\|u\|_{\tau,s}$ in place of $\|u\|_{\tau,H^s(\mathcal{N})}$. We use this notation also in the case $s < 0$ as no risk of confusion may arise: if we exclude the case of $\mathbb{R}^d$, we have not introduced any norm in the case of a large parameter for $s < 0$ up to now.

Note that we have

$$\|u\|_{\tau,s} \leq \|u\|_{H^s(\mathcal{N})}, \text{ for } s \leq 0, \qquad \|u\|_{H^s(\mathcal{N})} \leq \|u\|_{\tau,s}, \text{ for } s \geq 0. \tag{7.2.2}$$

In what follows, for $s \in \mathbb{R}$, we shall denote by $H^s_\tau(\mathcal{N})$ the space $H^s(\mathcal{N})$ if equipped with the norm $\|.\|_{\tau,s}$. With the definition of these norms for both $s \geq 0$ and $s < 0$, one can identify $H^{-s}_\tau(\mathcal{N})$ as the dual space of $H^s_\tau(\mathcal{N})$ as is done for $H^{-s}(\mathcal{N})$ and $H^s(\mathcal{N})$ in Sect. 18.5.

We have the following result.

PROPOSITION 7.1. *Let $s \in \mathbb{R}$ and $\mathcal{A} = (\mathcal{C}^i)_{1\leq i\leq N}$ be a finite atlas of $\mathcal{N}$ with $\mathcal{C}^i = (O^i, \kappa^i)$, and let $(\psi_i)_{1\leq i\leq N}$ be a smooth partition of unity subordinated to the open covering $(O^i)_{1\leq i\leq N}$. There exists $K = K_{\mathcal{A}} > 0$ such that*

$$K^{-1}\|u\|_{\tau,s} \leq \sum_{1\leq i\leq N} \|\Lambda^s_\tau(\psi_i u)^{\mathcal{C}^i}\|_{L^2(\mathbb{R}^d)} \leq K\|u\|_{\tau,s},$$

for $u \in \mathscr{C}^\infty(\mathcal{N})$. Moreover, if $i \in \{1,\ldots,N\}$, for some $K = K_i > 0$, we have

$$K^{-1}\|u\|_{\tau,s} \leq \|\Lambda^s_\tau u^{\mathcal{C}^i}\|_{L^2(\mathbb{R}^d)} \leq K\|u\|_{\tau,s},$$

if $u \in \mathscr{C}^\infty(\mathcal{N})$ and $\operatorname{supp}(u) \subset O^i$.

We recall that $\Lambda^s_\tau = \operatorname{Op}(\lambda^s_\tau)$ with $\lambda^2_\tau = \tau^2 + |\xi|^2$.

PROOF. In the case $s \geq 0$, with (7.2.1), one sees that the proof follows from the counterpart result in Proposition 18.21. Then, the case $s < 0$ is obtained as in the proof of Proposition 18.22. ■

Naturally, with the norms introduced above for $s \geq 0$ and $s < 0$, one has the following continuity result.

PROPOSITION 7.2. *Let Q_τ be a differential operator with a large parameter with smooth coefficients of order $k \in \mathbb{N}$ as in Sect. 16.3.4. Then, for $s \in \mathbb{R}$, there exists $C > 0$ such that*

$$\|Q_\tau u\|_{\tau,s} \leq C\|u\|_{\tau,s+k}, \qquad u \in \mathscr{C}^\infty(\mathcal{N}),\ \tau > 0.$$

7.3. Trace Norms

As in the above sections, consider ν to be the unique outward pointing vector field on $\mathcal{M}$ along $\partial\mathcal{M}$ such that, for all $m \in \partial\mathcal{M}$, $g_m(\nu_m, \nu_m) = 1$ and $g_m(\nu_m, v) = 0$ for all $v \in T_m\partial\mathcal{M}$. For a function u defined in $\mathcal{M}$, we set

$$\operatorname{tr}^k(u) = (u_{|\partial\mathcal{M}}, \partial_\nu u_{|\partial\mathcal{M}}, \ldots, \partial_\nu^k u_{|\partial\mathcal{M}}),$$

where $\partial_\nu^j u_{|\partial\mathcal{M}}$ is defined in (17.7.3). As shown in Proposition 17.25, in normal geodesic coordinates (to be used locally below), we have $\partial_\nu^j u_{|\partial\mathcal{M}} = (-\partial_{x_d})^j u_{|x_d=0^+}$.

With the norms introduced in the previous section, we then define the following norm, for $k \in \mathbb{N}$ and $r \in \mathbb{R}$,

$$|\operatorname{tr}^k(u)|^2_{\tau,k,r} = \sum_{j=0}^k |\partial_\nu^j u_{|\partial\mathcal{M}}|^2_{\tau,k-j+r}.$$

As there shall be no ambiguity, we shall write $|\operatorname{tr}(u)|_{\tau,k,r}$ instead of $|\operatorname{tr}^k(u)|_{\tau,k,r}$.

For $k, \ell \in \mathbb{N}$ and $r, s \geq 0$, if $k+r < \ell+s$ and $k \leq \ell$, we have $|\operatorname{tr}(u)|_{\tau,k,r} \ll |\operatorname{tr}(u)|_{\tau,\ell,s}$ for $\tau > 0$ chosen sufficiently large.

For a smooth function f defined on $\partial\mathcal{M}$, we shall write

$$|f \operatorname{tr}(u)|^2_{\tau,k,r} = \sum_{j=0}^k |f \partial_\nu^j u_{|\partial\mathcal{M}}|^2_{\tau,k-j+r}.$$

7.4. Carleman Weight Function and Sobolev-Norm Estimation

Our goal in this section is the proof of the following result.

PROPOSITION 7.3. *Let $s \in \mathbb{R}$, and let $\mathcal{N}$ be a smooth compact Riemannian manifold without boundary, Ω be an open set of $\mathcal{N}$, and $\varphi \in \mathscr{C}^\infty(\Omega)$. Let also K be a compact set of Ω, then there exists $C > 0$ such that for any $\tau \geq 1$, we have*

$$\|e^{\tau\varphi} u\|_{\tau,s} \leq C e^{\tau \max_K \varphi} \|u\|_{\tau,s}, \tag{7.4.1}$$

for all $u \in H^s_\tau(\mathcal{N})$ supported in K.

Note that the result of this proposition is clear in the case $s \in \mathbb{N}$. The proof we provide is thus of interest in the case s is not an integer.

As we shall see, the proof can be reduced to a similar estimate on $\mathbb{R}^d$ by means of Proposition 7.1. On the flat space, the argument relies on the introduction of a classical Littlewood–Paley decomposition yet here adapted to the Sobolev norms with the large parameter $\tau > 0$. Then, the product au with $a = e^{\tau\varphi}$ is split into different terms in the spirit of the paraproduct.

We pick $\phi \in \mathscr{C}_c^\infty(\mathbb{R})$, $0 \leq \phi \leq 1$, such that $\phi(t) = 1$ for $|t| \leq 1$ and $\phi(t) = 0$ for $|t| \geq 3/2$. We also set $\varphi(t) = \phi(t/2) - \phi(t)$; note that φ is supported on $\{t \in \mathbb{R};\ 1 \leq |t| \leq 3\}$ and $0 \leq \varphi \leq 1$. We further set

$\varphi_0 = \phi$ and, for $k \geq 1$, $\varphi_k(t) = \varphi(2^{1-k}t)$, the latter being supported in $\{t \in \mathbb{R};\ 2^{k-1} \leq |t| \leq 3 \cdot 2^{k-1}\}$. We have

$$\sum_{k=0}^{\infty} \varphi_k = 1, \tag{7.4.2}$$

with convergence in $\mathscr{C}^\infty(\mathbb{R})$ as the sum is locally finite. Note that any $t \in \mathbb{R}$ lies in the support of at most two functions in the summation.

For $u \in \mathscr{S}'(\mathbb{R}^d)$, and $\tau \geq 1$, we define $\Delta_{\tau,k}u$ for $k \in \mathbb{N}$ through their Fourier transform:

$$\mathscr{F}\Delta_{\tau,k}u(\xi) = \varphi_k(|\xi|/\tau)\hat{u}(\xi).$$

One thus has $u = \sum_{k\in\mathbb{N}} \Delta_{\tau,k}u$ with convergence in $\mathscr{S}'(\mathbb{R}^d)$. Observe that $\mathscr{F}\Delta_{\tau,k}u$ is supported in $\{\xi \in \mathbb{R}^d;\ 2^{k-1}\tau \leq |\xi| \leq 3\tau 2^{k-1}\}$ for $k \geq 1$. For $k \geq 3$, we define also

$$S_{\tau,k}u = \sum_{j=0}^{k-3} \Delta_{\tau,k}u.$$

One has $\mathscr{F}S_{\tau,k}u(\xi) = \phi(2^{3-k}\xi/\tau)\hat{u}(\xi)$. Note that $S_{\tau,k}u$ is supported in $\{\xi \in \mathbb{R}^d,\ |\xi| \leq 3\tau 2^{k-4}\}$. With their forms in the Fourier domain, one sees that the operators $\Delta_{\tau,k}$ and $S_{\tau,k}$ are uniformly bounded on $L^2(\mathbb{R}^d)$ since the functions ϕ and φ are bounded.

LEMMA 7.4. *Let $s \in \mathbb{R}$. One has $\|u\|^2_{\tau,s} \eqsim \sum_{k\in\mathbb{N}} (\tau 2^k)^{2s} \|\Delta_{\tau,k}u\|^2_{L^2(\mathbb{R}^d)}$.*

PROOF. Since $\varphi_k \geq 0$ and since any point meets at most two such functions, for all $t \in \mathbb{R}$, there exists $k \in \mathbb{N}$ such that $\varphi_k(t) \geq 1/2$. One thus finds

$$1/4 \leq \sum_{k\in\mathbb{N}} \varphi_k^2(t) \leq 2, \qquad t \in \mathbb{R}.$$

With this property, we first prove that one has

$$\|u\|^2_{\tau,s} \eqsim \sum_{k\in\mathbb{N}} \|\Delta_{\tau,k}u\|^2_{\tau,s}. \tag{7.4.3}$$

In fact, one writes

$$\begin{aligned}\|u\|^2_{\tau,s} = (\lambda_\tau^{2s}\hat{u}, \hat{u})_{L^2(\mathbb{R}^d)} &\eqsim \sum_{k\in\mathbb{N}} \big(\lambda_\tau^{2s}\varphi_k^2(|\xi|/\tau)\hat{u}, \hat{u}\big)_{L^2(\mathbb{R}^d)} \\ &= \sum_{k\in\mathbb{N}} \|\lambda_\tau^s \varphi_k(|\xi|/\tau)\hat{u}\|^2_{L^2(\mathbb{R}^d)},\end{aligned}$$

which is (7.4.3). We then write

$$\|\Delta_{\tau,k}u\|^2_{\tau,s} = \|\lambda_\tau^s \mathscr{F}\Delta_{\tau,k}u\|^2_{L^2(\mathbb{R}^d)} \eqsim \|(\tau 2^k)^s \mathscr{F}\Delta_{\tau,k}u\|^2_{L^2(\mathbb{R}^d)}, \tag{7.4.4}$$

since $\lambda_\tau \eqsim \tau 2^k$ on $\operatorname{supp}(\mathscr{F}\Delta_{\tau,k}u)$. ∎

We shall need the following estimates in what follows.

LEMMA 7.5. *There exists $C > 0$ such that*

$$\|\Delta_{\tau,k} a\|_{L^\infty(\mathbb{R}^d)} + \|S_{\tau,k} a\|_{L^\infty(\mathbb{R}^d)} \leq C\|a\|_{L^\infty(\mathbb{R}^d)}, \qquad a \in L^\infty(\mathbb{R}^d),\ k \in \mathbb{N}.$$

One also has

$$\|\Delta_{\tau,k} a\|_{L^\infty(\mathbb{R}^d)} \leq C\tau^{-1} 2^{-k} \|a\|_{W^{1,\infty}(\mathbb{R}^d)}, \qquad a \in W^{1,\infty}(\mathbb{R}^d),\ k \geq 1.$$

A proof is given below.

With Lemma 7.5, for two functions a and u in $\mathscr{S}(\mathbb{R}^d)$, one may write

$$\begin{aligned} au &= \sum_{k,j\in\mathbb{N}} \Delta_{\tau,j} a\, \Delta_{\tau,k} u \\ &= \sum_{k\geq j+3} \Delta_{\tau,j} a\, \Delta_{\tau,k} u + \sum_{j\geq k+3} \Delta_{\tau,j} a\, \Delta_{\tau,k} u + \sum_{|j-k|\leq 2} \Delta_{\tau,j} a\, \Delta_{\tau,k} u \\ &= \sum_{k=3}^{\infty} S_{\tau,k} a\, \Delta_{\tau,k} u + \sum_{k=3}^{\infty} S_{\tau,k} u\, \Delta_{\tau,k} a + \sum_{|j-k|\leq 2} \Delta_{\tau,j} a\, \Delta_{\tau,k} u, \end{aligned} \tag{7.4.5}$$

where convergence holds in $L^\infty(\mathbb{R}^d)$. With the support theorem (see Proposition 8.42 of Volume 1), we find

(7.4.6)

$$\operatorname{supp}\Big(\mathscr{F}\big(S_{\tau,k} a \Delta_{\tau,k} u\big)\Big) \subset \{\xi \in \mathbb{R}^d;\ 5\tau 2^{k-4} \leq |\xi| \leq 27\tau 2^{k-4}\}, \qquad k \geq 3$$

and $\Delta_{\tau,j}\big(S_{\tau,k} a \Delta_{\tau,k} u\big) = 0$ if $|j-k| \geq 3$.

One also finds

(7.4.7)

$$\operatorname{supp}\Big(\mathscr{F}\big(\Delta_{\tau,j} a \Delta_{\tau,k} u\big)\Big) \subset \{\xi \in \mathbb{R}^d;\ |\xi| \leq 3\tau 2^{\max(k,j)}\}, \qquad k,j \in \mathbb{N}.$$

and $\Delta_{\tau,n}\big(\Delta_{\tau,j} a \Delta_{\tau,k} u\big) = 0$ if $n \geq 3 + \max(k,j)$.

PROPOSITION 7.6. *Let $s \in \mathbb{R}$. There exists $C > 0$ such that*

$$\|au\|_{\tau,s} \leq C\Big(\sum_{|\alpha|\leq \lfloor |s| \rfloor + 1} \tau^{-|\alpha|} \|\partial^\alpha a\|_{L^\infty(\mathbb{R}^d)}\Big) \|u\|_{\tau,s}, \tag{7.4.8}$$

for $a \in W^{\lfloor |s| \rfloor + 1,\infty}(\mathbb{R}^d)$ and $u \in H^s_\tau(\mathbb{R}^d)$.

For $s \in \mathbb{N}$, a direct proof is obtained by computing $\tau^{s-|\alpha|}\partial^\alpha(au)$. In what follows, we shall assume $s \in \mathbb{R} \setminus \mathbb{N}$. To estimate the H^s_τ-norm of the product au, we need the following lemma, whose proof is given below.

LEMMA 7.7. *We have the following three estimations:*

(1) *There exists $C_0 > 0$ such that*

$$\|S_{\tau,k} a \Delta_{\tau,k} u\|_{L^2(\mathbb{R}^d)} \leq C_0 \|a\|_{L^\infty(\mathbb{R}^d)} \|\Delta_{\tau,k} u\|_{L^2(\mathbb{R}^d)}, \qquad k \geq 3, \tag{7.4.9}$$

for $a \in L^\infty(\mathbb{R}^d)$ and $u \in H^s_\tau(\mathbb{R}^d)$.

(2) *Let $s > 0$. There exists C_1 such that if*

$$\delta_n = (\tau 2^n)^s \Big\| \sum_{\substack{|j-k|\leq 2,\\ k\geq n}} \Delta_{\tau,j} a \Delta_{\tau,k} u \Big\|_{L^2(\mathbb{R}^d)}, \qquad n \in \mathbb{N},$$

then $\delta = (\delta_n)_{n\in\mathbb{N}} \in \ell^2(\mathbb{N})$ and

$$\|\delta\|_{\ell^2} \leq C_1 \|a\|_{L^\infty(\mathbb{R}^d)} \|u\|_{\tau,s}, \tag{7.4.10}$$

for $a \in L^\infty(\mathbb{R}^d)$ and $u \in H^s_\tau(\mathbb{R}^d)$.

(3) *Let $s \in (0,1)$. There exists C_2 such that*

$$(\tau 2^k)^s \|S_{\tau,k} u \Delta_{\tau,k} a\|_{L^2(\mathbb{R}^d)} \leq C\tau^{-1} \|a\|_{W^{1,\infty}(\mathbb{R}^d)} \|u\|_{\tau,s} 2^{-(1-s)k}, \tag{7.4.11}$$

for $a \in W^{1,\infty}(\mathbb{R}^d)$ and $u \in H^s_\tau(\mathbb{R}^d)$.

Proof of Proposition 7.6. We prove that each term in the r.h.s. of (7.4.5) fulfills the estimation in (7.4.8).

With the support of the Fourier transform of $S_{\tau,k} a \Delta_{\tau,k} u$ given in (7.4.6), we have

$$(\tau 2^n)^s \Big\| \Delta_{\tau,n} \big(\sum_{k=3}^{\infty} S_{\tau,k} a \Delta_{\tau,k} u \big) \Big\|_{L^2(\mathbb{R}^d)} \lesssim \sum_{|n-k|\leq 2} (\tau 2^k)^s \|S_{\tau,k} a \Delta_{\tau,k} u\|_{L^2(\mathbb{R}^d)}. \tag{7.4.12}$$

With Lemma 7.4 and (7.4.9), this implies $\| \sum_{k=3}^{\infty} S_{\tau,k} a \Delta_{\tau,k} u\|_{\tau,s} \lesssim \|a\|_{L^\infty(\mathbb{R}^d)} \|u\|_{\tau,s}$.

With the support of the Fourier transform of $\Delta_{\tau,j} a \Delta_{\tau,k} u$ given in (7.4.7), we have

$$\begin{aligned}(\tau 2^n)^s \Big\| \Delta_{\tau,n} \big(\sum_{|j-k|\leq 2} \Delta_{\tau,j} a \Delta_{\tau,k} u \big) \Big\|_{L^2(\mathbb{R}^d)} \\ \lesssim (\tau 2^n)^s \Big\| \sum_{\substack{|j-k|\leq 2\\ \max(j,k)+2\geq n}} \Delta_{\tau,j} a \Delta_{\tau,k} u \Big\|_{L^2(\mathbb{R}^d)} = \delta_n.\end{aligned}$$

If $s > 0$ with (7.4.10), one has $(\delta_n)_{n\in\mathbb{N}} \in \ell^2(\mathbb{N})$ and

$$\sum_{n\in\mathbb{N}} (\tau 2^n)^{2s} \Big\| \Delta_{\tau,n} \big(\sum_{|j-k|\leq 2} \Delta_{\tau,j} a \Delta_{\tau,k} u \big) \Big\|^2_{L^2(\mathbb{R}^d)} \lesssim \|a\|^2_{L^\infty(\mathbb{R}^d)} \|u\|^2_{\tau,s}.$$

This implies that $\| \sum_{|j-k|\leq 2} \Delta_{\tau,j}(a) \Delta_{\tau,k}(u)\|_{\tau,s} \lesssim \|a\|_{L^\infty(\mathbb{R}^d)} \|u\|_{\tau,s}$ by Lemma 7.4.

For $s \in (0,1)$, with the support of the Fourier transform of $S_{\tau,k} u \Delta_{\tau,k} a$ given in (7.4.6) and (7.4.11), we have

$$\begin{aligned}(\tau 2^n)^s \Big\| \Delta_{\tau,n} \big(\sum_{k=3}^{\infty} S_{\tau,k} u \Delta_{\tau,k} a \big) \Big\|_{L^2(\mathbb{R}^d)} &\lesssim \sum_{|n-k|\leq 2} (\tau 2^k)^s \|S_{\tau,k} u \Delta_{\tau,k} a\|_{L^2(\mathbb{R}^d)} \\ &\lesssim \tau^{-1} \|a\|_{W^{1,\infty}(\mathbb{R}^d)} \|u\|_{\tau,s} 2^{-(1-s)n}.\end{aligned}$$

With Lemma 7.4, we find $\| \sum_{k=3}^{\infty} S_{\tau,k} u \Delta_{\tau,k} a \|_{\tau,s} \lesssim \tau^{-1} \|a\|_{W^{1,\infty}(\mathbb{R}^d)} \|u\|_{\tau,s}$ for $s \in (0,1)$.

With the above estimations, the result of the proposition is proven for $s \in (0,1)$. We now prove the result for other values of s.

For $s > 0$, $s \notin \mathbb{N}$, we have $s = p + \sigma$ with $\sigma \in (0,1)$ and $p = \lfloor s \rfloor$. Applying the result of the proposition proven for $\sigma \in (0,1)$, we have

$$\begin{aligned}
\|au\|_{\tau,s} &\approx \sum_{|\alpha| \leq p} \tau^{p-|\alpha|} \|\partial^\alpha (au)\|_{\tau,\sigma} \approx \sum_{|\alpha|+|\beta| \leq p} \tau^{p-|\alpha|-|\beta|} \|\partial^\alpha a \partial^\beta u)\|_{\tau,\sigma} \\
&\approx \sum_{|\alpha|+|\beta| \leq p} \tau^{p-|\alpha|-|\beta|-1} \|\partial^\alpha a\|_{W^{1,\infty}(\mathbb{R}^d)} \|\partial^\beta u\|_{\tau,\sigma} \\
&\lesssim \Big(\sum_{|\alpha| \leq p+1} \tau^{-|\alpha|} \|\partial^\alpha a\|_{L^\infty(\mathbb{R}^d)} \Big) \|u\|_{\tau,s},
\end{aligned}$$

as $\tau^{p-|\beta|} \|\partial^\beta u\|_{\tau,\sigma} \lesssim \|u\|_{\tau,s}$. We thus obtain the result of the proposition for $s \geq 0$, recalling that the case $s \in \mathbb{N}$ follows from explicit computations.

To obtain the result for $s < 0$, we argue by duality. Let $\sigma = -s$, for $u \in H_\tau^{-\sigma}(\mathbb{R}^d)$, $v \in H_\tau^\sigma(\mathbb{R}^d)$, and $a \in W^{\lfloor \sigma \rfloor + 1}(\mathbb{R}^d)$. First note that $au \in H_\tau^{-\sigma}(\mathbb{R}^d)$. Second, we write

$$\begin{aligned}
|\langle au, v \rangle_{H_\tau^{-\sigma}(\mathbb{R}^d), H_\tau^\sigma(\mathbb{R}^d)}| &= |\langle u, av \rangle_{H_\tau^{-\sigma}(\mathbb{R}^d), H_\tau^\sigma(\mathbb{R}^d)}| \leq \|u\|_{\tau,-\sigma} \|av\|_{\tau,\sigma} \\
&\lesssim \|u\|_{\tau,-\sigma} \Big(\sum_{|\alpha| \leq \lfloor \sigma \rfloor + 1} \tau^{-|\alpha|} \|\partial^\alpha a\|_{L^\infty(\mathbb{R}^d)} \Big) \|v\|_{\tau,\sigma},
\end{aligned}$$

which implies the result for $s < 0$. ∎

We may now proceed with the proof of Proposition 7.3.

Proof of Proposition 7.3. Below we prove that (7.4.1) holds in the case of $\mathbb{R}^d$. This suffices for the conclusion in the case $\mathcal{N}$ is a smooth Riemannian manifold without boundary. In fact, with a finite atlas $\mathcal{A} = (\mathcal{C}^i)_{1 \leq i \leq N}$, with $\mathcal{C}^i = (\kappa^i, O^i)$, and a subordinated partition of unity $(\psi_i)_{1 \leq i \leq N}$, by applying Proposition 7.1 twice, one has

$$\begin{aligned}
\|e^{\tau\varphi} u\|_{\tau,s} &\lesssim \sum_{1 \leq i \leq N} \|\Lambda_\tau^s (\psi_i e^{\tau \varphi^{\mathcal{C}^i}} u^{\mathcal{C}^i})\|_{L^2(\mathbb{R}^d)} \\
&\lesssim \sum_{1 \leq i \leq N} e^{\tau \sup_{x \in \kappa(K \cap O^i)} \varphi^{\mathcal{C}^i}(x)} \|\Lambda_\tau^s (\psi_i u^{\mathcal{C}^i})\|_{L^2(\mathbb{R}^d)} \\
&\lesssim e^{\tau \max_K \varphi} \sum_{1 \leq i \leq N} \|\Lambda_\tau^s (\psi_i u^{\mathcal{C}^i})\|_{L^2(\mathbb{R}^d)} \\
&\lesssim e^{\tau \max_K \varphi} \|u\|_{\tau,s}.
\end{aligned}$$

In what follows, we can assume that Ω is an open set in $\mathbb{R}^d$ and we use the Euclidean distance. There exists $\chi_\tau \in \mathscr{C}_c^\infty(\Omega)$ such that $\chi_\tau \equiv 1$ in a neighborhood of K, $\chi_\tau(x) = 0$ if $\operatorname{dist}(x, K) \geq \tau^{-1}$, and for any multi-index α, one has $|\partial^\alpha \chi_\tau| \lesssim \tau^{|\alpha|}$. As $e^{\tau\varphi} u = e^{\tau\varphi} \chi_\tau u$, by Proposition 7.6

with $a = e^{\tau\varphi}\chi_\tau$ therein, we only need to estimate the L^∞-norm on $\mathbb{R}^d$ of $\tau^{-\alpha}\partial^\alpha(e^{\tau\varphi}\chi_\tau)$ for $|\alpha| = \lfloor |s| \rfloor + 1$. This term is a linear combination of terms of the form

$$\tau^{-|\alpha|+k} e^{\tau\varphi}(\partial^{\beta_1}\varphi)\cdots(\partial^{\beta_k}\varphi)\partial^{\beta'}\chi_\tau, \tag{7.4.13}$$

with $1 \le k \le |\alpha|$, $1 \le |\beta_i|$, $i = 1, \dots, k$, and $\beta_1 + \cdots + \beta_k + \beta' = \alpha$. In particular, $k + |\beta'| \le |\alpha|$. Hence, each term given by (7.4.13) has its L^∞-norm estimated by $C\|e^{\tau\varphi}\|_{L^\infty(K_\tau)}$, for some $C > 0$ independent of τ, and where $K_\tau = \{x \in \mathbb{R}^d;\ \mathrm{dist}(x, K) \le \tau^{-1}\}$.

The conclusion follows if we prove that $e^{\tau\varphi(y)} \lesssim e^{\tau \max_K \varphi}$, for all $y \in K_\tau$. We note that if $y \in K_\tau \setminus K$, there exists $x \in K$ such that $|x - y| \le \tau^{-1}$. We then write

$$\begin{aligned}\varphi(y) &\le \varphi(x) + |\varphi(x) - \varphi(y)| \le \varphi(x) + C|x - y| \\ &\le \max_K \varphi + C\tau^{-1}.\end{aligned}$$

We deduce that $e^{\tau\varphi(y)} \le e^{\tau \max_K \varphi + C}$, which gives the result. ∎

Proof of Lemma 7.5. Set $\tilde{\phi}$ and $\tilde{\varphi}$ as the inverse Fourier transforms of $\phi(|\xi|)$ and $\varphi(|\xi|)$, respectively. One has

$$\Delta_{\tau,0}a = \tau^d\tilde{\phi}(\tau x) * a \quad \text{and} \quad \Delta_{\tau,k}a = (\tau 2^{k-1})^d\tilde{\varphi}(\tau 2^{k-1}x) * a, \quad k \ge 1,$$

and

$$S_{\tau,k}a = (\tau 2^{k-3})^d\tilde{\phi}(\tau 2^{k-3}x) * a, \quad k \ge 3.$$

As $(\tau 2^n)^d\tilde{\phi}(2^n\tau x)$ and $(\tau 2^n)^d\tilde{\varphi}(2^n\tau x)$ are bounded in $L^1(\mathbb{R}^d)$ uniformly with respect to n and τ, we obtain the first inequality of the lemma.

For the second inequality, for $k \ge 1$, we write

$$\begin{aligned}\Delta_{\tau,k}a(x) &= (\tau 2^{k-1})^d \int_{\mathbb{R}^d} \tilde{\varphi}\big(\tau 2^{k-1}(x - y)\big)a(y)\,dy \\ &= (\tau 2^{k-1})^d \int_{\mathbb{R}^d} \tilde{\varphi}\big(\tau 2^{k-1}(x - y)\big)\big(a(y) - a(x)\big)dy,\end{aligned}$$

as $\tilde{\varphi}$ has a zero mean since $\varphi(0) = 0$. We deduce that

$$\begin{aligned}\|\Delta_{\tau,k}a\|_{L^\infty(\mathbb{R}^d)} &\lesssim (\tau 2^{k-1})^d\|a\|_{W^{1,\infty}(\mathbb{R}^d)} \int_{\mathbb{R}^d} \big|\tilde{\varphi}\big(\tau 2^{k-1}(x - y)\big)\big||x - y|\,dy \\ &\lesssim 2^{-k}\tau^{-1}\|a\|_{W^{1,\infty}(\mathbb{R}^d)},\end{aligned}$$

by a change of variables and using that $|x|\tilde{\varphi}(x) \in L^1(\mathbb{R}^d)$. ∎

Proof of Lemma 7.7. Let $a \in L^\infty(\mathbb{R}^d)$ and $u \in H^s_\tau(\mathbb{R}^d)$. From Lemma 7.5, we have

$$\|S_k(a)\Delta_k(u)\|_{L^2(\mathbb{R}^d)} \lesssim \|S_k(a)\|_{L^\infty}\|\Delta_k(u)\|_{L^2(\mathbb{R}^d)} \lesssim \|a\|_{L^\infty}\|\Delta_k(u)\|_{L^2(\mathbb{R}^d)}.$$

This gives (7.4.9).

Let now $s > 0$. We have by Lemma 7.5

$$\tau^s 2^{sn} \Big\| \sum_{\substack{|j-k|\leq 2,\\ k\geq n}} \Delta_{\tau,j} a \Delta_{\tau,k} u \Big\|_{L^2(\mathbb{R}^d)} \lesssim \|a\|_{L^\infty} \sum_{k\geq n} 2^{-s(k-n)} (2^{ks}\tau^s \|\Delta_k(u)\|_{L^2(\mathbb{R}^d)}).$$

As $(2^{-sn})_n$ is in $\ell^1(\mathbb{N})$ since $s > 0$ and $(2^{ks}\tau^s\|\Delta_k(u)\|_{L^2(\mathbb{R}^d)})_k$ is in $\ell^2(\mathbb{N})$ with norm estimated by $\|u\|_{\tau,s}$ by Lemma 7.4, by discrete convolution, we obtain that $\delta \in \ell^2$ with δ as in the statement and (7.4.10) follows.

Let now $s \in (0,1)$ and $k \geq 3$. As $S_{\tau,k}$ is uniformly bounded on $L^2(\mathbb{R}^d)$ and from Lemma 7.5, we have

$$\begin{aligned}
(\tau 2^k)^s \|S_{\tau,k} u \Delta_{\tau,k} a\|_{L^2(\mathbb{R}^d)} &\lesssim (\tau 2^k)^s \|u\|_{L^2(\mathbb{R}^d)} \|\Delta_{\tau,k} a\|_{L^\infty(\mathbb{R}^d)} \\
&\lesssim 2^{-k(1-s)} \tau^{s-1} \|u\|_{L^2(\mathbb{R}^d)} \|a\|_{W^{1,\infty}(\mathbb{R}^d)} \\
&\lesssim 2^{-k(1-s)} \tau^{-1} \|u\|_{\tau,s} \|a\|_{W^{1,\infty}(\mathbb{R}^d)},
\end{aligned}$$

which proves (7.4.11) ∎

7.5. Notes

The contents we present in this chapter are very classical and somewhat already used in the previous chapters, yet not for norms of orders that are integers. The present chapter is thus mainly written to make precise Sobolev norms with a large parameter at boundaries, in particular in the case of noninteger Sobolev orders (positive or negative). Trace inequalities are also written. This prepares for Chap. 8 where these norms are needed to derive more precise Carleman estimates at boundaries. The basic material for (usual) Sobolev spaces on a Riemannian manifold can be found in Chapter 18.

The Sobolev-norm estimation of the map $u \mapsto e^{\tau\varphi}u$ given in Proposition 7.3 is based on the analysis tools developed for the so-called paraproduct with extensions to paradifferential calculus. The reader is referred to the original article by J.-M. Bony [76] and the articles by Y. Meyer [248, 249]. See also the books of Y. Meyer and R.R. Coifman [250], M.E. Taylor [318], and H. Bahouri et al. [50].

CHAPTER 8

Estimates for General Boundary Conditions

Contents

8.1. Introduction

In Chap. 2, Lopatinskiĭ–Šapiro boundary conditions are introduced. Our goal is now to use these conditions and to generalize the Carleman estimates obtained in Chap. 5 in the case of Dirichlet boundary conditions on

J. Le Rousseau et al., *Elliptic Carleman Estimates and Applications to Stabilization and Controllability, Volume II*, PNLDE Subseries in Control 98, https://doi.org/10.1007/978-3-030-88670-7_8

a Riemannian manifold. The Lopatinskiĭ–Šapiro framework for the study of a boundary value problem is natural as it means that the elliptic system formed by the elliptic operator and the boundary operator is Fredholm; see Chap. 3.

We thus consider $(\mathcal{M}, g)$ a d-dimensional smooth compact Riemannian manifold. We refer to Chap. 15 for the notions of differential geometry that are needed in what follows. The elliptic operator under consideration is $P = -\Delta_g + R_1$ with R_1 a first-order differential operator on $\mathcal{M}$ with bounded coefficients. We also consider a differential operator B defined at (or in a neighborhood of) the boundary $\partial\mathcal{M}$. The elliptic problem under consideration is thus the following:

$$Pu = f \in L^2(\mathcal{M}), \qquad Bu_{|\partial\mathcal{M}} = f_\partial, \tag{8.1.1}$$

with f_∂ in a prescribed space depending on the nature of the operator B.

For a reader interested in a second-order elliptic operator in an open set of $\mathbb{R}^d$, we refer to Sect. 18.9 that exposes how the Riemannian point of view can be used.

As in Chap. 2, the boundary operator B is chosen of the form

$$B = B^k + B^{k-1}\partial_\nu, \tag{8.1.2}$$

one every connected component $\mathcal{N}$ of $\partial\mathcal{M}$, with B^k and B^{k-1} smooth differential operators on $\mathcal{N}$ of order k and $k-1$, respectively, with $k \leq \beta$. Naturally, if B is of order zero on $\mathcal{N}$, one has $B^{k-1} = 0$.

In Sect. 8.2, we show that the Lopatinskiĭ–Šapiro condition is robust under conjugation by a weight function, a key ingredient in the derivation of a Carleman estimate. In Sect. 8.3, we show how Carleman estimates can be obtained under the Lopatinskiĭ–Šapiro boundary condition. Finally, in Sect. 8.4, we take advantage of the analysis tools we develop to derive estimates in the case no boundary condition is imposed; yet the Dirichlet trace and a first-order trace need to be known. This allows us to refine the counterpart estimates derived in Chap. 5 (see Lemma 5.3 and Theorem 5.5).

8.2. Lopatinskiĭ–Šapiro Condition After Conjugation

We naturally use the same notation as in Sect. 2.2.

Let φ be a smooth function of $\mathcal{M}$. By Proposition 16.17, the operator $P_\varphi = e^{\tau\varphi} P e^{-\tau\varphi}$ is a second-order differential operator with a large parameter on $\mathcal{M}$, and its principal symbol is given by

$$\sigma(P_\varphi)(m, \omega, \tau) = p(m, \omega + i\tau d\varphi(m)) = (\omega + i\tau d\varphi(m), \omega + i\tau d\varphi(m))_g,$$

with $(.,.)_g$ the bilinear form for cotangent vectors associated with the metric at m, see (17.1.4) in Sect. 17.1. Following Sect. 2.2, for $(m, \omega') \in T^*\partial\mathcal{M}$, $z \in \mathbb{R}$, and $\tau \geq 0$, we set $\omega = \omega' - z n_m$ in the above symbol and define

$$\check{p}_\varphi(m, \omega', z, \tau) = p(m, \omega' - z n_m + i\tau d\varphi(m)).$$

We recall that n is the outward pointing unitary one-form on $\mathcal{M}$ along $\partial\mathcal{M}$, that is, $n = \nu^\flat$, with ν the unique outward pointing unitary vector field that is orthogonal to $T_m\partial\mathcal{M}$ in the sense of g.

Let $m \in \partial\mathcal{M}$. If $v \in T_m\mathcal{M}$, we have $v = t\nu_m + v'$ with $v' \in T_m\partial\mathcal{M}$. Going back to the definition of a tangent vector (see Sect. 15.3), we have $v'(\varphi) = v'(\varphi_{|\partial\mathcal{M}}) = d\varphi_{|\partial\mathcal{M}}(m)(v')$, where $d\varphi_{|\partial\mathcal{M}} = d(\varphi_{|\partial\mathcal{M}})$ is the differential on the submanifold $\partial\mathcal{M}$ of the restriction $\varphi_{|\partial\mathcal{M}}$. In particular, $d(\varphi_{|\partial\mathcal{M}})(m) \in T^*_m\partial\mathcal{M}$. We thus have

$$d\varphi(m)(v) = t d\varphi(m)(\nu_m) + d\varphi(m)(v') = t\partial_\nu\varphi(m) + d\varphi_{|\partial\mathcal{M}}(m)(v').$$

We thus write $d\varphi(m) = \partial_\nu\varphi(m)n_m + d\varphi_{|\partial\mathcal{M}}(m)$, which is the decomposition of $d\varphi(m)$ according to the orthogonal direct sum $N^*_m\partial\mathcal{M} \oplus T^*_m\partial\mathcal{M} = T^*_m\mathcal{M}$ (orthogonality to be understood with respect to g_m). Decomposing $\omega + i\tau d\varphi(m)$ accordingly, that is,

$$\omega + i\tau d\varphi(m) = \omega' + i\tau d\varphi_{|\partial\mathcal{M}}(m) + \big(-z + i\tau\partial_\nu\varphi(m)\big)n_m,$$

yields

$$\check{p}_\varphi(m,\omega',z,\tau) = \big(z - i\tau\partial_\nu\varphi(m)\big)^2 + \alpha^2, \tag{8.2.1}$$

with $\alpha = \alpha(m,\omega',\tau)$ such that $\operatorname{Re}\alpha \geq 0$ and

$$\begin{aligned} \alpha^2 &= p(m,\omega' + i\tau d\varphi_{|\partial\mathcal{M}}(m)) \\ &= |\omega'|^2_{g_\partial} - \tau^2|d\varphi_{|\partial\mathcal{M}}(m)|^2_{g_\partial} + 2i\tau(\omega', d\varphi_{|\partial\mathcal{M}}(m))_{g_\partial} \\ &= |\omega'|^2_{g_\partial} - \tau^2|d\varphi_{|\partial\mathcal{M}}(m)|^2_{g_\partial} + 2i\tau\omega'^\sharp(\varphi_{|\partial\mathcal{M}}), \end{aligned} \tag{8.2.2}$$

as we have, noticing that $\omega'^\sharp \in T_m\partial\mathcal{M}$,

$$(\omega', d\varphi(m))_g = \omega'^\sharp(\varphi) = \omega'^\sharp(\varphi_{|\partial\mathcal{M}}) = d\varphi_{|\partial\mathcal{M}}(m)(\omega'^\sharp) = (\omega', d\varphi_{|\partial\mathcal{M}}(m))_{g_\partial},$$

where g_∂ is the induced metric on $\partial\mathcal{M}$. We thus obtain

$$\check{p}_\varphi(m,\omega',z,\tau) = (z - \gamma_2)(z - \gamma_1),$$

with

$$\gamma_j = \gamma_j(m,\omega',\tau) = i\tau\partial_\nu\varphi(m) + i(-1)^j\alpha(m,\omega',\tau), \qquad j = 1,2. \tag{8.2.3}$$

As in Definition 2.2 (see also Remark 2.2), we define the following polynomial:

$$\check{p}^+_\varphi(m,\omega',z,\tau) = \prod_{\operatorname{Im}\gamma_j \geq 0} (z - \gamma_j),$$

with the convention that $\check{p}^+_\varphi(m,\omega',z,\tau) = 1$ if all the roots of $\check{p}_\varphi$ have negative imaginary parts. Depending on the sign of these roots, the polynomial $\check{p}^+_\varphi$ is thus of degree 0, 1, or 2.

We also define a conjugated version of the boundary operator B. Assume that m is in a connected component $\mathcal{N}$ of $\partial\mathcal{M}$ where B is of order $k \leq \beta$. We have $B = B^k + B^{k-1}\partial_\nu$, where B^k and B^{k-1} are smooth differential operators on $\mathcal{N}$ of order k and $k-1$, respectively. We denote by $b(m,\omega)$, $b^k(m,\omega')$, and $b^{k-1}(m,\omega')$ their respective principal symbols.

We set $B_\varphi = e^{\tau\varphi}Be^{-\tau\varphi}$, which is a differential operator with a large parameter of order k on a neighborhood of $\partial\mathcal{M}$, and its principal symbol is given by

$$\begin{aligned} b_\varphi(m,\omega,\tau) &= b\big(m,\omega + i\tau d\varphi(m)\big) \\ &= b^k\big(m,\omega' + i\tau d\varphi_{|\partial\mathcal{M}}(m)\big) \\ &\quad + ib^{k-1}\big(m,\omega' + i\tau d\varphi_{|\partial\mathcal{M}}(m)\big)\big(\omega_n + i\tau\partial_\nu\varphi(m)\big), \end{aligned}$$

where $\omega = \omega' + \omega_n n_m$ with $\omega' \in T^*_m\partial\mathcal{M}$. For $(m,\omega') \in T^*\partial\mathcal{M}$, and $\tau \geq 0$, we set

$$\begin{aligned} \check{b}_\varphi(m,\omega',z,\tau) &= b(m,\omega' - zn_m + i\tau d\varphi(m)) \\ &= b^k\big(m,\omega' + i\tau d\varphi_{|\partial\mathcal{M}}(m)\big) \\ &\quad + ib^{k-1}\big(m,\omega' + i\tau d\varphi_{|\partial\mathcal{M}}(m)\big)\big(i\tau\partial_\nu\varphi(m) - z\big). \end{aligned}$$

Note that $b^k\big(m,\omega' + i\tau d\varphi_{|\partial\mathcal{M}}(m)\big)$ and $b^{k-1}\big(m,\omega' + i\tau d\varphi_{|\partial\mathcal{M}}(m)\big)$ are polynomial functions in τ of degree less than or equal to k and $k-1$, respectively, by Proposition 16.13.

We may now state the Lopatinskiĭ–Šapiro condition for the conjugated operators, first microlocally, second at one point, and third locally.

DEFINITION 8.1 (Lopatinskiĭ–Šapiro Condition for Conjugated Operators). Let $(m,\omega') \in T^*\partial\mathcal{M}$ and $\tau \geq 0$ with $(\omega',\tau) \neq 0$. We say that the Lopatinskiĭ–Šapiro condition holds for (P,B,φ) at (m,ω',τ) if for any polynomial function $f(z)$ with complex coefficients, there exist $c \in \mathbb{C}$ and a polynomial function $\ell(z)$ with complex coefficients such that, for all $z \in \mathbb{C}$,

$$f(z) = c\check{b}_\varphi(m,\omega',z,\tau) + \ell(z)\check{p}^+_\varphi(m,\omega',z,\tau). \tag{8.2.4}$$

We say that the Lopatinskiĭ–Šapiro condition holds for (P,B,φ) at $m \in \partial\mathcal{M}$ if it holds at (m,ω',τ) for all $\omega' \in T^*\partial\mathcal{M}$ and $\tau \geq 0$ such that $(\omega',\tau) \neq 0$. If $\Gamma \subset \partial\mathcal{M}$, we say that the Lopatinskiĭ–Šapiro condition holds for (P,B,φ) in Γ if it holds at m for all $m \in \Gamma$.

REMARK 8.2. In Definition 8.1, we have $\tau \geq 0$. Note that the case $\tau = 0$ coincides with the Lopatinskiĭ–Šapiro condition for (P,B); see Definition 2.2.

REMARK 8.3. With the Euclidean division of polynomials, we see that it suffices to consider the polynomial function $f(z)$ to be of degree less than that of $\check{p}^+_\varphi(m,\omega',z,\tau)$ in (8.2.4). In any case, the degree of $f(z)$ can be chosen less than or equal to one.

REMARK 8.4. Similarly to what was observed in Remark 2.4, note that the Lopatinskiĭ–Šapiro property is written here without any use of local coordinates. It is thus a geometrical condition. As the principal parts of the conjugated operators P_φ and B_φ are geometrical objects (see Sect. 16.3.4), if a particular local chart is chosen, then the Lopatinskiĭ–Šapiro condition for (P, B, φ) can be equivalently expressed in this chart.

LEMMA 8.5. *Let* $(m, \omega') \in T^*\partial\mathcal{M}$ *and* $\tau \geq 0$ *with* $(\omega', \tau) \neq 0$. *The Lopatinskiĭ–Šapiro condition holds for* (P, B, φ) *at* (m, ω', τ) *if and only if*

(1) *either* $\check{p}_\varphi^+(m, \omega', z, \tau) = 1$

(2) *or* $\check{p}_\varphi^+(m, \omega', z, \tau) = z - \gamma$ *and* $\check{b}_\varphi(m, \omega', \gamma, \tau) \neq 0$.

PROOF. If $\check{p}_\varphi^+(m, \omega', z, \tau) = \check{p}_\varphi(m, \omega', z, \tau) = (z - \gamma_1)(z - \gamma_2)$, that is, both roots γ_1 and γ_2 have nonnegative imaginary parts, then condition (8.2.4) cannot hold, as by Remark 8.3, it means that the vector space of polynomials of degree less than or equal to one would be generated by the single polynomial $\check{b}_\varphi(m, \omega', z, \tau)$.

If $\check{p}_\varphi^+(m, \omega', z, \tau)$ is of degree one, then one of the roots, say γ_1, has a negative imaginary part and the second one, γ_2, has a nonnegative imaginary part. We have $\check{p}_\varphi^+(m, \omega', z, \tau) = z - \gamma_2$. Then, the Lopatinskiĭ–Šapiro condition holds at (m, ω', τ) if for any $f(z)$, the polynomial function $z \mapsto f(z) - c\check{b}(m, \omega', z, \tau)$ admits γ_2 for a root for some $c \in \mathbb{C}$. A necessary and sufficient condition is then $\check{b}(m, \omega', \gamma_2, \tau) \neq 0$. Observe that in this case, the root configuration is precisely that of the non-conjugated operators (compare with Proposition 2.3).

If finally $\check{p}_\varphi^+(m, \omega', z, \tau) = 1$, that is, both roots have negative imaginary parts, then the condition (8.2.4) holds trivially. ■

PROPOSITION 8.6. *Let* $(m^0, \omega^{0\prime}) \in T^*\partial\mathcal{M}$ *and* $\tau^0 \geq 0$ *with* $(\omega^{0\prime}, \tau^0) \neq 0$ *such that the Lopatinskiĭ–Šapiro condition holds for* (P, B, φ) *at* $(m^0, \omega^{0\prime}, \tau^0)$. *Then, there exists a conic neighborhood of* $(m, \omega^{0\prime}, \tau^0) \in T^*\partial\mathcal{M} \times \mathbb{R}_+$ *where the Lopatinskiĭ–Šapiro condition holds also.*

Let $m^0 \in \partial\mathcal{M}$ *be such that Lopatinskiĭ–Šapiro condition holds for* (P, B, φ) *at* m. *Then, there exists a neighborhood of* m^0 *in* $\partial\mathcal{M}$ *where the Lopatinskiĭ–Šapiro condition holds also.*

PROOF. The second part of the proof follows from the first part, using homogeneity and compacity.

If the Lopatinskiĭ–Šapiro condition holds at $(m^0, \omega^{0\prime}, \tau^0)$, by Lemma 8.5, one faces the following two cases.

Case 1: $\check{p}_\varphi^+(m^0, \omega^{0\prime}, z, \tau^0) = 1$, meaning that both roots γ_1 and γ_2 are such that $\operatorname{Im} \gamma_1 < 0$ and $\operatorname{Im} \gamma_2 < 0$. With the continuity of the roots as given by Lemma 6.27, one sees that $\check{p}_\varphi^+$ remains equal to one in a conic neighborhood.

Case 2: one root, say $\gamma_1(m^0, \omega^{0\prime}, \tau^0)$, has a nonnegative imaginary part. Then,

$$\check{p}_\varphi^+(m^0, \omega^{0\prime}, z, \tau^0) = z - \gamma_1(m^0, \omega^{0\prime}, \tau^0), \qquad \operatorname{Im} \gamma_2(m^0, \omega^{0\prime}, \tau^0) < 0$$

and

$$\check{b}_\varphi(m^0, \omega^{0\prime}, \gamma_1(m^0, \omega^{0\prime}, \tau^0), \tau^0) \neq 0.$$

Again, with the continuity of the roots, one has $\operatorname{Im}\gamma_2(m,\omega',\tau) < 0$ and

$$\check{b}_\varphi(m,\omega',\gamma_1(m,\omega',\tau),\tau) \neq 0 \tag{8.2.5}$$

for (m,ω',τ) in a conic neighborhood $\mathscr{U}$ of $(m^0,\omega^{0\prime},\tau^0)$.

For a point $(m,\omega',\tau) \in \mathscr{U}$, either $\operatorname{Im}\gamma_1(m,\omega',\tau) < 0$ yielding $\check{p}_\varphi^+(m,\omega', z,\tau) = 1$ or $\operatorname{Im}\gamma_1(m,\omega',\tau) \geq 0$ yielding $\check{p}_\varphi^+(m,\omega',z,\tau) = z - \gamma_1(m,\omega',\tau)$. In either case, Lemma 8.5 implies that the Lopatinskiĭ–Šapiro condition holds at (m,ω',τ). ∎

The following theorem states how the Lopatinskiĭ–Šapiro condition *before* conjugation of Definition 2.2 for (P,B) may or may not imply the Lopatinskiĭ–Šapiro condition *after* conjugation of Definition 8.1 for (P,B,φ) for some choices of φ.

THEOREM 8.7. *Let P, B, and φ be as above. Let $m \in \partial\mathcal{M}$.*

1. *If $\partial_\nu\varphi(m) \geq 0$, then the Lopatinskiĭ–Šapiro condition for (P,B,φ) does not hold at m.*
2. *Let the Lopatinskiĭ–Šapiro condition hold for (P,B) at m, and let φ be such that $\partial_\nu\varphi(m) < 0$. We have the following results:*
   (a) *If B is of order zero with nonvanishing principal symbol (Dirichlet boundary conditions), then the Lopatinskiĭ–Šapiro condition holds for (P,B,φ) at m.*
   (b) *Let B be of order $1 \leq k \leq \beta$ on $\mathcal{N}$ the connected component of $\partial\mathcal{M}$ that contains m. There exists $\mu_0 > 0$, depending only on the principal symbols p and b, such that the Lopatinskiĭ–Šapiro condition holds for (P,B,φ) at m if*

$$|d\varphi_{|\partial\mathcal{M}}(m)|_{g_\partial} \leq \mu_0 |\partial_\nu\varphi(m)|. \tag{8.2.6}$$

   By continuity, if the Lopatinskiĭ–Šapiro condition holds for (P,B) in a compact subset Γ of $\partial\mathcal{M}$, then there exists $\mu_0 > 0$ such that if (8.2.6) holds in Γ, then the Lopatinskiĭ–Šapiro condition holds for (P,B,φ) in Γ.
   (c) *Assume that B is of order $k = 1$ on $\mathcal{N}$ the connected component of $\partial\mathcal{M}$ that contains m with principal symbol of the form $b(m,\omega) = \omega_n + \langle\omega',t'_m\rangle$ for $\omega \in T_m^*\mathcal{M}$, $\omega = \omega_n n_m + \omega'$ with $\omega' \in T_m^*\partial\mathcal{M}$, and where t' is a real vector field on $\partial\mathcal{M}$ (see Example 2.5-(3)). If $d = 2$, the Lopatinskiĭ–Šapiro condition holds for (P,B,φ) at m. If $d \geq 3$, the Lopatinskiĭ–Šapiro condition holds for (P,B,φ) at m if and only if we have*

$$|\partial_\nu\varphi(m)| > \langle d\varphi_{|\partial\mathcal{M}}(m), t'_m\rangle. \tag{8.2.7}$$

   (d) *Assume that B is of order $k = 1$ on $\mathcal{N}$ the connected component of $\partial\mathcal{M}$ that contains m with principal symbol of the form $b(m,\omega) = \omega_n + i\langle\omega',v'_m\rangle$ for $\omega \in T_m^*\mathcal{M}$, $\omega = \omega_n n_m + \omega'$ with $\omega' \in T_m^*\partial\mathcal{M}$, and where v' is a real vector field on $\partial\mathcal{M}$. If*

$d=2$, necessarily $|v'_m| \neq 1$ and the Lopatinskiĭ–Šapiro condition holds for (P,B,φ) at m. If $d \geq 3$, necessarily $|v'_m|_{g_\partial} < 1$ and the Lopatinskiĭ–Šapiro condition holds for (P,B,φ) at m if and only if

$$(1-|v'_m|^2_{g_\partial})(\partial_\nu\varphi(m))^2 > |v'_m|^2_{g_\partial}|d\varphi_{|\partial\mathcal{M}}(m)|^2_{g_\partial} - \langle d\varphi_{|\partial\mathcal{M}}(m), v'_m\rangle^2. \tag{8.2.8}$$

The remainder of this section is devoted to the proof of this theorem.

REMARK 8.8. Observe that Item (2b) states that if the Lopatinskiĭ–Šapiro condition holds for (P,B) at m, if $\partial_\nu\varphi(m) < 0$, and if φ is constant on $\partial\mathcal{M}$, then the Lopatinskiĭ–Šapiro condition holds for (P,B,φ) at m.

By Item (2c), observe also that in the case of the Neumann boundary condition, we see that the Lopatinskiĭ–Šapiro condition holds for (P,B,φ) at m *if and only if* $\partial_\nu\varphi(m) < 0$.

Observe also that Item (2c) covers the case a boundary operator with principal symbol of the form $b(m,\omega) = c\langle\omega, v_m\rangle$ where $c \in \mathbb{C}^*$ and v is vector field in $\mathcal{M}$ along $\partial\mathcal{M}$ with $v^\nu_m = \langle n_m, v_m\rangle \neq 0$.

PROOF OF THEOREM 8.7.

Item (1). We prove that the equality $\check{p}^+_\varphi = \check{p}_\varphi$ occurs for some choices of ω' and τ. Then by Lemma 8.5, the Lopatinskiĭ–Šapiro condition cannot hold for (P,B,φ).

First, we assume that $\partial_\nu\varphi(m) = 0$, and then for $\omega' = 0$ and $\tau > 0$, we have $\alpha^2 \leq 0$ for α defined in (8.2.2) implying that $\operatorname{Re}\alpha = 0$. Consequently, for the roots of $z \mapsto \check{p}_\varphi(m,\omega',z,\tau)$ given in (8.2.3), we have $\operatorname{Im}\gamma_j = 0$ for $j=1,2$. Hence, $\check{p}^+_\varphi = \check{p}_\varphi$.

Second, we assume that $\partial_\nu\varphi(m) > 0$, and we pick some $\omega' \in T^*_m\partial\mathcal{M}$. Then, we write

$$\alpha(m,\omega',\tau)^2 = \tau^2\big(-|d\varphi_{|\partial\mathcal{M}}(m)|^2_g + |\omega'/\tau|^2_g + 2i\tau^{-1}\omega'^\sharp(\varphi_{|\partial\mathcal{M}})\big).$$

As $\tau\to\infty$, we have $(\alpha/\tau)^2 \to c \in \mathbb{C}$ with $\operatorname{Re} c \leq 0$, implying that $\operatorname{Re}\alpha/\tau \to 0$. As the roots of $z \mapsto \check{p}_\varphi(m,\omega',z,\tau)$ read

$$\gamma_j = i\tau\big(\partial_\nu\varphi(m) + (-1)^j\alpha(m,\omega',\tau)/\tau\big), \quad j=1,2,$$

we see that $\operatorname{Im}\gamma_j > 0$ for $\tau > 0$ chosen sufficiently large for both $j=1,2$. Hence, as above, $\check{p}^+_\varphi = \check{p}_\varphi$.

Item (2). As the Lopatinskiĭ–Šapiro condition holds for (P,B), it is sufficient to only consider $\tau > 0$ in the proof by Remark 8.2.

The condition $\partial_\nu\varphi(m) < 0$ implies $\operatorname{Im}\gamma_1 < 0$ with γ_1 given in (8.2.3). If $0 \leq \operatorname{Re}\alpha < \tau|\partial_\nu\varphi|$, we have $\operatorname{Im}\gamma_2 < 0$ and thus $\check{p}^+_\varphi = 1$; then the Lopatinskiĭ–Šapiro condition holds for (P,B,φ) at m by Lemma 8.5. If $\operatorname{Re}\alpha \geq \tau|\partial_\nu\varphi|$, we have $\operatorname{Im}\gamma_2 \geq 0$ and thus $\check{p}^+_\varphi = z-\gamma_2$; then by Lemma 8.5, the Lopatinskiĭ–Šapiro condition holds for (P,B,φ) at m if and only if $\check{b}_\varphi(m,\omega',\gamma_2,\tau) \neq 0$. The Lopatinskiĭ–Šapiro condition for (P,B,φ) at m

thus holds if and only if for any $\omega' \in T^*_m\partial\mathcal{M}$ and $\tau > 0$, we have

$$\operatorname{Re}\alpha(m,\omega',\tau) \geq \tau|\partial_\nu\varphi(m)| \quad \Rightarrow \quad \check{b}_\varphi(m,\omega',\gamma_2,\tau) \neq 0. \tag{8.2.9}$$

Item (2a). If B is of order zero, the principal symbol is such that $b(m,\omega) \neq 0$ is independent of ω. The condition $\check{b}_\varphi(m,\omega',\gamma_2,\tau) \neq 0$ thus holds.

Item (2b). We consider the case $\operatorname{Re}\alpha \geq \tau|\partial_\nu\varphi(m)| > 0$, that is, $\operatorname{Im}\gamma_2 \geq 0$. Lemma 8.11 below yields the equivalent condition

$$4\tau^2(\partial_\nu\varphi(m))^2 \operatorname{Re}\alpha^2 - 4\tau^4(\partial_\nu\varphi(m))^4 + (\operatorname{Im}\alpha^2)^2 \geq 0.$$

In turn, the definition of α^2 in (8.2.2) gives the equivalent condition

$$(\partial_\nu\varphi(m))^2\big(|\omega'|^2_{g_\partial} - \tau^2|d(\varphi_{|\partial\mathcal{M}})(m)|^2_g\big) - \tau^2(\partial_\nu\varphi(m))^4 + (\omega'^\sharp(\varphi_{|\partial\mathcal{M}}))^2 \geq 0,$$

which reads

$$(\tau\partial_\nu\varphi(m))^2\,|d\varphi(m)|^2_g \leq (\partial_\nu\varphi(m))^2|\omega'|^2_{g_\partial} + (\omega'^\sharp(\varphi_{|\partial\mathcal{M}}))^2. \tag{8.2.10}$$

In particular, we see that $\omega' \neq 0$ if $\operatorname{Im}\gamma_2 \geq 0$. With (8.2.3), and writing $d\varphi(m) = d\varphi_{|\partial\mathcal{M}}(m) + \partial_\nu\varphi(m)n_m$, we compute

$$\begin{aligned}\check{b}_\varphi(m,\omega',z=\gamma_2,\tau) &= b\big(m,\omega' - \gamma_2 n_m + i\tau d\varphi(m)\big)\\ &= b\big(m,\omega' - i|\omega'|_{g_\partial}n_m - i(\alpha - |\omega'|_{g_\partial})n_m\\ &\qquad + i\tau d\varphi_{|\partial\mathcal{M}}(m)\big).\end{aligned}$$

We now claim that for $\varepsilon > 0$, there exists $\mu_0 > 0$ such that

$$\begin{aligned}&|d\varphi_{|\partial\mathcal{M}}(m)|_{g_\partial} \leq \mu_0|\partial_\nu\varphi(m)| \text{ and } \operatorname{Im}\gamma_2 \geq 0\\ &\qquad\Rightarrow \quad |\alpha - |\omega'|_{g_\partial}| + \tau|(d\varphi_{|\partial\mathcal{M}})(m)|_{g_\partial} \leq \varepsilon|\omega'|_{g_\partial}.\end{aligned} \tag{8.2.11}$$

We then use that $|b(m,\omega' - i|\omega'|_{g_\partial}n_m)| \geq C_0|\omega'|^k_{g_\partial}$, for some $C_0 > 0$ as the Lopatinskiĭ–Šapiro condition holds for (P,B) at m, for $\omega' \neq 0$, and as b is homogeneous of degree k in ω here. With the mean value theorem and using that $d_\omega b$ is homogeneous of degree $k-1$ in ω, we then find that $|\check{b}_\varphi(m,\omega',z=\gamma_2,\tau)| \geq C|\omega'|^k_{g_\partial} > 0$ for $\mu_0 > 0$ chosen sufficiently small and φ satisfying the condition (8.2.6), using that in the case $\operatorname{Im}\gamma_2 \geq 0$, we have $\omega' \neq 0$, as seen above. This concludes the proof of the item (2b) of the theorem.

Proof of the Claim (8.2.11). Condition (8.2.10) and $|(d\varphi_{|\partial\mathcal{M}})(m)|_{g_\partial} \leq \mu_0|\partial_\nu\varphi(m)|$ imply, with the Cauchy–Schwarz inequality,

$$\begin{aligned}\tau^2(\partial_\nu\varphi(m))^4 &\leq (\partial_\nu\varphi(m))^2|\omega'|^2_{g_\partial} + |\omega'|^2_{g_\partial}|(d\varphi_{|\partial\mathcal{M}})(m)|^2_{g_\partial}\\ &\leq (1+\mu_0^2)(\partial_\nu\varphi(m))^2|\omega'|^2_{g_\partial},\end{aligned}$$

yielding $\tau|\partial_\nu\varphi(m)| \leq (1+\mu_0)|\omega'|_{g_\partial}$. We thus obtain

$$\tau|(d\varphi_{|\partial\mathcal{M}})(m)|_{g_\partial} \leq \mu_0(1+\mu_0)|\omega'|_{g_\partial}. \tag{8.2.12}$$

The form of α^2 given in (8.2.2) and inequality (8.2.12) give $\alpha^2 = |\omega'|^2_{g_\partial}(1 + \mathcal{O}(\mu_0))$. We thus obtain

$$\alpha = |\omega'|_{g_\partial}|(1 + \mathcal{O}(\mu_0)). \tag{8.2.13}$$

Inequalities (8.2.12) and (8.2.13) yield the claim (8.2.11).

Item (2c). Firstly, we consider the case $d = 2$. Then $T^*_m\partial\mathcal{M}$ is of dimension one. We pick e_m such that $\operatorname{span}(e_m) = T^*_m\partial\mathcal{M}$ and $|e_m|_{g_\partial} = 1$. Let $s \in \mathbb{R}$ be such that $\tau d\varphi_{|\partial\mathcal{M}}(m) = se_m$. For $\omega' = s'e_m$, we have

$$\alpha^2 = p(m, \omega' + i\tau d\varphi_{|\partial\mathcal{M}}(m)) = (s' + is)^2|e_m|^2_{g_\partial} = (s' + is)^2,$$

which gives $\alpha = \operatorname{sgn}(s')(s' + is)$. We thus find, by (8.2.3),

$$\begin{aligned}\check{b}_\varphi(m, \omega', z = \gamma_2, \tau) &= \langle\omega' + i\tau d\varphi_{|\partial\mathcal{M}}(m), t'_m\rangle - i\alpha\\ &= (s' + is)\big(\langle e_m, t'_m\rangle - i\operatorname{sgn}(s')\big).\end{aligned}$$

Assume that for some $\omega' \in T^*_m\partial\mathcal{M}$ and $\tau > 0$, we have

$$|s'| = \operatorname{Re}\alpha(m, \omega', \tau) \geq \tau|\partial_\nu\varphi(m)| \quad\text{and}\quad \check{b}_\varphi(m, \omega', z = \gamma_2, \tau) = 0.$$

Then, $s' \neq 0$, and thus $0 = \langle e_m, t'_m\rangle - i\operatorname{sgn}(s')$; we obtain a contradiction. The Lopatinskiĭ–Šapiro condition for (P, B, φ) thus holds at m.

Secondly, we treat the case $d \geq 3$. Assume that (8.2.7) holds. We have

$$\check{b}_\varphi(m, \omega', z, \tau) = \langle\omega' + i\tau d\varphi_{|\partial\mathcal{M}}(m), t'_m\rangle + i\tau\partial_\nu\varphi(m) - z,$$

yielding, by (8.2.3),

$$\check{b}_\varphi(m, \omega', z = \gamma_2, \tau) = \langle\omega' + i\tau d\varphi_{|\partial\mathcal{M}}(m), t'_m\rangle - i\alpha. \tag{8.2.14}$$

Assume that for some $\omega' \in T^*_m\partial\mathcal{M}$ and $\tau > 0$ we have

$$\operatorname{Re}\alpha(m, \omega', \tau) \geq \tau|\partial_\nu\varphi(m)| \quad\text{and}\quad \check{b}_\varphi(m, \omega', z = \gamma_2, \tau) = 0. \tag{8.2.15}$$

We then have

$$\tau\langle d\varphi_{|\partial\mathcal{M}}(m), t'_m\rangle = \operatorname{Re}\alpha \geq \tau|\partial_\nu\varphi(m)|.$$

We have thus reached a contradiction with (8.2.7).

Let us now assume that (8.2.7) does not hold at m, that is,

$$\langle d\varphi_{|\partial\mathcal{M}}(m), t'_m\rangle \geq |\partial_\nu\varphi(m)| > 0. \tag{8.2.16}$$

We wish to find $\omega' \in T^*_m\partial\mathcal{M}$ and $\tau > 0$ such that (8.2.15) holds. By (8.2.14), the condition $\check{b}_\varphi(m, \omega', z = \gamma_2, \tau) = 0$ reads

$$\alpha(m, \omega', \tau) = \langle\tau d\varphi_{|\partial\mathcal{M}}(m) - i\omega', t'_m\rangle, \tag{8.2.17}$$

yielding

$$\alpha(m, \omega', \tau)^2 = \langle\tau d\varphi_{|\partial\mathcal{M}}(m), t'_m\rangle^2 - \langle\omega', t'_m\rangle^2 - 2i\langle\tau d\varphi_{|\partial\mathcal{M}}(m), t'_m\rangle\langle\omega', t'_m\rangle.$$

With α^2 given by (8.2.2), considering the real and imaginary parts separately, we find

$$\tau^2\big(\langle d\varphi_{|\partial\mathcal{M}}(m), t'_m\rangle^2 + |d\varphi_{|\partial\mathcal{M}}(m)|^2_{g_\partial}\big) = \langle\omega', t'_m\rangle^2 + |\omega'|^2_{g_\partial}, \tag{8.2.18}$$

$$\langle d\varphi_{|\partial\mathcal{M}}(m), t'_m\rangle\langle\omega', t'_m\rangle = -\omega'^{\sharp}(\varphi_{|\partial\mathcal{M}}) = -\langle\omega', d\varphi_{|\partial\mathcal{M}}(m)^{\sharp}\rangle. \tag{8.2.19}$$

The solutions ω' of (8.2.19) form a linear subspace of $T^*_m\partial\mathcal{M}$ of dimension $d-2 \geq 1$. If we pick one of them different from zero, there exists $\tau > 0$ such that (8.2.18) holds also, as here $\langle d\varphi_{|\partial\mathcal{M}}(m), t'_m\rangle \neq 0$. We thus have

$$\alpha(m, \omega', \tau)^2 = \langle\tau d\varphi_{|\partial\mathcal{M}}(m) - i\omega', t'_m\rangle^2.$$

We may thus choose α as in (8.2.17), since our requirement is $\operatorname{Re}\alpha \geq 0$ (see above (8.2.2)) and here we have $\langle d\varphi_{|\partial\mathcal{M}}(m), t'_m\rangle \geq 0$ by (8.2.16). We thus obtain

$$\operatorname{Re}\alpha(m, \omega', \tau) = \langle\tau d\varphi_{|\partial\mathcal{M}}(m), t'_m\rangle.$$

By (8.2.16), we thus obtain that (8.2.15) holds for this choice of $\omega' \in T_m\partial\mathcal{M}$ and $\tau > 0$, meaning that the Lopatinskiĭ–Šapiro condition does not hold for (P, B, φ) at m. This concludes the proof of the item (2c) of the theorem.

Item (2d). Firstly, we consider the case $d = 2$. By Proposition 2.8, if the Lopatinskiĭ–Šapiro condition holds for (P, B), then $|v'_m|_{g_\partial} \neq 1$. Here $T^*_m\partial\mathcal{M}$ is of dimension one. We pick e_m such that $\operatorname{span}(e_m) = T^*_m\partial\mathcal{M}$ and $|e_m|_{g_\partial} = 1$. Let $s \in \mathbb{R}$ be such that $\tau d\varphi_{|\partial\mathcal{M}}(m) = s e_m$. For $\omega' = s' e_m$, we have

$$\alpha^2 = p(m, \omega' + i\tau d\varphi_{|\partial\mathcal{M}}(m)) = (s' + is)^2 |e_m|^2_{g_\partial} = (s' + is)^2,$$

which gives $\alpha = \operatorname{sgn}(s')(s' + is)$. We thus find, by (8.2.3),

$$\begin{aligned}\check{b}_\varphi(m, \omega', z = \gamma_2, \tau) &= i\langle\omega' + i\tau d\varphi_{|\partial\mathcal{M}}(m), v'_m\rangle - i\alpha \\ &= i(s' + is)\big(\langle e_m, v'_m\rangle - \operatorname{sgn}(s')\big).\end{aligned}$$

Assume that for some $\omega' \in T^*_m\partial\mathcal{M}$ and $\tau > 0$, we have

$$|s'| = \operatorname{Re}\alpha(m, \omega', \tau) \geq \tau|\partial_\nu\varphi(m)| \ \text{ and } \ \check{b}_\varphi(m, \omega', z = \gamma_2, \tau) = 0.$$

Then, $s' \neq 0$, and thus $0 = \langle e_m, v'_m\rangle - \operatorname{sgn}(s')$; we obtain a contradiction as $|v'_m|_{g_\partial} \neq 1$. The Lopatinskiĭ–Šapiro condition for (P, B, φ) thus holds at m.

Secondly, we treat the case $d \geq 3$. Since the Lopatinskiĭ–Šapiro condition holds for (P, B), then $|v'_m|_{g_\partial} < 1$ by Proposition 2.8. Assume that (8.2.8) holds. If $d\varphi_{|\partial\mathcal{M}}(m) = 0$, then the Lopatinskiĭ–Šapiro condition holds for (P, B, φ) at m by Item (2b) of the theorem that is proven above. If $v'_m = 0$, then the Lopatinskiĭ–Šapiro condition holds for (P, B, φ) at m by Item (2c) of the theorem that is proven above.

We may thus assume that $v'_m \neq 0$ and $d\varphi_{|\partial\mathcal{M}}(m) \neq 0$. We have

$$\check{b}_\varphi(m, \omega', z, \tau) = i\langle\omega' + i\tau d\varphi(m), v'_m\rangle + i\tau\partial_\nu\varphi(m) - z.$$

By (8.2.3), it yields

$$\check{b}_\varphi(m,\omega',z=\gamma_2,\tau) = i\langle \omega' + i\tau d\varphi_{|\partial\mathcal{M}}(m), v'_m\rangle - i\alpha(m,\omega',\tau). \tag{8.2.20}$$

Assume that for some $\omega' \in T^*_m\partial\mathcal{M}$ and $\tau > 0$, we have

$$\operatorname{Re}\alpha(m,\omega',\tau) \geq \tau|\partial_\nu\varphi(m)| \ \text{ and } \ \check{b}_\varphi(m,\omega',z=\gamma_2,\tau) = 0. \tag{8.2.21}$$

We then have

$$\langle \omega', v'_m\rangle = \operatorname{Re}\alpha \geq \tau|\partial_\nu\varphi(m)|. \tag{8.2.22}$$

Since $\check{b}_\varphi(m,\omega',z=\gamma_2,\tau)=0$, we have $\alpha(m,\omega',\tau) = \langle \omega' + i\tau d\varphi_{|\partial\mathcal{M}}(m), v'_m\rangle$ implying

$$\begin{aligned}\alpha(m,\omega',\tau)^2 = \langle \omega', v'_m\rangle^2 &- \tau^2\langle d\varphi_{|\partial\mathcal{M}}(m), v'_m\rangle^2 \\ &+ 2i\tau\langle d\varphi_{|\partial\mathcal{M}}(m), v'_m\rangle\langle \omega', v'_m\rangle.\end{aligned}$$

With α^2 given by (8.2.2), considering the real and imaginary parts separately, we find

$$\tau^2\Big(|d\varphi_{|\partial\mathcal{M}}(m)|^2_{g_\partial} - \langle d\varphi_{|\partial\mathcal{M}}(m), v'_m\rangle^2\Big) = |\omega'|^2_{g_\partial} - \langle \omega', v'_m\rangle^2 \tag{8.2.23}$$

and

$$(\omega', d\varphi_{|\partial\mathcal{M}}(m))_{g_\partial} = \langle d\varphi_{|\partial\mathcal{M}}(m), v'_m\rangle\langle \omega', v'_m\rangle. \tag{8.2.24}$$

If $\operatorname{rank}(d\varphi_{|\partial\mathcal{M}}(m)^\sharp, v'_m) \leq 1$, then $d\varphi_{|\partial\mathcal{M}}(m)^\sharp = av'_m$ with $a \neq 0$ since $d\varphi_{|\partial\mathcal{M}}(m) \neq 0$ here. Condition (8.2.24) leads to $a(1-|v'_m|^2_{g_\partial})\langle \omega', v'_m\rangle = 0$. Since $|v'_m|_{g_\partial} < 1$, we obtain $\langle \omega', v'_m\rangle = 0$, which contradicts (8.2.22).

Assume now that $\operatorname{rank}(d\varphi_{|\partial\mathcal{M}}(m)^\sharp, v'_m) = 2$ in addition to condition (8.2.8). Introduce

$$u' = d\varphi_{|\partial\mathcal{M}}(m)^\sharp - \langle d\varphi_{|\partial\mathcal{M}}(m), v'_m\rangle v'_m. \tag{8.2.25}$$

Condition (8.2.24) reads $\langle \omega', u'\rangle = 0$. Note that $\operatorname{rank}(u', v'_m) = 2$.

Set $Q = \operatorname{span}((v'_m)^\flat, d\varphi_{|\partial\mathcal{M}}(m)) = \operatorname{span}((v'_m)^\flat, (u')^\flat)$ and by ω'_0 the orthogonal projection of ω' onto Q. Introduce also p' the orthogonal projection of v'_m onto $\{u'\}^\perp$. Since $\operatorname{rank}(u', v'_m) = 2$, we have $p' \neq 0$ and $p' = v'_m - |u'|^{-2}_{g_\partial}(v'_m, u')_{g_\partial}u'$ and $(p')^\flat \in Q$. We have $Q = \operatorname{span}((p')^\flat, (u')^\flat)$.

Lemma 8.9. *We have* $\omega'_0 = \langle \omega', v'_m\rangle|p'|^{-2}_{g_\partial}(p')^\flat$.

Proof. Set $\tilde{\omega}' = \langle \omega', v'_m\rangle|p'|^{-2}_{g_\partial}(p')^\flat$. We have

$$\langle \omega' - \tilde{\omega}', p'\rangle = \langle \omega', p'\rangle - \langle \omega', v'_m\rangle = -|u'|^{-2}_{g_\partial}(v'_m, u')_{g_\partial}\langle \omega', u'\rangle = 0,$$

since $\langle \omega', u'\rangle = 0$. We also have $\langle \omega' - \tilde{\omega}', u'\rangle = 0$, as $\tilde{\omega}'$ is also orthogonal to u', since $\langle p', u'\rangle = 0$. Thus, $\omega' - \tilde{\omega}'$ is orthogonal to Q meaning that $\omega'_0 = \tilde{\omega}'$. ∎

Set $a \geq 1$ to be such that $\langle \omega', v'_m \rangle = a\tau |\partial_\nu \varphi(m)|$. We write $\omega' = \omega'_0 + \zeta'$, where ζ' is orthogonal to Q by the definition of ω'_0. Note that we have $\zeta' = 0$ in the case $d = 3$ since $T^*_m \partial \mathcal{M} = Q$.

We shall need the following lemma whose proof is given in Appendix 8.A.1.

LEMMA 8.10. *We have*

$$|p'|_{g_\partial}^{-2} - 1 = \big(1 - |v'_m|^2_{g_\partial}\big) \frac{\big(|d\varphi_{|\partial\mathcal{M}}(m)|^2_{g_\partial} - \langle d\varphi_{|\partial\mathcal{M}}(m), v'_m\rangle^2\big)}{|v'_m|^2_{g_\partial} |d\varphi_{|\partial\mathcal{M}}(m)|^2_{g_\partial} - \langle d\varphi_{|\partial\mathcal{M}}(m), v'_m\rangle^2}.$$

From condition (8.2.23), we have

$$\begin{aligned}
|\zeta'|^2 &= |\omega'|^2_{g_\partial} - |\omega'_0|^2_{g_\partial} \\
&= \langle \omega', v'_m\rangle^2 - |\omega'_0|^2_{g_\partial} + \tau^2\Big(|d\varphi_{|\partial\mathcal{M}}(m)|^2_{g_\partial} - \langle d\varphi_{|\partial\mathcal{M}}(m), v'_m\rangle^2\Big) \\
&= \tau^2\Big(a^2\big(1 - |p'|^{-2}_{g_\partial}\big)(\partial_\nu\varphi(m))^2 + |d\varphi_{|\partial\mathcal{M}}(m)|^2_{g_\partial} - \langle d\varphi_{|\partial\mathcal{M}}(m), v'_m\rangle^2\Big).
\end{aligned}$$

With Lemma 8.10, we obtain

$$\begin{aligned}
|\zeta'|^2 = M\Big(&a^2\big(|v'_m|^2_{g_\partial} - 1\big)(\partial_\nu\varphi(m))^2 + |v'_m|^2_{g_\partial}|d\varphi_{|\partial\mathcal{M}}(m)|^2_{g_\partial} \\
&- \langle d\varphi_{|\partial\mathcal{M}}(m), v'_m\rangle^2\Big),
\end{aligned}$$

with

$$M = \tau^2 \frac{|d\varphi_{|\partial\mathcal{M}}(m)|^2_{g_\partial} - \langle d\varphi_{|\partial\mathcal{M}}(m), v'_m\rangle^2}{|v'_m|^2_{g_\partial}|d\varphi_{|\partial\mathcal{M}}(m)|^2_{g_\partial} - \langle d\varphi_{|\partial\mathcal{M}}(m), v'_m\rangle^2} > 0,$$

since $|v'_m|_{g_\partial} < 1$. Necessarily, we have

$$\begin{aligned}
&\big(|v'_m|^2_{g_\partial} - 1\big)(\partial_\nu\varphi(m))^2 + |v'_m|^2_{g_\partial}|d\varphi_{|\partial\mathcal{M}}(m)|^2_{g_\partial} - \langle d\varphi_{|\partial\mathcal{M}}(m), v'_m\rangle^2 \\
&\quad \geq a^2\big(|v'_m|^2_{g_\partial} - 1\big)(\partial_\nu\varphi(m))^2 + |v'_m|^2_{g_\partial}|d\varphi_{|\partial\mathcal{M}}(m)|^2_{g_\partial} \\
&\qquad - \langle d\varphi_{|\partial\mathcal{M}}(m), v'_m\rangle^2 \\
&\quad = |\zeta'|^2/M \geq 0.
\end{aligned}$$

This gives a contradiction with (8.2.8). Having condition (8.2.8) thus implies that the Lopatinskiĭ–Šapiro condition holds for (P, B, φ) at m.

Let us now assume that (8.2.8) does not hold at m, that is,

$$|v'_m|^2_{g_\partial}|d\varphi_{|\partial\mathcal{M}}(m)|^2_{g_\partial} - \langle d\varphi_{|\partial\mathcal{M}}(m), v'_m\rangle^2 \geq \big(1 - |v'_m|^2_{g_\partial}\big)(\partial_\nu\varphi(m))^2. \tag{8.2.26}$$

Since $|v'_m|_{g_\partial} < 1$ here, observe that this implies that $\operatorname{rank}(d\varphi_{|\partial\mathcal{M}}(m)^\sharp, v'_m) = 2$. We wish to find $\omega' \in T^*_m\partial\mathcal{M}$ and $\tau > 0$ such that (8.2.21) holds. By (8.2.20), the condition $\check{b}_\varphi(m, \omega', z = \gamma_2, \tau) = 0$ reads

$$\alpha(m, \omega', \tau) = \langle \omega' + i\tau d\varphi_{|\partial\mathcal{M}}(m), v'_m\rangle. \tag{8.2.27}$$

If we find ω' and $\tau > 0$ such that $\alpha(m, \omega', \tau)^2 = \langle \omega' + i\tau d\varphi_{|\partial\mathcal{M}}(m), v'_m\rangle^2$ and $\langle \omega', v'_m\rangle \geq \tau|\partial_\nu\varphi(m)|$, then (8.2.27) holds since $\operatorname{Re}\alpha \geq 0$, and thus (8.2.21) is fulfilled meaning that the Lopatinskiĭ–Šapiro condition does not hold for

(P, B, φ) at m. We then conclude that condition (8.2.8) is necessary for the Lopatinskiĭ–Šapiro condition to hold for (P, B, φ) at m. The remainder of the proof thus concerns the construction of (ω', τ).

We recall that having $\alpha(m, \omega', \tau)^2 = \langle \omega' + i\tau d\varphi_{|\partial\mathcal{M}}(m), v'_m\rangle^2$ is equivalent to (8.2.23)–(8.2.24). With u' as given by (8.2.25), the second condition reads $\langle \omega', u'\rangle = 0$. We shall also use the notation introduced in the proof of the sufficiency of condition (8.2.8) given above.

Let $\tau > 0$ to be kept fixed and $a \geq 1$ to be determined below. Set

$$\omega'_0 = a|p'|_{g_\partial}^{-2}\tau|\partial_\nu\varphi(m)|(p')^\sharp \in Q,$$

and set $\omega' = \omega'_0 + \zeta'$ with ζ' orthogonal to Q. Note that in the case $d = 3$, we have $Q = T^*_m\partial\mathcal{M}$ meaning that $\zeta' = 0$. We also have $\langle \omega', v'_m\rangle = a\tau|\partial_\nu\varphi(m)|$.

We have $\langle \omega', u'\rangle = 0$ since $(u')^\flat \in Q$ and since p' is orthogonal to u'. Hence condition (8.2.24) is fulfilled. Now condition (8.2.23) reads

$$\tau^2\Big(|d\varphi_{|\partial\mathcal{M}}(m)|_{g_\partial}^2 - \langle d\varphi_{|\partial\mathcal{M}}(m), v'_m\rangle^2\Big) = |\omega'_0|_{g_\partial}^2 + |\zeta'|_{g_\partial}^2 - (a\tau\partial_\nu\varphi(m))^2.$$

As $|\omega'_0|_{g_\partial}^2 = (a\tau\partial_\nu\varphi(m))^2/|p'|_{g_\partial}^2$, we find

$$\begin{aligned}
|\zeta'|_{g_\partial}^2 &= (1 - |p'|_{g_\partial}^{-2})(a\tau\partial_\nu\varphi(m))^2 + \tau^2\Big(|d\varphi_{|\partial\mathcal{M}}(m)|_{g_\partial}^2 \\
&\qquad - \langle d\varphi_{|\partial\mathcal{M}}(m), v'_m\rangle^2\Big) \\
&= M\Big(a^2(|v'_m|_{g_\partial}^2 - 1)(\partial_\nu\varphi(m))^2 + |v'_m|_{g_\partial}^2|d\varphi_{|\partial\mathcal{M}}(m)|_{g_\partial}^2 \\
&\qquad - \langle d\varphi_{|\partial\mathcal{M}}(m), v'_m\rangle^2\Big).
\end{aligned}$$

with the computations made above. By (8.2.26), we have

$$\big(|v'_m|_{g_\partial}^2 - 1\big)(\partial_\nu\varphi(m))^2 + |v'_m|_{g_\partial}^2|d\varphi_{|\partial\mathcal{M}}(m)|_{g_\partial}^2 - \langle d\varphi_{|\partial\mathcal{M}}(m), v'_m\rangle^2 \geq 0.$$

Since $|v'_m|_{g_\partial} < 1$ for some $a \geq 1$, we obtain

$$a\big(|v'_m|_{g_\partial}^2 - 1\big)(\partial_\nu\varphi(m))^2 + |v'_m|_{g_\partial}^2|d\varphi_{|\partial\mathcal{M}}(m)|_{g_\partial}^2 - \langle d\varphi_{|\partial\mathcal{M}}(m), v'_m\rangle^2 = 0.$$

This means that for this value of $a \geq 1$, if we choose $\omega' = \omega'_0$, that is, $\zeta' = 0$, condition (8.2.23) is also fulfilled. Note that with the choice of $\zeta' = 0$, we cover the cases $d = 3$ and $d \geq 4$ together. We have thus obtained $\omega' \in T^*_m\partial\mathcal{M}$ and $\tau > 0$ such that $\alpha(m, \omega', \tau)^2 = \langle \omega' + i\tau d\varphi_{|\partial\mathcal{M}}(m), v'_m\rangle^2$ and $\langle \omega', v'_m\rangle \geq \tau|\partial_\nu\varphi(m)|$. This concludes the proof of the Item (2d) of the theorem. ■

LEMMA 8.11. *Let $t \in \mathbb{C}$ and $s = t^2$. We then have, for $r_0 > 0$,*

$$|\operatorname{Re} t| \lesseqqgtr r_0 \quad\Leftrightarrow\quad 4r_0^2\operatorname{Re} s - 4r_0^4 + (\operatorname{Im} s)^2 \lesseqqgtr 0.$$

PROOF. Let $t = x + iy$. We have $\operatorname{Re} s = x^2 - y^2$ and $\operatorname{Im} s = 2xy$, and we observe that

$$4r_0^2\operatorname{Re} s - 4r_0^4 + (\operatorname{Im} s)^2 = 4(r_0^2 + y^2)(x^2 - r_0^2),$$

which gives the result. ■

8.3. Carleman Estimate Under the Lopatinskiĭ–Šapiro Condition

The Sobolev norms used in the statements of the Carleman estimates are introduced in Chap. 7.

8.3.1. Statements.

In the neighborhood of a point m^0 of the boundary $\partial\mathcal{M}$, we prove the following estimate.

PROPOSITION 8.12. *Let $(\mathcal{M}, g)$ be a smooth compact Riemannian manifold with boundary, and let $P = -\Delta_g + R_1$ with R_1 a first-order differential operator with bounded coefficients on $\mathcal{M}$. Let $m^0 \in \partial\mathcal{M}$ and V^0 be an open neighborhood of m^0 in $\mathcal{M}$ that meets one connected component of $\partial\mathcal{M}$. Let $\varphi \in \mathscr{C}^\infty(\mathcal{M})$ be such that the pair (P, φ) has the sub-ellipticity property of Definition 5.1 in $\overline{V^0}$. Consider B a differential operator of order k in V^0 of the form of* (8.1.2), *Moreover, assume that (P, B, φ) satisfies the Lopatinskiĭ–Šapiro condition of Definition 8.1 at m^0. Then there exist a neighborhood V^1 of m^0 in $\mathcal{M}$ and two constants C and $\tau_* > 0$ such that*

$$\tau^{-1/2}\|e^{\tau\varphi}u\|_{\tau,2} + |e^{\tau\varphi_{|\partial\mathcal{M}}}\operatorname{tr}(u)|_{\tau,1,1/2} \leq C\Big(\|e^{\tau\varphi}Pu\|_{L^2(\mathcal{M})} + |e^{\tau\varphi}Bu_{|\partial\mathcal{M}}|_{\tau,3/2-k}\Big), \tag{8.3.1}$$

for all $u \in \mathscr{C}^\infty(\mathcal{M})$, with $\operatorname{supp}(u) \subset V^1$, and $\tau \geq \tau_*$.

REMARK 8.13. The Lopatinskiĭ–Šapiro condition for (P, B, φ) is only used in the estimate of Lemma 8.19 that represents a key point of the proof of Proposition 8.12. As pointed out in Remark 8.37, from the proof of Lemma 8.19 in Appendix 8.A.3, one can allow the boundary operator B to *continuously* depend on a parameter s that lies in a *compact* set S and such that the Lopatinskiĭ–Šapiro condition holds for (P, B, φ) for all values of $s \in S$. Note that the order of the operator should however remain constant as s varies. In such case, all the constants obtained in the Carleman estimate of Proposition 8.12 can be chosen uniform with respect to $s \in S$. This property can then be extended to all resulting Carleman estimates under Lopatinskiĭ–Šapiro conditions derived in this chapter.

Arguing as in Sect. 5.3, for Theorems 5.5 and 5.6, patching together estimates as above, we obtain the following result.

THEOREM 8.14. *Let $(\mathcal{M}, g)$ be a smooth compact Riemannian manifold with boundary, and let $P = -\Delta_g + R_1$ with R_1 a first-order differential operator with bounded coefficients on $\mathcal{M}$. Let V be an open set of $\mathcal{M}$ and set $V_\partial = V \cap \partial\mathcal{M}$. Let $\varphi \in \mathscr{C}^\infty(\mathcal{M})$ be such that the pair (P, φ) has the sub-ellipticity property of Definition 5.1 in $\overline{V}$. If $V_\partial \neq \emptyset$, consider B a differential operator of order β in V. For $0 \leq k \leq \beta$, denote by ${}^k\partial\mathcal{M}$ the union of the connected components of $\partial\mathcal{M}$ where B is of order k. Moreover, assume that (P, B, φ) satisfies the Lopatinskiĭ–Šapiro condition of Definition 8.1 at all* *points $m \in \overline{V_\partial}$. Then, there exist C and $\tau_* > 0$ such that*

$$\tau^{-1/2}\|e^{\tau\varphi}u\|_{\tau,2} + |e^{\tau\varphi_{|\partial\mathcal{M}}}\operatorname{tr}(u)|_{\tau,1,1/2} \leq C\Big(\|e^{\tau\varphi}Pu\|_{L^2(\mathcal{M})} + \sum_{1\leq k\leq\beta} |e^{\tau\varphi}Bu_{|^k\partial\mathcal{M}}|_{\tau,3/2-k}\Big), \tag{8.3.2}$$

for all $u \in \mathscr{C}^\infty(\mathcal{M})$, with $\operatorname{supp}(u) \subset V$, and $\tau \geq \tau_*$.

Note that the Lopatinskiĭ–Šapiro condition for (P, B, φ) is assumed on a closed set. By Proposition 8.6, the Lopatinskiĭ–Šapiro condition holds also in a neighborhood of $\overline{V_\partial}$. This is needed to have all the proper estimations in the patching procedure.

REMARK 8.15. With commutator arguments, as in Chapter 3 of Volume 1, we see that estimate (8.3.2) reads

$$\tau^{3/2}\|e^{\tau\varphi}u\|_{L^2(\mathcal{M})} + \tau^{1/2}\|e^{\tau\varphi}\,\mathsf{D}\,u\|_{L^2\Lambda^1(\mathcal{M})} + \tau^{-1/2}\|e^{\tau\varphi}\,\mathsf{H}\,u\|_{L^2\Lambda^2(\mathcal{M})}$$
$$+ |e^{\tau\varphi_{|\partial\mathcal{M}}}\operatorname{tr}(u)|_{\tau,1,1/2} \leq C\Big(\|e^{\tau\varphi}Pu\|_{L^2(\mathcal{M})} + \sum_{1\leq k\leq\beta} |e^{\tau\varphi}Bu_{|^k\partial\mathcal{M}}|_{\tau,3/2-k}\Big)$$

where $\mathsf{D}\,u$ stands for the covariant derivative on $\mathcal{M}$ of u and $\mathsf{H}\,u = \mathsf{D}^2\,u$ is the Hessian of u, defined by means of the Levi-Civita connection. These notions are recalled in Sects. 17.4 and 17.7.

We prove the result of Proposition 8.12 by means of local coordinates at the point of interest at the boundary. In a local chart, $\mathcal{C} = (O, \kappa)$ such that $m^0 \in O$, the boundary $\partial\mathcal{M}$ is given by $\{x_d = 0\}$ and $\mathcal{M} = \{x_d \geq 0\}$. More precisely, we have (see Sect. 15.1)

$$\kappa(\partial\mathcal{M} \cap O) = \{x_d = 0\} \cap \kappa(O), \qquad \kappa(\mathcal{M} \cap O) = \{x_d \geq 0\} \cap \kappa(O). \tag{8.3.3}$$

In this chart, the principal part of the Laplace-Beltrami operator takes the form

$$p^{\mathcal{C}}(x,\xi) = \sum_{1\leq i,j\leq d} g^{\mathcal{C},ij}(x)\xi_i\xi_j. \tag{8.3.4}$$

We denote the principal symbol of the operator B by $b(m,\omega)$ and by $b^{\mathcal{C}}(x,\xi)$ its local representative. In this local chart, the operator B is of order $0 \leq k \leq \beta$.

The form of the Carleman estimate of Proposition 8.12 is insensitive to lower order terms in the operator P following Remark 3.13. The same holds for the boundary operator B. With the local setting introduced above, proving Proposition 8.12 is thus equivalent to proving the following proposition.

PROPOSITION 8.16. *Let $P_0 = \operatorname{Op}(p^{\mathcal{C}})$, $B_0 = \operatorname{Op}(b^{\mathcal{C}})$. Let $m^0 \in \partial\mathcal{M}$ and V^0 be an open neighborhood of m^0 in $\mathcal{M}$ that meets one connected component of $\partial\mathcal{M}$. Let $x^0 = \kappa(m^0)$. Assume that $(P_0, \varphi^{\mathcal{C}})$ has the sub-ellipticity property of Definition 3.2 of Volume 1 in $\overline{U^0}$ with $U^0 = \kappa(V^0 \cap O)$. Moreover, assume that $(P_0, B_0, \varphi^{\mathcal{C}})$ satisfies the Lopatinskiĭ–Šapiro condition* *of Definition 8.1 at x^0. Then, there exist a bounded neighborhood U_+ of x^0 in $\overline{\mathbb{R}^d_+}$ such that $U_+ \subset U^0$ and two constants C and $\tau_* > 0$ such that*

$$\tau^{-1/2}\|e^{\tau\varphi^{\mathcal{C}}} u\|_{\tau,2} + |e^{\tau\varphi^{\mathcal{C}}_{|x_d=0^+}} \operatorname{tr}(u)|_{\tau,1,1/2} \leq C\Big(\|e^{\tau\varphi^{\mathcal{C}}} P_{\mathcal{O}} u\|_+ + |e^{\tau\varphi^{\mathcal{C}}} B_{\mathcal{O}} u_{|x_d=0^+}|_{\tau,3/2-k}\Big), \tag{8.3.5}$$

for all $u \in \overline{\mathscr{C}}^\infty_c(U_+)$ and $\tau \geq \tau_*$.

For some open set $W_+ = W \cap \mathbb{R}^d_+$ with W an open set of $\mathbb{R}^d$, the space $\overline{\mathscr{C}}^\infty_c(W_+)$ is defined in (3.4.11). By extension for U_+ an open set of $\overline{\mathbb{R}^d_+}$, we set

$$\overline{\mathscr{C}}^\infty_c(U_+) = \{u = v_{|\mathbb{R}^d_+};\, v \in \mathscr{C}^\infty_c(\mathbb{R}^d) \text{ and } \operatorname{supp} v \subset U\}, \tag{8.3.6}$$

where U is some open set of $\mathbb{R}^d$ such that $U_+ = U \cap \overline{\mathbb{R}^d_+}$. Note that the space $\overline{\mathscr{C}}^\infty_c(U_+)$ is independent of the choice of U.

The following sections are devoted to the proof of Proposition 8.16.

8.3.2. Choice of Local Coordinates. Choosing normal geodesic coordinates, near m^0, by possibly reducing the size of the open set O and changing the coordinate map κ (see Sects. 9.4 and 17.6), yields a local representative of the metric that satisfies

$$g^{\mathcal{C}}_{dd}(x) = 1, \quad \text{and } g^{\mathcal{C}}_{dj}(x) = 0 \quad \text{for } j = 1, \dots, d-1,$$

for $x \in \kappa(O)$. We moreover choose the map κ such that $0 = \kappa(m^0)$. We set $x = (x', x_d)$ with $x' = (x_1, \dots, x_{d-1})$ and accordingly $\xi = (\xi', \xi_d)$ with $\xi' = (\xi_1, \dots, \xi_{d-1})$.

The principal part of the Laplace–Beltrami operator then reads

$$p^{\mathcal{C}}(x, \xi) = \xi_d^2 + r(x, \xi'), \qquad r(x, \xi') = \sum_{1 \leq i,j \leq d-1} g^{\mathcal{C},ij}(x)\xi_i\xi_j \gtrsim |\xi'|^2.$$

Let $m \in \partial\mathcal{M}$ and $x = \kappa(m)$. In this chart if $\omega \in T^*_m\mathcal{M}$, we write $\xi = {}^t(T\kappa(m))^{-1}\omega$ (see Sect. 15.4). Then, $\omega \in T^*_m\partial\mathcal{M}$ if and only if $\xi_d = \omega^{\mathcal{C}}_d = 0$. An outward pointing vector field is simply given by its representative $\nu^{\mathcal{C}}_x = (0, \dots, 0, -1)$. It is the unique outward pointing vector field on O such that, for all $m \in \partial\mathcal{M} \cap O$, $g_m(\nu_m, \nu_m) = 1$ and $g_m(\nu_m, u) = 0$ for all $u \in T_m\partial\mathcal{M}$. Setting $n_m = \nu^\flat_m$ as in Sect. 2.2, we have $n^{\mathcal{C}}_x = (0, \dots, 0, -1)$ because of the form of the metric in the chosen normal geodesic coordinates. We also have $\partial_\nu\varphi(m) = -\partial_d\varphi^{\mathcal{C}}(x)$ for $x = \kappa(m)$.

In the chosen normal geodesic coordinates, the local representative of ∂_ν is $-\partial_d$. Hence, the representative of the boundary operator B of order k takes the form

$$B^{\mathcal{C}}(x, D) = B^{k,\mathcal{C}}(x, D') - iB^{k-1,\mathcal{C}}(x, D')D_d,$$

where $B^{k,\mathcal{C}}(x, D')$ and $B^{k-1,\mathcal{C}}(x, D')$ are the local representatives of B^k and B^{k-1}. We denote by $b^{k,\mathcal{C}}(x,\xi')$ and $b^{k-1,\mathcal{C}}(x,\xi')$ their respective principal symbols. With $b^{\mathcal{C}}(x,\xi)$ defined above, we have $b^{\mathcal{C}}(x,\xi) = b^{k,\mathcal{C}}(x,\xi') - ib^{k-1,\mathcal{C}}(x,\xi')\xi_d$ and $B_0 = \operatorname{Op}(b^{\mathcal{C}})$.

In the local chart, we define the conjugated operators of P_0 and B_0:

$$P_{0,\varphi} = e^{\tau\varphi^{\mathcal{C}}} P_0 e^{-\tau\varphi^{\mathcal{C}}}, \qquad B_{0,\varphi} = e^{\tau\varphi^{\mathcal{C}}} B_0 e^{-\tau\varphi^{\mathcal{C}}}.$$

The principal symbol of $P_{0,\varphi}$ and $B_{0,\varphi}$ is, respectively,

$$p_{0,\varphi}(x,\xi,\tau) = p^{\mathcal{C}}(x,\xi + i\tau d\varphi^{\mathcal{C}}(x)),$$

and

$$\begin{aligned} b_{0,\varphi}(x,\xi,\tau) &= b^{\mathcal{C}}_{\varphi}(x,\xi,\tau) = b^{\mathcal{C}}\big(x,\xi + i\tau d\varphi^{\mathcal{C}}(x)\big) \\ &= b^{k,\mathcal{C}}_{\varphi}(x,\xi',\tau) - ib^{k-1,\mathcal{C}}_{\varphi}(x,\xi',\tau)\big(\xi_d + i\tau\partial_d\varphi^{\mathcal{C}}\big), \end{aligned} \tag{8.3.7}$$

with

$$b^{j,\mathcal{C}}_{\varphi}(x,\xi',\tau) = b^{j,\mathcal{C}}\big(x,\xi' + i\tau d\varphi^{\mathcal{C}}_{|x_d=0^+}(x)\big), \qquad j = k, k-1.$$

By (8.2.1)–(8.2.2), for $m \in \partial\mathcal{M}$, $\omega' \in T^*_m\partial\mathcal{M}$, and $z \in \mathbb{R}$, we obtain

$$\begin{aligned} \check{p}_{\varphi}(m,\omega',z,\tau) &= p_{0,\varphi}(x,\xi',z,\tau) \\ &= p^{\mathcal{C}}(x,\xi' + i\tau d_{x'}\varphi^{\mathcal{C}}(x), z + i\tau\partial_d\varphi^{\mathcal{C}}(x)) \\ &= (z + i\tau\partial_d\varphi^{\mathcal{C}}(x))^2 + \alpha(x,\xi',\tau)^2, \end{aligned}$$

for $x = (x',0) = \kappa(m)$, with $(\xi',0) = (\omega')^{\mathcal{C}}$, and where

$$\begin{aligned} \alpha(x,\xi',\tau)^2 &= r(x,\xi' + i\tau d_{x'}\varphi^{\mathcal{C}}(x)) \\ &= r(x,\xi') - \tau^2 r(x,d_{x'}\varphi^{\mathcal{C}}(x)) + 2i\tau\tilde{r}(x,\xi',d_{x'}\varphi^{\mathcal{C}}(x)), \end{aligned} \tag{8.3.8}$$

with $\tilde{r}(x,.,.)$ the bilinear form associated with the quadratic form $r(x,.)$. As in Sect. 8.2, α is chosen such that $\operatorname{Re}\alpha \geq 0$. We thus have

$$\check{p}_{\varphi}(m,\omega',z,\tau) = p_{0,\varphi}(x,\xi',z,\tau) = (z-\gamma_1)(z-\gamma_2),$$

with

$$\gamma_j(x,\xi',\tau) = -i\tau\partial_d\varphi^{\mathcal{C}}(x) + i(-1)^j\alpha(x,\xi',\tau). \tag{8.3.9}$$

By the first result of Theorem 8.7, if the Lopatinskiĭ–Šapiro condition holds for (P,B,φ) at m^0, then $\partial_d\varphi^{\mathcal{C}}(x^0) > 0$, and locally we have $\partial_d\varphi^{\mathcal{C}}(x) > 0$, implying

$$\check{p}^+_{\varphi}(m,\omega',z,\tau) = \begin{cases} 1 & \text{if } \operatorname{Re}\alpha(x,\xi',\tau) < \tau\partial_d\varphi^{\mathcal{C}}(x), \\ z - \gamma_2(x,\xi',\tau) & \text{if } \operatorname{Re}\alpha(x,\xi',\tau) \geq \tau\partial_d\varphi^{\mathcal{C}}(x). \end{cases} \tag{8.3.10}$$

Moreover, by Lemma 8.5, if the Lopatinskiĭ–Šapiro condition holds at (m^0,ω',τ) with $\omega' \in T^*_{m^0}\partial\mathcal{M}$, $\tau \geq 0$ and $(\omega',\tau) \neq 0$, we have

$$\begin{aligned} &\operatorname{Re}\alpha(x^0,\xi',\tau) \geq \tau\partial_d\varphi^{\mathcal{C}}(x^0) \\ &\qquad\Rightarrow\quad b^{\mathcal{C}}(x^0,\xi' + i\tau d_{x'}\varphi^{\mathcal{C}}(x^0), \gamma_2 + i\tau\partial_{x_d}\varphi^{\mathcal{C}}(x^0)) \neq 0, \end{aligned} \tag{8.3.11}$$

as $\check{b}_\varphi(m^0,\omega',z) = b^{\mathcal{C}}(x^0,\xi' + i\tau d_{x'}\varphi^{\mathcal{C}}(x^0), z + i\tau\partial_{x_d}\varphi^{\mathcal{C}}(x^0))$; see the argumentation above (8.2.9). Here, as the Lopatinskiĭ–Šapiro condition holds at m^0, property (8.3.11) is valid for any $\omega' \in T^*_{m^0}\partial\mathcal{M}$ and $\tau \geq 0$ such that $(\omega',\tau) \neq 0$.

Observe that if we have $0 \leq \operatorname{Re}\alpha < \tau\partial_d\varphi^{\mathcal{C}}$, by Lemma 8.11, this reads

$$4(\tau\partial_d\varphi^{\mathcal{C}})^2 \operatorname{Re}\alpha^2 - 4(\tau\partial_d\varphi^{\mathcal{C}})^4 + (\operatorname{Im}\alpha^2)^2 < 0,$$

which, with (8.3.8), gives

$$(\partial_d\varphi^{\mathcal{C}})^2 r(x,\xi') + \tilde{r}^2(x,\xi',d_{x'}\varphi^{\mathcal{C}}) < (\tau\partial_d\varphi^{\mathcal{C}})^2\big(r(x,d_{x'}\varphi^{\mathcal{C}}) + (\partial_d\varphi^{\mathcal{C}})^2\big). \tag{8.3.12}$$

In particular, this yields $|\xi'| \lesssim \tau$, or rather

$$\lambda_{\mathsf{T},\tau} \lesssim \tau. \tag{8.3.13}$$

8.3.3. A Basic Estimate.

An argument based on the Gårding inequality for interior quadratic forms gives the following basic estimation. Note that the argument is in fact quite close to that used in Chapter 3 of Volume 1 and only relies on the sub-ellipticity property. A proof is given in Appendix 8.A.2.

LEMMA 8.17. *Let $P_0 = \operatorname{Op}(p^{\mathcal{C}})$. Assume that $(P_0,\varphi^{\mathcal{C}})$ has the sub-ellipticity property of Definition 3.2 of Volume 1 in $\overline{U^0}$ with $U^0 = \kappa(V^0 \cap O)$. Then, there exist two constants C and $\tau_* > 0$ such that*

$$\tau^{-1}\|w\|^2_{\tau,2} \leq C\big(\|P_{0,\varphi}w\|^2_+ + |\operatorname{tr}(w)|^2_{\tau,1,1/2}\big), \tag{8.3.14}$$

for $\tau \geq \tau_*$ and $w \in \overline{\mathscr{C}}^\infty_c(U^0)$.

The space $\overline{\mathscr{C}}^\infty_c(U^0)$ is defined in (8.3.6).

In this lemma, all traces appear on the right-hand side of the estimation: no assumption is made on the weight function at the boundary. This basic estimate is used in the following section as well as in Sect. 8.4.

8.3.4. A Microlocal Estimate Under the Lopatinskiĭ–Šapiro Condition.

In the local setting described above, we prove the following result.

PROPOSITION 8.18. *Let $x^0 = \kappa(m^0)$ be such that $x^0_d = 0$. Assume that $(P_0,\varphi^{\mathcal{C}})$ has the sub-ellipticity property of Definition 3.2 of Volume 1 in $\overline{U^0}$ with $U^0 = \kappa(V^0 \cap O)$. Let $\omega^{0\prime} \in T^*_{m^0}\partial\mathcal{M}$, with local representative $(\omega^{0\prime})^{\mathcal{C}} = \xi^{0\prime} \in \mathbb{R}^{d-1}$, and $\tau^0 \geq 0$ such that $(\xi^{0\prime},\tau^0) \neq 0$, and assume that the Lopatinskiĭ–Šapiro condition of Definition 8.1 holds at $(m^0,\omega^{0\prime},\tau^0)$. Then, there exists $\mathscr{U}$ a conic open neighborhood of $(x^0,\xi^{0\prime},\tau^0)$ in $U^0 \times \mathbb{R}^{d-1} \times \mathbb{R}_+$*

such that for $\chi \in S^0_{\mathsf{T},\tau}$*, homogeneous of degree* 0*, with* $\operatorname{supp}(\chi) \subset \mathscr{U}$*, there exist* $C > 0$ *and* $\tau_* > 0$ *such that*

$$\tau^{-1/2}\|\mathrm{Op}_{\mathsf{T}}(\chi)v\|_{\tau,2} + |\operatorname{tr}(\mathrm{Op}_{\mathsf{T}}(\chi)v)|_{\tau,1,1/2} \leq C\Big(\|P_{0,\varphi}v\|_+ + |B_{0,\varphi}v_{|x_d=0^+}|_{\tau,3/2-k} + \|v\|_{\tau,2,-1}\Big), \tag{8.3.15}$$

for $\tau \geq \tau_*$*,* $v \in \overline{\mathscr{S}}(\mathbb{R}^d_+)$*.*

For the proof of the proposition, we use some of the notation and techniques introduced in Chapter 3 of Volume 1. In what follows, to ease the notation, we shall write φ, p, b, and b_φ in place of $\varphi^{\mathcal{C}}$, $p^{\mathcal{C}}$, $b^{\mathcal{C}}$, and $b^{\mathcal{C}}_\varphi = b_{0,\varphi}$, respectively. With

$$P_2 = \frac{1}{2}\big(P_{0,\varphi} + P_{0,\varphi}^*\big) \in \mathscr{D}^2_\tau, \qquad P_1 = \frac{1}{2i}\big(P_{0,\varphi} - P_{0,\varphi}^*\big) \in \tau\mathscr{D}^1_\tau \tag{8.3.16}$$

see Sections 3.2.1 and 3.4.2), we write, for some $w \in \overline{\mathscr{C}}^\infty_c(U^0)$,

$$\|P_{0,\varphi}w\|_+^2 = Q(w) + 2\operatorname{Re}(P_2 w, iP_1 w)_+, \quad \text{with } Q(w) = \|P_2 w\|_+^2 + \|P_1 w\|_+^2. \tag{8.3.17}$$

The computations of Section 3.4.2 give, see (3.4.20)–(3.4.21),

$$2\operatorname{Re}(P_2 w, iP_1 w)_+ = \operatorname{Re}\big(i[P_2, P_1]w, w\big)_+ + \tau \operatorname{Re}\tilde{\mathscr{B}}(w), \tag{8.3.18}$$

with

$$\begin{aligned}\tilde{\mathscr{B}}(w) = {} & 2(\partial_d\varphi D_d w_{|x_d=0^+}, D_d w_{|x_d=0^+})_\partial \\ & + 2(\tilde{r}(x, D', d_{x'}\varphi)w_{|x_d=0^+}, D_d w_{|x_d=0^+})_\partial \\ & + 2\big(\tilde{r}(x, D', d_{x'}\varphi)D_d w_{|x_d=0^+}, w_{|x_d=0^+}\big)_\partial \\ & - 2\big(\partial_d\varphi\big(\mathrm{Op}_{\mathsf{T}}(r) - p(x, \tau d\varphi)\big)w_{|x_d=0^+}, w_{|x_d=0^+}\big)_\partial \\ & + (\mathrm{Op}(c_0)w_{|x_d=0^+}, D_d w_{|x_d=0^+})_\partial \\ & + \big(\mathrm{Op}(\tilde{c}_0)D_d + \mathrm{Op}(c_1)\big)w_{|x_d=0^+}, w_{|x_d=0^+}\big)_\partial,\end{aligned} \tag{8.3.19}$$

with $\mathrm{Op}(c_0), \mathrm{Op}(\tilde{c}_0) \in \mathscr{D}^0$ and $\mathrm{Op}(c_1) \in \mathscr{D}^1_{\mathsf{T},\tau}$. We have this estimation for $\tilde{\mathscr{B}}(w)$:

$$|\tilde{\mathscr{B}}(w)| \lesssim |\operatorname{tr}(w)|^2_{\tau,1,0}. \tag{8.3.20}$$

According to (8.3.10), we face two cases in the proof of Proposition 8.18: either one root is the upper complex half-plane, that is, $\operatorname{Im}\gamma_2 \geq 0$, implying $\check{p}^+_\varphi(m^0, \omega^{0\prime}, z) = z - \gamma_2$, or both roots γ_1 and γ_2 are in the lower complex half-plane, implying $\check{p}^+_\varphi(m^0, \omega^{0\prime}, z) = 1$.

8.3.4.1. *Case 1: One Root in the Upper Complex Half-Plane.* Here, we have

$$\check{p}_{\varphi}^{+}(m^0, \omega^{0\prime}, z, \tau^0) = z - \gamma_2(x^0, \xi^{0\prime}, \tau^0),$$

that is, $\operatorname{Im} \gamma_1(x^0, \xi^{0\prime}, \tau^0) < 0$ and $\operatorname{Im} \gamma_2(x^0, \xi^{0\prime}, \tau^0) \geq 0$. With the Lopatinskiĭ–Šapiro condition holding at the considered point, by (8.3.11), we have moreover

$$b_{\varphi}(x^0, \xi^{0\prime}, \xi_d = \gamma_2, \tau^0) = b(x^0, \xi^{0\prime} + i\tau^0 d_{x'}\varphi(x^0), \gamma_2 + i\tau^0 \partial_{x_d}\varphi(x^0)) \neq 0.$$

As the roots γ_1 and γ_2 are locally smooth with respect to (x, ξ', τ) and homogeneous of degree one in (ξ', τ) by Proposition 6.28, there exist $\mathscr{U}$ a conic open neighborhood of $(x^0, \xi^{0\prime}, \tau^0)$ in $U^0 \times \mathbb{R}^{d-1} \times \mathbb{R}_+$ and $C, C' > 0$ such that $\mathbb{S}_{\overline{\mathscr{U}}}$ is compact and

$$\gamma_1(\varrho') \neq \gamma_2(x, \xi', \tau), \ \operatorname{Im} \gamma_2(\varrho') \geq -C\lambda_{\mathsf{T},\tau}, \ \text{and} \ \ \operatorname{Im} \gamma_1(\varrho') \leq -C'\lambda_{\mathsf{T},\tau},$$

and

$$b_{\varphi}(x, \xi', \xi_d = \gamma_2(\varrho'), \tau) \neq 0, \tag{8.3.21}$$

if $\varrho' = (x, \xi', \tau) \in \overline{\mathscr{U}}$.

We let $\chi \in S^0_{\mathsf{T},\tau}$ be as in the statement and $\tilde{\chi} \in S^0_{\mathsf{T},\tau}$ be homogeneous of degree zero and be such that $\operatorname{supp}(\tilde{\chi}) \subset \mathscr{U}$ and $\tilde{\chi} \equiv 1$ on $\operatorname{supp}(\chi)$. From the smoothness and the homogeneity of the roots, we have $\tilde{\chi}\gamma_j \in S^1_{\mathsf{T},\tau}$, $j = 1, 2$. We set

$$P^+ = D_d - \operatorname{Op}_{\mathsf{T}}(\tilde{\chi}\gamma_2) \ \text{ and } \ P^- = D_d - \operatorname{Op}_{\mathsf{T}}(\tilde{\chi}\gamma_1).$$

LEMMA 8.19. *There exist $C > 0$ and $\tau_* > 0$ such that for any $N \in \mathbb{N}$, there exists $C_N > 0$ such that*

$$\begin{aligned} &|\operatorname{tr}(\operatorname{Op}_{\mathsf{T}}(\chi)v)|_{\tau,1,1/2} \\ &\quad \leq C\big(|B_{0,\varphi}\operatorname{Op}_{\mathsf{T}}(\chi)v_{|x_d=0^+}|_{\tau,3/2-k} + |P^+\operatorname{Op}_{\mathsf{T}}(\chi)v_{|x_d=0^+}|_{\tau,1/2}\big) \\ &\qquad + C_N|\operatorname{tr}(v)|_{\tau,1,-N}, \end{aligned}$$

for $\tau \geq \tau_*$, $v \in \overline{\mathscr{S}}(\mathbb{R}^d_+)$.

This lemma is the key point where the Lopatinskiĭ–Šapiro condition for (P, B, φ) is used. A proof is given in Appendix 8.A.3.

From Lemma 6.20 (with $s = 0$), we have for τ chosen sufficiently large,

$$\|\operatorname{Op}_{\mathsf{T}}(\chi)u\|_{\tau,1} + |\operatorname{Op}_{\mathsf{T}}(\chi)u_{|x_d=0^+}|_{\tau,1/2} \lesssim \|P^-\operatorname{Op}_{\mathsf{T}}(\chi)u\|_+ + \|u\|_{\tau,0,-N}, \tag{8.3.22}$$

for $u \in \overline{\mathscr{S}}(\mathbb{R}^d_+)$.

Let now $v \in \overline{\mathscr{S}}(\mathbb{R}^d_+)$. We apply estimate (8.3.22) for $u = P^+v$, yielding

$$\begin{aligned} |\operatorname{Op}_{\mathsf{T}}(\chi)P^+v_{|x_d=0^+}|_{\tau,1/2} &\lesssim \|P^-\operatorname{Op}_{\mathsf{T}}(\chi)P^+v\|_+ + \|v\|_{\tau,1,-N} \\ &\lesssim \|P_{0,\varphi}v\|_+ + \|v\|_{\tau,1}, \end{aligned}$$

using that $P^- \operatorname{Op}_{\mathsf{T}}(\chi) P^+ = \operatorname{Op}_{\mathsf{T}}(\chi) P^- P^+ \mod \Psi^{1,0}_{\tau,\mathrm{ph}} = \operatorname{Op}_{\mathsf{T}}(\chi) P_{0,\varphi} \mod \Psi^{1,0}_{\tau,\mathrm{ph}}$. We set $w = \operatorname{Op}_{\mathsf{T}}(\chi) v$. We observe that we have

$$
\begin{aligned}
|P^+ w_{|x_d=0^+}|_{\tau,1/2} &\lesssim |\operatorname{Op}_{\mathsf{T}}(\chi) P^+ v_{|x_d=0^+}|_{\tau,1/2} + |v_{|x_d=0^+}|_{\tau,1/2} \\
&\lesssim |\operatorname{Op}_{\mathsf{T}}(\chi) P^+ v_{|x_d=0^+}|_{\tau,1/2} + \|v\|_{\tau,1},
\end{aligned}
$$

using the trace inequality of Proposition 6.9. We thus obtain

$$
|P^+ w_{|x_d=0^+}|_{\tau,1/2} \lesssim \|P_{0,\varphi} v\|_+ + \|v\|_{\tau,1}.
$$

Together with Lemma 8.19, this yields the estimate

$$
|\operatorname{tr}(w)|_{\tau,1,1/2} \lesssim |B_{0,\varphi} w_{|x_d=0^+}|_{\tau,3/2-k} + \|P_{0,\varphi} v\|_+ + \|v\|_{\tau,2,-1}, \tag{8.3.23}
$$

using the trace inequality of Corollary 6.10.

By Lemma 8.17, we have

$$
\tau^{-1/2} \|w\|_{\tau,2} \lesssim \|P_{0,\varphi} w\|_+ + |\operatorname{tr}(w)|_{\tau,1,1/2},
$$

for $\tau > 0$ chosen sufficiently large. As $[P_{0,\varphi}, \operatorname{Op}_{\mathsf{T}}(\chi)] \in \Psi^{1,0}_{\tau}$, we obtain

$$
\tau^{-1/2} \|w\|_{\tau,2} \lesssim \|P_{0,\varphi} v\|_+ + \|v\|_1 + |\operatorname{tr}(w)|_{\tau,1,1/2}. \tag{8.3.24}
$$

Finally, with (8.3.23) and (8.3.24), we obtain

$$
\begin{aligned}
\tau^{-1/2} \|w\|_{\tau,2} + |\operatorname{tr}(w)|_{\tau,1,1/2} &\lesssim \|P_{0,\varphi} v\|_+ + |\operatorname{tr}(w)|_{\tau,1,1/2} + \|v\|_1 \\
&\lesssim \|P_{0,\varphi} v\|_+ + |B_{0,\varphi} w_{|x_d=0^+}|_{\tau,3/2-k} + \|v\|_{\tau,2,-1}.
\end{aligned}
$$

As $w = \operatorname{Op}_{\mathsf{T}}(\chi) v$, with a commutator argument, we have

$$
\begin{aligned}
|B_{0,\varphi} w_{|x_d=0^+}|_{\tau,3/2-k} &\lesssim |B_{0,\varphi} v_{|x_d=0^+}|_{\tau,3/2-k} + |\operatorname{tr}(v)|_{\tau,1,-1/2} \\
&\lesssim |B_{0,\varphi} v_{|x_d=0^+}|_{\tau,3/2-k} + \|v\|_{\tau,2,-1},
\end{aligned}
$$

using the trace inequality of Corollary 6.10. We then obtain (8.3.15), which concludes the proof for Case 1. ∎

8.3.4.2. *Case 2: Both Roots in the Lower Complex Half-Plane.* In this case, we have

$$
\begin{aligned}
\check{p}_{\varphi}^{-}(m^0, \omega^{0\prime}, z, \tau^0) &= \check{p}_{\varphi}(m^0, \omega^{0\prime}, z, \tau^0) \\
&= p^{\mathcal{C}}(x^0, \xi^{0\prime} + i\tau^0 d_{x'} \varphi^{\mathcal{C}}(x^0), z + i\tau^0 \partial_d \varphi^{\mathcal{C}}(x^0)) \\
&= (z - \gamma_1(x^0, \xi^{0\prime}, \tau^0))(z - \gamma_2(x^0, \xi^{0\prime}, \tau^0)).
\end{aligned}
$$

As the roots γ_1 and γ_2 depend continuously on the parameters (x, ξ', τ), there exists $\mathscr{U}$ a conic open neighborhood of $(x^0, \xi^{0\prime}, \tau^0)$ in $U^0 \times \mathbb{R}^{d-1} \times \mathbb{R}_+$ such that both $\operatorname{Im} \gamma_1 < 0$ and $\operatorname{Im} \gamma_2 < 0$ if $(x, \xi', \tau) \in \overline{\mathscr{U}}$. In particular, $\partial_d \varphi^{\mathcal{C}}_{|x_d=0^+} > 0$ in $\mathscr{U}$.

Letting $\chi \in S^0_{\mathsf{T},\tau}$ be as in the statement, we set $w = \operatorname{Op}_{\mathsf{T}}(\chi) v$. For $\varepsilon \in (0,1)$ to be set below, as $Q(w) \geq 0$, from (8.3.17), we write in fact

$$
\|P_{0,\varphi} w\|_+^2 \geq \varepsilon Q(w) + 2 \operatorname{Re}(P_2 w, i P_1 w)_+. \tag{8.3.25}
$$

First, the set

$$\mathcal{K} = \{(x,\xi,\tau);\, (x,\xi',\tau) \in \overline{\mathscr{U}},\ \xi_d \in \mathbb{R},\ |\xi|^2 + \tau^2 = 1\}$$

is a compact set recalling that $U^0 = \kappa(V^0 \cap O)$ is bounded since the manifold is compact here. On $\mathcal{K}$, we have $|p(x, \xi + i\tau d\varphi(x))| \geq C_0 > 0$, for $(x,\xi,\tau) \in \mathcal{K}$. By homogeneity, we obtain

$$|p(x, \xi + i\tau d\varphi(x))| \geq C_0 \lambda_\tau^2, \qquad (x,\xi',\tau) \in \mathscr{U}, \quad \xi_d \in \mathbb{R}. \tag{8.3.26}$$

Denoting by p_2 and p_1 the (real) principal symbols of P_2 and P_1, respectively, we have

$$q(x,\xi,\tau) = p_2^2(x,\xi,\tau) + p_1^2(x,\xi,\tau) \gtrsim \lambda_\tau^4, \qquad (x,\xi',\tau) \in \mathscr{U}, \quad \xi_d \in \mathbb{R}. \tag{8.3.27}$$

As the interior quadratic form Q has q for its principal symbol in the sense of Definition 6.13, by the Gårding inequality of Theorem 6.17, we obtain

$$\operatorname{Re} Q(w) \geq C\|w\|_{\tau,2}^2 - C'|\operatorname{tr}(w)|_{\tau,1,1/2}^2 - C_N\|v\|_{\tau,2,-N}^2, \tag{8.3.28}$$

for $\tau > 0$ chosen sufficiently large.

Second, as we have $[P_2, P_1] \in \tau \mathscr{D}_\tau^2$, we have

$$|\operatorname{Re}\big(i[P_2,P_1]w, w\big)_+| \lesssim \tau \|w\|_{\tau,2,-1}^2 \lesssim \tau^{-1}\|w\|_{\tau,2}^2 \tag{8.3.29}$$

(see Remark 6.8).

Third, we have the following lemma.

LEMMA 8.20. *There exist $C, C_N > 0$ and $\tau_* > 0$ such that*

$$\tau \operatorname{Re} \tilde{\mathscr{B}}(\operatorname{Op_T}(\chi)v) \geq C|\operatorname{tr}(\operatorname{Op_T}(\chi)v)|_{\tau,1,1/2}^2 - C_N|\operatorname{tr}(v)|_{\tau,1,-N}^2,$$

for $\tau \geq \tau_*$ and $v \in \overline{\mathscr{S}}(\mathbb{R}_+^d)$.

A proof is given in Appendix 8.A.4.

With this lemma, from (8.3.18) and (8.3.29), we obtain

$$\begin{aligned} 2\operatorname{Re}(P_2 w, iP_1 w)_+ &\geq C|\operatorname{tr}(w)|_{\tau,1,1/2}^2 - C'\tau\|w\|_{\tau,2,-1}^2 - C_N|\operatorname{tr}(v)|_{\tau,1,-N}^2 \\ &\geq C|\operatorname{tr}(w)|_{\tau,1,1/2}^2 - C'\tau\|w\|_{\tau,2,-1}^2 - C_N'\|v\|_{\tau,2,-N}^2, \end{aligned} \tag{8.3.30}$$

using the trace inequality of Corollary 6.10. With (8.3.25), (8.3.28), and (8.3.30), we obtain

$$\begin{aligned} C\varepsilon\|w\|_{\tau,2}^2 + (C - C'\varepsilon)|\operatorname{tr}(w)|_{\tau,1,1/2}^2 \leq C'\big(\|P_{0,\varphi}w\|_+^2 + \tau\|w\|_{\tau,2,-1}^2\big) \\ + C_N\|v\|_{\tau,2,-N}^2. \end{aligned}$$

Choosing $\tau > 0$ sufficiently large and $\varepsilon > 0$ sufficiently small allows one to obtain, for any $N \in \mathbb{N}$,

$$\|w\|_{\tau,2}^2 + |\operatorname{tr}(w)|_{\tau,1,1/2}^2 \lesssim \|P_{0,\varphi}w\|_+^2 + C_N\|v\|_{\tau,2,-N}^2.$$

As $w = \operatorname{Op_T}(\chi)v$, with a commutator argument, we then write

$$\|P_{0,\varphi}w\|_+ \lesssim \|P_{0,\varphi}v\|_+ + \|v\|_{\tau,2,-1},$$

yielding

$$\|w\|_{\tau,2} + |\operatorname{tr}(w)|_{\tau,1,1/2} \lesssim \|P_{0,\varphi}v\|_+ + \|v\|_{\tau,2,-1}, \tag{8.3.31}$$

which concludes the proof for Case 2. ■

REMARK 8.21. This is in fact a much better microlocal estimate than the sought estimate of Proposition 8.18. Observe, for instance, that this estimate is elliptic as the $\|\mathrm{Op}_{\mathsf{T}}(\chi)v\|_{\tau,2}$ is recovered from $\|P_{0,\varphi}\mathrm{Op}_{\mathsf{T}}(\chi)v\|_+$ and that no trace information is needed. Traces are in fact estimated from $\|P_{0,\varphi}\mathrm{Op}_{\mathsf{T}}(\chi)v\|_+$ also. The microlocal nature of this estimate is crucial with that respect. A similar estimate without the microlocal cutoff $\mathrm{Op}_{\mathsf{T}}(\chi)$ cannot hold.

8.3.5. Patching Microlocal Estimates Together. In the framework of the normal geodesic coordinates introduced in Sect. 8.3.2, we now deduce from the microlocal estimate of Proposition 8.18 the following local estimate.

PROPOSITION 8.22. *Under the assumption of Proposition 8.16, there exist a bounded neighborhood U_+ of x^0 in $\overline{\mathbb{R}^d_+}$ such that $U_+ \subset U^0$ and two constants C and $\tau_* > 0$ such that*

$$\tau^{-1/2}\|v\|_{\tau,2} + |\operatorname{tr}(v)|_{\tau,1,1/2} \leq C\Big(\|P_{0,\varphi}v\|_+ + |B_{0,\varphi}v_{|x_d=0^+}|_{\tau,3/2-k}\Big),$$

for all $v \in \overline{\mathscr{C}}^\infty_c(U_+)$ and $\tau \geq \tau_*$.

With the experience of Chapter 3 of Volume 1, it is now classical to deduce the result of Proposition 8.16 by setting $v = e^{\tau\varphi}u$, with a commutator argument between $e^{\tau\varphi}$ and D_d, to prove

$$|\operatorname{tr}(v)|_{\tau,1,1/2} \approx |e^{\tau\varphi_{|x_d=0^+}}\operatorname{tr}(u)|_{\tau,1,1/2}. \tag{8.3.32}$$

PROOF OF PROPOSITION 8.22. As was done in Sect. 8.3.4 above, in what follows, to ease the notation, we shall write φ and p in place of $\varphi^{\mathcal{C}}$ and $p^{\mathcal{C}}$, respectively.

With x^0 as in the statement of Proposition 8.16, the Lopatinskiĭ–Šapiro condition for (P_0, B_0, φ) holds at $(x^0, \xi^{0\prime}, \tau^0)$ for all $\xi^{0\prime} \in T^*_{x^0}\mathbb{R}^{d-1} \cong \mathbb{R}^{d-1}$ and $\tau^0 \geq 0$ such that $(\xi^{0\prime}, \tau^0) \neq 0$. It is in fact sufficient to consider $(\xi^{0\prime}, \tau^0)$ in the half-unit sphere $\mathbb{S}^{d-1}_+$, using the notation introduced in (1.7.3).

By Proposition 8.18 for all $(\xi^{0\prime}, \tau^0) \in \mathbb{S}^{d-1}_+$, there exists a conic open neighborhood $\mathscr{U}_{y^{0\prime}}$ of $\varrho^{0\prime} = (x^0, \xi^{0\prime}, \tau^0)$ in $U^0 \times \mathbb{R}^{d-1} \times \mathbb{R}_+$ such that the estimate (8.3.15) holds. In fact, by reducing $\mathscr{U}_{\varrho^{0\prime}}$, we can choose $\mathscr{U}_{\varrho^{0\prime}} = \mathcal{O}_{\varrho^{0\prime}} \times \Gamma_{\varrho^{0\prime}}$, where $\mathcal{O}_{\varrho^{0\prime}}$ is an open set in U^0 and $\Gamma_{\varrho^{0\prime}}$ is a conic open set in $\mathbb{R}^{d-1} \times \mathbb{R}_+$. With the compactness of $\mathbb{S}^{d-1}_+$, we can thus find finitely many such open sets $\mathscr{U}_j = \mathcal{O}_j \times \Gamma_j$, $j \in J$, such that $\mathbb{S}^{d-1}_+ \subset \cup_{j\in J}\Gamma_j$. We then set $\mathcal{O} = \cap_{j\in J}\mathcal{O}_j$ that is an open neighborhood of x^0 in U^0, and we set $\mathscr{V}_j = \mathcal{O} \times \Gamma_j \subset \mathscr{U}_j$. We also choose an open neighborhood U of x^0 in $\mathbb{R}^d$ such that $U_+ = U \cap U^0 \Subset \mathcal{O}$.

We then choose a smooth partition of unity, χ_j, $j \in J$, of the closed set $F = \overline{U_+} \times \mathbb{S}_+^{d-1}$ in the manifold $\mathcal{R} = \mathbb{R}^d \times \mathbb{S}_+^{d-1}$ subordinated to the covering by the open sets $\mathscr{V}_j \cap \mathcal{R}$ according to Theorem 15.14. We then extend each χ_j smoothly to $\mathbb{R}^d \times \mathbb{R}^{d-1} \times \mathbb{R}_+$ by homogeneity of degree 0 for $|(\xi', \tau)| \geq 1$. We have $\operatorname{supp}(\chi_j) \subset \mathscr{V}_j$ and

$$\sum_{j \in J} \chi_j(\varrho') = 1,$$

for $\varrho' = (x, \xi', \tau)$ in a conic neighborhood of $\overline{U_+} \times \mathbb{R}^{d-1} \times \mathbb{R}_+$ and $|(\xi', \tau)| \geq 10$. We also set $\underline{\chi} = 1 - \sum_{j \in J} \chi_j$

As $\operatorname{supp}(\chi_j) \subset \mathscr{U}_j$, we can apply the microlocal estimate of Proposition 8.18

$$\begin{aligned} \tau^{-1/2} \|\operatorname{Op}(\chi_j) v\|_{\tau,2} &+ |\operatorname{tr}(\operatorname{Op}(\chi_j) v)|_{\tau,1,1/2} \\ &\lesssim \|P_{0,\varphi} v\|_+ + |B_{0,\varphi} v_{|x_d=0^+}|_{\tau,3/2-k} + \|v\|_{\tau,2,-1}, \end{aligned} \tag{8.3.33}$$

for $\tau > 0$ chosen sufficiently large and for $v = w_{|\mathbb{R}^d_+}$ with $w \in \mathscr{C}_c^\infty(U)$.

Observe then that, for any $N \in \mathbb{N}$, using the support of v,

$$\begin{aligned} \|v\|_{\tau,2} &\leq \sum_{j \in J} \|\operatorname{Op}_{\mathsf{T}}(\chi_j) v\|_{\tau,2} + \|\operatorname{Op}_{\mathsf{T}}(\underline{\chi}) v\|_{\tau,2} \\ &\lesssim \sum_{j \in J} \|\operatorname{Op}_{\mathsf{T}}(\chi_j) v\|_{\tau,2} + \|v\|_{\tau,2,-N}, \end{aligned}$$

and

$$\begin{aligned} |\operatorname{tr}(v)|_{\tau,1,1/2} &\leq \sum_{j \in J} |\operatorname{tr}(\operatorname{Op}_{\mathsf{T}}(\chi_j) v)|_{\tau,1,1/2} + |\operatorname{tr}(\operatorname{Op}_{\mathsf{T}}(\underline{\chi}) v)|_{\tau,1,1/2} \\ &\lesssim \sum_{j \in J} |\operatorname{tr}(\operatorname{Op}_{\mathsf{T}}(\chi_j) v)|_{\tau,1,1/2} + |\operatorname{tr}(v)|_{\tau,1,-N} \\ &\lesssim \sum_{j \in J} |\operatorname{tr}(\operatorname{Op}_{\mathsf{T}}(\chi_j) v)|_{\tau,1,1/2} + \|v\|_{\tau,2,-N}. \end{aligned}$$

Summing estimates (8.3.33) for each $j \in J$, we thus obtain

$$\tau^{-1/2} \|v\|_{\tau,2} + |\operatorname{tr}(v)|_{\tau,1,1/2} \lesssim \|P_{0,\varphi} v\|_+ + |B_{0,\varphi} v_{|x_d=0^+}|_{\tau,3/2-k} + \|v\|_{\tau,2,-1}.$$

Choosing now $\tau > 0$ sufficiently large, we obtain the sought estimate. ■

We observe that the same proof allows us to deduce the following microlocal "semi-global" estimate.

PROPOSITION 8.23. *Let $P_0 = \operatorname{Op}(p^{\mathcal{C}})$, $B_0 = \operatorname{Op}(b^{\mathcal{C}})$. Assume that $(P_0, \varphi^{\mathcal{C}})$ has the sub-ellipticity property of Definition 3.2 of Volume 1 in $\overline{U^0}$ with $U^0 = \kappa(V^0 \cap O)$. Let K be a compact set of U^0. Let $\mathscr{V}$ be a conic set of $K \times \mathbb{R}^{d-1} \times \mathbb{R}_+$. Assume that $(P_0, B_0, \varphi^{\mathcal{C}})$ satisfies the Lopatinskiĭ–Šapiro condition of Definition 8.1 at all $(x^0, \xi^{0\prime}, \tau^0) \in \overline{\mathscr{V} \cap \{x_d = 0\}}$. Then, there exists $\mathscr{U}$ a conic open set of $\overline{\mathbb{R}^d_+} \times \mathbb{R}^{d-1} \times \mathbb{R}_+$ such that $\overline{\mathscr{V}} \subset \mathscr{U}$, and if*

$\chi \in S^0_\tau$ is homogeneous of degree 0 and such that $\operatorname{supp}(\chi) \subset \mathscr{U}$*, there exist two constants C and $\tau_* > 0$ such that*

$$\tau^{-1/2}\|\operatorname{Op_T}(\chi)v\|_{\tau,2} + |\operatorname{tr}(\operatorname{Op_T}(\chi)v)|_{\tau,1,1/2} \leq C\Big(\|P_{0,\varphi}v\|_+ + |B_{0,\varphi}v_{|x_d=0^+}|_{\tau,3/2-k} + \|v\|_{\tau,2,-1}\Big), \tag{8.3.34}$$

for $\tau \geq \tau_*$, $v \in \overline{\mathscr{S}}(\mathbb{R}^d_+)$. In particular, one can choose $\chi \equiv 1$ in a neighborhood of $\overline{\mathscr{V}}$.

8.3.6. A Shifted Estimate. With the estimations of Proposition 8.22, we can prove the following estimate, that is closer in form to those proven in Chapter 3 of Volume 1 at boundaries.

THEOREM 8.24. *Under the assumptions of Theorem 8.14, there exist C and $\tau_* > 0$ such that*

$$\tau^{1/2}\|e^{\tau\varphi}u\|_{\tau,1} + \tau^{1/2}|e^{\tau\varphi_{|\partial\mathcal{M}}}\operatorname{tr}(u)|_{\tau,1,0} \leq C\Big(\|e^{\tau\varphi}Pu\|_{L^2(\mathcal{M})} + \tau^{1/2}\sum_{1\leq k\leq\beta}|e^{\tau\varphi}Bu_{|^k\partial\mathcal{M}}|_{\tau,1-k}\Big),$$

for all $u \in \mathscr{C}^\infty(\mathcal{M})$, with $\operatorname{supp}(u) \subset V$*, and $\tau \geq \tau_*$.*

A proof is given in Appendix 8.A.5.

Weaker norms both in the interior of $\mathcal{M}$ or at the boundary $\partial\mathcal{M}$ are obtained as compared to the estimation in Theorem 8.14. However, the right-hand side of the estimate also shows weaker norms for the boundary operator. Note that

$$\begin{aligned}\tau^{1/2}|e^{\tau\varphi_{|\partial\mathcal{M}}}\operatorname{tr}(u)|_{\tau,1,0} \approx \tau^{3/2}&|e^{\tau\varphi_{|\partial\mathcal{M}}}u_{|\partial\mathcal{M}}|_{L^2(\partial\mathcal{M})}\\ &+ \tau^{1/2}|e^{\tau\varphi_{|\partial\mathcal{M}}}\,\mathsf{D}'\,u_{|\partial\mathcal{M}}|_{L^2\Lambda^1(\partial\mathcal{M})}\\ &+ \tau^{1/2}|e^{\tau\varphi_{|\partial\mathcal{M}}}\partial_\nu u_{|\partial\mathcal{M}}|_{L^2(\partial\mathcal{M})},\end{aligned}$$

where D' denotes the covariant derivative on $\partial\mathcal{M}$. As we also have

$$\tau^{1/2}\|e^{\tau\varphi}u\|_{\tau,1} \approx \tau^{3/2}\|e^{\tau\varphi}u\|_{L^2(\mathcal{M})} + \tau^{1/2}\|e^{\tau\varphi}\,\mathsf{D}\,u\|_{L^2\Lambda^1(\mathcal{M})},$$

we obtain the following estimation:

$$\tau^{3/2}\|e^{\tau\varphi}u\|_{L^2(\mathcal{M})} + \tau^{1/2}\|e^{\tau\varphi}\,\mathsf{D}\,u\|_{L^2\Lambda^1(\mathcal{M})} + \tau^{1/2}|e^{\tau\varphi_{|\partial\mathcal{M}}}\partial_\nu u_{|\partial\mathcal{M}}|_{L^2(\partial\mathcal{M})} \lesssim \|e^{\tau\varphi}Pu\|_{L^2(\mathcal{M})} + \tau^{1/2}\sum_{1\leq k\leq\beta}|e^{\tau\varphi}Bu_{|^k\partial\mathcal{M}}|_{\tau,1-k}.$$

Observe that in the case $\beta = 0$, that is, B is the Dirichlet boundary operator, we recover precisely the statement of Theorem 5.6.

8.3.7. A Basic Microlocal Elliptic Estimate. With a microlocalization procedure as in Sect. 8.3.4, we can improve the result of Lemma 8.17 in regions where roots are not real. This result is used in Chapter 14; the reader may thus skip this section.

We use the local notation of Sect. 8.3.2.

LEMMA 8.25. *Assume that* $(P_0, \varphi^{\mathsf{C}})$ *has the sub-ellipticity property of Definition 3.2 of Volume 1 in* $\overline{U^0}$ *with* $U^0 = \kappa(V^0 \cap O)$. *Let* $\chi \in S^0_{\mathsf{T},\tau}$ *be homogeneous of degree* 0 *and such that* $\operatorname{Im}\gamma_1$ *and* $\operatorname{Im}\gamma_2$ *do not vanish in* $\operatorname{supp}(\chi)$. *Then, there exist* $C > 0$ *and* $\tau_* > 0$ *such that*

$$\|\operatorname{Op}_{\mathsf{T}}(\chi)v\|_{\tau,2} \lesssim \|P_{0,\varphi}v\|_+ + |\operatorname{tr}(\operatorname{Op}_{\mathsf{T}}(\chi)v)|_{\tau,1,1/2} + \|v\|_{\tau,2,-1}, \tag{8.3.35}$$

for $\tau \geq \tau_*$, $v \in \overline{\mathscr{S}}(\mathbb{R}^d_+)$.

PROOF. We use the notations P_1 and P_2 as defined in (8.3.16) and denote by p_2 and p_1 their respective (real) principal symbols. Observe that $w \mapsto \|P_{0,\varphi}w\|_+^2$ is an interior quadratic form in the sense of Definition 6.13, with *principal* symbol $q = p_2^2 + p_1^2$.

Having $\operatorname{Im}\gamma_1 \neq 0$ and $\operatorname{Im}\gamma_2 \neq 0$ in $\operatorname{supp}(\chi)$, one obtains

$$q(x, \xi, \tau) \gtrsim \lambda_\tau^4, \qquad (x, \xi', \tau) \in \mathscr{U}, \tag{8.3.36}$$

for $\mathscr{U}$ a neighborhood of $\operatorname{supp}(\chi)$. Then, with the microlocal Gårding inequality of Theorem 6.17, one finds

$$\|\operatorname{Op}_{\mathsf{T}}(\chi)v\|_{\tau,2} \lesssim \|P_{0,\varphi}\operatorname{Op}_{\mathsf{T}}(\chi)v\|_+ + |\operatorname{tr}(\operatorname{Op}_{\mathsf{T}}(\chi)v)|_{\tau,1,1/2} + \|v\|_{\tau,2,-N}.$$

Finally, with $[P_{0,\varphi}, \operatorname{Op}_{\mathsf{T}}(\chi)] \in \Psi_\tau^{1,1}$, we obtain the result. ■

8.4. Estimates Without Any Prescribed Boundary Condition

We prove now estimates without assuming any boundary condition. Such a case is considered in Theorem 3.28 of Volume 1 in the case of an open set of $\mathbb{R}^d$ and in Theorem 5.5 in the case of a manifold with boundary. In those results, both the Dirichlet and Neumann traces are assumed known and allow one to obtain an estimation. In that case, no particular assumption is made on the weight function at the boundary.

Here, we wish to obtain bounds by means of norms on the Dirichlet trace $u_{|\partial\mathcal{M}}$ and a first-order trace $Bu_{|\partial\mathcal{M}}$, that is, where $B = \partial_\nu + B'$ is a first-order boundary operator. For this to make sense, one assumes B fulfills the Lopatinskiĭ–Šapiro condition of Definition 2.2 along with P. Such boundary operators are characterized in Proposition 2.8. However, no particular assumption on the weight function is made at the boundary. Naturally, as in the rest of this book, the sub-ellipticity property of the pair (P, φ) is assumed.

We first prove such an estimate with Sobolev norms as in Theorem 8.14 with an observation of all traces. Second, we improve upon this estimate, removing some the trace observations. Third, we provide a shifted estimate, similar to Theorems 3.28 and 5.5. The improvement made in the second step can be important in applications. The microlocal nature of the proof of the estimates in the present chapter is crucial with regard to this refinement.

8.4.1. A First Estimate.

PROPOSITION 8.26. *Let $(\mathcal{M}, g)$ be a smooth compact Riemannian manifold with boundary, and let $P = -\Delta_g + R_1$ with R_1 a first-order differential operator with bounded coefficients on $\mathcal{M}$. Let V be an open set of $\mathcal{M}$ such that $V \cap \partial\mathcal{M} \neq \emptyset$. Let $\varphi \in \mathscr{C}^\infty(\mathcal{M})$ be such that the pair (P, φ) has the sub-ellipticity property of Definition 5.1 in $\overline{V}$. Then, there exist C and $\tau_* > 0$ such that*

$$\tau^{-1/2}\|e^{\tau\varphi}u\|_{\tau,2} \leq C\Big(\|e^{\tau\varphi}Pu\|_{L^2(\mathcal{M})} + |e^{\tau\varphi}\operatorname{tr}(u)|_{\tau,1,1/2}\Big), \tag{8.4.1}$$

for all $u \in \mathscr{C}^\infty(\mathcal{M})$, with $\operatorname{supp}(u) \subset V$, *and $\tau \geq \tau_*$.*

This result simply follows from patching estimates in local charts as in Lemma 8.17 together.

This estimation is different from its counterparts in Chapter 3 of Volume 1, namely Theorem 3.28 in an open set and Theorem 5.5 on a manifold by the norms that appears on both the left-hand side and the right-hand side of the estimation. There is a shift by a half-tangential derivative. We refer to Sect. 8.3.6 where this issue is discussed.

8.4.2. A Refined Estimate. Let $B = \partial_\nu + B'$ be a first-order boundary operator such that (P, B) satisfies the Lopatinskiĭ–Šapiro condition. We prove an estimate with terms involving both the Dirichlet trace $u_{|\partial\mathcal{M}}$ and the trace $Bu_{|\partial\mathcal{M}}$.

THEOREM 8.27. *Let $(\mathcal{M}, g)$ be a smooth compact Riemannian manifold with boundary, and let $P = -\Delta_g + R_1$ with R_1 a first-order differential operator with bounded coefficients on $\mathcal{M}$. Let V be an open set of $\mathcal{M}$ such that $V_\partial = V \cap \partial\mathcal{M} \neq \emptyset$. Let B' be a differential operator of order one on $\partial\mathcal{M}$ such that (P, B) fulfills the Lopatinskiĭ–Šapiro condition of Definition 2.2 on $V \cap \partial\mathcal{M}$ for $B = \partial_\nu + B'$. Let $\varphi \in \mathscr{C}^\infty(\mathcal{M})$ be such that the pair (P, φ) has the sub-ellipticity property of Definition 5.1 in $\overline{V_\partial}$. Then, there exist C and $\tau_* > 0$ such that*

$$\begin{aligned}\tau^{-1/2}\|e^{\tau\varphi}u\|_{\tau,2} &+ |e^{\tau\varphi}\operatorname{tr}(u)|_{\tau,1,1/2}\\ &\leq C\Big(\|e^{\tau\varphi}Pu\|_{L^2(\mathcal{M})} + \tau^{3/2}|e^{\tau\varphi}u_{|\partial\mathcal{M}}|_{L^2(\partial\mathcal{M})} + |e^{\tau\varphi}Bu_{|\partial\mathcal{M}}|_{\tau,1/2}\Big),\end{aligned} \tag{8.4.2}$$

for all $u \in \mathscr{C}^\infty(\mathcal{M})$, with $\operatorname{supp}(u) \subset V$, *and $\tau \geq \tau_*$.*

REMARK 8.28.

(1) Note that one does not assume here that (P, B, φ) fulfills the Lopatinskiĭ–Šapiro condition of Definition 8.1, as opposed to Theorem 8.14.

(2) An important case is naturally $B = \partial_\nu$ since this operator fulfills the Lopatinskiĭ–Šapiro condition along with P; see example 2.5-(2). Note that the estimate one obtains in this case is stronger than that of Proposition 8.26 by replacing $|e^{\tau\varphi}u_{|\partial\mathcal{M}}|_{\tau,3/2}$ by $\tau^{3/2}$ $|e^{\tau\varphi}u_{|\partial\mathcal{M}}|_{L^2(\partial\mathcal{M})}$ on the right-hand side.

In the framework of the normal geodesic coordinates introduced in Sect. 8.3.2, we prove in fact the following proposition, that is, the counterpart of Proposition 8.22

Proposition 8.29. *Let $P_0 = \mathrm{Op}(p^{\mathcal{C}})$ and $B_0 = \mathrm{Op}(b^{\mathcal{C}})$. Let $x^0 = \kappa(m^0)$ be such that $x^0_d = 0$. Assume that (P_0, B_0) satisfies the Lopatinskiĭ–Šapiro condition of Definition 2.2 at x^0. Assume that $(P_0, \varphi^{\mathcal{C}})$ has the sub-ellipticity property of Definition 3.2 of Volume 1 in $\overline{U^0}$ with $U^0 = \kappa(V^0 \cap O)$. There exist a bounded neighborhood U_+ of x^0 in $\overline{\mathbb{R}^d_+}$ such that $U_+ \subset U^0$ and two constants C and $\tau_* > 0$ such that*

$$\begin{aligned}\tau^{-1/2}\|v\|_{\tau,2} &+ |\operatorname{tr}(v)|_{\tau,1,1/2}\\ &\leq C\Big(\|P_{0,\varphi}v\|_+ + \tau^{3/2}|v_{|x_d=0^+}|_\partial + |B_{0,\varphi}v_{|x_d=0^+}|_{\tau,1/2}\Big),\end{aligned}$$

for all $v \in \overline{\mathscr{C}}^\infty_c(U_+)$ and $\tau \geq \tau_*$.

With the experience of Chapter 3 of Volume 1, we deduce, with a commutator argument, a local version of Theorem 8.27 by setting $v = e^{\tau\varphi}u$. Arguing as in Sect. 5.3, for Theorems 5.5 and 5.6, patching together such local estimates, we then obtain the result of Theorem 8.27.

Proof. Let $\xi^{0\prime} \in \mathbb{R}^{d-1} \setminus \{0\}$. Observe that the Lopatinskiĭ–Šapiro condition for (P_0, B_0, φ) holds at $(x^0, \xi^{0\prime}, \tau^0 = 0)$ for any weight function φ, since the Lopatinskiĭ–Šapiro condition of Definition 8.1 coincides with the Lopatinskiĭ–Šapiro condition of Definition 2.2 at such a point. The Lopatinskiĭ–Šapiro condition thus holds on the conic set $\mathscr{V} = \{x^0\} \times \mathbb{R}^{d-1} \times \{\tau = 0\}$. By Proposition 8.23, there exists $\mathscr{U}$ a conic open set of $\overline{\mathbb{R}^d_+} \times \mathbb{R}^{d-1} \times \mathbb{R}_+$ such that $\overline{\mathscr{V}} \subset \mathscr{U}$ and for $\chi \in S^0_{\mathcal{T}}$, homogeneous of degree 0 and such that $\operatorname{supp}(\chi) \subset \mathscr{U}$ and $\chi \equiv 1$ in a neighborhood of $\mathscr{V}$, we have

$$\begin{aligned}(8.4.3)\quad \tau^{-1/2}\|\mathrm{Op}_{\mathsf{T}}(\chi)v\|_{\tau,2} &+ |\operatorname{tr}(\mathrm{Op}_{\mathsf{T}}(\chi)v)|_{\tau,1,1/2}\\ &\lesssim \|P_{0,\varphi}v\|_+ + |B_{0,\varphi}v_{|x_d=0^+}|_{\tau,1/2} + \|v\|_{\tau,2,-1},\end{aligned}$$

for $\tau > 0$ chosen sufficiently large.

We set $\tilde{\chi} = 1 - \chi \in S^0_{\mathcal{T}}$. By Lemma 8.17, we have

$$(8.4.4)\qquad \tau^{-1/2}\|\mathrm{Op}_{\mathsf{T}}(\tilde{\chi})v\|_{\tau,2} \lesssim \|P_{0,\varphi}\mathrm{Op}_{\mathsf{T}}(\tilde{\chi})v\|_+ + |\operatorname{tr}(\mathrm{Op}_{\mathsf{T}}(\tilde{\chi})v)|_{\tau,1,1/2},$$

for $\tau > 0$ chosen sufficiently large.

Observe that $\tau^{-3/2}\Lambda_{\mathsf{T},\tau}^{3/2}\operatorname{Op}_{\mathsf{T}}(\tilde{\chi}) \in \Psi_{\mathsf{T}}^{0}$ since $\tau \gtrsim \lambda_{\mathsf{T},\tau}$ in $\operatorname{supp}(\tilde{\chi})$. This yields

$$|\operatorname{Op}_{\mathsf{T}}(\tilde{\chi})v_{|x_d=0^+}|_{\tau,3/2} \lesssim \tau^{3/2}|v_{|x_d=0^+}|_{\partial}.$$

We thus find

$$\begin{aligned}|\operatorname{tr}(\operatorname{Op}_{\mathsf{T}}(\tilde{\chi})v)|_{\tau,1,1/2} &\eqsim |\operatorname{Op}_{\mathsf{T}}(\tilde{\chi})v_{|x_d=0^+}|_{\tau,3/2} + |D_d\operatorname{Op}_{\mathsf{T}}(\tilde{\chi})v_{|x_d=0^+}|_{\tau,1/2}\\ &\lesssim |\operatorname{Op}_{\mathsf{T}}(\tilde{\chi})v_{|x_d=0^+}|_{\tau,3/2} + |B_{0,\varphi}\operatorname{Op}_{\mathsf{T}}(\tilde{\chi})v_{|x_d=0^+}|_{\tau,1/2}\\ &\lesssim \tau^{3/2}|v_{|x_d=0^+}|_{\partial} + |B_{0,\varphi}v_{|x_d=0^+}|_{\tau,1/2}\\ &\quad + |[B_{0,\varphi},\operatorname{Op}_{\mathsf{T}}(\tilde{\chi})]v_{|x_d=0^+}|_{\tau,1/2},\end{aligned}$$

using that $B_{0,\varphi} - D_d \in \Psi_{\mathsf{T},\tau}^{1}$. Since $\tau^{-1/2}\Lambda_{\mathsf{T},\tau}^{1/2}[B_{0,\varphi},\operatorname{Op}_{\mathsf{T}}(\tilde{\chi})] \in \Psi^{0}$ with the same support argument as above, we obtain

$$|\operatorname{tr}(\operatorname{Op}_{\mathsf{T}}(\tilde{\chi})v)|_{\tau,1,1/2} \lesssim \tau^{3/2}|v_{|x_d=0^+}|_{\partial} + |B_{0,\varphi}v_{|x_d=0^+}|_{\tau,1/2}.$$

From (8.4.4), we find

$$\begin{aligned}(8.4.5)\quad \tau^{-1/2}\|\operatorname{Op}_{\mathsf{T}}(\tilde{\chi})v\|_{\tau,2} &+ |\operatorname{tr}(\operatorname{Op}_{\mathsf{T}}(\tilde{\chi})v)|_{\tau,1,1/2}\\ &\lesssim \|P_{0,\varphi}\operatorname{Op}_{\mathsf{T}}(\tilde{\chi})v\|_{+} + \tau^{3/2}|v_{|x_d=0^+}|_{\partial} + |B_{0,\varphi}v_{|x_d=0^+}|_{\tau,1/2}\\ &\lesssim \|P_{0,\varphi}v\|_{+} + \tau^{3/2}|v_{|x_d=0^+}|_{\partial} + |B_{0,\varphi}v_{|x_d=0^+}|_{\tau,1/2} + \|v\|_{\tau,1},\end{aligned}$$

as $[P_{0,\varphi},\operatorname{Op}_{\mathsf{T}}(\tilde{\chi})] \in \Psi_{\tau}^{1,0}$. With (8.4.3) and (8.4.5), one obtains

$$\begin{aligned}&\tau^{-1/2}\|v\|_{\tau,2} + |\operatorname{tr}(v)|_{\tau,1,1/2}\\ &\quad\lesssim \tau^{-1/2}\|\operatorname{Op}_{\mathsf{T}}(\chi)v\|_{\tau,2} + \tau^{-1/2}\|\operatorname{Op}_{\mathsf{T}}(\tilde{\chi})v\|_{\tau,2}\\ &\qquad + |\operatorname{tr}(\operatorname{Op}_{\mathsf{T}}(\chi)v)|_{\tau,1,1/2} + |\operatorname{tr}(\operatorname{Op}_{\mathsf{T}}(\tilde{\chi})v)|_{\tau,1,1/2}\\ &\quad\lesssim \|P_{0,\varphi}v\|_{+} + \tau^{3/2}|v_{|x_d=0^+}|_{\partial} + |B_{0,\varphi}v_{|x_d=0^+}|_{\tau,1/2} + \|v\|_{\tau,2,-1}.\end{aligned}$$

We then conclude the proof by taking $\tau > 0$ sufficiently large. ■

8.4.3. A Shifted Refined Estimate.

THEOREM 8.30. *Under the assumptions of Theorem 8.27, there exist C and $\tau_* > 0$ such that*

$$\begin{aligned}(8.4.6)\quad &\tau^{1/2}\|e^{\tau\varphi}u\|_{\tau,1} + \tau^{1/2}|e^{\tau\varphi}\operatorname{tr}(u)|_{\tau,1,0}\\ &\leq C\Big(\|e^{\tau\varphi}Pu\|_{L^2(\mathcal{M})} + \tau^{3/2}|e^{\tau\varphi}u_{|\partial\mathcal{M}}|_{L^2(\partial\mathcal{M})} + \tau^{1/2}|e^{\tau\varphi}Bu_{|\partial\mathcal{M}}|_{L^2(\partial\mathcal{M})}\Big),\end{aligned}$$

for all $u \in \mathscr{C}^\infty(\mathcal{M})$, with $\operatorname{supp}(u) \subset V$, *and $\tau \geq \tau_*$.*

A comparison with the statement of Theorem 5.5 shows that the trace norms on the right-hand side of the above estimate are weaker. No tangential derivative of the trace is needed here. In fact, this tangential derivative actually appears on the left-hand side of the estimate. A proof of Theorem 8.30 is given in Appendix 8.A.6.

REMARK 8.31. As pointed out in Remark 8.28, an important case is the Neumann boundary operator $B = \partial_\nu$.

8.5. Global Estimates

We consider the setting of Sect. 5.4, and we let $(\tilde{\mathcal{M}}, g)$ be a σ-compact Riemannian manifold (possibly not compact) with or without boundary and $\mathcal{M}$ be a bounded connected open set of $\tilde{\mathcal{M}}$. Let also Γ_0 be an open set of $\partial\mathcal{M}$ such that $\partial\mathcal{M}$ is smooth in a neighborhood of $\overline{\Gamma_0}$. Note that, whereas the boundary of $\tilde{\mathcal{M}}$ is smooth, we need not assume $\partial\mathcal{M}$ to be smooth everywhere. Let also ω_0 be an open subset of $\mathcal{M}$.

As for Theorem 5.8 using Theorem 8.14 (respectively, Theorem 8.24), one can adapt Section 3.6.1 and one obtains the following global estimates.

THEOREM 8.32 (Global Carleman Estimate—Inner Observation). *Let $P = P_0 + R_1$ with $P_0 = -\Delta_g$, where R_1 is a first-order differential operator with bounded coefficients. Let also W_0 be a neighborhood of $\partial\mathcal{M} \setminus \Gamma_0$ in $\tilde{\mathcal{M}}$. Let ω be an open set of $\mathcal{M}$ such that $\omega_0 \Subset \omega$. Let $\varphi \in \mathscr{C}^\infty(\overline{\mathcal{M}})$ be a global weight function adapted to Γ_0 and ω_0 in the sense of Definition 5.7.*

Consider B a differential operator of order β in Γ^0. For $0 \leq k \leq \beta$, denote by ${}^k\partial\mathcal{M}$ the union of the connected components of $\partial\mathcal{M}$ where B is of order k. Moreover, assume that (P, B, φ) satisfies the Lopatinskiĭ–Šapiro condition of Definition 8.1 at all points $m \in \Gamma_0$.

Then, there exist $\tau_* > 0$ and $C \geq 0$ such that

$$
\begin{aligned}
&\tau^{-1/2}\|e^{\tau\varphi}u\|_{\tau,2} + |e^{\tau\varphi_{|\partial\mathcal{M}}}\operatorname{tr}(u)|_{\tau,1,1/2} \\
&\leq C\big(\|e^{\tau\varphi}Pu\|_{L^2(\mathcal{M})} + \tau^{3/2}\|e^{\tau\varphi}u\|_{L^2(\omega)} + \sum_{1\leq k\leq\beta} |e^{\tau\varphi}Bu_{|\Gamma_0\cap{}^k\partial\mathcal{M}}|_{\tau,3/2-k}\big),
\end{aligned}
$$

for $\tau \geq \tau_*$ and $u \in \mathscr{C}^\infty(\overline{\mathcal{M}})$ vanishing in $W_0 \cap \mathcal{M}$. One also has the estimate

$$
\begin{aligned}
&\tau^{1/2}\|e^{\tau\varphi}u\|_{\tau,1} + \tau^{1/2}|e^{\tau\varphi_{|\partial\mathcal{M}}}\operatorname{tr}(u)|_{\tau,1,0} \\
&\leq C\big(\|e^{\tau\varphi}Pu\|_{L^2(\mathcal{M})} + \tau^{3/2}\|e^{\tau\varphi}u\|_{L^2(\omega)} + \tau^{1/2}\sum_{1\leq k\leq\beta} |e^{\tau\varphi}Bu_{|\Gamma_0\cap{}^k\partial\mathcal{M}}|_{\tau,1-k}\big).
\end{aligned}
$$

If $\Gamma_0 = \partial\mathcal{M}$, then one can write a corollary in the form of Corollary 5.9

As for Theorem 5.11 using Theorem 8.14 (respectively, 8.24) and Theorem 8.27 (respectively, Theorem 8.30), one can adapt Section 3.6.1 and one obtains the following global estimates.

THEOREM 8.33 (Global Carleman Estimate—Boundary Observation). *Let $P = P_0 + R_1$ with $P_0 = -\Delta_g$ where R_1 is a first-order differential operator with bounded coefficients. Let Γ_0 and Γ_{obs} be two nonempty open sets of $\partial\mathcal{M}$ such that $\Gamma_{\text{obs}} \setminus \overline{\Gamma_0} \neq \emptyset$ and such that $\partial\mathcal{M}$ is smooth in a neighborhood of $\overline{\Gamma_0} \cup \overline{\Gamma_{\text{obs}}}$. Let also W_0 be a neighborhood of $\partial\mathcal{M} \setminus (\Gamma_0 \cup \Gamma_{\text{obs}})$ in $\tilde{\mathcal{M}}$.*

Let $\varphi \in \mathscr{C}^\infty(\overline{\mathcal{M}})$ be a global weight function adapted to Γ_0 in the sense of Definition 5.10.

Consider B a differential operator of order β in Γ^0. For $0 \leq k \leq \beta$, denote by ${}^k\partial\mathcal{M}$ the union of the connected components of $\partial\mathcal{M}$ where B is of order k. Moreover, assume that (P, B, φ) satisfies the Lopatinskiĭ–Šapiro condition of Definition 8.1 at all points $m \in \Gamma_0$.

Consider also B' a first-order differential operator in Γ_{obs} and $\tilde{B} = \partial_\nu + B'$, and assume that $(P, \tilde{B})$ fulfills the Lopatinskiĭ–Šapiro condition of Definition 2.2 on Γ_{obs}.

Let also $\psi_0, \psi_{\text{obs}} \in \mathscr{C}^\infty(\overline{\mathcal{M}})$ be such that $\psi_{0|\partial\mathcal{M}}$ and $\psi_{\text{obs}|\partial\mathcal{M}}$ form a partition of unity of $F = \overline{\Gamma_0 \cup \Gamma_{\text{obs}}} \setminus W^0$ associated with the covering by Γ_0 and Γ_{obs}, that is,

$$\operatorname{supp}(\psi_{0|\partial\mathcal{M}}) \subset \Gamma_0, \quad \operatorname{supp}(\psi_{\text{obs}|\partial\mathcal{M}}) \subset \Gamma_{\text{obs}},$$

and $\psi_{0|\partial\mathcal{M}} + \psi_{\text{obs}|\partial\mathcal{M}} \equiv 1$ in a neighborhood of F.

Then, there exist $\tau_* > 0$ and $C \geq 0$ such that

$$\begin{aligned}
&\tau^{-1/2}\|e^{\tau\varphi}u\|_{\tau,2} + |e^{\tau\varphi_{|\partial\mathcal{M}}}\operatorname{tr}(u)|_{\tau,1,1/2} \\
&\quad \leq C\big(\|e^{\tau\varphi}Pu\|_{L^2(\mathcal{M})} + \tau^{3/2}|e^{\tau\varphi}u_{|\Gamma_{\text{obs}}}|_{L^2(\Gamma_{\text{obs}})} + |e^{\tau\varphi}\psi_{\text{obs}}\tilde{B}u_{|\Gamma_{\text{obs}}}|_{\tau,1/2} \\
&\qquad\qquad + \sum_{1\leq k\leq\beta} |e^{\tau\varphi}\psi_0 Bu_{|\Gamma_0\cap{}^k\partial\mathcal{M}}|_{\tau,3/2-k}\big),
\end{aligned}$$

for $\tau \geq \tau_*$ and $u \in \mathscr{C}^\infty(\overline{\mathcal{M}})$ vanishing in $W_0 \cap \mathcal{M}$.

One also has the estimate

$$\begin{aligned}
&\tau^{1/2}\|e^{\tau\varphi}u\|_{\tau,1} + \tau^{1/2}|e^{\tau\varphi_{|\partial\mathcal{M}}}\operatorname{tr}(u)|_{\tau,1,0} \\
&\quad \leq C\big(\|e^{\tau\varphi}Pu\|^2_{L^2(\mathcal{M})} + \tau^{3/2}|e^{\tau\varphi}u_{|\Gamma_{\text{obs}}}|_{L^2(\Gamma_{\text{obs}})} + \tau^{1/2}|e^{\tau\varphi}\tilde{B}u_{|\Gamma_{\text{obs}}}|_{L^2(\Gamma_{\text{obs}})} \\
&\qquad\qquad + \tau^{1/2}\sum_{1\leq k\leq\beta} |e^{\tau\varphi}Bu_{|\Gamma_0\cap{}^k\partial\mathcal{M}}|_{\tau,1-k}\big),
\end{aligned}$$

for $\tau \geq \tau_*$ and $u \in \mathscr{C}^\infty(\overline{\mathcal{M}})$ vanishing in $W_0 \cap \mathcal{M}$.

Remark 8.34. The cutoff functions ψ_0 and ψ_{obs} in the two estimates of Theorem 8.33 are useful for Sobolev norms of noninteger order, as we have not introduced properly Sobolev norms of fractional orders on a manifold with boundary like Γ_0 and Γ_{obs}.

If $\mathcal{M}$ is itself a smooth connected compact Riemannian manifold and $\Gamma_0 \cup \Gamma_{\text{obs}} = \partial\mathcal{M}$, then one can write a corollary in the form of Corollary 5.12.

8.6. Notes

In Sect. 8.2, we present how Lopatinskiĭ–Šapiro conditions need to be adapted to cover the case of a conjugated operator for the purpose of the derivation of a Carleman estimate. Such extended conditions can be found in the work of D. Tataru [314] and joint work with M. Bellassoued [70]. The proof of the Carleman estimate of Theorem 8.12 under such extended

Lopatinskiĭ–Šapiro conditions follows also from these two references. Here, we however make use of estimates for first-order factors as derived in Sect. 6.4. Note that the positivity argument of Sect. 8.3.4.2 exploits the fact that the roots of the symbol, viewed as polynomials in ξ_d, are microlocally located in the lower complex half-plane. This is generalized to operators of arbitrary order in [70].

The techniques we use yield estimates with optimal spaces for trace terms, for example, a Sobolev norm on $\partial\mathcal{M}$ of order 3/2 for the trace $u_{|\partial\mathcal{M}}$. Compare with the results in Chapter 3 of Volume 1: there we obtained a Sobolev norm on $\partial\mathcal{M}$ of order 1 for the trace $u_{|\partial\mathcal{M}}$ is obtained. We show in Sect. 8.3.6 that the estimates we find here imply estimates of the form of those obtained in Chapter 3.

In Sect. 8.4, with the techniques introduced for the treatment of the Lopatinskiĭ–Šapiro conditions, we revisit the case where no boundary condition is imposed as we done for instance[1] in Theorem 5.5, and we obtain a substantial improvement by removing the term $|e^{\tau\varphi} \mathsf{D}' u_{|\partial\mathcal{M}}|_{L^2\Lambda^1(\partial\mathcal{M})}$ from the right-hand side of the inequality. This improvement is crucial for the application to the logarithmic stabilization of the wave equation through a boundary damping treated in Chap. 10.

As mentioned above, Neumann conditions are a particular cases of Lopatinskiĭ–Šapiro type conditions. For the derivation of a Carleman estimate for a second-order elliptic operator, their treatment can be found in [219]. The Neumann and Robin conditions for the associated parabolic operator are treated in the work of A. Fursikov and O. Yu. Imanuvilov [156]. Mixed Zaremba type conditions are treated in a joint work with P. Cornilleau [108]. Ventcel boundary condition are treated by R. Buffe [92]. The Lopatinskiĭ–Šapiro conditions for higher order operators are treated in a joint work with M. Bellassoued [70]. Clamped boundary conditions for the bi-Laplace operator are treated in [215]. For the Lamé system, O. Yu. Imanuvilov and M. Yamamoto consider in [182] boundary conditions related to the stress associated with the solution.

Here, we only treat boundary conditions as given by differential operators, thus local conditions. Nonlocal boundary conditions are also of interest. The case of an integral condition for the Neumann trace can be found in the work of Q. Lu and Z. Yin [240].

The proofs of Carleman estimates for various boundary conditions were followed by the studies of interfaces associated with transmission conditions as in the works of A. Doubova et al. [127], M. Bellassoued [68] extended by Le Rousseau and Robbiano [213, 214], and some joint works with N. Lerner [211] and M. Léautaud [210]. For transmission problems for higher order

[1] See Theorem 3.28 of Volume 1 for the counterpart result in an open set of the Euclidean space.

elliptic operators, we refer to joint work with M. Bellassoued [71]. For transmission problems with regularity as low as Lipschitz in the principal part, we refer to the work of M. Di Cristo et al. [118] in the elliptic case and E. Francini and S. Vessella [151] in the parabolic case. The content of the present book and most of the above references concern scalar operators; estimates for elliptic systems like the Lamé system for elasticity are derived in the work of M. Bellassoued [67, 69].

Appendix

8.A. Some Technical Proofs

8.A.1. A Norm Computation. Here we prove Lemma 8.10.

We have $v' = p' + x'$ with $x' = |u'|_{g_\partial}^{-2}(v'_m, u')_{g_\partial} u'$. As p' is orthogonal to x', we obtain $|p'|_{g_\partial}^2 = |v'_m|_{g_\partial}^2 - |x'|_{g_\partial}^2$, yielding

$$|u'|_{g_\partial}^2 |p'|_{g_\partial}^2 = |v'_m|_{g_\partial}^2 |u'|_{g_\partial}^2 - (v'_m, u')_{g_\partial}^2.$$

Recalling the definition of u' in (8.2.25), we compute

$$|u'|_{g_\partial}^2 = |d\varphi_{|\partial\mathcal{M}}(m)|_{g_\partial}^2 + \langle d\varphi_{|\partial\mathcal{M}}(m), v'_m\rangle^2 |v'_m|_{g_\partial}^2 - 2\langle d\varphi_{|\partial\mathcal{M}}(m), v'_m\rangle^2,$$

and

$$(v'_m, u')_{g_\partial} = \langle d\varphi_{|\partial\mathcal{M}}(m), v'_m\rangle \big(1 - |v'_m|_{g_\partial}^2\big),$$

yielding, after algebraic simplifications,

$$|v'_m|_{g_\partial}^2 |u'|_{g_\partial}^2 - (v'_m, u')_{g_\partial}^2 = |v'_m|_{g_\partial}^2 |d\varphi_{|\partial\mathcal{M}}(m)|_{g_\partial}^2 - \langle d\varphi_{|\partial\mathcal{M}}(m), v'_m\rangle^2.$$

We thus find

$$|p'|_{g_\partial}^2 = \frac{|v'_m|_{g_\partial}^2 |d\varphi_{|\partial\mathcal{M}}(m)|_{g_\partial}^2 - \langle d\varphi_{|\partial\mathcal{M}}(m), v'_m\rangle^2}{|d\varphi_{|\partial\mathcal{M}}(m)|_{g_\partial}^2 + \langle d\varphi_{|\partial\mathcal{M}}(m), v'_m\rangle^2 |v'_m|_{g_\partial}^2 - 2\langle d\varphi_{|\partial\mathcal{M}}(m), v'_m\rangle^2},$$

and the result follows. ∎

8.A.2. The Classical Carleman Argument Revisited. To prove Lemma 8.17, we start with the following lemma whose proof is given below.

LEMMA 8.35. *Set $Q(w) = \|P_2 w\|_+^2 + \|P_1 w\|_+^2$. There exist $C_0, C > 0$, $\mu > 0$, and $\tau_* > 0$ such that*

$$\mu Q(w) + \tau \operatorname{Re}\big(i[P_2, P_1]w, w\big)_+ \geq C_0 \|w\|_{\tau,2}^2 - C|\operatorname{tr}(w)|_{\tau,1,1/2}^2,$$

for $\tau \geq \tau_*$ and $w \in \overline{\mathscr{C}}_c^\infty(U^0)$.

PROOF OF LEMMA 8.17. Let $\mu > 0$ be as given by Lemma 8.35, and let $\tau \geq \mu$. With this lemma, the identities (8.3.17)–(8.3.18), and the estimation (8.3.20), we then write

$$\begin{aligned}\|P_{0,\varphi}w\|_+^2 &\geq \mu\tau^{-1}Q(w) + 2\operatorname{Re}(P_2 w, iP_1 w)_+ \\ &= \mu\tau^{-1}Q(w) + \operatorname{Re}\big(i[P_2, P_1]w, w\big)_+ + \tau \operatorname{Re}\tilde{\mathscr{B}}(w) \\ &\geq C_0\tau^{-1}\|w\|_{\tau,2}^2 - C'\big(\tau^{-1}|\operatorname{tr}(w)|_{\tau,1,1/2}^2 + \tau|\operatorname{tr}(w)|_{\tau,1,0}^2\big) \\ &\geq C_0\tau^{-1}\|w\|_{\tau,2}^2 - C'|\operatorname{tr}(w)|_{\tau,1,1/2}^2,\end{aligned}$$

which gives (8.3.14). ■

PROOF OF LEMMA 8.35. We have $[P_2, P_1] \in \tau\mathscr{D}_\tau^2$. Writing

$$\tau\operatorname{Re}\big(i[P_2, P_1]w, w\big)_+ = \operatorname{Re}\big(i\tau^{-1}[P_2, P_1]w, \tau^2 w\big)_+,$$

we see that this is an interior quadratic form of type $(2, 0)$ (see Definition 6.13). We thus see that $K(w) = \mu Q(w) + \tau\operatorname{Re}\big(i[P_2, P_1]w, w\big)_+$ is also such an interior quadratic form with principal symbol

$$k(\varrho) = \mu|p_{0,\varphi}(\varrho)|^2 + \tau\{p_2, p_1\}(\varrho), \quad \varrho = (x, \xi, \tau).$$

The sub-ellipticity property of (P_0, φ) yields by Lemma 3.8 of Volume 1

$$k(\varrho) \gtrsim \lambda_\tau^4, \qquad \varrho \in \overline{U^0} \times \mathbb{R}^d \times [1, +\infty),$$

for $\mu > 0$ chosen sufficiently large. The Gårding inequality of Theorem 6.17 yields

$$K(w) \geq C\|w\|_{\tau,2}^2 - C'|\operatorname{tr}(w)|_{\tau,1,1/2}^2,$$

for some $C, C' > 0$ and for $\tau > 0$ chosen sufficiently large (no microlocalization is needed here and the estimation follows from (6.3.5) in Remark 6.18). ■

8.A.3. Proof of Lemma 8.19. The principal symbol of $B_{0,\varphi}(x, D, \tau)$ is given in (8.3.7) and is homogeneous of degree k in λ_τ. One sets $B_\varphi^j(x, D', \tau) = \operatorname{Op}_{\mathsf{T}}(b_\varphi^k)$, $j = k, k-1$, that is,

$$\begin{aligned}B_{0,\varphi}(x, D, \tau) &= B_\varphi^k(x, D', \tau) - iB_\varphi^{k-1}(x, D', \tau)\big(D_d + i\tau\partial_d\varphi(x)\big) \\ &= \tilde{B}_\varphi^k(x, D', \tau) - iB_\varphi^{k-1}(x, D', \tau)D_d,\end{aligned}$$

with

$$\tilde{B}_\varphi^k(x, D', \tau) = B_\varphi^k(x, D', \tau) + \tau B_\varphi^{k-1}(x, D', \tau)\partial_d\varphi(x) \in \mathscr{D}_\tau^k.$$

The principal symbol of $\tilde{B}_\varphi^k$ is given by $\tilde{b}_\varphi^k(x, \xi', \tau) = b_\varphi^k(x, \xi', \tau) + \tau b_\varphi^{k-1}(x, \xi', \tau)\partial_d\varphi(x)$ and is homogeneous of degree k in $\lambda_{\mathsf{T},\tau}$.

For $\sigma \in \mathbb{R}$, we introduce the operator

$$\mathsf{B}_\varphi^\sigma = \Big(\Lambda_{\mathsf{T},\tau}^{3/2-\sigma-k}\tilde{B}_\varphi^k(x, D', \tau) \quad -i\Lambda_{\mathsf{T},\tau}^{3/2-\sigma-k}B_\varphi^{k-1}(x, D', \tau)\Lambda_{\mathsf{T},\tau}^1\Big),$$

with principal symbol

$$\mathsf{b}^{\sigma}_{\varphi}(x,\xi',\tau) = \Big(\lambda_{\mathsf{T},\tau}^{3/2-\sigma-k}\tilde{b}^{k}_{\varphi}(x,\xi',\tau) \quad -i\lambda_{\mathsf{T},\tau}^{5/2-\sigma-k} b^{k-1}_{\varphi}(x,\xi',\tau)\Big).$$

We also introduce

$$\mathsf{P}^{\sigma,+} = \Big(-\Lambda_{\mathsf{T},\tau}^{1/2-\sigma}\mathrm{Op}_{\mathsf{T}}(\tilde{\gamma}_2) \quad \Lambda_{\mathsf{T},\tau}^{3/2-\sigma}\Big),$$

with principal symbol

$$\mathsf{p}^{\sigma,+} = \Big(-\lambda_{\mathsf{T},\tau}^{1/2-\sigma}\tilde{\chi}\gamma_2 \quad \lambda_{\mathsf{T},\tau}^{3/2-\sigma}\Big).$$

We set $\mathsf{M}^{\sigma} = (\mathsf{B}^{\sigma}_{\varphi})^{*}\,\mathsf{B}^{\sigma}_{\varphi} + (\mathsf{P}^{\sigma,+})^{*}\,\mathsf{P}^{\sigma,+}$. It is 2×2 matrix operator of order $3-2\sigma$. If $V = {}^{t}(v^0, v^1) \in \big(\mathscr{S}(\mathbb{R}^{d-1})\big)^2$, we have

$$(\mathsf{M}^{\sigma}V, V)_{(L^2(\mathbb{R}^{d-1}))^2} = |\mathsf{B}^{\sigma}_{\varphi}V|^2_{\partial} + |\mathsf{P}^{\sigma,+}V|^2_{\partial}.$$

We have the following lemma.

LEMMA 8.36. *Let $\mathscr{W}$ be an open conic set of $\mathbb{R}^d_+ \times \mathbb{R}^{d-1}\times\mathbb{R}_+$ such that $\mathbb{S}_{\overline{\mathscr{W}}}$ is compact and*

$$b_{\varphi}(x,\xi',\xi_d = \gamma_2(x,\xi',\tau),\tau) \neq 0, \quad (x,\xi',\tau)\in\mathscr{W}, \tag{8.A.1}$$

and such that $\tilde{\chi}\equiv 1$ on $\mathscr{W}$. Let $\hat{\chi}\in S^0_{\mathsf{T},\tau}$ be homogeneous of degree zero and be supported in $\mathscr{W}$. There exists $C>0$ such that for $N\in\mathbb{N}$, there exist $C_N>0$ and $\tau_*>0$ such that

$$(\mathsf{M}^{\sigma}\mathrm{Op}_{\mathsf{T}}(\hat{\chi})V, \mathrm{Op}_{\mathsf{T}}(\hat{\chi})V)_{(L^2(\mathbb{R}^{d-1}))^2} \geq C|\mathrm{Op}_{\mathsf{T}}(\hat{\chi})V|^2_{\tau,3/2-\sigma} - C_N|V|^2_{\tau,-N},$$

for $V = {}^{t}(v^0,v^1)\in\big(\mathscr{S}(\mathbb{R}^{d-1})\big)^2$ and $\tau\geq\tau_*$.

Here $|V|^2_{\tau,s} = |v^0|^2_{\tau,s} + |v^1|^2_{\tau,s}$. The proof of Lemma 8.36 can be found below.

We shall use this lemma in the case $\sigma = 0$ with $\mathscr{W}\subset\mathscr{U}$ an open conic neighborhood of $\mathrm{supp}(\chi)$ where $\tilde{\chi}\equiv 1$. We choose $\hat{\chi}\in S^0_{\mathsf{T},\tau}$ satisfying moreover $\hat{\chi}\equiv 1$ on $\mathrm{supp}(\chi)$. Condition (8.A.1) holds by (8.3.21) as $\mathscr{W}\subset\mathscr{U}$. Recall that this condition is a direct consequence of the Lopatinskiĭ–Šapiro condition.

For $u\in\overline{\mathscr{S}}(\mathbb{R}^d_+)$, we have

$$|B_{0,\varphi}u_{|x_d=0^+}|^2_{\tau,3/2-k} = |\Lambda_{\mathsf{T},\tau}^{3/2-k}B_{0,\varphi}u_{|x_d=0^+}|^2_{\partial} = |\mathsf{B}^0_{\varphi}U|^2_{\partial},$$

where $U = {}^{t}(u_{|x_d=0^+}, \Lambda_{\mathsf{T},\tau}^{-1}D_d u_{|x_d=0^+})$. Recalling that $P^+ = D_d - \mathrm{Op}_{\mathsf{T}}(\tilde{\gamma}_2)$ with $\tilde{\gamma}_2 = \tilde{\chi}\gamma_2$, we have

$$|P^+u_{|x_d=0^+}|^2_{\tau,1/2} = |\Lambda_{\mathsf{T},\tau}^{1/2}P^+u_{|x_d=0^+}|^2_{\partial} = |\mathsf{P}^{0,+}U|^2_{\partial}.$$

We thus obtain

$$|B_{0,\varphi}u_{|x_d=0^+}|^2_{\tau,3/2-k} + |P^+u_{|x_d=0^+}|^2_{\tau,1/2} = (\mathsf{M}^0U,U)_{L^2(\mathbb{R}^{d-1})^2}. \tag{8.A.2}$$

Lemma 8.36, for $\sigma = 0$ here, gives

$$(\mathsf{M}^0 \mathrm{Op}_\mathsf{T}(\hat{\chi})U, \mathrm{Op}_\mathsf{T}(\hat{\chi})U)_{(L^2(\mathbb{R}^{d-1}))^2} \geq C|\mathrm{Op}_\mathsf{T}(\hat{\chi})U|^2_{\tau,3/2} - C_N|U|^2_{\tau,-N},$$

for $N \in \mathbb{N}$ and $\tau \geq 1$ chosen sufficiently large. We now set $u = \mathrm{Op}_\mathsf{T}(\chi)v$, that is, $U = (\mathrm{Op}_\mathsf{T}(\chi)v_{|x_d=0^+}, \Lambda^{-1}_{\mathsf{T},\tau}D_d\mathrm{Op}_\mathsf{T}(\chi)v_{|x_d=0^+})$. Because of the support conditions of $\hat{\chi}$ and χ and from pseudo-differential calculus, we find, for any $N \in \mathbb{N}$,

$$(\mathsf{M}^0 U, U)_{(L^2(\mathbb{R}^{d-1}))^2} \geq \mathrm{Re}(\mathsf{M}^0 \mathrm{Op}_\mathsf{T}(\hat{\chi})U, \mathrm{Op}_\mathsf{T}(\hat{\chi})U)_{(L^2(\mathbb{R}^{d-1}))^2} - C_N|\operatorname{tr}(v)|^2_{\tau,1,-N},$$

and

$$|\mathrm{Op}_\mathsf{T}(\hat{\chi})U|^2_{\tau,3/2} \geq |U|^2_{\tau,3/2} - C_N|\operatorname{tr}(v)|^2_{\tau,1,-N},$$

yielding

$$\begin{aligned}(\mathsf{M}^0 U, U)_{(L^2(\mathbb{R}^{d-1}))^2} &\geq C|U|^2_{\tau,3/2} - C_N|\operatorname{tr}(v)|^2_{\tau,1,-N} \\ &= C|\operatorname{tr}(\mathrm{Op}_\mathsf{T}(\chi)v)|^2_{\tau,1,1/2} - C_N|\operatorname{tr}(v)|^2_{\tau,1,-N}.\end{aligned}$$

By (8.A.2), this concludes the proof of Lemma 8.19. ■

PROOF OF LEMMA 8.36. The matrix operator M^σ has the following 2×2 matrix principal symbol:

$$\ell(\varrho') = \big((\mathsf{b}^\sigma_\varphi)^*\mathsf{b}^\sigma_\varphi + (\mathsf{p}^{\sigma,+})^*\, \mathsf{p}^{\sigma,+}\big)(\varrho'), \qquad \varrho' = (x, \xi', \tau).$$

For $\mathsf{z} \in \mathbb{C}^2$, we have

$$(\ell(\varrho')\mathsf{z}, \mathsf{z})_{\mathbb{C}^2} = |\mathsf{b}^\sigma_\varphi(\varrho')\mathsf{z}|^2 + |\mathsf{p}^{\sigma,+}(\varrho')\mathsf{z}|^2.$$

Having $\mathsf{b}^\sigma_\varphi(\varrho')\mathsf{z} = \mathsf{p}^{\sigma,+}(\varrho')\mathsf{z} = 0$ for $\mathsf{z} \neq 0$ precisely means

$$\det\begin{pmatrix}\mathsf{b}^\sigma_\varphi(\varrho') \\ \mathsf{p}^{\sigma,+}(\varrho')\end{pmatrix} = 0,$$

that is, $0 = \lambda^{3-2\sigma-k}_{\mathsf{T},\tau}\big(\tilde{b}^k_\varphi(\varrho') - i b^{k-1}_\varphi(\varrho')\tilde{\gamma}_2\big)$. In $\mathscr{W}$, we have $\tilde{\chi} \equiv 1$ meaning $\tilde{\gamma}_2 = \gamma_2$ and then

$$b_\varphi(x, \xi', \xi_d = \gamma_2, \tau) = 0,$$

which is precisely excluded in $\mathscr{W}$ by (8.A.1). As a result, by homogeneity and the compactness of $\mathbb{S}_{\overline{\mathscr{W}}}$, we have

$$(\ell(\varrho')\mathsf{z}, \mathsf{z})_{\mathbb{C}^2} \geq C\lambda^{3-2\sigma}_{\mathsf{T},\tau}|\mathsf{z}|^2_{\mathbb{C}^2}, \tag{8.A.3}$$

for some $C > 0$ for $\varrho' \in \mathscr{W}$. The microlocal Gårding inequality for systems of Theorem 2.32 of Volume 1 then yields the result. ■

REMARK 8.37. Assume that the boundary operator B *continuously* depends on a parameter s that lies in a *compact* set S and such that the Lopatinskiĭ–Šapiro condition holds for all values of $s \in S$. Then because of the compactness of S and thus of $\mathbb{S}_{\overline{\mathscr{W}}} \times S$, the estimate in (8.A.3) holds uniformly with respect to $s \in S$, meaning that the result of Lemma 8.19 can be written uniformly with respect to $s \in S$. An inspection of the remainder of the proofs of Propositions 8.18 and 8.12 shows that their statements then follow with constants that are uniform with respect to $s \in S$.

8.A.4. Proof of Lemma 8.20. Setting $W = (\Lambda^1_{\mathsf{T},\tau} w_{|x_d=0^+}, D_d w_{|x_d=0^+})$, with (8.3.19), we write

$$\operatorname{Re} \tilde{\mathscr{B}}(w) \geq 2 \operatorname{Re}(\operatorname{Op}_{\mathsf{T}}(a)W, W)_\partial - C|\operatorname{tr}(w)_{|x_d=0^+}|^2_{\tau,1,-1/2} \tag{8.A.4}$$

with $a = (a^{ij})_{1\leq i,j\leq 2}$ given by

$$\begin{aligned}
\operatorname{Op}_{\mathsf{T}}(a^{11}) &= -\Lambda^{-1}_{\mathsf{T},\tau} \partial_d \varphi_{|x_d=0^+} \big(\operatorname{Op}_{\mathsf{T}}(r) - p(x, \tau d\varphi)\big)_{|x_d=0^+} \Lambda^{-1}_{\mathsf{T},\tau},\\
\operatorname{Op}_{\mathsf{T}}(a^{12}) &= \Lambda^{-1}_{\mathsf{T},\tau} \tilde{r}(x, D', d_{x'}\varphi)_{|x_d=0^+},\\
\operatorname{Op}_{\mathsf{T}}(a^{21}) &= \tilde{r}(x, D', d_{x'}\varphi)_{|x_d=0^+} \Lambda^{-1}_{\mathsf{T},\tau},\\
\operatorname{Op}_{\mathsf{T}}(a^{22}) &= \partial_d \varphi_{|x_d=0^+},
\end{aligned}$$

with principal parts in $S^0_{\mathsf{T},\tau}$ given by

$$\begin{aligned}
a_0^{11} &= -\lambda^{-2}_{\mathsf{T},\tau} \partial_d \varphi_{|x_d=0^+} \big(r(x,\xi') - p(x, \tau d\varphi)\big)_{|x_d=0^+},\\
a_0^{12} &= a_0^{21} = \lambda^{-1}_{\mathsf{T},\tau} \tilde{r}(x, \xi', d_{x'}\varphi)_{|x_d=0^+},\\
a_0^{22} &= \partial_d \varphi_{|x_d=0^+}.
\end{aligned}$$

We set $a_0 = (a_0^{ij})_{1\leq i,j\leq 2}$. It is symmetric. As we have $\partial_d \varphi_{|x_d=0^+} > 0$ in $\mathscr{U}$, we thus obtain, for $Z \in \mathbb{R}^2$,

$$\big(a_0(x,\xi',\tau)Z, Z\big)_{\mathbb{R}^2} \geq C\|Z\|^2_{\mathbb{R}^2}, \quad (x,\xi',\tau) \in \mathscr{U},\ \tau \in [1,+\infty),\ |(\xi',\tau)| \geq R, \tag{8.A.5}$$

for some C and R positive if and only if $\det a_0(x,\xi',\tau) > 0$ in $\mathscr{U}$, by compactness and homogeneity. We compute

$$\begin{aligned}
\lambda^2_{\mathsf{T},\tau} \det a_0(x,\xi',\tau) &= -(\partial_d\varphi)^2 \big(r(x,\xi') - p(x, \tau d\varphi)\big)_{|x_d=0^+}\\
&\quad - \tilde{r}^2(x, \xi', d_{x'}\varphi)_{|x_d=0^+}.
\end{aligned}$$

By (8.3.10), we have $0 \leq \operatorname{Re}\alpha < \tau \partial_d \varphi$. By (8.3.12), this reads

$$(\partial_d\varphi)^2 \big(r(x,\xi') - \tau^2 r(x, d_{x'}\varphi)\big) - \tau^2(\partial_d\varphi)^4 + \tilde{r}^2(x,\xi', d_{x'}\varphi) < 0,$$

which is precisely $\det a_0(x,\xi',\tau) > 0$.

By (8.A.4), (8.A.5), and Theorem 2.32, for any $N \in \mathbb{N}$, we have

$$\operatorname{Re} \tilde{\mathscr{B}}(w) \geq C|\operatorname{tr}(w)|^2_{\tau,1,0} - C'|\operatorname{tr}(w)|^2_{\tau,1,-1/2} - C_N|\operatorname{tr}(v)|^2_{\tau,1,-N}.$$

For $\tau > 0$ chosen sufficiently large, we obtain

$$\operatorname{Re} \tilde{\mathscr{B}}(w) \geq C|\operatorname{tr}(w)|^2_{\tau,1} - C_N|\operatorname{tr}(v)|^2_{\tau,1,-N}.$$

Observe that we have, for any $N \in \mathbb{N}$,

$$\tau^s|\operatorname{tr}(w)|_{\tau,1,s'} \geq C|\operatorname{tr}(w)|_{\tau,1,s+s'} - C'_N|\operatorname{tr}(v)|_{\tau,1,-N},$$

by the Gårding inequality of Theorem 2.29 of Volume 1, since, by (8.3.13), we have $\lambda_{\mathsf{T},\tau} \lesssim \tau$ in $\mathscr{U}$. We thus obtain the result. ■

8.A.5. Proof of Theorem 8.24. In the framework of the normal geodesic coordinates introduced in Sect. 8.3.2, we prove in fact the following proposition.

PROPOSITION 8.38. *Under the assumption of Proposition 8.16, there exist a bounded neighborhood U_+ of x^0 in $\overline{\mathbb{R}^d_+}$ such that $U_+ \subset U^0$ and two constants C and $\tau_* > 0$ such that*

$$\tau \|v\|^2_{\tau,1} + \tau |\operatorname{tr}(v)|^2_{\tau,1,0} \leq C\Big(\|P_{0,\varphi} v\|^2_+ + \tau |B_{0,\varphi} v_{|x_d=0^+}|^2_{\tau,1-k} \Big),$$

for all $v \in \overline{\mathscr{C}}^\infty_c(U_+)$ and $\tau \geq \tau_*$.

With the experience of Chapter 3 of Volume 1, it is now classical to deduce, with a commutator argument, a local version of Theorem 8.24 by setting $v = e^{\tau\varphi} u$. Arguing as in Sect. 5.3, for Theorems 5.5 and 5.6, patching together such local estimates, we then obtain the result of Theorem 8.24.

PROOF OF PROPOSITION 8.38. Let $\tilde{U}_+$ be the open set of U^0 given by Proposition 8.22, and let U_+ be a second open set of U^0 such that $U_+ \Subset \tilde{U}_+$ with $x_0 \subset U_+$. We choose $\chi \in \overline{\mathscr{C}}^\infty_c(\tilde{U}_+)$ be such that $\chi \equiv 1$ in a neighborhood of U_+.

We then set $w = \tau^{1/2} \chi \Lambda^{-1/2}_{\mathsf{T},\tau} v$, and we apply Proposition 8.22 to w:

$$\|\chi \Lambda^{-1/2}_{\mathsf{T},\tau} v\|^2_{\tau,2} + \tau |\operatorname{tr}(\chi \Lambda^{-1/2}_{\mathsf{T},\tau} v)|^2_{\tau,1,1/2} \lesssim \tau \|P_{0,\varphi} \chi \Lambda^{-1/2}_{\mathsf{T},\tau} v\|^2_+ + \tau |B_{0,\varphi} \chi \Lambda^{-1/2}_{\mathsf{T},\tau} v_{|x_d=0^+}|^2_{\tau,3/2-k}. \tag{8.A.6}$$

We then write, with Corollary 6.7,

$$\begin{aligned} \|v\|_{\tau,2,-1/2} = \|\chi v\|_{\tau,2,-1/2} &= \|\Lambda^{-1/2}_{\mathsf{T},\tau} \chi v\|_{\tau,2} \\ &\lesssim \|\chi \Lambda^{-1/2}_{\mathsf{T},\tau} v\|_{\tau,2} + \|[\Lambda^{-1/2}_{\mathsf{T},\tau}, \chi] v\|_{\tau,2} \\ &\lesssim \|\chi \Lambda^{-1/2}_{\mathsf{T},\tau} v\|_{\tau,2} + \|v\|_{\tau,2,-3/2}. \end{aligned}$$

Thus, for $\tau > 0$ chosen sufficiently large, we obtain

$$\|v\|_{\tau,2,-1/2} \lesssim \|\chi \Lambda^{-1/2}_{\mathsf{T},\tau} v\|_{\tau,2},$$

which yields

$$\tau^{1/2} \|v\|_{\tau,1} \lesssim \|v\|_{\tau,1,1/2} \lesssim \|v\|_{\tau,2,-1/2} \lesssim \|\chi \Lambda^{-1/2}_{\mathsf{T},\tau} v\|_{\tau,2}. \tag{8.A.7}$$

We now observe that $[D_d^k \chi, \Lambda_{\mathsf{T},\tau}^{-1/2}] \in \Psi_\tau^{k,-3/2}$, for $k = 0, 1$, which gives

$$
\begin{aligned}
|\operatorname{tr}(v)|_{\tau,1,0} &= |\Lambda_{\mathsf{T},\tau}^{-1/2} \operatorname{tr}(\chi v)|_{\tau,1,1/2} \\
&\asymp |\Lambda_{\mathsf{T},\tau}^{-1/2} \chi v_{|x_d=0^+}|_{\tau,3/2} + |\Lambda_{\mathsf{T},\tau}^{-1/2} D_{x_d}(\chi v)_{|x_d=0^+}|_{\tau,1/2} \\
&\lesssim |\chi \Lambda_{\mathsf{T},\tau}^{-1/2} v_{|x_d=0^+}|_{\tau,3/2} + |D_{x_d}(\chi \Lambda_{\mathsf{T},\tau}^{-1/2} v)_{|x_d=0^+}|_{\tau,1/2} \\
&\quad + |[\Lambda_{\mathsf{T},\tau}^{-1/2}, \chi] v_{|x_d=0^+}|_{\tau,3/2} + |[\Lambda_{\mathsf{T},\tau}^{-1/2}, D_{x_d}\chi] v_{|x_d=0^+}|_{\tau,1/2} \\
&\lesssim |\operatorname{tr}(\chi \Lambda_{\mathsf{T},\tau}^{-1/2} v)|_{\tau,1,1/2} + |\operatorname{tr}(v)|_{\tau,1,-1}.
\end{aligned}
$$

Thus, for $\tau > 0$ chosen sufficiently large, we obtain

$$
|\operatorname{tr}(v)|_{\tau,1,0} \lesssim |\operatorname{tr}(\chi \Lambda_{\mathsf{T},\tau}^{-1/2} v)|_{\tau,1,1/2}. \tag{8.A.8}
$$

As $\tau^{1/2}[P_{0,\varphi}, \chi \Lambda_{\mathsf{T},\tau}^{-1/2}] \in \tau^{1/2}\Psi_\tau^{1,-1/2} \subset \Psi_\tau^{1,0}$, we also have

(8.A.9)

$$
\begin{aligned}
\tau^{1/2}\|P_{0,\varphi}\chi \Lambda_{\mathsf{T},\tau}^{-1/2} v\|_+ &\lesssim \tau^{1/2}\|\chi \Lambda_{\mathsf{T},\tau}^{-1/2} P_{0,\varphi} v\|_+ + \tau^{1/2}\|[P_{0,\varphi}, \chi \Lambda_{\mathsf{T},\tau}^{-1/2}] v\|_+ \\
&\lesssim \|P_{0,\varphi} v\|_+ + \|v\|_{\tau,1}.
\end{aligned}
$$

Finally, we observe that $[B_{0,\varphi}\chi, \Lambda_{\mathsf{T},\tau}^{-1/2}] \in \Psi_{\mathsf{T},\tau}^{-3/2}$ if $k = 0$ and $[B_{0,\varphi}\chi, \Lambda_{\mathsf{T},\tau}^{-1/2}] \in \Psi_\tau^{1,k-5/2}$ if $k \geq 1$; thus, $[B_{0,\varphi}\chi, \Lambda_{\mathsf{T},\tau}^{-1/2}] \in \Psi_\tau^{1,k-5/2}$ for any value of k. We then find

(8.A.10)

$$
\begin{aligned}
|B_{0,\varphi}\chi \Lambda_{\mathsf{T},\tau}^{-1/2} v_{|x_d=0^+}|_{\tau,3/2-k} &\lesssim |B_{0,\varphi} v_{|x_d=0^+}|_{\tau,1-k} \\
&\quad + |[B_{0,\varphi}\chi, \Lambda_{\mathsf{T},\tau}^{-1/2}] v_{|x_d=0^+}|_{\tau,3/2-k} \\
&\lesssim |B_{0,\varphi} v_{|x_d=0^+}|_{\tau,1-k} + |\operatorname{tr}(v)_{|x_d=0^+}|_{\tau,1,-1}.
\end{aligned}
$$

With (8.A.6) and (8.A.7)–(8.A.10), we obtain

$$
\begin{aligned}
&\tau\|v\|_{\tau,1}^2 + \tau|\operatorname{tr}(v)|_{\tau,1,0}^2 \\
&\qquad \lesssim \|P_{0,\varphi} v\|_+^2 + \tau|B_{0,\varphi} v_{|x_d=0^+}|_{\tau,1-k}^2 + \|v\|_{\tau,1}^2 + \tau|\operatorname{tr}(v)_{|x_d=0^+}|_{\tau,1,-1}^2.
\end{aligned}
$$

We then obtain the result by choosing $\tau > 0$ sufficiently large. ∎

8.A.6. Proof of Theorem 8.30. In the framework of the normal geodesic coordinates introduced in Sect. 8.3.2, we prove in fact the following proposition.

PROPOSITION 8.39. *Under the assumption of Proposition 8.29, there exist a bounded neighborhood U_+ of x^0 in $\overline{\mathbb{R}^d_+}$ such that $U_+ \subset U^0$ and two constants C and $\tau_* > 0$ such that*

$$
\begin{aligned}
\tau^{1/2}\|v\|_{\tau,1} + \tau^{1/2}|\operatorname{tr}(v)|_{\tau,1,0} \leq C\Big(&\|P_{0,\varphi} v\|_+ + \tau^{3/2}|v_{|x_d=0^+}|_\partial \\
&+ \tau^{1/2}|B_{0,\varphi} v_{|x_d=0^+}|_\partial\Big),
\end{aligned}
$$

for all $v \in \overline{\mathscr{C}}_c^\infty(U_+)$ *and* $\tau \geq \tau_*$.

With the experience of Chapter 3 of Volume 1 we deduce, with a commutator argument, a local version of Theorem 8.30 by setting $v = e^{\tau\varphi}u$. Arguing as in Sect. 5.3, for Theorems 5.5 and 5.6, patching together such local estimates, we then obtain the result of Theorem 8.30.

PROOF OF PROPOSITION 8.39. Let $\tilde{U}_+$ be the open set of U^0 given by Proposition 8.29, and let U_+ be a second open set of U^0 such that $U_+ \Subset \tilde{U}_+$ with $x_0 \subset U_+$. We choose $\chi \in \overline{\mathscr{C}}_c^\infty(\tilde{U}_+)$ be such that $\chi \equiv 1$ in a neighborhood of U_+.

We then set $w = \tau^{1/2}\chi\Lambda_{\mathsf{T},\tau}^{-1/2}v$, and we apply Proposition 8.29 to w:

$$\begin{aligned} \|\chi\Lambda_{\mathsf{T},\tau}^{-1/2}v\|_{\tau,2} &+ \tau^{1/2}|\chi\Lambda_{\mathsf{T},\tau}^{-1/2}v_{|x_d=0^+}|_{\tau,3/2} \\ &\lesssim \tau^{1/2}\|P_{0,\varphi}\chi\Lambda_{\mathsf{T},\tau}^{-1/2}v\|_+ + \tau^2|\chi\Lambda_{\mathsf{T},\tau}^{-1/2}v_{|x_d=0^+}|_\partial \\ &\quad + \tau^{1/2}|B_{0,\varphi}(\chi\Lambda_{\mathsf{T},\tau}^{-1/2}v)_{|x_d=0^+}|_{\tau,1/2}. \end{aligned} \tag{8.A.11}$$

By (8.A.7), we have, for $\tau > 0$ sufficiently large,

$$\tau^{1/2}\|v\|_{\tau,1} \lesssim \|\chi\Lambda_{\mathsf{T},\tau}^{-1/2}v\|_{\tau,2}. \tag{8.A.12}$$

We now observe that $[\chi, \Lambda_{\mathsf{T},\tau}^{-1/2}] \in \Psi_{\mathsf{T},\tau}^{-3/2}$, which gives

$$\begin{aligned} |v_{|x_d=0^+}|_{\tau,1} &= |\Lambda_{\mathsf{T},\tau}^{-1/2}\chi v_{|x_d=0^+}|_{\tau,3/2} \\ &\lesssim |\chi\Lambda_{\mathsf{T},\tau}^{-1/2}v_{|x_d=0^+}|_{\tau,3/2} + |[\Lambda_{\mathsf{T},\tau}^{-1/2}, \chi]v_{|x_d=0^+}|_{\tau,3/2} \\ &\lesssim |\chi\Lambda_{\mathsf{T},\tau}^{-1/2}v_{|x_d=0^+}|_{\tau,3/2} + |v_{|x_d=0^+}|_\partial. \end{aligned}$$

Thus, for $\tau > 0$ chosen sufficiently large, we obtain

$$|v_{|x_d=0^+}|_{\tau,1} \lesssim |\chi\Lambda_{\mathsf{T},\tau}^{-1/2}v_{|x_d=0^+}|_{\tau,3/2}. \tag{8.A.13}$$

By (8.A.9), we have

$$\tau^{1/2}\|P_{0,\varphi}\chi\Lambda_{\mathsf{T},\tau}^{-1/2}v\|_+ \lesssim \|P_{0,\varphi}v\|_+ + \|v\|_{\tau,1}. \tag{8.A.14}$$

Finally, as $\tau^{1/2}\chi\Lambda_{\mathsf{T},\tau}^{-1/2} \in \Psi_{\mathsf{T},\tau}^0$ and $[B_{0,\varphi}, \chi\Lambda_{\mathsf{T},\tau}^{-1/2}] \in \Psi_\tau^{1,-3/2}$, we have

(8.A.15)

$$\begin{aligned} \tau^2|\chi\Lambda_{\mathsf{T},\tau}^{-1/2}v_{|x_d=0^+}|_\partial &+ \tau^{1/2}|B_{0,\varphi}(\chi\Lambda_{\mathsf{T},\tau}^{-1/2}v)_{|x_d=0^+}|_{\tau,1/2} \\ &\lesssim \tau^{3/2}|v_{|x_d=0^+}|_\partial + \tau^{1/2}|B_{0,\varphi}v_{|x_d=0^+}|_\partial + \tau^{1/2}|D_d v_{|x_d=0^+}|_{\tau,-1} \\ &\lesssim \tau^{3/2}|v_{|x_d=0^+}|_\partial + \tau^{1/2}|B_{0,\varphi}v_{|x_d=0^+}|_\partial, \end{aligned}$$

using that $B_{0,\varphi} - D_d \in \Psi^1_{\mathrm{T},\tau}$. With (8.A.11) and (8.A.12)–(8.A.15), we obtain

$$\tau^{1/2}\|v\|_{\tau,1} + \tau^{1/2}|v_{|x_d=0^+}|_{\tau,1} \lesssim \|P_{0,\varphi}v\|_+ + \tau^{3/2}|v_{|x_d=0^+}|_\partial + \tau^{1/2}|B_{0,\varphi}v_{|x_d=0^+}|_\partial + \|v\|^2_{\tau,1}.$$

By choosing $\tau > 0$ sufficiently large, we find

$$\tau^{1/2}\|v\|_{\tau,1} + \tau^{1/2}|v_{|x_d=0^+}|_{\tau,1} \lesssim \|P_{0,\varphi}v\|_+ + \tau^{3/2}|v_{|x_d=0^+}|_\partial + \tau^{1/2}|B_{0,\varphi}v_{|x_d=0^+}|_\partial.$$

Finally, we write

$$|D_d v_{|x_d=0^+}|_\partial \leq |v_{|x_d=0^+}|_{\tau,1} + |B_{0,\varphi}v_{|x_d=0^+}|_\partial,$$

and we obtain the sought estimate. ■

Part 3

Applications

CHAPTER 9

Quantified Unique Continuation on a Riemannian Manifold

Contents

In Chapter 5 of Volume 1, we saw how the Carleman estimates proven in Chapter 3 (also in Volume 1) for a second-order elliptic operator on an open set Ω of $\mathbb{R}^d$ could be used to provide unique continuation results. Moreover, a quantification of the unique continuation property could be expressed through an interpolation inequality of the form, for some $\delta \in (0, 1)$,

$$\|u\|_{H^1(U)} \leq C\|u\|_{H^1(\Omega)}^{1-\delta}\big(\|f\|_{L^2(\Omega)} + \|u\|_{L^2(\omega)}\big)^{\delta},$$

where U and ω are open subsets of an open set Ω of $\mathbb{R}^d$ and where $u \in H^2(\Omega)$ satisfies

$$Pu = f + g, \quad \text{with } f \in L^2(\Omega),$$
$$|g(x)| \leq C_0\big(|u(x)| + |Du(x)|\big) \text{ a.e. in } \Omega,$$

J. Le Rousseau et al., *Elliptic Carleman Estimates and Applications to Stabilization and Controllability, Volume II*, PNLDE Subseries in Control 98, https://doi.org/10.1007/978-3-030-88670-7_9

along with homogeneous Dirichlet boundary conditions. With the Carleman estimates of Chap. 5 obtained for a Riemannian manifold, away from and also at boundaries, the result of Chapter 5 can be extended to the manifold case along with homogeneous Dirichlet boundary conditions. The purpose of the present chapter is to generalize these results to the case of the general boundary conditions treated in Chaps. 2, 3, and 8, namely Lopatinskiĭ–Šapiro type boundary conditions.

We shall consider $(\mathcal{M}, g)$ a d-dimensional smooth Riemannian manifold, possibly noncompact. On $\mathcal{M}$, we consider the operator $P = -\Delta_g + R_1$ with R_1 a first-order differential operator on $\mathcal{M}$. When needed, we shall also consider a differential operator B of order β in a neighborhood of part of $\partial\mathcal{M}$ in the form introduced in Chap. 2.

9.1. Unique Continuation Estimate Away from Boundaries

We consider $\mathcal{U}$ and $\mathcal{V}$ two open sets of $\mathcal{M}$ such that $\mathcal{U}$ is connected, $\overline{\mathcal{U}} \cap \partial\mathcal{M} = \emptyset$, and $\mathcal{U} \Subset \mathcal{V} \Subset \mathcal{M}$; see Fig. 9.1. The following result, away from boundaries, is deduced from the results of Section 5.3.1 of Volume 1.

THEOREM 9.1 (Propagation of "smallness" Away from Boundaries). *Let ω be an open subset of $\mathcal{U}$. Let $C_0 > 0$. There exist $C > 0$ and $\delta \in (0,1)$ such that*

$$\|u\|_{H^1(\mathcal{U})} \leq C\|u\|_{H^1(\mathcal{V})}^{1-\delta}\big(\|f\|_{L^2(\mathcal{V})} + \|u\|_{L^2(\omega)}\big)^{\delta}, \tag{9.1.1}$$

for $u \in H^2(\mathcal{V})$ satisfying $Pu = f + \ell$ with $f \in L^2(\mathcal{V})$ and

$$|\ell(m)| \leq C_0\big(|u(m)| + |\mathsf{D}\, u(m)|_g\big) \quad a.e. \text{ in } \mathcal{V}.$$

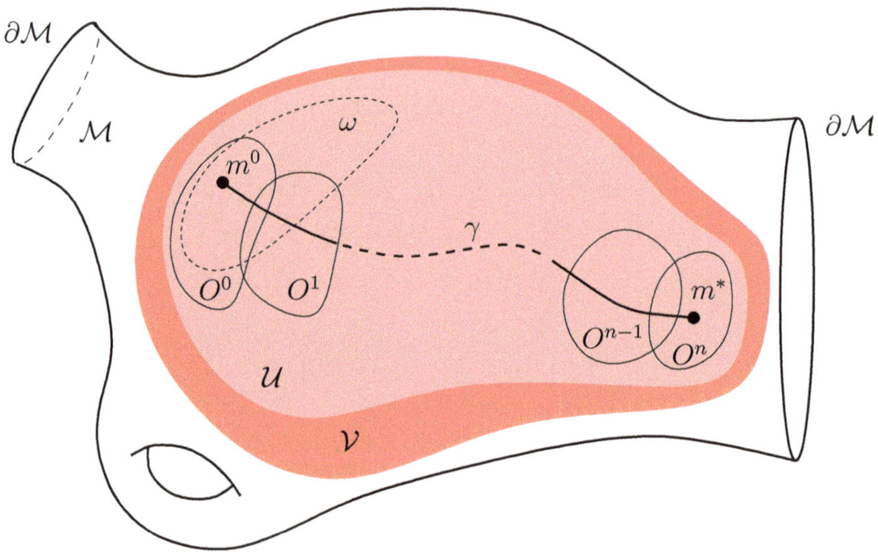

FIGURE 9.1. Propagation of smallness on a Riemannian manifold away from the boundary

Here, D denotes the Levi-Civita connection; see Sect. 17.4.

PROOF. We may assume that $\|f\|_{L^2(\mathcal{V})} \leq \|u\|_{H^1(\mathcal{V})}$, since otherwise the estimate we wish to prove is obvious.

Because of the compactness of $\overline{\mathcal{U}}$, it suffices to prove that for any $m^* \in \overline{\mathcal{U}}$, there exist a neighborhood W of m^* and $\delta \in (0,1)$ such that

$$\|u\|_{H^1(W)} \lesssim \|u\|_{H^1(\mathcal{V})}^{1-\delta} \big(\|f\|_{L^2(\mathcal{V})} + \|u\|_{L^2(\omega)}\big)^{\delta}.$$

As $\mathcal{U}$ is connected, there exists a continuous path $\gamma : [0,1] \to \mathcal{U}$ such that $\gamma(0) = m^0 \in \omega$ and $\gamma(1) = m^*$. Because of the compactness of $\gamma([0,1])$, we can pick a finite number of local charts $(\mathcal{C}^j)_{0\leq j\leq n}$, $\mathcal{C}^j = (\tilde{O}^j, \kappa^j)$, such that $\gamma([0,1]) \subset \cup_{0\leq j\leq n}\tilde{O}^j$. Replacing $\tilde{O}^j$ by $\tilde{O}^j \cap \mathcal{V}$ if needed, we may assume that $\tilde{O}^j \subset \mathcal{V}$. With Lemma 15.13, there exists an open set O^j such that $O^j \Subset \tilde{O}^j$, $j = 0, \dots, n$, and $\gamma([0,1]) \subset \cup_{0\leq j\leq n} O^j$. We order the local charts to have

$$m^0 \in O^0, \ m^* \in O^n, \quad \text{and} \quad O^j \cap O^{j-1} \neq \emptyset, \ j = 1, \dots, n.$$

See Fig. 9.1 for an illustration of this setting.

In the local chart $\mathcal{C}^0$, we set $V^0 = \kappa^0(O^0)$, $\tilde{V}^0 = \kappa^0(\tilde{O}^0)$ and $\omega^0 = \kappa^0(\omega \cap \tilde{O}^0)$. In $\tilde{V}^0$, the local representative $u^{\mathcal{C}^0}$ of u is in H^2 and satisfies

$$P^{\mathcal{C}^0} u^{\mathcal{C}^0} = f^{\mathcal{C}^0} + \ell^{\mathcal{C}^0},$$

where $P^{\mathcal{C}^0}$, $f^{\mathcal{C}^0}$, and $\ell^{\mathcal{C}^0}$ denote the local representatives of P, f, and ℓ, respectively, and we have

$$f^{\mathcal{C}^0} \in L^2(\tilde{V}^0), \quad |\ell^{\mathcal{C}^0}(x)| \lesssim \big(|u^{\mathcal{C}^0}(x)| + |du^{\mathcal{C}^0}(x)|\big) \ \text{ a.e. in } \tilde{V}^0.$$

Then, by Theorem 5.6 of Volume 1, we have, for some $\delta_0 \in (0,1)$,

$$\|u^{\mathcal{C}^0}\|_{H^1(V^0)} \lesssim \|u^{\mathcal{C}^0}\|_{H^1(\tilde{V}^0)}^{1-\delta_0} \big(\|f^{\mathcal{C}^0}\|_{L^2(\tilde{V}^0)} + \|u^{\mathcal{C}^0}\|_{L^2(\omega^0)}\big)^{\delta_0},$$

which gives, using the assumption made at the beginning of the proof,

$$\|u\|_{H^1(O^0)} + \|f\|_{L^2(\mathcal{V})} \lesssim \|u\|_{H^1(\mathcal{V})}^{1-\delta_0} \big(\|f\|_{L^2(\mathcal{V})} + \|u\|_{L^2(O^0\cap\omega)}\big)^{\delta_0}.$$

Similarly, for $j = 1, \dots, n$, we have, for some $\delta_j \in (0,1)$,

$$\|u\|_{H^1(O^j)} + \|f\|_{L^2(\mathcal{V})} \lesssim \|u\|_{H^1(\mathcal{V})}^{1-\delta_j} \big(\|f\|_{L^2(\mathcal{V})} + \|u\|_{L^2(O^j\cap O^{j-1})}\big)^{\delta_j}.$$

By induction, we then obtain

$$\|u\|_{H^1(O^n)} + \|f\|_{L^2(\mathcal{V})} \lesssim \|u\|_{H^1(\mathcal{V})}^{1-\delta} \big(\|f\|_{L^2(\mathcal{V})} + \|u\|_{L^2(\omega)}\big)^{\delta},$$

with $\delta = \prod_{0\leq j\leq n} \delta_j$. ■

9.2. Unique Continuation Estimates Up to Boundaries

We now consider $\mathcal{U}$ and $\mathcal{V}$ two open sets of $\mathcal{M}$ such that $\mathcal{U}$ is connected, and $\mathcal{U} \Subset \mathcal{V} \Subset \mathcal{M}$ (if $\mathcal{M}$ is connected and compact, we can choose $\mathcal{U} = \mathcal{V} = \mathcal{M}$). Here, we require $\overline{\mathcal{U}} \cap \partial\mathcal{M} \neq \emptyset$ as opposed to what was assumed in Sect. 9.1.

We first place ourselves at a point m^0 of $\partial\mathcal{M} \cap \overline{\mathcal{U}}$, and we prove how smallness can be propagated from regions of $\mathcal{V}$ away from the boundary up to a neighborhood of m^0. At the boundary, we consider Lopatinskiĭ–Šapiro type boundary conditions and the result we obtain follows from the Carleman estimate proven in Chap. 8.

The boundary operator B we consider is of order $\beta \in \mathbb{N}$, of the form introduced in Chap. 2, that is,

$$B = B^k + B^{k-1}\partial_\nu, \qquad 0 \leq k \leq \beta,$$

on every connected component $\mathcal{N}$ of $\partial\mathcal{M}$, with B^k and B^{k-1} smooth differential operators on $\mathcal{N}$ of order k and $k-1$, respectively. If B is of order zero on $\mathcal{N}$, one has $B^{k-1} = 0$. For each $k \in \{0, \dots, \beta\}$, we denote by ${}^k\partial\mathcal{M}$ the union of all the connected components of $\partial\mathcal{M}$ where the order of B is exactly k.

For $\hat{\mathcal{U}}$ an open set of $\mathcal{M}$, we define the following open set:

$$\hat{\mathcal{U}}_\varepsilon = \{m \in \hat{\mathcal{U}};\ \operatorname{dist}_g(m, \partial\mathcal{M}) > \varepsilon\}, \tag{9.2.1}$$

which is not empty for $\varepsilon > 0$ chosen sufficiently small. Here, $\operatorname{dist}_g(.,.)$ denotes the Riemannian distance (see Sect. 17.1).

LEMMA 9.2 (Local Interpolation Inequality at a Boundary). *Let $k \in \{0, \dots, \beta\}$, and assume that $\mathcal{V} \cap \partial\mathcal{M} \subset {}^k\partial\mathcal{M}$. Let $\hat{\mathcal{U}}$ be an open set of $\mathcal{M}$ such that $\mathcal{U} \Subset \hat{\mathcal{U}} \Subset \mathcal{V}$. The Lopatinskiĭ–Šapiro condition of Definition 2.2 is assumed to hold for (P, B) in $\mathcal{V} \cap \partial\mathcal{M}$. Let $m^0 \in \overline{\mathcal{U}} \cap \partial\mathcal{M}$, $C_0 > 0$, and $\chi_{\mathcal{U}} \in \mathscr{C}^\infty(\mathcal{V} \cap \partial\mathcal{M})$ such that $\chi_{\mathcal{U}} \equiv 1$ in a neighborhood of $\mathcal{U} \cap \partial\mathcal{M}$. There exist W an open neighborhood of m^0 in $\mathcal{V}$, $\varepsilon \in (0,1)$, $\delta \in (0,1)$, and $C > 0$, such that we have*

$$\|u\|_{H^1(W)} \leq C\|u\|_{H^1(\mathcal{V})}^{1-\delta}\Big(\|f\|_{L^2(\mathcal{V})} + |\chi_{\mathcal{U}} h|_{H^{1-k}({}^k\partial\mathcal{M})} + \|u\|_{H^1(\hat{\mathcal{U}}_\varepsilon)}\Big)^\delta, \tag{9.2.2}$$

for $u \in H^2(\mathcal{V})$ satisfying $Pu = f + \ell$ with $f \in L^2(\mathcal{V})$ and

$$|\ell(m)| \leq C_0\big(|u(m)| + |\mathsf{D}\, u(m)|_g\big) \quad \text{a.e. in } \mathcal{V},$$

and $Bu_{|\mathcal{V}\cap\partial\mathcal{M}} = h_{|\mathcal{V}\cap\partial\mathcal{M}} + \ell_\partial$ with $h \in H^{3/2-k}({}^k\partial\mathcal{M})$ and

$$\begin{cases} \ell_\partial \equiv 0 \quad \text{a.e. in } \mathcal{V} \cap \partial\mathcal{M} & \text{if } k = 0, \\ |\ell_\partial(m)| \leq C_0|u(m)| \quad \text{a.e. in } \mathcal{V} \cap \partial\mathcal{M} & \text{if } k = 1, \\ |\ell_\partial(m)| \leq C_0\big(|u(m)| + |\mathsf{D}\, u(m)|_g\big) \quad \text{a.e. in } \mathcal{V} \cap \partial\mathcal{M} & \text{if } k \geq 2. \end{cases}$$

The proof is evidently very similar to that of Lemma 5.10 of Volume 1.

REMARK 9.3. The proof is based on the Carleman estimate of Theorem 8.24 itself a consequence of Proposition 8.12 and Theorem 8.14. By Remark 8.13, if the boundary operator B *continuously* depends on a parameter s that lies in a *compact* set S and such that the Lopatinskiĭ–Šapiro condition holds for all values of $s \in S$, then all the Carleman estimates derived in Chap. 8 under these boundary conditions hold with constants that are uniform with respect to $s \in S$. Note that the order of the operator should however remain constant as s varies. Consequently, the quantified unique continuation result of Lemma 9.2 also follows with constants that uniform with respect to $s \in S$. The same holds for the results in Theorems 9.4 and 9.6 below.

PROOF. With the same argument as in the beginning of the proof of Proposition 5.1 of Volume 1, we may simply assume $P = P_0 = -\Delta_g$. We use a local chart $\mathcal{C} = (O, \kappa)$ such that $m^0 \in O \subset \hat{\mathcal{U}}$, $\chi_{\mathcal{U}} \equiv 1$ in a neighborhood of $\overline{O}$ and such that the associated local coordinates given by κ are normal geodesic coordinates; see Sections 9.4. With the coordinates $x = (x', x_d) = \kappa(m)$, the boundary $\kappa(O \cap \partial\mathcal{M})$ is given by $\kappa(O) \cap \{x_d = 0\}$ and $\kappa(O) \subset \{x_d \geq 0\}$. We assume moreover that $\kappa(m^0) = 0$. In this chart, a local representative of the metric that satisfies

$$g^{\mathcal{C}}_{dd}(x) = 1, \quad \text{and } g^{\mathcal{C}}_{dj}(x) = 0 \ \text{ for } j = 1, \dots, d-1,$$

for $x \in \kappa(O)$. In what follows, to ease the notation, we shall write φ, p, and g in place of $\varphi^{\mathcal{C}}$, $p^{\mathcal{C}}$, and $g^{\mathcal{C}}$, respectively.

Let $r > 0$ be chosen such that $B(x^{(1)}, 4r) \cap \mathbb{R}^d_+ \Subset \kappa(O)$ with $x^{(1)} = (0, 2r)$. We then set

$$\varphi(x) = e^{-\gamma |x - x^{(1)}|^2}.$$

By Lemma 3.5 of Volume 1, for $\gamma > 0$ chosen sufficiently large, the sub-ellipticity condition of Definition 3.2 of Volume 1 holds also in $\overline{V^0}$, open set in O, given by $\kappa(V^0) = \{x \in \overline{\mathbb{R}^d_+};\ r/2 < |x - x^{(1)}| < 4r\}$. We fix γ equal to this value.

Observe that $\partial_\nu \varphi(0) = -\partial_d \varphi(0) = -4\gamma r \varphi(0) < 0$. If B is of order $k = 0$, then the Lopatinskiĭ–Šapiro condition of Definition 8.1 holds for (P, B, φ) in a neighborhood of $x = \kappa(m^0) = 0$ by Theorem 8.7-(2a). If $k \geq 1$, let μ_0 be as given by Theorem 8.7-(2b). Observe that $\nabla_{x'} \varphi(0) = 0$ implying that condition (8.2.6) holds in a neighborhood of $x = \kappa(m^0) = 0$. In both cases, $k = 0$ or $k \geq 1$, there exists an open neighborhood $V^1 \subset V^0$ of m^0 such that the Lopatinskiĭ–Šapiro condition of Definition 8.1 holds for (P, B, φ) at all points of $\overline{V^1_\partial} = \overline{V^1 \cap \partial\mathcal{M}}$. In V^1, we may thus apply the Carleman estimate of Theorem 8.24.

Let $r_0 > 0$ such that $r_0 < r/4$ to be chosen below, and let $\chi_0 \in \mathscr{C}^\infty_c(\mathbb{R})$ be such that

$$\chi_0(s) = \begin{cases} 1 & \text{if } |s| < r_0, \\ 0 & \text{if } 2r_0 < |s|. \end{cases}$$

Let also $\chi_1 \in \mathscr{C}_c^\infty(B(x^{(1)}, 4r))$ be such that

$$\chi_1(x) = \begin{cases} 1 & \text{if } 3r/4 < |x - x^{(1)}| < r_1, \\ 0 & \text{if } |x - x^{(1)}| < 5r/8 \text{ or } r_1' < |x - x^{(1)}|, \end{cases}$$

where r_1 and r_1' are such that $2r < r_1 < r_1' < 3r$ and will be chosen below. We set $\chi(x) = \chi_0(x_d)\chi_1(x)$. As we have

$$\operatorname{supp}(\chi) \subset \{x \in \overline{\mathbb{R}^d_+}; x_d \leq 2r_0\} \cap \{x \in \overline{\mathbb{R}^d_+};\ |x - x^{(1)}| \leq r_1'\}.$$

Upon choosing $r_0 > 0$ and $r_1' - 2r > 0$ sufficiently small, we obtain $\operatorname{supp}(\chi) \subset \kappa(V^1)$. Below, we shall also write χ in place of $\chi \circ \kappa^{-1} \in \mathscr{C}^\infty(\mathcal{M})$ with support in V^1 as no confusion should arise.

The geometry associated with the functions we have just introduced is illustrated in Fig. 9.2. Note that we have chosen the notation to match that of the proof of Lemma 5.10 of Volume 1. Figure 9.2 is naturally a small variation of Figure 5.5.

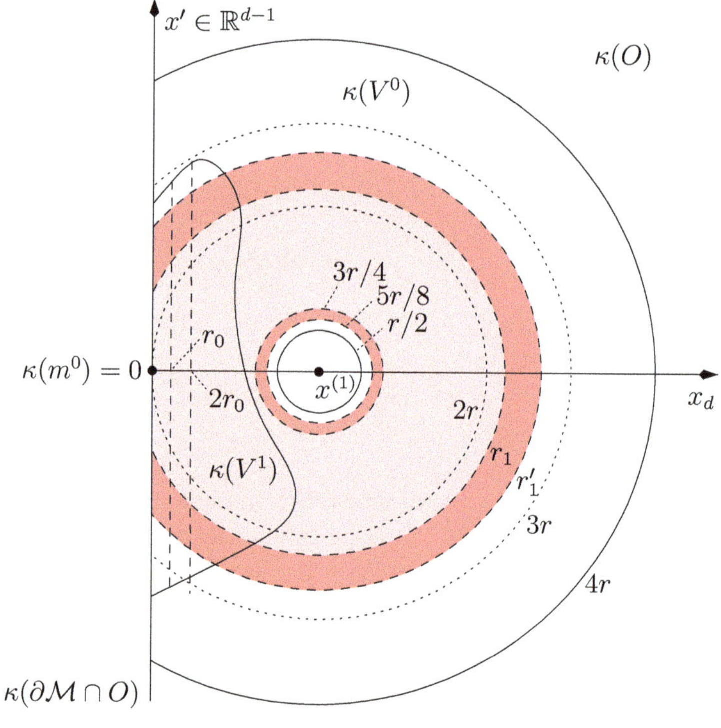

FIGURE 9.2. Geometry near the boundary for the application of the local Carleman estimate of Theorem 8.24 in the local chart $\mathcal{C} = (O, \kappa)$

Below we shall apply the Carleman estimate of Theorem 8.24 in V^1 neighborhood of m^0 as given above to the function $v = \chi u$. In both inequalities, we find the term $\|e^{\tau\varphi} P v\|_{L^2(\mathcal{M})}$ that we estimate as follows:

$$\|e^{\tau\varphi} P v\|_{L^2(\mathcal{M})} \lesssim \|e^{\tau\varphi} \chi f\|_{L^2(\mathcal{V})} + \|e^{\tau\varphi} \chi \ell\|_{L^2(\mathcal{V})} + \|e^{\tau\varphi} [P, \chi] u\|_{L^2(\mathcal{V})}.$$

By the assumption made on ℓ, we obtain

$$\begin{aligned}\|e^{\tau\varphi} \chi \ell\|_{L^2(\mathcal{V})} &\lesssim \|e^{\tau\varphi} \chi u\|_{L^2(\mathcal{V})} + \|e^{\tau\varphi} \chi \, \mathsf{D} \, u\|_{L^2\Lambda^1(\mathcal{V})} \\ &\lesssim \|e^{\tau\varphi} v\|_{L^2(\mathcal{V})} + \|e^{\tau\varphi} \, \mathsf{D} \, v\|_{L^2\Lambda^1(\mathcal{V})} + \|e^{\tau\varphi} [\mathsf{D}, \chi] u\|_{L^2(\mathcal{V})}.\end{aligned}$$

The operators $[P, \chi]$ and $[\mathsf{D}, \chi]$ are differential operators of order less than or equal to 1 with bounded coefficients and supported in $\operatorname{supp}(d\chi) \subset A_0 \cup A_1$ with

$$\begin{aligned}\kappa(A_0) &= \{x \in \overline{\mathbb{R}^d_+};\ r_0 \leq x_d \leq 2r_0 \text{ and } |x - x^{(1)}| \leq r_1'\}, \\ \kappa(A_1) &= \{x \in \overline{\mathbb{R}^d_+};\ 0 \leq x_d \leq \ 2r_0 \text{ and } r_1 \leq |x - x^{(1)}| \leq r_1'\}.\end{aligned}$$

We thus obtain

$$\begin{aligned}\|e^{\tau\varphi} P v\|_{L^2(\mathcal{V})} &\lesssim \|e^{\tau\varphi} \chi f\|_{L^2(\mathcal{V})} + \|e^{\tau\varphi} v\|_{L^2(\mathcal{V})} + \|e^{\tau\varphi} \, \mathsf{D} \, v\|_{L^2\Lambda^1(\mathcal{V})} \\ &\quad + \|e^{\tau\varphi} u\|_{L^2(A_0 \cup A_1)} + \|e^{\tau\varphi} \, \mathsf{D} \, u\|_{L^2\Lambda^1(A_0 \cup A_1)}.\end{aligned} \tag{9.2.3}$$

We set

$$C_1 = e^{-\gamma r_1^2}, \qquad C_2 = e^{-\gamma (r + r_1/2)^2}, \qquad C_3 = e^{-\gamma (2r - 2r_0)^2}. \tag{9.2.4}$$

We have $0 < C_1 < C_2 < C_3$. We also pick C_1' such that $C_1 < C_1' < C_2$. We have

$$\sup_{A_1} \varphi \leq C_1, \quad C_2 \leq \inf_{W} \varphi, \quad \sup_{\operatorname{supp}(\chi)} \varphi \leq C_3, \tag{9.2.5}$$

where the open set W in O is given by

$$\kappa(W) = \{x \in \mathbb{R}^d_+;\ 0 \leq x_d \leq r_0\} \cap \{x \in \mathbb{R}^d_+;\ |x - x^{(1)}| < r + r_1/2\}. \tag{9.2.6}$$

This is an open neighborhood of $m^0 = \kappa^{-1}(0)$ in $\mathcal{U}$ as $2r < r + r_1/2$. Since $r + r_1/2 < r_1$ and $x_d < r_0$ if $x \in \kappa(W)$, we have $\chi = \chi_0 \chi_1 \equiv 1$, and thus $u \equiv v$ in W.

We apply Theorem 8.24 to $v = \chi u$ in V^1, that is,

$$\begin{aligned}\tau^{1/2} \|e^{\tau\varphi} v\|_{\tau,1} + \tau^{1/2} |e^{\tau\varphi_{|\partial\mathcal{M}}} \operatorname{tr}(v)|_{\tau,1,0} &\lesssim \|e^{\tau\varphi} P v\|_{L^2(\mathcal{M})} \\ &\quad + \tau^{1/2} |e^{\tau\varphi} B v_{|\partial\mathcal{M}}|_{\tau,1-k},\end{aligned} \tag{9.2.7}$$

for $\tau \geq \tau_*$ for some $\tau_* > 0$. We write

$$\begin{aligned}|e^{\tau\varphi} B v_{|\partial\mathcal{M}}|_{\tau,1-k} &\lesssim |e^{\tau\varphi} \chi B u_{|\partial\mathcal{M}}|_{\tau,1-k} + |e^{\tau\varphi} [B, \chi] u_{|\partial\mathcal{M}}|_{\tau,1-k} \\ &\lesssim |e^{\tau\varphi} \chi_{|\partial\mathcal{M}} h|_{\tau,1-k} + |e^{\tau\varphi} \chi_{|\partial\mathcal{M}} \ell_\partial|_{\tau,1-k} \\ &\quad + |e^{\tau\varphi} [B, \chi] u_{|\partial\mathcal{M}}|_{\tau,1-k}.\end{aligned}$$

Note that

- $[B, \chi]$ vanishes if $k = 0$.
- $[B, \chi]_{|\partial\mathcal{M}}$ is a smooth function supported in $A_1 \cap \partial\mathcal{M}$ for $k = 1$.
- $[B, \chi] = B'^{k-1} + B'^{k-2}\partial_\nu$ with $B'^j \in \mathscr{D}^j(\partial\mathcal{M})$ supported in $\operatorname{supp}(\chi) \subset A_1 \cap \partial\mathcal{M}$, $j = k-1, k-2$, for $k \geq 2$.

Let $\tilde{\chi} \in \mathscr{C}^\infty(\partial\mathcal{M})$ be such that $\operatorname{supp}(\tilde{\chi}) \subset A_1 \cap \partial\mathcal{M}$ and $\tilde{\chi} \equiv 1$ on $\operatorname{supp}(d\chi) \cap \partial\mathcal{M}$. One has $B'^{k-1} = B'^{k-1}\tilde{\chi}$ and $B'^{k-2} = B'^{k-2}\tilde{\chi}$. We thus find

$$\begin{aligned}
\tau^{1/2}|e^{\tau\varphi}[B,\chi]u_{|\partial\mathcal{M}}|_{\tau,1-k} &\lesssim k\tau^{1/2}|e^{\tau\varphi}u_{|A_1\cap\partial\mathcal{M}}|_{L^2(\partial\mathcal{M})} \\
&\quad + k(k-1)\tau^{1/2}|e^{\tau\varphi}B'^{k-1}\tilde{\chi}u_{|\partial\mathcal{M}}|_{\tau,1-k} \\
&\quad + k(k-1)\tau^{1/2}|e^{\tau\varphi}B'^{k-2}\tilde{\chi}\partial_\nu u_{|\partial\mathcal{M}}|_{\tau,1-k} \\
&\lesssim \tau^{1/2}|e^{\tau\varphi}u_{|A_1\cap\partial\mathcal{M}}|_{L^2(\partial\mathcal{M})} + |e^{\tau\varphi}\tilde{\chi}\partial_\nu u_{|\partial\mathcal{M}}|_{\tau,-1/2}.
\end{aligned}$$

In the cases $k = 0, 1$, by the assumption made on ℓ_∂, one has

$$\tau^{1/2}|e^{\tau\varphi}\chi_{|\partial\mathcal{M}}\ell_\partial|_{\tau,1-k} \lesssim k\tau^{1/2}|e^{\tau\varphi}v_{|\partial\mathcal{M}}|_{L^2(\partial\mathcal{M})}.$$

In the case $k \geq 2$, since $3/2 - k \leq 0$, we write

$$\begin{aligned}
\tau^{1/2}|e^{\tau\varphi}\chi_{|\partial\mathcal{M}}\ell_\partial|_{\tau,1-k} &\lesssim |e^{\tau\varphi}\chi_{|\partial\mathcal{M}}\ell_\partial|_{\tau,3/2-k} \lesssim |e^{\tau\varphi}\chi_{|\partial\mathcal{M}}\ell_\partial|_{L^2(\partial\mathcal{M})} \\
&\lesssim |e^{\tau\varphi}v_{|\partial\mathcal{M}}|_{L^2(\partial\mathcal{M})} + |e^{\tau\varphi}|\,\mathsf{D}\,v|_{g\,|\partial\mathcal{M}}|_{L^2(\partial\mathcal{M})} \\
&\quad + |e^{\tau\varphi}[\mathsf{D},\chi]u_{|\partial\mathcal{M}}|_{L^2(\partial\mathcal{M})} \\
&\lesssim |e^{\tau\varphi}v_{|\partial\mathcal{M}}|_{L^2(\partial\mathcal{M})} + |e^{\tau\varphi}|\,\mathsf{D}\,v|_{g\,|\partial\mathcal{M}}|_{L^2(\partial\mathcal{M})} \\
&\quad + |e^{\tau\varphi}u_{|A_1\cap\partial\mathcal{M}}|_{L^2(\partial\mathcal{M})}.
\end{aligned}$$

We thus obtain

$$\begin{aligned}
&\tau^{1/2}|e^{\tau\varphi}Bv_{|\partial\mathcal{M}}|_{\tau,1-k} \\
&\quad\lesssim \tau^{1/2}|e^{\tau\varphi}\chi_{|\partial\mathcal{M}}h|_{\tau,1-k} + \tau^{1/2}|e^{\tau\varphi}u_{|A_1\cap\partial\mathcal{M}}|_{L^2(\partial\mathcal{M})} \\
&\qquad + |e^{\tau\varphi}\tilde{\chi}\partial_\nu u_{|\partial\mathcal{M}}|_{\tau,-1/2} \\
&\qquad + \tau^{1/2}|e^{\tau\varphi}v_{|\partial\mathcal{M}}|_{L^2(\partial\mathcal{M})} + |e^{\tau\varphi}|\,\mathsf{D}\,v|_{g\,|\partial\mathcal{M}}|_{L^2(\partial\mathcal{M})}.
\end{aligned}$$

With (9.2.3) and (9.2.7), for $\tau > 0$ chosen sufficiently large, we thus obtain

$$\begin{aligned}
&\tau^{1/2}\|e^{\tau\varphi}v\|_{\tau,1} + \tau^{1/2}|e^{\tau\varphi_{|\partial\mathcal{M}}}\operatorname{tr}(v)|_{\tau,1,0} \\
&\quad\lesssim \|e^{\tau\varphi}\chi f\|_{L^2(\mathcal{V})} + \tau^{1/2}|e^{\tau\varphi}\chi_{|\partial\mathcal{M}}h|_{\tau,1-k} + \|e^{\tau\varphi}u\|_{L^2(A_0\cup A_1)} \\
&\qquad + \|e^{\tau\varphi}\,\mathsf{D}\,u\|_{L^2\Lambda^1(A_0\cup A_1)} + \tau^{1/2}|e^{\tau\varphi}u_{|A_1\cap\partial\mathcal{M}}|_{L^2(\partial\mathcal{M})} \\
&\qquad + |e^{\tau\varphi}\tilde{\chi}\partial_\nu u_{|\partial\mathcal{M}}|_{\tau,-1/2}.
\end{aligned}$$

With the constant C_3 defined in (9.2.4)–(9.2.5), we write, by Proposition 7.3,

$$\tau^{1/2}|e^{\tau\varphi}\chi_{|\partial\mathcal{M}}h|_{\tau,1-k} \lesssim \tau^{1/2}e^{\tau\sup_{\operatorname{supp}(\chi_{|\partial\mathcal{M}})}\varphi}|\chi_{|\partial\mathcal{M}}h|_{\tau,1-k} \lesssim e^{\tau C_3}|\chi_{|\partial\mathcal{M}}h|_{H^{1-k}(\partial\mathcal{M})}.$$

With the constant C_1 also defined in (9.2.4)–(9.2.5) and Proposition 7.3, we obtain

$$\begin{aligned}\tau^{1/2}\|e^{\tau\varphi}v\|_{\tau,1} &+ \tau^{1/2}|e^{\tau\varphi_{|\partial\mathcal{M}}}\operatorname{tr}(v)|_{\tau,1,0} \\ &\lesssim e^{\tau C_3}\big(\|f\|_{L^2(\mathcal{V})} + |\chi u h|_{H^{1-k}(\partial\mathcal{M})} + \|u\|_{H^1(\hat{\mathcal{U}}_\varepsilon)}\big) \\ &\quad + e^{\tau C_1'}\big(\|u\|_{H^1(\mathcal{V})} + |u_{|A_1\cap\partial\mathcal{M}}|_{L^2(\mathcal{U}\cap\partial\mathcal{M})} + |\tilde{\chi}\partial_\nu u_{|\partial\mathcal{M}}|_{\tau,-1/2}\big),\end{aligned}\tag{9.2.8}$$

where $0 < \varepsilon < r_0$, implying $A_0 \subset \hat{\mathcal{U}}_\varepsilon$, recalling the definition of the set $\hat{\mathcal{U}}_\varepsilon$ in (9.2.1).

Let $\mathcal{W} \subset \mathcal{V}$ be a *smooth* bounded open set[1] such that $\overline{\mathcal{U}} \cap \partial\mathcal{M} \Subset \partial\mathcal{W} \cap \partial\mathcal{M}$. Since $u_{|\mathcal{W}} \in H^2(\mathcal{W})$, by Lemma 18.42, we have

$$|\tilde{\chi}\partial_\nu u_{|\partial\mathcal{M}}|_{H^{-1/2}(\partial\mathcal{M})} \lesssim |\partial_\nu u_{|\partial\mathcal{W}}|_{H^{-1/2}(\partial\mathcal{W})} \lesssim \|u\|_{H^1(\mathcal{W})} + \|Pu\|_{L^2(\mathcal{W})} \lesssim \|u\|_{H^1(\mathcal{V})} + \|f\|_{L^2(\mathcal{V})},$$

using that $Pu = f + \ell$ and the assumed pointwise estimation of ℓ. From (9.2.8), we thus obtain, as $C_1' < C_3$,

$$\begin{aligned}\tau^{1/2}\|e^{\tau\varphi}v\|_{\tau,1} &+ \tau^{1/2}|e^{\tau\varphi_{|\partial\mathcal{M}}}\operatorname{tr}(v)|_{\tau,1,0} \\ &\lesssim e^{\tau C_3}\big(\|f\|_{L^2(\mathcal{V})} + |\chi u h|_{H^{1-k}(\partial\mathcal{M})} + \|u\|_{H^1(\hat{\mathcal{U}}_\varepsilon)}\big) \\ &\quad + e^{\tau C_1'}\big(\|u\|_{H^1(\mathcal{V})} + |u_{|A_1\cap\partial\mathcal{M}}|_{L^2(\mathcal{U}\cap\partial\mathcal{M})}\big). \\ &\lesssim e^{\tau C_3}\big(\|f\|_{L^2(\mathcal{V})} + |\chi u h|_{H^{1-k}(\partial\mathcal{M})} + \|u\|_{H^1(\hat{\mathcal{U}}_\varepsilon)}\big) + e^{\tau C_1'}\|u\|_{H^1(\mathcal{V})},\end{aligned}$$

using the trace inequality 18.24 of Proposition 18.24.

If we restrict the first term of the l.h.s. of (9.2.8) to W defined in (9.2.6), as on this set we have $\varphi \geq C_2$, we obtain

$$\tau^{1/2}\|e^{\tau\varphi}v\|_{\tau,1} \gtrsim \tau^{3/2}\|e^{\tau\varphi}u\|_{L^2(W)} + \tau^{1/2}\|e^{\tau\varphi}\,\mathsf{D}\,u\|_{L^2\Lambda^1(W)} \gtrsim e^{\tau C_2}\|u\|_{H^1(W)}.$$

We have thus obtained

$$e^{\tau C_2}\|u\|_{H^1(W)} \lesssim e^{\tau C_3}\big(\|f\|_{L^2(\mathcal{V})} + |\chi u h|_{H^{1-k}(\partial\mathcal{M})} + \|u\|_{H^1(\hat{\mathcal{U}}_\varepsilon)}\big) + e^{\tau C_1'}\|u\|_{H^1(\mathcal{V})},$$

which we write

$$\|u\|_{H^1(W)} \lesssim e^{\tau(C_3-C_2)}\big(\|f\|_{L^2(\mathcal{V})} + |\chi u h|_{H^{1-k}(\partial\mathcal{M})} + \|u\|_{H^1(\hat{\mathcal{U}}_\varepsilon)}\big) + e^{-\tau(C_2-C_1')}\|u\|_{H^1(\mathcal{V})}.$$

[1]The open set $\mathcal{W}$ is introduced since the two open sets $\mathcal{U}$ and $\mathcal{V}$ are not assumed smooth.

Recalling that $C_1' < C_2 < C_3$, by Lemma 5.4, we obtain the local interpolation inequalities (9.2.2) at the boundary. ■

As in Chapter 5 of Volume 1, with the local results of Lemma 9.2 at the boundary and the result of Theorem 9.1, we deduce the following quantifications of the propagation of "smallness" from a location away from the boundary up to the boundary.

THEOREM 9.4 (Propagation of "smallness" Up to Boundaries). *Let ω be an open subset of $\mathcal{U}$. The Lopatinskiĭ–Šapiro condition of Definition 2.2 is assumed to hold for (P, B) in $\mathcal{V} \cap \partial\mathcal{M}$. Let $C_0 > 0$ and $\chi_\mathcal{U} \in \mathscr{C}^\infty(\mathcal{V} \cap \partial\mathcal{M})$ such that $\chi_\mathcal{U} \equiv 1$ in a neighborhood of $\mathcal{U} \cap \partial\mathcal{M}$. There exist $C > 0$ and $\delta \in (0,1)$ such that*

(9.2.9)

$$\|u\|_{H^1(\mathcal{U})} \leq C \|u\|_{H^1(\mathcal{V})}^{1-\delta} \Big(\|f\|_{L^2(\mathcal{V})} + \sum_{0 \leq k \leq \beta} |\chi_\mathcal{U} h|_{H^{1-k}({}^k\partial\mathcal{M})} + \|u\|_{L^2(\omega)} \Big)^\delta,$$

for $u \in H^2(\mathcal{V})$ satisfying $Pu = f + \ell$ with $f \in L^2(\mathcal{V})$ and

$$|\ell(m)| \leq C_0 \big(|u(m)| + |\operatorname{D} u(m)|_g \big) \quad a.e.\ in\ \mathcal{V},$$

and $Bu_{|\mathcal{V}\cap\partial\mathcal{M}} = h_{|\mathcal{V}\cap\partial\mathcal{M}} + \ell_\partial$ with $h_{|{}^k\partial\mathcal{M}} \in H^{3/2-k}({}^k\partial\mathcal{M})$ and, for all $k = 0, \dots, \beta$,

$$\begin{cases} \ell_\partial \equiv 0 \quad a.e.\ in\ \mathcal{V} \cap {}^k\partial\mathcal{M} & if\ k = 0, \\ |\ell_\partial(m)| \leq C_0 |u(m)| \quad a.e.\ in\ \mathcal{V} \cap {}^k\partial\mathcal{M} & if\ k = 1, \\ |\ell_\partial(m)| \leq C_0 \big(|u(m)| + |\operatorname{D} u(m)|_g \big) \quad a.e.\ in\ \mathcal{V} \cap {}^k\partial\mathcal{M} & if\ k \geq 2. \end{cases}$$

The geometry, here different from that of Theorem 9.1, is illustrated in Fig. 9.3.

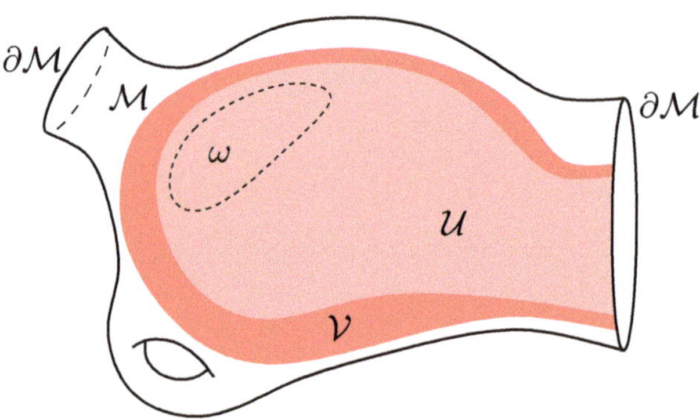

FIGURE 9.3. Propagation of smallness on a Riemannian manifold up to a part of the boundary. Here, $\mathcal{V}$ only meets one connected component of $\partial\mathcal{M}$

PROOF. We may assume that

$$\|f\|_{L^2(\mathcal{V})} + \sum_{0\le k\le\beta} |\chi u h|_{H^{1-k}(^k\partial\mathcal{M})} \le \|u\|_{H^1(\mathcal{V})}, \tag{9.2.10}$$

since otherwise inequality (9.2.9) is obvious.

Let $\hat{\mathcal{U}}$ be an open set of $\mathcal{M}$ such that $\mathcal{U} \Subset \hat{\mathcal{U}} \Subset \mathcal{V}$. Recall the definition of $\hat{\mathcal{U}}_\varepsilon$ given in (9.2.1). With a compactness argument and by Lemma 9.2, we can find a finite number of open sets W_j of $\mathcal{V}\cap\hat{\mathcal{U}}$, $j\in J$, such that

$$\partial\mathcal{M}\cap\overline{\mathcal{U}} \subset \cup_{j\in J} W_j, \qquad W_j\cap\partial\mathcal{M}\subset {}^k\partial\mathcal{M},$$
$$\text{for a unique } k = k(j)\in\{0,\dots,\beta\},$$

and such that, for some values $\delta=\delta_j\in(0,1)$ and $\varepsilon_j>0$, we have

$$\begin{aligned}\|u\|_{H^1(W_j)} &\lesssim \|u\|_{H^1(\mathcal{V})}^{1-\delta_j}\Big(\|f\|_{L^2(\mathcal{V})} + |\chi u h|_{H^{1-k(j)}(^{k(j)}\partial\mathcal{M})} + \|u\|_{H^1(\hat{\mathcal{U}}_{\varepsilon_j})}\Big)^{\delta_j}\\ &\lesssim \|u\|_{H^1(\mathcal{V})}^{1-\delta_j}\Big(\|f\|_{L^2(\mathcal{V})} + \sum_{0\le k\le\beta}|\chi u h|_{H^{1-k}(^k\partial\mathcal{M})} + \|u\|_{H^1(\hat{\mathcal{U}}_{\varepsilon_j})}\Big)^{\delta_j}.\end{aligned} \tag{9.2.11}$$

For $\varepsilon\in(0,1)$, we set

$$_\varepsilon\mathcal{U} = \{m\in\mathcal{U}, \operatorname{dist}_g(m,\partial\mathcal{M})<\varepsilon\}.$$

There exists $\varepsilon\in(0,1)$ such that $_\varepsilon\mathcal{U}\subset(\cup_{j\in J}W_j)$. Applying (9.2.11) for each $j\in J$, using now

$$\delta' = \min_{j\in J}\delta_j\in(0,1), \quad \text{and} \quad \varepsilon' < \min\big(\min_{j\in J}\varepsilon_j, \varepsilon\big)\in(0,1),$$

in place of δ_j and ε_j (note that the set $\hat{\mathcal{U}}_\varepsilon$ increases as ε decreases), we obtain

$$\|u\|_{H^1(_\varepsilon\mathcal{U})} \lesssim \|u\|_{H^1(\mathcal{V})}^{1-\delta'}\Big(\|f\|_{L^2(\mathcal{V})} + \sum_{0\le k\le\beta}|\chi u h|_{H^{1-k}(^k\partial\mathcal{M})} + \|u\|_{H^1(\hat{\mathcal{U}}_{\varepsilon'})}\Big)^{\delta'}. \tag{9.2.12}$$

By Theorem 9.1, with $\mathcal{U}=\hat{\mathcal{U}}_{\varepsilon'}$ therein, there exists $\delta_0\in(0,1)$ such that

$$\|u\|_{H^1(\hat{\mathcal{U}}_{\varepsilon'})} \lesssim \|u\|_{H^1(\mathcal{V})}^{1-\delta}\Big(\|f\|_{L^2(\mathcal{V})} + \|u\|_{L^2(\omega)}\Big)^{\delta}, \quad \delta\in(0,\delta_0]. \tag{9.2.13}$$

By (9.2.10) and (9.2.13), we have

$$\begin{aligned}&\|f\|_{L^2(\mathcal{V})} + \sum_{0\le k\le\beta}|\chi u h|_{H^{1-k}(^k\partial\mathcal{M})} + \|u\|_{H^1(\hat{\mathcal{U}}_{\varepsilon'})}\\ &\qquad\lesssim \|u\|_{H^1(\mathcal{V})}^{1-\delta_0}\Big(\|f\|_{L^2(\mathcal{V})} + \sum_{0\le k\le\beta}|\chi u h|_{H^{1-k}(^k\partial\mathcal{M})} + \|u\|_{L^2(\omega)}\Big)^{\delta_0}.\end{aligned} \tag{9.2.14}$$

Then, estimates (9.2.12) and (9.2.14) give

$$\|u\|_{H^1(_\varepsilon\mathcal{U})} \lesssim \|u\|_{H^1(\mathcal{V})}^{1-\delta}\Big(\|f\|_{L^2(\mathcal{V})} + \sum_{0\le k\le\beta}|\chi u h|_{H^{1-k}(^k\partial\mathcal{M})} + \|u\|_{L^2(\omega)}\Big)^{\delta}, \tag{9.2.15}$$

for $\delta = \delta'\delta_0$. Since $\varepsilon' \in (0, \varepsilon)$, we have $\mathcal{U} \subset {}_\varepsilon\mathcal{U} \cup \hat{\mathcal{U}}_{\varepsilon'}$. Thus, estimate (9.2.13) together with (9.2.15) yields the result. ∎

9.3. Boundary Initiated Unique Continuation: Improved Estimates

Here, we extend the result of Section 5.4 of Volume 1. As above, we consider $\mathcal{U}$ and $\mathcal{V}$ two open sets of $\mathcal{M}$ such that $\mathcal{U}$ is connected, and $\mathcal{U} \Subset \mathcal{V} \Subset \mathcal{M}$.

One part of the boundary $\partial\mathcal{V}$ we consider a first-order differential operator operator $\tilde{B}'$ and set $\tilde{B} = \partial_\nu + \tilde{B}'$. We assume that $(P, \tilde{B})$ fulfills the Lopatinskiĭ–Šapiro condition of Definition 2.2. Using the improved Carleman estimate of Theorem 8.30 we obtain the following local result. Compare with Lemma 5.12 of Volume 1.

LEMMA 9.5 (Local Interpolation Inequality with Boundary Observations). *Let $m^0 \in \partial\mathcal{V}$ and V^0 be an open neighborhood of m^0 where the boundary $\partial\mathcal{V}$ is smooth and where $(P, \tilde{B})$ fulfills the Lopatinskiĭ–Šapiro condition of Definition 2.2. Let $C_0 > 0$. There exist W^0 a neighborhood of m^0 in $\mathcal{V}$, $C > 0$, and $\delta \in (0, 1)$ such that*

$$\|u\|_{H^1(W^0)} \leq C\|u\|_{H^1(\mathcal{V})}^{1-\delta} \times \Big(\|f\|_{L^2(\mathcal{V})} + |u_{|V^0\cap\partial\mathcal{V}}|_{L^2(V^0\cap\partial\mathcal{V})} + |\tilde{B}u_{|V^0\cap\partial\mathcal{V}}|_{L^2(V^0\cap\partial\mathcal{V})}\Big)^\delta, \tag{9.3.1}$$

for $u \in H^2(\mathcal{V})$ satisfying $Pu = f + \ell$ with $f \in L^2(\mathcal{V})$ and

$$|\ell(m)| \leq C_0\big(|u(m)| + |Du(m)|\big) \quad \text{a.e. in } \mathcal{V}.$$

The proof can be written *mutatis mutandis* from that of Lemma 5.12, with an adaptation to the manifold case, by replacing the Carleman estimate of Theorem 3.28 of Volume 1 by that of Theorem 8.30. We then deduce the following result.

THEOREM 9.6 (Propagation of "smallness" Initiated from the Boundary). *Let $\chi_\mathcal{U} \in \mathscr{C}^\infty(\mathcal{V} \cap \partial\mathcal{M})$ such that $\chi_\mathcal{U} \equiv 1$ in a neighborhood of $\mathcal{U} \cap \partial\mathcal{M}$. Let also $m^0 \in \partial\mathcal{V}$ and V^0 be an open neighborhood of m^0 where the boundary $\partial\mathcal{V}$ is smooth. The Lopatinskiĭ–Šapiro condition of Definition 2.2 is assumed to hold for (P, B) in $\mathcal{V} \cap \partial\mathcal{M}$ and for $(P, \tilde{B})$ in $V^0 \cap \partial\mathcal{V}$. Let $C_0 > 0$. There exist $C > 0$ and $\delta \in (0, 1)$ such that*

$$\|u\|_{H^1(\mathcal{U})} \leq C\|u\|_{H^1(\mathcal{V})}^{1-\delta}\Big(\|f\|_{L^2(\mathcal{V})} + \sum_{0\leq k\leq\beta} |\chi_\mathcal{U} h|_{H^{1-k}({}^k\partial\mathcal{M})} + |u_{|V^0\cap\partial\mathcal{V}}|_{L^2(V^0\cap\partial\mathcal{V})} + |\tilde{B}u_{|V^0\cap\partial\mathcal{V}}|_{L^2(V^0\cap\partial\mathcal{V})}\Big)^\delta, \tag{9.3.2}$$

for $u \in H^2(\mathcal{V})$ *satisfying* $Pu = f + \ell$ *with* $f \in L^2(\mathcal{V})$ *and*

$$|\ell(m)| \leq C_0\big(|u(m)| + |\operatorname{D} u(m)|_g\big) \quad a.e.\ in\ \mathcal{V},$$

and $Bu_{|\mathcal{V}\cap\partial\mathcal{M}} = h_{|\mathcal{V}\cap\partial\mathcal{M}} + \ell_\partial$ *with* $h_{|{}^k\partial\mathcal{M}} \in H^{3/2-k}({}^k\partial\mathcal{M})$ *and, for all* $k = 0, \dots, \beta$,

$$\begin{cases} \ell_\partial \equiv 0 \quad a.e.\ in\ \mathcal{V} \cap {}^k\partial\mathcal{M} & if\ k = 0, \\ |\ell_\partial(m)| \leq C_0|u(m)| \quad a.e.\ in\ \mathcal{V} \cap {}^k\partial\mathcal{M} & if\ k = 1, \\ |\ell_\partial(m)| \leq C_0\big(|u(m)| + |\operatorname{D} u(m)|_g\big) \quad a.e.\ in\ \mathcal{V} \cap {}^k\partial\mathcal{M} & if\ k \geq 2. \end{cases}$$

The proof simply follows that of Theorem 5.13 of Volume 1 using Theorem 9.4 and Lemma 9.5.

REMARK 9.7.
(1) In the above results, $m^0 \in \partial\mathcal{V}$. This allows one to choose m^0 in $\mathcal{V} \cap \partial\mathcal{M}$ or, if needed, in $\partial\mathcal{V} \setminus \partial\mathcal{M}$. Various choices will be done in applications in Chaps. 10, 11, and 12.
(2) A possible and often natural choice for $\tilde{B}$ is naturally $\tilde{B} = \partial_\nu$; see Remarks 8.28 and 8.31.

The following non-quantified unique continuation result can be simply proven with Lemma 9.5 and an adaptation of Theorem 5.2 of Volume 1 to the manifold case.

THEOREM 9.8. *Let* $u \in H^2_{\text{loc}}(\mathcal{M})$ *be such that* $Pu = \ell$ *with*

$$|\ell(m)| \leq C_0\big(|u(m)| + |\operatorname{D} u(m)|_g\big) \quad a.e.\ in\ \mathcal{V}.$$

If Γ *is a bounded open set of* $\partial\mathcal{M}$ *and both* $\partial_\nu u_{|\partial\mathcal{M}}$ *and* $u_{|\partial\mathcal{M}}$ *vanish on* Γ, *then* $u = 0$.

Note that for this last result, no particular boundary condition is assumed away from Γ.

9.4. Notes

This chapter generalizes results obtained in Chapter 5 of Volume 1 in the case of Dirichlet boundary conditions. A result of quantification of the unique continuation property in the case of Neumann boundary conditions can be found in [219]. There it was derived for the purpose of understanding the stabilization of a boundary-damped wave equation. This topic is covered in the next two chapters. The generalization we provide here encompasses all boundary operators that fit the framework given by the Lopatinskiĭ–Šapiro conditions recalled in Chap. 2. We do not believe this can be found in the existing literature with the same level of generality.

The proof of the results presented in this chapter is based on the Carleman estimates derived in Chap. 8. Once a Carleman estimate is obtained at a boundary, one can apply the method used in the proof of Lemma 9.2 to obtain a quantification of the unique continuation property up to the boundary under such conditions. The mixed Zaremba type conditions are not covered by the analysis we do here. However, with the Carleman estimate proven in [108], a quantification of the unique continuation property as in Lemma 9.2 is derived therein.

Here, we consider a boundary operator than can be of arbitrary order in the tangential direction. Ventcel boundary conditions are of order 2 tangentially, and the result we present can be found in [92] for these conditions.

CHAPTER 10

Stabilization of Waves Under Neumann Boundary Damping

Contents

10.1. Setting

For $(\mathcal{M}, g)$ a smooth d-dimensional connected compact Riemannian manifold with boundary we set $P_0 = -\Delta_g$. With a smooth function $\alpha \geq 0$ on $\partial\mathcal{M}$, we consider the following wave equation

$$
\begin{cases}
\partial_t^2 y + P_0 y = 0 & \text{in } (0, +\infty) \times \mathcal{M}, \\
\partial_\nu y + \alpha \partial_t y = 0 & \text{on } (0, +\infty) \times \partial\mathcal{M}, \\
y_{|t=0} = y^0, \ \partial_t y_{|t=0} = y^1 & \text{in } \mathcal{M},
\end{cases}
\tag{10.1.1}
$$

where we denote by ∂_ν the outward pointing normal derivative in the sense of the metric g (see (5.3.1) in Sect. 5.3 and (17.4.12) in Sect. 17.4).

J. Le Rousseau et al., *Elliptic Carleman Estimates and Applications to Stabilization and Controllability, Volume II*, PNLDE Subseries in Control 98, https://doi.org/10.1007/978-3-030-88670-7_10

In the case $\alpha \equiv 0$ the solution exists for $t \in \mathbb{R}$ with an energy that remains constant with respect to t. Here, we shall assume that $\alpha > 0$ on a nonempty open subset $\Gamma \subset \partial\mathcal{M}$ and we shall prove that in such case the energy decays to *zero* as $t \to +\infty$. The boundary term $\alpha\partial_t y$ thus acts as a damping and we refer to (10.1.1) as to the *boundary-damped* wave equation.

Similarly to the result of Chapter 6 of Volume 1, without any assumption made on the open set Γ, we prove that the decay rate is at least logarithmic. The outline of this chapter in naturally very similar to that of Chapter 6.

10.2. Preliminaries on the Boundary-Damped Wave Equation

To properly define solutions to (10.1.1) we need to provide the proper functional framework, in particular to account for the boundary condition $\partial_\nu y + \alpha\partial_t y = 0$. We address first strong solutions and second weak solution.

10.2.1. Strong Solutions of the Boundary-Damped Wave Equation. We define the boundary operator B_∂ by

$$B_\partial : H^2(\mathcal{M}) \times H^1(\mathcal{M}) \to H^{1/2}(\partial\mathcal{M}) \qquad (10.2.1)$$
$$(v^0, v^1) \mapsto \partial_\nu v^0{}_{|\partial\mathcal{M}} + \alpha v^1{}_{|\partial\mathcal{M}}.$$

This operator is well defined by the trace formula of Proposition 18.24.

We have the following existence and uniqueness result.

THEOREM 10.1 (Strong Solutions). *For $(y^0, y^1) \in H^2(\mathcal{M}) \times H^1(\mathcal{M})$ such that $B_\partial(y^0, y^1) = 0$, there exists a unique*

$$y \in \mathscr{C}^2\big([0,+\infty); L^2(\mathcal{M})\big) \cap \mathscr{C}^1\big([0,+\infty);\, H^1(\mathcal{M})\big) \cap \mathscr{C}^0\big([0,+\infty); H^2(\mathcal{M})\big)$$

such that

$$\partial_t^2 y + P_0 y = 0 \quad \text{in } L^\infty([0,+\infty); L^2(\mathcal{M})), \qquad y_{|t=0} = y^0,\ \partial_t y_{|t=0} = y^1, \qquad (10.2.2)$$

and $B_\partial(y(t), \partial_t y(t)) = 0$ in $L^\infty([0,+\infty); H^{1/2}(\partial\mathcal{M}))$. Moreover, there exists $C > 0$ such that

$$\|y(t)\|_{H^2(\mathcal{M})} + \|\partial_t y(t)\|_{H^1(\mathcal{M})} \le C\big(\|y^0\|_{H^2(\mathcal{M})} + \|y^1\|_{H^1(\mathcal{M})}\big), \qquad t \ge 0. \qquad (10.2.3)$$

Solutions given in Theorem 10.1 are called *strong* solutions of the boundary-damped wave equation.

To prove the results of Theorem 10.1 it is convenient to cast the boundary-damped wave equation into a semigroup formalism. We refer to Chapter 12 of Volume 1 for some elements of semigroup theory. We set

$$A = \begin{pmatrix} 0 & -1 \\ P_0 & 0 \end{pmatrix}, \qquad Y(t) = \begin{pmatrix} y(t) \\ \partial_t y(t) \end{pmatrix}, \qquad Y^0 = \begin{pmatrix} y^0 \\ y^1 \end{pmatrix}, \qquad (10.2.4)$$

and we consider the Hilbert sum

$$\mathscr{H} = H^1(\mathcal{M}) \oplus L^2(\mathcal{M}),$$

naturally endowed with the inner product

$$(U, U')_{\mathscr{H}} = (u^0, u'^0)_{H^1(\mathcal{M})} + (u^1, u'^1)_{L^2(\mathcal{M})}, \quad U = (u^0, u^1), \ U' = (u'^0, u'^1),$$

and norm

$$\|U\|_{\mathscr{H}}^2 = \|u^0\|_{H^1(\mathcal{M})}^2 + \|u^1\|_{L^2(\mathcal{M})}^2, \quad U = (u^0, u^1).$$

If y is a strong solution to (10.2.2) we have

$$Y(t) \in \mathscr{C}^0([0, +\infty); H^2(\mathcal{M}) \oplus H^1(\mathcal{M})) \cap \mathscr{C}^1([0, +\infty); \mathscr{H}), \tag{10.2.5}$$

and

$$\frac{d}{dt} Y(t) + AY(t) = 0 \text{ for } t \geq 0 \ \text{ and } Y_{|t=0} = Y^0, \tag{10.2.6}$$

with moreover $B_\partial(Y(t)) = 0$ for $t \geq 0$.

Accordingly, we define the unbounded operator A given by (10.2.4) on $\mathscr{H}$ with domain

$$D(A) = \{V = {}^t(v^0, v^1) \in H^2(\mathcal{M}) \times H^1(\mathcal{M});\ B_\partial(V) = 0\}.$$

By Proposition 10.14 and Corollary 10.19 in Appendix 10.A the operator $(A, D(A))$ is closed on $\mathscr{H}$ and moreover $D(A)$ is dense in $\mathscr{H}$.

We endow $D(A)$ with the graph norm $\|V\|_{D(A)}^2 = \|V\|_{\mathscr{H}}^2 + \|AV\|_{\mathscr{H}}^2$ for $V \in D(A)$. In fact, this norm is equivalent to the norm on $D(A)$ inherited from that of $H^2(\mathcal{M}) \oplus H^1(\mathcal{M})$ by Corollary 10.26.

THEOREM 10.2. *The unbounded operator $(A, D(A))$ generates a bounded C_0-semigroup $S(t)$ on $\mathscr{H}$.*

This result is proven in Appendix 10.A: see Theorem 10.23.

PROPOSITION 10.3. *Let $Y^0 = {}^t(y^0, y^1) \in D(A)$ and set $Y(t) = S(t)Y^0$. Then, $\frac{d}{dt}Y(t) + AY(t) = 0$, for $t \geq 0$. If $Y(t) = {}^t(y(t), z(t))$, then*

$$y \in \mathscr{C}^2([0, +\infty); L^2(\mathcal{M})) \cap \mathscr{C}^1([0, +\infty); H^1(\mathcal{M})) \cap \mathscr{C}^0([0, +\infty); H^2(\mathcal{M})),$$

$z(t) = \partial_t y(t)$, and y is the unique solution of the boundary damped wave equation in the sense of Theorem 10.1.

Note that we obtain $\frac{d}{dt}Y(t) + AY(t) = 0$, for $t \geq 0$, by applying Proposition 12.2 of Volume 1. The proof of Proposition 10.3 is actually contained in that of Theorem 10.1. Using the uniqueness part of Theorem 10.1, Proposition 10.3 shows that the solution of the boundary-damped wave equation is simply given by the first component of $Y(t) = S(t)Y^0$.

Proof of Theorem 10.1. If y is a solution as described in the statement of the theorem, then $Y(t) = {}^t(y(t), \partial_t y(t))$ satisfies (10.2.5) and (10.2.6). Moreover $Y(t) \in D(A)$ for $t \geq 0$ since $B_\partial\big(Y(t)\big) = 0$. As the graph norm on $D(A)$ is equivalent to the norm on $D(A)$ inherited from that of $H^2(\mathcal{M}) \oplus H^1(\mathcal{M})$ by Corollary 10.26, we obtain

$$Y(t) \in \mathscr{C}^0\big([0,+\infty); D(A)\big) \cap \mathscr{C}^1\big([0,+\infty); \mathscr{H}\big).$$

Then, the second part of Proposition 12.2 of Volume 1 yields $Y(t)=S(t)Y(0)$. This implies the uniqueness of the strong solution.

Conversely, let $(y^0, y^1) \in H^2(\mathcal{M}) \times H^1(\mathcal{M})$ such that $B_\partial(y^0, y^1) = 0$. We have $Y^0 = {}^t(y^0, y^1) \in D(A)$. If we set $Y(t) = S(t)Y^0$, by Proposition 12.2 we have

$$Y(t) \in \mathscr{C}^0([0,+\infty); D(A)) \cap \mathscr{C}^1([0,+\infty); \mathscr{H}) \quad \text{and } Y(0) = Y^0.$$

If we write $Y(t) = {}^t(y(t), y^1(t))$ we obtain

$$y(t) \in \mathscr{C}^0([0,+\infty); H^2(\mathcal{M})) \cap \mathscr{C}^1([0,+\infty); H^1(\mathcal{M})),$$

and

$$y^1(t) \in \mathscr{C}^0([0,+\infty); H^1(\mathcal{M})) \cap \mathscr{C}^1([0,+\infty); L^2(\mathcal{M})),$$

and in addition $B_\partial\big(y(t), y^1(t)\big) = 0$ for $t \geq 0$. Moreover, the first part of Proposition 12.2 yields $\frac{d}{dt}Y(t) + AY(t) = 0$ for $t \geq 0$. From the form of A we find $\partial_t y = y^1$, which gives

$$\begin{aligned} y \in \mathscr{C}^2\big([0,+\infty); L^2(\mathcal{M})\big) \cap \mathscr{C}^1\big([0,+\infty);\\ H^1(\mathcal{M})\big) \cap \mathscr{C}^0\big([0,+\infty); H^2(\mathcal{M})\big). \end{aligned}$$

Finally, $\frac{d}{dt}Y(t) + AY(t) = 0$ for $t \geq 0$ and $Y(0) = Y^0$ precisely imply the content of (10.2.2). We have thus proven the existence of a strong solution.

Finally, we prove estimate (10.2.3). Let $Y^0 \in D(A)$ and $Y(t) = S(t)Y^0 \in D(A)$. From the boundedness of the semigroup $S(t)$ on $\mathscr{H}$ we have

$$\|Y(t)\|_{\mathscr{H}} \lesssim \|Y^0\|_{\mathscr{H}}. \tag{10.2.7}$$

In addition, with Proposition 12.2 we deduce,

$$\|AY(t)\|_{\mathscr{H}} = \|S(t)AY^0\|_{\mathscr{H}} \lesssim \|AY^0\|_{\mathscr{H}}. \tag{10.2.8}$$

If $Y^0 = {}^t(y^0, y^1)$ we have $Y(t) = {}^t(y(t), \partial_t y(t))$ and we thus obtain, from Corollary 10.26,

$$\begin{aligned} \|y(t)\|_{H^2(\mathcal{M})} + \|\partial_t y(t)\|_{H^1(\mathcal{M})} &\eqsim \|Y(t)\|_{H^2(\mathcal{M})\oplus H^1(\mathcal{M})} \eqsim \|Y(t)\|_{D(A)} \\ &\lesssim \|AY(t)\|_{\mathscr{H}} + \|Y(t)\|_{\mathscr{H}}. \end{aligned}$$

With (10.2.7) and (10.2.8) we find

$$\begin{aligned} \|y(t)\|_{H^2(\mathcal{M})} + \|\partial_t y(t)\|_{H^1(\mathcal{M})} &\lesssim \|AY^0\|_{\mathscr{H}} + \|Y^0\|_{\mathscr{H}} \\ &\eqsim \|Y^0\|_{D(A)} \eqsim \|y^0\|_{H^2(\mathcal{M})} + \|y^1\|_{H^1(\mathcal{M})}, \end{aligned}$$

which is precisely (10.2.3). ∎

For a solution to the boundary-damped wave equation we introduce the natural energy:

$$\mathcal{E}(y)(t) = E\big(y(t), \partial_t y(t)\big) = \frac{1}{2}\Big(\|\partial_t y(t)\|^2_{L^2(\mathcal{M})} + \|\nabla_g y(t)\|^2_{L^2V(\mathcal{M})}\Big). \tag{10.2.9}$$

If we have

$$y \in \mathscr{C}^2\big([0,+\infty); L^2(\mathcal{M})\big) \cap \mathscr{C}^1\big([0,+\infty);\\ H^1(\mathcal{M})\big) \cap \mathscr{C}^0\big([0,+\infty); H^2(\mathcal{M})\big),$$

we compute

$$\frac{d}{dt}\mathcal{E}(y)(t) = \operatorname{Re}(\partial_t^2 y(t), \partial_t y(t))_{L^2(\mathcal{M})} + \operatorname{Re}\big(\nabla_g y(t), \nabla_g \partial_t y(t)\big)_{L^2V(\mathcal{M})}.$$

With (18.5.5) consequence of the divergence formula of Proposition 18.28 we obtain

$$\begin{aligned}\big(\nabla_g y(t), \nabla_g \partial_t y(t)\big)_{L^2V(\mathcal{M})} &= \big(P_0 y(t), \partial_t y(t)\big)_{L^2(\mathcal{M})}\\ &\quad + (\partial_\nu y(t)_{|\partial\mathcal{M}}, \partial_t y(t)_{|\partial\mathcal{M}})_{L^2(\partial\mathcal{M})}\\ &= \big(P_0 y(t), \partial_t y(t)\big)_{L^2(\mathcal{M})}\\ &\quad - (\alpha\partial_t y(t)_{|\partial\mathcal{M}}, \partial_t y(t)_{|\partial\mathcal{M}})_{L^2(\partial\mathcal{M})},\end{aligned}$$

where each term is well defined by the regularity property of $y(t)$ using in particular the trace formula of Theorem 18.25. If moreover we have $\partial_t^2 y + P_0 y = 0$ in $L^\infty([0,+\infty); L^2(\mathcal{M}))$ we obtain, as $\alpha \geq 0$,

$$\frac{d}{dt}\mathcal{E}(y)(t) = -|\alpha^{1/2}\partial_t y(t)_{|\partial\mathcal{M}}|^2_{L^2(\partial\mathcal{M})},$$

or rather

$$\mathcal{E}(y)(t') = \mathcal{E}(y)(t) - \int_t^{t'} |\alpha^{1/2}\partial_t y(\sigma)_{|\partial\mathcal{M}}|^2_{L^2(\partial\mathcal{M})}\, d\sigma,$$

this implies the decay of the energy $\mathcal{E}(y)(t)$, viz.,

$$0 \leq \mathcal{E}(y)(t') \leq \mathcal{E}(y)(t), \qquad \text{for } 0 \leq t \leq t'. \tag{10.2.10}$$

Observe, however, that the energy $E(y^0, y^1)$ does not provide a norm on $\mathscr{H}$ and, thus, it cannot be used to obtain the uniqueness of strong solutions. Compare with Section 6.2 of Volume 1.

10.2.2. Weak Solution of the Boundary-Damped Wave Equation. Weak solutions of the boundary-damped wave equation are solutions that lie in

$$\mathscr{C}^0\big([0,+\infty); H^1(\mathcal{M})\big) \cap \mathscr{C}^1\big([0,+\infty); L^2(\mathcal{M})\big),$$

that is, $Y(t) = {}^t(y(t), \partial_t y(t)) \in \mathscr{C}^0([0,+\infty); \mathscr{H})$. Here, as opposed to the setting in Chapter 6 of Volume 1, the functional setting does not express explicitly the boundary condition. No boundary information is provided by the space $\mathscr{H}$. For strong solutions described in Sect. 10.2.1, the boundary condition that carries the damping term is explicitly given by the domain of the semigroup generator A. Weak solutions cannot just be defined by requiring the wave equation to hold in the sense of distribution, as opposed to what is done in Chapter 6.

We start by considering y a strong solution of the boundary-damped wave equation as given by Theorem 10.1, with $y_{|t=0} = y^0$ and $\partial_t y_{|t=0} = y^1$ with the requirement $Y^0 = {}^t(y^0, y^1) \in D(A)$. For $T > 0$, we pick

$$z \in \mathscr{C}^2\big([0,T]; L^2(\mathcal{M})\big) \cap \mathscr{C}^1\big([0,T]; H^1(\mathcal{M})\big) \cap \mathscr{C}^0\big([0,T]; H^2(\mathcal{M})\big),$$

by integration by parts, we compute

$$\begin{aligned}(\partial_t^2 y, z)_{L^2((0,T)\times\mathcal{M})} = (y, \partial_t^2 z)_{L^2((0,T)\times\mathcal{M})} &+ (\partial_t y(T), z(T))_{L^2(\mathcal{M})} \\ &- (y(T), \partial_t z(T))_{L^2(\mathcal{M})} \\ &- \big((y^1, z(0))_{L^2(\mathcal{M})} - (y^0, \partial_t z(0))_{L^2(\mathcal{M})}\big),\end{aligned}$$

and, with the Green formula of Proposition 18.30, we write, for $t \geq 0$,

$$\begin{aligned}(P_0 y(t), z(t))_{L^2(\mathcal{M})} = (y(t), P_0 z(t))_{L^2(\mathcal{M})} &\\ - (\partial_\nu y(t)_{|\partial\mathcal{M}}, z(t)_{|\partial\mathcal{M}})_{L^2(\partial\mathcal{M})}&\\ + (y(t)_{|\partial\mathcal{M}}, \partial_\nu z(t)_{|\partial\mathcal{M}})_{L^2(\partial\mathcal{M})}&.\end{aligned}$$

We thus find

$$\begin{aligned}0 = (y, (\partial_t^2 + P_0)z)_{L^2((0,T)\times\mathcal{M})} &+ (\partial_t y(T), z(T))_{L^2(\mathcal{M})} \\ - (y(T), \partial_t z(T))_{L^2(\mathcal{M})}& \\ - \big((y^1, z(0))_{L^2(\mathcal{M})} &- (y^0, \partial_t z(0))_{L^2(\mathcal{M})}\big) \\ - (\partial_\nu y(t)_{|\partial\mathcal{M}}, z(t)_{|\partial\mathcal{M}})_{L^2((0,T)\times\partial\mathcal{M})} &+ (y(t)_{|\partial\mathcal{M}}, \partial_\nu z(t)_{|\partial\mathcal{M}})_{L^2((0,T)\times\partial\mathcal{M})}.\end{aligned}$$

Since $B_\partial(y(t), \partial_t(y(t)) = 0$ we have $\partial_\nu y(t)_{|\partial\mathcal{M}} = -\alpha \partial_t y(t)_{|\partial\mathcal{M}}$, yielding, by integration by parts,

$$\begin{aligned}(\partial_\nu y(t)_{|\partial\mathcal{M}}, z(t)_{|\partial\mathcal{M}})_{L^2((0,T)\times\partial\mathcal{M})} = (y(t)_{|\partial\mathcal{M}}, \alpha\partial_t z(t)_{|\partial\mathcal{M}})_{L^2((0,T)\times\partial\mathcal{M})}&\\ - (\alpha y(T)_{|\partial\mathcal{M}}, z(T)_{|\partial\mathcal{M}})_{L^2(\partial\mathcal{M})}&\\ + (\alpha y^0_{|\partial\mathcal{M}}, z(0)_{|\partial\mathcal{M}})_{L^2(\partial\mathcal{M})}&.\end{aligned}$$

The above computations thus give the identity

$$\begin{aligned}0 = (y, (\partial_t^2 &+ P_0)z)_{L^2((0,T)\times\mathcal{M})} \\ &+ (y(t)_{|\partial\mathcal{M}}, \partial_\nu z(t)_{|\partial\mathcal{M}} - \alpha\partial_t z(t)_{|\partial\mathcal{M}})_{L^2((0,T)\times\partial\mathcal{M})} \\ &+ (\partial_t y(T), z(T))_{L^2(\mathcal{M})} - (y(T), \partial_t z(T))_{L^2(\mathcal{M})} \\ &+ (\alpha y(T)_{|\partial\mathcal{M}}, z(T)_{|\partial\mathcal{M}})_{L^2(\partial\mathcal{M})} \\ &- \big((y^1, z(0))_{L^2(\mathcal{M})} - (y^0, \partial_t z(0))_{L^2(\mathcal{M})} + (\alpha y^0_{|\partial\mathcal{M}}, z(0)_{|\partial\mathcal{M}})_{L^2(\partial\mathcal{M})}\big).\end{aligned} \tag{10.2.11}$$

Observe now that this last formula is in fact well defined for functions y that lie in $\mathscr{C}^0\big([0,+\infty); H^1(\mathcal{M})\big) \cap \mathscr{C}^1\big([0,+\infty); L^2(\mathcal{M})\big)$. We are thus led to the following definition of weak solutions of the boundary-damped wave equation.

DEFINITION 10.4 (Weak Solutions). Let $(y^0, y^1) \in H^1(\mathcal{M}) \times L^2(\mathcal{M})$. A function

$$y \in \mathscr{C}^0([0, +\infty); H^1(\mathcal{M})) \cap \mathscr{C}^1([0, +\infty); L^2(\mathcal{M}))$$

such that $y(0) = y^0$ and $\partial_t y(0) = y^1$ is said to be a weak solution to the boundary-damped equation (10.1.1) if identity (10.2.11) holds for all $T > 0$ and all

$$z \in \mathscr{C}^2([0, T]; L^2(\mathcal{M})) \cap \mathscr{C}^1([0, T]; H^1(\mathcal{M})) \cap \mathscr{C}^0([0, T]; H^2(\mathcal{M})).$$

REMARK 10.5. The computations above show that a strong solution as given by Theorem 10.1 is also a weak solution. In such a case $(y^0, y^1) \in H^2(\mathcal{M}) \times H^1(\mathcal{M})$ and $B_\partial(y^0, y^1) = 0$.

The following result further shows that there is existence and uniqueness of a weak solution in the case we wish to consider, that is, $(y^0, y^1) \in H^1(\mathcal{M}) \times L^2(\mathcal{M})$.

THEOREM 10.6 (Weak Solutions). *Let $(y^0, y^1) \in H^1(\mathcal{M}) \times L^2(\mathcal{M})$. There exists a unique weak solution*

$$y \in \mathscr{C}^0([0, +\infty); H^1(\mathcal{M})) \cap \mathscr{C}^1([0, +\infty); L^2(\mathcal{M}))$$

to the boundary-damped wave equation in the sense of Definition 10.4. Moreover, there exists $C > 0$ such that

$$\|y(t)\|_{H^1(\mathcal{M})} + \|\partial_t y(t)\|_{L^2(\mathcal{M})} \leq C\big(\|y^0\|_{H^1(\mathcal{M})} + \|y^1\|_{L^2(\mathcal{M})}\big), \qquad t \geq 0. \tag{10.2.12}$$

If we set $Y(t) = S(t)Y^0$ with $Y^0 = {}^t(y^0, y^1)$, then $Y(t) = {}^t(y(t), \partial_t y(t))$. Finally, for such a weak solution we observe decay of the energy function defined in (10.2.9), *that is,*

$$0 \leq \mathcal{E}(y)(t') \leq \mathcal{E}(y)(t), \qquad \text{for } 0 \leq t \leq t'. \tag{10.2.13}$$

REMARK 10.7. Because of the uniqueness of weak solution and by Remark 10.5, we see that if $(y^0, y^1) \in H^2(\mathcal{M}) \times H^1(\mathcal{M})$, then the weak solution given by Theorem 10.6 is in fact a strong solution in the sense of Theorem 10.1.

PROOF. First, we address uniqueness. We thus consider the case $y^0 = 0$ and $y^1 = 0$. For $\psi^1 \in \mathscr{C}_c^\infty(\Omega)$, by Theorem 10.1 we consider

$$\psi \in \mathscr{C}^2([0, +\infty); L^2(\mathcal{M})) \cap \mathscr{C}^1([0, +\infty); H^1(\mathcal{M})) \cap \mathscr{C}^0([0, +\infty); H^2(\mathcal{M}))$$

such that

$$\partial_t^2 \psi + P_0 \psi = 0 \quad \text{in } L^\infty([0, +\infty); L^2(\mathcal{M})), \qquad \psi_{|t=0} = 0, \ \partial_t \psi_{|t=0} = \psi^1,$$

and $B_\partial(\psi(t), \partial_t \psi(t)) = 0$ in $L^\infty([0, +\infty); H^{1/2}(\partial\mathcal{M}))$.

For $T \geq 0$ we set $z(t) = \psi(T-t)$ and we have

$$\partial_t^2 z + P_0 z = 0 \quad \text{in } L^\infty([0,T]; L^2(\mathcal{M})), \qquad z_{|t=T} = 0, \ \partial_t z_{|t=T} = -\psi^1,$$

and $\partial_\nu z(t) - \alpha \partial_t z(t) = 0$ in $L^\infty([0,T]; H^{1/2}(\partial\mathcal{M}))$. Identity (10.2.11) then reads

$$(y(T), \psi^1)_{L^2(\mathcal{M})} = 0.$$

As $\psi^1 \in \mathscr{C}_c^\infty(\Omega)$ and $T \geq 0$ are both arbitrary, we conclude that $y \equiv 0$.

Second, we address existence. Setting $Y^0 = {}^t(y^0, y^1)$ we have $Y^0 \in \mathscr{H}$ and we set $Y(t) = S(t)Y^0$. We also pick a sequence $Y^{0,n} \subset D(A)$ such that $Y^{0,n} \to Y^0$ as $n \to +\infty$, using the density of the domain of the semigroup generator, and set $Y^n(t) = S(t)Y^{0,n}$. By Proposition 10.3, we have $Y^n(t) = {}^t(y^n(t), \partial_t y^n(t))$ and $y^n(t)$ is a strong solution of the boundary-damped wave equation in the sense of Theorem 10.1.

The boundedness of the semigroup stated in Theorem 10.2 implies

$$\|Y(t) - Y^n(t)\|_{\mathscr{H}} \lesssim \|Y^0 - Y^{n,0}\|_{\mathscr{H}}, \qquad t \geq 0,$$

that is, uniform convergence in $\mathscr{C}^0([0,+\infty); \mathscr{H})$. If $Y(t) = {}^t(y(t), z(t))$ this gives $z = \partial_t y$. We thus write

(10.2.14)
$$\|y(t) - y^n(t)\|_{H^1(\mathcal{M})} + \|\partial_t y(t) - \partial_t y^n(t)\|_{L^2(\mathcal{M})} \lesssim \|Y^0 - Y^{n,0}\|_{\mathscr{H}}, \qquad t \geq 0.$$

By remark 10.5, we see that (10.2.11) holds for the strong solution $y^n(t)$. Yet, all terms in (10.2.11) converge according to (10.2.14). We thus obtain that $y(t)$ is a weak solution.

For strong solution we have obtained the decay of the energy function $\mathcal{E}(y)(t)$ given in (10.2.10). Yet, the uniform convergence written in (10.2.14) implies that the energy associated with $y^n(t)$ converges to that of $y(t)$. We thus obtain (10.2.13). ■

The weak formulation hides the damping phenomenon that is expressed by $B_\partial(y(t), \partial_t y(t)) = 0$ in the case a strong solution. For such a solution this equality holds in $L^\infty([0,+\infty); H^{1/2}(\partial\mathcal{M}))$. As $L^\infty([0,+\infty); H^{1/2}(\partial\mathcal{M})) \subset H^{-1/2}((0,T) \times \partial\mathcal{M})$ we define the maps

$$\begin{aligned} T_1 : D(A) &\to H^{-1/2}((0,T) \times \partial\mathcal{M}) \\ Y^0 &\mapsto \partial_\nu y_{|(0,T)\times\partial\mathcal{M}}, \end{aligned} \qquad \begin{aligned} T_1' : D(A) &\to H^{-1/2}((0,T) \times \partial\mathcal{M}) \\ Y^0 &\mapsto (\partial_t y)_{|(0,T)\times\partial\mathcal{M}}, \end{aligned}$$

where ${}^t(y(t), \partial_t y(t)) = S(t)Y^0$. In the case of a weak solution we have the following result.

PROPOSITION 10.8. *The maps T_1, T_1' can be uniquely extended to $\mathscr{H}$ and for a weak solution of the boundary-damped wave equation ${}^t(y(t), \partial_t y(t)) = S(t)Y^0$, with $Y^0 \in \mathscr{H}$, we have*

$$T_1(Y^0) + \alpha T_1'(Y^0) = 0. \tag{10.2.15}$$

PROOF. Let $Y^0 \in \mathscr{H}$. As $y \in \mathscr{C}^1([0,+\infty), L^2(\mathcal{M})) \cap \mathscr{C}^0([0,+\infty), H^1(\mathcal{M}))$ we have $H^1((0,T) \times \mathcal{M})$ and the trace formula of Theorem 18.25 gives $y_{|(0,T)\times\partial\mathcal{M}} \in H^{1/2}((0,T) \times \partial\mathcal{M})$. A weak derivation then gives $\partial_t (y_{|(0,T)\times\partial\mathcal{M}}) \in H^{-1/2}((0,T) \times \partial\mathcal{M})$. The map $T_1'' : Y^0 \mapsto \partial_t(y_{|(0,T)\times\partial\mathcal{M}})$ is thus bounded from $\mathscr{H}$ into $H^{-1/2}((0,T) \times \partial\mathcal{M})$.

Let now $(Y^{0,n})_{n\in\mathbb{N}} \subset D(A)$ be such that $Y^{0,n} \to Y^0$ in $\mathscr{H}$.

For ${}^t(y^n(t), \partial_t y^n(t)) = S(t)Y^{0,n}$ the damping condition (10.2.15) holds. Observe that we have

$$T_1' Y^{0,n} = (\partial_t y^n)_{|(0,T)\times\partial\mathcal{M}} = \partial_t \big(y^n_{|(0,T)\times\partial\mathcal{M}}\big) = T_1'' Y^{0,n},$$

since $y^n \in \mathscr{C}^1([0,+\infty), H^1(\mathcal{M}))$. The boundedness of T_1'' on $\mathscr{H}$ shows that T_1' extends uniquely to $\mathscr{H}$ and that this extension is precisely T_1''.

Now since $T_1 Y^{0,n} = -\alpha T_1' Y^{0,n}$ and since α is smooth we find that T_1 also uniquely extends to $\mathscr{H}$ and that (10.2.15) holds for the two extended maps. ∎

10.3. Reduced Functional Space and Generator

Here, we introduce a linear space $\dot{\mathscr{H}}$ of $\mathscr{H}$ of codimension one, on which the norm inherited by $\mathscr{H}$ is equivalent to the square root of the energy function

$$E(U) = \frac{1}{2}\Big(\|u^1\|^2_{L^2(\mathcal{M})} + \|\nabla_g u^0\|^2_{L^2V(\mathcal{M})}\Big), \qquad U = {}^t(u^0, u^1). \tag{10.3.1}$$

The generator of the boundary-damped wave semigroup $(A, D(A))$ yields an unbounded operator $(\dot{A}, D(\dot{A}))$ that generates a C_0-semigroup of contraction on $\dot{\mathscr{H}}$. In the next section, a resolvent estimate for $(\dot{A}, D(\dot{A}))$ is derived leading to the stabilization property of the boundary-damped wave equation.

Here, we only define $\dot{\mathscr{H}}$ and $(\dot{A}, D(\dot{A}))$ and provide their properties and the connection with the semigroup generated by $(A, D(A))$. Details are provided in Appendix 10.A. The reader will also note that the results of the previous section rely in fact on the analysis carried on in this appendix. For clarity, in the main text of this chapter, we presented first $(A, D(A))$ and $S(t)$ as they are related directly to the boundary-damped wave equation. However, their analysis relies essentially on the understanding of the unbounded operator $(\dot{A}, D(\dot{A}))$ and the semigroup $\dot{S}(t)$ it generates as one can see in Appendix 10.A.

Below, to ease reading, we provide precise references to the different results of Appendix 10.A.

We introduce the continuous linear form

$$\begin{aligned} F_\alpha : \mathscr{H} &\to \mathbb{C} \\ \begin{pmatrix} v^0 \\ v^1 \end{pmatrix} &\mapsto |\alpha|^{-1}_{L^1(\partial\mathcal{M})}\Big(\int_{\partial\mathcal{M}} \alpha v^0{}_{|\partial\mathcal{M}} \mu_{g_\partial} + \int_{\mathcal{M}} v^1 \mu_g\Big), \end{aligned} \tag{10.3.2}$$

(recall that we assume that $\alpha \geq 0$ is positive on some open subset of the boundary) and the space

$$\dot{\mathscr{H}} = \ker(F_\alpha) = \{V \in \mathscr{H};\ F_\alpha(V) = 0\}.$$

We set $\Pi_{\mathcal{N}} : \mathscr{H} \to \mathscr{H}$ and $\Pi_{\dot{\mathscr{H}}} : \mathscr{H} \to \mathscr{H}$ as the bounded linear maps

$$\Pi_{\mathcal{N}} V = F_\alpha(V)\Theta, \qquad \Pi_{\dot{\mathscr{H}}} = \mathrm{Id}_{\mathscr{H}} - \Pi_{\mathcal{N}}, \tag{10.3.3}$$

with $\Theta = {}^t(1,0)$. They are projectors onto $\mathcal{N} = \operatorname{span}\{\Theta\}$ and $\dot{\mathscr{H}}$, respectively, according to the direct sum

$$\mathscr{H} = \dot{\mathscr{H}} \oplus \mathcal{N}. \tag{10.3.4}$$

In particular note that $\dot{\mathscr{H}}$ and $\mathcal{N}$ are *not orthogonal* in $\mathscr{H}$. Note that the continuity of F_α implies that of $\Pi_{\dot{\mathscr{H}}}$ and $\Pi_{\mathcal{N}}$.

By Lemma 10.16, we defined the following norm on $\dot{\mathscr{H}}$

$$\|V\|^2_{\dot{\mathscr{H}}} := \|\nabla_g v^0\|^2_{L^2V(\mathcal{M})} + \|v^1\|^2_{L^2(\mathcal{M})}, \quad V = {}^t(v^0, v^1),$$

and we have the following norm equivalence

$$\|V\|_{\dot{\mathscr{H}}} \eqsim \|V\|_{\mathscr{H}}, \quad V \in \dot{\mathscr{H}}.$$

Moreover, for the energy function recalled in (10.3.1), we have, for $U \in \mathscr{H}$,

$$E(U) = E(U - \Pi_{\mathcal{N}} U) = E(\Pi_{\dot{\mathscr{H}}} U) = \frac{1}{2}\|\Pi_{\dot{\mathscr{H}}} U\|^2_{\dot{\mathscr{H}}}. \tag{10.3.5}$$

The operator A has the following properties: $\operatorname{Ran}(A) \subset \dot{\mathscr{H}}$ and $\ker(A) = \mathcal{N}$, by Lemma 10.15 and (10.A.4). In particular, if $Y(t) \in \mathscr{C}^0([0,+\infty); D(A)) \cap \mathscr{C}^1([0,+\infty); \mathscr{H})$ is a strong solution of the semigroup equation $\frac{d}{dt}Y(t) + AY(t) = 0$, then we find that $t \mapsto F_\alpha(Y(t))$ is constant for $t \in [0,+\infty)$; see (10.A.7).

One is thus led to introduce the unbounded operator $\dot{A}$ on $\dot{\mathscr{H}}$ given by the following domain

$$\begin{aligned} D(\dot{A}) &= \big\{V \in H^2(\mathcal{M}) \times H^1(\mathcal{M});\ B_\partial(V) = 0,\ F_\alpha(V) = 0\big\} \\ &= \Pi_{\dot{\mathscr{H}}}\big(D(A)\big) \subset D(A), \end{aligned}$$

and such that $\dot{A}V = AV$ for $V \in D(\dot{A})$. We then have $A = \dot{A} \circ \Pi_{\dot{\mathscr{H}}}$; see the diagram above (10.A.9). Below, we shall refer to $(\dot{A}, D(\dot{A}))$ as to the reduced generator of the boundary-damped wave equation.

The unbounded operator $(\dot{A}, D(\dot{A}))$ generates a C_0-semigroup of contraction on $\dot{\mathscr{H}}$ by Theorem 10.22. We denote this semigroup $\dot{S}(t)$ and we have

$$S(t) = \dot{S}(t) \circ \Pi_{\dot{\mathscr{H}}} + \Pi_{\mathcal{N}}.$$

where $S(t)$ is the bounded C_0-semigroup[1] generated by $(A, D(A))$.

---

[1]Note in fact, that this is the actual definition we give of $S(t)$ in Appendix 10.A; see (10.A.24).

If $Y^0 \in D(A)$, the solution of the semigroup equation $\frac{d}{dt}Y(t)+AY(t)=0$ reads

$$Y(t) = S(t)Y^0 = \dot{S}(t) \circ \Pi_{\dot{\mathscr{H}}} Y^0 + \Pi_{\mathcal{N}} Y^0.$$

We set $\dot{Y}(t) = \Pi_{\dot{\mathscr{H}}} Y(t) = \dot{S}(t) \circ \Pi_{\dot{\mathscr{H}}} Y^0$. We have

$$\Pi_{\mathcal{N}} Y^0 = F_\alpha(Y^0)\Theta = \begin{pmatrix} F_\alpha(Y^0) \\ 0 \end{pmatrix}.$$

Hence, by (10.3.5) we find that the energy of $Y(t)$ is precisely that of $\dot{Y}(t)$:

$$E\big(Y(t)\big) = E\big(\Pi_{\dot{H}} Y(t)\big) = E\big(\dot{Y}(t)\big) = \frac{1}{2}\|\dot{Y}(t)\|^2_{\dot{\mathscr{H}}}. \tag{10.3.6}$$

Understanding the stabilization properties of the semigroup $S(t)$ can thus be done through that of the semigroup $\dot{S}(t)$. This is the subject of the next section. As in Chapter 6 of Volume 1 the analysis relies on the derivation of a resolvent estimate for the semigroup generator.

10.4. Resolvent Estimate and Stabilization Result

For the reduced generator $(\dot{A}, D(\dot{A}))$ of the boundary-damped wave equation we have the following resolvent estimate.

PROPOSITION 10.9. *Let Γ be a nonempty open subset of $\partial\mathcal{M}$ and $\alpha \in \mathscr{C}^\infty(\partial\mathcal{M})$ be such that $\alpha > 0$ on Γ. Then, the unbounded operator $i\sigma \operatorname{Id}_{\dot{\mathscr{H}}} - \dot{A}$ is invertible on $\dot{\mathscr{H}}$ for all $\sigma \in \mathbb{R}$ and there exist $K > 0$ and $\sigma_0 > 0$ such that*

$$\|(i\sigma \operatorname{Id}_{\dot{\mathscr{H}}} - \dot{A})^{-1}\|_{\mathscr{L}(\dot{\mathscr{H}},\dot{\mathscr{H}})} \leq K e^{K|\sigma|}, \qquad \sigma \in \mathbb{R},\ |\sigma| \geq \sigma_0.$$

The proof of Proposition 10.9 is given below in Sect. 10.5. The main consequences of this estimate, with the result of Theorem 6.6 of Volume 1, are the following result.

THEOREM 10.10. *Let Γ be a nonempty open subset of $\partial\mathcal{M}$ and $\alpha \in \mathscr{C}^\infty(\partial\mathcal{M})$ be such that $\alpha > 0$ on Γ. Let $k \in \mathbb{N}$. Then, there exists $C > 0$ such that, if the initial condition Y^0 is in the domain of the operator $\dot{A}^k$, then*

$$\|\dot{S}(t)Y^0\|_{\dot{\mathscr{H}}} \leq \frac{C}{\big(\log(2+t)\big)^k}\|\dot{A}^k Y^0\|_{\dot{\mathscr{H}}}. \tag{10.4.1}$$

Observing that $D(\dot{A}^k) = \Pi_{\dot{\mathscr{H}}} D(A^k)$ as $A = \dot{A} \circ \Pi_{\dot{\mathscr{H}}}$ and using (10.3.6) we obtain the following stabilization result for the boundary-damped wave equation (10.1.1).

COROLLARY 10.11. *Let Γ be a nonempty open subset of $\partial\mathcal{M}$ and $\alpha \in \mathscr{C}^\infty(\partial\mathcal{M})$ be such that $\alpha > 0$ on Γ. Let $k \in \mathbb{N}$. Then, there exists $C > 0$ such* *that, if the initial condition $Y^0 = (y^0, y^1)$ is in the domain of the operator A^k, the energy of the strong solution $y(t)$ to (10.2.2) satisfies*

$$\mathcal{E}(y)(t) \leq \frac{C}{\big(\log(2+t)\big)^{2k}} \|A^k Y^0\|^2_{\mathscr{H}}. \tag{10.4.2}$$

By Proposition 6.12 of volume 1, with Theorem 10.10 we obtain

COROLLARY 10.12. *The energy of any weak solution to the boundary-damped wave equation (10.1.1) goes to 0 as $t \to +\infty$.*

10.5. Proof of the Resolvent Estimate

Here, we give the proof of the resolvent estimate of Proposition 10.9.

First, for $\sigma \in \mathbb{R}$, the resolvent operator $(i\sigma \operatorname{Id}_{\dot{\mathscr{H}}} - \dot{A})^{-1}$ is well defined and continuous on $\dot{\mathscr{H}}$ as the spectrum of $(\dot{A}, D(\dot{A}))$ is contained in $\{z \in \mathbb{C};\ \operatorname{Re} z > 0\}$ by Proposition 10.27. We shall now estimate the operator norm of this resolvent.

Let thus $U \in D(\dot{A})$ and $F \in \dot{\mathscr{H}}$ be such that

$$(i\sigma \operatorname{Id}_{\dot{\mathscr{H}}} - \dot{A})U = F, \quad U = {}^t(u^0, u^1), \quad F = {}^t(f^0, f^1). \tag{10.5.1}$$

Our goal is to find an estimate of the form $\|U\|_{\dot{\mathscr{H}}} \leq K e^{K|\sigma|} \|F\|_{\dot{\mathscr{H}}}$.

The resolvent equation (10.5.1) reads

$$i\sigma u^0 + u^1 = f^0, \quad i\sigma u^1 - P_0 u^0 = f^1, \qquad (\partial_\nu u^0 + \alpha u^1)_{|\partial\mathcal{M}} = 0,$$

which we write, with $f = i\sigma f^0 - f^1$,

$$i\sigma u^0 + u^1 = f^0, \quad (-\sigma^2 + P_0)u^0 = f, \quad \partial_\nu u^0{}_{|\partial\mathcal{M}} = (i\sigma\alpha u^0 - \alpha f^0)_{|\partial\mathcal{M}}. \tag{10.5.2}$$

Multiplying the second equation by $\overline{u}^0$ and an integration over $\mathcal{M}$ give

$$((-\sigma^2 + P_0)u^0, u^0)_{L^2(\mathcal{M})} = (f, u^0)_{L^2(\mathcal{M})}.$$

An integration by parts and the boundary equation in (10.5.2) yield

$$\begin{aligned}(f, u^0)_{L^2(\mathcal{M})} &= -\|\sigma u^0\|^2_{L^2(\mathcal{M})} + \|\nabla_g u^0\|^2_{L^2V(\mathcal{M})} - (\partial_\nu u^0{}_{|\partial\mathcal{M}}, u^0{}_{|\partial\mathcal{M}})_{L^2(\partial\mathcal{M})} \\ &= -\|\sigma u^0\|^2_{L^2(\mathcal{M})} + \|\nabla_g u^0\|^2_{L^2V(\mathcal{M})} \\ &\quad - i\sigma|\alpha^{1/2} u^0{}_{|\partial\mathcal{M}}|^2_{L^2(\partial\mathcal{M})} + (\alpha f^0{}_{|\partial\mathcal{M}}, u^0{}_{|\partial\mathcal{M}})_{L^2(\partial\mathcal{M})},\end{aligned} \tag{10.5.3}$$

using that $\alpha \geq 0$. Computing the imaginary part of (10.5.3) we obtain

$$\sigma|\alpha^{1/2} u^0{}_{|\partial\mathcal{M}}|^2_{L^2(\partial\mathcal{M})} = \operatorname{Im}(\alpha f^0{}_{|\partial\mathcal{M}}, u^0{}_{|\partial\mathcal{M}})_{L^2(\partial\mathcal{M})} - \operatorname{Im}(f, u^0)_{L^2(\mathcal{M})}.$$

We then write, for $|\sigma| \geq \sigma_0 > 0$,

$$\begin{aligned}|\alpha^{1/2} u^0{}_{|\partial\mathcal{M}}|^2_{L^2(\partial\mathcal{M})} &\lesssim |\sigma|^{-1}(\|f^0\|_{H^1(\mathcal{M})} + \|f\|_{L^2(\mathcal{M})})\|u^0\|_{H^1(\mathcal{M})} \\ &\lesssim \|F\|_{\dot{\mathscr{H}}} \|u^0\|_{H^1(\mathcal{M})},\end{aligned} \tag{10.5.4}$$

using the trace inequality of Proposition 18.24. From the properties of the damping function α on $\partial\mathcal{M}$, there exist Γ_0 an open subset of Γ and $\delta > 0$ such that $\alpha \geq \delta > 0$ in Γ_0. This yields

$$\delta \, |u^0{}_{|\partial\mathcal{M}}|^2_{L^2(\Gamma_0)} \lesssim \|F\|_{\dot{\mathscr{H}}} \|u^0\|_{H^1(\mathcal{M})}, \qquad |\sigma| \geq \sigma_0. \tag{10.5.5}$$

The key estimate is given by the following observation lemma.

LEMMA 10.13. *For Γ_0 as introduced above, there exists $C > 0$ such that*

$$\|u^0\|_{H^1(\mathcal{M})} \leq C e^{C|\sigma|} \big(\|F\|_{\dot{\mathscr{H}}} + |u^0{}_{|\partial\mathcal{M}}|_{L^2(\Gamma_0)} \big).$$

The proof is given below.

Then estimate (10.5.5) yields

$$\|u^0\|_{H^1(\mathcal{M})} \lesssim e^{C|\sigma|} \big(\|F\|_{\dot{\mathscr{H}}} + \|u^0\|^{\frac{1}{2}}_{H^1(\mathcal{M})} \|F\|^{\frac{1}{2}}_{\dot{\mathscr{H}}} \big),$$

and with the Young inequality we obtain

$$\|u^0\|_{H^1(\mathcal{M})} \lesssim e^{C|\sigma|} \|F\|_{\dot{\mathscr{H}}}.$$

Finally, as $u^1 = f^0 - i\sigma u^0$ we obtain

$$\|u^0\|_{H^1(\mathcal{M})} + \|u^1\|_{L^2(\mathcal{M})} \lesssim e^{C|\sigma|} \|F\|_{\dot{\mathscr{H}}},$$

yielding the resolvent estimate of Proposition 10.9. ■

PROOF OF LEMMA 10.13. The proof we provide is based on the results on the quantification of the unique continuation property that were obtained in Chap. 9.

We set $\tilde{\mathcal{M}} = \mathbb{R} \times \mathcal{M}$ and we denote by $n = (s, m)$ a point of $\tilde{\mathcal{M}}$ with $s \in \mathbb{R}$ and $m \in \mathcal{M}$. We have $\partial\tilde{\mathcal{M}} = \mathbb{R} \times \partial\mathcal{M}$. On $\tilde{\mathcal{M}}$ we consider the metric $g_{\tilde{\mathcal{M}}} = ds \otimes ds + g$, where g is the metric on $\mathcal{M}$. On $\tilde{\mathcal{M}}$ we consider the operator $Q = D_s^2 + P_0$, that is, the Laplace–Beltrami operator for $g_{\tilde{\mathcal{M}}}$. The function $u(s, m) = e^{\sigma s} u^0(m)$ is then a solution to

$$Qu = h \ \text{ in } \tilde{\mathcal{M}}, \qquad Bu = h_\partial \ \text{ on } \partial\tilde{\mathcal{M}},$$

where B is the Neumann boundary operator on $\partial\tilde{\mathcal{M}}$, that is, $Bu = \partial_\nu u_{|\partial\tilde{\mathcal{M}}}$ and

$$h(n) = e^{\sigma s} f(m), \quad n = (s, m) \in \tilde{\mathcal{M}},$$
$$h_\partial(n) = \alpha(m) e^{\sigma s} (i\sigma u^0 - f^0)_{|\partial\mathcal{M}}(m), \quad n = (s, m) \in \partial\tilde{\mathcal{M}}.$$

In particular, by Example 2.5, the Lopatinskiĭ–Šapiro condition of Definition 2.2 holds for (Q, B) at $\partial\tilde{\mathcal{M}}$. We set $\mathcal{U} = (-1, 1) \times \mathcal{M}$ and $\mathcal{V} = (-2, 2) \times \mathcal{M}$ to fit the setting of Chap. 9. The geometry is illustrated in Fig. 10.1. We consider $m^0 \in \Gamma_0 \subset \partial\mathcal{M}$, with Γ_0 as introduced above the statement of Lemma 10.13, $n^0 = (0, m^0)$ and V^0 a neighborhood of n^0 in $\mathcal{U}$ such that $V^0 \cap \partial\tilde{\mathcal{M}} \subset (-1, 1) \times \Gamma_0$. We then apply Theorem 9.6 to the

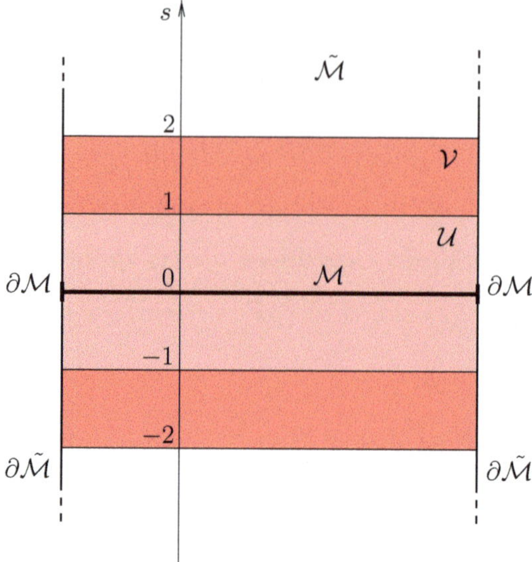

FIGURE 10.1. Geometry for the application of the quantified unique continuation property of Theorem 9.6

function u with $\tilde{\mathcal{M}}$ in place of $\mathcal{M}$ and $B = \tilde{B} = \partial_\nu$ therein, yielding, for some $\delta \in (0,1)$,

(10.5.6)

$$\begin{aligned}\|u\|_{H^1(\mathcal{U})} &\lesssim \|u\|^{1-\delta}_{H^1(\mathcal{V})}\Big(\|h\|_{L^2(\mathcal{V})} + |h_\partial|_{L^2(\mathcal{V}\cap\partial\tilde{\mathcal{M}})} + |u_{|V^0\cap\partial\tilde{\mathcal{M}}}|_{L^2(V^0\cap\partial\tilde{\mathcal{M}})} \\ &\qquad\qquad + |\partial_\nu u_{|V^0\cap\partial\tilde{\mathcal{M}}}|_{L^2(V^0\cap\partial\tilde{\mathcal{M}})}\Big)^\delta \\ &\lesssim \|u\|^{1-\delta}_{H^1(\mathcal{V})}\Big(\|h\|_{L^2(\mathcal{V})} + |h_\partial|_{L^2(\mathcal{V}\cap\partial\tilde{\mathcal{M}})} + |u_{|V^0\cap\partial\tilde{\mathcal{M}}}|_{L^2(V^0\cap\partial\tilde{\mathcal{M}})}\Big)^\delta,\end{aligned}$$

using that $Bu_{|\partial\tilde{\mathcal{M}}} = \partial_\nu u_{|\partial\tilde{\mathcal{M}}} = h_\partial$ and that $V^0 \subset \mathcal{U}$.

We have

$$\begin{aligned}&\|u\|_{H^1(\mathcal{V})} \lesssim e^{2|\sigma|}\|u^0\|_{H^1(\mathcal{M})},\\ &\|h\|_{L^2(\mathcal{V})} \leq e^{2|\sigma|}\|f\|_{L^2(\mathcal{M})} \lesssim e^{2|\sigma|}\|F\|_{\mathscr{H}},\\ &|u_{|\tilde{\mathcal{M}}}|_{L^2(V^0\cap\partial\tilde{\mathcal{M}})} \lesssim e^{|\sigma|}|u^0{}_{|\partial\mathcal{M}}|_{L^2(\Gamma_0)},\end{aligned}$$

and

$$\|u\|_{H^1(\mathcal{U})} \geq \|u\|_{L^2(-1,1;H^1(\mathcal{M}))} \gtrsim e^{-|\sigma|}\|u^0\|_{H^1(\mathcal{M})}.$$

We also write

$$\begin{aligned}|h_\partial|_{L^2(\mathcal{V}\cap\partial\tilde{\mathcal{M}})} &\lesssim e^{3|\sigma|}\big(|\alpha u^0{}_{|\partial\mathcal{M}}|_{L^2(\partial\mathcal{M})} + |f^0{}_{|\partial\mathcal{M}}|_{L^2(\partial\mathcal{M})}\big)\\ &\lesssim e^{3|\sigma|}\big(|\alpha^{1/2}u^0{}_{|\partial\mathcal{M}}|_{L^2(\partial\mathcal{M})} + \|f^0\|_{H^1(\mathcal{M})}\big),\end{aligned}$$

using that α is bounded and the trace inequality of Proposition 18.24. With (10.5.4) we obtain

$$|h_\partial|_{L^2(\mathcal{V}\cap\partial\tilde{\mathcal{M}})} \lesssim e^{3|\sigma|}\big(\|F\|_{\dot{\mathscr{H}}}^{1/2}\,\|u^0\|_{H^1(\mathcal{M})}^{1/2} + \|F\|_{\dot{\mathscr{H}}}\big).$$

Collecting the above estimations, with (10.5.6) we find

$$\|u^0\|_{H^1(\mathcal{M})} \lesssim e^{C|\sigma|}\big(\|F\|_{\dot{\mathscr{H}}} + |u^0{}_{|\partial\mathcal{M}}|_{L^2(\Gamma_0)} + \|F\|_{\dot{\mathscr{H}}}^{1/2}\,\|u^0\|_{H^1(\mathcal{M})}^{1/2}\big).$$

The Young inequality yields the sought result. ■

10.6. Notes

The results of the present chapter are generalized in Chap. 11 in the case of a damping associated with Lopatinskiĭ–Šapiro boundary conditions. Thus, the present chapter can very well be omitted. Yet, the generalization provided in Chap. 11 requires the reading of the material of Part 1. Since Neumann boundary conditions are natural in many models, it appeared to us that writing the proof of the logarithmic stabilization in this particular case was sensible. Moreover, the presentation of Chap. 11 follows very much that of the present chapter, which eases the reading of that latter chapter.

We refer to Section 6.7 of Volume 1 for references on stabilization. The result proven here is that of [219]. The method is based on [217] (where logarithmic stabilization is proven in the case of an inner damping and Dirichlet boundary conditions).

In Sect. 10.5 where the key resolvent estimate of Proposition 10.9 is proven we use a quantified unique continuation property of Chap. 9. One could also proceed as in Section 6.5.2 and obtain the resolvent estimate from the derivation of a global Carleman estimate directly without going through an interpolation type estimate. This is left to the reader and requires the construction of a global Carleman weight function. As explained in Section 6.6 of Volume 1, another approach lies also in the derivation of a Carleman estimate for the operator $P_\sigma = P_0 - \sigma^2$ with the boundary operator $\partial_\nu - i\sigma\alpha$.

Observe that we could also treat the case of an inner damping with homogeneous Neumann boundary conditions:

$$\begin{cases} \partial_t^2 y + P_0 y + \alpha\partial_t y = 0 & \text{in } (0,+\infty)\times\mathcal{M},\\ \partial_\nu y = 0 & \text{on } (0,+\infty)\times\partial\mathcal{M},\\ y_{|t=0} = y^0,\ \partial_t y_{|t=0} = y^1 & \text{in } \mathcal{M}, \end{cases}$$

with $\alpha \geq 0$. Constant functions are solutions to this damped wave equation. The C_0-semigroup associated with this damped wave equation acts on $\mathscr{H} = H^1(\mathcal{M}) \oplus L^2(\mathcal{M})$ and is given by

$$A = \begin{pmatrix} 0 & -1 \\ P_0 & \alpha \end{pmatrix}$$

with domain $D(A) = \{{}^t(v^0, v^1) \in H^2(\mathcal{M}) \times H^1(\mathcal{M}); \partial_\nu v^0{}_{|\partial\mathcal{M}} = 0\}$. Setting

$$\begin{aligned} F_\alpha : \mathscr{H} &\to \mathbb{C} \\ \begin{pmatrix} v^0 \\ v^1 \end{pmatrix} &\mapsto |\alpha|_{L^1(\mathcal{M})}^{-1} \int_{\mathcal{M}} (\alpha v^0 + v^1)\mu_g, \end{aligned}$$

and $\dot{\mathscr{H}} = \ker(F_\alpha)$ we have $\mathscr{H} = \dot{\mathscr{H}} \oplus \mathscr{N}$ with $\mathscr{N} = \operatorname{span}\{\Theta\}$, $\Theta = {}^t(1, 0)$. The space $\dot{\mathscr{H}}$ is invariant by A and is the counterpart of the reduced functional space introduced in Sect. 10.3. This setting allows one to adapt the analysis of the present chapter and the analysis carried out in Chapter 6 of Volume 1.

Appendix

10.A. The Generator of the Boundary-Damped Wave Semigroup

The boundary-damped wave equation associated with the operator $P_0 = -\Delta_g$ reads in a system form (see Sect. 10.2)

$$\partial_t V + AV = 0, \quad V(0) = V^0 \in \mathscr{H} = H^1(\mathcal{M}) \oplus L^2(\mathcal{M}), \tag{10.A.1}$$

with

$$\begin{aligned} A &= \begin{pmatrix} 0 & -1 \\ P_0 & 0 \end{pmatrix}, \\ D(A) &= \big\{V = {}^t(v^0, v^1) \in H^2(\mathcal{M}) \times H^1(\mathcal{M}); B_\partial(V) = 0\big\}, \end{aligned}$$

with the operator B_∂ defined in (10.2.1).

The different results given in this section rely a lot on the analysis of Neumann problem for the Laplace–Beltrami operator P_0 on the Riemannian manifold $(\mathcal{M}, g)$ whose treatment is given in Sect. 18.7. In particular, some of the notations of that section are used here.

PROPOSITION 10.14. *The unbounded operators $(A, D(A))$ is a closed operator on $\mathscr{H}$.*

PROOF. Let $(V^n)_n \subset D(A)$, $V^n = {}^t(v^{0,n}, v^{1,n})$, be such that both $V^n \to V = {}^t(v^0, v^1)$ and $AV^n \to W = {}^t(w^0, w^1)$ in $\mathscr{H}$ as $n \to +\infty$.

We thus have $(v^{0,n})_n \subset H^2(\mathcal{M})$, $(v^{1,n})_n \subset H^1(\mathcal{M})$ and

$$v^{0,n} \to v^0 \text{ in } H^1(\mathcal{M}), \qquad v^{1,n} \to v^1 \text{ in } L^2(\mathcal{M}).$$

As $AV^n = {}^t(-v^{1,n}, P_0 v^{0,n})$ we have

$$-v^{1,n} \to w^0 \text{ in } H^1(\mathcal{M}), \qquad P_0 v^{0,n} \to w^1 \text{ in } L^2(\mathcal{M}).$$

We thus obtain $v^1 = -w^0 \in H^1(\mathcal{M})$ and in fact $v^{1,n} \to v^1$ in $H^1(\mathcal{M})$. It thus remains to prove that $v^0 \in H^2(\mathcal{M})$, $P_0 v^0 = w^1$ and $B_\partial V = 0$.

Set $w^{1,n} = P_0 v^{0,n} \in L^2(\mathcal{M})$. We have $w^{1,n} \to w^1$ in $L^2(\mathcal{M})$. With the divergence formula of Proposition 18.28 we have

$$\int_{\mathcal{M}} w^{1,n} \mu_g = -\int_{\partial\mathcal{M}} \partial_\nu v^{0,n}{}_{|\partial\mathcal{M}}\, \mu_{g_\partial} = -\int_{\partial\mathcal{M}} h^n \mu_{g_\partial}, \qquad h^n = -\alpha v^{1,n}{}_{|\partial\mathcal{M}},$$

as $B_\partial(V^n) = 0$.

We recall that for a function $w \in H^k(\mathcal{M})$, $[w]$ denotes its class in $H^k(\mathcal{M})/\mathbb{C}$ and $\underline{w} = \underline{\Pi} w$ is the unique representative of $[w]$ with a vanishing average. See the discussion below (18.7.7). Here, we have $P_0 \underline{v}^{0,n} = w^{1,n}$ and $\partial_\nu \underline{v}^{0,n} = h^n$.

As $v^{1,n} \to v^1$ in $H^1(\mathcal{M})$ we find that $h^n \to h = -\alpha v^1{}_{|\partial\mathcal{M}}$ in $H^{1/2}(\partial\mathcal{M})$ by Proposition 18.24. In particular,

$$\int_{\mathcal{M}} w^1 \mu_g + \int_{\partial\mathcal{M}} h \mu_{g_\partial} = \lim_{n\to+\infty} \Big(\int_{\mathcal{M}} w^{1,n} \mu_g + \int_{\partial\mathcal{M}} h^n \mu_{g_\partial} \Big) = 0.$$

Then, by Proposition 18.48 and Corollary 18.49 there exists a unique $[u] \in H^2(\mathcal{M})/\mathbb{C}$ solution to the Neumann problem, in the sense that $P_0 u = w^1$ and $\partial_\nu u_{|\partial\mathcal{M}} = h$, for any $u \in [u]$, in particular $\underline{u}$. Moreover we have

$$\begin{aligned} \|\underline{v}^{0,n} - \underline{u}\|_{H^2(\mathcal{M})} &\lesssim \|w^{1,n} - w^1\|_{L^2(\mathcal{M})} + |h^n - h|_{H^{1/2}(\partial\mathcal{M})} \\ &\lesssim \|w^{1,n} - w^1\|_{L^2(\mathcal{M})} + \|v^{1,n} - v^1\|_{H^1(\mathcal{M})}, \end{aligned}$$

by the trace inequality of Proposition 18.24. Hence, $\underline{v}^{0,n} \to \underline{u}$ in $\underline{H}^2(\mathcal{M})$, implying, $[v^{0,n}] \to [u]$ in $H^2(\mathcal{M})/\mathbb{C}$, as the map $\underline{\Phi} : \underline{H}^2(\mathcal{M}) \to H^2(\mathcal{M})/\mathbb{C}$ is an isometry by Lemma 18.47. Since $v^{0,n} \to v^0$ in $H^1(\mathcal{M})$ we have $v^0 \in [u]$ by the continuity of the map $\Phi : H^2(\mathcal{M}) \to H^2(\mathcal{M})/\mathbb{C}$ introduced below (18.7.7). This gives $v^0 \in H^2(\mathcal{M})$, $P_0 v^0 = w^1$, and $\partial_\nu v^0{}_{|\partial\mathcal{M}} = h = -\alpha v^1{}_{|\partial\mathcal{M}}$ meaning precisely that $B_\partial(V) = 0$. ∎

Constant functions (in both time and space) are solutions to the boundary-damped wave equation (10.1.1). For such solution the energy (10.2.9) is identically zero. As a result, the energy does not stand as a proper norm to carry out the analysis of the solutions.

In the semigroup framework, a constant solution $y(t) \equiv \text{Cst}$ is identified to $V(t) \equiv \text{Cst}\Theta$ with $\Theta = {}^t(1,0)$. One observes that indeed $\Theta \in D(A)$ and $A\Theta = 0$ and thus $V(t) = \Theta$ is a trivial solution to (10.A.1). We set $\mathcal{N} = \text{span}\{\Theta\} = \mathsf{C} \times \{0\}$, with C the space of constant functions on $\mathcal{M}$ as introduced above (18.7.7).

Observe that by Lemma 18.46, we have

$$\text{in } \mathscr{H} = H^1(\mathcal{M}) \oplus L^2(\mathcal{M}), \quad \mathcal{N}^\perp = \underline{H}^1(\mathcal{M}) \oplus L^2(\mathcal{M}),$$

and

$$\text{in } H^2(\mathcal{M}) \oplus H^1(\mathcal{M}), \quad \mathcal{N}^\perp = \underline{H}^2(\mathcal{M}) \oplus H^1(\mathcal{M}),$$

with the spaces $\underline{H}^1(\mathcal{M})$ and $\underline{H}^2(\mathcal{M})$ as given in (18.7.8). However, even though A vanishes on $\mathcal{N}$ this does not imply that it maps $\underline{H}^2(\mathcal{M}) \oplus H^1(\mathcal{M})$ into $\underline{H}^1(\mathcal{M}) \oplus L^2(\mathcal{M})$, i.e., A does not preserve the orthogonality with respect to $\mathcal{N}$, the reason being that A is simply not symmetric (and thus not selfadjoint). The orthogonal spaces of $\mathcal{N}$ in $\mathscr{H}$ and $H^2(\mathcal{M}) \oplus H^1(\mathcal{M})$ are thus not the proper spaces to consider if one wishes to ignore constant functions in the analysis.

An important observation is the following result.

LEMMA 10.15. *We have* $\ker(A) = \mathcal{N}$ *and if* $V \in D(A)$ *and* $AV = {}^t(u^0, u^1)$ *the following relation holds*

$$\int_{\partial\mathcal{M}} \alpha u^0{}_{|\partial\mathcal{M}} \mu_{g_\partial} + \int_{\mathcal{M}} u^1 \mu_g = 0.$$

PROOF. Above we saw that $\mathcal{N} \subset \ker(A)$. If now $V = {}^t(v^0, v^1) \in \ker(A)$, then $V \in D(A)$ and $AV = 0$. This reads $v^1 = 0$ and

$$v^0 \in H^2(\mathcal{M}),\ \partial_\nu v^0{}_{|\partial\mathcal{M}} = 0, \quad P_0 v^0 = 0.$$

From Proposition 18.48 and Corollary 18.49 we have $v^0 \equiv \text{Cst}$, which gives $V \in \mathcal{N}$.

Let now $V = {}^t(v^0, v^1) \in D(A)$. We have $U = {}^t(u^0, u^1) = AV = {}^t(-v^1, P_0 v^0) \in \mathscr{H}$. With the divergence formula of Proposition 18.28, we compute

$$\begin{aligned}\int_{\partial\mathcal{M}} \alpha u^0{}_{|\partial\mathcal{M}} \mu_{g_\partial} + \int_{\mathcal{M}} u^1 \mu_g &= - \int_{\partial\mathcal{M}} \alpha v^1{}_{|\partial\mathcal{M}} \mu_{g_\partial} + \int_{\mathcal{M}} P_0 v^0 \mu_g \\ &= - \int_{\partial\mathcal{M}} \big(\alpha v^1{}_{|\partial\mathcal{M}} + \partial_\nu v^0{}_{|\partial\mathcal{M}}\big) \mu_{g_\partial} = 0,\end{aligned}$$

since $B_\partial(V) = 0$. ■

With the above lemma we are naturally led to introduce the following linear form

$$\begin{aligned} F_\alpha : \mathscr{H} &\to \mathbb{C} \\ \begin{pmatrix} v^0 \\ v^1 \end{pmatrix} &\mapsto |\alpha|^{-1}_{L^1(\partial\mathcal{M})} \Big(\int_{\partial\mathcal{M}} \alpha v^0{}_{|\partial\mathcal{M}} \mu_{g_\partial} + \int_{\mathcal{M}} v^1 \mu_g \Big), \end{aligned} \tag{10.A.2}$$

and the space

$$\dot{\mathscr{H}} = \ker(F_\alpha) = \{V \in \mathscr{H};\ F_\alpha(V) = 0\}. \tag{10.A.3}$$

The second part of Lemma 10.15 thus reads

$$\operatorname{Ran}(A) \subset \dot{\mathscr{H}}. \tag{10.A.4}$$

The form F_α is bounded by the trace inequality of Proposition 18.24 and we note that $F_\alpha(\Theta) = 1$ with $\Theta = {}^t(1, 0)$. We set $\Pi_{\mathcal{N}} : \mathscr{H} \to \mathscr{H}$ and $\Pi_{\dot{\mathscr{H}}} : \mathscr{H} \to \mathscr{H}$ as the bounded linear maps

$$\Pi_{\mathcal{N}} V = F_\alpha(V)\Theta, \qquad \Pi_{\dot{\mathscr{H}}} = \operatorname{Id}_{\mathscr{H}} - \Pi_{\mathcal{N}}. \tag{10.A.5}$$

We see that $\operatorname{Ran}(\Pi_{\mathcal{N}}) = \ker(\Pi_{\dot{\mathscr{H}}}) = \mathcal{N}$ and $\operatorname{Ran}(\Pi_{\dot{\mathscr{H}}}) = \ker(\Pi_{\mathcal{N}}) = \dot{\mathscr{H}}$, that both maps $\Pi_{\mathcal{N}}$ and $\Pi_{\dot{\mathscr{H}}}$ are projectors and we have

$$\mathscr{H} = \dot{\mathscr{H}} \oplus \mathcal{N}, \qquad \mathcal{N} = \operatorname{span}\{\Theta\}. \tag{10.A.6}$$

Note that this direct sum is to be understood in the sense of vector spaces only and not in the sense of Hilbert spaces. The space $\mathcal{N}$ is not orthogonal to $\dot{\mathscr{H}}$.

From the point of view of the semigroup formulation of the boundary-damped wave equation, the introduction of the form F_α and the space $\dot{\mathscr{H}}$ also appear natural with the following argument. Let us assume for now that we know that A generates a C_0-semigroup on $\mathscr{H}$ (a proof is given below). For a strong solution $U(t) = {}^t(u^0(t), u^1(t))$, that is $U(t) \in \mathscr{C}^0(\mathbb{R}_+; D(A)) \cap \mathscr{C}^1(\mathbb{R}_+; \mathscr{H})$ such that $\frac{d}{dt}U(t) + AU(t) = 0$, we observe that $t \mapsto F_\alpha(U(t))$ is constant. In fact, we write

$$\frac{d}{dt}F_\alpha(U(t)) = F_\alpha\Big(\frac{d}{dt}U(t)\Big) = -F_\alpha\big(AU(t)\big) = 0, \tag{10.A.7}$$

by Lemma 10.15. Then, the strong solution associated with the initial condition $\Pi_{\dot{\mathscr{H}}}U(0) \in D(A) \cap \dot{\mathscr{H}}$ remains in $\dot{\mathscr{H}}$ for all $t > 0$.

From the above analysis, we have $A \circ \Pi_{\dot{\mathscr{H}}} = A$. If we define the unbounded map $(\dot{A}, D(\dot{A}))$ with

$$\begin{aligned} D(\dot{A}) = \Pi_{\dot{\mathscr{H}}}\big(D(A)\big) = D(A) \cap \dot{\mathscr{H}} &= \big\{V \in D(A);\ F_\alpha(V) = 0\big\} \\ &= \big\{V \in H^2(\mathcal{M}) \times H^1(\mathcal{M});\ B_\partial(V) = 0,\ F_\alpha(V) = 0\big\}, \end{aligned} \tag{10.A.8}$$

(we recall that the operator B_∂ is defined in (10.2.1) and the operator F_α in (10.A.2)) by $\dot{A}U = AU$ for $U \in D(\dot{A})$, we then have $A = \dot{A} \circ \Pi_{\dot{\mathscr{H}}}$. This is summarized in the following diagram

$$\begin{array}{ccc} \mathscr{H} \supset D(A) & \xrightarrow{\ A\ } & \mathscr{H} \\ \big\downarrow \Pi_{\dot{\mathscr{H}}} \quad \big\downarrow \Pi_{\dot{\mathscr{H}}} & \nearrow \dot{A} & \\ \dot{\mathscr{H}} \supset D(\dot{A}) & & \end{array}$$

Note also that since $\mathcal{N} \subset D(A)$ we have

$$D(A) = D(\dot{A}) \oplus \mathcal{N}. \tag{10.A.9}$$

In particular, with (10.A.6), this gives

$$\Pi_{\dot{\mathscr{H}}}V \in D(\dot{A}) \quad \Leftrightarrow \quad V \in D(A), \qquad \text{for } V \in \mathscr{H}. \tag{10.A.10}$$

The space $\dot{\mathscr{H}}$ inherits the norm of $\mathscr{H} = H^1(\mathcal{M}) \oplus L^2(\mathcal{M})$. However, the following equivalent norm will be more adapted to our purpose here.

LEMMA 10.16. *The norm* $\|.\|_{\dot{\mathscr{H}}}$ *defined by*

$$\begin{aligned} \|V\|_{\dot{\mathscr{H}}}^2 &:= \|\nabla_g v^0\|_{L^2V(\mathcal{M})}^2 + \|v^1\|_{L^2(\mathcal{M})}^2 \\ &= \|\mathsf{D}\, v^0\|_{L^2\Lambda^1(\mathcal{M})}^2 + \|v^1\|_{L^2(\mathcal{M})}^2, \quad V = {}^t(v^0, v^1), \end{aligned}$$

is equivalent to $\|.\|_{\mathscr{H}}$ *on* $\dot{\mathscr{H}}$.

PROOF. It suffices to prove the following adapted Poincaré inequality

$$\|v^0\|_{L^2(\mathcal{M})} \lesssim \|V\|_{\dot{\mathscr{H}}}, \quad V \in \dot{\mathscr{H}},$$

to reach the conclusion. The proof is similar to that of the Poincaré inequality of Proposition 18.9. Assume that the inequality does not hold. Then, there exists a sequence $V^n = {}^t(v^{0,n}, v^{1,n})_{n\in\mathbb{N}} \subset \dot{\mathscr{H}}$ such that

$$\|v^{0,n}\|_{L^2(\mathcal{M})} = 1, \qquad \|\nabla_g v^{0,n}\|_{L^2V(\mathcal{M})} + \|v^{1,n}\|_{L^2(\mathcal{M})} \to 0.$$

In particular, $v^{0,n}$ is bounded in $H^1(\mathcal{M})$ and, thus, there exists $v^0 \in H^1(\mathcal{M})$ such that $v^{0,n} \rightharpoonup v^0$ in $H^1(\mathcal{M})$ up to a subsequence.

On the one hand, we have $\nabla_g v^{0,n} \rightharpoonup \nabla_g v^0$ in $L^2V(\mathcal{M})$, implying that $\nabla_g v^0 = 0$. Because of the form of ∇_g given in (17.2.2) this implies that $v^0 \equiv \mathrm{Cst}$ in every chart and it thus constant on the whole $\mathcal{M}$ since $\mathcal{M}$ is connected.

On the other hand, from the continuity of the trace map (see Proposition 18.24) we have $v^{0,n}_{|\partial\mathcal{M}} \rightharpoonup v^0_{|\partial\mathcal{M}}$ in $H^{1/2}(\partial\mathcal{M})$ (apply, for instance, Proposition 35.8 in [320] or Theorem 3.10 in [90]). Consequently

$$\int_{\partial\mathcal{M}} \alpha v^{0,n}_{|\partial\mathcal{M}} \mu_{g_\partial} \to \int_{\partial\mathcal{M}} \alpha v^0_{|\partial\mathcal{M}} \mu_{g_\partial}.$$

As $\int_{\mathcal{M}} v^{1,n}\mu_g \to 0$ and $F_\alpha(V^n) = 0$ we obtain in the limit $\int_{\partial\mathcal{M}} \alpha v^0_{|\partial\mathcal{M}}\mu_{g_\partial} = 0$.

Above we obtained $v^0 \equiv \mathrm{Cst}$; since $\int_{\partial\mathcal{M}} \alpha\, \mu_{g_\partial} > 0$, this gives $v^0 \equiv 0$, which yields a contradiction as $\|v^0\|_{L^2(\mathcal{M})} = \lim \|v^{0,n}\|_{L^2(\mathcal{M})} = 1$ since $v^{0,n} \to v^0$ in $L^2(\mathcal{M})$ from the compact injection $\iota : H^1(\mathcal{M}) \to L^2(\mathcal{M})$ by the Rellich–Kondrachov theorem (Theorem 18.7). ∎

Naturally, associated with the norm $\|.\|_{\dot{\mathscr{H}}}$ is the following inner product, for $U, V \in \dot{\mathscr{H}}$,

$$(V,U)_{\dot{\mathscr{H}}} := (\nabla_g v^0, \nabla_g u^0)_{L^2V(\mathcal{M})} + (v^1, u^1)_{L^2(\mathcal{M})} = (\mathsf{D}\, v^0, \mathsf{D}\, u^0)_{L^2\Lambda^1(\mathcal{M})} + (v^1, u^1)_{L^2(\mathcal{M})}. \tag{10.A.11}$$

PROPOSITION 10.17. *The unbounded operator $(\dot{A}, D(\dot{A}))$ on $\dot{\mathscr{H}}$ is closed. Moreover, $D(\dot{A})$ is dense in $\dot{\mathscr{H}}$ and the injection of $D(\dot{A})$ into $\dot{\mathscr{H}}$ is compact.*

In the proof of Proposition 10.17 we shall need the following lemma.

LEMMA 10.18. *The map*

$$D(\dot{A}) \to H^1(\mathcal{M})$$
$$(u^0, u^1) \mapsto u^1$$

is surjective. Moreover, there exists a bounded linear map $\Psi : H^1(\mathcal{M}) \to H^2(\mathcal{M})$ such that $(\Psi w, w) \in D(\dot{A})$ for all $w \in H^1(\mathcal{M})$.

PROOF OF LEMMA 10.18. Let $w \in H^1(\mathcal{M})$ and set $h = -\alpha w_{|\partial\mathcal{M}} \in H^{1/2}(\partial\mathcal{M})$ using the trace property of Proposition 18.24. Then, with the trace theorem (Theorem 18.25), there exists $v \in H^2(\mathcal{M})$ obtained linearly and continuously with respect to h such that $v_{|\partial\mathcal{M}} = 0$ and $\partial_\nu v_{|\partial\mathcal{M}} = h$ and

$$\|v\|_{H^2(\mathcal{M})} \lesssim |h|_{H^{1/2}(\partial\mathcal{M})} \lesssim \|w\|_{H^1(\mathcal{M})}.$$

Since $\int_{\partial\mathcal{M}} \alpha\, \mu_{g_\partial} \neq 0$, we can pick $c^0 \in \mathbb{C}$ such that $c^0 \int_{\partial\mathcal{M}} \alpha\mu_{g_\partial} + \int_{\mathcal{M}} w\mu_g = 0$. We then set $\Psi w = v + c^0$. Note that the map $w \mapsto \Psi w$ is linear and that we have $V = {}^t(\Psi w, w) \in D(\dot{A})$. Moreover, we see that

$$\|c^0\|_{L^2(\mathcal{M})} \lesssim \|w\|_{L^1(\mathcal{M})} \lesssim \|w\|_{H^1(\mathcal{M})}.$$

Consequently $\|\Psi w\|_{H^2(\mathcal{M})} \lesssim \|w\|_{H^1(\mathcal{M})}$. The proof is complete. ■

PROOF OF PROPOSITION 10.17. Let $(u^n)_n \subset D(\dot{A})$ be such that both $(u^n)_n$ and $(\dot{A}u^n)_n$ converge in $\dot{\mathscr{H}}$. Denote by u and v the respective limits. Recall that $D(\dot{A}) \subset D(A)$ and $\dot{A}u^n = Au^n$. As $(A, D(A))$ is closed, and as the topology of $\dot{\mathscr{H}}$ is that of a subspace of $\mathscr{H}$, we find that $u \in D(A) \cap \dot{\mathscr{H}} = D(\dot{A})$ and $v = Au = \dot{A}u$. The operator $(\dot{A}, D(\dot{A}))$ is thus closed on $\dot{\mathscr{H}}$.

Let now $U \in \dot{\mathscr{H}}$ be such that $(U, V)_{\dot{\mathscr{H}}} = 0$ for all $V \in D(\dot{A})$. Let us prove that $U = 0$, which will assert that $D(\dot{A})$ is dense in $\dot{\mathscr{H}}$.

Let $v \in H^2(\mathcal{M})$ be such that $\partial_\nu v_{|\partial\mathcal{M}} = 0$. We then set $v^0 = \underline{v} + c^0$ with c^0 chosen so that $\int_{\partial\mathcal{M}} \alpha v^0{}_{|\partial\mathcal{M}} \mu_{g_\partial} = 0$, which is possible as $\int_{\partial\mathcal{M}} \alpha\, \mu_{g_\partial} \neq 0$. Observe then that $V = {}^t(v^0, 0) \in D(\dot{A})$ and, with (18.5.5) consequence of the divergence formula of Proposition 18.28, we obtain

$$0 = (U, V)_{\dot{\mathscr{H}}} = (\nabla_g u^0, \nabla_g v^0)_{L^2V(\mathcal{M})} = (u^0, P_0 v^0)_{L^2(\mathcal{M})} = (u^0, P_0 v)_{L^2(\mathcal{M})}.$$

By Lemma 18.51, this implies that $u^0 \equiv \mathrm{Cst}$. We thus find that

$$0 = (U, V)_{\dot{\mathscr{H}}} = (u^1, v^1)_{L^2(\mathcal{M})}, \qquad V = {}^t(v^0, v^1) \in D(\dot{A}). \tag{10.A.12}$$

Let $w \in H^1(\mathcal{M})$. By Lemma 10.18, there exists $v \in H^2(\mathcal{M})$ such that $V = {}^t(v, w) \in D(\dot{A})$. Then, (10.A.12) yields $(u^1, w)_{L^2(\mathcal{M})} = 0$. As $w \in H^1(\mathcal{M})$ is arbitrary, this gives $u^1 = 0$ as $H^1(\mathcal{M})$ is dense in $L^2(\mathcal{M})$. Thus, $U = {}^t(u^0, 0) \in \dot{\mathscr{H}} \cap \mathcal{N}$ since $u^0 \equiv \mathrm{Cst}$. By (10.A.6) we finally obtain $U = 0$. This concludes the proof of the density property.

The compact property of the injection of $D(\dot{A})$ into $\dot{\mathscr{H}}$ follows from the definitions of $D(\dot{A})$ and $\dot{\mathscr{H}}$ (see (10.A.3) and (10.A.8)) and the Rellich–Kondrachov theorem (Theorem 18.7). ■

From (10.A.6) and (10.A.9), and the previous proposition we obtain the following result.

COROLLARY 10.19. *The domain $D(A)$ is dense in $\mathscr{H}$.*

We now make explicit the operator $\dot{A}^*$ adjoint of $\dot{A}$ with respect to the inner product on $\dot{\mathscr{H}}$ introduced in (10.A.11). We recall that the domain of the adjoint is defined by

$$D(\dot{A}^*) = \{V \in \dot{\mathscr{H}};\ \exists C > 0,\ \forall U \in D(\dot{A}),\ |(V, \dot{A}U)_{\dot{\mathscr{H}}}| \leq C\|U\|_{\dot{\mathscr{H}}}\}.$$

PROPOSITION 10.20. *We have*

(10.A.13)
$$D(\dot{A}^*)=\{V={}^t(v^0,v^1)\in \mathscr{H}; v^0\in H^2(\mathcal{M}),\ v^1\in H^1(\mathcal{M});\ B^*_\partial(V)=0\}$$

with $B^*_\partial(V)=-\partial_\nu v^0+\alpha v^1$. *For* $V={}^t(v^0,v^1)\in D(\dot{A}^*)$ *we have*

$$\dot{A}^*V=\begin{pmatrix} v^1\\ -P_0v^0\end{pmatrix},\quad \textit{that is,}\quad \dot{A}^*=\begin{pmatrix}0 & 1\\ -P_0 & 0\end{pmatrix}. \tag{10.A.14}$$

PROOF. Let $V\in \big(H^2(\mathcal{M})\times H^1(\mathcal{M})\big)\cap\mathscr{H}$ be such that $B^*_\partial(V)=0$. If $U={}^t(u^0,u^1)\in D(\dot{A})$, with (18.5.5) consequence of the divergence formula of Proposition 18.28 we compute

$$\begin{aligned}(\dot{A}U,V)_{\dot{\mathscr{H}}} &= -(\nabla_g u^1,\nabla_g v^0)_{L^2V(\mathcal{M})}+(P_0u^0,v^1)_{L^2(\mathcal{M})}\\ &= -(u^1,P_0v^0)_{L^2(\mathcal{M})}+(\nabla_g u^0,\nabla_g v^1)_{L^2V(\mathcal{M})}\\ &\quad -(u^1{}_{|\partial\mathcal{M}},\partial_\nu v^0{}_{|\partial\mathcal{M}})_{L^2(\partial\mathcal{M})}-(\partial_\nu u^0{}_{|\partial\mathcal{M}},v^1{}_{|\partial\mathcal{M}})_{L^2(\partial\mathcal{M})}.\end{aligned}$$

Using that $B_\partial(U)=0$, that is, $(\partial_\nu u^0+\alpha u^1)_{|\partial\mathcal{M}}=0$, for the two boundary terms, we obtain

$$\begin{aligned}&(u^1{}_{|\partial\mathcal{M}},\partial_\nu v^0{}_{|\partial\mathcal{M}})_{L^2(\partial\mathcal{M})}+(\partial_\nu u^0{}_{|\partial\mathcal{M}},v^1{}_{|\partial\mathcal{M}})_{L^2(\partial\mathcal{M})}\\ &\quad=(u^1{}_{|\partial\mathcal{M}},\partial_\nu v^0{}_{|\partial\mathcal{M}})_{L^2(\partial\mathcal{M})}-(\alpha u^1{}_{|\partial\mathcal{M}},v^1{}_{|\partial\mathcal{M}})_{L^2(\partial\mathcal{M})}\\ &\quad=(u^1{}_{|\partial\mathcal{M}},(\partial_\nu v^0-\alpha v^1)_{|\partial\mathcal{M}})_{L^2(\partial\mathcal{M})}\\ &\quad=0,\end{aligned}$$

since $B^*_\partial(V)=(\partial_\nu v^0-\alpha v^1)_{|\partial\mathcal{M}}=0$. We thus obtain

$$(\dot{A}U,V)_{\dot{\mathscr{H}}}=-(u^1,P_0v^0)_{L^2(\mathcal{M})}+(\nabla_g u^0,\nabla_g v^1)_{L^2V(\mathcal{M})}, \tag{10.A.15}$$

which yields, as $v^0\in H^2(\mathcal{M})$ and $v^1\in H^1(\mathcal{M})$,

$$|(\dot{A}U,V)_{\dot{\mathscr{H}}}|\lesssim \|u^1\|_{L^2(\mathcal{M})}+\|\nabla_g u^0\|_{L^2V(\mathcal{M})}\eqsim\|U\|_{\dot{\mathscr{H}}}.$$

Thus $V\in D(\dot{A}^*)$ and (10.A.15) precisely yields the form of $\dot{A}^*V$ given in (10.A.14).

Conversely, let $V={}^t(v^0,v^1)\in D(\dot{A}^*)$. If $U={}^t(u^0,u^1)\in D(\dot{A})$ we have

$$(\dot{A}U,V)_{\dot{\mathscr{H}}}=-(\nabla_g u^1,\nabla_g v^0)_{L^2V(\mathcal{M})}+(P_0u^0,v^1)_{L^2(\mathcal{M})}, \tag{10.A.16}$$

and

$$|(\dot{A}U,V)_{\dot{\mathscr{H}}}|\lesssim\|U\|_{\dot{\mathscr{H}}}\eqsim\|\nabla_g u^0\|_{L^2V(\mathcal{M})}+\|u^1\|_{L^2(\mathcal{M})}. \tag{10.A.17}$$

By the Riesz representation theorem, there exists $W={}^t(w^0,w^1)\in\mathscr{H}$ such that

$$(\dot{A}U,V)_{\dot{\mathscr{H}}}=(U,W)_{\dot{\mathscr{H}}}=(\nabla_g u^0,\nabla_g w^0)_{L^2V(\mathcal{M})}+(u^1,w^1)_{L^2(\mathcal{M})}. \tag{10.A.18}$$

Let $u\in H^2(\mathcal{M})$ be such that $\partial_\nu u_{|\partial\mathcal{M}}=0$. We then pick $u^0=u+c^0$ with c^0 chosen so that $\int_{\partial\mathcal{M}}\alpha u^0{}_{|\partial\mathcal{M}}\mu_{g_\partial}=0$, which is possible as $\int_{\partial\mathcal{M}}\alpha\,\mu_{g_\partial}\neq 0$.

Observe that we have $U = {}^t(u^0, 0) \in D(\dot{A})$. With such a choice, we have on the one hand

$$(\dot{A}U, V)_{\dot{\mathscr{H}}} = (P_0 u^0, v^1)_{L^2(\mathcal{M})},$$

and on the other hand

$$(\dot{A}U, V)_{\dot{\mathscr{H}}} = (\nabla_g u^0, \nabla_g w^0)_{L^2V(\mathcal{M})} = (P_0 u^0, w^0)_{L^2(\mathcal{M})},$$

by (18.5.5) consequence of the divergence formula of Proposition 18.28. Consequently,

$$(P_0 u, v^1 - w^0)_{L^2(\mathcal{M})} = (P_0 u^0, v^1 - w^0)_{L^2(\mathcal{M})} = 0,$$

for all $u \in H^2(\mathcal{M})$ such that $\partial_\nu u_{|\partial\mathcal{M}} = 0$. By Lemma 18.51, this implies $v^1 = w^0 + \text{Cst}$. In particular, $v^1 \in H^1(\mathcal{M})$.

From (10.A.16), with $U = {}^t(u^0, u^1) \in D(\dot{A})$, we may now write, by (18.5.5),

$$\begin{aligned}(\dot{A}U, V)_{\dot{\mathscr{H}}} &= -(\nabla_g u^1, \nabla_g v^0)_{L^2V(\mathcal{M})} - (\partial_\nu u^0{}_{|\partial\mathcal{M}}, v^1{}_{|\partial\mathcal{M}})_{L^2(\partial\mathcal{M})} \\ &\quad + (\nabla_g u^0, \nabla_g v^1)_{L^2V(\mathcal{M})} \\ &= -(\nabla_g u^1, \nabla_g v^0)_{L^2V(\mathcal{M})} + (\alpha u^1{}_{|\partial\mathcal{M}}, v^1{}_{|\partial\mathcal{M}})_{L^2(\partial\mathcal{M})} \\ &\quad + (\nabla_g u^0, \nabla_g w^0)_{L^2V(\mathcal{M})},\end{aligned} \tag{10.A.19}$$

using that $B_\partial(U) = 0$ and $v^1 = w^0 + \text{Cst}$. Note that the boundary L^2-inner product is well defined since both $\partial_\nu u^0{}_{|\partial\mathcal{M}}$ and $v^1{}_{|\partial\mathcal{M}}$ are in $H^{1/2}(\partial\mathcal{M})$ the trace formula of Theorem 18.25.

With (10.A.18) we conclude that

$$(\nabla_g u^1, \nabla_g v^0)_{L^2V(\mathcal{M})} - (u^1{}_{|\partial\mathcal{M}}, \alpha v^1{}_{|\partial\mathcal{M}})_{L^2(\partial\mathcal{M})} = -(u^1, w^1)_{L^2(\mathcal{M})}. \tag{10.A.20}$$

Let $w \in H^1(\mathcal{M})$. From Lemma 10.18 there exists $z \in H^2(\mathcal{M})$ such that $(z, w) \in D(\dot{A})$. With (10.A.20) we find that

$$(\nabla_g w, \nabla_g v^0)_{L^2V(\mathcal{M})} = -(w, w^1)_{L^2(\mathcal{M})} + (w_{|\partial\mathcal{M}}, \alpha v^1{}_{|\partial\mathcal{M}})_{L^2(\partial\mathcal{M})}, \tag{10.A.21}$$

for all $w \in H^1(\mathcal{M})$. Note that choosing $w = 1$, yields

$$-\int_{\mathcal{M}} w^1 \mu_g + \int_{\partial\mathcal{M}} \alpha v^1{}_{|\partial\mathcal{M}}\, \mu_{g_\partial} = 0. \tag{10.A.22}$$

Then, Proposition 18.48 applies to the (variational) Eq. (10.A.21) under condition (10.A.22), yielding a unique solution $[v] \in H^2(\mathcal{M})/\mathbb{C}$. As a result $v^0 \in [v]$ implying that $v^0 \in H^2(\mathcal{M})$ and $\partial_\nu v^0 = \alpha v^1$ on $\partial\mathcal{M}$. As $V = {}^t(v^0, v^1) \in \mathscr{H}$ we have $F_\alpha(V) = 0$. We have therefore obtained that $D(\dot{A}^*)$ is as described in (10.A.13). The proof is complete. ■

We now proceed toward the proof that $\dot{A}$ generates a C_0-semigroup on $\dot{\mathscr{H}}$. To that purpose we need following lemma that provides a first resolvent estimate.

LEMMA 10.21. *Let $B = \dot{A}$ or $\dot{A}^*$ and $z \in \mathbb{C}$ be such that* $\operatorname{Re} z < 0$. *We have*

$$\|(z \operatorname{Id}_{\dot{\mathscr{H}}} - B)U\|_{\dot{\mathscr{H}}} \geq |\operatorname{Re} z| \, \|U\|_{\dot{\mathscr{H}}}, \quad U \in D(B).$$

PROOF. Let $U = {}^t(u^0, u^1) \in D(\dot{A})$. We write

$$\begin{aligned}
((z \operatorname{Id}_{\dot{\mathscr{H}}} - \dot{A})U, U)_{\dot{\mathscr{H}}} &= \left(\begin{pmatrix} zu^0 + u^1 \\ zu^1 - P_0 u^0 \end{pmatrix}, \begin{pmatrix} u^0 \\ u^1 \end{pmatrix} \right)_{\dot{\mathscr{H}}} \\
&= z\|U\|^2_{\dot{\mathscr{H}}} + (\nabla_g u^1, \nabla_g u^0)_{L^2V(\mathcal{M})} - (P_0 u^0, u^1)_{L^2(\mathcal{M})} \\
&= z\|U\|^2_{\dot{\mathscr{H}}} + 2i \operatorname{Im}(\nabla_g u^1, \nabla_g u^0)_{L^2V(\mathcal{M})} \\
&\quad + (\partial_\nu u^0{}_{|\partial\mathcal{M}}, u^1{}_{|\partial\mathcal{M}})_{L^2(\partial\mathcal{M})}.
\end{aligned}$$

We find, computing the real part and using that $B_\partial(U) = 0$,

$$-\operatorname{Re}((z \operatorname{Id}_{\dot{\mathscr{H}}} - \dot{A})U, U)_{\dot{\mathscr{H}}} = -\operatorname{Re}(z)\|U\|^2_{\dot{\mathscr{H}}} + \int_{\partial\mathcal{M}} \alpha |u^1|^2_{|\partial\mathcal{M}} \, \mu_{g_\partial}. \tag{10.A.23}$$

As $\alpha \geq 0$ and $\operatorname{Re} z < 0$, we find

$$|\operatorname{Re}((z \operatorname{Id}_{\dot{\mathscr{H}}} - \dot{A})U, U)_{\dot{\mathscr{H}}}| \geq |\operatorname{Re}(z)| \, \|U\|^2_{\dot{\mathscr{H}}},$$

which yields the conclusion for $B = \dot{A}$. In the case $B = \dot{A}^*$ we similarly have, for $U \in D(\dot{A}^*)$,

$$((z \operatorname{Id}_{\dot{\mathscr{H}}} - \dot{A}^*)U, U)_{\dot{\mathscr{H}}} = z\|U\|^2_{\dot{\mathscr{H}}} - (\nabla_g u^1, \nabla_g u^0)_{L^2V(\mathcal{M})} + (P_0 u^0, u^1)_{L^2(\mathcal{M})}$$

yielding

$$\begin{aligned}
((z \operatorname{Id}_{\dot{\mathscr{H}}} - \dot{A}^*)U, U)_{\dot{\mathscr{H}}} &= z\|U\|^2_{\dot{\mathscr{H}}} - (\nabla_g u^1, \nabla_g u^0)_{L^2V(\mathcal{M})} \\
&\quad + (\nabla_g u^0, \nabla_g u^1)_{L^2V(\mathcal{M})} \\
&\quad - (\partial_\nu u^0{}_{|\partial\mathcal{M}}, u^1{}_{|\partial\mathcal{M}})_{L^2(\partial\mathcal{M})}.
\end{aligned}$$

As $B^*_\partial(U) = -\partial_\nu u^0 + \alpha u^1 = 0$ we find

$$-\operatorname{Re}((z \operatorname{Id}_{\dot{\mathscr{H}}} - \dot{A}^*)U, U)_{\dot{\mathscr{H}}} = -\operatorname{Re}(z)\|U\|^2_{\dot{\mathscr{H}}} + \int_{\partial\mathcal{M}} \alpha |u^1|^2_{|\partial\mathcal{M}} \, \mu_{g_\partial},$$

which also yields the conclusion in this case. ■

We may now state and prove that $\dot{A}$ generates a C_0-semigroup.

THEOREM 10.22. *The unbounded operator $(\dot{A}, D(\dot{A}))$ generates a C_0-semigroup of contraction $\dot{S}(t) = e^{-t\dot{A}}$ on $\dot{\mathscr{H}}$.*

PROOF. By Lemma 10.21 , for $\lambda < 0$, the operator $\lambda \operatorname{Id}_{\dot{\mathscr{H}}} - \dot{A}$ is invertible with a bounded inverse and moreover we have $\|(\lambda \operatorname{Id}_{\dot{\mathscr{H}}} - \dot{A})^{-1}\|_{\mathscr{L}(\dot{\mathscr{H}})} \leq |\lambda|^{-1}$. We then obtain the result by the Hille-Yosida theorem (see Theorem 12.6 of Volume 1) as $(\dot{A}, D(\dot{A}))$ is closed and $D(\dot{A})$ is dense in $\dot{\mathscr{H}}$ by Proposition 10.17. ■

Recalling the diagram illustrating the definition of $(\dot{A}, D(\dot{A}))$ above (10.A.9), and that A vanishes on $\mathscr{N} = \operatorname{span}\{\theta\}$, we define the following operator on $\mathscr{H}$

$$S(t) := \dot{S}(t) \circ \Pi_{\dot{\mathscr{H}}} + \Pi_{\mathscr{N}}, \tag{10.A.24}$$

with $t \geq 0$ as a parameter, and where $\Pi_{\dot{\mathscr{H}}}$ and $\Pi_{\mathscr{N}}$ are the projectors defined in (10.A.5). Observe that we have the following commutative diagram

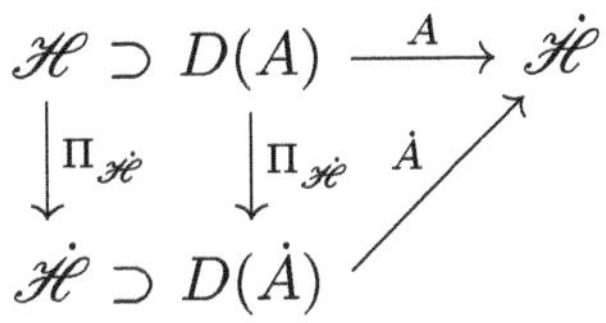

$$\begin{array}{ccc} \mathscr{H} \supset D(A) & \xrightarrow{\ A\ } & \dot{\mathscr{H}} \\ \big\downarrow \Pi_{\dot{\mathscr{H}}} \quad \big\downarrow \Pi_{\dot{\mathscr{H}}} & \nearrow \dot{A} & \\ \dot{\mathscr{H}} \supset D(\dot{A}) & & \end{array}$$

The operator $S(t)$ coincides with $\dot{S}(t)$ on $\dot{\mathscr{H}}$ and coincides with the identity map on its complement $\mathscr{N}$.

THEOREM 10.23. *The map $S(t)$ is a bounded C_0-semigroup on $\mathscr{H}$; its generator is the unbounded operator $(A, D(A))$.*

PROOF. First, we see that $S(0) = \mathrm{Id}_{\dot{\mathscr{H}}} \circ \Pi_{\dot{\mathscr{H}}} + \Pi_{\mathscr{N}} = \mathrm{Id}_{\mathscr{H}}$. Second, from the strong continuity of $\dot{S}(t)$ at $t = 0^+$ we deduce that of $S(t)$. Third, we write, using the above commutative diagram, (10.A.5) and the semigroup property of $\dot{S}(t)$, for $t, t' \geq 0$, and $U \in \mathscr{H}$.

$$\begin{aligned} S(t') \circ S(t)U &= \dot{S}(t') \circ \Pi_{\dot{\mathscr{H}}} \circ S(t)U + \Pi_{\mathscr{N}} S(t)U \\ &= \dot{S}(t') \circ \dot{S}(t) \circ \Pi_{\dot{\mathscr{H}}} U + \Pi_{\mathscr{N}} U \\ &= \dot{S}(t' + t) \circ \Pi_{\dot{\mathscr{H}}} U + \Pi_{\mathscr{N}} U \\ &= S(t' + t)U. \end{aligned}$$

These three properties show that $S(t)$ is a C_0-semigroup on $\mathscr{H}$.

For the boundedness of $S(t)$ we write

$$\|S(t)U\|_{\mathscr{H}} \leq \|\dot{S}(t) \circ \Pi_{\dot{\mathscr{H}}} U\|_{\mathscr{H}} + \|\Pi_{\mathscr{N}} U\|_{\mathscr{H}}.$$

As $\Pi_{\mathscr{N}}$ is continuous we have $\|\Pi_{\mathscr{N}} U\|_{\mathscr{H}} \lesssim \|U\|_{\mathscr{H}}$. Next, since $\dot{S}(t) \circ \Pi_{\dot{\mathscr{H}}} U \in \dot{\mathscr{H}}$ we have

$$\|\dot{S}(t) \circ \Pi_{\dot{\mathscr{H}}} U\|_{\mathscr{H}} \lesssim \|\dot{S}(t) \circ \Pi_{\dot{\mathscr{H}}} U\|_{\dot{\mathscr{H}}} \lesssim \|\Pi_{\dot{\mathscr{H}}} U\|_{\dot{\mathscr{H}}} \lesssim \|\Pi_{\dot{\mathscr{H}}} U\|_{\mathscr{H}} \lesssim \|U\|_{\mathscr{H}},$$

by Lemma 10.16 and the boundedness of $\dot{S}(t)$ on $\dot{H}$ and that of $\Pi_{\dot{\mathscr{H}}}$ on $\mathscr{H}$. Gathered together, the last three inequalities yield $\|S(t)U\|_{\mathscr{H}} \lesssim \|U\|_{\mathscr{H}}$.

Finally, to determine the generator $(B, D(B))$ of $S(t)$, for $t > 0$ and $U \in \mathscr{H}$, we compute

$$\frac{1}{t}\big(U - S(t)U\big) = \frac{1}{t}\big(\mathrm{Id}_{\dot{\mathscr{H}}} - \dot{S}(t)\big)\Pi_{\dot{\mathscr{H}}} U,$$

which allows one to conclude that $U \in D(B)$ if and only if $\Pi_{\dot{\mathscr{H}}} U \in D(\dot{A})$, that is, $U \in D(A)$ by (10.A.10). In such a case, we have

$$BU = \lim_{t \to 0+} \frac{1}{t}\big(U - S(t)U\big) = \lim_{t \to 0+} \frac{1}{t}\big(\mathrm{Id}_{\dot{\mathscr{H}}} - \dot{S}(t)\big)\Pi_{\dot{\mathscr{H}}} U = \dot{A}\Pi_{\dot{\mathscr{H}}} U = AU,$$

as $A = \dot{A} \circ \Pi_{\dot{\mathscr{H}}}$; see the commutative diagram above (10.A.10). As a conclusion the generator of $S(t)$ is precisely $(A, D(A))$. ■

REMARK 10.24. Note that Theorem 10.23 provides also a proof of Corollary 10.19 by Corollary 1.2.5 in [270].

PROPOSITION 10.25. *On $D(\dot{A})$ the norm inherited from that of $H^2(\mathcal{M}) \oplus H^1(\mathcal{M})$ is equivalent to the norm $U \mapsto \|\dot{A}U\|_{\dot{\mathscr{H}}} \eqsim \|\dot{A}U\|_{\mathscr{H}}$.*

PROOF. Let $U = {}^t(u^0, u^1) \in D(\dot{A}) \subset H^2(\mathcal{M}) \oplus H^1(\mathcal{M})$. We have $\dot{A}U = {}^t(-u^1, P_0 u^0) \in \dot{\mathscr{H}}$ by Lemma 10.15 and as $A = \dot{A} \circ \Pi_{\dot{\mathscr{H}}}$; see the commutative diagram above (10.A.10). Since the norm $\|.\|_{\dot{\mathscr{H}}}$ is equivalent to the norm $\|.\|_{\mathscr{H}}$ on $\dot{\mathscr{H}}$ by Lemma 10.16 we have

$$\|\dot{A}U\|_{\dot{\mathscr{H}}} \eqsim \|\dot{A}U\|_{\mathscr{H}} \eqsim \|u^1\|_{H^1(\mathcal{M})} + \|P_0 u^0\|_{L^2(\mathcal{M})}.$$

As we have $\|P_0 u^0\|_{L^2(\mathcal{M})} \lesssim \|u^0\|_{H^2(\mathcal{M})}$ by Proposition 18.12, to conclude, it remains to prove that we have

$$\|u^0\|_{H^2(\mathcal{M})} \lesssim \|u^1\|_{H^1(\mathcal{M})} + \|P_0 u^0\|_{L^2(\mathcal{M})}. \tag{10.A.25}$$

Since $B_\partial(U) = 0$ we have

$$\partial_\nu u^0{}_{|\partial\mathcal{M}} = -\alpha u^1{}_{|\partial\mathcal{M}} \in H^{1/2}(\partial\mathcal{M}), \quad \text{with } \|\alpha u^1{}_{|\partial\mathcal{M}}\|_{H^{1/2}(\partial\mathcal{M})} \lesssim \|u^1\|_{H^1(\mathcal{M})},$$

by the trace inequality of Proposition 18.24. With the divergence formula of Proposition 18.28, we have

$$\int_{\mathcal{M}} P_0 u^0 \mu_g + \int_{\partial\mathcal{M}} \partial_\nu u^0{}_{|\partial\mathcal{M}}\, \mu_{g_\partial} = 0.$$

Then, by the elliptic regularity result of Proposition 18.48 and Corollary 18.49 for the Neumann problem we obtain

$$\|\underline{u}^0\|_{H^2(\mathcal{M})} \lesssim \|P_0 u^0\|_{L^2(\mathcal{M})} + \|u^1\|_{H^1(\mathcal{M})}, \tag{10.A.26}$$

where $\underline{u}^0 = u^0 - \fint_{\mathcal{M}} u^0 \mu_g$. We thus have

$$\|u^0\|_{H^2(\mathcal{M})} \lesssim \|\underline{u}^0\|_{H^2(\mathcal{M})} + \Big|\fint_{\mathcal{M}} u^0 \mu_g\Big| \operatorname{Vol_g}(\mathcal{M})^{1/2}. \tag{10.A.27}$$

We set $\underline{U} = {}^t(\underline{u}^0, u^1) = U - (\fint_{\mathcal{M}} u^0 \mu_g)\Theta \in \mathscr{H}$. Since $U \in D(\dot{A})$ one has

$$\Pi_{\mathcal{N}} \underline{U} = -\Big(\fint_{\mathcal{M}} u^0 \mu_g\Big)\Pi_{\mathcal{N}}\Theta = -\Big(\fint_{\mathcal{M}} u^0 \mu_g\Big)\Theta.$$

From the continuity of $\Pi_{\mathcal{N}}$ we obtain

$$\Big|\fint_{\mathcal{M}} u^0 \mu_g\Big| \lesssim \|\underline{U}\|_{\mathscr{H}} \lesssim \|\underline{u}^0\|_{H^1(\mathcal{M})} + \|u^1\|_{L^2(\mathcal{M})}.$$

With (10.A.26) and (10.A.27), we thus obtain (10.A.25), which concludes the proof. ■

COROLLARY 10.26. *On $D(A)$ the norm inherited from that of $H^2(\mathcal{M}) \oplus H^1(\mathcal{M})$ is equivalent to the graph norm $\|.\|_{D(A)}$ given by $\|V\|^2_{D(A)} = \|V\|^2_{\mathscr{H}} + \|AV\|^2_{\mathscr{H}}$.*

PROOF. Let $U = {}^t(u^0, u^1) \in D(A)$. Using the projectors $\Pi_{\dot{\mathscr{H}}}$ and $\Pi_{\mathcal{N}}$ defined above (10.A.6), by (10.A.9) we have

$$U = \Pi_{\dot{\mathscr{H}}} U + \Pi_{\mathcal{N}} U,$$

with $\Pi_{\dot{\mathscr{H}}} U \in D(\dot{A})$ and $\Pi_{\mathcal{N}} U \in \mathcal{N} = \ker(A)$, using Lemma 10.15. By Proposition 10.25 we obtain

$$\begin{aligned}\|U\|_{H^2(\mathcal{M})\oplus H^1(\mathcal{M})} &\lesssim \|\Pi_{\dot{\mathscr{H}}} U\|_{H^2(\mathcal{M})\oplus H^1(\mathcal{M})} + \|\Pi_{\mathcal{N}} U\|_{H^2(\mathcal{M})\oplus H^1(\mathcal{M})} \\ &\lesssim \|\dot{A} \circ \Pi_{\dot{\mathscr{H}}} U\|_{\dot{\mathscr{H}}} + \|\Pi_{\mathcal{N}} U\|_{H^2(\mathcal{M})\oplus H^1(\mathcal{M})}.\end{aligned}$$

As $AU = A \circ \Pi_{\dot{\mathscr{H}}} U = \dot{A} \circ \Pi_{\dot{\mathscr{H}}} U$ we thus have

$$\|U\|_{H^2(\mathcal{M})\oplus H^1(\mathcal{M})} \lesssim \|AU\|_{\mathscr{H}} + \|\Pi_{\mathcal{N}} U\|_{H^2(\mathcal{M})\oplus H^1(\mathcal{M})}.$$

Since $\dim \mathcal{N} = 1$ one has

$$\|\Pi_{\mathcal{N}} U\|_{H^2(\mathcal{M})\oplus H^1(\mathcal{M})} \lesssim \|\Pi_{\mathcal{N}} U\|_{\mathscr{H}} \lesssim \|U\|_{\mathscr{H}},$$

by the continuity of $\Pi_{\mathcal{N}}$ on $\mathscr{H}$. We have thus obtained

$$\|U\|_{H^2(\mathcal{M})\oplus H^1(\mathcal{M})} \lesssim \|AU\|_{\mathscr{H}} + \|U\|_{\mathscr{H}}.$$

As the opposite inequality is clear, the proof is complete. ■

The study of the stabilization properties of the damped wave equation requires to prove that the imaginary axis $\{\operatorname{Re} z = 0\}$ is in the resolvent set of the operator $\dot{A}$. This is the subject of the following proposition.

PROPOSITION 10.27. *The spectrum of $(\dot{A}, D(\dot{A}))$ is contained in $\{z \in \mathbb{C}; \operatorname{Re}(z) > 0\}$.*

PROOF. Let $z \in \mathbb{C}$. We consider the two cases.

Case 1: $\operatorname{Re} z < 0$. By Lemma 10.21 $z \operatorname{Id}_{\dot{\mathscr{H}}} - \dot{A}$ is injective. Moreover, as its adjoint $\bar{z} \operatorname{Id}_{\dot{\mathscr{H}}} - \dot{A}^*$ is injective and satisfies $\|(\bar{z} \operatorname{Id}_{\dot{\mathscr{H}}} - \dot{A}^*)U\|_{\dot{\mathscr{H}}} \gtrsim \|U\|_{\dot{\mathscr{H}}}$ for $U \in D(\dot{A})$ by Lemma 10.21, using $B = \dot{A}^*$ therein, the map $z \operatorname{Id}_{\dot{\mathscr{H}}} - \dot{A}$ is surjective (see e.g. [90, Theorem 2.20]). The estimation of Lemma 10.21, for $B = \dot{A}$, then gives the continuity of the operator $(z \operatorname{Id}_{\dot{\mathscr{H}}} - \dot{A})^{-1}$ on $\dot{\mathscr{H}}$.

Case 2: $\operatorname{Re} z = 0$. We start by proving the injectivity of $z \operatorname{Id}_{\dot{\mathscr{H}}} - \dot{A}$. Let thus $U = {}^t(u^0, u^1) \in D(\dot{A})$ be such that $zU - \dot{A}U = 0$. This gives

$$zu^0 + u^1 = 0, \quad -P_0 u^0 + zu^1 = 0. \tag{10.A.28}$$

First, if $z = 0$ one has $u^1 = 0$, $P_0 u^0 = 0$, and $\partial_\nu u^0{}_{|\partial\mathcal{M}} = 0$ as $B_\partial(U) = 0$. Thus $u^0 \equiv \mathrm{Cst}$ by Corollary 18.49. As moreover $F_\alpha(U) = 0$, we find $\int_{\partial\mathcal{M}} \alpha u^0 \mu_{g_\partial} = 0$ meaning that $u^0 = 0$ as $\int_{\partial\mathcal{M}} \alpha \mu_{g_\partial} > 0$.

Second, if now $z \neq 0$, using (10.A.23) we obtain

$$0 = \operatorname{Re}((z \operatorname{Id}_{\dot{\mathscr{H}}} - \dot{A})U, U)_{\dot{\mathscr{H}}} = - \int_{\partial\mathcal{M}} \alpha |u^1|^2_{|\partial\mathcal{M}} \, \mu_{g_\partial}.$$

As $\alpha |u^1|^2_{|\partial\mathcal{M}} \geq 0$, this implies that $\alpha |u^1|^2_{|\partial\mathcal{M}}$ vanishes a.e. on $\partial\mathcal{M}$. Then, on the one hand $u^0{}_{|\partial\mathcal{M}}$ vanishes a.e. on $\operatorname{supp}(\alpha)$ by computing the trace of the first equality in (10.A.28) and, on the other hand, $\partial_\nu u^0{}_{|\partial\mathcal{M}} = 0$ as $B_\partial(U) = 0$. We observe that we have

$$P_0 u^0 = zu^1 = -z^2 u^0.$$

With the unique continuation property, initiated from the boundary, as stated in Theorem 9.6, we obtain that u^0 vanishes in $\mathcal{M}$ and u^1 as well.

If we now prove that $z \operatorname{Id}_{\dot{\mathscr{H}}} - \dot{A}$ is surjective, the result then follows from the closed graph theorem as $\dot{A}$ is a closed operator. We write $z \operatorname{Id}_{\dot{\mathscr{H}}} - \dot{A} = T + \operatorname{Id}_{\dot{\mathscr{H}}}$ with $T = (z-1) \operatorname{Id}_{\dot{\mathscr{H}}} - \dot{A}$. By the first part of the proof, T is invertible with a bounded inverse. The operator T is unbounded on $\dot{\mathscr{H}}$. We denote by $\tilde{T}$ the restriction of T to $D(\dot{A})$ equipped with the graph norm associated with $\dot{A}$. The operator $\tilde{T}$ is bounded by Proposition 10.25. It is also invertible. It is thus a bounded Fredholm operator of index $\operatorname{ind} \tilde{T} = 0$ (see Definition 11.6 of Volume 1). Similarly, we denote by ι the injection of $D(\dot{A})$ into $\dot{\mathscr{H}}$ and $\tilde{A}$ the restriction of $\dot{A}$ on $D(\dot{A})$ viewed as a bounded operator. We have $z\iota - \tilde{A} = \tilde{T} + \iota$. Since ι is a compact operator by Proposition 10.17, we obtain that $z\iota - \tilde{A}$ is also a bounded Fredholm operator of index 0 by Theorem 11.15 of Volume 1. Hence, $z\iota - \tilde{A}$ is surjective since $z \operatorname{Id}_{\dot{\mathscr{H}}} - \dot{A}$ is injective as proven above. Consequently, $z \operatorname{Id}_{\dot{\mathscr{H}}} - \dot{A}$ is surjective. ■

CHAPTER 11

Stabilization of Waves Under General Boundary Damping

Contents

11.1. Setting

For $(\mathcal{M}, g)$ a smooth d-dimensional connected compact Riemannian manifold with boundary, with $d \geq 2$, we consider a second-order differential operator P on $\mathcal{M}$ given by $P = -\Delta_g + R_1$, where R_1 is a first-order differential operator. We also consider a differential boundary operator B defined in a neighborhood of $\partial\mathcal{M}$ of order 0 or 1 on each connected component of $\partial\mathcal{M}$. On the part of $\partial\mathcal{M}$ where the order is 0, denoted by ${}^0\partial\mathcal{M}$, we consider the Dirichlet boundary operator, that is, $Bu_{|{}^0\partial\mathcal{M}} = u_{|{}^0\partial\mathcal{M}}$. We assume that the part of $\partial\mathcal{M}$ where the order is 1, denoted by ${}^1\partial\mathcal{M}$, is nonempty, meaning that there exists a connected component of $\partial\mathcal{M}$ where B is a genuine first-order operator.

J. Le Rousseau et al., *Elliptic Carleman Estimates and Applications to Stabilization and Controllability, Volume II*, PNLDE Subseries in Control 98, https://doi.org/10.1007/978-3-030-88670-7_11

We introduce the unbounded operator P on $L^2(\mathcal{M})$ with domain

$$D(\mathsf{P}) = \{u \in L^2(\mathcal{M});\ Pu \in L^2(\mathcal{M}) \text{ and } Bu_{|\partial\mathcal{M}} = 0\}.$$

To ensure Fredholm properties, we require that (P, B) satisfies the Lopatinskiĭ–Šapiro conditions; see Theorem 3.1. Consequently, as observed in the beginning of Chap. 4, one has

$$D(\mathsf{P}) = \{u \in H^2(\mathcal{M});\ Bu_{|\partial\mathcal{M}} = 0\}.$$

We also require the operator to be selfadjoint; by Theorem 4.3, this implies that R_1 and B take the following forms:

(11.1.1)
$$R_1 = iV + f \text{ in } \mathcal{M} \quad \text{and} \quad B = \partial_\nu + B' \text{ with } B' = iX' + h \text{ in } {}^1\partial\mathcal{M},$$

where we denote by ∂_ν the outward pointing normal derivative in the sense of the metric g (see (5.3.1) in Sect. 5.3 and (17.4.12) in Sect. 17.4), V a smooth real vector field on $\mathcal{M}$, f a complex valued function on $\mathcal{M}$, X' a smooth real vector field on ${}^1\partial\mathcal{M}$, and h a complex valued function on $\partial\mathcal{M}$, with the additional properties

(11.1.2)
$$\operatorname{Im} f = \operatorname{div}_g V/2 \text{ in } \mathcal{M} \quad \text{and} \quad 2\operatorname{Im} h - \operatorname{div}_g X' + g(V,\nu)_{|{}^1\partial\mathcal{M}} = 0 \text{ in } {}^1\partial\mathcal{M}.$$

Moreover, in the case $d = 2$, one has $|X'|_{g_\partial} \neq 1$, and in the case $d \geq 3$, we moreover have $|X'|_{g_\partial} < 1$.

In the present chapter, we shall furthermore require $|X'|_{g_\partial} < 1$ independently of the dimension. Consequently, the spectrum of P is composed of a sequence of eigenvalues of finite multiplicities that grows to $+\infty$; see Theorems 4.9 and 4.12. If one does not assume $|X'|_{g_\partial} < 1$, in dimension $d = 2$, the case of eigenvalues that go to $-\infty$ can occur (see Proposition 4.13). This would imply that the associated wave equation is not well posed, having its "generator" with unbounded positive eigenvalues.

Finally, if we denote by $\mu_0 \leq \mu_1 \leq \cdots \leq \mu_n \leq \cdots$ the real eigenvalues of the selfadjoint operator $(\mathsf{P}, D(\mathsf{P}))$, we moreover assume that $\mu_0 \geq 0$.

With a smooth function $\alpha \geq 0$ given on ${}^1\partial\mathcal{M}$, we consider the following wave equation:

$$\begin{cases} \partial_t^2 y + Py = 0 & \text{in } (0,+\infty) \times \mathcal{M}, \\ y = 0 & \text{on } (0,+\infty) \times {}^0\partial\mathcal{M}, \\ By + \alpha\partial_t y = 0 & \text{on } (0,+\infty) \times {}^1\partial\mathcal{M}, \\ y_{|t=0} = y^0,\ \partial_t y_{|t=0} = y^1 & \text{in } \mathcal{M}. \end{cases} \tag{11.1.3}$$

We assume that $\alpha > 0$ on a nonempty open subset $\Gamma \subset {}^1\partial\mathcal{M}$. Below we define a proper energy for solutions of this equation, and we prove that the energy decays to *zero* as $t \to +\infty$. The present chapter thus generalizes the result of Chap. 10 obtained in the case of a Neumann boundary damping.

The content of the present chapter is thus very close and adapted from that of Chap. 10.

11.2. Strong Solutions and Energy

We define the boundary operator B_∂ by

$$(11.2.1) \qquad \begin{aligned} B_\partial : H^2(\mathcal{M}) \times H^1(\mathcal{M}) &\to H^{1/2}({}^1\partial\mathcal{M}) \\ (v^0, v^1) &\mapsto B v^0{}_{|{}^1\partial\mathcal{M}} + \alpha v^1{}_{|{}^1\partial\mathcal{M}}. \end{aligned}$$

This operator is well defined by the trace formula of Proposition 18.24.

We recall the notation

$$H^1_{\mathsf{D}}(\mathcal{M}) = \{u \in H^1(\mathcal{M});\ u_{|{}^0\partial\mathcal{M}} = 0\},$$

and we introduce the space $H^2_{\mathsf{D}}(\mathcal{M}) = H^2(\mathcal{M}) \cap H^1_{\mathsf{D}}(\mathcal{M})$. On $H^1_{\mathsf{D}}(\mathcal{M})$, we use the usual H^1-inner product and H^1-norm. On $H^2_{\mathsf{D}}(\mathcal{M})$, we use the usual H^2-inner product and H^2-norm.

We have the following existence and uniqueness result.

THEOREM 11.1 (Strong Solutions). *For $(y^0, y^1) \in H^2_{\mathsf{D}}(\mathcal{M}) \times H^1_{\mathsf{D}}(\mathcal{M})$ such that $B_\partial(y^0, y^1) = 0$, there exists a unique*

$$\begin{aligned} y \in \mathscr{C}^2\big([0,+\infty); L^2(\mathcal{M})\big) \cap \mathscr{C}^1\big([0,+\infty); \\ H^1_{\mathsf{D}}(\mathcal{M})\big) \cap \mathscr{C}^0\big([0,+\infty); H^2_{\mathsf{D}}(\mathcal{M})\big) \end{aligned}$$

such that

(11.2.2)

$$\partial_t^2 y + P y = 0 \quad \text{in } L^\infty([0,+\infty); L^2(\mathcal{M})), \qquad y_{|t=0} = y^0,\ \partial_t y_{|t=0} = y^1,$$

and

$$(11.2.3) \qquad B_\partial(y(t), \partial_t y(t)) = 0 \text{ in } L^\infty\big([0,+\infty); H^{1/2}({}^1\partial\mathcal{M})\big).$$

Moreover, there exists $C > 0$ such that

(11.2.4)

$$\|y(t)\|_{H^2(\mathcal{M})} + \|\partial_t y(t)\|_{H^1(\mathcal{M})} \leq C\big(\|y^0\|_{H^2(\mathcal{M})} + \|y^1\|_{H^1(\mathcal{M})}\big), \qquad t \geq 0.$$

Solutions given in Theorem 11.1 are called *strong* solutions of the boundary-damped wave equation (11.1.3).

To prove the results of Theorem 11.1 it is convenient to cast the boundary-damped wave equation into a semigroup formalism. Theorem 11.1 then follows from Theorem 11.2 and Proposition 11.3 below similarly to what is done in Chapter 6 of Volume 1 and Chap. 10 (see, for instance, the proof of Theorem 10.1).

We refer to Chapter 12 for some elements of semigroup theory. We set

$$(11.2.5) \qquad A = \begin{pmatrix} 0 & -1 \\ P & 0 \end{pmatrix}, \qquad Y(t) = \begin{pmatrix} y(t) \\ \partial_t y(t) \end{pmatrix}, \qquad Y^0 = \begin{pmatrix} y^0 \\ y^1 \end{pmatrix},$$

and we consider the Hilbert sum

$$\mathscr{H} = H^1_{\mathsf{D}}(\mathcal{M}) \oplus L^2(\mathcal{M}),$$

naturally endowed with the inner product

$$(U, U')_{\mathscr{H}} = (u^0, u'^0)_{H^1(\mathcal{M})} + (u^1, u'^1)_{L^2(\mathcal{M})}, \quad U = (u^0, u^1), \ U' = (u'^0, u'^1),$$

and norm

$$\|U\|^2_{\mathscr{H}} = \|u^0\|^2_{H^1(\mathcal{M})} + \|u^1\|^2_{L^2(\mathcal{M})}, \quad U = (u^0, u^1).$$

If y is a strong solution to (11.2.2) and (11.2.3), we have

$$Y(t) \in \mathscr{C}^0\big([0, +\infty); H^2_{\mathsf{D}}(\mathcal{M}) \oplus H^1_{\mathsf{D}}(\mathcal{M})\big) \cap \mathscr{C}^1\big([0, +\infty); \mathscr{H}\big), \tag{11.2.6}$$

and

$$\frac{d}{dt} Y(t) + AY(t) = 0 \text{ for } t \geq 0 \ \text{ and } Y_{|t=0} = Y^0, \tag{11.2.7}$$

with moreover $B_\partial(Y(t)) = 0$ for $t \geq 0$.

We define the unbounded operator A given by (11.2.5) on $\mathscr{H}$ with domain

$$D(A) = \big\{V = {}^t(v^0, v^1) \in H^2_{\mathsf{D}}(\mathcal{M}) \times H^1_{\mathsf{D}}(\mathcal{M});\ B_\partial(V) = 0\big\}.$$

By Proposition 11.10 and Corollary 11.14 in Appendix 11.A, the operator $(A, D(A))$ is closed on $\mathscr{H}$, and moreover $D(A)$ is dense in $\mathscr{H}$.

We endow $D(A)$ with the graph norm $\|V\|^2_{D(A)} = \|V\|^2_{\mathscr{H}} + \|AV\|^2_{\mathscr{H}}$ for $V \in D(A)$. In fact, this norm is equivalent to the norm on $D(A)$ inherited from that of $H^2_{\mathsf{D}}(\mathcal{M}) \oplus H^1_{\mathsf{D}}(\mathcal{M})$ by Corollary 11.20.

THEOREM 11.2. *The unbounded operator* $(A, D(A))$ *generates a bounded* C_0*-semigroup* $S(t)$ *on* $\mathscr{H}$.

This result is proven in Appendix 11.A; see Theorem 11.18.

PROPOSITION 11.3. *Let* $Y^0 = {}^t(y^0, y^1) \in D(A)$ *and set* $Y(t) = S(t)Y^0$. *Then,* $\frac{d}{dt}Y(t) + AY(t) = 0$, *for* $t \geq 0$. *If* $Y(t) = {}^t(y(t), z(t))$, *then*

$$y \in \mathscr{C}^2\big([0, +\infty); L^2(\mathcal{M})\big) \cap \mathscr{C}^1\big([0, +\infty); H^1_{\mathsf{D}}(\mathcal{M})\big) \cap \mathscr{C}^0\big([0, +\infty); H^2_{\mathsf{D}}(\mathcal{M})\big),$$

$z(t) = \partial_t y(t)$, *and* y *is the unique strong solution* (11.2.2) *and* (11.2.3).

Note that we obtain $\frac{d}{dt}Y(t) + AY(t) = 0$, for $t \geq 0$, by applying Proposition 12.2 of Volume 1. The proof of Proposition 11.3 is actually contained in that of Theorem 11.1. Using the uniqueness part of Theorem 11.1, Proposition 11.3 shows that the solution of (11.2.2) and (11.2.3) is simply given by the first component of $Y(t) = S(t)Y^0$.

To analyze stabilization, we need to introduce a proper energy. We set

$$\mathcal{E}(y)(t) = E\big(y(t), \partial_t y(t)\big) = \frac{1}{2}\Big(\|\partial_t y(t)\|^2_{L^2(\mathcal{M})} + N(y(t))\Big), \tag{11.2.8}$$

with $N(.)$ given by

$$
\begin{aligned}
(11.2.9)\quad N(u) &= \|\nabla_g u\|^2_{L^2V(\mathcal{M})} \\
&\quad + (R_1 u, u)_{L^2(\mathcal{M})} + \langle B'\gamma^{\mathsf{D}}(u), \gamma^{\mathsf{D}}(\bar{u})\rangle_{H^{-1/2}({}^1\partial\mathcal{M}),H^{1/2}({}^1\partial\mathcal{M})},
\end{aligned}
$$

as in (4.6.6). This is motivated by the density of $D(\mathsf{P})$ in $H^1_{\mathsf{D}}(\mathcal{M})$ and the fact that $N(u) = (Pu, u)_{L^2(\mathcal{M})} \geq 0$ if $u \in D(\mathsf{P})$; hence $N(u) \geq 0$ on $H^1_{\mathsf{D}}(\mathcal{M})$. Note that this definition of the energy coincides with that given in a Neumann boundary damping in Chap. 10; see (10.2.9).

Considering $y(t)$ a strong solution and observing that $t \mapsto \mathcal{E}(y)(t)$ is differentiable, we find

$$
\frac{d}{dt}\mathcal{E}(y)(t) = \operatorname{Re}(\partial_t^2 y(t), \partial_t y(t))_{L^2(\mathcal{M})} + \operatorname{Re}\tilde{N}\big(\partial_t y(t), y(t)\big),
$$

with $\tilde{N}(.,.)$ defined in (4.6.5). We then compute

$$
\begin{aligned}
\operatorname{Re}\tilde{N}\big(\partial_t y(t), y(t)\big) &= (\nabla_g \partial_t y(t), \nabla_g y(t))_{L^2V(\mathcal{M})} \\
&\quad + (R_1 \partial_t y(t), y(t))_{L^2(\mathcal{M})} \\
&\quad + \langle B'\partial_t y(t)_{|{}^1\partial\mathcal{M}}, \bar{y}(t)_{|{}^1\partial\mathcal{M}}\rangle_{H^{-1/2}({}^1\partial\mathcal{M}),H^{1/2}({}^1\partial\mathcal{M})} \\
&= (\partial_t y(t), -\Delta_g y(t))_{L^2(\mathcal{M})} \\
&\quad + (\partial_t y(t)_{|{}^1\partial\mathcal{M}}, \partial_\nu y(t)_{|{}^1\partial\mathcal{M}})_{L^2({}^1\partial\mathcal{M})} \\
&\quad + (\partial_t y(t), R_1 y(t))_{L^2(\mathcal{M})} \\
&\quad + \big(\partial_t y(t)_{|{}^1\partial\mathcal{M}}, B' y(t)_{|{}^1\partial\mathcal{M}}\big)_{L^2({}^1\partial\mathcal{M})},
\end{aligned}
$$

with Lemma 4.7. We then obtain

$$
\frac{d}{dt}\mathcal{E}(y)(t) = -|\alpha^{1/2}\partial_t y(t)_{|{}^1\partial\mathcal{M}}|^2_{L^2({}^1\partial\mathcal{M})},
$$

using that $\partial_\nu y(t) + B'y(t) + \alpha\partial_t y(t) = 0$ on ${}^1\partial\mathcal{M}$. This implies the decay of the energy $\mathcal{E}(y)(t)$, viz.,

$$
(11.2.10)\qquad 0 \leq \mathcal{E}(y)(t') \leq \mathcal{E}(y)(t), \qquad \text{for } 0 \leq t \leq t'.
$$

11.3. Reduced Functional Space and Generator

Set $\mathcal{N} = \ker(A)$, and observe that $U \in \mathcal{N}$ if and only if $U = {}^t(\varphi, 0)$ with $\varphi \in \ker(\mathsf{P})$, that is, $P\varphi = 0$, $u_{|{}^0\partial\mathcal{M}} = 0$, and $B\varphi_{|{}^1\partial\mathcal{M}} = 0$.

If $\varphi \in \ker(\mathsf{P})$ and if $(\alpha\varphi_{|{}^1\partial\mathcal{M}}, \varphi_{|{}^1\partial\mathcal{M}})_{L^2({}^1\partial\mathcal{M})} = 0$, then, as $\alpha > 0$ on some open set Γ of ${}^1\partial\mathcal{M}$, one finds

$$
\varphi_{|\Gamma} = 0 \quad \text{and} \quad \partial_\nu\varphi_{|\Gamma} = 0,
$$

using the form of B on ${}^1\partial\mathcal{M}$. Then, the unique continuation result of Theorem 9.8 applies yielding $\varphi = 0$ in $\mathcal{M}$. Hence, we have the following result.

LEMMA 11.4. *The bilinear map* $(\varphi, \psi) \mapsto (\alpha\varphi_{|{}^1\partial\mathcal{M}}, \psi_{|{}^1\partial\mathcal{M}})_{L^2({}^1\partial\mathcal{M})}$ *is an inner product on the finite dimensional space* $\ker(\mathsf{P})$.

We now consider an orthonormal basis $\{\varphi_0, \dots, \varphi_{n-1}\}$ of $\ker(\mathsf{P})$ associated with this inner product.

For $\varphi \in \ker(\mathsf{P})$, $\varphi \neq 0$, we introduce the linear form

$$
\begin{aligned}
(11.3.1) \qquad & F_{\alpha,\varphi} : \mathscr{H} \to \mathbb{C} \\
& \begin{pmatrix} v^0 \\ v^1 \end{pmatrix} \mapsto (\alpha\varphi_{|^1\partial\mathcal{M}}, \varphi_{|^1\partial\mathcal{M}})^{-1}_{L^2(^1\partial\mathcal{M})} \\
& \qquad \big((\alpha v^0_{|^1\partial\mathcal{M}}, \varphi_{|^1\partial\mathcal{M}})_{L^2(^1\partial\mathcal{M})} + (v^1, \varphi)_{L^2(\mathcal{M})}\big),
\end{aligned}
$$

and we set $\Theta_\varphi = {}^t(\varphi, 0) \in \ker(A)$. Note that $F_{\alpha,\varphi}$ is a bounded map. We then set

$$
\dot{\mathscr{H}} = \bigcap_{\varphi \in \ker(\mathsf{P})} \ker(F_{\alpha,\varphi}) = \bigcap_{j=0}^{n-1} \ker(F_{\alpha,\varphi_j}).
$$

Note that for $j = 0, \dots, n-1$, one simply has $F_{\alpha,\varphi_j}(V) = (\alpha v^0_{|^1\partial\mathcal{M}}, \varphi_{|^1\partial\mathcal{M}})_{L^2(^1\partial\mathcal{M})} + (v^1, \varphi)_{L^2(\mathcal{M})}$. Setting $\Theta_j = {}^t(\varphi_j, 0)$, $j = 0, \dots, n-1$, one has $F_{\alpha,\varphi_j}(\Theta_j) = 1$. We define

$$
\Pi_{\mathcal{N}} V = \sum_{j=0}^{n-1} F_{\alpha,\varphi_j}(V)\Theta_j, \qquad V \in \mathscr{H},
$$

and $\Pi_{\dot{\mathscr{H}}} = \operatorname{Id}_{\mathscr{H}} - \Pi_{\mathcal{N}}$, and we obtain that $\Pi_{\mathcal{N}}$ and $\Pi_{\dot{\mathscr{H}}}$ are *continuous* projectors associated with the direct sum

$$
(11.3.2) \qquad \mathscr{H} = \dot{\mathscr{H}} \oplus \mathcal{N} \quad \text{and} \quad \dot{\mathscr{H}} = \ker(\Pi_{\mathcal{N}}).
$$

Note that $\dot{\mathscr{H}}$ and $\mathcal{N}$ are *not orthogonal* in $\mathscr{H}$.

LEMMA 11.5. *We have* $\operatorname{Ran}(A) \subset \dot{\mathscr{H}}$.

PROOF. Let $U = {}^t(u^0, u^1) = AV$ with $V = {}^t(v^0, v^1) \in D(A)$. One has $u^0 = -v^1 \in H^1_{\mathsf{D}}(\mathcal{M})$ and $u^1 = Pv^0 \in L^2(\mathcal{M})$. If $\varphi \in \ker(\mathsf{P})$, one computes

$$
\begin{aligned}
(\alpha\varphi_{|^1\partial\mathcal{M}}, \varphi_{|^1\partial\mathcal{M}})_{L^2(^1\partial\mathcal{M})} F_{\alpha,\varphi}(U) &= -(\alpha v^1_{|^1\partial\mathcal{M}}, \varphi_{|^1\partial\mathcal{M}})_{L^2(^1\partial\mathcal{M})} \\
&\quad + (Pv^0, \varphi)_{L^2(\mathcal{M})} \\
&= -(\alpha v^1_{|^1\partial\mathcal{M}}, \varphi_{|^1\partial\mathcal{M}})_{L^2(^1\partial\mathcal{M})} \\
&\quad - (Bv^1_{|^1\partial\mathcal{M}}, \varphi_{|^1\partial\mathcal{M}})_{L^2(^1\partial\mathcal{M})} \\
&= 0,
\end{aligned}
$$

applying the Green formula (4.5.3), using that $B_\partial V = 0$. ■

The space $\dot{\mathscr{H}}$ inherits the norm of $\mathscr{H} = H^1_{\mathsf{D}}(\mathcal{M}) \oplus L^2(\mathcal{M})$. By Lemma 11.11, the norm $\|.\|_{\dot{\mathscr{H}}}$ defined by

$$
\|V\|^2_{\dot{\mathscr{H}}} := N(v^0) + \|v^1\|^2_{L^2(\mathcal{M})}, \quad V = {}^t(v^0, v^1),
$$

is equivalent to $\|.\|_{\mathscr{H}}$ on $\dot{\mathscr{H}}$. This latter norm is more adapted to our purpose here.

The energy of a strong solution is given in Sect. 11.2 by $\mathcal{E}(y)(t) = E\big(Y(t)\big)$ with

$$E(U) = \frac{1}{2}\Big(N(u^0) + \|u^1\|^2_{L^2(\mathcal{M})}\Big), \qquad U = {}^t(u^0, u^1), \tag{11.3.3}$$

and $Y = {}^t(y, \partial_t y)$. We note that

$$E(U) = E(\Pi_{\dot{\mathscr{H}}} U) = \frac{1}{2}\|\Pi_{\dot{\mathscr{H}}} U\|^2_{\dot{\mathscr{H}}}. \tag{11.3.4}$$

In fact, we have $\Pi_{\dot{\mathscr{H}}} U = U + \Theta_\varphi$ for some $\varphi \in \ker(\mathsf{P})$. We then have

$$2E(U + \Theta_\varphi) = N(u^0 + \varphi) + \|u^1\|^2_{L^2(\mathcal{M})},$$

and

$$N(u^0 + \varphi) = N(u^0) + N(\varphi) + 2\operatorname{Re}\tilde{N}(\varphi, u^0).$$

By (4.6.7), we have $N(\varphi) = (P\varphi, \varphi) = 0$ and $\tilde{N}(\varphi, u^0) = (P\varphi, u^0) = 0$, yielding (11.3.4).

Below we shall consider the semigroup generated by the restriction of A on the reduced space $\dot{\mathscr{H}}$. A proper norm is then $\|\Pi_{\dot{\mathscr{H}}} U\|_{\dot{\mathscr{H}}}$ and (11.3.4) shows that the chosen energy of the associated solution of the damped wave equation is precisely the norm in the space where evolution occurs.

Following Chap. 10, we introduce the unbounded operator $\dot{A}$ on $\dot{\mathscr{H}}$ given by the following domain:

$$\begin{aligned} D(\dot{A}) &= D(A) \cap \dot{\mathscr{H}} \\ &= \big\{V \in H^2_{\mathsf{D}}(\mathcal{M}) \times H^1_{\mathsf{D}}(\mathcal{M}); \\ &\qquad\qquad B_\partial(V) = 0,\ F_{\alpha,\varphi_j}(V) = 0,\ j = 0, \dots, n-1\big\}, \end{aligned}$$

and such that $\dot{A}V = AV$ for $V \in D(\dot{A})$. We then have $A = \dot{A} \circ \Pi_{\dot{\mathscr{H}}}$. We say that $(\dot{A}, D(\dot{A}))$ is the reduced generator of the damped wave equation we consider here.

Observe that $D(\dot{A}) = \Pi_{\dot{\mathscr{H}}}\big(D(A)\big)$ since $\mathcal{N} = \ker(A) \subset D(A)$. Thus, one has

$$D(A) = D(\dot{A}) \oplus \mathcal{N}. \tag{11.3.5}$$

With (11.3.2) and (11.3.5), one has

$$\Pi_{\dot{\mathscr{H}}} V \in D(\dot{A}) \quad \Leftrightarrow \quad V \in D(A), \qquad \text{for } V \in \mathscr{H}. \tag{11.3.6}$$

The unbounded operator $(\dot{A}, D(\dot{A}))$ generates a C_0-semigroup of contraction on $\dot{\mathscr{H}}$ by Theorem 11.17. We denote this semigroup $\dot{S}(t)$, and we have

$$S(t) = \dot{S}(t) \circ \Pi_{\dot{\mathscr{H}}} + \Pi_{\mathcal{N}},$$

where $S(t)$ is the bounded C_0-semigroup generated by $(A, D(A))$.

If $Y^0 \in D(A)$, the solution of the semigroup equation $\frac{d}{dt}Y(t)+AY(t)=0$ reads

$$Y(t) = S(t)Y^0 = \dot{S}(t) \circ \Pi_{\dot{\mathscr{H}}} Y^0 + \Pi_{\mathcal{N}} Y^0.$$

We set $\dot{Y}(t) = \Pi_{\dot{\mathscr{H}}} Y(t) = \dot{S}(t) \circ \Pi_{\dot{\mathscr{H}}} Y^0$. By (11.3.4), we find that the energy of $Y(t)$ is precisely that of $\dot{Y}(t)$:

$$E\big(Y(t)\big) = E\big(\Pi_{\dot{\mathscr{H}}} Y(t)\big) = E\big(\dot{Y}(t)\big) = \frac{1}{2}\|\dot{Y}(t)\|^2_{\dot{\mathscr{H}}}. \tag{11.3.7}$$

The stabilization properties of the semigroup $S(t)$ are thus analyzed by that of the semigroup $\dot{S}(t)$ in the next section.

11.4. Resolvent Estimate and Stabilization Result

For the reduced generator $(\dot{A}, D(\dot{A}))$ of the boundary-damped wave equation, we have the following resolvent estimate.

PROPOSITION 11.6. *Let Γ be a nonempty open subset of ${}^1\partial\mathcal{M}$, and let $\alpha \in \mathscr{C}^\infty({}^1\partial\mathcal{M})$ be such that $\alpha > 0$ on Γ. Then, the unbounded operator $i\sigma \operatorname{Id}_{\dot{\mathscr{H}}} - \dot{A}$ is invertible on $\dot{\mathscr{H}}$ for all $\sigma \in \mathbb{R}$, and there exist $K > 0$ and $\sigma_0 > 0$ such that*

$$\|(i\sigma \operatorname{Id}_{\dot{\mathscr{H}}} - \dot{A})^{-1}\|_{\mathscr{L}(\dot{\mathscr{H}},\dot{\mathscr{H}})} \leq K e^{K|\sigma|}, \qquad \sigma \in \mathbb{R},\ |\sigma| \geq \sigma_0.$$

The proof of Proposition 11.6 is given below in Sect. 11.5. Proposition 11.6 is the counterpart of Proposition 10.9.

By Theorem 6.6 of Volume 1, a consequence is the logarithmic asymptotic stability of the semigroup $\dot{S}(t)$. The statement of this result is the same as in Theorem 10.10, from which one deduces the stabilization result for the damped wave equation (11.1.3).

THEOREM 11.7. *Let Γ be a nonempty open subset of ${}^1\partial\mathcal{M}$, and let $\alpha \in \mathscr{C}^\infty({}^1\partial\mathcal{M})$ be such that $\alpha > 0$ on Γ. Let $k \in \mathbb{N}$. Then, there exists $C > 0$ such that, if the initial condition $Y^0 = (y^0, y^1)$ is in the domain of the operator A^k, the energy of the strong solution $y(t)$ to (11.2.2) and (11.2.3) satisfies*

$$\mathcal{E}(y)(t) \leq \frac{C}{\big(\log(2+t)\big)^{2k}} \|A^k Y^0\|^2_{\mathscr{H}}. \tag{11.4.1}$$

If one were to introduce weak solutions, one would then obtain the counterpart result of Corollary 10.12. This is left to the reader.

11.5. Proof of the Resolvent Estimate

Here, we give the proof of the resolvent estimate of Proposition 11.6.

First, for $\sigma \in \mathbb{R}$, the resolvent operator $(i\sigma \operatorname{Id}_{\dot{\mathscr{H}}} - \dot{A})^{-1}$ is well defined and continuous on $\dot{\mathscr{H}}$ as the spectrum of $(\dot{A}, D(\dot{A}))$ is contained in $\{z \in \mathbb{C};\ \operatorname{Re} z > 0\}$ by Proposition 11.21.

Let $U \in D(\dot{A})$ and $F \in \dot{\mathscr{H}}$ be such that

$$(i\sigma \operatorname{Id}_{\dot{\mathscr{H}}} - \dot{A})U = F, \quad U = {}^t(u^0, u^1), \quad F = {}^t(f^0, f^1). \tag{11.5.1}$$

The resolvent equation (11.5.1) reads

$$i\sigma u^0 + u^1 = f^0, \quad i\sigma u^1 - Pu^0 = f^1, \qquad (Bu^0 + \alpha u^1)_{|{}^1\partial\mathcal{M}} = 0,$$

which we write, with $f = i\sigma f^0 - f^1$,

$$i\sigma u^0 + u^1 = f^0, \quad (-\sigma^2 + P)u^0 = f, \quad Bu^0{}_{|{}^1\partial\mathcal{M}} = \alpha(i\sigma u^0 - f^0)_{|{}^1\partial\mathcal{M}}. \tag{11.5.2}$$

Writing

$$((-\sigma^2 + P)u^0, u^0)_{L^2(\mathcal{M})} = (f, u^0)_{L^2(\mathcal{M})},$$

with the integration by parts formula (4.6.4), we find

$$\begin{aligned}
(f, u^0)_{L^2(\mathcal{M})} &= \|\nabla_g u^0\|^2_{L^2V(\mathcal{M})} + (R_1 u^0, u^0)_{L^2(\mathcal{M})} \\
&\quad - (\partial_\nu u^0{}_{|{}^1\partial\mathcal{M}}, u^0{}_{|{}^1\partial\mathcal{M}})_{L^2({}^1\partial\mathcal{M})} - \|\sigma u^0\|^2_{L^2(\mathcal{M})} \\
&= \|\nabla_g u^0\|^2_{L^2V(\mathcal{M})} + (R_1 u^0, u^0)_{L^2(\mathcal{M})} \\
&\quad + (B' u^0_{|{}^1\partial\mathcal{M}}, u^0_{|{}^1\partial\mathcal{M}})_{L^2({}^1\partial\mathcal{M})} \\
&\quad - i\sigma |\alpha^{1/2} u^0{}_{|{}^1\partial\mathcal{M}}|^2_{L^2({}^1\partial\mathcal{M})} + (\alpha f^0{}_{|{}^1\partial\mathcal{M}}, u^0{}_{|{}^1\partial\mathcal{M}})_{L^2({}^1\partial\mathcal{M})} \\
&\quad - \|\sigma u_0\|^2_{L^2(\mathcal{M})} \\
&= N(u^0) - i\sigma |\alpha^{1/2} u^0{}_{|{}^1\partial\mathcal{M}}|^2_{L^2({}^1\partial\mathcal{M})} \\
&\quad + (\alpha f^0{}_{|{}^1\partial\mathcal{M}}, u^0{}_{|{}^1\partial\mathcal{M}})_{L^2({}^1\partial\mathcal{M})} - \|\sigma u_0\|^2_{L^2(\mathcal{M})},
\end{aligned} \tag{11.5.3}$$

using that $\alpha \geq 0$, where $N(.)$ as given in (11.2.9). Recall that $N(.)$ is real (and nonnegative) on $H^1_{\mathsf{D}}(\mathcal{M})$. Computing the imaginary part of (11.5.3), we obtain

$$\sigma |\alpha^{1/2} u^0{}_{|{}^1\partial\mathcal{M}}|^2_{L^2({}^1\partial\mathcal{M})} = \operatorname{Im}(\alpha f^0{}_{|{}^1\partial\mathcal{M}}, u^0{}_{|{}^1\partial\mathcal{M}})_{L^2({}^1\partial\mathcal{M})} - \operatorname{Im}(f, u^0)_{L^2(\mathcal{M})}.$$

We then write, for $|\sigma| \geq \sigma_0 > 0$,

$$\begin{aligned}
|\alpha^{1/2} u^0{}_{|{}^1\partial\mathcal{M}}|^2_{L^2({}^1\partial\mathcal{M})} &\lesssim |\sigma|^{-1} (\|f^0\|_{H^1(\mathcal{M})} + \|f\|_{L^2(\mathcal{M})}) \|u^0\|_{H^1(\mathcal{M})} \\
&\lesssim \|F\|_{\dot{\mathscr{H}}} \|u^0\|_{H^1(\mathcal{M})},
\end{aligned} \tag{11.5.4}$$

using the trace inequality of Proposition 18.24. From the properties of the damping function α on ${}^1\partial\mathcal{M}$, there exist Γ_0 an open subset of Γ and $\delta > 0$ such that $\alpha \geq \delta > 0$ in Γ_0. This yields

$$\delta\, |u^0{}_{|{}^1\partial\mathcal{M}}|^2_{L^2(\Gamma_0)} \lesssim \|F\|_{\dot{\mathscr{H}}} \|u^0\|_{H^1(\mathcal{M})}, \qquad |\sigma| \geq \sigma_0. \tag{11.5.5}$$

The key estimate is given by the following observation lemma.

LEMMA 11.8. *For Γ_0 as introduced above, there exists $C > 0$ such that*

$$\|u^0\|_{H^1(\mathcal{M})} \leq Ce^{C|\sigma|}\big(\|F\|_{\dot{\mathscr{H}}} + |u^0{}_{|^1\partial\mathcal{M}}|_{L^2(\Gamma_0)}\big).$$

The proof is given below.

Then estimate (11.5.5) yields

$$\|u^0\|_{H^1(\mathcal{M})} \lesssim e^{C|\sigma|}\big(\|F\|_{\dot{\mathscr{H}}} + \|F\|_{\dot{\mathscr{H}}}^{\frac{1}{2}}\|u^0\|_{H^1(\mathcal{M})}^{\frac{1}{2}}\big),$$

and with the Young inequality, we obtain

$$\|u^0\|_{H^1(\mathcal{M})} \lesssim e^{C|\sigma|}\|F\|_{\dot{\mathscr{H}}}.$$

Finally, as $u^1 = f^0 - i\sigma u^0$, we obtain

$$\|u^0\|_{H^1(\mathcal{M})} + \|u^1\|_{L^2(\mathcal{M})} \lesssim e^{C|\sigma|}\|F\|_{\dot{\mathscr{H}}},$$

yielding the resolvent estimate of Proposition 11.6. ∎

PROOF OF LEMMA 11.8. We set $\tilde{\mathcal{M}} = \mathbb{R} \times \mathcal{M}$, and we denote by $n = (s, m)$ a point of $\tilde{\mathcal{M}}$ with $s \in \mathbb{R}$ and $m \in \mathcal{M}$. We have $\partial\tilde{\mathcal{M}} = \mathbb{R} \times \partial\mathcal{M}$. We set ${}^0\partial\tilde{M} = \mathbb{R} \times {}^0\partial\mathcal{M}$ and ${}^1\partial\tilde{M} = \mathbb{R} \times {}^1\partial\mathcal{M}$.

On $\tilde{\mathcal{M}}$, we consider the metric $\tilde{g} = ds \otimes ds + g$, where g is the metric on $\mathcal{M}$. On $\tilde{\mathcal{M}}$, we consider the operator $Q = D_s^2 + P$. The function $u(s, m) = e^{\sigma s}u^0(m)$ is then a solution to

$$Qu = h \text{ in } \tilde{\mathcal{M}}, \quad u = 0 \text{ on } {}^0\partial\tilde{M}, \quad \text{and } \; Bu_{|^1\partial\mathcal{M}} = h_\partial \text{ on } {}^1\partial\tilde{M},$$

where

$$h(n) = e^{\sigma s} f(m), \quad n = (s, m) \in \tilde{\mathcal{M}},$$
$$h_\partial(n) = \alpha(m)e^{\sigma s}(i\sigma u^0 - f^0)_{|^1\partial\mathcal{M}}(m), \quad n = (s, m) \in {}^1\partial\tilde{M}.$$

The principal symbol of Q is $q(s, m, \sigma, \omega) = \sigma^2 + |\omega|_{g_m}^2 = |(\sigma, \omega)|_{\tilde{g}_m}^2$. Let $s \in \mathbb{R}$, $\sigma \in \mathbb{R}$, and $(m, \omega') \in T^*\partial\mathcal{M}$. Following Sect. 2.2, we set

$$\check{q}(s, m, \sigma, \omega', z) = |(\sigma, \omega') - zn_m|_{\tilde{g}_m}^2.$$

As we have $(n_m, (\sigma, \omega'))_{\tilde{g}_m} = 0$, we find

$$\check{q}(m, s, \omega', \sigma, z) = z^2 + |(\sigma, \omega')|_{\tilde{g}_m}^2 = (z - i|(\sigma, \omega')|_{\tilde{g}_m})(z + i|(\sigma, \omega')|_{\tilde{g}_m}).$$

The Lopatinskiĭ–Šapiro condition holds on ${}^0\partial\mathcal{M}$ by Example 2.5–(1). With $\check{b}(m, \omega', z) = b(m, \omega' - zn_m)$, on the one hand, having the Lopatinskiĭ–Šapiro condition for (P, B) on ${}^1\partial\mathcal{M}$ reads $\check{b}(m, \omega', i|\omega'|_{g_\partial}) \neq 0$ for all $(m, \omega') \in T^{*1}\partial\mathcal{M}$ with $\omega' \neq 0$ by Proposition 2.3. This condition is part of the assumption formulated here. On the other hand, the Lopatinskiĭ–Šapiro condition for (Q, B) on $\mathbb{R} \times {}^1\partial\mathcal{M}$ reads $\check{b}(m, \omega', i|(\sigma, \omega')|_{\tilde{g}_m}) \neq 0$ for all $(s, m, \sigma, \omega') \in T^{*1}\partial\tilde{M}$ with $(\sigma, \omega') \neq (0, 0)$ also by applying Proposition 2.3. We wish to prove that this latter condition holds too.

On ${}^1\partial\mathcal{M}$, we have $Bu_{|{}^1\partial\mathcal{M}} = \partial_\nu u_{|{}^1\partial\mathcal{M}} + (iX' + h)u_{|{}^1\partial\mathcal{M}}$, where X' is a real valued vector field. This gives $b(m, \omega' - zn_m) = -iz - \langle X'_m, \omega'\rangle$. In the case $d \geq 3$, we have $|X'|_{g_\partial} < 1$ on ${}^1\partial\mathcal{M}$ by Proposition 4.6. In the case $d = 2$, this property is assumed here. Because of the latter property, if $\omega' \neq 0$, observe that we have

$$\check{b}(m, \omega', i|(\sigma, \omega')|_{\tilde{g}_m}) = |(\sigma, \omega')|_{\tilde{g}_m} - \langle X'_m, \omega'\rangle \geq |\omega'|_{g_\partial} - \langle X'_m, \omega'\rangle > 0.$$

If $\omega' = 0$ and $\sigma \neq 0$, then $\check{b}(m, \omega', i|(\sigma, \omega')|_{\tilde{g}_m}) = |\sigma| \neq 0$. Consequently, the Lopatinskiĭ–Šapiro condition holds for (Q, B) on $\mathbb{R} \times {}^1\partial\mathcal{M}$.

We set $\mathcal{U} = (-1, 1)\times\mathcal{M}$ and $\mathcal{V} = (-2, 2)\times\mathcal{M}$ to fit the setting of Chap. 9. The geometry is the same as in Fig. 10.1. We consider $m^0 \in \Gamma_0 \subset {}^1\partial\mathcal{M}$, with Γ_0 as introduced above the statement of Lemma 11.8, $n^0 = (0, m^0)$ and V^0 a neighborhood of n^0 in $\mathcal{U}$ such that $V^0 \cap \partial\tilde{\mathcal{M}} \subset (-1, 1)\times\Gamma_0 \subset \mathcal{U} \cap \partial\tilde{\mathcal{M}}$. We then apply Theorem 9.6 to the function u with $\tilde{\mathcal{M}}$ in place of $\mathcal{M}$ and $\tilde{B} = B$ therein, yielding, for some $\delta \in (0, 1)$,

$$\begin{aligned}\|u\|_{H^1(\mathcal{U})} &\lesssim \|u\|_{H^1(\mathcal{V})}^{1-\delta}\Big(\|h\|_{L^2(\mathcal{V})} + |h_\partial|_{L^2(\mathcal{V}\cap\partial\tilde{\mathcal{M}})} + |u_{|V^0\cap\partial\tilde{\mathcal{M}}}|_{L^2(V^0\cap\partial\tilde{\mathcal{M}})}\\ &\qquad + |Bu_{|V^0\cap\partial\tilde{\mathcal{M}}}|_{L^2(V^0\cap\partial\tilde{\mathcal{M}})}\Big)^\delta\\ &\lesssim \|u\|_{H^1(\mathcal{V})}^{1-\delta}\Big(\|h\|_{L^2(\mathcal{V})} + |h_\partial|_{L^2(\mathcal{V}\cap\partial\tilde{\mathcal{M}})} + |u_{|V^0\cap\partial\tilde{\mathcal{M}}}|_{L^2(V^0\cap\partial\tilde{\mathcal{M}})}\Big)^\delta.\end{aligned}$$

With this estimate, we conclude following the end of the proof of Lemma 10.13 starting from estimate (10.5.6). ■

REMARK 11.9. Here, in the case $d = 2$, we have assumed that $|X'|_{g_\partial} < 1$. A reason associated with the nature of the spectrum of the generator of the wave semigroup is put forward in the introductory Sect. 11.1.

In the proof of Lemma 11.8, the condition $|X'|_{g_\partial} < 1$ also appears in a crucial way for the Lopatinskiĭ–Šapiro condition to hold for the pair (Q, B) if it holds for the pair (P, B). We further discuss this point here.

In fact, assume now that $|X'|_{g_\partial} > 1$. If one assumes that P, along with the boundary condition B, yields a selfadjoint operator, this is the alternative case as shown in Theorem 4.3 in the case $d = 2$.

In the proof of Lemma 11.8, the Lopatinskiĭ–Šapiro condition holds for (Q, B) at m if and only if

$$\check{b}(m, \omega', i|(\sigma, \omega')|_{\tilde{g}_m}) = |(\sigma, \omega')|_{\tilde{g}_m} - \langle X'_m, \omega'\rangle \neq 0,$$

for $(\sigma, \omega') \neq (0, 0)$. This follows from Proposition 2.3. Here, ω' and X'_m both lie in a one-dimensional space. Thus, one can choose $\omega' \neq 0$ such that $\langle X'_m, \omega'\rangle = -|X'_m|_{g_\partial}|\omega'|_{g_\partial}$. Then, the condition $\check{b}(m, \omega', i|(\sigma, \omega')|_{\tilde{g}_m}) = 0$ reads

$$\sigma^2 + |\omega'|^2_{g_\partial} = |X'_m|^2_{g_\partial}|\omega'|^2_{g_\partial}.$$

Since $|X'_m|^2_{g_\partial} - 1 > 0$, if one chooses $\sigma = \pm\big(|X'_m|^2_{g_\partial} - 1\big)^{1/2}|\omega'|_{g_\partial}$, one obtains $\check{b}(m, \omega', i|(\sigma, \omega')|_{\tilde{g}_m}) = 0$ meaning that the Lopatinskiĭ–Šapiro condition does not hold for (Q, B) at m. The proof based on the unique continuation results of Chap. 9 cannot be carried out, since the Lopatinskiĭ–Šapiro condition is assumed at the boundary therein. One thus sees that the condition $|X'_m|_{g_\partial} < 1$ is also an important technical assumption in the proof scheme we carry out.

11.6. Notes

The present chapter generalizes the result of Chap. 10 obtained in the case of a Neumann boundary damping. Chapter 10 could have thus been omitted. Yet the Neumann boundary condition appears as quite classical and their independent treatment seemed important for that respect. The generalization we provide here encompasses all boundary operators that fit the framework given by the Lopatinskiĭ–Šapiro conditions recalled in Chap. 2 with the additional requirement that the Laplace–Beltrami operator be selfadjoint. Up to our knowledge, this cannot be found in the existing literature with the same level of generality.

Here damping occurs at the boundary. Observe that we could also treat the case of an inner damping

$$\begin{cases} \partial_t^2 y + P_0 y + \alpha \partial_t y = 0 & \text{in } (0, +\infty) \times \mathcal{M}, \\ By = 0 & \text{on } (0, +\infty) \times \partial\mathcal{M}, \\ y_{|t=0} = y^0, \ \partial_t y_{|t=0} = y^1 & \text{in } \mathcal{M}, \end{cases}$$

with some $\alpha \geq 0$. Such a result relies on the analysis used in the Chapter 6 of Volume 1 and the present chapter.

Appendix

11.A. The Generator of the Boundary-Damped Wave Semigroup

PROPOSITION 11.10. *The unbounded operators $(A, D(A))$ are a closed operator on $\mathscr{H}$.*

PROOF. Let $(V^n)_n \subset D(A)$, $V^n = {}^t(v^{0,n}, v^{1,n})$, be such that both $V^n \to V = {}^t(v^0, v^1)$ and $AV^n \to W = {}^t(w^0, w^1)$ in $\mathscr{H}$ as $n \to +\infty$.

We thus have $(v^{0,n})_n \subset H^2_{\mathsf{D}}(\mathcal{M})$, $(v^{1,n})_n \subset H^1_{\mathsf{D}}(\mathcal{M})$ and

$$v^{0,n} \to v^0 \text{ in } H^1_{\mathsf{D}}(\mathcal{M}), \qquad v^{1,n} \to v^1 \text{ in } L^2(\mathcal{M}).$$

As $AV^n = {}^t(-v^{1,n}, Pv^{0,n})$, we have

$$-v^{1,n} \to w^0 \text{ in } H^1_{\mathsf{D}}(\mathcal{M}), \qquad Pv^{0,n} \to w^1 \text{ in } L^2(\mathcal{M}).$$

We thus obtain $v^1 = -w^0 \in H^1_{\mathsf{D}}(\mathcal{M})$ and in fact $v^{1,n} \to v^1$ in $H^1_{\mathsf{D}}(\mathcal{M})$. It thus remains to prove that $v^0 \in H^2_{\mathsf{D}}(\mathcal{M})$, $Pv^0 = w^1$, and $B_\partial V = 0$.

Set $w^{1,n} = Pv^{0,n} \in L^2(\mathcal{M})$. We have $w^{1,n} \to w^1$ in $L^2(\mathcal{M})$. With the Green formula (4.5.3), for $\varphi \in \ker(\mathsf{P})$, one writes

$$(w^{1,n}, \varphi)_{L^2(\mathcal{M})} = -(Bv^{0,n}_{|^1\partial\mathcal{M}}, \varphi_{|^1\partial\mathcal{M}})_{L^2(^1\partial\mathcal{M})} = (\alpha v^{1,n}_{|^1\partial\mathcal{M}}, \varphi_{|^1\partial\mathcal{M}})_{L^2(^1\partial\mathcal{M})},$$

using that $v^{0,n}_{|^0\partial\mathcal{M}} = 0$ and $B_\partial(V^n) = 0$. We set $h^n = -\alpha v^{1,n}_{|^1\partial\mathcal{M}} \in H^{1/2}(^1\partial\mathcal{M})$.

As $v^{1,n} \to v^1$ in $H^1_{\mathsf{D}}(\mathcal{M})$, we find that $h^n \to h = -\alpha v^1{}_{|^1\partial\mathcal{M}}$ in $H^{1/2}(^1\partial\mathcal{M})$ by the trace inequality of Proposition 18.24. In particular,

$$\begin{aligned}(w^1, \varphi)_{L^2(\mathcal{M})} &+ (h, \varphi_{|^1\partial\mathcal{M}})_{L^2(^1\partial\mathcal{M})} \\ &= \lim_{n\to+\infty} \big((w^{1,n}, \varphi)_{L^2(\mathcal{M})} + (h^n, \varphi_{|^1\partial\mathcal{M}})_{L^2(^1\partial\mathcal{M})}\big) = 0,\end{aligned}$$

if $\varphi \in \ker(\mathsf{P})$. Then, by Theorem 4.18, there exists a unique $u \in \underline{H}^2(\mathcal{M}) \cap H^1_{\mathsf{D}}(\mathcal{M})$ solution to $Pu = w^1$ and $Bu_{|^1\partial\mathcal{M}} = h$. Moreover, one has

$$\begin{aligned}\|\underline{v}^{0,n} - u\|_{H^2(\mathcal{M})} &\lesssim \|w^{1,n} - w^1\|_{L^2(\mathcal{M})} + |h^n - h|_{H^{1/2}(^1\partial\mathcal{M})} \\ &\lesssim \|w^{1,n} - w^1\|_{L^2(\mathcal{M})} + \|v^{1,n} - v^1\|_{H^1(\mathcal{M})},\end{aligned}$$

by Proposition 18.24; see (4.7.3) for the definition of $\underline{H}^2(\mathcal{M})$.

Since $v^{0,n} \to v^0$ in $H^1_{\mathsf{D}}(\mathcal{M})$, we have $\underline{v}^{0,n} \to \underline{v}^0$ in $H^1_{\mathsf{D}}(\mathcal{M})$ implying that $\underline{v}^0 = u$, that is, $v^0 - u \in \ker(\mathsf{P})$. This gives $v^0 \in H^2_{\mathsf{D}}(\mathcal{M})$, $Pv^0 = Pu = w^1$, and $Bv^0{}_{|^1\partial\mathcal{M}} = Bu_{|^1\partial\mathcal{M}} = h = -\alpha v^1{}_{|^1\partial\mathcal{M}}$ meaning precisely that $B_\partial(V) = 0$. ∎

LEMMA 11.11. *The norm $\|.\|_{\dot{\mathscr{H}}}$ defined by*

$$\|V\|^2_{\dot{\mathscr{H}}} := N(v^0) + \|v^1\|^2_{L^2(\mathcal{M})}, \quad V = {}^t(v^0, v^1),$$

is equivalent to the norm $\|.\|_{\mathscr{H}}$ on $\dot{\mathscr{H}}$.

PROOF. The proof follows that of Lemma 10.16 by proving that

$$\|v^0\|_{L^2(\mathcal{M})} \lesssim \|V\|_{\dot{\mathscr{H}}}, \quad V = {}^t(v^0, v^1) \in \dot{\mathscr{H}}.$$

Assume that the inequality does not hold. Then, there exists a sequence $V^n = {}^t(v^{0,n}, v^{1,n})_{n\in\mathbb{N}} \subset \dot{\mathscr{H}}$ such that

$$\|v^{0,n}\|_{L^2(\mathcal{M})} = 1, \qquad N(v^{0,n})^{1/2} + \|v^{1,n}\|_{L^2(\mathcal{M})} \to 0, \quad \text{as } n \to \infty.$$

For $\lambda \in \mathbb{R}$ well chosen, $N_\lambda(.)^{1/2}$ is a norm on $H^1_{\mathsf{D}}(M)$ by Proposition 4.14 with $N_\lambda(.) = N(.) + \lambda\|.\|^2_{L^2(\mathcal{M})}$. Thus the sequence $(v^{0,n})_{n\in\mathbb{N}}$ is bounded in $H^1_{\mathsf{D}}(\mathcal{M})$, and, hence, there exists $v^0 \in H^1_{\mathsf{D}}(\mathcal{M})$ such that $v^{0,n} \rightharpoonup v^0$ in $H^1_{\mathsf{D}}(\mathcal{M})$ up to a subsequence. If $u \in H^1_{\mathsf{D}}(\mathcal{M})$, the map $v \mapsto \tilde{N}(v, u)$ is a continuous form. We thus find $\tilde{N}(v^{0,n}, u) \to \tilde{N}(v^0, u)$ as $n \to +\infty$.

Yet by Cauchy–Schwarz like inequality (4.6.8), one finds

$$\tilde{N}(v^{0,n}, u) \to 0, \quad \text{as } n \to +\infty,$$

since $N(v^{0,n}) \to 0$. One thus obtains

$$\tilde{N}(v^0, u) = 0, \quad \text{for all } u \in H^1_{\mathsf{D}}(\mathcal{M}).$$

Then, by Theorems 4.18 and 4.19, one has $v^0 \in \ker(\mathsf{P})$.

Since $V^n \subset \dot{\mathscr{H}}$, one has $F_{\alpha,v^0}(V^n) = 0$, which reads

$$(\alpha v^{0,n}_{|{}^1\partial\mathcal{M}}, v^0_{|{}^1\partial\mathcal{M}})_{L^2({}^1\partial\mathcal{M})} + (v^{1,n}, v^0)_{L^2(\mathcal{M})} = 0.$$

Since $v^{0,n}_{|{}^1\partial\mathcal{M}} \rightharpoonup v^0_{|{}^1\partial\mathcal{M}}$ in $H^{1/2}({}^1\partial\mathcal{M})$ and $v^{1,n} \to 0$ in $L^2(\mathcal{M})$, one obtains

$$(\alpha v^0_{|{}^1\partial\mathcal{M}}, v^0_{|{}^1\partial\mathcal{M}})_{L^2({}^1\partial\mathcal{M})} = 0.$$

By Lemma 11.4, we conclude that $v^0 = 0$ since $v^0 \in \ker(\mathsf{P})$. This is in contradiction with having $\|v^{0,n}\|_{L^2(\mathcal{M})} = 1$ since $v^{0,n} \to v^0$ in $L^2(\mathcal{M})$. ■

Associated with the norm $\|.\|_{\dot{\mathscr{H}}}$ is the following inner product, for U, $V \in \dot{\mathscr{H}}$,

$$(V, U)_{\dot{\mathscr{H}}} := \tilde{N}(v^0, u^0) + (v^1, u^1)_{L^2(\mathcal{M})}. \tag{11.A.1}$$

PROPOSITION 11.12. *The unbounded operator $(\dot{A}, D(\dot{A}))$ on $\dot{\mathscr{H}}$ is closed. Moreover, $D(\dot{A})$ is dense in $\dot{\mathscr{H}}$ and the injection of $D(\dot{A})$ into $\dot{\mathscr{H}}$ is compact.*

In the proof of Proposition 11.12, we shall need the following lemma.

LEMMA 11.13. *The map*

$$D(\dot{A}) \to H^1_{\mathsf{D}}(\mathcal{M})$$
$$(u^0, u^1) \mapsto u^1$$

is surjective. Moreover, there exists a bounded linear map $\Psi : H^1_{\mathsf{D}}(\mathcal{M}) \to H^2_{\mathsf{D}}(\mathcal{M})$ such that $(\Psi w, w) \in D(\dot{A})$ for all $w \in H^1_{\mathsf{D}}(\mathcal{M})$.

PROOF OF LEMMA 11.13. Let $w \in H^1_{\mathsf{D}}(\mathcal{M})$ and set $h = -\alpha w_{|{}^1\partial\mathcal{M}} \in H^{1/2}({}^1\partial\mathcal{M})$ using the trace property of Proposition 18.24. Pick then $f = \sum_{j=0}^{n-1} f_j \phi_j \in \ker(\mathsf{P})$ such that $(f, \phi_j)_{L^2(\mathcal{M})} + (h, \phi_{j|{}^1\partial\mathcal{M}})_{L^2({}^1\partial\mathcal{M})} = 0$, $j = 0, \ldots, n-1$. The map $w \mapsto f$ is linear, and we have

$$\|f\|_{L^2(\mathcal{M})} \lesssim |h|_{L^2({}^1\partial\mathcal{M})} \lesssim \|w\|_{H^1(\mathcal{M})}.$$

By Theorem 4.18, there exists $v \in \underline{H}^2(\mathcal{M}) \cap H^1_{\mathsf{D}}(\mathcal{M})$ such that $Pv = f$ and $Bv_{|{}^1\partial\mathcal{M}} = h$. Moreover, we have

$$\|v\|_{H^2(\mathcal{M})} \lesssim \|f\|_{L^2(\mathcal{M})} + |h|_{H^{1/2}({}^1\partial\mathcal{M})} \lesssim \|w\|_{H^1(\mathcal{M})}.$$

With a Hilbert basis $\{\varphi_0, \ldots, \varphi_{n-1}\}$ of $\ker(\mathsf{P})$ associated with the inner product of Lemma 11.4, we set $a_j = (\alpha v_{|{}^1\partial\mathcal{M}}, \varphi_{j|{}^1\partial\mathcal{M}})_{L^2({}^1\partial\mathcal{M})} + (w_{|{}^1\partial\mathcal{M}}, \varphi_{j|{}^1\partial\mathcal{M}})_{L^2({}^1\partial\mathcal{M})}$ and $\Psi w = v - \sum_{j=0}^{n-1} a_j \varphi_j$. We then have $(\alpha \Psi w_{|{}^1\partial\mathcal{M}}, \varphi_{j|{}^1\partial\mathcal{M}})_{L^2({}^1\partial\mathcal{M})} + (w_{|{}^1\partial\mathcal{M}}, \varphi_{j|{}^1\partial\mathcal{M}})_{L^2({}^1\partial\mathcal{M})} = 0$. The map $w \mapsto \Psi w$ is linear and $V = {}^t(\Psi w, w) \in D(\dot{A})$. Moreover, we see that

$$\begin{aligned}\Big\| \sum_{j=0}^{n-1} a_j \varphi_j \Big\|_{L^2(\mathcal{M})} &\lesssim \|v_{|{}^1\partial\mathcal{M}}\|_{L^2({}^1\partial\mathcal{M})} + \|w_{|{}^1\partial\mathcal{M}}\|_{L^2({}^1\partial\mathcal{M})} \\ &\lesssim \|v\|_{H^2(\mathcal{M})} + \|w\|_{H^1(\mathcal{M})} \lesssim \|w\|_{H^1(\mathcal{M})}.\end{aligned}$$

Consequently, $\|\Psi w\|_{H^2(\mathcal{M})} \lesssim \|w\|_{H^1(\mathcal{M})}$. The proof is complete. ∎

PROOF OF PROPOSITION 11.12. To prove that $(\dot{A}, D(\dot{A}))$ is closed, one can write the proof given for Proposition 10.17 without any change. The argument for the compactness of the injection of $D(\dot{A})$ into $\dot{\mathscr{H}}$ is also the same.

We now address the question of the density of $D(\dot{A})$ in $\dot{\mathscr{H}}$. We consider $U \in \dot{\mathscr{H}}$ such that $(U, V)_{\dot{\mathscr{H}}} = 0$ for all $V \in D(\dot{A})$, and we prove that $U = 0$, which asserts that $D(\dot{A})$ is dense in $\dot{\mathscr{H}}$.

Let $f \in \underline{L}^2(\mathcal{M}) = \ker(\mathsf{P})^\perp$. By Proposition 4.17, one has $f \in \operatorname{Ran}(\mathsf{P})$. Thus there exists $v \in D(\mathsf{P})$ such that $\mathsf{P}v = f$. Note that $v \in H^2_{\mathsf{D}}(\mathcal{M})$ and $Bv_{|^1\partial\mathcal{M}} = 0$. With a Hilbert basis $\{\varphi_0, \dots, \varphi_{n-1}\}$ of $\ker(\mathsf{P})$ associated with the inner product of Lemma 11.4, we set $a_j = (\alpha v_{|^1\partial\mathcal{M}}, \varphi_{j|^1\partial\mathcal{M}})_{L^2(^1\partial\mathcal{M})}$ and $v^0 = v - \sum_{j=0}^{n-1} a_j \varphi_j \in D(\mathsf{P})$. Observe that $(\alpha v^0_{|^1\partial\mathcal{M}}, \varphi_{j|^1\partial\mathcal{M}})_{L^2(^1\partial\mathcal{M})} = 0$ for $j = 0, \dots, n-1$.

Set $V = {}^t(v^0, 0)$ and observe that $V \in D(\dot{A})$. Then, with (4.6.7),

$$0 = (U, V)_{\dot{\mathscr{H}}} = \tilde{N}(u^0, v^0) = (u^0, \mathsf{P}v^0)_{L^2(\mathcal{M})} = (u^0, f)_{L^2(\mathcal{M})}.$$

Since $f \in \ker(\mathsf{P})^\perp$ is arbitrary, one concludes that $u^0 \in \ker(\mathsf{P})$. We thus find that

$$\begin{aligned} 0 = (U, V)_{\dot{\mathscr{H}}} &= \tilde{N}(u^0, v^0) + (u^1, v^1)_{L^2(\mathcal{M})} \\ &= (\mathsf{P}u^0, v^0)_{L^2(\mathcal{M})} + (u^1, v^1)_{L^2(\mathcal{M})} \\ &= (u^1, v^1)_{L^2(\mathcal{M})}, \end{aligned} \tag{11.A.2}$$

for $V = {}^t(v^0, v^1) \in D(\dot{A})$. Let $w \in H^1_{\mathsf{D}}(\mathcal{M})$. By Lemma 11.13, there exists $v \in H^2_{\mathsf{D}}(\mathcal{M})$ such that $V = {}^t(v, w) \in D(\dot{A})$. Then, (11.A.2) yields $(u^1, w)_{L^2(\mathcal{M})} = 0$. As $w \in H^1_{\mathsf{D}}(\mathcal{M})$ is arbitrary, this gives $u^1 = 0$ as $H^1_{\mathsf{D}}(\mathcal{M})$ is dense in $L^2(\mathcal{M})$. Thus, $U = {}^t(u^0, 0) \in \dot{\mathscr{H}} \cap \mathcal{N}$ since $u^0 \in \ker(\mathsf{P})$. By (11.3.2), we finally obtain $U = 0$. This concludes the proof of the density property. ∎

From (11.3.2), (11.3.5), and Proposition 11.12, we obtain the following result.

COROLLARY 11.14. *The domain $D(A)$ is dense in $\mathscr{H}$.*

We now make explicit the operator $\dot{A}^*$ adjoint of $\dot{A}$ with respect to the inner product on $\dot{\mathscr{H}}$ introduced in (11.A.1). We recall that the domain of the adjoint is defined by

$$D(\dot{A}^*) = \{V \in \dot{\mathscr{H}};\ \exists C > 0,\ \forall U \in D(\dot{A}),\ |(V, \dot{A}U)_{\dot{\mathscr{H}}}| \le C\|U\|_{\dot{\mathscr{H}}}\}.$$

PROPOSITION 11.15. *We have*

$$D(\dot{A}^*) = \{V = {}^t(v^0, v^1) \in \dot{\mathscr{H}}; v^0 \in H^2_{\mathsf{D}}(\mathcal{M}),\ v^1 \in H^1_{\mathsf{D}}(\mathcal{M});\ B^*_\partial(V) = 0\} \tag{11.A.3}$$

with $B_\partial^*(V) = -Bv^0 + \alpha v^1$. *For* $V = {}^t(v^0, v^1) \in D(\dot{A}^*)$, *we have*

$$\dot{A}^* V = \begin{pmatrix} v^1 \\ -Pv^0 \end{pmatrix}, \quad \textit{that is,} \quad \dot{A}^* = \begin{pmatrix} 0 & 1 \\ -P & 0 \end{pmatrix}. \tag{11.A.4}$$

PROOF. Let $V \in \big(H^2_{\mathsf{D}}(\mathcal{M}) \times H^1_{\mathsf{D}}(\mathcal{M})\big) \cap \mathscr{H}$ be such that $B_\partial^*(V) = 0$. If $U = {}^t(u^0, u^1) \in D(\dot{A})$, we have

$$(\dot{A}U, V)_{\mathscr{H}} = -\tilde{N}(u^1, v^0) + (Pu^0, v^1)_{L^2(\mathcal{M})}.$$

With the definition of $\tilde{N}$ in (4.6.5), Lemma 4.7, and formula (18.5.5), and the form of B, we compute

$$\begin{aligned}
\tilde{N}(u^1, v^0) &= (\nabla_g u^1, \nabla_g v^0)_{L^2V(\mathcal{M})} + (u^1, R_1 v^0)_{L^2(\mathcal{M})} \\
&\quad + ({u^1}_{|{}^1\partial\mathcal{M}}, B' {v^0}_{|{}^1\partial\mathcal{M}})_{L^2({}^1\partial\mathcal{M})} \\
&= (u^1, -\Delta_g v^0)_{L^2(\mathcal{M})} + (u^1, R_1 v^0)_{L^2(\mathcal{M})} \\
&\quad + ({u^1}_{|{}^1\partial\mathcal{M}}, \partial_\nu {v^0}_{|{}^1\partial\mathcal{M}})_{L^2({}^1\partial\mathcal{M})} \\
&\quad + ({u^1}_{|{}^1\partial\mathcal{M}}, B' {v^0}_{|{}^1\partial\mathcal{M}})_{L^2({}^1\partial\mathcal{M})} \\
&= (u^1, Pv^0)_{L^2(\mathcal{M})} + ({u^1}_{|{}^1\partial\mathcal{M}}, B {v^0}_{|{}^1\partial\mathcal{M}})_{L^2({}^1\partial\mathcal{M})}.
\end{aligned}$$

Similarly,

$$(Pu^0, v^1)_{L^2(\mathcal{M})} = \tilde{N}(u^0, v^1) - (B{u^0}_{|{}^1\partial\mathcal{M}}, {v^1}_{|{}^1\partial\mathcal{M}})_{L^2({}^1\partial\mathcal{M})},$$

yielding

$$\begin{aligned}
(\dot{A}U, V)_{\mathscr{H}} &= -(u^1, Pv^0)_{L^2(\mathcal{M})} + \tilde{N}(u^0, v^1) \\
&\quad - ({u^1}_{|{}^1\partial\mathcal{M}}, B{v^0}_{|{}^1\partial\mathcal{M}})_{L^2({}^1\partial\mathcal{M})}. - (B{u^0}_{|{}^1\partial\mathcal{M}}, {v^1}_{|{}^1\partial\mathcal{M}})_{L^2({}^1\partial\mathcal{M})}.
\end{aligned}$$

Using that $B_\partial(U) = 0$, that is, $(Bu^0 + \alpha u^1)_{|{}^1\partial\mathcal{M}} = 0$, for the two boundary terms, we obtain

$$\begin{aligned}
&({u^1}_{|{}^1\partial\mathcal{M}}, B{v^0}_{|{}^1\partial\mathcal{M}})_{L^2({}^1\partial\mathcal{M})} + (B{u^0}_{|{}^1\partial\mathcal{M}}, {v^1}_{|{}^1\partial\mathcal{M}})_{L^2({}^1\partial\mathcal{M})} \\
&\qquad = ({u^1}_{|{}^1\partial\mathcal{M}}, B{v^0}_{|{}^1\partial\mathcal{M}})_{L^2({}^1\partial\mathcal{M})} - (\alpha {u^1}_{|{}^1\partial\mathcal{M}}, {v^1}_{|{}^1\partial\mathcal{M}})_{L^2({}^1\partial\mathcal{M})} \\
&\qquad = ({u^1}_{|{}^1\partial\mathcal{M}}, (Bv^0 - \alpha v^1)_{|{}^1\partial\mathcal{M}})_{L^2({}^1\partial\mathcal{M})} \\
&\qquad = 0,
\end{aligned}$$

since $B_\partial^*(V) = (Bv^0 - \alpha v^1)_{|{}^1\partial\mathcal{M}} = 0$. We thus have

$$(\dot{A}U, V)_{\mathscr{H}} = -(u^1, Pv^0)_{L^2(\mathcal{M})} + \tilde{N}(u^0, v^1), \tag{11.A.5}$$

which yields, as $v^0 \in H^2_{\mathsf{D}}(\mathcal{M})$ and $v^1 \in H^1_{\mathsf{D}}(\mathcal{M})$,

$$|(\dot{A}U, V)_{\mathscr{H}}| \lesssim \|u^1\|_{L^2(\mathcal{M})} + N(u^0)^{1/2} \eqsim \|U\|_{\mathscr{H}},$$

by (4.6.8) and the form of the norm $\|.\|_{\mathscr{H}}$ defined in Lemma (11.11). Thus $V \in D(\dot{A}^*)$ and (11.A.5) precisely yields the form of $\dot{A}^*V$ given in (11.A.4).

Conversely, let $V = {}^t(v^0, v^1) \in D(\dot{A}^*)$. If $U = {}^t(u^0, u^1) \in D(\dot{A})$, we have

$$(\dot{A}U, V)_{\dot{\mathscr{H}}} = -\tilde{N}(u^1, v^0) + (Pu^0, v^1)_{L^2(\mathcal{M})}, \tag{11.A.6}$$

and

$$|(\dot{A}U, V)_{\dot{\mathscr{H}}}| \lesssim \|U\|_{\dot{\mathscr{H}}} \eqsim N(u^0)^{1/2} + \|u^1\|_{L^2(\mathcal{M})}. \tag{11.A.7}$$

By the Riesz representation theorem, there exists $W = {}^t(w^0, w^1) \in \dot{\mathscr{H}}$ such that

$$(\dot{A}U, V)_{\dot{\mathscr{H}}} = (U, W)_{\dot{\mathscr{H}}} = \tilde{N}(u^0, w^0) + (u^1, w^1)_{L^2(\mathcal{M})}. \tag{11.A.8}$$

Let $f \in \operatorname{Ran}(\mathsf{P})$, and let $u \in H^2_{\mathsf{D}}(\mathcal{M})$ be such that $Bu_{|^1\partial\mathcal{M}} = 0$ and $Pu = f$ by Theorem 4.18. With the orthonormal basis $\{\varphi_0, \dots, \varphi_{n-1}\}$ of $\ker(\mathsf{P})$ introduced in Sect. 11.3, we set $u^0 = u + \sum_{j=0}^{n-1} \nu_j \varphi_j$, with ν_j chosen such that

$$(u^0_{|^1\partial\mathcal{M}}, \varphi_{j|^1\partial\mathcal{M}})_{L^2(^1\partial\mathcal{M})} = 0, \quad j = 0, \dots, n-1.$$

Observe that we have $u^0 \in D(\mathsf{P})$ with $Pu^0 = f$ and moreover $U = {}^t(u^0, 0) \in D(\dot{A})$. With such a choice, we have on the one hand

$$(\dot{A}U, V)_{\dot{\mathscr{H}}} = (Pu^0, v^1)_{L^2(\mathcal{M})},$$

and on the other hand,

$$(\dot{A}U, V)_{\dot{\mathscr{H}}} = \tilde{N}(u^0, w^0) = (Pu^0, w^0)_{L^2(\mathcal{M})},$$

by (4.6.7). Consequently,

$$(f, v^1 - w^0)_{L^2(\mathcal{M})} = (Pu^0, v^1 - w^0)_{L^2(\mathcal{M})} = 0,$$

for all $f \in \operatorname{Ran}(\mathsf{P})$ implying that $v^1 - w^0 \in \operatorname{Ran}(\mathsf{P})^\perp = \ker(\mathsf{P})$ by Proposition 4.17. In particular, $v^1 \in H^1_{\mathsf{D}}(\mathcal{M})$.

From (11.A.6), with $U = {}^t(u^0, u^1) \in D(\dot{A})$, we may now write, by (4.6.4),

$$\begin{aligned}
(\dot{A}U, V)_{\dot{\mathscr{H}}} &= -\tilde{N}(u^1, v^0) + (Pu^0, v^1)_{L^2(\mathcal{M})} \\
&= -\tilde{N}(u^1, v^0) - (\partial_\nu u^0{}_{|^1\partial\mathcal{M}}, v^1{}_{|^1\partial\mathcal{M}})_{L^2(^1\partial\mathcal{M})} \\
&\quad + (\nabla_g u^0, \nabla_g v^1)_{L^2V(\mathcal{M})} + (R_1 u^0, v^1)_{L^2(\mathcal{M})} \\
&= -\tilde{N}(u^1, v^0) + (\alpha u^1{}_{|^1\partial\mathcal{M}}, v^1{}_{|^1\partial\mathcal{M}})_{L^2(^1\partial\mathcal{M})} + \tilde{N}(u^0, v^1),
\end{aligned}$$

using that $B_\partial(U) = 0$ and the definition of $\tilde{N}(.,.)$ in (4.6.5). We then find

$$(\dot{A}U, V)_{\dot{\mathscr{H}}} = -\tilde{N}(u^1, v^0) + (\alpha u^1{}_{|^1\partial\mathcal{M}}, v^1{}_{|^1\partial\mathcal{M}})_{L^2(^1\partial\mathcal{M})} + \tilde{N}(u^0, w^0), \tag{11.A.9}$$

since $\tilde{N}(z, z') = 0$ for all $z \in H^1_{\mathsf{D}}(\mathcal{M})$ if $z' \in \ker(\mathsf{P})$; see (4.6.7).

With (11.A.8), we conclude that

$$-\tilde{N}(u^1, v^0) + (\alpha u^1{}_{|^1\partial\mathcal{M}}, v^1{}_{|^1\partial\mathcal{M}})_{L^2(^1\partial\mathcal{M})} = (u^1, w^1)_{L^2(\mathcal{M})}. \tag{11.A.10}$$

Let $w \in H^1_{\mathsf{D}}(\mathcal{M})$. From Lemma 11.13, there exists $z \in H^2_{\mathsf{D}}(\mathcal{M})$ such that ${}^t(z, w) \in D(A)$. Then, from (11.A.10), one deduces

(11.A.11)
$$-\tilde{N}(w, v^0) + (w_{|{}^1\partial\mathcal{M}}, \alpha v^1{}_{|{}^1\partial\mathcal{M}})_{L^2({}^1\partial\mathcal{M})} = (w, w^1)_{L^2(\mathcal{M})}, \qquad w \in H^1_{\mathsf{D}}(\mathcal{M}).$$

Note that choosing $w \in \ker(\mathsf{P})$, one finds

$$-(w, w^1)_{L^2(\mathcal{M})} + (w_{|{}^1\partial\mathcal{M}}, \alpha v^1{}_{|{}^1\partial\mathcal{M}})_{L^2({}^1\partial\mathcal{M})} = 0. \tag{11.A.12}$$

Then, Theorem 4.19 applies to the variational equation (11.A.11) under condition (11.A.12), yielding a unique solution $v \in \underline{H}^2(\mathcal{M}) \cap H^1_{\mathsf{D}}(\mathcal{M})$ (use also Theorem 4.18). As a result, $v^0 - v \in \ker(\mathsf{P})$ implying that $v^0 \in H^2_{\mathsf{D}}(\mathcal{M})$ and $Bv^0 = \alpha v^1$ on ${}^1\partial\mathcal{M}$. As $V = {}^t(v^0, v^1) \in \dot{\mathscr{H}}$, we have $F_\alpha(V) = 0$. We have therefore obtained that $D(\dot{A}^*)$ is as described in (11.A.3). The proof is complete. ■

The resolvent estimate of the following lemma is key in the proof that $\dot{A}$ generates a C_0-semigroup on $\dot{\mathscr{H}}$.

LEMMA 11.16. *Let $B = \dot{A}$ or $\dot{A}^*$ and $z \in \mathbb{C}$ be such that $\operatorname{Re} z < 0$. We have*

$$\|(z \operatorname{Id}_{\dot{\mathscr{H}}} - B)U\|_{\dot{\mathscr{H}}} \geq |\operatorname{Re} z|\, \|U\|_{\dot{\mathscr{H}}}, \quad U \in D(B).$$

PROOF. Let $U = {}^t(u^0, u^1) \in D(\dot{A})$. We write

$$\begin{aligned}((z \operatorname{Id}_{\dot{\mathscr{H}}} - \dot{A})U, U)_{\dot{\mathscr{H}}} &= \left(\begin{pmatrix} zu^0 + u^1 \\ zu^1 - P_0 u^0 \end{pmatrix}, \begin{pmatrix} u^0 \\ u^1 \end{pmatrix} \right)_{\dot{\mathscr{H}}} \\ &= z\|U\|^2_{\dot{\mathscr{H}}} + \tilde{N}(u^1, u^0) - (Pu^0, u^1)_{L^2(\mathcal{M})} \\ &= z\|U\|^2_{\dot{\mathscr{H}}} + \tilde{N}(u^1, u^0) - (R_1 u^0, u^1)_{L^2(\mathcal{M})} \\ &\quad - (\nabla_g u^0, \nabla_g u^1)_{L^2V(\mathcal{M})} + (\partial_\nu u^0{}_{|\partial\mathcal{M}}, u^1{}_{|\partial\mathcal{M}})_{L^2(\partial\mathcal{M})},\end{aligned}$$

using formula (18.5.5). Since we have $B_\partial(U) = (\partial_\nu + B')u^0{}_{|\partial\mathcal{M}} + \alpha u^1{}_{|\partial\mathcal{M}} = 0$, we obtain

$$((z \operatorname{Id}_{\dot{\mathscr{H}}} - \dot{A})U, U)_{\dot{\mathscr{H}}} = z\|U\|^2_{\dot{\mathscr{H}}} + 2i \operatorname{Im} \tilde{N}(u^1, u^0) - (\alpha u^0{}_{|\partial\mathcal{M}}, u^1{}_{|\partial\mathcal{M}})_{L^2(\partial\mathcal{M})},$$

recalling the definition of $\tilde{N}(.,.)$ in (4.6.5). Computing the real part of the above equality, one obtains

$$-\operatorname{Re}((z \operatorname{Id}_{\dot{\mathscr{H}}} - \dot{A})U, U)_{\dot{\mathscr{H}}} = -\operatorname{Re}(z)\|U\|^2_{\dot{\mathscr{H}}} + \int_{\partial\mathcal{M}} \alpha |u^1|^2_{|\partial\mathcal{M}}\, \mu_{g_\partial}. \tag{11.A.13}$$

As $\alpha \geq 0$ and $\operatorname{Re} z < 0$, we find

$$|\operatorname{Re}((z \operatorname{Id}_{\dot{\mathscr{H}}} - \dot{A})U, U)_{\dot{\mathscr{H}}}| \geq |\operatorname{Re}(z)|\, \|U\|^2_{\dot{\mathscr{H}}},$$

which yields the conclusion for $B = \dot{A}$.

The conclusion follows similarly in the case $B = \dot{A}^*$. ■

We may now state and prove that $\dot{A}$ generates a C_0-semigroup.

THEOREM 11.17. *The unbounded operator* $(\dot{A}, D(\dot{A}))$ *generates a* C_0-*semigroup of contraction* $\dot{S}(t) = e^{-t\dot{A}}$ *on* $\dot{\mathscr{H}}$.

The argument, based on the Hille–Yosida theorem, is that of the proof of Theorem 10.22 without modification.

We define the following bounded operator on $\mathscr{H}$:

$$S(t) := \dot{S}(t) \circ \Pi_{\dot{\mathscr{H}}} + \Pi_{\mathcal{N}}, \tag{11.A.14}$$

with $t \geq 0$ as a parameter, and where $\Pi_{\dot{\mathscr{H}}}$ and $\Pi_{\mathcal{N}}$ are the projectors associated with the direct sum in (11.3.2). Observe that we have the following commutative diagram:

$$\begin{array}{ccc} \mathscr{H} & \xrightarrow{S(t)} & \mathscr{H} \\ \downarrow \Pi_{\dot{\mathscr{H}}} & & \downarrow \Pi_{\dot{\mathscr{H}}} \\ \dot{\mathscr{H}} & \xrightarrow{\dot{S}(t)} & \dot{\mathscr{H}} \end{array}$$

The operator $S(t)$ coincides with $\dot{S}(t)$ on $\dot{\mathscr{H}}$ and coincides with the identity map on its complement $\mathcal{N}$.

THEOREM 11.18. *The map* $S(t)$ *is a bounded* C_0-*semigroup on* $\mathscr{H}$; *its generator is the unbounded operator* $(A, D(A))$.

The proof is *mutatis mutandis* that of Theorem 10.23.

PROPOSITION 11.19. *On* $D(\dot{A})$ *the norm inherited from that of* $H^2(\mathcal{M}) \oplus H^1(\mathcal{M})$ *is equivalent to the norm* $U \mapsto \|\dot{A}U\|_{\dot{\mathscr{H}}} \eqsim \|\dot{A}U\|_{\mathscr{H}}$.

PROOF. Let $U = {}^t(u^0, u^1) \in D(\dot{A}) \subset H^2(\mathcal{M}) \oplus H^1(\mathcal{M})$. We have $\dot{A}U = {}^t(-u^1, Pu^0) \in \dot{\mathscr{H}}$ by Lemma 11.5 and as $A = \dot{A} \circ \Pi_{\dot{\mathscr{H}}}$. Following the proof of the counterpart Proposition 10.25, it suffices to prove that we have

$$\|u^0\|_{H^2(\mathcal{M})} \lesssim \|u^1\|_{H^1(\mathcal{M})} + \|Pu^0\|_{L^2(\mathcal{M})}. \tag{11.A.15}$$

Since $B_\partial(U) = 0$ we have

$$\begin{aligned} Bu^0{}_{|{}^1\partial\mathcal{M}} &= -\alpha u^1{}_{|{}^1\partial\mathcal{M}} \in H^{1/2}({}^1\partial\mathcal{M}), \\ &\text{with } \|\alpha u^1{}_{|{}^1\partial\mathcal{M}}\|_{H^{1/2}({}^1\partial\mathcal{M})} \lesssim \|u^1\|_{H^1(\mathcal{M})}, \end{aligned}$$

by the trace inequality of Proposition 18.24. If $\varphi \in \ker(\mathsf{P})$ with the identity (4.5.3), one has

$$(Pu^0, \varphi)_{L^2(\mathcal{M})} + (Bu^0_{|{}^1\partial\mathcal{M}}, \varphi_{|{}^1\partial\mathcal{M}})_{L^2({}^1\partial\mathcal{M})} = 0.$$

Then, by the elliptic regularity result of Theorem 4.18, one has

$$\|\underline{u}^0\|_{H^2(\mathcal{M})} \lesssim \|Pu^0\|_{L^2(\mathcal{M})} + \|u^1\|_{H^1(\mathcal{M})}, \tag{11.A.16}$$

where $\underline{u}^0 = u^0 - \Pi_{\ker(\mathsf{P})} u^0$, where $\Pi_{\ker(\mathsf{P})}$ is the orthogonal projection onto $\ker(\mathsf{P})$ in $L^2(\mathcal{M})$. We thus have

$$\|u^0\|_{H^2(\mathcal{M})} \lesssim \|\underline{u}^0\|_{H^2(\mathcal{M})} + \|\Pi_{\ker(\mathsf{P})} u^0\|_{H^2(\mathcal{M})}. \tag{11.A.17}$$

We set $\underline{U} = {}^t(\underline{u}^0, u^1) = U - {}^t(\Pi_{\ker(\mathsf{P})} u^0, 0) \in \mathscr{H}$. Since $U \in D(\dot{A})$, one has

$$\Pi_{\mathcal{N}} \underline{U} = -{}^t(\Pi_{\ker(\mathsf{P})} u^0, 0),$$

since ${}^t(\Pi_{\ker(\mathsf{P})} u^0, 0) \in \mathcal{N}$.

Since $\dim \ker(\mathsf{P}) < \infty$ one has

$$\begin{aligned} \|\Pi_{\ker(\mathsf{P})} u^0\|_{H^2(\mathcal{M})} &\lesssim \|\Pi_{\ker(\mathsf{P})} u^0\|_{H^1(\mathcal{M})} \lesssim \|\Pi_{\mathcal{N}} \underline{U}\|_{\mathscr{H}} \\ &\lesssim \|\underline{U}\|_{\mathscr{H}} \lesssim \|\underline{u}^0\|_{H^1(\mathcal{M})} + \|u^1\|_{L^2(\mathcal{M})}, \end{aligned}$$

from the continuity of $\Pi_{\mathcal{N}}$ on $\mathscr{H}$. With (11.A.16) and (11.A.17), we thus obtain (11.A.15), which concludes the proof. ■

From Proposition 11.19 arguing as in the proof of Corollary 10.26, one obtains the following result.

COROLLARY 11.20. *On $D(A)$, the norm inherited from that of $H^2(\mathcal{M}) \oplus H^1(\mathcal{M})$ is equivalent to the graph norm $\|.\|_{D(A)}$ given by $\|V\|^2_{D(A)} = \|V\|^2_{\mathscr{H}} + \|AV\|^2_{\mathscr{H}}$.*

The study of the stabilization properties of the damped wave equation requires to prove that the imaginary axis $\{\operatorname{Re} z = 0\}$ is in the resolvent set of the operator $\dot{A}$. This is the subject of the following proposition.

PROPOSITION 11.21. *The spectrum of $(\dot{A}, D(\dot{A}))$ is contained in*

$$\{z \in \mathbb{C};\ \operatorname{Re}(z) > 0\}.$$

PROOF. Let $z \in \mathbb{C}$. We consider the two cases.

Case 1: $\operatorname{Re} z < 0$. Replacing Lemma 10.21 by Lemma 11.16, one concludes that z is in the resolvent set of $\dot{A}$ as in the counterpart case in the proof of Proposition 10.27.

Case 2: $\operatorname{Re} z = 0$. We start by proving the injectivity of $z \operatorname{Id}_{\dot{\mathscr{H}}} - \dot{A}$. Let thus $U = {}^t(u^0, u^1) \in D(\dot{A})$ be such that $zU - \dot{A}U = 0$. This gives

$$z u^0 + u^1 = 0, \quad -P u^0 + z u^1 = 0.$$

First, if $z = 0$, one has $u^1 = 0$, $P u^0 = 0$, and $B u^0{}_{|\partial\mathcal{M}} = 0$ as $B_\partial(U) = 0$. Thus $u^0 \in \ker(\mathsf{P})$, meaning that $U = {}^t(u^0, 0) \in \mathcal{N}$. Yet $\Pi_{\mathcal{N}} U = 0$ since $U \in \dot{\mathscr{H}} = \ker(\Pi_{\mathcal{N}})$. Thus $U = 0$.

Second, if $z \neq 0$, using (11.A.13), we obtain

$$0 = \operatorname{Re}((z \operatorname{Id}_{\dot{\mathscr{H}}} - \dot{A})U, U)_{\dot{\mathscr{H}}} = - \int_{\partial\mathcal{M}} \alpha |u^1|^2_{|\partial\mathcal{M}} \, \mu_{g_\partial}.$$

With the same argument as that used in the counterpart case in the proof of Proposition 10.27, one obtains

$$P_0 u^0 = -z^2 u^0, \quad \text{and} \quad u^0_{|{}^1\partial\mathcal{M}} = \partial_\nu u^0_{|{}^1\partial\mathcal{M}} = 0 \text{ on } \Gamma,$$

where Γ is an open set of ${}^1\partial\mathcal{M}$ where $\alpha > 0$. With the unique continuation property, initiated from the boundary, as stated in Theorem 9.8, we obtain that u^0 vanishes in $\mathcal{M}$ and u^1 as well.

The proof of the surjectivity of $z \operatorname{Id}_{\dot{\mathscr{H}}} - \dot{A}$ is the same as in the end of the proof of Proposition 10.27; it is based on the Fredholm index arguments. The result then follows from the closed graph theorem using that $\dot{A}$ is a closed operator. ∎

CHAPTER 12

Spectral Inequality for General Boundary Conditions and Application

Contents

12.1. Setting

For $(\mathcal{M}, g)$ a smooth d-dimensional connected compact Riemannian manifold with boundary, with $d \geq 2$, we consider a second-order differential operator P on $\mathcal{M}$ given by $P = -\Delta_g + R_1$ where R_1 is a first-order differential operator. We also consider a differential boundary operator B defined in a neighborhood of $\partial\mathcal{M}$ of order 0 or 1 on each connected component of $\partial\mathcal{M}$. On the part of $\partial\mathcal{M}$ where the order is 0, denoted by ${}^0\partial\mathcal{M}$, we consider the Dirichlet boundary operator, that is, $Bu_{|{}^0\partial\mathcal{M}} = u_{|{}^0\partial\mathcal{M}}$.

J. Le Rousseau et al., *Elliptic Carleman Estimates and Applications to Stabilization and Controllability, Volume II*, PNLDE Subseries in Control 98, https://doi.org/10.1007/978-3-030-88670-7_12

We introduce the unbounded operator P on $L^2(\mathcal{M})$ with domain

$$D(\mathsf{P}) = \{u \in L^2(\mathcal{M});\ Pu \in L^2(\mathcal{M}) \text{ and } Bu_{|\partial\mathcal{M}} = 0\}.$$

As in the previous chapter we consider the case where

- the map $u \mapsto (Pu, Bu)$ has Fredholm properties;
- the unbounded operator $(\mathsf{P}, D(\mathsf{P}))$ is selfadjoint;
- the spectrum of P is composed of a sequence of eigenvalues

$$\mu_0 \le \mu_1 \le \cdots \le \mu_n \le \cdots,$$

each of finite multiplicity, that grows to $+\infty$.

Those properties are fulfilled under assumptions made on R_1 and B.

Fredholm properties are equivalent to having (P, B) satisfying the Lopatinskiĭ–Šapiro conditions; see Theorem 3.1. In such case, as observed in the beginning of Chap. 4 one has

$$D(\mathsf{P}) = \{u \in H^2(\mathcal{M});\ Bu_{|\partial\mathcal{M}} = 0\}.$$

By Theorem 4.3 the selfadjointness of $(\mathsf{P}, D(\mathsf{P}))$ is equivalent to having

$$R_1 = iV + f \text{ in } \mathcal{M} \quad \text{and} \quad B = \partial_\nu + B' \text{ with } B' = iX' + h \text{ in } {}^1\partial\mathcal{M},$$

where we denote by ∂_ν the outward pointing normal derivative in the sense of the metric g (see (5.3.1) in Sect. 5.3 and (17.4.12) in Sect. 17.4), V a smooth real vector field on $\mathcal{M}$, f a complex-valued function on $\mathcal{M}$, X' a smooth real vector field on ${}^1\partial\mathcal{M}$, and h a complex-valued function on $\partial\mathcal{M}$, with the additional properties

$$\operatorname{Im} f = \operatorname{div}_g V/2 \text{ in } \mathcal{M} \quad \text{and} \quad 2\operatorname{Im} h - \operatorname{div}_g X' + g(V, \nu)_{|{}^1\partial\mathcal{M}} = 0 \text{ in } {}^1\partial\mathcal{M}.$$

Moreover, in the case $d = 2$, one has $|X'|_{g_\partial} \neq 1$ and in the case $d \geq 3$, we moreover have $|X'|_{g_\partial} < 1$.

We furthermore require $|X'|_{g_\partial} < 1$ independently of the dimension. Consequently, the spectrum of P is composed of a sequence of eigenvalues of finite multiplicities that grows to $+\infty$; see Theorems 4.9 and 4.12. If one does not assume $|X'|_{g_\partial} < 1$, in dimension $d = 2$ the case of eigenvalues that go to $-\infty$ can occur (see Proposition 4.13). The operator P could not be the generator of a C_0-semigroup in such case. This case is not of relevance for us as we aim to study the null-controllability of the associated parabolic equation.

In Sect. 12.2 we prove a spectral inequality associated with the spectral decomposition of P of the form of the spectral inequality proven in Chapter 7 of Volume 1. In Sect. 12.3 we define the parabolic semigroup associated with P and weak solution of the nonhomogeneous Cauchy problem that are needed for the definition of the controlled parabolic equation and its solutions

$$
\begin{cases} \partial_t y + Py = \mathbf{1}_\omega v & \text{in } (0,T)\times\mathcal{M}, \\ By = 0 & \text{on } (0,T)\times\partial\mathcal{M}, \\ y(0) = y^0 & \text{in } \mathcal{M}, \end{cases} \tag{12.1.1}
$$

where v is the control function that only acts in an open region ω of $\mathcal{M}$. Finally, in Sect. 12.4 we prove the null-controllability of (12.1.1).

The proof schemes are similar to those given in Chapter 7 of Volume 1. Arguments are adapted to take into account the more general boundary conditions and operator involved here.

12.2. Spectral Inequality

One of the main results of this chapter is the following spectral inequality, counterpart of Theorem 7.10 of Volume 1.

THEOREM 12.1 (Spectral Inequality). *Let P and B be such that the Lopatinskiĭ–Šapiro condition holds on $\partial\mathcal{M}$ and $(\mathsf{P}, D(\mathsf{P}))$ is selfadjoint. In the case $d=2$ assume furthermore that X' is such that $|X'|_{g_\partial} < 1$ on ${}^1\partial\mathcal{M}$.*

Let $(\phi_j)_{j\in\mathbb{N}}$ be a Hilbert basis of $L^2(\mathcal{M})$ made of eigenfunctions of $(\mathsf{P}, D(\mathsf{P}))$ associated with the sequence $\mu_0 \leq \mu_1 \leq \cdots \leq \mu_n \leq \cdots$ of eigenvalues. Let ω be an open set in $\mathcal{M}$. There exists $K>0$ such that for all $\mu \geq \max(\mu_0, 0)$ one has

$$
\|w\|_{L^2(\mathcal{M})} \leq K e^{K\sqrt{\mu}} \|w\|_{L^2(\omega)}, \qquad w \in \operatorname{span}\{\phi_j;\ \mu_j \leq \mu\}. \tag{12.2.1}
$$

PROOF. The proof is along the lines of that of Theorem 7.10, yet it uses the unique continuation property of Chap. 9 consequence of the Carleman estimate of Chap. 8 derived under the Lopatinskiĭ–Šapiro condition.

If $\mu_0 \leq 0$, we replace P by $\mathsf{P}_\lambda = \mathsf{P} + \lambda\,\mathrm{Id}$ with $\lambda > -\mu_0$. Then, the eigenvalues of P_λ are simply $0 < \mu_0 + \lambda \leq \mu_1 + \lambda \leq \cdots$. Observe that if one derives a spectral inequality as (12.2.1) for P_λ, then one obtains also such an inequality for P, since $(\mu+\lambda)^{1/2} \lesssim \mu^{1/2}$ for $\mu \geq 0$. We may therefore assume that $\mu_0 > 0$.

Let $S_0 > 0$ and $\alpha \in (0, S_0/2)$ and set $\mathcal{V} = (0, S_0) \times \mathcal{M}$ and $\mathcal{U} = (\alpha, S_0 - \alpha) \times \mathcal{M}$. Set also $z = (s,x)$ with $s \in (0, S_0)$ and $x \in \mathcal{M}$. We define the following augmented elliptic operator $Q := D_s^2 + P$ in $\mathcal{V}$. We equip $\tilde{\mathcal{M}} = \mathbb{R} \times \mathcal{M}$ with the metric $\tilde{g} = ds \otimes ds + g$.

With the same argument as in the proof of Lemma 11.8 in Sect. 11.5 we find that the Lopatinskiĭ–Šapiro condition holds for (Q, B) on $\mathbb{R} \times \partial\mathcal{M} = \partial\tilde{\mathcal{M}}$.

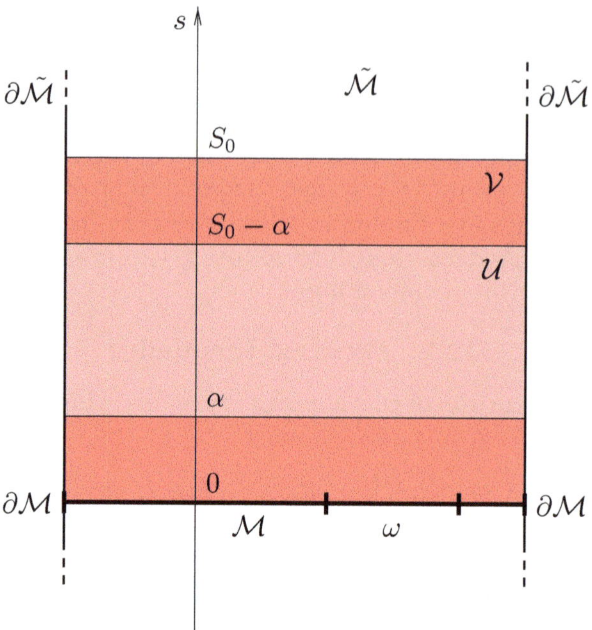

FIGURE 12.1. Geometry for the application of the quantified unique continuation property of Theorem 9.6

Since ω is an open set in $\mathcal{M}$, from Theorem 9.6 there exist $C > 0$ and $\delta \in (0, 1)$ such that for $u \in H^2(\mathcal{V})$ that satisfies $Bu_{|\partial\tilde{\mathcal{M}}} = 0$ for $s \in (0, S_0)$ one has

(12.2.2)

$$\|u\|_{H^1(\mathcal{U})} \leq C\|u\|_{H^1(\mathcal{V})}^{1-\delta} \left(\|Qu\|_{L^2(\mathcal{V})} + |u_{|s=0}|_{H^1(\omega)} + |\partial_s u_{|s=0}|_{L^2(\omega)}\right)^{\delta}.$$

The geometry associated with the application of Theorem 9.6 is illustrated in Fig. 12.1.

Let $\mu \geq \mu_0$ and $w \in \text{span}(\phi_j;\ \mu_j \leq \mu)$, meaning that $w = \sum_{\mu_j \leq \mu} \alpha_j \phi_j$. We set

$$u(s, .) = \sum_{\mu_j \leq \mu} \frac{\alpha_j}{\mu_j^{1/2}} \sinh\left(s\mu_j^{1/2}\right)\phi_j.$$

It is in $H^2(\mathcal{V})$ and

$$Qu = 0, \qquad u_{|s=0} = 0.$$

Inequality (12.2.2) applied to u reads

$$\|u\|_{H^1(\mathcal{U})} \leq C\|u\|_{H^1(\mathcal{V})}^{1-\delta} |\partial_s u_{|s=0}|_{L^2(\omega)}^{\delta}.$$

As $\partial_s u_{|s=0} = \sum_{\mu_j \leq \mu} \alpha_j \phi_j = w$, we have

$$|\partial_s u_{|s=0}|_{L^2(\omega)} = \|w\|_{L^2(\omega)}. \tag{12.2.3}$$

Note that

$$\mu_j^{1/2}|\sinh\big(s\mu_j^{1/2}\big)| + |\cosh\big(s\mu_j^{1/2}\big)| \lesssim e^{C\sqrt{\mu}}. \tag{12.2.4}$$

This gives

$$\|u\|^2_{L^2(\mathcal{V})} + \|\partial_s u\|^2_{L^2(\mathcal{V})} \lesssim e^{C\sqrt{\mu}} \sum_{\mu_j\le\mu} |\alpha_j|^2.$$

We also write with Corollary 3.3

$$\|\nabla_g u(s,.)\|_{L^2V(\mathcal{M})} \lesssim \|u(s,.)\|_{H^2(\mathcal{M})} \lesssim \|Pu(s,.)\|_{L^2(\mathcal{M})} + \|u(s,.)\|_{L^2(\mathcal{M})},$$

With (12.2.4) one has

$$\int_0^{S_0} \|Pu(s,.)\|^2_{L^2(\mathcal{M})} ds = \sum_{\mu_j\le\mu} |\alpha_j|^2 \int_0^{S_0} \mu_j |\sinh(s\mu_j^{1/2})|^2 ds \lesssim e^{C\sqrt{\mu}} \sum_{\mu_j\le\mu} |\alpha_j|^2,$$

yielding

$$\int_0^{S_0} \|\nabla_g u(s,.)\|^2_{L^2V(\mathcal{M})} ds \lesssim e^{C\sqrt{\mu}} \sum_{\mu_j\le\mu} |\alpha_j|^2.$$

We thus conclude that

$$\|u\|^2_{H^1(\mathcal{V})} \lesssim e^{C\sqrt{\mu}} \sum_{\mu_j\le\mu} |\alpha_j|^2. \tag{12.2.5}$$

Observe now that for $s \ge \alpha$ we have $\sinh(s\mu_j^{1/2})/\mu_j^{1/2} \ge \alpha$. Then, for some $\kappa > 0$ we have

$$\begin{aligned} \kappa \sum_{\mu_j\le\mu} |\alpha_j|^2 &\le \sum_{0<\mu_j\le\mu} |\alpha_j|^2 \int_\alpha^{S_0-\alpha} \mu_j^{-1} |\sinh(s\mu_j^{1/2})|^2 ds \\ &= \|w\|^2_{L^2(\mathcal{U})} \le \|w\|^2_{H^1(\mathcal{U})}. \end{aligned} \tag{12.2.6}$$

With (12.2.2), (12.2.3), (12.2.5), and (12.2.6) we obtain

$$\sum_{\mu_j\le\mu} |\alpha_j|^2 \lesssim e^{C\sqrt{\mu}} \|w\|^2_{L^2(\omega)},$$

which is the result. ∎

12.3. The Parabolic Semigroup

As in the proof of the spectral inequality in the previous section, if $\mu_0 \le 0$ we consider the operator $\mathsf{P}_\lambda = \mathsf{P} + (\lambda - \mu_0)\,\mathrm{Id}$, with the same domain $D(\mathsf{P})$. If we find that P_λ generates a C_0-semigroup $S_\lambda(t)$ on $L^2(\mathcal{M})$, then $e^{(\lambda-\mu_0)t} S_\lambda(t) = S_\lambda(t) e^{(\lambda-\mu_0)t}$ is a C_0-semigroup generated by P. We may thus assume that $\mu_0 > 0$.

Under the above assumption, we have the following results. Proofs are the same as in the classical case of homogeneous Dirichlet boundary conditions presented in Section 10.2 of Volume 1, using the analysis of the elliptic and spectral properties of $(\mathsf{P}, D(\mathsf{P}))$ in Sects. 4.6 and 4.7.

THEOREM 12.2. *Assume that* $\mu_0 > 0$. *The operator* $(\mathsf{P}, D(\mathsf{P}))$ *is maximal monotone. Let* $T \in \mathbb{R}_+ \cup \{+\infty\}$. *The operator* P *generates* C_0*-semigroup of contraction* $S(t) = e^{-t\mathsf{P}}$ *on* $L^2(\mathcal{M})$. *If* $y^0 \in D(\mathsf{P})$, *then* $y(t) = S(t)y^0$ *is the unique solution in*

$$\mathscr{C}^0([0,T]; D(\mathsf{P})) \cap \mathscr{C}^1([0,T]; L^2(\mathcal{M})),$$

such that $y(0) = y^0$ *and*

$$\frac{d}{dt}y(t) + \mathsf{P}y(t) = 0$$

holds in $L^2(\mathcal{M})$ *for all* $0 \le t \le T$.

The definition of monotone operators is recalled in Definition 12.22 of Volume 1 in the case of Hilbert spaces. That of maximal monotone operators is given in Definition 12.10. Here, $[0,T]$ means $[0,+\infty)$ if $T = +\infty$.

12.3.1. Spectral Representation and Sobolev Scale. The spectral representation of the semigroup $S(t)$ is the same as in Section 10.2.1 of Volume 1 and is based on the spectral characterization of $L^2(\mathcal{M})$ by means of a Hilbert basis $(\phi_j)_{j\in\mathbb{N}}$ of eigenfunction of $(\mathsf{P}, D(\mathsf{P}))$, and that of $H^1_{\mathsf{D}}(\mathcal{M})$ and $D(\mathsf{P})$ given in Proposition 4.16.

If $y^0 \in L^2(\mathcal{M})$ and we set $y_j(t) = e^{-t\mu_j}(y^0, \phi_j)_{L^2(\mathcal{M})}$, for $t \ge 0$ and $j \in \mathbb{N}$, then $(y_j(t))_j \in \mathscr{C}^0([0,T], \ell^2(\mathbb{C}))$ and

$$S(t)y^0 = \sum_{j\in\mathbb{N}} y_j(t)\phi_j, \quad t \ge 0,$$

with convergence in $L^2(\mathcal{M})$.

As in Section 10.1.3 of Volume 1 we can define the following spaces associated with the spectral family: for $s \ge 0$, we set

$$K^s(\mathcal{M}) = \{u = \sum_{j\in\mathbb{N}} u_j\phi_j;\ (\mu_j^{s/2}u_j)_j \in \ell^2(\mathbb{C})\}.$$

We have $K^2(\mathcal{M}) = D(\mathsf{P})$, $K^1(\mathcal{M}) = H^1_{\mathsf{D}}(\mathcal{M})$; see Proposition 4.16. Moreover, one has $K^s = D(\mathsf{P}^{s/2})$. For $s \ge 0$ we also set $K^{-s}(\mathcal{M}) = \big(K^s(\mathcal{M})\big)'$ and we find that if $u \in K^{-s}(\mathcal{M})$, then

$$\big(\mu_j^{-s/2}\langle u, \phi_j\rangle_{K^{-s}(\mathcal{M}),K^s(\mathcal{M})}\big)_j \in \ell^2(\mathbb{C}).$$

Using $L^2(\mathcal{M})$ as a *pivot* space, we thus identify

$$K^{-s}(\mathcal{M}) = \{u = \sum_{j\in\mathbb{N}} u_j\phi_j;\ (\mu_j^{-s/2}u_j)_j \in \ell^2(\mathbb{C})\}.$$

Similarly to Theorem 10.22 of Volume 1 we have the following result.

THEOREM 12.3. *Let* $T \in \mathbb{R}_+ \cup \{+\infty\}$. *The semigroup* $S(t)$ *is analytic and for* $y^0 \in L^2(\mathcal{M})$, *the function* $y(t) = S(t)y^0$ *is in*

$$\mathscr{C}^0([0,T]; L^2(\mathcal{M})) \cap \mathscr{C}^\infty((0,T]; K^s(\mathcal{M})), \quad s \in \mathbb{R},$$

and is such that

(12.3.1)

$$y(0) = y^0 \quad \text{and} \quad \frac{d}{dt}y(t) + \mathsf{P}y(t) = 0 \text{ holds in } L^2(\mathcal{M}) \text{ for } 0 < t \leq T.$$

Moreover, $y(t) = S(t)y^0$ *is the unique solution of* (12.3.1) *in*

$$\mathscr{C}^0([0,T]; L^2(\mathcal{M})) \cap \mathscr{C}^1((0,T]; L^2(\mathcal{M})) \cap \mathscr{C}^0((0,T]; D(\mathsf{P})).$$

For $r \in \mathbb{R}$ and $s \geq 0$, we also define the unbounded operator $\mathsf{P}_r^s : K^r(\mathcal{M}) \to K^r(\mathcal{M})$ with domain $K^{r+2s}(\mathcal{M})$ given by

$$\mathsf{P}_r^s u = \sum_{j\in\mathbb{N}} \mu_j^s u_j \phi_j, \quad u = \sum_{j\in\mathbb{N}} u_j \phi_j \in K^{r+2s}(\mathcal{M}).$$

For $s = 1$ we write P_r instead of P_r^s. We have $\mathsf{P}_0 = \mathsf{P}$. If $r \geq 0$, the operator P_r^s is a restriction of P_0^s to $K^r(\mathcal{M}) \subset K^0(\mathcal{M}) = L^2(\mathcal{M})$. If $r < 0$, the operator P_r^s is an extension of P_0^s to $K^r(\mathcal{M}) \supset L^2(\mathcal{M})$.

For any $r \in \mathbb{R}$ the operator P_r generates a C_0-semigroup on $K^r(\mathcal{M})$. For these operators we have counterpart results to Lemma 10.28 and Theorem 10.29 of Volume 1.

For the nonhomogeneous parabolic Cauchy problem we can prove a result counterpart to Theorem 10.44 (also of Volume 1).

THEOREM 12.4. *Let* $T \in \mathbb{R}^+ \cup \{+\infty\}$ *and* $r \in \mathbb{R}$. *Let* $f \in L^2(0,T; K^r(\mathcal{M}))$ *and* $y^0 \in K^{r+1}(\mathcal{M})$. *There exists a unique function* $y \in \mathscr{C}^0([0,T]; K^{r+1}(\mathcal{M})) \cap L^2(0,T; K^{r+2}(\mathcal{M})) \cap H^1(0,T; K^r(\mathcal{M}))$ *that is solution of the parabolic equation*

$$\frac{d}{dt}y + \mathsf{P}_r y = f$$

in $L^2(0,T; K^r(\mathcal{M}))$ *and satisfies moreover* $y(0) = y^0$. *The solution is given by*

$$y(t) = S(t)y^0 + \int_0^t S(t-\sigma) f(\sigma) d\sigma.$$

Moreover, there exists $C > 0$ *such that*

$$\|y\|_{L^\infty(0,T;K^{r+1}(\mathcal{M}))} + \|y\|_{L^2(0,T;K^{r+2}(\mathcal{M}))} + \left\|\frac{d}{dt}y\right\|_{L^2(0,T;K^r(\mathcal{M}))} \leq C\Big(\|y^0\|_{K^{r+1}(\mathcal{M})} + \|f\|_{L^2(0,T;K^r(\mathcal{M}))}\Big).$$

For $r = 0$ we refer to strong solution in the case $y^0 \in K^1(\mathcal{M})$ and $f \in L^2((0,T) \times \mathcal{M})$ or in the case $y^0 \in D(\mathsf{P})$ and $f \in L^2(0,T; D(\mathsf{P}_0))$. Statements of Corollaries 10.46 and 10.47 of Volume 1 can be adapted.

12.3.2. Weak Solutions of the Nonhomogeneous Parabolic Cauchy Problem.

Weak solutions in the case of Lopatinskiĭ–Šapiro type boundary condition can be introduced similarly to the more classical case of homogeneous Dirichlet conditions, as presented in Section 10.3.4 of Volume 1.

DEFINITION 12.5. Let $T \in \mathbb{R}_+ \cup \{+\infty\}$. Let $y^0 \in L^2(\mathcal{M})$ and $f \in L^2(0,T;K^{-1}(\mathcal{M}))$. One says that $y \in \mathscr{C}^0([0,T];L^2(\mathcal{M})) \cap L^2(0,T;K^1(\mathcal{M}))$ is a weak solution to the parabolic equation

$$\frac{d}{dt}y + \mathsf{P}y = f, \qquad y(0) = y^0,$$

if we have

$$(y(t),\psi)_{L^2(\mathcal{M})} + (y,\mathsf{P}\psi)_{L^2((0,t)\times\mathcal{M})} = (y^0,\psi)_{L^2(\mathcal{M})} + \int_0^t \langle f(\sigma),\overline{\psi}\rangle_{K^{-1}(\mathcal{M}),K^1(\mathcal{M})}\, d\sigma,$$

for all $\psi \in D(\mathsf{P})$ and for all $t \in [0,T]$.

An inspection of the proof of Theorem 10.49 of Volume 1 shows the following result can be proven *mutatis mutandis.*

THEOREM 12.6. *Let $T \in \mathbb{R}_+ \cup \{+\infty\}$. Let $y^0 \in L^2(\mathcal{M})$ and $f \in L^2(0,T;K^{-1}(\mathcal{M}))$. There exists a unique weak solution y to the parabolic equation*

$$\frac{d}{dt}y + \mathsf{P}y = f, \qquad y(0) = y^0,$$

in the sense of Definition 12.5. It coincides with the solution of the semigroup equation

$$\frac{d}{dt}y + \mathsf{P}_{-1}y = f, \qquad y(0) = y^0,$$

given by Theorem 12.4 in the case $r = -1$. In particular we have

$$y \in \mathscr{C}^0([0,T];L^2(\mathcal{M})) \cap L^2(0,T;K^1(\mathcal{M})) \cap H^1(0,T;K^{-1}(\mathcal{M})).$$

Naturally, uniqueness implies that a strong solution is also a weak solution.

12.4. Null-Controllability of the Associated Parabolic Equation

For some $T > 0$ we consider the null-controllability problem of the parabolic problem associated with (P,B) with an inner control function v that acts on an open subset ω of $\mathcal{M}$,

$$\begin{cases} \partial_t y + Py = \mathbf{1}_\omega v & \text{in } (0,T)\times\mathcal{M}, \\ By = 0 & \text{on } (0,T)\times\partial\mathcal{M}, \\ y(0) = y^0 & \text{in } \mathcal{M}. \end{cases} \tag{12.4.1}$$

The technique we use is that of Chapter 7 of Volume 1, based on the spectral inequality derived in Sect. 12.2.

For $y^0 \in L^2(\mathcal{M})$ and $v \in L^2((0,T)\times\mathcal{M})$ solutions to the inner-controlled parabolic equation (12.4.1) are of weak type and given by Definition 12.5 and Theorem 12.6. In particular, such a solution y is unique in

$$\mathscr{C}^0([0,T];L^2(\mathcal{M})) \cap L^2(0,T;K^1(\mathcal{M})) \cap H^1(0,T;K^{-1}(\mathcal{M}))$$

and is given by the Duhamel formula

$$y(t) = S(t)y^0 + \int_0^t S(t-\sigma)\mathbf{1}_\omega v(\sigma)d\sigma, \tag{12.4.2}$$

with the semigroup defined in Sect. 12.3.

DEFINITION 12.7. We say that the system in (12.4.1) is null-controllable in $L^2(\mathcal{M})$ at time $T>0$ if for all $y^0 \in L^2(\mathcal{M})$ one can choose $v \in L^2\big((0,T)\times\mathcal{M}\big)$ such that the weak solution given by (12.4.2) satisfies $y(T)=0$.

Above we saw that the C_0-semigroup $S_\lambda(t)$ generated by $\mathsf{P}_\lambda = \mathsf{P} + (\lambda - \mu_0)\,\mathrm{Id}$ is given by $S_\lambda(t) = e^{-(\lambda-\mu_0)t}S(t)$. Hence, with (12.4.2) one sees that the parabolic equation associated with P is null-controllable *if and only if* the parabolic equation associated with P_λ is null-controllable. We may thus assume as above that $\mu_0 > 0$.

Null-controllability is equivalent to an observability inequality according to Proposition 7.7 of Volume 1. Here, it reads

$$\|S(T)z\|_{L^2(\mathcal{M})} \le C_{\mathrm{obs}}\|\mathbf{1}_\omega S(t)z\|_{L^2((0,T)\times\mathcal{M})}, \qquad z \in L^2(\mathcal{M}). \tag{12.4.3}$$

We prove this observability inequality with the dual technique of Section 7.8 of Volume 1.

LEMMA 12.8. *Let $T>0$. Let $f:[0,T]\to\mathbb{R}$ be given by*

$$f(t) = \frac{t}{4}\exp\big(-K(K+\sqrt{K^2+2t\ln 6})/t\big),$$

where K is the constant in the spectral inequality of Theorem 12.1. We then have

$$f(t)\|S(t)z\|^2_{L^2(\mathcal{M})} - f(t/2)\|z\|^2_{L^2(\mathcal{M})} \le \int_0^t \|\mathbf{1}_\omega S(\sigma)z\|^2_{L^2(\mathcal{M})}d\sigma.$$

The proof is exactly that of Lemma 7.20 of Volume 1 replacing the spectral inequality of Theorem 7.10 by that of Theorem 12.1; see Remark 7.21 (recall that $(\mathsf{P}, D(\mathsf{P}))$ is such that $\mu_0 > 0$ here).

With Lemma 7.19 of Volume 1 we conclude that the observability inequality holds.

THEOREM 12.9 (Observability Under General Boundary Conditions). *Let $T>0$. There exists C_T such that,*

$$\|S(T)z^0\|_{L^2(\mathcal{M})} \le C_T\|\mathbf{1}_\omega S(t)z^0\|_{L^2((0,T)\times\mathcal{M})}, \qquad z^0 \in L^2(\mathcal{M}).$$

Equivalently, we have the following null-controllability result.

THEOREM 12.9′ (Null-Controllability Under General Boundary Conditions). *Let $T > 0$. There exists $C_T > 0$ such that for all initial conditions $y^0 \in L^2(\mathcal{M})$, there exists $v \in L^2((0,T) \times \mathcal{M})$, with $\|v\|_{L^2((0,T)\times\mathcal{M})} \leq C_T \|y^0\|_{L^2(\mathcal{M})}$, such that the weak solution $y(t) \in \mathscr{C}^0([0,T]; L^2(\mathcal{M})) \cap L^2(0,T; K^1(\mathcal{M})) \cap H^1(0,T; K^{-1}(\mathcal{M}))$ to*

$$\frac{d}{dt}y + \mathsf{P}y = \mathbf{1}_\omega v, \qquad y(0) = y^0,$$

satisfies $y(T) = 0$.

12.5. Notes

In the case $d = 2$, let $P = -\Delta_g + iV + f$ be as in the main text and $B = \partial_\nu + iX' + h$. Assume that the conditions of Theorem 4.3 hold meaning that $(\mathsf{P}, D(\mathsf{P}))$ is selfadjoint. Assume that $|X'|_{g_\partial} > 1$ on some connected component $\mathcal{N}$ of ${}^1\partial\mathcal{M}$, which is permitted by Lemma 4.5. Such an example is given in Proposition 4.13. Such an operator does not fit the framework of Theorem 12.1. If a spectral inequality were to hold for $(\mathsf{P}, D(\mathsf{P}))$, in the case of some discrete spectrum going to $-\infty$ the inequality should naturally take the following form

$$\|w\|_{L^2(\mathcal{M})} \leq K e^{K\sqrt{\mu}} \|w\|_{L^2(\omega)}, \qquad w \in \operatorname{span}\{\phi_j;\ |\mu_j| \leq \mu\}. \tag{12.5.1}$$

While the Lopatinskiĭ–Šapiro condition holds for (P, B) on $\mathcal{N}$ one can check with Proposition 2.8 that the same condition fails to hold for (Q, B) with the operator Q given by $Q = D_s^2 + P$. Note that the latter pair of operators is to be considered in dimension three as far as the Lopatinskiĭ–Šapiro condition is concerned. The proof of Theorem 12.1 does not apply to this case. To our knowledge having a spectral inequality for the operator $(\mathsf{P}, D(\mathsf{P}))$ in the form of (12.5.1) is an open question.

Here, we have only considered first-order boundary conditions. There exist higher-order boundary conditions such that together with the Laplace–Beltrami operator, the Lopatinskiĭ–Šapiro condition holds and one obtains a selfadjoint operator. An example is the so-called Ventcel boundary condition $B = \partial_\nu - h\Delta_{g_{\partial\mathcal{M}}}$ for $h > 0$, see Example 2.5-(4). With the results of R. Buffe in [92] obtained for a proof of logarithmic stabilization of the wave equation, it is very likely that a spectral inequality as in Theorem 12.1 can be derived and the null-controllability of the associated parabolic equation can be achieved.

Part 4

Further Aspects of Carleman Estimates

CHAPTER 13

Carleman Estimates with Source Terms of Weaker Regularity

Contents

J. Le Rousseau et al., *Elliptic Carleman Estimates and Applications to Stabilization and Controllability, Volume II*, PNLDE Subseries in Control 98, https://doi.org/10.1007/978-3-030-88670-7_13

13.1. Setting and Main Result

As in Chap. 8, on a d-dimensional compact Riemannian manifold $(\mathcal{M}, g)$, we consider the following elliptic problem

$$Pu = f, \qquad Bu_{|\partial\mathcal{M}} = f_\partial,$$

where $P = -\Delta_g + R_1$ with R_1 a first-order differential operator on $\mathcal{M}$ and B is differential operator of order at most $k \in \mathbb{N}$ in a neighborhood of a point of $\partial\mathcal{M}$, where we assume that P and B satisfy the Lopatinskiĭ–Šapiro condition (see Definition 2.2). Here, the novelty lies in the regularity assumed for both f and f_∂. We shall in fact assume that f takes the form

$$f = F_0 + \operatorname{div}_g F, \qquad F_0 \in L^2(\mathcal{M}),\ F \in L^2V(\mathcal{M}), \tag{13.1.1}$$

that is, $f \in H^{-1}(\mathcal{M})$. We recall that $L^2V(\mathcal{M})$ is the space of L^2-vector fields on $\mathcal{M}$ (see Sect. 18.1). Then, one cannot expect the solution to have a regularity higher than $H^1(\mathcal{M})$ in general. In fact, away from the boundary we have the following local result.

Proposition 13.1. *Let $(\mathcal{M}, g)$ be a compact Riemannian manifold with boundary and let $P = -\Delta_g + R_1$ with R_1 a first-order differential operator with bounded coefficients on $\mathcal{M}$. Let $m^0 \in \mathcal{M} \setminus \partial\mathcal{M}$, V^0 be an open neighborhood of m^0 in $\mathcal{M}$ such that $V^0 \cap \partial\mathcal{M} = \emptyset$, and let $\varphi \in \mathscr{C}^\infty(\mathcal{M})$ be such that the pair (P, φ) has the sub-ellipticity property of Definition 5.1 in $\overline{V^0}$. Then there exist an open neighborhood V^1 of m^0 in $\mathcal{M}$ and two constants C and $\tau_* > 0$ such that*

$$\tau^{1/2}\|e^{\tau\varphi}u\|_{\tau,1} \leq C\big(\|e^{\tau\varphi}F_0\|_{L^2(\mathcal{M})} + \tau\|e^{\tau\varphi}F\|_{L^2V(\mathcal{M})}\big), \tag{13.1.2}$$

for $\tau \geq \tau_*$ and $u \in H^1(\mathcal{M})$, $F_0 \in L^2(\mathcal{M})$, $F \in L^2V(\mathcal{M})$ such that

$$Pu = F_0 + \operatorname{div}_g F, \quad \text{and} \quad \operatorname{supp}(u) \cup \operatorname{supp}(F_0) \cup \operatorname{supp}(F) \subset \overline{V^1}.$$

A proof is given below.

If $u \in H^1(\mathcal{M})$, if a Carleman estimate is derived at the boundary, the norms associated with the boundary terms need to be adapted. Naturally, for $u \in H^1(\mathcal{M})$ the Dirichlet trace $u_{|\partial\mathcal{M}} \in H^{1/2}(\partial\mathcal{M})$. The Neumann trace $\partial_\nu u_{|\partial\mathcal{M}}$ is not defined in general. Here, ν is the outward pointing unit vector field at $\partial\mathcal{M}$. However, one has the following result.

Proposition 13.2. *Let $u \in H^1(\mathcal{M})$ be such that $Pu = F_0 + \operatorname{div}_g F$ with $F_0 \in L^2(\mathcal{M})$ and $F \in L^2V(\mathcal{M})$. Then, $\partial_\nu u + g(F, \nu)$ admits a trace at $\partial\mathcal{M}$ that lies in $H^{-1/2}(\partial\mathcal{M})$. Moreover, there exists $C > 0$ such that*

$$\big|(\partial_\nu u + g(F,\nu))_{|\partial\mathcal{M}}\big|_{H^{-1/2}(\partial\mathcal{M})} \leq C\big(\|F_0\|_{L^2(\mathcal{M})} + \|F\|_{L^2V(\mathcal{M})} + \|u\|_{H^1(\mathcal{M})}\big).$$

With the expression "admits a trace" one means that the trace map well defined for u and F smooth extends uniquely for the regularity we have here. A proof is given in Appendix 13.A.1.

As in Chaps. 2 and 8, we consider a boundary operator B of order at most one in the direction normal to the boundary, that is, if B is of order $k \in \mathbb{N}$ we have

$$Bu_{|\partial\mathcal{M}} = B^{k-1}\partial_\nu u_{|\partial\mathcal{M}} + B^k u_{|\partial\mathcal{M}},$$

where B^{k-1} and B^k are smooth differential operators on $\partial\mathcal{M}$ of order $k-1$ and k, respectively. If $k=0$, that is, $Bu_{|\partial\mathcal{M}}$ is the Dirichlet trace, we have $B^{k-1}=0$ and $B^k=1$.

Here, we also define the boundary operator

$$\mathsf{B}(u,h) = B^{k-1}(\partial_\nu u + h)_{|\partial\mathcal{M}} + B^k u_{|\partial\mathcal{M}}, \tag{13.1.3}$$

for a function h such that the trace $(\partial_\nu u + h)_{|\partial\mathcal{M}}$ makes sense. Proposition 13.2 indicates that one can use $h = g(F,\nu)$.

In the neighborhood of a point m^0 of the boundary $\partial\mathcal{M}$, we prove the following estimate.

PROPOSITION 13.3. *Let $(\mathcal{M},g)$ be a compact Riemannian manifold with boundary and let $P = -\Delta_g + R_1$ with R_1 a first-order differential operator with bounded coefficients on $\mathcal{M}$. Let $m^0 \in \partial\mathcal{M}$, V^0 be an open set of $\mathcal{M}$ such that $m^0 \in V^0$, and let $\varphi \in \mathscr{C}^\infty(\mathcal{M})$ be such that the pair (P,φ) has the sub-ellipticity property of Definition 5.1 in $\overline{V^0}$. Consider B a differential operator of order $k \in \mathbb{N}$ in V^0. Moreover, assume that (P,B,φ) satisfies the Lopatinskiĭ–Šapiro condition at m^0 (Definition 8.1). Then there exist an open neighborhood V^1 of m^0 in $\mathcal{M}$ and two constants C and $\tau_* > 0$ such that*

$$\begin{aligned} &\tau^{1/2}\|e^{\tau\varphi}u\|_{\tau,1} + \tau|e^{\tau\varphi}u_{|\partial\mathcal{M}}|_{\tau,1/2} + \tau|e^{\tau\varphi}(\partial_\nu u + g(F,\nu))_{|\partial\mathcal{M}}|_{\tau,-1/2} \\ &\quad\leq C\big(\|e^{\tau\varphi}F_0\|_{L^2(\mathcal{M})} + \tau\|e^{\tau\varphi}F\|_{L^2V(\mathcal{M})} + \tau|e^{\tau\varphi_{|\partial\mathcal{M}}}\mathsf{B}(u,g(F,\nu))|_{\tau,1/2-k}\big), \end{aligned} \tag{13.1.4}$$

for $\tau \geq \tau_*$ and $u \in H^1(\mathcal{M})$, $F_0 \in L^2(\mathcal{M})$, $F \in L^2V(\mathcal{M})$ such that

$$Pu = F_0 + \operatorname{div}_g F, \quad \textit{and} \quad \operatorname{supp}(u) \cup \operatorname{supp}(F_0) \cup \operatorname{supp}(F) \subset \overline{V^1}.$$

The norms used in the theorem are the Sobolev norms with a large parameter on $\mathcal{M}$ and $\partial\mathcal{M}$ introduced in Chap. 7.

Following Chap. 8, the proof of this proposition is carried out by first the introduction of local coordinates in Sect. 13.2, second the derivation of microlocal estimates in Sect. 13.3 and, third the patching of these estimates in Sect. 13.4.1.

The main result of this chapter follows from Propositions 13.1 and 13.3 by a patching argument in Sect. 13.4.2 already used in Section 3.5.

THEOREM 13.4. *Let $(\mathcal{M}, g)$ be a compact Riemannian manifold with boundary and let $P = -\Delta_g + R_1$ with R_1 a first-order differential operator on $\mathcal{M}$. Let V be an open set of $\mathcal{M}$ and set $V_\partial = V \cap \partial\mathcal{M}$. Let $\varphi \in \mathscr{C}^\infty(\mathcal{M})$ be such that the pair (P, φ) has the sub-ellipticity property of Definition 5.1 in $\overline{V}$. If $V_\partial \neq \emptyset$, consider B a differential operator of order $k \in \mathbb{N}$ in V and assume moreover that (P, B, φ) satisfies the Lopatinskiĭ–Šapiro condition at all points $m \in \overline{V_\partial}$ (Definition 8.1). Then, there exist C and $\tau_* > 0$ such that*

(13.1.5)
$$\tau^{1/2}\|e^{\tau\varphi}u\|_{\tau,1} + \tau|e^{\tau\varphi}u_{|\partial\mathcal{M}}|_{\tau,1/2} + \tau|e^{\tau\varphi}(\partial_\nu u + g(F,\nu))_{|\partial\mathcal{M}}|_{\tau,-1/2}$$
$$\leq C\big(\|e^{\tau\varphi}F_0\|_{L^2(\mathcal{M})} + \tau\|e^{\tau\varphi}F\|_{L^2V(\mathcal{M})} + \tau|e^{\tau\varphi_{|\partial\mathcal{M}}}\mathsf{B}(u, g(F,\nu))|_{\tau,1/2-k}\big),$$

for $\tau \geq \tau_*$ and $u \in H^1(\mathcal{M})$, $F_0 \in L^2(\mathcal{M})$, $F \in L^2V(\mathcal{M})$ such that

$$Pu = F_0 + \operatorname{div}_g F, \quad \textit{and} \quad \operatorname{supp}(u) \cup \operatorname{supp}(F_0) \cup \operatorname{supp}(F) \subset \overline{V}.$$

REMARK 13.5. With commutator arguments, as in Chapter 3 of Volume 1, we have

$$\|e^{\tau\varphi}u\|_{\tau,1} \eqsim \tau\|e^{\tau\varphi}u\|_{L^2(\mathcal{M})} + \|e^{\tau\varphi}\,\mathsf{D}\,u\|_{L^2\Lambda^1(\mathcal{M})},$$

where $\mathsf{D}\,u$ stands for the covariant derivative on $\mathcal{M}$ of u defined by means of the Levi-Civita connection. These notions are recalled in Sect. 17.4.

REMARK 13.6. The estimate of Theorem 13.4 is not a global estimate in the sense of Sections 3.6 and 5.4.1: the open set V cannot be equal to $\mathcal{M}$ since the sub-ellipticity condition on the whole $\mathcal{M}$ and the Lopatinskiĭ–Šapiro condition on the whole boundary $\partial\mathcal{M}$ are incompatible. The argument is similar to that given at the beginning of Section 3.6. If one has a global estimate, then Theorem 4.5 of Volume 1 used in all local charts implies that $|d\varphi| \geq C > 0$ in $\mathcal{M}$. Hence φ reaches its maximum on $\partial\mathcal{M}$, say at $m^0 \in \partial\mathcal{M}$. Then, $\partial_\nu\varphi(m^0) \geq 0$, which is compatible with having the Lopatinskiĭ–Šapiro condition holding at m^0 by the first item of Theorem 8.7.

A global estimate requires the introduction of an observation term as in Sections 3.6 and 5.4.1. Result of this type are given in Sect. 13.7.

REMARK 13.7. Compared with Theorem 8.14 the norm

$$\tau^{-1/2}\|e^{\tau\varphi}u\|_{\tau,2} \eqsim \tau^{3/2}\|e^{\tau\varphi}u\|_{L^2(\mathcal{M})} + \tau^{1/2}\|e^{\tau\varphi}\,\mathsf{D}\,u\|_{L^2\Lambda^1(\mathcal{M})}$$
$$+ \tau^{-1/2}\|e^{\tau\varphi}\,\mathsf{H}\,u\|_{L^2\Lambda^2(\mathcal{M})}$$

is replaced by

$$\tau^{1/2}\|e^{\tau\varphi}u\|_{\tau,1} \eqsim \tau^{3/2}\|e^{\tau\varphi}u\|_{L^2(\mathcal{M})} + \tau^{1/2}\|e^{\tau\varphi}\,\mathsf{D}\,u\|_{L^2\Lambda^1(\mathcal{M})}.$$

This is consistent with the smoothness of the function u that is limited to H^1 because of the nature of the source term $F_0 + \operatorname{div}_g F \in H^{-1}(\mathcal{M})$.

Similarly, for the estimation of the trace $u_{|\partial\mathcal{M}}$ the norm $|e^{\tau\varphi}\operatorname{tr}(u)|_{\tau,3/2}$ is replaced by $\tau|e^{\tau\varphi}u_{|\partial\mathcal{M}}|_{\tau,1/2}$ since here $u_{|\partial\mathcal{M}} \in H^{1/2}(\partial\mathcal{M})$.

We finish this introductory section by providing the proof of the Carleman estimate away from the boundary. The proof is rather simple; in fact, the main difficulties arise at the boundary and do not appear here. They are treated in the sections below where Proposition 13.3 is proven.

PROOF OF PROPOSITION 13.1. Consider $\mathcal{C} = (O, \kappa)$ a local chart such that $m^0 \in O$. It is sufficient to prove an estimate in an open neighborhood U of $x^0 = \kappa(m^0)$ in $\mathbb{R}^d$ with $U \subset \kappa(O)$. In fact, we take U an open set of $\mathbb{R}^d$ such that $U \Subset \kappa(O \cap V)$.

Here, we simply write φ and P for the local representatives of the weight function and the operator. The conjugated operator is $P_\varphi = e^{\tau\varphi}Pe^{-\tau\varphi}$. Let $u \in H_0^1(U)$ such that $Pu = F_0 + \operatorname{div}_g F$ where $F_0 \in L^2(\mathbb{R}^d)$ and $F \in L^2V(\mathbb{R}^2)$ are such that $\operatorname{supp}(F_0) \cup \operatorname{supp}(F) \subset \overline{U}$. We set $v = e^{\tau\varphi}u$ and $G = e^{\tau\varphi}F$. We then have $P_\varphi v = G_0 + \operatorname{div}_g G$ with

$$G_0 = e^{\tau\varphi}F_0 - \tau e^{\tau\varphi}g(\nabla_g\varphi, F),$$

using (17.2.6). Since U is bounded we have $G_0 \in L^2(U)$ and $G \in L^2V(U)$.

Let $\psi \in \mathscr{C}_c^\infty(\kappa(O \cap V))$ be such that $\psi \equiv 1$ in a neighborhood of $\overline{U}$. Setting $w = \psi\Lambda_\tau^{-1}v$, we write

$$\begin{aligned} P_\varphi w = \psi\Lambda_\tau^{-1}P_\varphi v + [P_\varphi, \psi\Lambda_\tau^{-1}]v &= \psi\Lambda_\tau^{-1}(G_0 + \operatorname{div}_g G) \\ &\quad + [P_\varphi, \psi\Lambda_\tau^{-1}]v \in L^2(U), \end{aligned}$$

and we find

$$\|P_\varphi w\|_{L^2(U)} \lesssim \tau^{-1}\|G_0\|_{L^2(U)} + \|G\|_{L^2V(U)} + \|v\|_{L^2(U)}. \tag{13.1.6}$$

Because of the respective supports of ψ and v, we write $w = \Lambda_\tau^{-1}v + [\psi, \Lambda_\tau^{-1}]v$. This yields

$$\|v\|_{\tau,1} = \|\Lambda_\tau^{-1}v\|_{\tau,2} \lesssim \|w\|_{\tau,2} + \|v\|_{L^2(U)}. \tag{13.1.7}$$

Since φ and P have the sub-ellipticity property in $\overline{U}$. As a result, Theorem 3.11 of Volume 1 applies. We rather use the equivalent form (3.3.3), that is, $\tau^{-1/2}\|w\|_{\tau,2} \lesssim \|P_\varphi w\|_{L^2(U)}$ that, together with (13.1.6) and (13.1.7), implies

$$\tau^{-1/2}\|v\|_{\tau,1} \lesssim \tau^{-1}\|G_0\|_{L^2(U)} + \|G\|_{L^2V(U)} + \|v\|_{L^2(U)}.$$

Taking $\tau > 0$ sufficiently large and multiplying by τ we obtain

$$\tau^{1/2}\|v\|_{\tau,1} \lesssim \|G_0\|_{L^2(U)} + \tau\|G\|_{L^2V(U)} \lesssim \|e^{\tau\varphi}F_0\|_{L^2(U)} + \tau\|e^{\tau\varphi}F\|_{L^2V(U)},$$

which concludes the proof with $V^1 = \kappa^{-1}(U)$. ∎

13.2. Local Setting at the Boundary

13.2.1. Choice of Coordinates. At a point $m^0 \in \partial\mathcal{M}$, we use the local chart $\mathcal{C} = (O, \kappa)$ associated with normal geodesic coordinates introduced in Sect. 8.3.2; see also Sections 9.4 and 17.6. We have

$$\kappa(\partial\mathcal{M} \cap O) = \{x_d = 0\} \cap \kappa(O), \qquad \kappa(\mathcal{M} \cap O) = \{x_d \geq 0\} \cap \kappa(O).$$

Set $x^0 = \kappa(m^0)$. In this chart, the principal symbol of the Laplace–Beltrami operator takes the form

$$p^{\mathcal{C}}(x, \xi) = \xi_d^2 + r(x, \xi'), \qquad r(x, \xi') = \sum_{1 \leq i,j \leq d-1} g^{\mathcal{C},ij}(x)\xi_i\xi_j \gtrsim |\xi'|^2.$$

As $\partial_\nu = -iD_d$, the local representative of the principal symbol of B is given by

$$b^{\mathcal{C}}(x, \xi) = b^k(x, \xi') - ib^{k-1}(x, \xi')\xi_d,$$

where $b^j(x, \xi')$, $j = k, k-1$ are polynomial and homogeneous of degree j in ξ'. We set

$$P_0(x, D) = \operatorname{Op}(p^{\mathcal{C}}) = D_d^2 + R_0(x, D'), \quad \text{with } R_0 = \operatorname{Op}_{\mathsf{T}}(r),$$

and $B_0 = \operatorname{Op}(b^{\mathcal{C}}) = \operatorname{Op}_{\mathsf{T}}(b^k) - i\operatorname{Op}_{\mathsf{T}}(b^{k-1})D_d$. In agreement with (13.1.3), we then define

$$\mathsf{B}_0(u, h) = \operatorname{Op}_{\mathsf{T}}(b^k)u_{|x_d=0^+} - i\operatorname{Op}_{\mathsf{T}}(b^{k-1})(D_d u + ih)_{|x_d=0^+}. \tag{13.2.1}$$

Consider $u \in H^1(\mathbb{R}^d_+)$ that satisfies

$$P_0 u = F_0 + \operatorname{div}_g F, \quad \text{with } F_0 \in L^2(\mathbb{R}^d_+),\ F \in L^2V(\mathbb{R}^d_+). \tag{13.2.2}$$

From Proposition 13.2 one has

$$(D_d u - iF^d)_{|x_d=0^+} = i(\partial_\nu u + g(F, \nu)) \in H^{-1/2}(\mathbb{R}^{d-1}), \tag{13.2.3}$$

implying $\mathsf{B}_0(u, -F^d) \in H^{1/2-k}(\mathbb{R}^{d-1})$.

The following result is equivalent to Proposition 13.3.

PROPOSITION 13.8. *Let $P_0 = \operatorname{Op}(p^{\mathcal{C}})$, $B_0 = \operatorname{Op}(b^{\mathcal{C}})$. Let $m^0 \in \partial\mathcal{M}$, V^0 be an open neighborhood of m^0 in $\mathcal{M}$ that meets one connected component of $\partial\mathcal{M}$. Assume that $(P_0, \varphi^{\mathcal{C}})$ has the sub-ellipticity property of Definition 3.2 of Volume 1 in $\overline{U^0}$ with $U^0 = \kappa(V^0 \cap O)$. Moreover, assume that $(P_0, B_0, \varphi^{\mathcal{C}})$ satisfies the Lopatinskiĭ–Šapiro condition at x^0 (Definition 8.1). Then, there exists a bounded neighborhood U_+ of x^0 in $\overline{\mathbb{R}^d_+}$ such that $U_+ \subset U^0$ and two constants C and $\tau_* > 0$ such that*

(13.2.4)

$$\tau^{1/2}\|e^{\tau\varphi^{\mathcal{C}}} u\|_{\tau,1} + \tau|e^{\tau\varphi^{\mathcal{C}}} u_{|x_d=0^+}|_{\tau,1/2} + \tau|e^{\tau\varphi^{\mathcal{C}}}(D_d u - iF^d)_{|x_d=0^+}|_{\tau,-1/2}$$
$$\leq C\Big(\|e^{\tau\varphi^{\mathcal{C}}} F_0\|_+ + \tau\|e^{\tau\varphi^{\mathcal{C}}} F\|_+ + \tau|e^{\tau\varphi^{\mathcal{C}}} \mathsf{B}_0(u, -F^d)|_{\tau,1/2-k}\Big),$$

for $\tau \geq \tau_*$ *and* $u \in H^1(\mathbb{R}^d_+)$, $F_0 \in L^2(\mathbb{R}^d_+)$, $F \in L^2V(\mathbb{R}^d_+)$ *such that*

$$P_0 u = F_0 + \operatorname{div}_g F, \quad \textit{and} \quad \operatorname{supp}(u) \cup \operatorname{supp}(F_0) \cup \operatorname{supp}(F) \subset U_+.$$

13.2.2. Conjugation. In the local chart $\mathcal{C}$, we define the conjugated operators of P_0 and B_0:

$$P_{0,\varphi} = e^{\tau\varphi^{\mathcal{C}}} P_0 e^{-\tau\varphi^{\mathcal{C}}}, \qquad B_{0,\varphi} = e^{\tau\varphi^{\mathcal{C}}} B_0 e^{-\tau\varphi^{\mathcal{C}}}.$$

In what follows, to ease notation we shall write φ and p in place of $\varphi^{\mathcal{C}}$ and $p^{\mathcal{C}}$, respectively.

We have

$$\begin{aligned} B_{0,\varphi} &= e^{\tau\varphi}\operatorname{Op}_{\mathsf{T}}(b^k)e^{-\tau\varphi} - ie^{\tau\varphi}\operatorname{Op}_{\mathsf{T}}(b^{k-1})e^{-\tau\varphi}(D_d + i\tau\partial_d\varphi) \\ &= \tilde{B}^k_\varphi(x, D', \tau) - iB^{k-1}_\varphi(x, D', \tau)D_d, \end{aligned}$$

with

$$\begin{aligned} B^{k-1}_\varphi(x, D', \tau) &= e^{\tau\varphi}\operatorname{Op}_{\mathsf{T}}(b^{k-1})e^{-\tau\varphi}, \\ \tilde{B}^k_\varphi(x, D', \tau) &= e^{\tau\varphi}\operatorname{Op}_{\mathsf{T}}(b^k)e^{-\tau\varphi} + \tau B^{k-1}_\varphi(x, D', \tau)\partial_d\varphi, \end{aligned}$$

with respective principal symbols

$$\begin{aligned} b^{k-1}_\varphi(\varrho') &= b^{k-1}(x, \xi' + i\tau d_{x'}\varphi(x)), \\ \tilde{b}^k_\varphi(\varrho') &= b^k(x, \xi' + i\tau d_{x'}\varphi(x)) + \tau b^{k-1}_\varphi(\varrho')\partial_d\varphi(x), \end{aligned}$$

with $\varrho' = (x, \xi', \tau)$. Note that $\tilde{b}^k_\varphi(\varrho')$ and $b^{k-1}_\varphi(\varrho')$ are polynomial and homogeneous of degree k and $k-1$ in (ξ', τ), respectively.

In agreement with (13.2.1) we define

(13.2.5)
$$\mathsf{B}_{0,\varphi}(u, h) = \tilde{B}^k_\varphi(x, D', \tau)u_{|x_d=0^+} - iB^{k-1}_\varphi(x, D', \tau)(D_d u + ih)_{|x_d=0^+}.$$

With u satisfying (13.2.2), setting $v = e^{\tau\varphi}u$ and

$$G = e^{\tau\varphi}F, \qquad \tilde{G}_0 = e^{\tau\varphi}\big(F_0 - \tau g(\nabla_g\varphi, F)\big),$$

with (17.2.6) we have $P_{0,\varphi}v = \tilde{G}_0 + \operatorname{div}_g G$. From the form of the divergence in (17.2.5) we find $P_{0,\varphi}v = G_0 + \sum_{1\leq i\leq d}\partial_j G^j$, with

$$G_0 = \tilde{G}_0 + \frac{1}{2\det g_x}\sum_{1\leq i\leq d}(\partial_j \det g_x)G^j.$$

With u, F_0, F as in Proposition 13.8 we have $v \in H^1(\mathbb{R}^d_+)$, $G_0 \in L^2(\mathbb{R}^d_+)$, $G \in L^2V(\mathbb{R}^d_+)$ with supports in $\overline{U_+}$. From (13.2.3) we have

$$(D_d v - iG^d)_{|x_d=0^+} = e^{\tau\varphi}\big(D_d u - iF^d - i\tau\partial_d\varphi u\big)_{|x_d=0^+} \in H^{-1/2}(\mathbb{R}^{d-1}),$$

yielding

$$\begin{aligned} |e^{\tau\varphi^{\mathcal{C}}} u_{|x_d=0^+}|_{\tau,1/2} &+ |e^{\tau\varphi^{\mathcal{C}}_{|x_d=0^+}}(D_d u - iF^d)_{|x_d=0^+}|_{\tau,-1/2} \\ &\asymp |v_{|x_d=0^+}|_{\tau,1/2} + |(D_d v - iG^d)_{|x_d=0^+}|_{\tau,-1/2}. \end{aligned}$$

We shall prove the following proposition that is equivalent to Proposition 13.8.

PROPOSITION 13.9. *With the same setting and assumption as in Proposition 13.8, there exist there exists a bounded neighborhood U_+ of x^0 in $\overline{\mathbb{R}^d_+}$ such that $U_+ \subset U^0$, $C > 0$, and $\tau_* > 0$ such that*

$$\tau^{1/2}\|v\|_{\tau,1} + \tau|v_{|x_d=0^+}|_{\tau,1/2} + \tau|(D_d v - iG^d)_{|x_d=0^+}|_{\tau,-1/2} \leq C\Big(\|G_0\|_+ + \tau\|G\|_+ + \tau|\mathsf{B}_{0,\varphi}(v, -G^d)|_{\tau,1/2-k}\Big), \tag{13.2.6}$$

for $\tau \geq \tau_*$ and $v \in H^1(\mathbb{R}^d_+)$, $G_0 \in L^2(\mathbb{R}^d_+)$, $G \in L^2V(\mathbb{R}^d_+)$ such that

$$P_{0,\varphi}v = G_0 + \sum_{1\leq i\leq d} \partial_j G^j, \qquad \operatorname{supp}(v) \cup \operatorname{supp}(G_0) \cup \operatorname{supp}(G) \subset U_+.$$

In the next section we prove microlocal versions of this result. The proof of Proposition 13.9 is given in Sect. 13.4.1 with a patching argument.

13.3. A Microlocal Estimate

Here, we shall only assume that the Lopatinskiĭ–Šapiro condition holds at one point in the cotangent bundle $T^*\partial\mathcal{M} \times [0, +\infty)$. Introducing a cutoff function $\chi \in S^0_{\mathsf{T},\tau}$ in phase space near this point we prove an estimate for the microlocalized function $\operatorname{Op}_{\mathsf{T}}(\chi)v$.

PROPOSITION 13.10. *Let $U^0 = \kappa(V^0 \cap O)$. Assume that (P_0, φ) has the sub-ellipticity property of Definition 3.2 of Volume 1 in $\overline{U^0}$. Let $\omega^{0\prime} \in T^*_{m^0}\partial\mathcal{M}$, with local representative $(\omega^{0\prime})^{\mathcal{C}} = \xi^{0\prime} \in \mathbb{R}^{d-1}$, and $\tau^0 \geq 0$ such that $(\xi^{0\prime}, \tau^0) \neq 0$, and assume that the Lopatinskiĭ–Šapiro condition of Definition 8.1 holds at $(m^0, \omega^{0\prime}, \tau^0)$. Then, there exists $\mathscr{U}$ a conic open neighborhood of $\varrho^{0\prime} = (x^0, \xi^{0\prime}, \tau^0)$ in $U^0 \times \mathbb{R}^{d-1} \times \mathbb{R}_+$ such that for $\chi \in S^0_{\mathsf{T},\tau}$, homogeneous of degree zero, with $\operatorname{supp}(\chi) \subset \mathscr{U}$, there exist $C > 0$ and $\tau_* > 0$ such that*

$$\begin{aligned}\tau^{1/2}\|\operatorname{Op}_{\mathsf{T}}(\chi)v\|_{\tau,1} &+ \tau|(D_d\operatorname{Op}_{\mathsf{T}}(\chi)v - i\operatorname{Op}_{\mathsf{T}}(\chi)G^d)_{|x_d=0^+}|_{\tau,-1/2} \\ &+ \tau|\operatorname{Op}_{\mathsf{T}}(\chi)v_{|x_d=0^+}|_{\tau,1/2} \leq C\Big(\|G_0\|_+ + \tau\|G\|_+ + \tau|\mathsf{B}_{0,\varphi}(v, -G^d)|_{\tau,1/2-k} \\ &+ \|v\|_{\tau,1} + |(D_d v - iG^d)_{|x_d=0^+}|_{\tau,-3/2}\Big),\end{aligned} \tag{13.3.1}$$

for $\tau \geq \tau_*$ and $v \in H^1(\mathbb{R}^d_+)$, $G_0 \in L^2(\mathbb{R}^d_+)$, $G \in L^2V(\mathbb{R}^d_+)$ such that

$$P_{0,\varphi}v = G_0 + \sum_{1\leq i\leq d} \partial_j G^j.$$

With the normal geodesic coordinates we introduced locally, we may use some of the notation of Sect. 8.3.2 that we recall now. We have

$$P_{0,\varphi} = (D_d + i\tau\partial_d\varphi(x))^2 + R_{0,\varphi}, \tag{13.3.2}$$

where the principal symbol of $R_{0,\varphi} = e^{\tau\varphi} R_0 e^{-\tau\varphi} \in \Psi^2_{\mathsf{T},\tau}$ is precisely

$$r_\varphi(\varrho') = r(x, \xi' + i\tau d_{x'}\varphi(x)) \qquad \varrho' = (x, \xi', \tau),$$

We introduced $\alpha(\varrho')$ such that $\operatorname{Re}\alpha(\varrho') \geq 0$ and $\alpha^2(\varrho') = r_\varphi(\varrho')$ and we set

$$\gamma_j(\varrho') = -i\tau\partial_d\varphi(x) + i(-1)^j\alpha(\varrho'), \qquad \varrho' = (x, \xi', \tau).$$

This yields

$$p_\varphi(\varrho) = \check{p}_\varphi(m, \omega', \tau, z) = \big(z - \gamma_1(\varrho')\big)\big(z - \gamma_2(\varrho')\big), \;\; \varrho = (x, \xi', \xi_d = z, \tau).$$

By the first result of Theorem 8.7, if the Lopatinskiĭ–Šapiro condition holds for (P, B, φ) at m^0, then $\partial_d\varphi(x^0) > 0$. This implies that we have $\operatorname{Im}\gamma_1(\varrho^{0\prime}) < 0$. For the proof of Proposition 13.10 we shall treat differently the situations $\gamma_2(\varrho^{0\prime}) \neq \gamma_1(\varrho^{0\prime})$ and $\gamma_2(\varrho^{0\prime}) = \gamma_1(\varrho^{0\prime})$. Following (8.3.10), we prove the estimate of Proposition 13.10 in the following three (exhaustive) cases:

(1) either $\operatorname{Im}\gamma_2(\varrho^{0\prime}) \geq 0$;
(2) or $\operatorname{Im}\gamma_2(\varrho^{0\prime}) < 0$ and $\gamma_2(\varrho^{0\prime}) \neq \gamma_1(\varrho^{0\prime})$;
(3) or $\operatorname{Im}\gamma_2(\varrho^{0\prime}) < 0$ and $\gamma_2(\varrho^{0\prime}) = \gamma_1(\varrho^{0\prime})$.

In case (1), we have $\check{p}^+_\varphi(m^0, \omega^{0\prime}, z, \tau^0) = z - \gamma_2(x^0, \xi^{0\prime}, \tau^0)$, while in cases (2) and (3), we have $\check{p}^+_\varphi(m^0, \omega^{0\prime}, z, \tau^0) = 1$.

In Sect. 8.3.4.2, Cases (2) and (3) were treated together. However, because of the low regularity of the source terms, the proof in this case (2) require to factorize the principal symbol of P_φ in a smooth way and this can only be done in the case where roots do not cross. This explains the different treatments of Cases (2) and (3).

13.3.1. Case 1: One Root in the Upper Complex Half-Plane.

Here, we have

$$\check{p}^+_\varphi(m^0, \omega^{0\prime}, z, \tau^0) = z - \gamma_2(x^0, \xi^{0\prime}, \tau^0),$$

that is, $\operatorname{Im}\gamma_1(x^0, \xi^{0\prime}, \tau^0) < 0$ and $\operatorname{Im}\gamma_2(x^0, \xi^{0\prime}, \tau^0) \geq 0$. With the Lopatinskiĭ–Šapiro condition holding for (P, B, φ) at the considered point, by (8.3.11), we have moreover

$$b_\varphi(x^0, \xi^{0\prime}, \xi_d = \gamma_2, \tau^0) = b(x^0, \xi^{0\prime} + i\tau^0 d_{x'}\varphi(x^0), \gamma_2 + i\tau^0\partial_d\varphi(x^0)) \neq 0.$$

As the roots γ_1 and γ_2 are locally smooth with respect to $\varrho' = (x, \xi', \tau)$ and homogeneous of degree one in (ξ', τ) by Proposition 6.28, there exists $\mathscr{U}$ a conic open neighborhood of $(x^0, \xi^{0\prime}, \tau^0)$ in $U^0 \times \mathbb{R}^{d-1} \times \mathbb{R}_+$ and $C, C' > 0$ such that

$$\gamma_1(\varrho') \neq \gamma_2(\varrho'), \; \operatorname{Im}\gamma_2(\varrho') \geq -C\lambda_{\mathsf{T},\tau}, \text{ and } \operatorname{Im}\gamma_1(\varrho') \leq -C'\lambda_{\mathsf{T},\tau}, \tag{13.3.3}$$

and

$$b_\varphi(x, \xi', \xi_d = \gamma_2(x, \xi', \tau), \tau) \neq 0, \tag{13.3.4}$$

if $\varrho' = (x, \xi', \tau) \in \overline{\mathscr{U}}$. Without any loss of generality we may moreover assume that $\mathbb{S}_{\overline{\mathscr{U}}}$ is compact.

We let $\chi \in S^0_{\mathsf{T},\tau}$ be as in the statement and $\tilde{\chi} \in S^0_{\mathsf{T},\tau}$ be homogeneous of degree zero and be such that $\operatorname{supp}(\tilde{\chi}) \subset \mathscr{U}$ and $\tilde{\chi} \equiv 1$ on $\operatorname{supp}(\chi)$. From the smoothness and the homogeneity of the roots, we have $\tilde{\chi}\gamma_j \in S^1_{\mathsf{T},\tau}$, $j = 1, 2$. We set

$$P^+ = D_d - \operatorname{Op}_{\mathsf{T}}(\tilde{\chi}\gamma_2) \quad \text{and} \quad P^- = D_d - \operatorname{Op}_{\mathsf{T}}(\tilde{\chi}\gamma_1).$$

Since $\operatorname{Op}_{\mathsf{T}}(\chi)P_{0,\varphi} = \operatorname{Op}_{\mathsf{T}}(\chi)P^-P^+ + R$ with $R \in \Psi^{1,0}_{\tau}$, we write

$$\begin{aligned}\operatorname{Op}_{\mathsf{T}}(\chi)P^-P^+v = \operatorname{Op}_{\mathsf{T}}(\chi)P_{0,\varphi}v - Rv &= \operatorname{Op}_{\mathsf{T}}(\chi)G_0 \\ &+ \operatorname{Op}_{\mathsf{T}}(\chi) \sum_{1\leq j\leq d} \partial_j G^j - Rv,\end{aligned}$$

and we obtain

$$\begin{aligned}&\operatorname{Op}_{\mathsf{T}}(\chi)P^-(P^+v - iG^d) \\ &\quad = \operatorname{Op}_{\mathsf{T}}(\chi)G_0 + \operatorname{Op}_{\mathsf{T}}(\chi) \sum_{1\leq j\leq d-1} \partial_j G^j + i\operatorname{Op}_{\mathsf{T}}(\chi)\operatorname{Op}_{\mathsf{T}}(\tilde{\chi}\gamma_1)G^d - Rv,\end{aligned}$$

which yields

$$\begin{aligned}\|\operatorname{Op}_{\mathsf{T}}(\chi)P^-(P^+v - iG^d)\|_{\tau,0,-1} &\lesssim \|G_0\|_{\tau,0,-1} + \|G\|_+ + \|v\|_{\tau,1,-1} \\ &\lesssim \tau^{-1}\|G_0\|_+ + \|G\|_+ + \|v\|_{\tau,1,-1},\end{aligned} \tag{13.3.5}$$

using Remark 6.8.

From Lemma 6.20 (with $s = -1$), we have for τ chosen sufficiently large,

$$\|\operatorname{Op}_{\mathsf{T}}(\chi)z\|_+ + |\operatorname{Op}_{\mathsf{T}}(\chi)z_{|x_d=0^+}|_{\tau,-1/2} \lesssim \|P^-\operatorname{Op}_{\mathsf{T}}(\chi)z\|_{\tau,0,-1} + \|z\|_{\tau,0,-N}, \tag{13.3.6}$$

for $z \in \overline{\mathscr{S}}(\mathbb{R}^d_+)$. With a density argument, estimate (13.3.6) applies to $z = P^+v - iG^d$, yielding

$$\begin{aligned}&\|\operatorname{Op}_{\mathsf{T}}(\chi)(P^+v - iG^d)\|_+ + |\operatorname{Op}_{\mathsf{T}}(\chi)(P^+v - iG^d)_{|x_d=0^+}|_{\tau,-1/2} \\ &\qquad \lesssim \|P^-\operatorname{Op}_{\mathsf{T}}(\chi)(P^+v - iG^d)\|_{\tau,0,-1} + \|P^+v - iG^d\|_{\tau,0,-N} \\ &\qquad \lesssim \|\operatorname{Op}_{\mathsf{T}}(\chi)P^-(P^+v - iG^d)\|_{\tau,0,-1} + \|v\|_{\tau,1,-1} + \|G^d\|_+,\end{aligned}$$

using that $P^-\operatorname{Op}_{\mathsf{T}}(\chi) = \operatorname{Op}_{\mathsf{T}}(\chi)P^- \mod \Psi^0_{\mathsf{T},\tau}$. With (13.3.5), we then obtain

$$\begin{aligned}&\|\operatorname{Op}_{\mathsf{T}}(\chi)(P^+v - iG^d)\|_+ + |\operatorname{Op}_{\mathsf{T}}(\chi)(P^+v - iG^d)_{|x_d=0^+}|_{\tau,-1/2} \\ &\qquad \lesssim \tau^{-1}\|G_0\|_+ + \|G\|_+ + \|v\|_{\tau,1,-1}.\end{aligned}$$

To lighten notation we set $w = \mathrm{Op}_\mathsf{T}(\chi)v$ and we write

$$\begin{aligned}
&\|P^+ w\|_+ + |(P^+ w - i\mathrm{Op}_\mathsf{T}(\chi)G^d)_{|x_d=0^+}|_{\tau,-1/2} \\
&\quad\lesssim \|\mathrm{Op}_\mathsf{T}(\chi)P^+ v\|_+ + |\mathrm{Op}_\mathsf{T}(\chi)(P^+ v - iG^d)_{|x_d=0^+}|_{\tau,-1/2} \\
&\qquad + \|v\|_+ + |v_{|x_d=0^+}|_{\tau,-1/2} \\
&\quad\lesssim \|\mathrm{Op}_\mathsf{T}(\chi)(P^+ v - iG^d)\|_+ + |\mathrm{Op}_\mathsf{T}(\chi)(P^+ v - iG^d)_{|x_d=0^+}|_{\tau,-1/2} \\
&\qquad + \|G^d\|_+ + \|v\|_{\tau,1,-1},
\end{aligned}$$

using the trace inequality of Corollary 6.10 (with $m = 0$ and $s = -1/2$). We thus obtain

$$\begin{aligned}
(13.3.7) \qquad &\|P^+ w\|_+ + |(P^+ w - i\mathrm{Op}_\mathsf{T}(\chi)G^d)_{|x_d=0^+}|_{\tau,-1/2} \\
&\qquad\lesssim \tau^{-1}\|G_0\|_+ + \|G\|_+ + \|v\|_{\tau,1,-1}.
\end{aligned}$$

We now use the following lemma.

LEMMA 13.11. *There exist $C > 0$ and $\tau_* > 0$ such that for any $N \in \mathbb{N}$, there exists $C_N > 0$ such that*

$$\begin{aligned}
&|\mathrm{Op}_\mathsf{T}(\chi)\ell|_{\tau,1/2} + |\mathrm{Op}_\mathsf{T}(\chi)\ell'|_{\tau,-1/2} \\
&\quad\leq C\Big(|\tilde{B}^k_\varphi \mathrm{Op}_\mathsf{T}(\chi)\ell - iB^{k-1}_\varphi \mathrm{Op}_\mathsf{T}(\chi)\ell'|_{\tau,1/2-k} \\
&\qquad + |\mathrm{Op}_\mathsf{T}(\chi)\ell' - \mathrm{Op}_\mathsf{T}(\tilde{\chi}\gamma_2)\mathrm{Op}_\mathsf{T}(\chi)\ell|_{\tau,-1/2}\Big) + C_N\big(|\ell|_{\tau,-N} + |\ell'|_{\tau,-N}\big),
\end{aligned}$$

for $\tau \geq \tau_*$, $\ell \in H^{1/2}(\mathbb{R}^{d-1})$, and $\ell' \in H^{-1/2}(\mathbb{R}^{d-1})$.

A proof is given in Appendix 13.A.2. This is precisely where the Lopatinskiĭ–Šapiro condition is used.

We set $\ell = v_{|x_d=0^+} \in H^{1/2}(\mathbb{R}^{d-1})$ and $\ell' = (D_d v - iG^d)_{|x_d=0^+} \in H^{-1/2}(\mathbb{R}^{d-1})$. As $[D_d, \mathrm{Op}_\mathsf{T}(\chi)] \in \Psi^0_{\mathsf{T},\tau}$, we note that

$$\begin{aligned}
|(D_d\mathrm{Op}_\mathsf{T}(\chi)v - i\mathrm{Op}_\mathsf{T}(\chi)G^d)_{|x_d=0^+}|_{\tau,-1/2} &\lesssim |\mathrm{Op}_\mathsf{T}(\chi)\ell'_{|x_d=0^+}|_{\tau,-1/2} \\
&\quad + |v_{|x_d=0^+}|_{\tau,-1/2},
\end{aligned}$$

and

$$\begin{aligned}
&|\tilde{B}^k_\varphi \mathrm{Op}_\mathsf{T}(\chi)\ell - iB^{k-1}_\varphi \mathrm{Op}_\mathsf{T}(\chi)\ell'|_{\tau,1/2-k} \\
&\quad\lesssim |\tilde{B}^k_\varphi \mathrm{Op}_\mathsf{T}(\chi)v_{|x_d=0^+} - iB^{k-1}_\varphi (D_d\mathrm{Op}_\mathsf{T}(\chi)v - i\mathrm{Op}_\mathsf{T}(\chi)G^d)_{|x_d=0^+}|_{\tau,1/2-k} \\
&\qquad + |v_{|x_d=0^+}|_{\tau,-1/2} \\
&\quad\lesssim |\mathsf{B}_{0,\varphi}(w, -\mathrm{Op}_\mathsf{T}(\chi)G^d)|_{\tau,1/2-k} + |v_{|x_d=0^+}|_{\tau,-1/2},
\end{aligned}$$

recalling the form of $\mathsf{B}_{0,\varphi}$ given in (13.2.5). We also have

$$\begin{aligned}
&|\mathrm{Op}_{\mathsf{T}}(\chi)\ell' - \mathrm{Op}_{\mathsf{T}}(\tilde{\chi}\gamma_2)\mathrm{Op}_{\mathsf{T}}(\chi)\ell|_{\tau,-1/2} \\
&\quad\lesssim |(D_d w - \mathrm{Op}_{\mathsf{T}}(\tilde{\chi}\gamma_2)w)_{|x_d=0^+} - i\mathrm{Op}_{\mathsf{T}}(\chi)G^d_{|x_d=0^+}|_{\tau,-1/2} + |v_{|x_d=0^+}|_{\tau,-1/2} \\
&\quad\lesssim |P^+ w_{|x_d=0^+} - i\mathrm{Op}_{\mathsf{T}}(\chi)G^d_{|x_d=0^+}|_{\tau,-1/2} + |v_{|x_d=0^+}|_{\tau,-1/2}.
\end{aligned}$$

With these observations, the result of Lemma 13.11 reads

(13.3.8)
$$\begin{aligned}
&|w_{|x_d=0^+}|_{\tau,1/2} + |(D_d w - i\mathrm{Op}_{\mathsf{T}}(\chi)G^d)_{|x_d=0^+}|_{\tau,-1/2} \\
&\quad\lesssim |\mathsf{B}_{0,\varphi}(w, -\mathrm{Op}_{\mathsf{T}}(\chi)G^d)|_{\tau,1/2-k} + |(P^+ w - i\mathrm{Op}_{\mathsf{T}}(\chi)G^d)_{|x_d=0^+}|_{\tau,-1/2} \\
&\qquad + \|v\|_{\tau,1,-1} + |(D_d v - iG^d)_{|x_d=0^+}|_{\tau,-N},
\end{aligned}$$

for $\tau \geq 1$ chosen sufficiently large, using the trace inequality of Corollary 6.10 (with $m = 0$ and $s = -1/2$). With a commutator argument we find

$$\begin{aligned}
&|\mathsf{B}_{0,\varphi}(w, -\mathrm{Op}_{\mathsf{T}}(\chi)G^d)|_{\tau,1/2-k} \\
&\quad\lesssim |\mathsf{B}_{0,\varphi}(v, -G^d)|_{\tau,1/2-k} + |v_{|x_d=0^+}|_{\tau,-1/2} + |(D_d v - iG^d)_{|x_d=0^+}|_{\tau,-3/2},
\end{aligned}$$

and we write

$$|v_{|x_d=0^+}|_{\tau,-1/2} \lesssim \tau^{-1}|v_{|x_d=0^+}|_{\tau,1/2} \lesssim \tau^{-1}\|v\|_{\tau,1},$$

with the trace inequality of Proposition 6.9. Estimate (13.3.8) thus becomes

(13.3.9)
$$\begin{aligned}
&|w_{|x_d=0^+}|_{\tau,1/2} + |(D_d w - i\mathrm{Op}_{\mathsf{T}}(\chi)G^d)_{|x_d=0^+}|_{\tau,-1/2} \\
&\quad\lesssim |\mathsf{B}_{0,\varphi}(w, -\mathrm{Op}_{\mathsf{T}}(\chi)G^d)|_{\tau,1/2-k} + |(P^+ w - i\mathrm{Op}_{\mathsf{T}}(\chi)G^d)_{|x_d=0^+}|_{\tau,-1/2} \\
&\qquad + \tau^{-1}\|v\|_{\tau,1} + |(D_d v - iG^d)_{|x_d=0^+}|_{\tau,-3/2}.
\end{aligned}$$

With (13.3.7) we thus obtain

$$\begin{aligned}
\text{(13.3.10)}\quad &\|P^+ w\|_+ + |w_{|x_d=0^+}|_{\tau,1/2} + |(D_d w - i\mathrm{Op}_{\mathsf{T}}(\chi)G^d)_{|x_d=0^+}|_{\tau,-1/2} \\
&\quad\lesssim \tau^{-1}\|G_0\|_+ + \|G\|_+ + |\mathsf{B}_{0,\varphi}(w, -\mathrm{Op}_{\mathsf{T}}(\chi)G^d)|_{\tau,1/2-k} \\
&\qquad + \tau^{-1}\|v\|_{\tau,1} + |(D_d v - iG^d)_{|x_d=0^+}|_{\tau,-3/2}.
\end{aligned}$$

We recall that $\tilde{\gamma}_2 = \tilde{\chi}\gamma_2 \in S^1_{\mathsf{T},\tau}$ and $P^+ = D_d - \mathrm{Op}_{\mathsf{T}}(\tilde{\gamma}_2)$. We write $P^+ = P_2^+ + iP_1^+$ with

$$\begin{aligned}
P_2^+ &= \frac{1}{2}(P^+ + (P^+)^*) = D_d - \frac{1}{2}(\mathrm{Op}_{\mathsf{T}}(\tilde{\gamma}_2) + \mathrm{Op}_{\mathsf{T}}(\tilde{\gamma}_2)^*), \\
P_1^+ &= \frac{1}{2i}(P^+ - (P^+)^*) = -\frac{1}{2i}(\mathrm{Op}_{\mathsf{T}}(\tilde{\gamma}_2) - \mathrm{Op}_{\mathsf{T}}(\tilde{\gamma}_2)^*).
\end{aligned}$$

Observe that $P_2^+ \in \Psi_\tau^{1,0}$ and $P_1^+ \in \Psi_{\mathsf{T},\tau}^1$ are both formally selfadjoint. Their respective principal symbols are

$$p_2^+(\varrho) = \xi_d - \operatorname{Re}\tilde{\gamma}_2(\varrho') \in S_\tau^{1,0}, \qquad p_1^+(\varrho') = -\operatorname{Im}\tilde{\gamma}_2(\varrho') \in S_{\mathsf{T},\tau}^1.$$

The principal symbol of P^+ is $p^+(\varrho) = p_2^+(\varrho) + ip_1^+(\varrho')$.

LEMMA 13.12. *There exists a conic open neighborhood $\mathscr{W}$ of* $\operatorname{supp}(\chi)$ *in* $\mathbb{R}_+^d \times \mathbb{R}^{d-1} \times \mathbb{R}_+$ *such that* $\tilde{\chi} \equiv 1$ *in* $\mathscr{W}$ *and*

$$p^+(\varrho) = 0 \;\Rightarrow\; \frac{1}{2i}\{\overline{p^+}, p^+\}(\varrho') = \{p_2^+, p_1^+\}(\varrho') > 0, \;\; \varrho' \in \overline{\mathscr{W}}, \varrho = (\varrho', \xi_d).$$

A proof is given in Appendix 13.A.3.

With Lemma 13.12, applying Lemma 6.22 with a density argument, for $\tau \geq 1$ sufficiently large, we obtain

$$\tau^{-1/2}\|w\|_{\tau,1} \lesssim \|P^+ w\|_+ + |w_{|x_d=0^+}|_{\tau,1/2} + \|v\|_{\tau,0,-N},$$

yielding, with (13.3.10),

$$\begin{aligned}\tau^{-1/2}\|w\|_{\tau,1} + |w_{|x_d=0^+}|_{\tau,1/2} + |(D_d w - i\operatorname{Op}_{\mathsf{T}}(\chi)G^d)_{|x_d=0^+}|_{\tau,-1/2} \\ \lesssim \tau^{-1}\|G_0\|_+ + \|G\|_+ + |\mathsf{B}_{0,\varphi}(w, -\operatorname{Op}_{\mathsf{T}}(\chi)G^d)|_{\tau,1/2-k} \\ + \tau^{-1}\|v\|_{\tau,1} + |(D_d v - iG^d)_{|x_d=0^+}|_{\tau,-3/2}.\end{aligned}$$

Upon multiplication by τ we obtain (13.3.1). This concludes the proof of Proposition 13.10 in Case 1. ∎

13.3.2. Case 2: Roots Are Different and Both in the Open Lower Complex Half-Plane. Here, we have $\operatorname{Im}\gamma_1(\varrho^{0\prime}) < 0$ and $\operatorname{Im}\gamma_2(\varrho^{0\prime}) < 0$, and $\gamma_1(\varrho^{0\prime}) \neq \gamma_2(\varrho^{0\prime})$, where $\varrho^{0\prime} = (x^0, \xi^{0\prime}, \tau^0)$.

As the roots γ_1 and γ_2 are locally smooth with respect to $\varrho' = (x, \xi', \tau)$ and homogeneous of degree one in (ξ', τ) by Proposition 6.28, there exists $\mathscr{U}$ a conic open neighborhood of $(x^0, \xi^{0\prime}, \tau^0)$ in $U^0 \times \mathbb{R}^{d-1} \times \mathbb{R}_+$ and $C > 0$ such that

$$\gamma_1(\varrho') \neq \gamma_2(\varrho'), \; \operatorname{Im}\gamma_2(\varrho') \leq -C\lambda_{\mathsf{T},\tau}, \;\text{ and }\; \operatorname{Im}\gamma_1(\varrho') \leq -C\lambda_{\mathsf{T},\tau},$$

if $\varrho' \in \overline{\mathscr{U}}$. Without any loss of generality we may moreover assume that $\mathbb{S}_{\overline{\mathscr{U}}}$ is compact.

We let $\chi \in S_{\mathsf{T},\tau}^0$ be as in the statement and $\tilde{\chi} \in S_{\mathsf{T},\tau}^0$ be homogeneous of degree zero and be such that $\operatorname{supp}(\tilde{\chi}) \subset \mathscr{U}$ and $\tilde{\chi} \equiv 1$ on $\operatorname{supp}(\chi)$. From the smoothness and the homogeneity of the roots, we have $\tilde{\chi}\gamma_j \in S_{\mathsf{T},\tau}^1$, $j = 1, 2$. We set

$$P^+ = D_d - \operatorname{Op}_{\mathsf{T}}(\tilde{\chi}\gamma_2) \;\text{ and }\; P^- = D_d - \operatorname{Op}_{\mathsf{T}}(\tilde{\chi}\gamma_1).$$

Since $\mathrm{Op_T}(\chi)P_{0,\varphi} = \mathrm{Op_T}(\chi)P^-P^+ + R$ with $R \in \Psi_\tau^{1,0}$, we write

$$\mathrm{Op_T}(\chi)P^-P^+v = \mathrm{Op_T}(\chi)P_{0,\varphi}v - Rv = \mathrm{Op_T}(\chi)G_0 + \mathrm{Op_T}(\chi)\sum_{1\leq j\leq d}\partial_j G^j - Rv,$$

and we obtain

$$\mathrm{Op_T}(\chi)P^-(P^+v - iG^d) = \mathrm{Op_T}(\chi)G_0 + \mathrm{Op_T}(\chi)\sum_{1\leq j\leq d-1}\partial_j G^j + i\mathrm{Op_T}(\chi)\mathrm{Op_T}(\tilde{\chi}\gamma_1)G^d - Rv,$$

which yields

$$\|\mathrm{Op_T}(\chi)P^-(P^+v - iG^d)\|_{\tau,0,-1} \lesssim \|G_0\|_{\tau,0,-1} + \|G\|_+ + \|v\|_{\tau,1,-1} \lesssim \tau^{-1}\|G_0\|_+ + \|G\|_+ + \|v\|_{\tau,1,-1}, \tag{13.3.11}$$

using Remark 6.8.

From Lemma 6.20 (with $s = -1$), we have for τ chosen sufficiently large,

$$\|\mathrm{Op_T}(\chi)z\|_+ + |\mathrm{Op_T}(\chi)z_{|x_d=0^+}|_{\tau,-1/2} \lesssim \|P^-\mathrm{Op_T}(\chi)z\|_{\tau,0,-1} + \|z\|_{\tau,0,-N}, \tag{13.3.12}$$

for $z \in \overline{\mathscr{S}}(\mathbb{R}^d_+)$.

With a density argument we apply estimate (13.3.12) for $z = P^+v - iG^d$, yielding

$$\begin{aligned}\|\mathrm{Op_T}(\chi)(P^+v - iG^d)\|_+ &+ |\mathrm{Op_T}(\chi)(P^+v - iG^d)_{|x_d=0^+}|_{\tau,-1/2} \\ &\lesssim \|P^-\mathrm{Op_T}(\chi)(P^+v - iG^d)\|_{\tau,0,-1} + \|P^+v - iG^d\|_{\tau,0,-N} \\ &\lesssim \|\mathrm{Op_T}(\chi)P^-(P^+v - iG^d)\|_{\tau,0,-1} + \|v\|_{\tau,1,-1} + \|G^d\|_+,\end{aligned}$$

using that $P^-\mathrm{Op_T}(\chi) = \mathrm{Op_T}(\chi)P^- \mod \Psi^0_{\mathsf{T},\tau}$. With (13.3.11), we then obtain

$$\begin{aligned}\|\mathrm{Op_T}(\chi)(P^+v - iG^d)\|_+ &+ |\mathrm{Op_T}(\chi)(P^+v - iG^d)_{|x_d=0^+}|_{\tau,-1/2} \\ &\lesssim \tau^{-1}\|G_0\|_+ + \|G\|_+ + \|v\|_{\tau,1,-1}.\end{aligned}$$

We set $w = \mathrm{Op_T}(\chi)v$ and we write

$$\begin{aligned}\|P^+w\|_+ &+ |(P^+w - i\mathrm{Op_T}(\chi)G^d)_{|x_d=0^+}|_{\tau,-1/2} \\ &\lesssim \|\mathrm{Op_T}(\chi)P^+v\|_+ + |\mathrm{Op_T}(\chi)(P^+v - iG^d)_{|x_d=0^+}|_{\tau,-1/2} \\ &\quad + \|v\|_{\tau,0} + |v_{|x_d=0^+}|_{\tau,-1/2} \\ &\lesssim \|\mathrm{Op_T}(\chi)(P^+v - iG^d)\|_+ + |\mathrm{Op_T}(\chi)(P^+v - iG^d)_{|x_d=0^+}|_{\tau,-1/2} \\ &\quad + \|G^d\|_+ + \|v\|_{\tau,1,-1},\end{aligned}$$

using the trace inequality of Corollary 6.10 (with $m = 0$ and $s = -1/2$). We thus obtain

$$\|P^+ w\|_+ + |(P^+ w - i\mathrm{Op}_\mathsf{T}(\chi) G^d)_{|x_d=0^+}|_{\tau,-1/2} \lesssim \tau^{-1}\|G_0\|_+ + \|G\|_+ + \|v\|_{\tau,1,-1}.$$

Next, from Lemma 6.20 (with $s = 0$), with a density argument, we have for τ chosen sufficiently large,

$$\|w\|_{\tau,1} + |w_{|x_d=0^+}|_{\tau,1/2} \lesssim \|P^+ w\|_+ + \|v\|_{\tau,0,-N}.$$

This yields

$$\|w\|_{\tau,1} + |w_{|x_d=0^+}|_{\tau,1/2} + |(P^+ w - i\mathrm{Op}_\mathsf{T}(\chi) G^d)_{|x_d=0^+}|_{\tau,-1/2} \lesssim \tau^{-1}\|G_0\|_+ + \|G\|_+ + \|v\|_{\tau,1,-1}.$$

Observing that

$$|(D_d w - i\mathrm{Op}_\mathsf{T}(\chi) G^d)_{|x_d=0^+}|_{\tau,-1/2} \lesssim |w_{|x_d=0^+}|_{\tau,1/2} + |(P^+ w - i\mathrm{Op}_\mathsf{T}(\chi) G^d)_{|x_d=0^+}|_{\tau,-1/2},$$

we finally obtain

$$\tau\Big(\|w\|_{\tau,1} + |w_{|x_d=0^+}|_{\tau,1/2} + |(D_d w - i\mathrm{Op}_\mathsf{T}(\chi) G^d)_{|x_d=0^+}|_{\tau,-1/2}\Big) \lesssim \|G_0\|_+ + \tau\|G\|_+ + \tau\|v\|_{\tau,1,-1},$$

which concludes the proof of Proposition 13.10 in Case 2. ∎

REMARK 13.13. We observe that this microlocal estimate is in fact stronger than the estimate in the statement of Proposition 13.10. One has $\tau\|w\|_{\tau,1}$ instead of $\tau^{1/2}\|w\|_{\tau,1}$ on the left-hand side and the term $\tau|\mathsf{B}_{0,\varphi}(v, -G^d)|_{\tau,1/2-k}$ is not present in the right-hand side. This is consistent with the result obtained in Sect. 8.3.4.2 where the roots are also in the lower complex half-plane; see Remark 8.21. This is natural as this implies a perfectly elliptic estimate, meaning that all relevant traces can be microlocally estimated. Here, a particular treatment is needed because of the low regularity of the source terms; the smooth factorization $p_\varphi = p^+ p^-$ in the neighborhood of $\mathrm{supp}(\chi)$ is crucial here; it was needed in Sect. 8.3.4.2. The need for this smooth factorization forces us to consider separately the case of roots that coincide. This is the subject of the next section.

13.3.3. Case 3: Roots Coincide in the Open Lower Complex Half-Plane. Here, we consider the case $\mathrm{Im}\,\gamma_2(\varrho^{0\prime}) < 0$ and $\gamma_2(\varrho^{0\prime}) = \gamma_1(\varrho^{0\prime})$, which means $\alpha(\varrho^{0\prime}) = 0$. We recall that α is defined through its square, viz.,

$$\alpha(\varrho')^2 = r(x, \xi' + i\tau d_{x'}\varphi(x)), \quad \varrho' = (x, \xi', \tau).$$

At the set $\{\alpha = 0\}$ roots may not be smooth (whereas they are smooth away from this set by Proposition 6.28). Here, we shall concentrate our analysis close that set.

Note that $\operatorname{Re}\alpha(\varrho')^2 = r(x,\xi') - \tau^2 r(x, d_{x'}\varphi(x))$, implying

$$\lambda_{\mathsf{T},\tau}^2 \lesssim \tau^2 + r(x,\xi') \lesssim \tau^2(1 + r(x, d_{x'}\varphi(x)) + \operatorname{Re}\alpha(\varrho')^2 \lesssim \tau^2 + |\alpha(\varrho')|^2.$$

Thus, for $\varepsilon_0 > 0$ chosen sufficiently small, if $|\alpha^2(\varrho')| \leq \varepsilon_0 \lambda_{\mathsf{T},\tau}^2$, then

$$\lambda_{\mathsf{T},\tau} \eqsim \tau. \tag{13.3.13}$$

Hence, there exists $\mathscr{U}^0$ a conic open neighborhood of $\varrho^{0\prime}$ in $U^0 \times \mathbb{R}^{d-1} \times \mathbb{R}_+$ where

$$\partial_d \varphi \geq C > 0 \quad \text{and} \quad |\alpha^2(\varrho')| \leq \varepsilon_0 \lambda_{\mathsf{T},\tau}^2,$$

implying that (13.3.13) holds there. Without any loss of generality we may moreover assume that $\mathbb{S}_{\overline{\mathscr{U}^0}}$ is compact.

From the form of the operator $P_{0,\varphi}$, since we consider a microlocal region where $|\alpha|$ is small, we can foresee that the estimation we shall obtain near that region is almost given by an estimate for the conjugated operator

$$Q_\varphi = (D_d + i\tau\partial_d\varphi(x))^2.$$

LEMMA 13.14. *Let $\tilde{\chi} \in S_{\mathsf{T},\tau}^0$, homogeneous of degree zero, be such that $\operatorname{supp}(\tilde{\chi}) \subset \mathscr{U}^0$. There exist $C > 0$ and $\tau_* > 0$ such that*

$$\begin{aligned} \tau\big(\|\operatorname{Op}_{\mathsf{T}}(\tilde{\chi})v\|_{\tau,1} + |(D_d \operatorname{Op}_{\mathsf{T}}(\tilde{\chi})v - i\operatorname{Op}_{\mathsf{T}}(\tilde{\chi})H^d)_{|x_d=0^+}|_{\tau,-1/2} \\ + |\operatorname{Op}_{\mathsf{T}}(\tilde{\chi})v_{|x_d=0^+}|_{\tau,1/2}\big) \leq C\Big(\|H_0\|_+ + \tau\|H\|_+ + \tau\|v\|_{\tau,1,-1}\Big), \end{aligned} \tag{13.3.14}$$

for $\tau \geq \tau_*$ and $v \in H^1(\mathbb{R}_+^d)$, $H_0 \in L^2(\mathbb{R}_+^d)$, $H \in L^2V(\mathbb{R}_+^d)$ such that

$$Q_\varphi v = H_0 + \sum_{1 \leq i \leq d} \partial_j H^j.$$

A proof is given below.

For $\varepsilon > 0$ to be fixed below, we let $\mathscr{U}_\varepsilon$ be the conic open set of $U^0 \times \mathbb{R}^{d-1} \times \mathbb{R}_+$ given by

$$\mathscr{U}_\varepsilon = \{\varrho' \in \mathscr{U}^0; |\alpha^2(\varrho')| < \varepsilon \lambda_{\mathsf{T},\tau}^2\}.$$

We take $0 < \varepsilon \leq \varepsilon_0/2$ allowing us to choose $\chi_\varepsilon \in S_{\mathsf{T},\tau}^0$ homogeneous of degree zero such that $\operatorname{supp}(\chi_\varepsilon) \subset \mathscr{U}_\varepsilon$. In $\mathscr{U}_\varepsilon \subset \mathscr{U}^0$ one has $\lambda_{\mathsf{T},\tau} \lesssim \tau$. With Proposition 6.11, there exists $C_1 > 0$ such that

$$\|\operatorname{Op}_{\mathsf{T}}(\chi_\varepsilon)v\|_{\tau,0,2} \leq C_1\tau^2\|\operatorname{Op}_{\mathsf{T}}(\chi_\varepsilon)v\|_+ + C_{\varepsilon,N}\|v\|_{\tau,0,-N},$$

for any $N \in \mathbb{N}$, with some $C_{\varepsilon,N} > 0$.

As $|\alpha^2(\varrho')| \leq \varepsilon \lambda^2_{\mathsf{T},\tau}$ in $\mathscr{U}_\varepsilon$, one has by Corollary 2.51, for $C'_0 > 1$ and $C'_1 = C'_0 C_1$,

$$\|R_{0,\varphi}\mathrm{Op}_{\mathsf{T}}(\chi_\varepsilon)v\|_+ \leq C'_0 \varepsilon \|\mathrm{Op}_{\mathsf{T}}(\chi_\varepsilon)v\|_{\tau,0,2} + C'_{\varepsilon,N}\|v\|_{\tau,0,-N} \leq 2\varepsilon C'_1 \tau^2 \|\mathrm{Op}_{\mathsf{T}}(\chi_\varepsilon)v\|_+ + C''_{\varepsilon,N}\|v\|_{\tau,0,-N}. \tag{13.3.15}$$

Recalling (13.3.2), we write

$$\begin{aligned} Q_\varphi \mathrm{Op}_{\mathsf{T}}(\chi_\varepsilon)v &= P_{0,\varphi}\mathrm{Op}_{\mathsf{T}}(\chi_\varepsilon)v - R_{0,\varphi}\mathrm{Op}_{\mathsf{T}}(\chi_\varepsilon)v \\ &= \mathrm{Op}_{\mathsf{T}}(\chi_\varepsilon)P_{0,\varphi}v + [P_{0,\varphi}, \mathrm{Op}_{\mathsf{T}}(\chi_\varepsilon)]v - R_{0,\varphi}\mathrm{Op}_{\mathsf{T}}(\chi_\varepsilon)v \\ &= H_0 + \sum_{1\leq j\leq d} \partial_j H^j, \end{aligned}$$

with $H = \mathrm{Op}_{\mathsf{T}}(\chi_\varepsilon)G$ and

$$\begin{aligned} H_0 = \mathrm{Op}_{\mathsf{T}}(\chi_\varepsilon)G_0 + \sum_{1\leq j\leq d} [\mathrm{Op}_{\mathsf{T}}(\chi_\varepsilon), \partial_j]G^j + [P_{0,\varphi}, \mathrm{Op}_{\mathsf{T}}(\chi_\varepsilon)]v \\ - R_{0,\varphi}\mathrm{Op}_{\mathsf{T}}(\chi_\varepsilon)v. \end{aligned}$$

We have $\|H\|_+ \lesssim \|G\|_+$ and, with (13.3.15),

$$\|H_0\|_+ \leq C_\varepsilon \big(\|G_0\|_+ + \|G\|_+ + \|v\|_{\tau,1}\big) + C\varepsilon\tau^2\|\mathrm{Op}_{\mathsf{T}}(\chi_\varepsilon)v\|_+.$$

We choose $\tilde{\chi} \in S^0_{\mathsf{T},\tau}$ homogeneous of degree zero such that $\mathrm{supp}(\tilde{\chi}) \subset \mathscr{U}^0$ and $\tilde{\chi} \equiv 1$ in a neighborhood of $\mathscr{U}_{\varepsilon_0/2}$. Applying Lemma 13.14 we thus obtain, with $\tilde{\chi}_\varepsilon = \tilde{\chi} \circ \chi_\varepsilon$,

$$\begin{aligned} \tau\Big(\|\mathrm{Op}_{\mathsf{T}}(\tilde{\chi}_\varepsilon)v\|_{\tau,1} + |(D_d \mathrm{Op}_{\mathsf{T}}(\tilde{\chi}_\varepsilon)v - i\mathrm{Op}_{\mathsf{T}}(\tilde{\chi}_\varepsilon)G^d)_{|x_d=0^+}|_{\tau,-1/2} \\ + |\mathrm{Op}_{\mathsf{T}}(\tilde{\chi}_\varepsilon)v_{|x_d=0^+}|_{\tau,1/2}\Big) \leq C_\varepsilon\Big(\|G_0\|_+ + \tau\|G\|_+ + \|v\|_{\tau,1}\Big) \\ + C\varepsilon\tau^2\|\mathrm{Op}_{\mathsf{T}}(\chi_\varepsilon)v\|_+. \end{aligned}$$

Since $\tilde{\chi} \equiv 1$ in a neighborhood of $\mathrm{supp}(\chi_\varepsilon)$ we have $\tilde{\chi}_\varepsilon = \chi_\varepsilon \mod S^{-N}_{\mathsf{T},\tau}$ for any $N \in \mathbb{N}$ by pseudo-differential calculus. We then find

$$\begin{aligned} &|(D_d \mathrm{Op}_{\mathsf{T}}(\chi_\varepsilon)v - i\mathrm{Op}_{\mathsf{T}}(\chi_\varepsilon)G^d)_{|x_d=0^+}|_{\tau,-1/2} \\ &\quad\leq |\mathrm{Op}_{\mathsf{T}}(\chi_\varepsilon)(D_d v - iG^d)_{|x_d=0^+}|_{\tau,-1/2} + |[D_d, \mathrm{Op}_{\mathsf{T}}(\chi_\varepsilon)]v_{|x_d=0^+}|_{\tau,-1/2} \\ &\quad\leq |\mathrm{Op}_{\mathsf{T}}(\tilde{\chi}_\varepsilon)(D_d v - iG^d)_{|x_d=0^+}|_{\tau,-1/2} \\ &\qquad + C_\varepsilon\big(|(D_d v - iG^d)_{|x_d=0^+}|_{\tau,-N} + \|v\|_{\tau,1,-1}\big) \\ &\quad\leq |(D_d \mathrm{Op}_{\mathsf{T}}(\tilde{\chi}_\varepsilon)v - i\mathrm{Op}_{\mathsf{T}}(\tilde{\chi}_\varepsilon)G^d)_{|x_d=0^+}|_{\tau,-1/2} \\ &\qquad + |[D_d, \mathrm{Op}_{\mathsf{T}}(\tilde{\chi}_\varepsilon)]v_{|x_d=0^+}|_{\tau,-1/2} \\ &\qquad + C_\varepsilon\big(|(D_d v - iG^d)_{|x_d=0^+}|_{\tau,-N} + \|v\|_{\tau,1,-1}\big) \\ &\quad\leq |(D_d \mathrm{Op}_{\mathsf{T}}(\tilde{\chi}_\varepsilon)v - i\mathrm{Op}_{\mathsf{T}}(\tilde{\chi}_\varepsilon)G^d)_{|x_d=0^+}|_{\tau,-1/2} \\ &\qquad + C_\varepsilon\big(|(D_d v - iG^d)_{|x_d=0^+}|_{\tau,-N} + \|v\|_{\tau,1,-1}\big), \end{aligned}$$

using the trace inequality of Corollary 6.10 (with $m = 0$ and $s = -1/2$). Arguing similarly the term $|\mathrm{Op}_{\mathsf{T}}(\tilde{\chi}_\varepsilon)v_{|x_d=0^+}|_{\tau,1/2}$, and writing

$$\|\mathrm{Op}_{\mathsf{T}}(\chi_\varepsilon)v\|_{\tau,1} \lesssim \|\mathrm{Op}_{\mathsf{T}}(\tilde{\chi}_\varepsilon)v\|_{\tau,1} + C_\varepsilon \|v\|_{\tau,1,-N},$$

we then obtain

$$\begin{aligned}\tau\Big(\|\mathrm{Op}_{\mathsf{T}}(\chi_\varepsilon)v\|_{\tau,1} &+ |(D_d \mathrm{Op}_{\mathsf{T}}(\chi_\varepsilon)v - i\mathrm{Op}_{\mathsf{T}}(\chi_\varepsilon)G^d)_{|x_d=0^+}|_{\tau,-1/2} \\ &+ |\mathrm{Op}_{\mathsf{T}}(\chi_\varepsilon)v_{|x_d=0^+}|_{\tau,1/2}\Big) \leq C\varepsilon\tau^2 \|\mathrm{Op}_{\mathsf{T}}(\chi_\varepsilon)v\|_+ \\ &+ C_\varepsilon\Big(\|G_0\|_+ + \tau\|G\|_+ + \|v\|_{\tau,1} + |(D_d v - iG^d)_{|x_d=0^+}|_{\tau,-N}\Big).\end{aligned}$$

Choosing $\varepsilon > 0$ sufficiently small, to be kept fixed and setting $\chi = \chi_\varepsilon$ we finally obtain

$$\begin{aligned}\tau\Big(\|\mathrm{Op}_{\mathsf{T}}(\chi)v\|_{\tau,1} &+ |(D_d \mathrm{Op}_{\mathsf{T}}(\chi)v - i\mathrm{Op}_{\mathsf{T}}(\chi)G^d)_{|x_d=0^+}|_{\tau,-1/2} \\ &+ |\mathrm{Op}_{\mathsf{T}}(\chi)v_{|x_d=0^+}|_{\tau,1/2}\Big) \\ &\lesssim \|G_0\|_+ + \tau\|G\|_+ + \|v\|_{\tau,1} + |(D_d v - iG^d)_{|x_d=0^+}|_{\tau,-N},\end{aligned} \tag{13.3.16}$$

which concludes the proof of Proposition 13.10 in Case 3. ■

PROOF OF LEMMA 13.14. We set $L = D_d + i\tau\partial_d\varphi(x)$. We have $\mathrm{Op}_{\mathsf{T}}(\tilde{\chi})L^2 v = \mathrm{Op}_{\mathsf{T}}(\tilde{\chi})H_0 + \sum_{1\leq j\leq d} \mathrm{Op}_{\mathsf{T}}(\tilde{\chi})\partial_j H^j$, which we write

$$\begin{aligned}\mathrm{Op}_{\mathsf{T}}(\tilde{\chi})L(Lv - iH^d) = \mathrm{Op}_{\mathsf{T}}(\tilde{\chi})H_0 &+ \sum_{1\leq j\leq d-1} \mathrm{Op}_{\mathsf{T}}(\tilde{\chi})\partial_j H^j \\ &+ \tau\mathrm{Op}_{\mathsf{T}}(\tilde{\chi})\partial_d\varphi H^d.\end{aligned}$$

This yields

$$\|\mathrm{Op}_{\mathsf{T}}(\tilde{\chi})L(Lv - iH^d)\|_{\tau,0,-1} \lesssim \|H_0\|_{\tau,0,-1} + \|H\|_+ \lesssim \tau^{-1}\|H_0\|_+ + \|H\|_+, \tag{13.3.17}$$

using Remark 6.8.

Since $\tau\partial_d\varphi \gtrsim \lambda_{\mathsf{T},\tau}$ in $\mathscr{U}^0$, from Lemma 6.20 (with $s = -1$), we have for τ chosen sufficiently large,

$$\|\mathrm{Op}_{\mathsf{T}}(\tilde{\chi})z\|_+ + |\mathrm{Op}_{\mathsf{T}}(\tilde{\chi})z_{|x_d=0^+}|_{\tau,-1/2} \lesssim \|L\mathrm{Op}_{\mathsf{T}}(\tilde{\chi})z\|_{\tau,0,-1} + \|z\|_{\tau,0,-N}, \tag{13.3.18}$$

for $z \in \overline{\mathscr{S}}(\mathbb{R}^d_+)$. We apply estimate (13.3.18) for $z = Lv - H^d$, with a density argument, yielding

$$\begin{aligned}\|\mathrm{Op}_{\mathsf{T}}(\tilde{\chi})(Lv - iH^d)\|_+ &+ |\mathrm{Op}_{\mathsf{T}}(\tilde{\chi})(Lv - iH^d)_{|x_d=0^+}|_{\tau,-1/2} \\ &\lesssim \|L\mathrm{Op}_{\mathsf{T}}(\tilde{\chi})(Lv - iH^d)\|_{\tau,0,-1} + \|Lv - iH^d\|_{\tau,0,-N} \\ &\lesssim \|\mathrm{Op}_{\mathsf{T}}(\tilde{\chi})L(Lv - iH^d)\|_{\tau,0,-1} + \|v\|_{\tau,1,-1} + \|H^d\|_+,\end{aligned}$$

using that $[L, \mathrm{Op}_{\mathsf{T}}(\tilde{\chi})] \in \Psi^0_{\mathsf{T},\tau}$. With (13.3.17), we then obtain

$$\|\mathrm{Op}_{\mathsf{T}}(\tilde{\chi})(Lv - iH^d)\|_+ + |\mathrm{Op}_{\mathsf{T}}(\tilde{\chi})(Lv - iH^d)_{|x_d=0^+}|_{\tau,-1/2}$$
$$\lesssim \tau^{-1}\|H_0\|_+ + \|H\|_+ + \|v\|_{\tau,1,-1}.$$

We then write

$$\begin{aligned}\|L\mathrm{Op}_{\mathsf{T}}(\tilde{\chi})v\|_+ &+ |(L\mathrm{Op}_{\mathsf{T}}(\tilde{\chi})v - i\mathrm{Op}_{\mathsf{T}}(\tilde{\chi})H^d)_{|x_d=0^+}|_{\tau,-1/2}\\ &\lesssim \|\mathrm{Op}_{\mathsf{T}}(\tilde{\chi})Lv\|_+ + |\mathrm{Op}_{\mathsf{T}}(\tilde{\chi})(Lv - iH^d)_{|x_d=0^+}|_{\tau,-1/2}\\ &\quad + \|v\|_+ + |v|_{\tau,-1/2}\\ &\lesssim \|\mathrm{Op}_{\mathsf{T}}(\tilde{\chi})(Lv - iH^d)\|_+ + |\mathrm{Op}_{\mathsf{T}}(\tilde{\chi})(Lv - iH^d)_{|x_d=0^+}|_{\tau,-1/2}\\ &\quad + \|H^d\|_+ + \|v\|_{\tau,1,-1},\end{aligned}$$

using the trace inequality of Corollary 6.10 (with $m = 0$ and $s = -1/2$). We thus obtain

$$\|L\mathrm{Op}_{\mathsf{T}}(\tilde{\chi})v\|_+ + |(L\mathrm{Op}_{\mathsf{T}}(\tilde{\chi})v - i\mathrm{Op}_{\mathsf{T}}(\tilde{\chi})H^d)_{|x_d=0^+}|_{\tau,-1/2}$$
$$\lesssim \tau^{-1}\|H_0\|_+ + \|H\|_+ + \|v\|_{\tau,1,-1}.$$

Next, invoking Lemma 6.20 a second time (with $s = 0$ this time), we have for τ chosen sufficiently large,

$$\|\mathrm{Op}_{\mathsf{T}}(\tilde{\chi})v\|_{\tau,1} + |\mathrm{Op}_{\mathsf{T}}(\tilde{\chi})v_{|x_d=0^+}|_{\tau,1/2} \lesssim \|L\mathrm{Op}_{\mathsf{T}}(\tilde{\chi})v\|_+ + \|v\|_{\tau,0,-N}.$$

This yields

$$\begin{aligned}\|\mathrm{Op}_{\mathsf{T}}(\tilde{\chi})v\|_{\tau,1} &+ |\mathrm{Op}_{\mathsf{T}}(\tilde{\chi})v_{|x_d=0^+}|_{\tau,1/2}\\ &+ |(L\mathrm{Op}_{\mathsf{T}}(\tilde{\chi})v - \mathrm{Op}_{\mathsf{T}}(\tilde{\chi})H^d)_{|x_d=0^+}|_{\tau,-1/2}\\ &\qquad\qquad \lesssim \tau^{-1}\|H_0\|_+ + \|H\|_+ + \|v\|_{\tau,1,-1}.\end{aligned}$$

Observing that

$$\begin{aligned}|(D_d\mathrm{Op}_{\mathsf{T}}(\tilde{\chi})v &- \mathrm{Op}_{\mathsf{T}}(\tilde{\chi})H^d)_{|x_d=0^+}|_{\tau,-1/2}\\ &\lesssim |\mathrm{Op}_{\mathsf{T}}(\tilde{\chi})v_{|x_d=0^+}|_{\tau,1/2} + |(L\mathrm{Op}_{\mathsf{T}}(\tilde{\chi})v - \mathrm{Op}_{\mathsf{T}}(\tilde{\chi})H^d)_{|x_d=0^+}|_{\tau,-1/2},\end{aligned}$$

which concludes the proof. ■

13.4. Patching Estimates Together

First, we show how microlocal estimates given by Proposition 13.10 can be patched together to form the local estimate of Proposition 13.9. Second, we show how such local estimate can be patched to obtain the result of Theorem 13.4.

13.4.1. Patching Microlocal Estimates. Here, we prove Proposition 13.9 by means of the microlocal result obtained in Sect. 13.3. Equivalently, this gives the result of Propositions 13.3 and 13.8.

We use the argument and the notation of the proof of Proposition 8.22. We recall the setting introduced therein and we only detail the argument for the terms that only occur in the present case.

With x^0 as in the statement of Propositions 13.8 and 13.9 the Lopatinskiĭ–Šapiro condition for (P_0, B_0, φ) holds at $(x^0, \xi^{0\prime}, \tau^0)$ for all $\xi^{0\prime} \in T^*_{x^0}\mathbb{R}^{d-1} \cong \mathbb{R}^{d-1}$ and $\tau^0 \geq 0$ such that $(\xi^{0\prime}, \tau^0) \neq 0$. It is in fact sufficient to consider $(\xi^{0\prime}, \tau^0)$ in the half-unit sphere $\mathbb{S}^{d-1}_+$, using the notation introduced in (1.7.3).

By Proposition 13.10 for all $(\xi^{0\prime}, \tau^0) \in \mathbb{S}^{d-1}_+$ there exists a conic open neighborhood $\mathscr{U}_{y^{0\prime}}$ of $\varrho^{0\prime} = (x^0, \xi^{0\prime}, \tau^0)$ in $U^0 \times \mathbb{R}^{d-1} \times \mathbb{R}_+$ such that estimate (13.3.1) holds. In fact, by reducing $\mathscr{U}_{\varrho^{0\prime}}$ we can choose $\mathscr{U}_{\varrho^{0\prime}} = \mathscr{O}_{\varrho^{0\prime}} \times \Gamma_{\varrho^{0\prime}}$ where $\mathscr{O}_{\varrho^{0\prime}}$ is an open set in U^0 and $\Gamma_{\varrho^{0\prime}}$ is a conic open set in $\mathbb{R}^{d-1} \times \mathbb{R}_+$. With the compactness of $\mathbb{S}^{d-1}_+$ we can thus find finitely many such open sets $\mathscr{U}_j = \mathscr{O}_j \times \Gamma_j$, $j \in J$, such that $\mathbb{S}^{d-1}_+ \subset \cup_{j\in J}\Gamma_j$. We then set $\mathscr{O} = \cap_{j\in J}\mathscr{O}_j$ that is an open neighborhood of x_0 in U^0 and we set $\mathscr{V}_j = \mathscr{O} \times \Gamma_j \subset \mathscr{U}_j$. We also choose an open neighborhood U of x_0 in $\mathbb{R}^d$ such that $U_+ = U \cap U^0 \Subset \mathscr{O}$.

We then choose a smooth partition of unity, χ_j, $j \in J$, of the closed set $F = \overline{U_+} \times \mathbb{S}^{d-1}_+$ in the manifold $\mathcal{N} = \mathbb{R}^d \times \mathbb{S}^{d-1}_+$ subordinated to the covering by the open sets $\mathscr{V}_j \cap \mathcal{N}$ according to Theorem 15.14. We then extend each χ_j smoothly to $\mathbb{R}^d \times \mathbb{R}^{d-1} \times \mathbb{R}_+$ by homogeneity of order 0 for $|(\xi', \tau)| \geq 1 > 0$. We have $\operatorname{supp}(\chi_j) \subset \mathscr{V}_j$ and

$$\sum_{j\in J} \chi_j(\varrho') = 1,$$

for $\varrho' = (x, \xi', \tau)$ in a conic neighborhood of $\overline{U_+} \times \mathbb{R}^{d-1} \times \mathbb{R}_+$ and $|(\xi', \tau)| \geq 1$. We also set $\underline{\chi} = 1 - \sum_{j\in J} \chi_j$

Let $v \in H^1(\mathbb{R}^d_+)$, $G_0 \in L^2(\mathbb{R}^d_+)$, and $G \in L^2V(\mathbb{R}^d_+)$ be such that $\operatorname{supp}(v) \cup \operatorname{supp}(G_0) \cup \operatorname{supp}(G) \subset \overline{U_+}$ and

$$P_{0,\varphi} v = G_0 + \sum_{1\leq i\leq d} \partial_j G^j.$$

As $\operatorname{supp}(\chi_j) \subset \mathscr{U}_j$, we can apply the microlocal estimate of Proposition 13.10

$$\begin{aligned} \tau^{1/2} \|\mathrm{Op}_{\mathsf{T}}(\chi_j) v\|_{\tau,1} + \tau |(D_d \mathrm{Op}_{\mathsf{T}}(\chi_j) v - i\mathrm{Op}_{\mathsf{T}}(\chi_j) G^d)_{|x_d=0^+}|_{\tau,-1/2} \\ + \tau |\mathrm{Op}_{\mathsf{T}}(\chi_j) v_{|x_d=0^+}|_{\tau,1/2} \lesssim \|G_0\|_+ + \tau \|G\|_+ + \tau |\mathsf{B}_{0,\varphi}(v, -G^d)|_{\tau,1/2-k} \\ + \|v\|_{\tau,1} + |(D_d v - iG^d)_{|x_d=0^+}|_{\tau,-N}, \end{aligned}$$

for $\tau > 0$ chosen sufficiently large.

Arguing as in the proof of Proposition 8.22 we have

$$\|v\|_{\tau,1} \lesssim \sum_{j\in J} \|\mathrm{Op}_{\mathsf{T}}(\chi_j)v\|_{\tau,1} + \|v\|_{\tau,1,-N},$$

and

$$|v_{|x_d=0^+}|_{\tau,1/2} \lesssim \sum_{j\in J} |\mathrm{Op}_{\mathsf{T}}(\chi_j)v_{|x_d=0^+}|_{\tau,1/2} + \|v\|_{\tau,1,-N}.$$

We now consider the term that is specific to the estimates proven in this chapter:

$$\begin{aligned}|(D_d v - iG^d)_{|x_d=0^+}|_{\tau,-1/2} &\lesssim \sum_{j\in J} |(D_d \mathrm{Op}_{\mathsf{T}}(\chi_j)v - i\mathrm{Op}_{\mathsf{T}}(\chi_j)G^d)_{|x_d=0^+}|_{\tau,-1/2}\\ &\quad + |(D_d \mathrm{Op}_{\mathsf{T}}(\underline{\chi})v - i\mathrm{Op}_{\mathsf{T}}(\underline{\chi})G^d)_{|x_d=0^+}|_{\tau,-1/2}.\end{aligned}$$

We then write

$$\begin{aligned}&|(D_d \mathrm{Op}_{\mathsf{T}}(\underline{\chi})v - i\mathrm{Op}_{\mathsf{T}}(\underline{\chi})G^d)_{|x_d=0^+}|_{\tau,-1/2}\\ &\qquad\lesssim |\mathrm{Op}_{\mathsf{T}}(\underline{\chi})(D_d v - iG^d)_{|x_d=0^+}|_{\tau,-1/2} + |[D_d, \mathrm{Op}_{\mathsf{T}}(\underline{\chi})]v_{|x_d=0^+}|_{\tau,-1/2}\\ &\qquad\lesssim |(D_d v - iG^d)_{|x_d=0^+}|_{\tau,-N} + |v_{|x_d=0^+}|_{\tau,-N}\\ &\qquad\lesssim |(D_d v - iG^d)_{|x_d=0^+}|_{\tau,-N} + \|v\|_{\tau,1,-N},\end{aligned}$$

since $\underline{\chi}$ is supported away from U_+ and using the trace inequality of Corollary 6.10.

We thus obtain

$$\begin{aligned}\tau^{1/2}\|v\|_{\tau,1} &+ \tau|v_{|x_d=0^+}|_{\tau,1/2} + \tau|(D_d v - iG^d)_{|x_d=0^+}|_{\tau,-1/2}\\ &\lesssim \|G_0\|_+ + \tau\|G\|_+ + \tau|\mathsf{B}_{0,\varphi}(v, -G^d)|_{\tau,1/2-k}\\ &\quad + |(D_d v - iG^d)_{|x_d=0^+}|_{\tau,-N} + \|v\|_{\tau,1}.\end{aligned}$$

We conclude the proof of Proposition 13.9 by choosing $\tau > 0$ sufficiently large. ■

13.4.2. Patching Local Estimates. Here we show how the patching method first described in Section 3.5 allows one to obtain our main result, Theorem 13.4, from Proposition 13.3.

For all $m \in \overline{V}$, by Lemma 3.9 of Volume 1 used in a local chart (upon a smooth extension of the coefficients and the weight function locally in $\mathbb{R}^d_-$ if needed), there exists an open neighborhood V_m^0 of m where φ and P have the sub-ellipticity property and where either the result of Proposition 13.1 or Proposition 13.3 applies: there exists an open neighborhood $V_m^1 \subset V_m^0$, with $m \in V_m^1$, where either the local estimate (13.1.2) or (13.1.4) holds.

From the covering of the compact set $\overline{V}$ by the open sets V_m^1, $m \in \overline{V}$, we can extract a *finite* covering $(V_i)_{i\in\mathcal{I}}$, such that for all $i \in \mathcal{I}$ the Carleman estimate in V_i holds for $\tau \geq \tau_i > 0$, $C = C_i > 0$:

(13.4.1)
$$\begin{aligned}\tau^{1/2}\|e^{\tau\varphi}w\|_{\tau,1} + \tau|e^{\tau\varphi}w_{|\partial\mathcal{M}}|_{\tau,1/2} + \tau|e^{\tau\varphi_{|\partial\mathcal{M}}}(\partial_\nu w + g(H,\nu))_{|\partial\mathcal{M}}|_{\tau,-1/2}\\ \leq C\big(\|e^{\tau\varphi}H_0\|_{L^2(\mathcal{M})} + \tau\|e^{\tau\varphi}H\|_{L^2V(\mathcal{M})}\\ + \tau|e^{\tau\varphi_{|\partial\mathcal{M}}}\mathsf{B}(w, g(H,\nu))|_{\tau,1/2-k}\big),\end{aligned}$$

for $w \in H^1(\mathcal{M})$, $H_0 \in L^2(\mathcal{M})$, $H \in L^2V(\mathcal{M})$ such that

$$Pw = H_0 + \operatorname{div}_g H, \quad \text{and} \quad \operatorname{supp}(w) \cup \operatorname{supp}(H_0) \cup \operatorname{supp}(H) \subset \overline{V_i}.$$

Note that the boundary terms on the l.h.s. and the r.h.s. vanish if $\overline{V_i} \cap \partial\mathcal{M} = \emptyset$.

Let $(\chi^{(i)})_{i\in\mathcal{I}}$ be a partition of unity of $\overline{V}$ subordinated to the open covering V_i, $i \in \mathcal{I}$, as given by Theorem 15.14, that is, $\chi^{(i)} \in \mathscr{C}^\infty(\mathcal{M})$, with

$$\operatorname{supp}(\chi^{(i)}) \subset V_i, \quad 0 \leq \chi^{(i)} \leq 1, \quad i \in \mathcal{I},$$

and $\chi = \sum_{i\in\mathcal{I}} \chi^{(i)} \equiv 1$ in a neighborhood of $\overline{V}$.

Let now u, F_0, and F be as in the statement. For all $i \in \mathcal{I}$, we set $u^{(i)} = \chi^{(i)}u \in H^1(\mathcal{M})$ with $\operatorname{supp}(u^{(i)}) \subset V_i$. We have $Pu^{(i)} = H_0^{(i)} + \operatorname{div}_g F^{(i)}$ with

$$H_0^{(i)} = \chi^{(i)}F_0 - [P, \chi^{(i)}]u - g(\nabla_g\chi^{(i)}, F), \ \ F^{(i)} = \chi^{(i)}F.$$

We have

$$\begin{aligned}\|e^{\tau\varphi}H_0^{(i)}\|_{L^2(\mathcal{M})} + \tau\|e^{\tau\varphi}F^{(i)}\|_{L^2V(\mathcal{M})}\\ \lesssim \|e^{\tau\varphi}F_0\|_{L^2(\mathcal{M})} + \tau\|e^{\tau\varphi}F\|_{L^2V(\mathcal{M})} + \|e^{\tau\varphi}u\|_{L^2(\mathcal{M})} + \|e^{\tau\varphi}\nabla_g u\|_{L^2V(\mathcal{M})}\\ \lesssim \|e^{\tau\varphi}F_0\|_{L^2(\mathcal{M})} + \tau\|e^{\tau\varphi}F\|_{L^2V(\mathcal{M})} + \|e^{\tau\varphi}u\|_{\tau,1},\end{aligned}$$

and

$$\begin{aligned}|e^{\tau\varphi_{|\partial\mathcal{M}}}\mathsf{B}(u^{(i)}, g(F^{(i)},\nu))|_{\tau,1/2-k}\\ \lesssim |e^{\tau\varphi_{|\partial\mathcal{M}}}\mathsf{B}(u, g(F,\nu))|_{\tau,1/2-k} + |e^{\tau\varphi}u_{|\partial\mathcal{M}}|_{\tau,-1/2}\\ \lesssim |e^{\tau\varphi_{|\partial\mathcal{M}}}\mathsf{B}(u, g(F,\nu))|_{\tau,1/2-k} + \tau^{-1}\|e^{\tau\varphi}u\|_{\tau,1}.\end{aligned}$$

As $u = \chi u = \sum_{i\in\mathcal{I}} u^{(i)}$, $F = \chi F = \sum_{i\in\mathcal{I}} F^{(i)}$ we find

$$\begin{aligned}\tau^{1/2}\|e^{\tau\varphi}u\|_{\tau,1} + \tau|e^{\tau\varphi}u_{|\partial\mathcal{M}}|_{\tau,1/2} + \tau|e^{\tau\varphi_{|\partial\mathcal{M}}}(\partial_\nu u + g(F,\nu))_{|\partial\mathcal{M}}|_{\tau,-1/2}\\ \lesssim \sum_{i\in\mathcal{I}}\big(\tau^{1/2}\|e^{\tau\varphi}u^{(i)}\|_{\tau,1} + \tau|e^{\tau\varphi}u^{(i)}_{|\partial\mathcal{M}}|_{\tau,1/2}\\ + \tau|e^{\tau\varphi_{|\partial\mathcal{M}}}(\partial_\nu u^{(i)} + g(F^{(i)},\nu))_{|\partial\mathcal{M}}|_{\tau,-1/2}\big).\end{aligned}$$

Since, for each $u^{(i)}$ we have a local Carleman estimate of the form given by (13.4.1), we obtain

$$\begin{aligned}\tau^{1/2}\|e^{\tau\varphi}u\|_{\tau,1} + \tau|e^{\tau\varphi}u_{|\partial\mathcal{M}}|_{\tau,1/2} + \tau|e^{\tau\varphi_{|\partial\mathcal{M}}}(\partial_\nu u + g(F,\nu))_{|\partial\mathcal{M}}|_{\tau,-1/2} \\ \leq C\big(\|e^{\tau\varphi}F_0\|_{L^2(\mathcal{M})} + \tau\|e^{\tau\varphi}F\|_{L^2V(\mathcal{M})} \\ + \tau|e^{\tau\varphi_{|\partial\mathcal{M}}}\mathsf{B}(u, g(F,\nu))|_{\tau,1/2-k} + \|e^{\tau\varphi}u\|_{\tau,1}\big).\end{aligned}$$

We conclude the proof of Theorem 13.4 by choosing $\tau > 0$ sufficiently large. ∎

13.5. Shifted Estimate

Similarly to Theorem 8.24 we have the following shifted estimate.

THEOREM 13.15. *Let $(\mathcal{M}, g)$ be a compact Riemannian manifold with boundary and let $P = -\Delta_g + R_1$ with R_1 a first-order differential operator on $\mathcal{M}$. Let V be an open set of $\mathcal{M}$ and set $V_\partial = V \cap \partial\mathcal{M}$. Let $\varphi \in \mathscr{C}^\infty(\mathcal{M})$ be such that the pair (P, φ) has the sub-ellipticity property of Definition 5.1 in $\overline{V}$. If $V_\partial \neq \emptyset$, consider B a differential operator of order $k \in \mathbb{N}$ in V and assume moreover that (P, B, φ) satisfies the Lopatinskiĭ–Šapiro condition at all points $m \in \overline{V_\partial}$ (Definition 8.1). Then, there exist C and $\tau_* > 0$ such that*

(13.5.1)
$$\begin{aligned}&\tau\|e^{\tau\varphi}u\|_{\tau,0,1/2} + \tau^{3/2}|e^{\tau\varphi}u_{|\partial\mathcal{M}}|_{L^2(\partial\mathcal{M})} \\ &\quad + \tau^{3/2}|e^{\tau\varphi_{|\partial\mathcal{M}}}(\partial_\nu u + g(F,\nu))_{|\partial\mathcal{M}}|_{\tau,-1} \\ &\quad \leq C\big(\|e^{\tau\varphi}F_0\|_{L^2(\mathcal{M})} + \tau\|e^{\tau\varphi}F\|_{L^2V(\mathcal{M})} + \tau^{3/2}|e^{\tau\varphi_{|\partial\mathcal{M}}}\mathsf{B}(u, g(F,\nu))|_{\tau,-k}\big),\end{aligned}$$

for $\tau \geq \tau_*$ and $u \in H^1(\mathcal{M})$, $F_0 \in L^2(\mathcal{M})$, $F \in L^2V(\mathcal{M})$ such that

$$Pu = F_0 + \operatorname{div}_g F, \quad \textit{and} \quad \operatorname{supp}(u) \cup \operatorname{supp}(F_0) \cup \operatorname{supp}(F) \subset \overline{V}.$$

A proof based on Proposition 13.9 is given in Appendix 13.A.4.

13.6. Estimates Without Prescribed Boundary Conditions

Let $B = \partial_\nu + B'$ be a first-order boundary operator such that (P, B) satisfies the Lopatinskiĭ–Šapiro condition. As in Sect. 8.4.2 we prove estimates with terms involving both the Dirichlet trace $u_{|\partial\mathcal{M}}$ and the trace $Bu_{|\partial\mathcal{M}}$. Yet, one does not impose any condition on the weight function at the boundary, in particular in connection with B.

Firsts, we consider the case a weight function φ such that $\partial_\nu\varphi \neq 0$. We discuss the case where $\partial_\nu\varphi$ can vanish in Sect. 13.6.2 below.

13.6.1. Estimate for a Weight Function with Nonvanishing Neumann Trace. Here we prove an estimate in an open set V that meets the boundary and where $\partial_\nu\varphi$ does not vanish.

THEOREM 13.16. *Let $(\mathcal{M}, g)$ be a compact Riemannian manifold with boundary and let $P = -\Delta_g + R_1$ with R_1 a first-order differential operator with bounded coefficients on $\mathcal{M}$. Let V be an open set of $\mathcal{M}$ such that $V_\partial = V \cap \partial\mathcal{M} \neq \emptyset$. Let B' be a differential operator of order one on $\partial\mathcal{M}$ such that (P, B) fulfills the Lopatinskiĭ–Šapiro condition of Definition 2.2 on $\overline{V_\partial}$ for $B = \partial_\nu + B'$. Let $\varphi \in \mathscr{C}^\infty(\mathcal{M})$ be such that the pair (P, φ) has the sub-ellipticity property of Definition 5.1 in $\overline{V}$ and such that $\partial_\nu\varphi \neq 0$ in $\overline{V_\partial}$. Then, there exist C and $\tau_* > 0$ such that*

(13.6.1)
$$\tau^{1/2}\|e^{\tau\varphi}u\|_{\tau,1} + \tau|e^{\tau\varphi}u_{|\partial\mathcal{M}}|_{\tau,1/2} + \tau|e^{\tau\varphi_{|\partial\mathcal{M}}}(\partial_\nu u + g(F,\nu))_{|\partial\mathcal{M}}|_{\tau,-1/2}$$
$$\leq C\Big(\|e^{\tau\varphi}F_0\|_{L^2(\mathcal{M})} + \tau\|e^{\tau\varphi}F\|_{L^2V(\mathcal{M})}$$
$$+ \tau^{3/2}|e^{\tau\varphi}u_{|\partial\mathcal{M}}|_{L^2(\partial\mathcal{M})} + \tau|e^{\tau\varphi_{|\partial\mathcal{M}}}\mathsf{B}(u, g(F,\nu))|_{\tau,-1/2}\Big),$$

for $\tau \geq \tau_*$ and $u \in H^1(\mathcal{M})$, $F_0 \in L^2(\mathcal{M})$, $F \in L^2V(\mathcal{M})$ such that

$$Pu = F_0 + \operatorname{div}_g F, \quad \textit{and} \quad \operatorname{supp}(u) \cup \operatorname{supp}(F_0) \cup \operatorname{supp}(F) \subset \overline{V}.$$

Note that one does not assume that (P, B, φ) fulfills the Lopatinskiĭ–Šapiro condition of Definition 8.1. The same is done in Sect. 8.4.2.

We shall prove a microlocal version of this estimate. With the notation of Sect. 13.2, in a local chart $\mathcal{C} = (O, \kappa)$ near a point of $\partial\mathcal{M}$ we set $P_0 = \operatorname{Op}(p^{\mathcal{C}})$, where $p^{\mathcal{C}}$ is the local representative of the principal symbol of the Laplace–Beltrami operator. The associated conjugated operator is $P_{0,\varphi} = e^{\tau\varphi^{\mathcal{C}}}P_0 e^{-\tau\varphi^{\mathcal{C}}}$.

We also set $B_0' = \operatorname{Op}_{\mathsf{T}}(b'^{\mathcal{C}})$ where $b'^{\mathcal{C}}$ is the local representative of the principal symbol of B'. Here, $\mathsf{B}_{0,\varphi}(v, h)$ takes the form

$$\mathsf{B}_{0,\varphi}(v,h) = \tilde{B}_\varphi u_{|x_d=0^+} - i(D_d u + ih)_{|x_d=0^+}, \tag{13.6.2}$$

where $\tilde{B}_\varphi = B'_{0,\varphi} + \tau\partial_d\varphi^{\mathcal{C}}$ with $B'_{0,\varphi} = e^{\tau\varphi^{\mathcal{C}}}B_0' e^{-\tau\varphi^{\mathcal{C}}} \in \mathscr{D}^1_{\mathsf{T},\tau}$. This is consistent with the definition of $\mathsf{B}_{0,\varphi}(v,h)$ given in (13.2.5).

PROPOSITION 13.17. *Let $m^0 \in \partial\mathcal{M}$ and V^0 be an open neighborhood of m^0 in $\mathcal{M}$. Let $U^0 = \kappa(V^0 \cap O)$ and $x^0 = \kappa(m^0)$. Assume that (P_0, φ) has the sub-ellipticity property of Definition 3.2 of Volume 1 in $\overline{U^0}$ and $\partial_d\varphi^{\mathcal{C}} \neq 0$ in $\overline{U^0} \cap \{x_d = 0\}$. Let B' be a differential operator of order one on $\partial\mathcal{M} \cap U^0$ and set $B = \partial_\nu + B'$. Let $\omega^{0\prime} \in T^*_{m^0}\partial\mathcal{M}$, with local representative $(\omega^{0\prime})^{\mathcal{C}} = \xi^{0\prime} \in \mathbb{R}^{d-1}$, and $\tau^0 \geq 0$ such that $(\xi^{0\prime}, \tau^0) \neq 0$. If $\tau^0 = 0$ assume moreover that the Lopatinskiĭ–Šapiro condition of Definition 2.2 holds at $(m^0, \omega^{0\prime})$ for $(P_0, B_0, \varphi^{\mathcal{C}})$. Then, there exists $\mathscr{U}$ a conic open neighborhood of $\varrho^{0\prime} = (x^0, \xi^{0\prime}, \tau^0)$ in $U^0 \times \mathbb{R}^{d-1} \times \mathbb{R}_+$ such that for $\chi \in S^0_{\mathsf{T},\tau}$, homogeneous* *of degree zero, with* $\operatorname{supp}(\chi) \subset \mathscr{U}$, *there exist* $C > 0$ *and* $\tau_* > 0$ *such that*

$$\begin{aligned}(13.6.3)\quad &\tau^{1/2}\|\mathrm{Op_T}(\chi)v\|_{\tau,1} + \tau|\mathrm{Op_T}(\chi)v_{|x_d=0^+}|_{\tau,1/2} \\ &\quad + \tau|(D_d\mathrm{Op_T}(\chi)v - i\mathrm{Op_T}(\chi)G^d)_{|x_d=0^+}|_{\tau,-1/2} \\ &\leq C\Big(\|G_0\|_+ + \tau\|G\|_+ + \tau^{3/2}|v_{|x_d=0^+}|_\partial + \tau|\mathsf{B}_{0,\varphi}(v,-G^d)|_{\tau,-1/2} \\ &\qquad + \|v\|_{\tau,1} + |(D_dv - iG^d)_{|x_d=0^+}|_{\tau,-3/2}\Big),\end{aligned}$$

for $\tau \geq \tau_*$ *and* $v \in H^1(\mathbb{R}^d_+)$, $G_0 \in L^2(\mathbb{R}^d_+)$, $G \in L^2V(\mathbb{R}^d_+)$ *such that*

$$P_{0,\varphi}v = G_0 + \sum_{1\leq i\leq d} \partial_j G^j.$$

Estimates of this sort can be patched together as in Sect. 13.4.1 yielding a local version of Theorem 13.16. Local such versions can then be patched together as in Sect. 13.4.2 yielding the result of Theorem 13.16.

With the normal geodesic coordinates we introduced in Sect. 13.2, we may use some of the notation of Sect. 8.3.2 that we recall now. In particular, to ease notation we shall write φ in place of $\varphi^{\mathcal{C}}$, for example, in what follows, with a similar simplification for other symbols.

We have

$$(13.6.4)\qquad P_{0,\varphi} = (D_d + i\tau\partial_d\varphi(x))^2 + R_{0,\varphi},$$

where the principal symbol of $R_{0,\varphi} = e^{\tau\varphi}R_0e^{-\tau\varphi} \in \Psi^2_{\mathsf{T},\tau}$ is precisely

$$(13.6.5)\qquad r_\varphi(\varrho') = r(x,\xi' + i\tau d_{x'}\varphi(x)) \qquad \varrho' = (x,\xi',\tau).$$

We introduced $\alpha(\varrho')$ such that $\operatorname{Re}\alpha(\varrho') \geq 0$ and $\alpha^2(\varrho') = r_\varphi(\varrho')$ and we set

$$(13.6.6)\qquad \gamma_j(\varrho') = -i\tau\partial_d\varphi(x) + i(-1)^j\alpha(\varrho'), \qquad \varrho' = (x,\xi',\tau).$$

This yields, for the principal symbol of $P_{0,\varphi}$,

$$p_\varphi(\varrho) = \check{p}_\varphi(m,\omega',\tau,z) = \big(z - \gamma_1(\varrho')\big)\big(z - \gamma_2(\varrho')\big),\ \varrho = (x,\xi',\xi_d = z,\tau).$$

Note that if $\tau = 0$ and $\xi' \neq 0$, then $r_\varphi(\varrho') = r(x,\xi') > 0$. With $\gamma_j(\varrho') = i(-1)^j r(x,\xi')^{1/2}$ in this case one finds that having $\gamma_1(\varrho') = \gamma_2(\varrho')$ and $(\tau,\xi') \neq 0$ excludes $\tau = 0$.

Below, we prove the estimate of Proposition 13.17 in the following three (exhaustive) cases:

(1) $\tau^0 = 0$;
(2) $\tau^0 \neq 0$ and $\gamma_2(\varrho^{0\prime}) \neq \gamma_1(\varrho^{0\prime})$;
(3) $\gamma_2(\varrho^{0\prime}) = \gamma_1(\varrho^{0\prime})$ (and necessarily $\tau^0 \neq 0$).

13.6.1.1. *Case 1: Vanishing Carleman Parameter.* Here, $\tau^0 = 0$. The Lopatinskiĭ–Šapiro condition of Definition 2.2 holds at $(m^0,\omega^{0\prime})$ for (P_0,φ), meaning that the Lopatinskiĭ–Šapiro condition of Definition 8.1 holds at $(m^0,\omega^{0\prime},\tau^0)$. Hence, Proposition 13.10 applies and yields an estimation stronger than that of Proposition 13.17.

13.6.1.2. *Case 2:* $\boldsymbol{\tau^0 \neq 0}$ *and* $\boldsymbol{\gamma_2(\varrho^{0\prime}) \neq \gamma_1(\varrho^{0\prime})}$. As the roots γ_1 and γ_2 are locally smooth with respect to $\varrho' = (x, \xi', \tau)$ and homogeneous of degree one in (ξ', τ) by Proposition 6.28, there exists $\mathscr{U}$ a conic open neighborhood of $(x^0, \xi^{0\prime}, \tau^0)$ in $U^0 \times \mathbb{R}^{d-1} \times \mathbb{R}_+$ and $C > 0$ such that

$$\gamma_1(\varrho') \neq \gamma_2(\varrho') \text{ and } \tau \geq C\lambda_{\mathsf{T},\tau}, \quad \varrho' = (x, \xi', \tau) \in \overline{\mathscr{U}}. \tag{13.6.7}$$

Without any loss of generality we may moreover assume that $\mathbb{S}_{\overline{\mathscr{U}}}$ is compact.

We let $\chi \in S^0_{\mathsf{T},\tau}$ be as in the statement and $\tilde{\chi} \in S^0_{\mathsf{T},\tau}$ be homogeneous of degree zero and be such that $\operatorname{supp}(\tilde{\chi}) \subset \mathscr{U}$ and $\tilde{\chi} \equiv 1$ on $\operatorname{supp}(\chi)$. From the smoothness and the homogeneity of the roots, we have $\tilde{\chi}\gamma_j \in S^1_{\mathsf{T},\tau}$, $j = 1, 2$. We set

$$Q_2 = D_d - \operatorname{Op}_{\mathsf{T}}(\tilde{\chi}\gamma_2) \text{ and } Q_1 = D_d - \operatorname{Op}_{\mathsf{T}}(\tilde{\chi}\gamma_1).$$

We set $z_2 = Q_1 v$ and $z_1 = Q_2 v$. Since $\operatorname{Op}_{\mathsf{T}}(\chi) P_{0,\varphi} = \operatorname{Op}_{\mathsf{T}}(\chi) Q_1 Q_2 + R_1 = \operatorname{Op}_{\mathsf{T}}(\chi) Q_2 Q_1 + R_2$ with $R_1, R_2 \in \Psi^{1,0}_{\tau}$, we write

$$\begin{aligned} \operatorname{Op}_{\mathsf{T}}(\chi) Q_1 z_1 = \operatorname{Op}_{\mathsf{T}}(\chi) P_{0,\varphi} v - R_1 v &= \operatorname{Op}_{\mathsf{T}}(\chi) G_0 \\ &+ \operatorname{Op}_{\mathsf{T}}(\chi) \sum_{1 \leq j \leq d} \partial_j G^j - R_1 v, \end{aligned}$$

and

$$\begin{aligned} \operatorname{Op}_{\mathsf{T}}(\chi) Q_2 z_2 = \operatorname{Op}_{\mathsf{T}}(\chi) P_{0,\varphi} v - R_2 v &= \operatorname{Op}_{\mathsf{T}}(\chi) G_0 \\ &+ \operatorname{Op}_{\mathsf{T}}(\chi) \sum_{1 \leq j \leq d} \partial_j G^j - R_2 v. \end{aligned}$$

Writing $D_d G^d = Q_j G^d + \operatorname{Op}_{\mathsf{T}}(\tilde{\chi}\gamma_j) G^d$, one obtains

$$\begin{aligned} &\|\operatorname{Op}_{\mathsf{T}}(\chi) Q_1 (z_1 - iG^d)\|_+ + \|\operatorname{Op}_{\mathsf{T}}(\chi) Q_2 (z_2 - iG^d)\|_{\tau,+} \\ &\quad \lesssim \|G_0\|_+ + \sum_{j=1,2} \|\operatorname{Op}_{\mathsf{T}}(\chi) \operatorname{Op}_{\mathsf{T}}(\tilde{\chi}\gamma_j) G^d\|_+ \\ &\qquad + \sum_{1 \leq j \leq d-1} \|\operatorname{Op}_{\mathsf{T}}(\chi) \partial_j G^j\|_+ + \|v\|_{\tau,1} \\ &\quad \lesssim \|G_0\|_+ + \sum_{j=1,2} \|\operatorname{Op}_{\mathsf{T}}(\tilde{\chi}\gamma_j) \operatorname{Op}_{\mathsf{T}}(\chi) G^d\|_+ \\ &\qquad + \sum_{1 \leq j \leq d-1} \|\partial_j (\operatorname{Op}_{\mathsf{T}}(\chi) G^j)\|_+ + \|G\|_+ + \|v\|_{\tau,1}, \end{aligned}$$

as $[\operatorname{Op}_{\mathsf{T}}(\tilde{\chi}\gamma_j), \operatorname{Op}_{\mathsf{T}}(\chi)] \in \Psi^0_{\mathsf{T},\tau}$ and $[\partial_j, \operatorname{Op}_{\mathsf{T}}(\chi)] \in \Psi^0_{\mathsf{T},\tau}$. Since $\tau \gtrsim \lambda_{\mathsf{T},\tau}$ in $\overline{\mathscr{U}}$, with Proposition 6.12 we obtain

$$\begin{aligned} &\|\operatorname{Op}_{\mathsf{T}}(\chi) Q_1 (z_1 - iG^d)\|_+ + \|\operatorname{Op}_{\mathsf{T}}(\chi) Q_2 (z_2 - iG^d)\|_{\tau,+} \\ &\quad \lesssim \|G_0\|_+ + \tau \|G\|_+ + \|v\|_{\tau,1} \end{aligned}$$

Since $[Q_j, \mathrm{Op}_{\mathsf{T}}(\chi)] \in \Psi^0_{\mathsf{T},\tau}$ we obtain

(13.6.8)
$$\begin{aligned}\|Q_1\mathrm{Op}_{\mathsf{T}}(\chi)(z_1 - iG^d)\|_+ &+ \|Q_2\mathrm{Op}_{\mathsf{T}}(\chi)(z_2 - iG^d)\|_+\\ &\lesssim \|G_0\|_+ + \tau\|G\|_+ + \|v\|_{\tau,1} + \|(z_1 - iG^d)\|_+ + \|(z_2 - iG^d)\|_+\\ &\lesssim \|G_0\|_+ + \tau\|G\|_+ + \|v\|_{\tau,1}.\end{aligned}$$

We choose a conic open neighborhood $\mathscr{W}$ of $\operatorname{supp}(\chi)$ such that $\overline{\mathscr{W}} \subset \{\tilde{\chi} \equiv 1\}$. We then have $\overline{\mathscr{W}} \subset \mathscr{U}$. We denote by p_φ, q_2, q_1 the respective principal symbols of $P_{0,\varphi}, Q_2, Q_1$. We have $q_j(\varrho) = \xi_d - \tilde{\chi}\gamma_j(\varrho')$, $j = 1, 2$. For $\varrho' \in \overline{\mathscr{W}}$, as $\tilde{\chi}(\varrho') \equiv 1$, we have $p_\varphi(\varrho) = q_2 q_1(\varrho)$, for $\varrho = (\varrho', \xi_d)$.

By (13.6.7) one has $\gamma_2(\varrho') \neq \gamma_1(\varrho')$ in $\overline{\mathscr{W}}$ and thus Lemma 13.25 applies:

$$q_j(\varrho) = 0 \quad \Rightarrow \quad \frac{1}{2i}\{\overline{q_j}, q_j\}(\varrho') > 0, \qquad \varrho' \in \overline{\mathscr{V}},\ \varrho = (\varrho', \xi_d),\ \ j = 1, 2,$$

that is, both factors Q_2 and Q_1 are sub-elliptic in $\mathscr{W}$. Thus, Lemma 6.22 applies, yielding

$$\begin{aligned}\tau^{-1/2}\|\mathrm{Op}_{\mathsf{T}}(\chi)(z_j - iG^d)\|_{\tau,1} \lesssim\ & \|Q_j\mathrm{Op}_{\mathsf{T}}(\chi)(z_j - iG^d)\|_+\\ &+ |\mathrm{Op}_{\mathsf{T}}(\chi)(z_j - iG^d)_{|x_d=0^+}|_{\tau,1/2}\\ &+ \|(z_j - iG^d)\|_{\tau,0,-N},\end{aligned}$$

for $j = 1, 2$. With (13.6.8) we obtain

(13.6.9)
$$\begin{aligned}\tau^{-1/2} &\sum_{j=1,2} \|\mathrm{Op}_{\mathsf{T}}(\chi)(z_j - iG^d)\|_{\tau,1}\\ &\lesssim \|G_0\|_+ + \tau\|G\|_+ + \sum_{j=1,2} |\mathrm{Op}_{\mathsf{T}}(\chi)(z_j - iG^d)_{|x_d=0^+}|_{\tau,1/2} + \|v\|_{\tau,1}.\end{aligned}$$

Next, we observe that

$$\begin{aligned}\mathrm{Op}_{\mathsf{T}}(\chi)(z_1 - z_2) &= \mathrm{Op}_{\mathsf{T}}(\chi)(Q_2 - Q_1)v = \mathrm{Op}_{\mathsf{T}}(\chi)\mathrm{Op}_{\mathsf{T}}\big(\tilde{\chi}(\gamma_1 - \gamma_2)\big)v\\ &= \mathrm{Op}_{\mathsf{T}}\big(\tilde{\chi}(\gamma_1 - \gamma_2)\big)\mathrm{Op}_{\mathsf{T}}(\chi)v + R_{12}v,\end{aligned}$$

where $R_{12} = [\mathrm{Op}_{\mathsf{T}}(\chi), \mathrm{Op}_{\mathsf{T}}\big(\tilde{\chi}(\gamma_1 - \gamma_2)] \in \Psi^0_{\mathsf{T},\tau}$. We also write

$$\begin{aligned}D_d\mathrm{Op}_{\mathsf{T}}(\chi)(z_1 - z_2) &= D_d\mathrm{Op}_{\mathsf{T}}\big(\tilde{\chi}(\gamma_1 - \gamma_2)\big)\mathrm{Op}_{\mathsf{T}}(\chi)v + D_d R_{12}v\\ &= \mathrm{Op}_{\mathsf{T}}\big(\tilde{\chi}(\gamma_1 - \gamma_2)\big)\mathrm{Op}_{\mathsf{T}}(\chi)D_d v + R'_{12}v,\end{aligned}$$

with $R'_{12} \in \Psi^{1,0}_\tau$.

Since $\tilde{\chi}(\gamma_1 - \gamma_2) \in S^1_{\mathsf{T},\tau}$ is elliptic in a neighborhood of $\operatorname{supp}(\chi)$, with Proposition 2.37 of volume 1 (and Remark 2.38 both adapted to tangential operators) we then obtain

$$\|\mathrm{Op}_{\mathsf{T}}(\chi)w\|_{\tau,0,s+1} \lesssim \|\mathrm{Op}_{\mathsf{T}}\big(\tilde{\chi}(\gamma_1 - \gamma_2)\big)\mathrm{Op}_{\mathsf{T}}(\chi)w\|_{\tau,0,s} + \|w\|_{\tau,0,-N},$$

and

$$|\mathrm{Op}_{\mathsf{T}}(\chi) w_{|x_d=0^+}|_{\tau,s+1} \lesssim |\mathrm{Op}_{\mathsf{T}}\big(\tilde{\chi}(\gamma_1-\gamma_2)\big)\mathrm{Op}_{\mathsf{T}}(\chi) w_{|x_d=0^+}|_{\tau,s} + |w_{|x_d=0^+}|_{\tau,-N},$$

for $s \in \mathbb{R}$, $N \in \mathbb{N}$. It yields

$$\begin{aligned}
&\|\mathrm{Op}_{\mathsf{T}}(\chi) v\|_{\tau,0,2} + \|\mathrm{Op}_{\mathsf{T}}(\chi) D_d v\|_{\tau,0,1} \\
&\quad\lesssim \|\mathrm{Op}_{\mathsf{T}}\big(\tilde{\chi}(\gamma_1-\gamma_2)\big)\mathrm{Op}_{\mathsf{T}}(\chi) v\|_{\tau,0,1} \\
&\qquad + \|\mathrm{Op}_{\mathsf{T}}\big(\tilde{\chi}(\gamma_1-\gamma_2)\big)\mathrm{Op}_{\mathsf{T}}(\chi) D_d v\|_+ + \|v\|_{\tau,1,-N} \\
&\quad\lesssim \|\mathrm{Op}_{\mathsf{T}}(\chi)(z_1-z_2)\|_{\tau,1} + \|v\|_{\tau,1} \\
&\quad\lesssim \sum_{j=1,2} \|\mathrm{Op}_{\mathsf{T}}(\chi)(z_j - iG^d)\|_{\tau,1} + \|v\|_{\tau,1},
\end{aligned}$$

and similarly

$$\begin{aligned}
|\mathrm{Op}_{\mathsf{T}}(\chi) v_{|x_d=0^+}|_{\tau,3/2} &\lesssim |\mathrm{Op}_{\mathsf{T}}\big(\tilde{\chi}(\gamma_1-\gamma_2)\big)\mathrm{Op}_{\mathsf{T}}(\chi) v_{|x_d=0^+}|_{\tau,1/2} \\
&\quad + |v_{|x_d=0^+}|_{\tau,-N} \\
&\lesssim |\mathrm{Op}_{\mathsf{T}}(\chi)(z_1-z_2)_{|x_d=0^+}|_{\tau,1/2} + |v_{|x_d=0^+}|_{\tau,1/2} \\
&\lesssim \sum_{j=1,2} |\mathrm{Op}_{\mathsf{T}}(\chi)(z_j - iG^d)_{|x_d=0^+}|_{\tau,1/2} + |v_{|x_d=0^+}|_{\tau,1/2}.
\end{aligned}$$

We then write

$$\begin{aligned}
\tau\|\mathrm{Op}_{\mathsf{T}}(\chi) v\|_{\tau,1} &\lesssim \|\mathrm{Op}_{\mathsf{T}}(\chi) v\|_{\tau,1,1} \\
&\lesssim \|\mathrm{Op}_{\mathsf{T}}(\chi) v\|_{\tau,0,2} + \|D_d \mathrm{Op}_{\mathsf{T}}(\chi) v\|_{\tau,0,1} \\
&\lesssim \|\mathrm{Op}_{\mathsf{T}}(\chi) v\|_{\tau,0,2} + \|\mathrm{Op}_{\mathsf{T}}(\chi) D_d v\|_{\tau,0,1} + \|v\|_{\tau,0,1} \\
&\lesssim \sum_{j=1,2} \|\mathrm{Op}_{\mathsf{T}}(\chi)(z_j - iG^d)\|_{\tau,1} + \|v\|_{\tau,1}.
\end{aligned}$$

With (13.6.9) we obtain

$$\begin{aligned}
&\tau^{1/2}\|\mathrm{Op}_{\mathsf{T}}(\chi) v\|_{\tau,1} + |\mathrm{Op}_{\mathsf{T}}(\chi) v_{|x_d=0^+}|_{\tau,3/2} \\
&\quad\lesssim \tau^{-1/2} \sum_{j=1,2} \|\mathrm{Op}_{\mathsf{T}}(\chi)(z_j - iG^d)\|_{\tau,1} \\
&\qquad + \sum_{j=1,2} |\mathrm{Op}_{\mathsf{T}}(\chi)(z_j - iG^d)_{|x_d=0^+}|_{\tau,1/2} \\
&\qquad + \tau^{-1/2}\|v\|_{\tau,1} + |v_{|x_d=0^+}|_{\tau,1/2} \\
&\quad\lesssim \|G_0\|_+ + \tau\|G\|_+ + \sum_{j=1,2} |\mathrm{Op}_{\mathsf{T}}(\chi)(z_j - iG^d)_{|x_d=0^+}|_{\tau,1/2} + \|v\|_{\tau,1},
\end{aligned}$$

using the trace inequality of Corollary 6.10 (with $m = 0$ and $s = 1/2$). Because of the form of z_j we have, with commutator arguments as above,

$$\begin{aligned}\tau^{1/2}\|\mathrm{Op}_{\mathsf{T}}(\chi)v\|_{\tau,1} &+ |\mathrm{Op}_{\mathsf{T}}(\chi)v_{|x_d=0^+}|_{\tau,3/2}\\ &+ |\mathrm{Op}_{\mathsf{T}}(\chi)(D_d v - iG^d)_{|x_d=0^+}|_{\tau,1/2}\\ &\lesssim \|G_0\|_+ + \tau\|G\|_+ + \sum_{j=1,2} |\mathrm{Op}_{\mathsf{T}}(\chi)(z_j - iG^d)_{|x_d=0^+}|_{\tau,1/2} + \|v\|_{\tau,1},\end{aligned}$$

yielding, again with commutator arguments,

$$\begin{aligned}\tau^{1/2}\|\mathrm{Op}_{\mathsf{T}}(\chi)v\|_{\tau,1} &+ |\mathrm{Op}_{\mathsf{T}}(\chi)v_{|x_d=0^+}|_{\tau,3/2}\\ &+ |(D_d\mathrm{Op}_{\mathsf{T}}(\chi)v - i\mathrm{Op}_{\mathsf{T}}(\chi)G^d)_{|x_d=0^+}|_{\tau,1/2}\\ &\lesssim \|G_0\|_+ + \tau\|G\|_+ + \sum_{j=1,2} |\mathrm{Op}_{\mathsf{T}}(\chi)(z_j - iG^d)_{|x_d=0^+}|_{\tau,1/2} + \|v\|_{\tau,1}.\end{aligned} \tag{13.6.10}$$

With Proposition 6.12 as $\tau \gtrsim \lambda_{\mathsf{T},\tau}$ in $\overline{U}$ and with the form of z_j and that of $\mathsf{B}_{0,\varphi}(v,h)$ given in (13.6.2) we have

$$\begin{aligned}\sum_{j=1,2} &|\mathrm{Op}_{\mathsf{T}}(\chi)(z_j - iG^d)_{|x_d=0^+}|_{\tau,1/2}\\ &\lesssim \tau^{3/2}\sum_{j=1,2} |\mathrm{Op}_{\mathsf{T}}(\chi)(z_j - iG^d)_{|x_d=0^+}|_{\tau,-1}\\ &\quad + \sum_{j=1,2} |(z_j - iG^d)_{|x_d=0^+}|_{\tau,-N}\\ &\lesssim \tau^{3/2}\big(|(D_d v - iG^d + i\tilde{B}_\varphi v)_{|x_d=0^+}|_{\tau,-1}\\ &\quad + \sum_{j=1,2} \big|\big(i\tilde{B}_\varphi + \mathrm{Op}_{\mathsf{T}}(\tilde{\chi}\gamma_j)\big)v_{|x_d=0^+}|_{\tau,-1}\big)\\ &\quad + |(D_d v - iG^d)_{|x_d=0^+}|_{\tau,-N} + |v_{|x_d=0^+}|_{\tau,-N}\\ &\lesssim \tau|\mathsf{B}_{0,\varphi}(v,-G^d)|_{\tau,-1/2} + \tau^{3/2}|v_{|x_d=0^+}|_{\partial}\\ &\quad + |(D_d v - iG^d)_{|x_d=0^+}|_{\tau,-N}.\end{aligned}$$

With (13.6.10) we thus obtain an estimate that is stronger than the sought result.

13.6.1.3. *Case 3:* $\boldsymbol{\gamma_2(\varrho^{0\prime}) = \gamma_1(\varrho^{0\prime})}$. If $\gamma_2(\varrho^{0\prime}) = \gamma_1(\varrho^{0\prime})$, then by (13.6.5)–(13.6.6) one has $r_\varphi(\varrho^{\prime 0}) = 0$ and

$$\gamma_2(\varrho^{0\prime}) = \gamma_1(\varrho^{0\prime}) = -i\tau\partial_d\varphi(x^0).$$

As $\partial_d\varphi(x^0) \neq 0$ this double root is not real here. Hence, near $\varrho^{0\prime}$ the operator $P_{0,\varphi}$ is elliptic.

Let $\varepsilon_0 > 0$. Since $\tau^0 \neq 0$, there exists $\mathscr{U}^0$ a conic open neighborhood of $\varrho^{0\prime}$ in $U^0 \times \mathbb{R}^{d-1} \times \mathbb{R}_+$ where

$$|r_\varphi(\varrho')| \leq \varepsilon_0\lambda^2_{\mathsf{T},\tau}, \quad \text{and} \quad |\partial_d\varphi(x)| \geq C > 0, \quad \text{and} \quad \lambda_{\mathsf{T},\tau} \lesssim \tau.$$

Without any loss of generality we may moreover assume that $\mathbb{S}_{\overline{\mathscr{U}^0}}$ is compact.

From the form of the operator $P_{0,\varphi}$, since we consider a microlocal region where $|\alpha|$ is small, we can foresee that the estimation we shall obtain near that region is almost given by an estimate for the conjugated operator

$$Q_\varphi = (D_d + i\tau\partial_d\varphi(x))^2.$$

In the case $\partial_d\varphi(x) \geq C > 0$ an perfect elliptic estimate for the operator Q_φ is given by Lemma 13.14.

In the case $\partial_d\varphi(x) \leq -C < 0$, the counterpart result is given by the following lemma.

LEMMA 13.18. *Let $\tilde{\chi} \in S^0_{\mathsf{T},\tau}$, homogeneous of degree zero, be such that $\operatorname{supp}(\tilde{\chi}) \subset \mathscr{U}^0$. There exist $C > 0$ and $\tau_* > 0$ such that*

$$\begin{aligned}(13.6.11)\quad \tau\|\operatorname{Op}_{\mathsf{T}}(\tilde{\chi})v\|_{\tau,1} \leq C\Big(&\|H_0\|_+ + \tau\|H\|_+ \\ &+ \tau|(D_d\operatorname{Op}_{\mathsf{T}}(\tilde{\chi})v - i\operatorname{Op}_{\mathsf{T}}(\tilde{\chi})H^d)_{|x_d=0^+}|_{\tau,-1/2} \\ &+ \tau|\operatorname{Op}_{\mathsf{T}}(\tilde{\chi})v_{|x_d=0^+}|_{\tau,1/2} + \tau\|v\|_{\tau,1,-1}\Big),\end{aligned}$$

for $\tau \geq \tau_*$ and $v \in H^1(\mathbb{R}^d_+)$, $H_0 \in L^2(\mathbb{R}^d_+)$, $H \in L^2V(\mathbb{R}^d_+)$ such that

$$Q_\varphi v = H_0 + \sum_{1\leq i\leq d} \partial_j H^j.$$

A proof is given below.

Note that the estimate given by Lemma 13.14 in the case $\partial_d\varphi \geq C > 0$ implies (13.6.11). Here, in any case only an estimate as in (13.6.11) is needed. We shall thus apply this estimate either in both the cases $\partial_d\varphi \geq C > 0$ and $\partial_d\varphi \leq -C < 0$.

For $\varepsilon > 0$ to be fixed below, we let $\mathscr{U}_\varepsilon$ be the conic open set of $U^0 \times \mathbb{R}^{d-1} \times \mathbb{R}_+$ given by

$$\mathscr{U}_\varepsilon = \{\varrho' \in \mathscr{U}^0; |\alpha^2(\varrho')| < \varepsilon\lambda^2_{\mathsf{T},\tau}\}.$$

We take $0 < \varepsilon \leq \varepsilon_0/2$ allowing us to choose $\chi_\varepsilon \in S^0_{\mathsf{T},\tau}$ homogeneous of degree zero such that $\operatorname{supp}(\chi_\varepsilon) \subset \mathscr{U}_\varepsilon$. In $\mathscr{U}_\varepsilon \subset \mathscr{U}^0$ one has $\lambda_{\mathsf{T},\tau} \lesssim \tau$. With Proposition 6.11, there exists $C_1 > 0$ such that

$$\|\operatorname{Op}_{\mathsf{T}}(\chi_\varepsilon)v\|_{\tau,0,2} \leq C_1\tau^2\|\operatorname{Op}_{\mathsf{T}}(\chi_\varepsilon)v\|_+ + C_{\varepsilon,N}\|v\|_{\tau,0,-N},$$

for any $N \in \mathbb{N}$, with some $C_{\varepsilon,N} > 0$.

As $|\alpha^2(\varrho')| \leq \varepsilon\lambda^2_{\mathsf{T},\tau}$ in $\mathscr{U}_\varepsilon$, one has by Corollary 2.51, for $C_0' > 1$ and $C_1' = C_0'C_1$,

$$\begin{aligned}(13.6.12)\qquad \|R_{0,\varphi}\operatorname{Op}_{\mathsf{T}}(\chi_\varepsilon)v\|_+ &\leq C_0'\varepsilon\|\operatorname{Op}_{\mathsf{T}}(\chi_\varepsilon)v\|_{\tau,0,2} + C'_{\varepsilon,N}\|v\|_{\tau,0,-N} \\ &\leq 2\varepsilon C_1'\tau^2\|\operatorname{Op}_{\mathsf{T}}(\chi_\varepsilon)v\|_+ + C''_{\varepsilon,N}\|v\|_{\tau,0,-N}.\end{aligned}$$

Recalling (13.6.4), we write

$$\begin{aligned} Q_\varphi \mathrm{Op}_{\mathsf{T}}(\chi_\varepsilon)v &= P_{0,\varphi}\mathrm{Op}_{\mathsf{T}}(\chi_\varepsilon)v - R_{0,\varphi}\mathrm{Op}_{\mathsf{T}}(\chi_\varepsilon)v \\ &= \mathrm{Op}_{\mathsf{T}}(\chi_\varepsilon)P_{0,\varphi}v + [P_{0,\varphi}, \mathrm{Op}_{\mathsf{T}}(\chi_\varepsilon)]v - R_{0,\varphi}\mathrm{Op}_{\mathsf{T}}(\chi_\varepsilon)v \\ &= H_0 + \sum_{1\le j\le d} \partial_j H^j, \end{aligned}$$

with $H = \mathrm{Op}_{\mathsf{T}}(\chi_\varepsilon)G$ and

$$\begin{aligned} H_0 = \mathrm{Op}_{\mathsf{T}}(\chi_\varepsilon)G_0 &+ \sum_{1\le j\le d} [\mathrm{Op}_{\mathsf{T}}(\chi_\varepsilon), \partial_j]G^j + [P_{0,\varphi}, \mathrm{Op}_{\mathsf{T}}(\chi_\varepsilon)]v \\ &- R_{0,\varphi}\mathrm{Op}_{\mathsf{T}}(\chi_\varepsilon)v. \end{aligned}$$

We have $\|H\|_+ \lesssim \|G\|_+$ and, with (13.6.12),

$$\|H_0\|_+ \le C_\varepsilon\big(\|G_0\|_+ + \|G\|_+ + \|v\|_{\tau,1}\big) + C\varepsilon\tau^2\|\mathrm{Op}_{\mathsf{T}}(\chi_\varepsilon)v\|_+.$$

We choose $\tilde\chi \in S^0_{\mathsf{T},\tau}$ homogeneous of degree zero such that $\mathrm{supp}(\tilde\chi) \subset \mathscr{U}^0$ and $\tilde\chi \equiv 1$ in a neighborhood of $\mathscr{U}_{\varepsilon_0/2}$. Applying Lemma 13.18 we thus obtain, with $\tilde\chi_\varepsilon = \tilde\chi \circ \chi_\varepsilon$,

$$\begin{aligned} \tau\|\mathrm{Op}_{\mathsf{T}}(\tilde\chi_\varepsilon)v\|_{\tau,1} \le C_\varepsilon\Big(&\|G_0\|_+ + \tau\|G\|_+ + \|v\|_{\tau,1} + \tau|\mathrm{Op}_{\mathsf{T}}(\tilde\chi_\varepsilon)v_{|x_d=0^+}|_{\tau,1/2} \\ &+ \tau|(D_d\mathrm{Op}_{\mathsf{T}}(\tilde\chi_\varepsilon)v - i\mathrm{Op}_{\mathsf{T}}(\tilde\chi_\varepsilon)G^d)_{|x_d=0^+}|_{\tau,-1/2}\Big) + C\varepsilon\tau^2\|\mathrm{Op}_{\mathsf{T}}(\chi_\varepsilon)v\|_+. \end{aligned}$$

We write

$$\begin{aligned} &|(D_d\mathrm{Op}_{\mathsf{T}}(\tilde\chi_\varepsilon)v - i\mathrm{Op}_{\mathsf{T}}(\tilde\chi_\varepsilon)G^d)_{|x_d=0^+}|_{\tau,-1/2} \\ &\quad\le |\mathrm{Op}_{\mathsf{T}}(\tilde\chi_\varepsilon)(D_dv - iG^d)_{|x_d=0^+}|_{\tau,-1/2} + |[D_d, \mathrm{Op}_{\mathsf{T}}(\tilde\chi_\varepsilon)]v_{|x_d=0^+}|_{\tau,-1/2} \\ &\quad\le C|\mathrm{Op}_{\mathsf{T}}(\chi_\varepsilon)(D_dv - iG^d)_{|x_d=0^+}|_{\tau,-1/2} + C_\varepsilon\|v\|_{\tau,1,-1} \\ &\quad\le C\big(|(D_d\mathrm{Op}_{\mathsf{T}}(\chi_\varepsilon)v - i\mathrm{Op}_{\mathsf{T}}(\chi_\varepsilon)G^d)_{|x_d=0^+}|_{\tau,-1/2} \\ &\qquad + |[D_d, \mathrm{Op}_{\mathsf{T}}(\chi_\varepsilon)]v_{|x_d=0^+}|_{\tau,-1/2}\big) + C_\varepsilon\|v\|_{\tau,1,-1} \\ &\quad\le C|(D_d\mathrm{Op}_{\mathsf{T}}(\chi_\varepsilon)v - i\mathrm{Op}_{\mathsf{T}}(\chi_\varepsilon)G^d)_{|x_d=0^+}|_{\tau,-1/2} + C_\varepsilon\|v\|_{\tau,1,-1}, \end{aligned}$$

using the trace inequality of Corollary 6.10 (with $m = 0$ and $s = -1/2$). We also have

$$|\mathrm{Op}_{\mathsf{T}}(\tilde\chi_\varepsilon)v_{|x_d=0^+}|_{\tau,1/2} \lesssim |\mathrm{Op}_{\mathsf{T}}(\chi_\varepsilon)v_{|x_d=0^+}|_{\tau,1/2}.$$

Since $\tilde\chi \equiv 1$ in a neighborhood of $\mathrm{supp}(\chi_\varepsilon)$ we have $\tilde\chi_\varepsilon = \chi_\varepsilon \mod S^{-N}_{\mathsf{T},\tau}$ for any $N \in \mathbb{N}$ by pseudo-differential calculus. We then find

$$\|\mathrm{Op}_{\mathsf{T}}(\chi_\varepsilon)v\|_{\tau,1} \le \|\mathrm{Op}_{\mathsf{T}}(\tilde\chi_\varepsilon)v\|_{\tau,1} + C_\varepsilon\|v\|_{\tau,1,-N}.$$

We have thus obtained

$$\begin{aligned}\tau\|\mathrm{Op_T}(\chi_\varepsilon)v\|_{\tau,1} \leq C_\varepsilon\Big(&\|G_0\|_+ + \tau\|G\|_+\\ &+ |(D_d\mathrm{Op_T}(\chi_\varepsilon)v - i\mathrm{Op_T}(\chi_\varepsilon)G^d)_{|x_d=0^+}|_{\tau,-1/2}\\ &+ |\mathrm{Op_T}(\chi_\varepsilon)v_{|x_d=0^+}|_{\tau,1/2} + \|v\|_{\tau,1}\Big)\\ &+ C\varepsilon\tau^2\|\mathrm{Op_T}(\chi_\varepsilon)v\|_+.\end{aligned}$$

Choosing $\varepsilon > 0$ sufficiently small, to be kept fixed and setting $\chi = \chi_\varepsilon$ we finally obtain

$$\begin{aligned}\tau\|\mathrm{Op_T}(\chi)v\|_{\tau,1} \lesssim{}& \|G_0\|_+ + \tau\|G\|_+\\ &+ |(D_d\mathrm{Op_T}(\chi)v - i\mathrm{Op_T}(\chi)G^d)_{|x_d=0^+}|_{\tau,-1/2}\\ &+ |\mathrm{Op_T}(\chi)v_{|x_d=0^+}|_{\tau,1/2} + \|v\|_{\tau,1},\end{aligned}$$

which concludes the proof of Proposition 13.10 in Case 3. ∎

PROOF OF LEMMA 13.18. We set $L = D_d + i\tau\partial_d\varphi(x)$. We have $\mathrm{Op_T}(\tilde\chi)L^2 v = \mathrm{Op_T}(\tilde\chi)H_0 + \sum_{1\leq j\leq d}\mathrm{Op_T}(\tilde\chi)\partial_j H^j$, which we write

$$\begin{aligned}\mathrm{Op_T}(\tilde\chi)L(Lv - iH^d) = \mathrm{Op_T}(\tilde\chi)H_0 + &\sum_{1\leq j\leq d-1}\mathrm{Op_T}(\tilde\chi)\partial_j H^j\\ &+ \tau\mathrm{Op_T}(\tilde\chi)\partial_d\varphi H^d.\end{aligned}$$

This yields

(13.6.13)
$$\|\mathrm{Op_T}(\tilde\chi)L(Lv - iH^d)\|_{\tau,0,-1} \lesssim \|H_0\|_{\tau,0,-1} + \|H\|_+ \lesssim \tau^{-1}\|H_0\|_+ + \|H\|_+,$$

using Remark 6.8.

Since $\tau\partial_d\varphi \lesssim -\lambda_{\mathrm{T},\tau}$ in $\mathscr{U}^0$, from Lemma 6.21 (with $s = -1$), we have for τ chosen sufficiently large,

(13.6.14)
$$\|\mathrm{Op_T}(\tilde\chi)z\|_+ \lesssim \|L\mathrm{Op_T}(\tilde\chi)z\|_{\tau,0,-1} + |\mathrm{Op_T}(\tilde\chi)z_{|x_d=0^+}|_{\tau,-1/2} + \|z\|_{\tau,0,-N},$$

for $z \in \overline{\mathscr{S}}(\mathbb{R}^d_+)$. We apply estimate (13.6.14) for $z = Lv - H^d$, with a density argument, yielding

$$\begin{aligned}&\|\mathrm{Op_T}(\tilde\chi)(Lv - iH^d)\|_+\\ &\quad\lesssim \|L\mathrm{Op_T}(\tilde\chi)(Lv - iH^d)\|_{\tau,0,-1} + |\mathrm{Op_T}(\tilde\chi)(Lv - iH^d)_{|x_d=0^+}|_{\tau,-1/2}\\ &\qquad + \|Lv - iH^d\|_{\tau,0,-N}\\ &\quad\lesssim \|\mathrm{Op_T}(\tilde\chi)L(Lv - iH^d)\|_{\tau,0,-1} + |\mathrm{Op_T}(\tilde\chi)(Lv - iH^d)_{|x_d=0^+}|_{\tau,-1/2}\\ &\qquad + \|v\|_{\tau,1,-1} + \|H^d\|_+,\end{aligned}$$

using that $[L, \mathrm{Op}_\mathsf{T}(\tilde{\chi})] \in \Psi^0_{\mathsf{T},\tau}$. With (13.6.13), we then obtain

$$\begin{aligned}&\|\mathrm{Op}_\mathsf{T}(\tilde{\chi})(Lv - iH^d)\|_+ \\ &\quad \lesssim \tau^{-1}\|H_0\|_+ + |\mathrm{Op}_\mathsf{T}(\tilde{\chi})(Lv - iH^d)_{|x_d=0^+}|_{\tau,-1/2} + \|H\|_+ + \|v\|_{\tau,1,-1}.\end{aligned}$$

We then write

$$\begin{aligned}\|L\mathrm{Op}_\mathsf{T}(\tilde{\chi})v\|_+ &\lesssim \|\mathrm{Op}_\mathsf{T}(\tilde{\chi})Lv\|_+ + \|v\|_+ \\ &\lesssim \|\mathrm{Op}_\mathsf{T}(\tilde{\chi})(Lv - iH^d)\|_+ + \|H^d\|_+ + \|v\|_+,\end{aligned}$$

and

$$\begin{aligned}&|\mathrm{Op}_\mathsf{T}(\tilde{\chi})(Lv - iH^d)_{|x_d=0^+}|_{\tau,-1/2} \\ &\quad \lesssim |(L\mathrm{Op}_\mathsf{T}(\tilde{\chi})v - i\mathrm{Op}_\mathsf{T}(\tilde{\chi})H^d)_{|x_d=0^+}|_{\tau,-1/2} + |v_{|x_d=0^+}|_{\tau,-1/2} \\ &\quad \lesssim |(L\mathrm{Op}_\mathsf{T}(\tilde{\chi})v - i\mathrm{Op}_\mathsf{T}(\tilde{\chi})H^d)_{|x_d=0^+}|_{\tau,-1/2} + \|v\|_{\tau,1,-1}\end{aligned}$$

using the trace inequality of Corollary 6.10 (with $m = 0$ and $s = -1/2$).

We thus obtain

$$\begin{aligned}\|L\mathrm{Op}_\mathsf{T}(\tilde{\chi})v\|_+ &\lesssim \tau^{-1}\|H_0\|_+ + \|H\|_+ \\ &\quad + |(L\mathrm{Op}_\mathsf{T}(\tilde{\chi})v - i\mathrm{Op}_\mathsf{T}(\tilde{\chi})H^d)_{|x_d=0^+}|_{\tau,-1/2} + \|v\|_{\tau,1,-1}.\end{aligned}$$

Next, invoking Lemma 6.21 a second time (with $s = 0$ this time), we have for τ chosen sufficiently large,

$$\|\mathrm{Op}_\mathsf{T}(\tilde{\chi})v\|_{\tau,1} \lesssim \|L\mathrm{Op}_\mathsf{T}(\tilde{\chi})v\|_+ + |\mathrm{Op}_\mathsf{T}(\tilde{\chi})v_{|x_d=0^+}|_{\tau,1/2} + \|v\|_{\tau,0,-N}.$$

This yields

$$\begin{aligned}\|\mathrm{Op}_\mathsf{T}(\tilde{\chi})v\|_{\tau,1} &\lesssim \tau^{-1}\|H_0\|_+ + \|H\|_+ + |\mathrm{Op}_\mathsf{T}(\tilde{\chi})v_{|x_d=0^+}|_{\tau,1/2} \\ &\quad + |(L\mathrm{Op}_\mathsf{T}(\tilde{\chi})v - \mathrm{Op}_\mathsf{T}(\tilde{\chi})H^d)_{|x_d=0^+}|_{\tau,-1/2} + \|v\|_{\tau,1,-1}.\end{aligned}$$

Observing that

$$\begin{aligned}&|(L\mathrm{Op}_\mathsf{T}(\tilde{\chi})v - \mathrm{Op}_\mathsf{T}(\tilde{\chi})H^d)_{|x_d=0^+}|_{\tau,-1/2} \\ &\quad \lesssim |\mathrm{Op}_\mathsf{T}(\tilde{\chi})v_{|x_d=0^+}|_{\tau,1/2} + |(D_d\mathrm{Op}_\mathsf{T}(\tilde{\chi})v - \mathrm{Op}_\mathsf{T}(\tilde{\chi})H^d)_{|x_d=0^+}|_{\tau,-1/2},\end{aligned}$$

which concludes the proof. ■

13.6.2. Estimate for a Weight Function with a Vanishing Neumann Trace. Here, we place ourselves in the neighborhood of a point of $\partial\mathcal{M}$ where $\partial_\nu\varphi$ may vanish.

THEOREM 13.19. *Let $(\mathcal{M}, g)$ be a compact Riemannian manifold with boundary and let $P = -\Delta_g + R_1$ with R_1 a first-order differential operator with bounded coefficients on $\mathcal{M}$. Let V be an open set of $\mathcal{M}$ such that $V_\partial = V \cap \partial\mathcal{M} \neq \emptyset$. Let B' be a differential operator of order one on $\partial\mathcal{M}$ such that (P, B) fulfills the Lopatinskiĭ–Šapiro condition of Definition 2.2 on $\overline{V_\partial}$ for $B = \partial_\nu + B'$. Let $\varphi \in \mathscr{C}^\infty(\mathcal{M})$ be such that the pair (P, φ) has the sub-ellipticity property of Definition 5.1 in $\overline{V}$.*

Assume that $\partial_\nu \varphi_{|\partial\mathcal{M}}$ vanishes in some point of V_∂ and let $V' \subset V$ be a neighborhood in $\mathcal{M}$ of such points.

Then, there exist C and $\tau_* > 0$ such that

(13.6.15)

$$\begin{aligned}\tau^{1/2}\|e^{\tau\varphi}u\|_{\tau,1} + \tau|e^{\tau\varphi}u_{|\partial\mathcal{M}}|_{\tau,1/2} + \tau|e^{\tau\varphi_{|\partial\mathcal{M}}}(\partial_\nu u + g(F,\nu))_{|\partial\mathcal{M}}|_{\tau,-1/2} \\ \leq C\Big(\|e^{\tau\varphi}F_0\|_{L^2(\mathcal{M})} + \tau\|e^{\tau\varphi}F\|_{L^2V(\mathcal{M})} + \tau^{3/2}\|e^{\tau\varphi}g(\nu,F)\|_{L^2V(\mathcal{M})} \\ + \tau^{3/2}|e^{\tau\varphi}u_{|\partial\mathcal{M}}|_{L^2(\partial\mathcal{M})} + \tau|e^{\tau\varphi_{|\partial\mathcal{M}}}\mathsf{B}(u,g(F,\nu))|_{\tau,-1/2}\Big),\end{aligned}$$

for $\tau \geq \tau_*$ and $u \in H^1(\mathcal{M})$, $F_0 \in L^2(\mathcal{M})$, $F \in L^2V(\mathcal{M})$ such that

$$Pu = F_0 + \operatorname{div}_g F, \quad \textit{and} \quad \operatorname{supp}(u) \cup \operatorname{supp}(F_0) \cup \operatorname{supp}(F) \subset \overline{V'}.$$

The difference with Theorem 13.16 lies in the term $\|e^{\tau\varphi}g(\nu,F)\|_{L^2V(\mathcal{M})}$ on the right-hand side on the estimate. Here, it appears with the power $\tau^{3/2}$ of the large parameter. In Theorem 13.16 one only has τ implying that the present estimate is weaker.

For the proof of Theorem 13.19 one can proceed as in Sect. 13.6.1. Near each point $\partial\mathcal{M} \cap V'$ one proves a local estimate. For that, one uses a local chart $\mathcal{C} = (\kappa, O)$ as above. The local estimate is itself based on microlocal estimates. For a point x^0 where $\partial_d\varphi(x^0) \neq 0$ one uses the microlocal estimate of Proposition 13.17. Here, it suffices to consider points x^0 such that $\partial_d\varphi(x^0) = 0$. Since the sub-ellipticity property holds, in this case one has $d_{x'}\varphi(x^0) \neq 0$. This property holds locally.

We thus consider $\varrho^{0\prime} = (x^0, \xi^{0\prime}, \tau^0)$ with $(\xi^{0\prime}, \tau^0) \in \mathbb{R}^{d-1} \times [0,+\infty)$. If $\tau^0 = 0$, then Case 1 in the proof of Proposition 13.17 applies.

We may now assume that $\tau^0 > 0$. One writes $p_\varphi(\varrho) = \big(\xi_d + i\tau\partial_d\varphi(x)\big)^2 + r_\varphi(\varrho')$. One has $p_\varphi(\varrho^{0\prime}, \xi_d) = \xi_d^2 + r_\varphi(\varrho^{0\prime})$. If $r_\varphi(\varrho^{0\prime}) \neq 0$, then in a conic neighborhood of $\varrho^{0\prime}$ one can write $p_\varphi(\varrho) = \big(\xi_d - \gamma_1(\varrho')\big)\big(\xi_d - \gamma_2(\varrho')\big)$ with $\gamma_1(\varrho') \neq \gamma_2(\varrho')$. Then, one observes that Case 2 in the proof of Proposition 13.17 applies.

Then, the only remaining case to consider here is $r_\varphi(\varrho^{0\prime}) = 0$. We recall that

$$r_\varphi(\varrho') = r(x, \xi' + i\tau d_{x'}\varphi(x)),$$

with $r(x,\xi') \gtrsim |\xi'|^2$. If $\tilde{r}(x,.,.)$ is the associated bilinear form, $r_\varphi(\varrho')$ vanishes if and only if

$$r(x,\xi') = \tau^2 r(x, d_{x'}\varphi(x)) \text{ and } \tilde{r}(x,\xi', d_{x'}\varphi(x)) = 0. \tag{13.6.16}$$

If $d = 2$, then $\xi' \in \mathbb{R}$ and $r_\varphi(\varrho') = 0$ cannot occur if $(\xi',\tau) \neq (0,0)$ as $d_{x'}\varphi(x) \neq 0$ locally. However, if $d \geq 3$, with $\xi' \in \mathbb{R}^{d-1}$ there always exists $(\xi',\tau) \neq (0,0)$ such that (13.6.16) holds. In the following proposition we give a microlocal estimate in a conic neighborhood of such a point. Note that in such case $\xi' = 0$ if $\tau = 0$.

The following proposition yields the additional microlocal estimate near $\varrho^{0\prime}$ to complete the proof of Theorem 13.19.

PROPOSITION 13.20. *Let $m^0 \in \partial\mathcal{M}$ and V^0 be an open neighborhood of m^0 in $\mathcal{M}$. Let $U^0 = \kappa(V^0 \cap O)$ and $x^0 = \kappa(m^0)$. Assume that (P_0, φ) has the sub-ellipticity property of Definition 3.2 of Volume 1 in $\overline{U^0}$ and $\partial_d\varphi(x^0) = 0$. Let B' be a differential operator of order one on $\partial\mathcal{M}\cap U^0$ and set $B = \partial_\nu + B'$. Let $\omega^{0\prime} \in T^*_{m^0}\partial\mathcal{M}$, with local representative $(\omega^{0\prime})^{\mathcal{C}} = \xi^{0\prime} \in \mathbb{R}^{d-1}$, and $\tau^0 > 0$ such that $(\xi^{0\prime}, \tau^0) \neq 0$ and $r_\varphi(x^0, \xi^{0\prime}, \tau^0) = 0$. Then, there exists $\mathscr{U}$ a conic open neighborhood of $\varrho^{0\prime} = (x^0, \xi^{0\prime}, \tau^0)$ in $U^0 \times \mathbb{R}^{d-1} \times \mathbb{R}_+$ such that for $\chi \in S^0_{\mathsf{T},\tau}$, homogeneous of degree zero, with* $\operatorname{supp}(\chi) \subset \mathscr{U}$, *there exist $C > 0$ and $\tau_* > 0$ such that*

$$
\begin{aligned}
(13.6.17)\quad & \tau^{1/2}\|\mathrm{Op}_{\mathsf{T}}(\chi)v\|_{\tau,1} + \tau|\mathrm{Op}_{\mathsf{T}}(\chi)v_{|x_d=0^+}|_{\tau,1/2} \\
& \qquad + \tau|(D_d\mathrm{Op}_{\mathsf{T}}(\chi)v - i\mathrm{Op}_{\mathsf{T}}(\chi)G^d)_{|x_d=0^+}|_{\tau,-1/2} \\
& \leq C\Big(\|G_0\|_+ + \tau\|G'\|_+ + \tau^{3/2}\|G^d\|_+ + \tau^{3/2}|v_{|x_d=0^+}|_\partial + \tau|\mathsf{B}_{0,\varphi}(v, -G^d)|_{\tau,-1/2} \\
& \qquad + \|v\|_{\tau,1} + |(D_dv - iG^d)_{|x_d=0^+}|_{\tau,-3/2}\Big),
\end{aligned}
$$

for $\tau \geq \tau_*$ and $v \in H^1(\mathbb{R}^d_+)$, $G_0 \in L^2(\mathbb{R}^d_+)$, $G \in L^2V(\mathbb{R}^d_+)$, and $G' = (G^1, \dots, G^{d-1})$, such that

$$P_{0,\varphi}v = G_0 + \sum_{1\leq i\leq d} \partial_j G^j.$$

REMARK 13.21. Note that this estimate only differs from that of Proposition 13.17 by the term $\tau^{3/2}\|G^d\|_+$ that replaces the term $\tau\|G^d\|_+$.

PROOF. We first observe that $r_\varphi(\varrho')$ satisfies a strong sub-elliptic property for $\varrho' = (x, \xi', \tau')$ in a conic neighborhood of $\varrho^{0\prime} = (x^0, \xi^{0\prime}, \tau^0)$. Indeed, one has $r_\varphi(\varrho^{0\prime}) = 0$. With $\varrho^0 = (\varrho^{0\prime}, \xi_d = 0)$ one has $p_\varphi(\varrho^0) = 0$ and thus, since (P_0, φ) has the sub-ellipticity property one has $\{\overline{p_\varphi}, p_\varphi\}(\varrho^0)/i > 0$. Observe that

$$
\begin{aligned}
\{\overline{p_\varphi}, p_\varphi\}(\varrho) &= 2\big(\xi_d - i\tau\partial_d\varphi(x)\big)\{\xi_d - i\tau\partial_d\varphi, p_\varphi\}(\varrho) \\
&\quad + 2\big(\xi_d + i\tau\partial_d\varphi(x)\big)\{\overline{r_\varphi}, \xi_d + i\tau\partial_d\varphi\} + \{\overline{r_\varphi}, r_\varphi\},
\end{aligned}
$$

which yields

$$0 < \frac{1}{i}\{\overline{p_\varphi}, p_\varphi\}(\varrho^0) = \frac{1}{i}\{\overline{r_\varphi}, r_\varphi\}(\varrho^{0\prime}).$$

By continuity and homogeneity there exists a conic neighborhood $\mathscr{V}$ of $\varrho^{0\prime}$ such that

$$\frac{1}{i}\{\overline{r_\varphi}, r_\varphi\}(\varrho') \geq C_0\lambda^3_{\mathsf{T},\tau} \quad\text{and}\quad \tau \geq C_0\lambda_{\mathsf{T},\tau}, \tag{13.6.18}$$

for some $C_0 > 0$. Without any loss of generality we may moreover assume that $\mathbb{S}_{\overline{\mathscr{V}}}$ is compact. Let $0 < \varepsilon < 1$; there exists a conic neighborhood

$\mathscr{U}_\varepsilon \subset \mathscr{V}$ of $\varrho^{0\prime}$ such that

$$|r_\varphi(\varrho')| \leq \varepsilon \lambda_{\mathsf{T},\tau}^2, \quad \text{and} \quad |\partial_d \varphi(x)| \leq \varepsilon, \qquad \varrho' = (x, \xi', \tau) \in \mathscr{U}_\varepsilon. \tag{13.6.19}$$

We consider $\chi \in S_\tau^0$ supported in $\mathscr{U}_\varepsilon$. The value of ε be set below.

We define

$$R_2 = \frac{1}{2}(R_{0,\varphi} + R_{0,\varphi}^*) \quad \text{and}$$
$$R_1 = \frac{1}{2i}(R_{0,\varphi} - R_{0,\varphi}^*),$$

with respective principal symbols

$$r_2(\varrho') = r(x, \xi') - \tau^2 r(x, d_{x'}\varphi) \in S_\tau^2, \quad \text{and}$$
$$r_1(\varrho') = 2\tau \tilde{r}(x, \xi', d_{x'}\varphi) \in \tau S_\tau^1 \subset S_\tau^2.$$

Note that (13.6.18) reads

$$\{r_2, r_1\}(\varrho') \gtrsim \lambda_{\mathsf{T},\tau}^3 \approx \tau^3, \qquad \varrho' \in \mathscr{V}, \tag{13.6.20}$$

using that $\tau \gtrsim \lambda_{\mathsf{T},\tau}$ in $\mathscr{V}$.

From $P_{0,\varphi} v = G_0 + \sum_{1 \leq j \leq d} \partial_j G^j$, we write

$$P_{0,\varphi} \mathrm{Op}_\mathsf{T}(\chi) v = \mathrm{Op}_\mathsf{T}(\chi) G_0 + \mathrm{Op}_\mathsf{T}(\chi) \sum_{1 \leq j \leq d} \partial_j G^j + [P_{0,\varphi}, \mathrm{Op}_\mathsf{T}(\chi)] v,$$

yielding

$$w_2 + i w_1 = H_0 + \sum_{1 \leq j \leq d-1} \partial_j H^j,$$

with

$$w_2 = D_d(D_d \mathrm{Op}_\mathsf{T}(\chi) v - i \mathrm{Op}_\mathsf{T}(\chi) G^d) - (\tau \partial_d \varphi)^2 \mathrm{Op}_\mathsf{T}(\chi) v + R_2 \mathrm{Op}_\mathsf{T}(\chi) v,$$
$$w_1 = 2\tau \partial_d \varphi \big(D_d \mathrm{Op}_\mathsf{T}(\chi) v - i \mathrm{Op}_\mathsf{T}(\chi) G^d\big) + R_1 \mathrm{Op}_\mathsf{T}(\chi) v$$

and

$$H_0 = \mathrm{Op}_\mathsf{T}(\chi) G_0 + \sum_{1 \leq j \leq d} [\mathrm{Op}_\mathsf{T}(\chi), \partial_j] G^j + [P_{0,\varphi}, \mathrm{Op}_\mathsf{T}(\chi)] v$$
$$+ 2\tau(\partial_d \varphi) \mathrm{Op}_\mathsf{T}(\chi) G^d - \tau \partial_d^2 \varphi \mathrm{Op}_\mathsf{T}(\chi) v,$$

and $H^j = \mathrm{Op}_\mathsf{T}(\chi) G^j$. One has

$$\|H_0\|_+ \leq C_\varepsilon \big(\|G_0\|_+ + \|G'\|_+ + \tau \|G^d\|_+ + \|v\|_{\tau,1}\big),$$

and

$$\|\partial_j H^j\|_+ \leq C_\varepsilon \tau \|G^j\|_+,$$

using Proposition 6.12 since $\tau \gtrsim \lambda_{\mathsf{T},\tau}$ in $\mathscr{V}$.

Exploiting the standard Carleman approach we then write

$$\|w_2\|_+^2 + \|w_1\|_+^2 + 2\,\mathrm{Re}(w_2, i w_1)_+ \leq C_\varepsilon \big(\|G_0\|_+^2 + \tau^2 \|G\|_+^2 + \|v\|_{\tau,1}^2\big). \tag{13.6.21}$$

We focus our attention on the computation of the term $\operatorname{Re}(w_2, iw_1)_+$ in (13.6.21) that we write as a sum of six terms I_{ij}, $1 \le i \le 3$, $1 \le j \le 2$, where I_{ij} is the inner product of the ith term in the expression of w_2 and the jth term in the expression of iw_1.

Term I_{11}. With an integration by parts, we have

$$\begin{aligned}(\tau)^{-1} I_{11} &= 2\operatorname{Re}\big(D_d(D_d\mathrm{Op}_\mathsf{T}(\chi)v - i\mathrm{Op}_\mathsf{T}(\chi)G^d), i\partial_d\varphi\big(D_d\mathrm{Op}_\mathsf{T}(\chi)v \\ &\qquad\qquad - i\mathrm{Op}_\mathsf{T}(\chi)G^d\big)\big)_+ \\ &= -\int_{\mathbb{R}^d_+} \partial_d\varphi\partial_d\big|(D_d\mathrm{Op}_\mathsf{T}(\chi)v - i\mathrm{Op}_\mathsf{T}(\chi)G^d)\big|^2 dx \\ &= \int_{\mathbb{R}^{d-1}} \partial_d\varphi\big|(D_d\mathrm{Op}_\mathsf{T}(\chi)v - i\mathrm{Op}_\mathsf{T}(\chi)G^d)\big|^2_{|x_d=0^+}\, dx' \\ &\quad + \int_{\mathbb{R}^d_+} \partial_d^2\varphi\big|(D_d\mathrm{Op}_\mathsf{T}(\chi)v - i\mathrm{Op}_\mathsf{T}(\chi)G^d)\big|^2 dx,\end{aligned}$$

yielding

$$\begin{aligned}|I_{11}| &\lesssim \tau|(D_d\mathrm{Op}_\mathsf{T}(\chi)v - i\mathrm{Op}_\mathsf{T}(\chi)G^d)_{|x_d=0^+}|^2_\partial \\ &\quad + \tau\|D_d\mathrm{Op}_\mathsf{T}(\chi)v\|_+^2 + \tau\|G^d\|_+^2.\end{aligned} \tag{13.6.22}$$

Term I_{12}. With an integration by parts, we have

$$\begin{aligned}I_{12} &= \operatorname{Re}(D_d(D_d\mathrm{Op}_\mathsf{T}(\chi)v - i\mathrm{Op}_\mathsf{T}(\chi)G^d), iR_1\mathrm{Op}_\mathsf{T}(\chi)v)_+ \\ &= \operatorname{Re}((D_d\mathrm{Op}_\mathsf{T}(\chi)v - i\mathrm{Op}_\mathsf{T}(\chi)G^d)_{|x_d=0^+}, R_1\mathrm{Op}_\mathsf{T}(\chi)v_{|x_d=0^+})_\partial \\ &\quad + \operatorname{Re}((D_d\mathrm{Op}_\mathsf{T}(\chi)v - i\mathrm{Op}_\mathsf{T}(\chi)G^d), iD_dR_1\mathrm{Op}_\mathsf{T}(\chi)v)_+\end{aligned}$$

With $R_1 \in \tau\Psi^1_{\mathsf{T},\tau}$ we find

$$\begin{aligned}&\big|\operatorname{Re}((D_d\mathrm{Op}_\mathsf{T}(\chi)v - i\mathrm{Op}_\mathsf{T}(\chi)G^d)_{|x_d=0^+}, R_1\mathrm{Op}_\mathsf{T}(\chi)v_{|x_d=0^+})_\partial\big| \\ &\qquad \lesssim \tau|(D_d\mathrm{Op}_\mathsf{T}(\chi)v - i\mathrm{Op}_\mathsf{T}(\chi)G^d)_{|x_d=0^+}|^2_\partial + \tau|\mathrm{Op}_\mathsf{T}(\chi)v_{|x_d=0^+}|^2_{\tau,1}.\end{aligned}$$

We also write

$$\begin{aligned}&\operatorname{Re}(D_d\mathrm{Op}_\mathsf{T}(\chi)v - i\mathrm{Op}_\mathsf{T}(\chi)G^d, iD_dR_1\mathrm{Op}_\mathsf{T}(\chi)v)_+ \\ &\qquad = -\operatorname{Re}(i\mathrm{Op}_\mathsf{T}(\chi)G^d, iR_1D_d\mathrm{Op}_\mathsf{T}(\chi)v)_+ \\ &\qquad\quad + \operatorname{Re}(D_d\mathrm{Op}_\mathsf{T}(\chi)v - i\mathrm{Op}_\mathsf{T}(\chi)G^d, i[D_d, R_1]\mathrm{Op}_\mathsf{T}(\chi)v)_+,\end{aligned}$$

using that iR_1 is formally anti-adjoint. Since $R_1 \in \tau\Psi^1_\tau$, $[D_d, R_1] \in \tau\Psi^0_{\mathsf{T},\tau}$ one finds

$$\begin{aligned}&\big|\operatorname{Re}((D_d\mathrm{Op}_\mathsf{T}(\chi)v - i\mathrm{Op}_\mathsf{T}(\chi)G^d), iD_dR_1\mathrm{Op}_\mathsf{T}(\chi)v)_+\big| \\ &\qquad \lesssim \tau\|D_d\mathrm{Op}_\mathsf{T}(\chi)v\|_+^2 + \tau\|\mathrm{Op}_\mathsf{T}(\chi)G^d\|^2_{\tau,1} + \tau\|\mathrm{Op}_\mathsf{T}(\chi)v\|_+^2.\end{aligned}$$

Consequently, one has

$$I_{12} \lesssim \tau \|D_d \mathrm{Op}_{\mathsf{T}}(\chi)v\|_+^2 + \tau \|\mathrm{Op}_{\mathsf{T}}(\chi)G^d\|_{\tau,1}^2 + \tau \|\mathrm{Op}_{\mathsf{T}}(\chi)v\|_+^2 \\ + \tau |(D_d \mathrm{Op}_{\mathsf{T}}(\chi)v - i\mathrm{Op}_{\mathsf{T}}(\chi)G^d)_{|x_d=0^+}|_\partial^2 + \tau |\mathrm{Op}_{\mathsf{T}}(\chi)v_{|x_d=0^+}|_{\tau,1}^2.$$

Term I_{21}. We write

$$\begin{aligned}\tau^{-3} I_{21} &= -2\,\mathrm{Re}\,\big((\partial_d\varphi)^3 \mathrm{Op}_{\mathsf{T}}(\chi)v, i\big(D_d \mathrm{Op}_{\mathsf{T}}(\chi)v - i\mathrm{Op}_{\mathsf{T}}(\chi)G^d\big)\big)_+ \\ &= -\int_{\mathbb{R}^d_+} (\partial_d\varphi)^3 \partial_d |\mathrm{Op}_{\mathsf{T}}(\chi)v|^2 \, dx \\ &\quad - 2\,\mathrm{Re}\,\big((\partial_d\varphi)^3 \mathrm{Op}_{\mathsf{T}}(\chi)v, \mathrm{Op}_{\mathsf{T}}(\chi)G^d\big)_+.\end{aligned}$$

With an integration by parts one has

$$\begin{aligned}-\int_{\mathbb{R}^d_+} (\partial_d\varphi)^3 \partial_d |\mathrm{Op}_{\mathsf{T}}(\chi)v|^2 \, dx &= 3 \int_{\mathbb{R}^d_+} (\partial_d\varphi)^2 (\partial_d^2\varphi) |\mathrm{Op}_{\mathsf{T}}(\chi)v|^2 \, dx \\ &\quad + \int_{\mathbb{R}^{d-1}} (\partial_d\varphi)^3 |\mathrm{Op}_{\mathsf{T}}(\chi)v|^2_{|x_d=0^+} \, dx'.\end{aligned}$$

We thus find

$$|I_{21}| \lesssim \varepsilon\tau^3 \big(\|\mathrm{Op}_{\mathsf{T}}(\chi)v\|_+^2 + |\mathrm{Op}_{\mathsf{T}}(\chi)v_{|x_d=0^+}|_\partial^2 + \|G^d\|_+^2\big),$$

since $|\partial_d\varphi| \leq \varepsilon < 1$ in $\mathrm{supp}(\chi)$.

Term I_{22}. We write, using that R_1 is formally selfadjoint,

$$\begin{aligned}I_{22} &= -\tau^2\,\mathrm{Re}\,\big((\partial_d\varphi)^2 \mathrm{Op}_{\mathsf{T}}(\chi)v, iR_1 \mathrm{Op}_{\mathsf{T}}(\chi)v\big)_+ \\ &= -\tau^2 \big(i[(\partial_d\varphi)^2, R_1] \mathrm{Op}_{\mathsf{T}}(\chi)v, \mathrm{Op}_{\mathsf{T}}(\chi)v\big)_+.\end{aligned}$$

The principal symbol of $i[(\partial_d\varphi)^2, R_1]$ is given by $s_1 = \partial_d\varphi\{\partial_d\varphi, r_1\} \in \tau\partial_d\varphi S^0_{\mathsf{T},\tau}$ yielding $|s_1| \lesssim \varepsilon\tau$ in $\mathscr{U}_\varepsilon$. Thus, Corollary 2.51 yields

$$|I_{22}| \lesssim \varepsilon\tau^3 \|\mathrm{Op}_{\mathsf{T}}(\chi)v\|_+^2 + \|v\|_{0,-N}^2,$$

for any $N \in \mathbb{N}$.

Term I_{31}. We write

$$I_{31} = 2\tau\,\mathrm{Re}\,\big(R_2 \mathrm{Op}_{\mathsf{T}}(\chi)v, i(\partial_d\varphi)(D_d \mathrm{Op}_{\mathsf{T}}(\chi)v - i\mathrm{Op}_{\mathsf{T}}(\chi)G^d)\big)_+$$

First, one has

$$\begin{aligned}\big|2\tau\,\mathrm{Re}\,\big(R_2 \mathrm{Op}_{\mathsf{T}}(\chi)v, (\partial_d\varphi)\mathrm{Op}_{\mathsf{T}}(\chi)G^d\big)_+\big| &\lesssim \varepsilon\tau \|\mathrm{Op}_{\mathsf{T}}(\chi)v\|_{\tau,0,1}^2 \\ &\quad + \tau \|\mathrm{Op}_{\mathsf{T}}(\chi)G^d\|_{\tau,0,1}^2,\end{aligned}$$

since $|\partial_d \varphi| \leq \varepsilon$ in $\operatorname{supp}(\chi)$. Second, one computes

$$\begin{aligned}\operatorname{Re}\big(R_2 \operatorname{Op}_{\mathsf T}(\chi)v, i(\partial_d\varphi)D_d\operatorname{Op}_{\mathsf T}(\chi)v\big)_+ &= \big((\partial_d\varphi)R_2\operatorname{Op}_{\mathsf T}(\chi)v, iD_d\operatorname{Op}_{\mathsf T}(\chi)v\big)_+ \\ &\quad + \big(i(\partial_d\varphi)D_d\operatorname{Op}_{\mathsf T}(\chi)v, R_2\operatorname{Op}_{\mathsf T}(\chi)v\big)_+ \\ &= -\big((\partial_d\varphi)R_2\operatorname{Op}_{\mathsf T}(\chi)v_{|x_d=0^+}, \operatorname{Op}_{\mathsf T}(\chi)v_{|x_d=0^+}\big)_\partial \\ &\quad + \big(iA\operatorname{Op}_{\mathsf T}(\chi)v, \operatorname{Op}_{\mathsf T}(\chi)v\big)_+,\end{aligned}$$

with $A = R_2(\partial_d\varphi)D_d - D_d(\partial_d\varphi)R_2$ with principal symbol given by $a_2 + a_1\xi_d$ with $a_2 \in S^2_{\mathsf T,\tau}$ and $a_1 \in S^1_{\mathsf T,\tau}$ with the additional property $a_2 \lesssim \varepsilon\lambda^2_{\mathsf T,\tau}$ in $\mathscr U_\varepsilon$ where $|\partial_d\varphi| + |r_2|\lambda^{-2}_{\mathsf T,\tau} \lesssim \varepsilon$. Consequently, with the Young inequality, one obtains

$$\begin{aligned}\Big|2\tau \operatorname{Re}\big(R_2\operatorname{Op}_{\mathsf T}(\chi)v, i(\partial_d\varphi)D_d\operatorname{Op}_{\mathsf T}(\chi)v\big)_+\Big| &\lesssim \tau|\operatorname{Op}_{\mathsf T}(\chi)v_{|x_d=0^+}|^2_{\tau,1} + (\varepsilon+\varepsilon^{1/2})\tau\|\operatorname{Op}_{\mathsf T}(\chi)v\|^2_{\tau,0,1} \\ &\quad + \varepsilon^{-1/2}\tau\|D_d\operatorname{Op}_{\mathsf T}(\chi)v\|^2_+,\end{aligned}$$

yielding

$$\begin{aligned}|I_{31}| &\lesssim \tau|\operatorname{Op}_{\mathsf T}(\chi)v_{|x_d=0^+}|^2_{\tau,1} + (\varepsilon+\varepsilon^{1/2})\tau\|\operatorname{Op}_{\mathsf T}(\chi)v\|^2_{\tau,0,1} \\ &\quad + \varepsilon^{-1/2}\tau\|D_d\operatorname{Op}_{\mathsf T}(\chi)v\|^2_+ + \tau\|\operatorname{Op}_{\mathsf T}(\chi)G^d\|^2_{\tau,0,1}.\end{aligned}$$

Term I_{32}. Since R_1 and R_2 are both tangential and formally selfadjoint we have

$$I_{32} = \operatorname{Re}(R_2\operatorname{Op}_{\mathsf T}(\chi)v, iR_1\operatorname{Op}_{\mathsf T}(\chi)v)_+ = (i[R_2,R_1]\operatorname{Op}_{\mathsf T}(\chi)v, \operatorname{Op}_{\mathsf T}(\chi)v)_+,$$

where $i[R_2, R_1] \in \Psi^3_{\mathsf T,\tau}$ has $\{r_2, r_1\}$ for principal symbol. With (13.6.20) and the microlocal Gårding inequality of Theorem 2.50 of Volume 1 one finds

$$I_{32} \geq C\|\operatorname{Op}_{\mathsf T}(\chi)v\|^2_{\tau,0,3/2} - C_N\|v\|^2_{\tau,0,-N},$$

for any $N \in \mathbb N$.

With (13.6.21) and collecting the estimations obtained for the six terms we find

$$\begin{aligned}\|\operatorname{Op}_{\mathsf T}(\chi)v\|_{\tau,0,3/2} + \|w_2\|_+ &\leq C_\varepsilon\big(\|G_0\|_+ + \tau\|G\|_+ + \|v\|_{\tau,1}\big) \\ &\quad + C\tau^{1/2}\Big(|(D_d\operatorname{Op}_{\mathsf T}(\chi)v - i\operatorname{Op}_{\mathsf T}(\chi)G^d)_{|x_d=0^+}|_\partial \\ &\quad + |\operatorname{Op}_{\mathsf T}(\chi)v_{|x_d=0^+}|_{\tau,1} + (1+\varepsilon^{-1/4})\|D_d\operatorname{Op}_{\mathsf T}(\chi)v\|_+ \\ &\quad + (1+\varepsilon^{1/2}\tau)\|\operatorname{Op}_{\mathsf T}(\chi)v\|_+ + (\varepsilon^{1/2}+\varepsilon^{1/4})\|\operatorname{Op}_{\mathsf T}(\chi)v\|_{\tau,0,1} \\ &\quad + \|\operatorname{Op}_{\mathsf T}(\chi)G^d\|_{\tau,1}\Big).\end{aligned}$$

For $\varepsilon > 0$ chosen sufficiently small and for $\tau \geq 1$ chosen sufficiently large we find

$$\begin{aligned}
&\|\mathrm{Op}_\mathsf{T}(\chi)v\|_{\tau,0,3/2} + \|w_2\|_+ \\
&\qquad \leq C_\varepsilon\big(\|G_0\|_+ + \tau\|G\|_+ + \|v\|_{\tau,1}\big) \\
&\qquad\quad + C\tau^{1/2}\Big(|(D_d\mathrm{Op}_\mathsf{T}(\chi)v - i\mathrm{Op}_\mathsf{T}(\chi)G^d)_{|x_d=0^+}|_\partial \\
&\qquad\quad + |\mathrm{Op}_\mathsf{T}(\chi)v_{|x_d=0^+}|_{\tau,1} + (1+\varepsilon^{-1/4})\|D_d\mathrm{Op}_\mathsf{T}(\chi)v\|_+ \\
&\qquad\quad + \|\mathrm{Op}_\mathsf{T}(\chi)G^d\|_{\tau,1}\Big).
\end{aligned}$$

The following lemma explains how the L^2-norm of $D_d\mathrm{Op}_\mathsf{T}(\chi)v$ can be recovered from that of $\mathrm{Op}_\mathsf{T}(\chi)v$ and w_2.

LEMMA 13.22. *One has*

$$\begin{aligned}
\tau^{1/2}\|D_d\mathrm{Op}_\mathsf{T}(\chi)v\|_+ \lesssim{}& \varepsilon^{1/2}\tau^{3/2}\|\mathrm{Op}_\mathsf{T}(\chi)v\|_+ + (\varepsilon\tau)^{-1/2}\|w_2\|_+ + \tau^{1/2}\|G^d\|_+ \\
&+ \tau^{1/2}|(D_d\mathrm{Op}_\mathsf{T}(\chi)v - i\mathrm{Op}_\mathsf{T}(\chi)G^d)_{|x_d=0^+}|_\partial \\
&+ \tau^{1/2}|\mathrm{Op}_\mathsf{T}(\chi)v_{|x_d=0^+}|_\partial + C_{\varepsilon,N}\|v\|_{\tau,0,-N}.
\end{aligned}$$

A proof is given below.

With the lemma, for $\tau \geq \varepsilon^{-2}$, we obtain

$$\begin{aligned}
&\varepsilon^{-1/2}\tau^{1/2}\|D_d\mathrm{Op}_\mathsf{T}(\chi)v\|_+ + \|\mathrm{Op}_\mathsf{T}(\chi)v\|_{\tau,0,3/2} + \|w_2\|_+ \\
&\quad \leq C_\varepsilon\Big(\|G_0\|_+ + \tau\|G\|_+ + \|v\|_{\tau,1} \\
&\qquad + \tau^{1/2}|(D_d\mathrm{Op}_\mathsf{T}(\chi)v - i\mathrm{Op}_\mathsf{T}(\chi)G^d)_{|x_d=0^+}|_\partial \\
&\qquad + \tau^{1/2}|\mathrm{Op}_\mathsf{T}(\chi)v_{|x_d=0^+}|_{\tau,1}\Big) \\
&\quad + C\tau^{1/2}\Big((1+\varepsilon^{-1/4})\|D_d\mathrm{Op}_\mathsf{T}(\chi)v\|_+ + \|\mathrm{Op}_\mathsf{T}(\chi)G^d\|_{\tau,1}\Big).
\end{aligned}$$

For $\varepsilon > 0$ chosen sufficiently small one obtains

$$\begin{aligned}
&\tau^{1/2}\|D_d\mathrm{Op}_\mathsf{T}(\chi)v\|_+ + \|\mathrm{Op}_\mathsf{T}(\chi)v\|_{\tau,0,3/2} + \|w_2\|_+ \\
&\quad \lesssim \|G_0\|_+ + \tau\|G\|_+ + \tau^{1/2}\|\mathrm{Op}_\mathsf{T}(\chi)G^d\|_{\tau,1} + \|v\|_{\tau,1} \\
&\qquad + \tau^{1/2}|(D_d\mathrm{Op}_\mathsf{T}(\chi)v - i\mathrm{Op}_\mathsf{T}(\chi)G^d)_{|x_d=0^+}|_\partial + \tau^{1/2}|\mathrm{Op}_\mathsf{T}(\chi)v_{|x_d=0^+}|_{\tau,1}.
\end{aligned}$$

With the form of $\mathsf{B}_{0,\varphi}$ given in (13.6.2) we then obtain

(13.6.23)

$$\begin{aligned}
&\tau^{1/2}\|D_d\mathrm{Op}_\mathsf{T}(\chi)v\|_+ + \|\mathrm{Op}_\mathsf{T}(\chi)v\|_{\tau,0,3/2} + \|w_2\|_+ \\
&\quad \lesssim \|G_0\|_+ + \tau\|G\|_+ + \tau^{1/2}\|\mathrm{Op}_\mathsf{T}(\chi)G^d\|_{\tau,1} + \|v\|_{\tau,1} \\
&\qquad + \tau^{1/2}|\mathsf{B}_{0,\varphi}(\mathrm{Op}_\mathsf{T}(\chi)v, \mathrm{Op}_\mathsf{T}(\chi)G^d)|_\partial + \tau^{1/2}|\mathrm{Op}_\mathsf{T}(\chi)v_{|x_d=0^+}|_{\tau,1}.
\end{aligned}$$

We write

$$\begin{aligned}
&\mathsf{B}_{0,\varphi}(\mathrm{Op}_{\mathsf{T}}(\chi)v, -\mathrm{Op}_{\mathsf{T}}(\chi)G^d)\\
&\quad = \tilde{B}_\varphi \mathrm{Op}_{\mathsf{T}}(\chi)v_{|x_d=0^+} - i(D_d\mathrm{Op}_{\mathsf{T}}(\chi)v - i\mathrm{Op}_{\mathsf{T}}(\chi)G^d)_{|x_d=0^+}\\
&\quad = \mathrm{Op}_{\mathsf{T}}(\chi)\mathsf{B}_{0,\varphi}(v, -G^d) + \big([\tilde{B}_\varphi, \mathrm{Op}_{\mathsf{T}}(\chi)] - i[D_d, \mathrm{Op}_{\mathsf{T}}(\chi)]\big)v.
\end{aligned}$$

With Proposition 6.11, since $\tau \gtrsim \lambda_{\mathsf{T},\tau}$ in $\mathscr{V}$, one finds

$$\begin{aligned}
\tau^{1/2}&|\mathsf{B}_{0,\varphi}(\mathrm{Op}_{\mathsf{T}}(\chi)v, -\mathrm{Op}_{\mathsf{T}}(\chi)G^d)|_\partial\\
&\lesssim \tau|\mathsf{B}_{0,\varphi}(v, -G^d)|_{\tau,-1/2} + \tau^{1/2}|v_{|x_d=0^+}|_\partial\\
&\lesssim \tau|\mathsf{B}_{0,\varphi}(v, -G^d)|_{\tau,-1/2} + \|v\|_1,
\end{aligned}$$

using the trace inequality of Corollary 6.10. With the same arguments applied to other terms in the right-hand side of (13.6.23) yielding

$$\begin{aligned}
\tau^{1/2}\|\mathrm{Op}_{\mathsf{T}}(\chi)G^d\|_{\tau,1} &\lesssim \tau^{3/2}\|G^d\|_+,\\
\tau^{1/2}|\mathrm{Op}_{\mathsf{T}}(\chi)v_{|x_d=0^+}|_{\tau,1} &\lesssim \tau^{3/2}|v_{|x_d=0^+}|_\partial,
\end{aligned}$$

we then obtain the sought result. ∎

Proof of Lemma 13.22. We write

$$D_d(D_d\mathrm{Op}_{\mathsf{T}}(\chi)v - i\mathrm{Op}_{\mathsf{T}}(\chi)G^d) = w_2 + (\tau\partial_d\varphi)^2\mathrm{Op}_{\mathsf{T}}(\chi)v - R_2\mathrm{Op}_{\mathsf{T}}(\chi)v.$$

We estimate

$$\begin{aligned}
\tau\big|&\big(w_2 + (\tau\partial_d\varphi)^2\mathrm{Op}_{\mathsf{T}}(\chi)v - R_2\mathrm{Op}_{\mathsf{T}}(\chi)v, \mathrm{Op}_{\mathsf{T}}(\chi)v\big)_+\big|\\
&\lesssim (\varepsilon\tau)^{-1}\|w_2\|_+^2 + \varepsilon\tau^3\|\mathrm{Op}_{\mathsf{T}}(\chi)v\|_+^2 + C_{\varepsilon,N}\|v\|_{\tau,0,-N}^2,
\end{aligned}$$

using Corollary 2.51 as $|r_2| \lesssim \varepsilon\lambda_{\mathsf{T},\tau}^2$ and $|\partial_d\varphi| \le \varepsilon$ in $\mathrm{supp}(\chi)$. We compute

$$\begin{aligned}
\tau(&D_d(D_d\mathrm{Op}_{\mathsf{T}}(\chi)v - i\mathrm{Op}_{\mathsf{T}}(\chi)G^d), \mathrm{Op}_{\mathsf{T}}(\chi)v)_+\\
&= i\tau((D_d\mathrm{Op}_{\mathsf{T}}(\chi)v - i\mathrm{Op}_{\mathsf{T}}(\chi)G^d)_{|x_d=0^+}, \mathrm{Op}_{\mathsf{T}}(\chi)v_{|x_d=0^+})_\partial\\
&\quad + \tau\|D_d\mathrm{Op}_{\mathsf{T}}(\chi)v\|_+^2 - i\tau(\mathrm{Op}_{\mathsf{T}}(\chi)G^d, D_d\mathrm{Op}_{\mathsf{T}}(\chi)v)_+.
\end{aligned}$$

This yields

$$\begin{aligned}
\tau&\|D_d\mathrm{Op}_{\mathsf{T}}(\chi)v\|_+^2\\
&\lesssim \tau\big|(D_d(D_d\mathrm{Op}_{\mathsf{T}}(\chi)v - i\mathrm{Op}_{\mathsf{T}}(\chi)G^d), \mathrm{Op}_{\mathsf{T}}(\chi)v)_+\big|^2 + \tau\|G^d\|_+^2\\
&\quad + \tau|(D_d\mathrm{Op}_{\mathsf{T}}(\chi)v - i\mathrm{Op}_{\mathsf{T}}(\chi)G^d)_{|x_d=0^+}|_\partial^2 + \tau|\mathrm{Op}_{\mathsf{T}}(\chi)v_{|x_d=0^+}|_\partial^2,
\end{aligned}$$

which gives the result. ∎

13.7. Global Estimates

13.7.1. Global Estimate with an Inner Observation. With the same arguments as in Chapters 3 and 5 we have the following global estimate with an inner observation.

THEOREM 13.23. *Let $\mathcal{M}$ be a smooth compact connected Riemannian manifold. Let $P = -\Delta_g + R_1$ where R_1 is a first-order differential operator with bounded coefficients. Let ω, ω_0 be two nonempty open subsets of $\mathcal{M}$ such that $\omega_0 \Subset \omega$.*

Let φ be a global weight function adapted to $\Gamma_0 = \partial\mathcal{M}$ and ω_0 in the sense of Definition 5.7. Consider also B a differential operator of order $k \in \mathbb{N}$ in $\partial\mathcal{M}$ and assume moreover that (P, B, φ) satisfies the Lopatinskiĭ–Šapiro condition on $\partial\mathcal{M}$ (Definition 8.1).

Then, there exist $\tau_* > 0$ and $C \geq 0$ such that

$$\tau^{1/2}\|e^{\tau\varphi}u\|_{\tau,1} + \tau|e^{\tau\varphi}u_{|\partial\mathcal{M}}|_{\tau,1/2} + \tau|e^{\tau\varphi}(\partial_\nu u + g(F,\nu))_{|\partial\mathcal{M}}|_{\tau,-1/2}$$
$$\leq C\big(\|e^{\tau\varphi}F_0\|_{L^2(\mathcal{M})} + \tau\|e^{\tau\varphi}F\|_{L^2V(\mathcal{M})}$$
$$+ \tau|e^{\tau\varphi_{|\partial\mathcal{M}}}\mathsf{B}(u, g(F,\nu))|_{\tau,1/2-k} + \tau^{3/2}\|e^{\tau\varphi}u\|_{L^2(\omega)}\big),$$

for $\tau \geq \tau_*$ and $u \in H^1(\mathcal{M})$, $F_0 \in L^2(\mathcal{M})$, $F \in L^2V(\mathcal{M})$ such that

$$Pu = F_0 + \operatorname{div}_g F.$$

13.7.2. Global Estimate with a Boundary Observation. Consider a neighborhood O of $\partial\mathcal{M}$ where one has global normal geodesic coordinates as given by Theorem 17.22 using that $\mathcal{M}$ is compact here: for some $z_0 > 0$ one has a diffeomorphism Φ, such that

$$\begin{aligned}\Phi : \partial\mathcal{M} \times [0, z_0) &\to O \\ (m', z) &\mapsto \Phi(m', z),\end{aligned} \tag{13.7.1}$$

allowing one to use (m', z) to parameterize O, and the pullback of g takes the form $g'_{m'}(z) \otimes 1_z + d_z \otimes d_z$. In O we denote by ν the image of $-\partial_z$ by the tangent map $T\Phi(m', z)$ (see Sect. 15.3.2). At the boundary ν is precisely the outward pointing unit vector field. With ν thus introduced we can give a precise meaning of $g(F, \nu)$ for a vector field F corresponding to the *normal* part of F near the boundary.

With the results of Theorems 13.4 and 13.16 and Proposition 13.20 we deduce the following global estimates.

THEOREM 13.24. *Let $P = -\Delta_g + R_1$ where R_1 is a first-order differential operator with bounded coefficients. Let Γ_0, Γ_{obs} be two nonempty open subsets of $\partial\mathcal{M}$ such that $\Gamma_0 \cup \Gamma_{\text{obs}} = \partial\mathcal{M}$ and $\Gamma_{\text{obs}} \setminus \overline{\Gamma_0} \neq \emptyset$.*

Let φ be a global weight function adapted to Γ_0 in the sense of Definition 5.10. Consider also B a differential operator of order $k \in \mathbb{N}$ in Γ_0 and assume moreover that (P, B, φ) satisfies the Lopatinskiĭ–Šapiro condition on Γ_0 (Definition 8.1).

Consider also B' a first-order differential operator on Γ_{obs} and $\tilde{B} = \partial_\nu + B'$ and assume that $(P, \tilde{B})$ fulfills the Lopatinskiĭ–Šapiro condition of Definition 2.2 on Γ_{obs}.

Let also $\chi_0, \chi_{\text{obs}} \in \mathscr{C}^\infty(\overline{\mathcal{M}})$ be such that $\chi_{0|\partial\mathcal{M}}$ and $\chi_{\text{obs}|\partial\mathcal{M}}$ form a partition of unity of $\partial\mathcal{M}$ associated with the covering by Γ_0 and Γ_{obs}, that is,

$$\operatorname{supp}(\chi_{0|\partial\mathcal{M}}) \subset \Gamma_0, \quad \operatorname{supp}(\chi_{\text{obs}|\partial\mathcal{M}}) \subset \Gamma_{\text{obs}},$$

and $\chi_{0|\partial\mathcal{M}} + \chi_{\text{obs}|\partial\mathcal{M}} \equiv 1$ in $\partial\mathcal{M}$. Moreover, we assume that χ_{obs} is supported in O (neighborhood of $\partial\mathcal{M}$ introduced above).

Then, there exist $\tau_* > 0$ and $C \geq 0$ such that

$$\begin{aligned}
\tau^{1/2}\|e^{\tau\varphi}u\|_{\tau,1} &+ \tau|e^{\tau\varphi}u_{|\partial\mathcal{M}}|_{\tau,1/2} + \tau|e^{\tau\varphi}(\partial_\nu u + g(F,\nu))_{|\partial\mathcal{M}}|_{\tau,-1/2} \\
&\leq C\big(\|e^{\tau\varphi}F_0\|_{L^2(\mathcal{M})} + \tau\|e^{\tau\varphi}F\|_{L^2V(\mathcal{M})} + \tau^{3/2}\|\chi_{\text{obs}}e^{\tau\varphi}g(\nu,F)\|_{L^2(\mathcal{M})} \\
&\quad + \tau|\chi_0 e^{\tau\varphi_{|\partial\mathcal{M}}}\mathsf{B}(u, g(F,\nu))|_{\tau,1/2-k} + \tau^{3/2}|\chi_{\text{obs}}e^{\tau\varphi}u_{|\partial\mathcal{M}}|_{L^2(\partial\mathcal{M})} \\
&\qquad + \tau|\chi_{\text{obs}}e^{\tau\varphi_{|\partial\mathcal{M}}}\tilde{\mathsf{B}}(u, g(F,\nu))|_{\tau,-1/2}\big),
\end{aligned}$$

for $\tau \geq \tau_*$ and $u \in H^1(\mathcal{M})$, $F_0 \in L^2(\mathcal{M})$, $F \in L^2V(\mathcal{M})$ such that

$$Pu = F_0 + \operatorname{div}_g F.$$

In the theorem statement, $\tilde{\mathsf{B}}(.,.)$ has the same definition as $\mathsf{B}(.,.)$ given in (13.1.3) with B replaced by $\tilde{B}$.

Note that results in the form given in Theorems 5.8 and 5.11 could also be written. Imposing limitation on the supports of the functions then allows for nonsmooth parts of the boundary. This is, for instance, used in Sections 6.5.2 and 7.5.2 for the derivation of a resolvent estimate and a spectral inequality.

If one compares the result of Theorems 13.23 and 13.24 one sees that the term $\|\chi_{\text{obs}}e^{\tau\varphi}g(\nu,F)\|_{L^2(\mathcal{M})}$ appears in the latter with a factor $\tau^{3/2}$ instead of τ. The reason lies in the observation that the chosen weight function φ is such that $\partial_\nu\varphi = 0$ at places in Γ_{obs}. Indeed, if it is not the case, then $\partial_\nu\varphi < 0$ on $\partial\mathcal{M}$ and necessarily φ reaches a maximum in $\mathcal{M} \setminus \partial\mathcal{M}$ which is excluded for a Carleman estimate to hold by Theorems 4.5 and 4.7. It would be satisfying to be able to obtain the factor τ, in particular if needed in applications.

13.8. Notes

A Carleman estimate for a parabolic operator with a source term in H^{-1} appears in the work of O. Yu. Imanuvilov and M. Yamamoto [181]. There, homogeneous Dirichlet boundary conditions are considered.

The case of an elliptic operator, also with homogeneous Dirichlet boundary conditions, can be found in the work of O. Yu. Imanuvilov and J.-P. Puel [179]. In the work of E. Fernández-Cara et al. [144], Neumann boundary conditions are considered for a parabolic operator. Motivations for these works are inverse problems and controllability of Navier-Stokes equations.

Here, we prove estimation with a source term in H^{-1} for any boundary operator for which the Lopatinskiĭ–Šapiro condition holds. As in Chap. 8 the proofs we give are based on the analysis of the conjugated operator viewed in certain microlocal regions as a product of two first-order operators. This allows us to circumvent the difficulty raised by not having a well defined Neumann trace of the solution; only a modified version of the Neumann trace make sense; see Proposition 13.2. This modified Neumann trace appears in the results proven here; see Theorem 13.4.

Appendix

13.A. Some Technical Proofs

13.A.1. A Trace Result. Here, we prove Proposition 13.2.

Since $Pu = F_0 + \operatorname{div}_g F$ and $P = -\Delta_g + R_1$ one has

$$\operatorname{div}_g(\nabla_g u + F) = -F_0 + R_1 u \in L^2(\mathcal{M}).$$

This implies that $V = \nabla_g u + F$ is in the space of L^2-vector fields that have a L^2-divergence. This space is $H(\operatorname{div}_g, \mathcal{M})$. By Lemma 18.43, the trace of $\big(\partial_\nu u + g(F,\nu)\big)_{|\partial\mathcal{M}} = g(V,\nu)_{|\partial\mathcal{M}}$ makes sense and is in $H^{-1/2}(\mathcal{M})$. Moreover, one has

$$\begin{aligned}&\big|\big(\partial_\nu u + g(F,\nu)\big)_{|\partial\mathcal{M}}\big|_{H^{-1/2}(\partial\mathcal{M})}\\ &\quad\lesssim \|V\|_{H(\operatorname{div}_g,\mathcal{M})} \lesssim \|V\|_{L^2V(\mathcal{M})} + \|\operatorname{div}_g V\|_{L^2(\mathcal{M})}.\end{aligned}$$

One has

$$\|V\|_{L^2V(\mathcal{M})} \lesssim \|\nabla_g u\|_{L^2V(\mathcal{M})} + \|F\|_{L^2V(\mathcal{M})} \lesssim \|u\|_{H^1(\mathcal{M})} + \|F\|_{L^2V(\mathcal{M})},$$

and

$$\begin{aligned}\|\operatorname{div}_g V\|_{L^2(\mathcal{M})} &\lesssim \|F_0\|_{L^2(\mathcal{M})} + \|R_1 u\|_{L^2(\mathcal{M})}\\ &\lesssim \|F_0\|_{L^2(\mathcal{M})} + \|u\|_{H^1(\mathcal{M})}.\end{aligned}$$

We thus obtain the expected estimation for $\big|\big(\partial_\nu u + g(F,\nu)\big)_{|\partial\mathcal{M}}\big|_{H^{-1/2}(\partial\mathcal{M})}$. ■

13.A.2. Trace Estimate Under the Lopatinskiĭ–Šapiro Condition. Here, we prove Lemma 13.11 by adapting the proof of Lemma 8.19 in Sect. 8.A.3 and using its notation. In particular, with $\sigma = 1$ therein, we introduce

$$\begin{aligned}\mathsf{B}^1_\varphi &= \Big(\Lambda^{1/2-k}_{\mathsf{T},\tau}\tilde{B}^k_\varphi(x,D',\tau) \quad -i\Lambda^{1/2-k}_{\mathsf{T},\tau}B^{k-1}_\varphi(x,D',\tau)\Lambda^1_{\mathsf{T},\tau}\Big),\\ \mathsf{P}^{1,+} &= \Big(-\Lambda^{-1/2}_{\mathsf{T},\tau}\operatorname{Op}_{\mathsf{T}}(\tilde{\chi}\gamma_2) \quad \Lambda^{1/2}_{\mathsf{T},\tau}\Big),\end{aligned}$$

with respective principal symbols

$$\mathsf{b}_{\varphi}^{1}(\varrho') = \begin{pmatrix} \lambda_{\mathsf{T},\tau}^{1/2-k} \tilde{b}_{\varphi}^{k}(\varrho') & -i\lambda_{\mathsf{T},\tau}^{3/2-k} b_{\varphi}^{k-1}(\varrho') \end{pmatrix}, \quad \mathsf{p}^{1,+} = \begin{pmatrix} -\lambda_{\mathsf{T},\tau}^{-1/2} \tilde{\chi}\gamma_2 & \lambda_{\mathsf{T},\tau}^{1/2} \end{pmatrix}.$$

We also set $\mathsf{M}^1 = (\mathsf{B}_{\varphi}^1)^* \ \mathsf{B}_{\varphi}^1 + (\mathsf{P}^{1,+})^* \ \mathsf{P}^{1,+}$. It is 2×2 matrix operator of order one. Let $u_1 \in H^{1/2}(\mathbb{R}^{d-1})$ and $u_2 \in H^{-1/2}(\mathbb{R}^{d-1})$. Setting $U = {}^t(u_1, \Lambda_{\mathsf{T},\tau}^{-1} u_2) \in H^{1/2}(\mathbb{R}^{d-1})^2$, we have

$$\begin{aligned} \langle \mathsf{M}^1 U, U \rangle_{H^{-1/2}(\mathbb{R}^{d-1})^2, H^{1/2}(\mathbb{R}^{d-1})^2} &= |\mathsf{B}_{\varphi}^1 U|_{\partial}^2 + |\mathsf{P}_{\varphi}^1 U|_{\partial}^2 \\ &= |\tilde{B}_{\varphi}^k(x, D', \tau) u_1 - i B_{\varphi}^{k-1}(x, D', \tau) u_2|_{\tau,1/2-k}^2 \\ &\quad + |u_2 - \operatorname{Op_T}(\tilde{\chi}\gamma_2) u_1|_{\tau,-1/2}. \end{aligned} \tag{13.A.1}$$

Let $\mathscr{W} \subset \mathscr{U}$ be an open conic neighborhood of $\operatorname{supp}(\chi)$ where $\tilde{\chi} \equiv 1$. Let $\hat{\chi} \in S_{\mathsf{T},\tau}^0$ be homogeneous of degree zero and be supported in $\mathscr{W}$ and such that $\hat{\chi} \equiv 1$ on $\operatorname{supp}(\chi)$. As $\mathscr{W} \subset \mathscr{U}$ we have

$$b_{\varphi}(x, \xi', \xi_d = \gamma_2(x, \xi', \tau), \tau) \neq 0, \quad \varrho' = (x, \xi', \tau) \in \mathscr{W}.$$

Lemma 8.36, for $\sigma = 1$ and the order of the boundary operator equal to k therein, gives with a density argument

$$\begin{aligned} &\langle \mathsf{M}^1 \operatorname{Op_T}(\hat{\chi}) U, \operatorname{Op_T}(\hat{\chi}) U \rangle_{H^{-1/2}(\mathbb{R}^{d-1})^2, H^{1/2}(\mathbb{R}^{d-1})^2} \\ &\quad \geq C |\operatorname{Op_T}(\hat{\chi}) U|_{\tau,1/2}^2 - C_N |U|_{\tau,-N}^2, \end{aligned}$$

for $N \in \mathbb{N}$ and $\tau \geq 1$ chosen sufficiently large.

We now set $u_1 = \operatorname{Op_T}(\chi)\ell$ and $u_2 = \operatorname{Op_T}(\chi)\ell'$, that is $U = (\operatorname{Op_T}(\chi)\ell, \Lambda_{\mathsf{T},\tau}^{-1} \operatorname{Op_T}(\chi)\ell')$. Because of the support conditions of $\hat{\chi}$ and χ and from pseudo-differential calculus, we find, for any $N \in \mathbb{N}$,

$$\begin{aligned} &\langle \mathsf{M}^1 U, U \rangle_{H^{-1/2}(\mathbb{R}^{d-1})^2, H^{1/2}(\mathbb{R}^{d-1})^2} \\ &\quad \geq \operatorname{Re} \langle \mathsf{M}^1 \operatorname{Op_T}(\hat{\chi}) U, \operatorname{Op_T}(\hat{\chi}) U \rangle_{H^{-1/2}(\mathbb{R}^{d-1})^2, H^{1/2}(\mathbb{R}^{d-1})^2} \\ &\qquad - C_N \big(|\ell|_{\tau,-N}^2 + |\ell'|_{\tau,-N}^2 \big), \end{aligned}$$

and

$$|\operatorname{Op_T}(\hat{\chi}) U|_{\tau,1/2}^2 \geq |U|_{\tau,1/2}^2 - C_N \big(|\ell|_{\tau,-N}^2 + |\ell'|_{\tau,-N}^2 \big),$$

yielding

$$\begin{aligned} \langle \mathsf{M}^1 U, U \rangle_{H^{-1/2}(\mathbb{R}^{d-1})^2, H^{1/2}(\mathbb{R}^{d-1})^2} &\geq C |U|_{\tau,1/2}^2 - C_N \big(|\ell|_{\tau,-N}^2 + |h|_{\tau,-N}^2 \big) \\ &= C \big(|\operatorname{Op_T}(\chi)\ell|_{\tau,1/2}^2 + |\operatorname{Op_T}(\chi)\ell'|_{\tau,-1/2}^2 \big) \\ &\quad - C_N \big(|\ell|_{\tau,-N}^2 + |\ell'|_{\tau,-N}^2 \big). \end{aligned}$$

By (13.A.1), this concludes the proof of Lemma 13.11. ∎

13.A.3. Sub-ellipticity of a First-Order Factor. Here, we prove the sub-ellipticity property of the first-order operator P^+ stated in Lemma 13.12.

We start with the following lemma

LEMMA 13.25. *Let $\mathscr{V}$ be a conic open set of $U^0 \times \mathbb{R}^{d-1} \times \mathbb{R}_+$. Assume that p_φ read $p_\varphi = q_1 q_2$ in $\overline{\mathscr{V}}$ with q_1 and q_2 in the form $q_j(\varrho) = \xi_d - r_j(\varrho')$, $j = 1, 2$, where $r_1, r_2(\varrho') \in S^1_{\mathsf{T},\tau}$ and are such that $r_1(\varrho') \neq r_2(\varrho')$ for $\varrho' \in \overline{\mathscr{V}}$. If the sub-ellipticity property holds for (P, φ) in $\overline{U^0}$, then one has*

$$q_j(\varrho) = 0 \quad \Rightarrow \quad \frac{1}{2i}\{\overline{q_j}, q_j\}(\varrho') > 0, \qquad \varrho' \in \overline{\mathscr{V}},\ \varrho = (\varrho', \xi_d),\ \ j = 1, 2.$$

PROOF. We treat the case $j = 1$. We compute

$$\{\overline{p_\varphi}, p_\varphi\} = |q_1|^2\{\overline{q_2}, q_2\} + |q_2|^2\{\overline{q_1}, q_1\} + \overline{q_1} q_2\{\overline{q_2}, q_1\} + \overline{q_2} q_1\{\overline{q_1}, q_2\}.$$

We recall that the sub-ellipticity property reads (see Definition 3.2 of Volume 1)

$$p_\varphi(\varrho) = 0 \quad \Rightarrow \quad \frac{1}{2i}\{\overline{p_\varphi}, p_\varphi\}(\varrho) > 0.$$

Let $\varrho = (\varrho', \xi_d)$ with $\varrho' \in \overline{\mathscr{V}}$ be such that $q_1(\varrho) = 0$. Then $p_\varphi(\varrho') = 0$ and $q_2(\varrho) \neq 0$ since $r_1(\varrho') \neq r_2(\varrho')$. We thus find

$$\frac{1}{2i}\{\overline{q_1}, q_1\}(\varrho') > 0,$$

since $\{\overline{p_\varphi}, p_\varphi\}(\varrho) = |q_2|^2(\varrho)\{\overline{q_1}, q_1\}(\varrho')$. ■

We now proceed with the proof of Lemma 13.12.

We choose the conic open neighborhood $\mathscr{W}$ of $\operatorname{supp}(\chi)$ such that $\overline{\mathscr{W}} \subset \{\tilde{\chi} \equiv 1\}$. We then have $\overline{\mathscr{W}} \subset \mathscr{U}$. We denote by p_φ, p^- the respective principal symbols of $P_{0,\varphi}, P^-$. We have $p^-(\varrho) = \xi_d - \tilde{\gamma}_1(\varrho')$. We recall that $p^+(\varrho) = \xi_d - \tilde{\gamma}_2(\varrho')$. For $\varrho' \in \overline{\mathscr{W}}$, as $\tilde{\chi}(\varrho') \equiv 1$, we have $\tilde{\gamma}_2 = \gamma_2$ and $\tilde{\gamma}_1 = \gamma_1$ and thus $p_\varphi(\varrho) = p^+ p^-(\varrho)$, for $\varrho = (\varrho', \xi_d)$.

By (13.3.3) one has $\tilde{\gamma}_2(\varrho') \neq \tilde{\gamma}_1(\varrho')$ in $\overline{\mathscr{W}}$ and thus Lemma 13.25 applies. ■

13.A.4. Proof of the Shifted Estimate. Here, we prove Theorem 13.15.

In the framework of the normal geodesic coordinates introduced in Sect. 13.2.1 we prove in fact the following proposition.

PROPOSITION 13.26. *With the same setting and assumption as in Proposition 13.8, there exist U a bounded open neighborhood of x^0 in U^0, $C > 0$, and $\tau_* > 0$ such that*

$$\begin{aligned}(13.A.2)\quad \tau\|v\|_{\tau,1,-1/2} &+ \tau^{3/2}|v_{|x_d=0^+}|_\partial + \tau^{3/2}|(D_d v - iG^d)_{|x_d=0^+}|_{\tau,-1}\\ &\leq C\Big(\|G_0\|_+ + \tau\|G\|_+ + \tau^{3/2}|\mathsf{B}_{0,\varphi}(v, -G^d)|_{\tau,-k}\Big),\end{aligned}$$

for $\tau \geq \tau_*$ *and* $v \in H^1(\mathbb{R}^d_+)$, $G_0 \in L^2(\mathbb{R}^d_+)$, $G \in L^2V(\mathbb{R}^d_+)$ *such that*

$$P_{0,\varphi} v = G_0 + \sum_{1 \leq i \leq d} \partial_j G^j, \qquad \operatorname{supp}(v) \cup \operatorname{supp}(G_0) \cup \operatorname{supp}(G) \subset \overline{U_+}.$$

It implies the estimate

(13.A.3)

$$\begin{aligned}\tau \|e^{\tau\varphi} u\|_{\tau,1,-1/2} &+ \tau^{3/2} |e^{\tau\varphi} u_{|\partial\mathcal{M}}|_\partial + \tau^{3/2} |e^{\tau\varphi_{|\partial\mathcal{M}}} (\partial_\nu u + g(F,\nu))_{|\partial\mathcal{M}}|_{\tau,-1} \\ &\leq C\big(\|e^{\tau\varphi} F_0\|_{L^2(\mathcal{M})} + \tau \|e^{\tau\varphi} F\|_{L^2V(\mathcal{M})} + \tau^{3/2} |e^{\tau\varphi_{|\partial\mathcal{M}}} \mathsf{B}(u, g(F,\nu))|_{\tau,-k}\big),\end{aligned}$$

for $\tau \geq \tau_*$ and $u \in H^1(\mathcal{M})$, $F_0 \in L^2(\mathcal{M})$, $F \in L^2V(\mathcal{M})$ such that

$$Pu = F_0 + \operatorname{div}_g F, \quad \text{and} \quad \operatorname{supp}(u) \cup \operatorname{supp}(F_0) \cup \operatorname{supp}(F) \subset \overline{V^1}.$$

Then, following the patching argument of Sect. 13.4.2 we deduce Theorem 13.15.

PROOF OF PROPOSITION 13.26. Let $\tilde{U}$ be the open set of $\mathbb{R}^d$ subset of U^0 given by Proposition 13.9 and let U be a second open set of $\mathbb{R}^d$ such that $U \Subset \tilde{U}$. We choose $\chi \in \overline{\mathscr{C}}^\infty_c(\tilde{U}_+)$ be such that $\chi \equiv 1$ in a neighborhood of U_+.

We set $w = \tau^{1/2} \chi \Lambda^{-1/2}_{\mathsf{T},\tau} v$. One has

$$\begin{aligned} P_\varphi w &= \tau^{1/2} \big(\chi \Lambda^{-1/2}_{\mathsf{T},\tau} P_\varphi + [P_\varphi, \chi \Lambda^{-1/2}_{\mathsf{T},\tau}]\big) v \\ &= \tau^{1/2} \big(\chi \Lambda^{-1/2}_{\mathsf{T},\tau} G_0 + \sum_{1 \leq i \leq d} \chi \partial_j (\Lambda^{-1/2}_{\mathsf{T},\tau} G^j) + [P_\varphi, \chi \Lambda^{-1/2}_{\mathsf{T},\tau}] v\big) \\ &= \tilde{G}_0 + \sum_{1 \leq i \leq d} \tilde{G}^j, \end{aligned}$$

with

$$\tilde{G}_0 = \tau^{1/2} \big(\chi \Lambda^{-1/2}_{\mathsf{T},\tau} G_0 + \sum_{1 \leq i \leq d} [\chi, \partial_j] (\Lambda^{-1/2}_{\mathsf{T},\tau} G^j) + [P_\varphi, \chi \Lambda^{-1/2}_{\mathsf{T},\tau}] v\big),$$

and $\tilde{G}^j = \tau^{1/2} \chi \Lambda^{-1/2}_{\mathsf{T},\tau} G^j$. We apply Proposition 13.9 to w:

$$\begin{aligned} \text{(13.A.4)} \quad \tau \|\chi \Lambda^{-1/2}_{\mathsf{T},\tau} v\|_{\tau,1} &+ \tau^{3/2} |\chi \Lambda^{-1/2}_{\mathsf{T},\tau} v_{|x_d=0^+}|_{\tau,1/2} \\ &+ \tau^{3/2} |(D_d(\chi \Lambda^{-1/2}_{\mathsf{T},\tau} v)_{|x_d=0^+} - i \chi \Lambda^{-1/2}_{\mathsf{T},\tau} G^d)_{|x_d=0^+}|_{\tau,-1/2} \\ &\leq C\Big(\|\tilde{G}_0\|_+ + \tau \|\tilde{G}\|_+ + \tau^{3/2} |\mathsf{B}_{0,\varphi}(\chi \Lambda^{-1/2}_{\mathsf{T},\tau} v, -\chi \Lambda^{-1/2}_{\mathsf{T},\tau} G^d)|_{\tau,1/2-k}\Big), \end{aligned}$$

We write

$$\begin{aligned} \|v\|_{\tau,1,-1/2} = \|\chi v\|_{\tau,1,-1/2} &= \|\Lambda^{-1/2}_{\mathsf{T},\tau} \chi v\|_{\tau,1} \\ &\lesssim \|\chi \Lambda^{-1/2}_{\mathsf{T},\tau} v\|_{\tau,1} + \|[\Lambda^{-1/2}_{\mathsf{T},\tau}, \chi] v\|_{\tau,1} \\ &\lesssim \|\chi \Lambda^{-1/2}_{\mathsf{T},\tau} v\|_{\tau,1} + \|v\|_{\tau,1,-3/2}. \end{aligned}$$

Thus, for $\tau > 0$ chosen sufficiently large we obtain

$$\|v\|_{\tau,1,-1/2} \lesssim \|\chi \Lambda_{\mathsf{T},\tau}^{-1/2} v\|_{\tau,1}. \tag{13.A.5}$$

Similarly one has

$$|v_{|x_d=0^+}|_{\partial} \lesssim |\chi \Lambda_{\mathsf{T},\tau}^{-1/2} v|_{\tau,1/2}. \tag{13.A.6}$$

Next, one writes

$$\begin{aligned} |(D_d v - iG^d)_{|x_d=0^+}|_{\tau,-1} &= |(D_d(\chi v) - i\chi G^d)_{|x_d=0^+}|_{\tau,-1} \\ &= |\Lambda_{\mathsf{T},\tau}^{-1/2}(D_d(\chi v) - i\chi G^d)_{|x_d=0^+}|_{\tau,-1/2}, \end{aligned}$$

and

$$\begin{aligned} &\Lambda_{\mathsf{T},\tau}^{-1/2}(D_d(\chi v) - i\chi G^d) \\ &\quad = D_d(\chi \Lambda_{\mathsf{T},\tau}^{-1/2} v) - i\chi \Lambda_{\mathsf{T},\tau}^{-1/2} G^d + D_d[\Lambda_{\mathsf{T},\tau}^{-1/2}, \chi] v - i[\Lambda_{\mathsf{T},\tau}^{-1/2}, \chi] G^d \\ &\quad = D_d(\chi \Lambda_{\mathsf{T},\tau}^{-1/2} v) - i\chi \Lambda_{\mathsf{T},\tau}^{-1/2} G^d + [\Lambda_{\mathsf{T},\tau}^{-1/2}, \chi](D_d v - iG^d) + \big[D_d, [\Lambda_{\mathsf{T},\tau}^{-1/2}, \chi]\big] v. \end{aligned}$$

Since $[\Lambda_{\mathsf{T},\tau}^{-1/2}, \chi]$ and $\big[D_d, [\Lambda_{\mathsf{T},\tau}^{-1/2}, \chi]\big]$ are in $\Psi_{\mathsf{T},\tau}^{-3/2}$, it yields

$$\begin{aligned} |(D_d v - iG^d)_{|x_d=0^+}|_{\tau,-1} \lesssim{}& \big|\big(D_d(\chi \Lambda_{\mathsf{T},\tau}^{-1/2} v) - i\chi \Lambda_{\mathsf{T},\tau}^{-1/2} G^d\big)_{|x_d=0^+}\big|_{\tau,-1/2} \\ &+ |(D_d v - iG^d)_{|x_d=0^+}|_{\tau,-2} + |v_{|x_d=0^+}|_{\tau,-2}, \end{aligned}$$

and with (13.A.6) and for $\tau > 0$ chosen sufficiently large we obtain

(13.A.7)

$$\begin{aligned} &|v_{|x_d=0^+}|_{\partial} + |(D_d v - iG^d)_{|x_d=0^+}|_{\tau,-1} \\ &\qquad \lesssim |\chi \Lambda_{\mathsf{T},\tau}^{-1/2} v|_{\tau,1/2} + \big|\big(D_d(\chi \Lambda_{\mathsf{T},\tau}^{-1/2} v) - i\chi \Lambda_{\mathsf{T},\tau}^{-1/2} G^d\big)_{|x_d=0^+}\big|_{\tau,-1/2}. \end{aligned}$$

Combining (13.A.4), (13.A.5) and (13.A.7) we obtain

$$\begin{aligned} &\tau \|v\|_{\tau,1,-1/2} + \tau^{3/2}\big(|v_{|x_d=0^+}|_{\partial} + |(D_d v - iG^d)_{|x_d=0^+}|_{\tau,-1}\big) \\ &\qquad \leq C\Big(\|\tilde{G}_0\|_+ + \tau \|\tilde{G}\|_+ + \tau^{3/2} |\mathsf{B}_{0,\varphi}(\chi \Lambda_{\mathsf{T},\tau}^{-1/2} v, -\chi \Lambda_{\mathsf{T},\tau}^{-1/2} G^d)|_{\tau,1/2-k}\Big). \end{aligned} \tag{13.A.8}$$

Next, we estimate $\|\tilde{G}_0\|_+$. As one has $[P_\varphi, \chi \Lambda_{\mathsf{T},\tau}^{-1/2}] \in \Psi_\tau^{1,-1/2}$ one obtains

$$\|\tilde{G}_0\|_+ \lesssim \|G_0\|_+ + \sum_{1 \leq i \leq d} \|G\|_+ + \tau^{1/2} \|v\|_{\tau,1,-1/2}. \tag{13.A.9}$$

We also have

$$\|\tilde{G}^j\|_+ \lesssim \|G^j\|_+ \lesssim \|G\|_+. \tag{13.A.10}$$

With the definition of $\mathsf{B}_{0,\varphi}$ in (13.2.5) one has

$$
\begin{aligned}
&\mathsf{B}_{0,\varphi}(\chi\Lambda_{\mathsf{T},\tau}^{-1/2}v, -\chi\Lambda_{\mathsf{T},\tau}^{-1/2}G^d)\\
&\quad = \tilde{B}_\varphi^k(\chi\Lambda_{\mathsf{T},\tau}^{-1/2}v)_{|x_d=0^+} - iB_\varphi^{k-1}\big(D_d(\chi\Lambda_{\mathsf{T},\tau}^{-1/2}v) - i\chi\Lambda_{\mathsf{T},\tau}^{-1/2}G^d\big)_{|x_d=0^+}.\\
&\quad = \Lambda_{\mathsf{T},\tau}^{-1/2}\Big(\tilde{B}_\varphi^k v_{|x_d=0^+} - iB_\varphi^{k-1}\big(D_d v - iG^d\big)_{|x_d=0^+}\Big)\\
&\qquad + [\tilde{B}_\varphi^k\chi, \Lambda_{\mathsf{T},\tau}^{-1/2}]v_{|x_d=0^+}\\
&\qquad - i[B_\varphi^{k-1}D_d\chi, \Lambda_{\mathsf{T},\tau}^{-1/2}]v_{|x_d=0^+} - [B_\varphi^{k-1}\chi, \Lambda_{\mathsf{T},\tau}^{-1/2}]G^d{}_{|x_d=0^+}\\
&\quad = \Lambda_{\mathsf{T},\tau}^{-1/2}\mathsf{B}_{0,\varphi}(v, -G^d) + [\tilde{B}_\varphi^k\chi, \Lambda_{\mathsf{T},\tau}^{-1/2}]v_{|x_d=0^+}\\
&\qquad - i\Big([B_\varphi^{k-1}D_d\chi, \Lambda_{\mathsf{T},\tau}^{-1/2}]v_{|x_d=0^+} - [B_\varphi^{k-1}\chi, \Lambda_{\mathsf{T},\tau}^{-1/2}]D_d\Big)v_{|x_d=0^+}\\
&\qquad - i[B_\varphi^{k-1}\chi, \Lambda_{\mathsf{T},\tau}^{-1/2}]\big(D_d v - iG^d\big)_{|x_d=0^+}
\end{aligned}
$$

Observe that

$$
\begin{aligned}
&[B_\varphi^{k-1}D_d\chi, \Lambda_{\mathsf{T},\tau}^{-1/2}]v_{|x_d=0^+} - [B_\varphi^{k-1}\chi, \Lambda_{\mathsf{T},\tau}^{-1/2}]D_d v_{|x_d=0^+}\\
&\quad = B_\varphi^{k-1}[D_d, \chi]\Lambda_{\mathsf{T},\tau}^{-1/2}v_{|x_d=0^+}.
\end{aligned}
$$

If fact, in a neighborhood of $\operatorname{supp}(v)$, one has $\chi \equiv 1$ implying that $\chi D_d v_{|x_d=0^+} = D_d\chi v_{|x_d=0^+} = D_d v_{|x_d=0^+}$ yielding

$$
\begin{aligned}
&[B_\varphi^{k-1}D_d\chi, \Lambda_{\mathsf{T},\tau}^{-1/2}]v_{|x_d=0^+} - [B_\varphi^{k-1}\chi, \Lambda_{\mathsf{T},\tau}^{-1/2}]D_d v_{|x_d=0^+}\\
&\quad = B_\varphi^{k-1}D_d\chi\Lambda_{\mathsf{T},\tau}^{-1/2}v_{|x_d=0^+} - \Lambda_{\mathsf{T},\tau}^{-1/2}B_\varphi^{k-1}D_d\chi v_{|x_d=0^+}\\
&\qquad - B_\varphi^{k-1}\chi\Lambda_{\mathsf{T},\tau}^{-1/2}D_d v_{|x_d=0^+} + \Lambda_{\mathsf{T},\tau}^{-1/2}B_\varphi^{k-1}\chi D_d v_{|x_d=0^+}\\
&\quad = B_\varphi^{k-1}D_d\chi\Lambda_{\mathsf{T},\tau}^{-1/2}v_{|x_d=0^+} - B_\varphi^{k-1}\chi\Lambda_{\mathsf{T},\tau}^{-1/2}D_d v_{|x_d=0^+}\\
&\quad = B_\varphi^{k-1}[D_d, \chi]\Lambda_{\mathsf{T},\tau}^{-1/2}v_{|x_d=0^+}.
\end{aligned}
$$

Since $[\tilde{B}_\varphi^k\chi, \Lambda_{\mathsf{T},\tau}^{-1/2}] \in \Psi_{\mathsf{T},\tau}^{k-3/2}$, $B_\varphi^{k-1}[D_d, \chi]\Lambda_{\mathsf{T},\tau}^{-1/2} \in \Psi_{\mathsf{T},\tau}^{k-3/2}$, and $[B_\varphi^{k-1}\chi, \Lambda_{\mathsf{T},\tau}^{-1/2}] \in \Psi_{\mathsf{T},\tau}^{k-5/2}$ one obtains

(13.A.11)

$$
\begin{aligned}
&|\mathsf{B}_{0,\varphi}(\chi\Lambda_{\mathsf{T},\tau}^{-1/2}v, -\chi\Lambda_{\mathsf{T},\tau}^{-1/2}G^d)|_{\tau,1/2-k}\\
&\quad \lesssim |\mathsf{B}_{0,\varphi}(v, -G^d)|_{\tau,-k} + |v_{|x_d=0^+}|_{\tau,-1} + |\big(D_d v - iG^d\big)_{|x_d=0^+}|_{\tau,-2}.
\end{aligned}
$$

Combining (13.A.8)–(13.A.11) we obtain

$$
\begin{aligned}
\text{(13.A.12)}\quad &\tau\|v\|_{\tau,1,-1/2} + \tau^{3/2}\big(|v_{|x_d=0^+}|_\partial + |(D_d v - iG^d)_{|x_d=0^+}|_{\tau,-1}\big)\\
&\leq C\Big(\|G_0\|_+ + \tau\|G\|_+ + \tau^{3/2}|\mathsf{B}_{0,\varphi}(v, -G^d)|_{\tau,-k}\Big)\\
&\quad + \tau^{1/2}\|v\|_{\tau,1,-1/2} + \tau^{3/2}\big(|v_{|x_d=0^+}|_{\tau,-1} + |\big(D_d v - iG^d\big)_{|x_d=0^+}|_{\tau,-2}\big).
\end{aligned}
$$

We then obtain the result by choosing $\tau > 0$ sufficiently large. ∎

CHAPTER 14

Optimal Estimates at the Boundary

Contents

J. Le Rousseau et al., *Elliptic Carleman Estimates and Applications to Stabilization and Controllability, Volume II*, PNLDE Subseries in Control 98, https://doi.org/10.1007/978-3-030-88670-7_14

In this chapter, we provide a Carleman estimate near the boundary under Lopatinskiĭ–Šapiro conditions. Compared to the results obtained in Chap. 8, the estimate we obtain improves at the level of the boundary terms. We also prove that this improvement is in fact optimal.

14.1. Statement and Proof Scheme

Let $(\mathcal{M}, g)$ be a smooth compact Riemannian manifold with boundary, and let $P = -\Delta_g + R_1$ with R_1 a first-order differential operator with bounded coefficients on $\mathcal{M}$.

THEOREM 14.1. *Let V be an open set of $\mathcal{M}$ and set $V_\partial = V \cap \partial\mathcal{M}$. Let $\varphi \in \mathscr{C}^\infty(\mathcal{M})$ be such that the pair (P, φ) has the sub-ellipticity property of Definition 5.1 in $\overline{V}$. If $V_\partial \neq \emptyset$, consider B a differential operator of order β in V. For $0 \leq k \leq \beta$, denote by ${}^k\partial\mathcal{M}$ the union of the connected components of $\partial\mathcal{M}$ where B is of order k. Moreover, assume that (P, B, φ) satisfies the Lopatinskiĭ–Šapiro condition at all points $m \in \overline{V_\partial}$ (Definition 8.1). Then, there exist C and $\tau_* > 0$ such that*

$$\tau^{-1/2}\|e^{\tau\varphi}u\|_{\tau,2} + \tau^{-1/4}|e^{\tau\varphi_{|\partial\mathcal{M}}}\operatorname{tr}(u)|_{\tau,1,1/2} \leq C\Big(\|e^{\tau\varphi}Pu\|_{L^2(\mathcal{M})} + \tau^{-1/4}\sum_{1\leq k\leq\beta}|e^{\tau\varphi}Bu_{|{}^k\partial\mathcal{M}}|_{\tau,3/2-k}\Big), \tag{14.1.1}$$

for all $u \in \mathscr{C}^\infty(\mathcal{M})$, with $\operatorname{supp}(u) \subset V$, and $\tau \geq \tau_*$.

If one compares the estimates of Theorem 14.1 to that of Theorem 8.14, one finds an improvement of a factor $\tau^{-1/4}$ associated with the trace terms on the r.h.s. of the estimates in the present results. Naturally, with such a factor on the r.h.s., the same factor occurs for the trace term on the l.h.s.

We prove in Sect. 14.5 that the estimate in Theorem 14.1 is optimal with respect to the boundary terms. Note that optimality with respect to the volume terms is proven in Section 4.1.2 of Volume 1.

Arguing as for the proof of Theorem 8.24, we deduce from Theorem 14.1 the following shifted estimate.

THEOREM 14.2. *Under the assumptions of Theorem 14.1, there exist C and $\tau_* > 0$ such that*

$$\tau^{1/2}\|e^{\tau\varphi}u\|_{\tau,1} + \tau^{1/4}|e^{\tau\varphi_{|\partial\mathcal{M}}}\operatorname{tr}(u)|_{\tau,1,0} \leq C\Big(\|e^{\tau\varphi}Pu\|_{L^2(\mathcal{M})} + \tau^{1/4}\sum_{1\leq k\leq\beta}|e^{\tau\varphi}Bu_{|^k\partial\mathcal{M}}|_{\tau,1-k}\Big),$$

for all $u \in \mathscr{C}^\infty(\mathcal{M})$, with $\operatorname{supp}(u) \subset V$, *and $\tau \geq \tau_*$.*

14.1.1. Local Estimates and Patching. As in Chap. 8, we prove a similar estimate in the neighborhood of a point m^0 of the boundary $\partial\mathcal{M}$.

PROPOSITION 14.3. *Let $m^0 \in \partial\mathcal{M}$ and V^0 be an open neighborhood of m^0 in $\mathcal{M}$ that meets one connected component of $\partial\mathcal{M}$. Let $\varphi \in \mathscr{C}^\infty(\mathcal{M})$ be such that the pair (P,φ) has the sub-ellipticity property of Definition 5.1 in $\overline{V^0}$. Consider B a differential operator of order k in V^0 of the form of (8.1.2). Moreover, assume that (P,B,φ) satisfies the Lopatinskiĭ–Šapiro condition at m^0 (Definition 8.1). Then there exist a neighborhood V^1 of m^0 in $\mathcal{M}$ and two constants C and $\tau_* > 0$ such that*

$$\tau^{-1/2}\|e^{\tau\varphi}u\|_{\tau,2} + \tau^{-1/4}|e^{\tau\varphi_{|\partial\mathcal{M}}}\operatorname{tr}(u)|_{\tau,1,1/2} \leq C\Big(\|e^{\tau\varphi}Pu\|_{L^2(\mathcal{M})} + \tau^{-1/4}|e^{\tau\varphi}Bu_{|\partial\mathcal{M}}|_{\tau,3/2-k}\Big), \tag{14.1.2}$$

for all $u \in \mathscr{C}^\infty(\mathcal{M})$, with $\operatorname{supp}(u) \subset V^1$, *and $\tau \geq \tau_*$.*

Then, arguing as in Sect. 5.3, for Theorems 5.5 and 5.6, patching together estimates as above, we obtain the result of Theorem 14.1.

We follow the setting and notation of Sect. 8.3. We consider $m^0 \in \partial\mathcal{M}$, V^0 an open neighborhood of m^0 in $\mathcal{M}$ where the sub-ellipticity property for the pair (P,φ) holds, and $\mathcal{C} = (O,\kappa)$ a local chart at the boundary associated with coordinates (x', x_d) as described by (8.3.3), with $m^0 \in O$. With the local representative of the principal symbols $p^{\mathcal{C}}$, $b^{\mathcal{C}}$ of P, and B as given in (8.3.4), it suffices to prove the following estimate.

PROPOSITION 14.4. *Let $P_0 = \operatorname{Op}(p^{\mathcal{C}})$ and $B_0 = \operatorname{Op}(b^{\mathcal{C}})$. Let $m^0 \in \partial\mathcal{M}$ and V^0 be an open neighborhood of m^0 in $\mathcal{M}$ that meets one connected component of $\partial\mathcal{M}$. Set $x^0 = \kappa(m^0)$. Assume that $(P_0, \varphi^{\mathcal{C}})$ has the sub-ellipticity property of Definition 3.2 of Volume 1 in $\overline{U^0}$ with $U^0 = \kappa(V^0 \cap O)$. Moreover, assume that $(P_0, B_0, \varphi^{\mathcal{C}})$ satisfies the Lopatinskiĭ–Šapiro condition at x^0 (Definition 8.1). Then, there exist a bounded open neighborhood U_+ of x^0 in $\overline{\mathbb{R}^d_+}$ such that $U_+ \subset U^0$ and two constants C and $\tau_* > 0$ such that*

$$\tau^{-1/2}\|e^{\tau\varphi^{\mathcal{C}}}u\|_{\tau,2} + \tau^{-1/4}|e^{\tau\varphi^{\mathcal{C}}_{|x_d=0^+}}\operatorname{tr}(u)|_{\tau,1,1/2} \leq C\Big(\|e^{\tau\varphi^{\mathcal{C}}}P_0u\|_+ + \tau^{-1/4}|e^{\tau\varphi^{\mathcal{C}}}B_0u_{|x_d=0^+}|_{\tau,3/2-k}\Big), \tag{14.1.3}$$

for all $u \in \overline{\mathscr{C}}_c^\infty(U_+)$ *and* $\tau \geq \tau_*$.

We recall that for an open subset U_+ of $\overline{\mathbb{R}^d_+}$, the space $\overline{\mathscr{C}}_c^\infty(U_+)$ is introduced in (8.3.6).

The proof of Proposition 14.4 relies on a microlocal version of the estimate that we state below and it follows the proof scheme of Proposition 8.16 by patching together such microlocal estimates as in Sect. 8.3.5.

14.1.2. Microlocal Estimate. Using the local setting and notation of Sect. 8.3, we prove the following result. In particular, in the considered local chart, we define the conjugated operators of P_0 and B_0:

$$P_{0,\varphi} = e^{\tau\varphi^{\mathcal{C}}} P_0 e^{-\tau\varphi^{\mathcal{C}}}, \qquad B_{0,\varphi} = e^{\tau\varphi^{\mathcal{C}}} B_0 e^{-\tau\varphi^{\mathcal{C}}},$$

with principal symbols $p_{0,\varphi}$ and $b_{0,\varphi}$, respectively.

PROPOSITION 14.5. *Let* $x^0 = \kappa(m^0)$ *be such that* $x^0_d = 0$. *Assume that* $(P_0, \varphi^{\mathcal{C}})$ *has the sub-ellipticity property of Definition 3.2 of Volume 1 in* $\overline{U^0}$ *with* $U^0 = \kappa(V^0 \cap O)$. *Let* $\omega^{0\prime} \in T^*_{m^0}\partial\mathcal{M}$, *with local representative* $(\omega^{0\prime})^{\mathcal{C}} = \xi^{0\prime} \in \mathbb{R}^{d-1}$, *and* $\tau^0 \geq 0$ *such that* $(\xi^{0\prime}, \tau^0) \neq 0$, *and assume that the Lopatinskiĭ–Šapiro condition of Definition 8.1 holds at* $(m^0, \omega^{0\prime}, \tau^0)$. *Then, there exists* $\mathscr{U}$ *a conic open neighborhood of* $(x^0, \xi^{0\prime}, \tau^0)$ *in* $U^0 \times \mathbb{R}^{d-1} \times \mathbb{R}_+$ *such that for* $\chi \in S^0_{\mathsf{T},\tau}$, *homogeneous of degree* 0, *with* $\operatorname{supp}(\chi) \subset \mathscr{U}$, *there exist* $C > 0$ *and* $\tau_* > 0$ *such that*

$$\tag{14.1.4} \begin{aligned} &\tau^{-1/2}\|\operatorname{Op}_{\mathsf{T}}(\chi)v\|_{\tau,2} + \tau^{-1/4}|\operatorname{tr}(\operatorname{Op}_{\mathsf{T}}(\chi)v)|_{\tau,1,1/2} \\ &\qquad \leq C\Big(\|P_{0,\varphi}v\|_+ + \tau^{-1/4}|B_{0,\varphi}v_{|x_d=0^+}|_{\tau,3/2-k} + \|v\|_{\tau,2,-1}\Big), \end{aligned}$$

for $\tau \geq \tau_*$, $v \in \overline{\mathscr{S}}(\mathbb{R}^d_+)$.

The proof of Proposition 14.5 is given in Sect. 14.4. In some microlocal regions, the key aspect of the proof lies in the use of an estimate for a first-order operator that improves upon what can be found in Sect. 6.4.

14.1.3. An Improved Estimate for a First-Order Sub-elliptic Factor. Let $f(\varrho') \in S^1_{\mathsf{T},\tau}$ be homogeneous of degree one and $L = D_d - \operatorname{Op}_{\mathsf{T}}(f)$ with principal symbol $\ell(\varrho) = \xi_d - f(\varrho')$. We introduce $\operatorname{Re} F = \frac{1}{2}(\operatorname{Op}_{\mathsf{T}}(f) + \operatorname{Op}_{\mathsf{T}}(f)^*)$ and $\operatorname{Im} F = \frac{1}{2i}(\operatorname{Op}_{\mathsf{T}}(f) - \operatorname{Op}_{\mathsf{T}}(f)^*)$ both formally selfadjoint and

$$L_2 = \frac{1}{2}(L + L^*) = D_d - \operatorname{Re} F \in \Psi^{1,0}_\tau,$$
$$L_1 = \frac{1}{2i}(L - L^*) = -\operatorname{Im} F \in \Psi^1_{\mathsf{T},\tau}$$

also (formally) selfadjoint. Their respective principal symbols are

$$\ell_2(\varrho) = \operatorname{Re}\ell(\varrho) = \xi_d - \operatorname{Re} f(\varrho') \in S^{1,0}_\tau,$$
$$\ell_1(\varrho') = \operatorname{Im}\ell(\varrho) = -\operatorname{Im} f(\varrho') \in S^1_{\mathsf{T},\tau}.$$

Let $\mathscr{U}$ be a conic open set of $\mathbb{R}^d_+ \times \mathbb{R}^{d-1} \times \mathbb{R}_+$. We assume that we have, for some $C > 0$,

$$\lambda_{\mathsf{T},\tau}(\varrho') \le C\tau, \qquad \varrho' \in \mathscr{U}, \tag{14.1.5}$$

and that the following sub-ellipticity property holds:

$$\ell(\varrho) = 0 \quad \Rightarrow \quad \frac{1}{2i}\{\overline{\ell}, \ell\}(\varrho') = \{\ell_2, \ell_1\}(\varrho') > 0, \qquad \varrho' \in \overline{\mathscr{U}},\ \varrho = (\varrho', \xi_d). \tag{14.1.6}$$

As in Sect. 6.4.3, this latter property reads

$$\ell_1(\varrho') = 0 \quad \Rightarrow \quad \{\ell_2, \ell_1\}(\varrho') > 0, \quad \varrho' \in \overline{\mathscr{U}}. \tag{14.1.7}$$

We have the following estimate.

PROPOSITION 14.6. *Let $\mathscr{U}$ be as above with the additional assumption that $\mathbb{S}_{\overline{\mathscr{U}}}$ is compact. Assume that* (14.1.5) *and* (14.1.7) *hold. Let $\chi \in S^0_{\mathsf{T},\tau}$, homogeneous of degree* 0*, be such that* $\operatorname{supp}(\chi) \subset \mathscr{U}$. *Let $s \in \mathbb{R}$. There exist $C > 0$ and $\tau_* > 0$ such that for any $N \in \mathbb{N}$, there exists $C_N > 0$ such that*

$$\begin{aligned} &\tau^{-1/2} \|\mathrm{Op}_{\mathsf{T}}(\chi) u\|_{\tau,1,s} \\ &\quad \le C\Big(\|L \mathrm{Op}_{\mathsf{T}}(\chi) u\|_{\tau,0,s} + \tau^{-1/4} |\mathrm{Op}_{\mathsf{T}}(\chi) u_{|x_d=0^+}|_{\tau,s+1/2} \Big) + C_N \|u\|_{\tau,0,-N}, \end{aligned} \tag{14.1.8}$$

for $\tau \ge \tau_*$ and $u \in \overline{\mathscr{S}}(\mathbb{R}^d_+)$.

If one compares with the result of Lemma 6.22, in the case $s = 0$, one observes the improvement by the factor $\tau^{-1/4}$ for the trace term. This improvement is optimal; we give an indirect explanation of this optimality in Remark 14.28.

The proof of Proposition 14.6 is quite involved and is given in Sect. 14.3 below. It relies on general and sharp pseudo-differential methods introduced by L. Hörmander.

14.2. Some Elements of Hörmander Calculus

Most of the material presented here is adapted from Sections 18.4 to 18.6 in the book by L. Hörmander [175]. The calculus of pseudo-differential operators introduced therein is very general. For example, the calculus of operators with a large operators presented in Chapter 2 of Volume 1 and used throughout both Volumes 1 and 2 can be expressed in terms of the Hörmander calculus. Note, however, that, because of its complexity, it would have been a bit awkward to introduce this calculus at an early stage of this book.

14.2.1. Metrics and Order Functions. As in the previous chapters, we write $\varrho = (x, \xi, \tau) \in \mathbb{R}^{2d} \times [1, +\infty)$ and $\varrho' = (x, \xi', \tau) \in \mathbb{R}^{2d-1} \times [1, +\infty)$.

Recalling that $\lambda_{\mathsf{T},\tau}(\varrho) = \lambda_{\mathsf{T},\tau}(\varrho') = (|\xi'|^2 + \tau^2)^{1/2}$, we introduce some notation. We define the following metrics in phase space:

$$g_\varrho = |dx|^2 + \lambda_{\mathsf{T},\tau}^{-2}(\varrho')|d\xi|^2 \ \text{ and } \ g^\gamma_\varrho = \tau\gamma^{-1}|dx|^2 + \tau\gamma^{-1}\lambda_{\mathsf{T},\tau}^{-2}(\varrho')|d\xi|^2,$$

for $\gamma \geq 1$. For the first metric, this reads

$$g_\varrho(z,\zeta) = |z|^2 + \lambda_{\mathsf{T},\tau}^{-2}(\varrho')|\zeta|^2, \qquad (z,\zeta) \in \mathbb{R}^{2d}.$$

In proofs below, the parameter γ will be chosen large. The parameter τ will also be chosen large as in the other chapters with $\tau \geq \gamma$. Note that one has

$$g_\varrho \leq g_\varrho^\gamma.$$

For such a metric $\mathbf{g}_\varrho$, the associated dual metric is defined through

$$\mathbf{g}_\varrho^\sigma(y,\eta) = \sup_{(z,\zeta)} \frac{\big((y,\eta)\cdot(z,\zeta)\big)^2}{g_\varrho\big(A(z,\zeta)\big)},$$

with $A = \begin{pmatrix} 0 & I_d \\ -I_d & 0 \end{pmatrix}$ the $(2d)\times(2d)$ matrix associated with the symplectic 2-form σ on $\mathbb{R}^{2d}$ (see Sect. 15.7.1), that is,

$$\sigma\big((z,\zeta),(y,\eta)\big) = \big(A(z,\zeta)\big)\cdot(y,\eta) = \zeta\cdot y - z\cdot\eta.$$

The dual metrics g_ϱ^σ and $(g^\gamma)_\varrho^\sigma$ are given by

(14.2.1)
$$g_\varrho^\sigma = \lambda_{\mathsf{T},\tau}^2(\varrho')|dx|^2 + |d\xi|^2 \ \text{ and } \ (g^\gamma)_\varrho^\sigma = \lambda_{\mathsf{T},\tau}^2(\varrho')\tau^{-1}\gamma|dx|^2 + \tau^{-1}\gamma|d\xi|^2,$$

and one defines $h(\varrho)$ and $h^\gamma(\varrho)$ by

$$h(\varrho)^2 = \sup_{(y,\eta')} (g/g^\sigma)_\varrho(y,\eta) = \lambda_{\mathsf{T},\tau}^{-2}(\varrho') \leq 1,$$
$$h^\gamma(\varrho)^2 = \sup_{(y,\eta)} (g^\gamma/(g^\gamma)^\sigma)_\varrho(y,\eta) = (\tau/\gamma)^2\lambda_{\mathsf{T},\tau}^{-2}(\varrho') \leq 1/\gamma^2 \leq 1.$$

Note that

$$(g^\gamma)_\varrho^\sigma \leq g_\varrho^\sigma.$$

As we shall see below, the metric g (respectively, g^γ) and its dual metric g^σ (respectively, $(g^\gamma)^\sigma$) satisfy the necessary properties given by L. Hörmander [175, Sections 18.4-18.5] allowing one to define symbol classes and the associated pseudo-differential operators and calculus. With the metric g, one recovers in fact the tangential symbol classes $S_{\mathsf{T},\tau}^m$ introduced in Section 2.10 of Volume 1 and the associated operators (see below).

Let $\mathbf{g}_\varrho$ be a metric acting on $\mathbb{R}^{2d}$ as above. Following Definition 18.4.1 in [175], it is said to be slowly varying if there exist $\varepsilon > 0$ and $C > 0$ such that

(14.2.2)
$$\mathbf{g}_\varrho(y,\eta) < \varepsilon \ \Rightarrow \ \mathbf{g}_{\tilde\varrho} \leq C\mathbf{g}_\varrho, \qquad \text{for } \tilde\varrho = (x+y,\xi+\eta,\tau),$$
$$\forall\ \varrho = (x,\xi,\tau) \in \mathbb{R}^{2d}\times[1,+\infty),\ (y,\eta),\in \mathbb{R}^{2d}.$$

Following Definition 18.4.7 in [175], it is said to be σ-temperate if there exist $C > 0$ and $N > 0$ such that

$$\text{(14.2.3)} \quad \mathsf{g}_\varrho \le C\big(1 + \mathsf{g}^\sigma_\varrho(x - \tilde{x}, \xi - \tilde{\xi})\big)^N \mathsf{g}_{\tilde{\varrho}}, \quad \forall\ \varrho = (x, \xi, \tau),\ \tilde{\varrho} = (\tilde{x}, \tilde{\xi}, \tau) \in \mathbb{R}^{2d} \times [1, +\infty).$$

PROPOSITION 14.7. *The metrics g and g^γ are slowly varying and σ-temperate.*

We refer to Sect. 14.A.1 for a proof.

Let g be a slowly varying and σ-temperate metric on $\mathbb{R}^{2d}$, here g or g^γ. A positive function $m(\varrho)$ (possibly depending on the parameter γ) is said to be g-continuous if there exist $\varepsilon > 0$ and $C > 0$ such that

$$\text{(14.2.4)} \quad \mathsf{g}_\varrho(y, \eta) < \varepsilon \quad \Rightarrow \quad m(\varrho)/C \le m(\tilde{\varrho}) \le C m(\varrho), \quad \text{with } \tilde{\varrho} = (x + y, \xi + \eta, \tau), \quad \forall\ \varrho \in \mathbb{R}^{2d} \times [1, +\infty),\ (y, \eta) \in \mathbb{R}^{2d}.$$

REMARK 14.8. Since $g_\varrho \le g^\gamma_\varrho$, a function that is g-continuous is also g^γ-continuous.

Note also that if $m(\varrho)$ is independent of γ and is g^γ-continuous, then it is g-continuous, as can be seen by taking $\gamma = \tau$.

A positive function $m(\varrho)$ is said to be σ, g-temperate if there exist $C > 0$ and $N > 0$ such that

$$\text{(14.2.5)} \quad m(\varrho) \le C\big(1 + \mathsf{g}^\sigma_\varrho(x - \tilde{x}, \xi - \tilde{\xi})\big)^N m(\tilde{\varrho}), \quad \forall\ \varrho = (x, \xi, \tau),\ \tilde{\varrho} = (\tilde{x}, \tilde{\xi}, \tau) \in \mathbb{R}^{2d} \times [1, +\infty).$$

REMARK 14.9. Since $(g^\gamma)^\sigma_\varrho \le g^\sigma_\varrho$, a function that is σ, g^γ-temperate is also σ, g-temperate.

A function $m(\varrho)$ that is both g-continuous and σ, g-temperate will be called a g-admissible order function or simply an order function if there is no ambiguity with respect to the used metric.

PROPOSITION 14.10. *Let $\mathsf{g} = g$ or g^γ. Let $r \in \mathbb{R}$. The functions $\lambda_\tau(\varrho)^r$ and $\lambda_{\mathsf{T},\tau}(\varrho')^r$ are both g-admissible order functions.*

If $m_1(\varrho)$ and $m_2(\varrho)$ are g-admissible order functions, then so is $m_1 m_2(\varrho)$.

We refer to Sect. 14.A.2 for a proof.

In what follows, we shall mainly use $\tau^a \gamma^b \lambda_{\mathsf{T},\tau}(\varrho')^c$, with $a, b, c \in \mathbb{R}$, as an order function.

14.2.2. Symbols and Pseudo-Differential Operators.

DEFINITION 14.11 (Symbols for the Metrics g and g^γ). Consider $m(\varrho)$ a positive g-admissible order function. Recall that $\varrho = (x, \xi, \tau)$. Let $a(\varrho) \in$

$\mathscr{C}^\infty(\mathbb{R}^{2d})$ with $\tau \in [1,+\infty)$ acting as a parameter. One says that $a(\varrho) \in S(m,g)$ if one has, for $\alpha, \beta \in \mathbb{N}^d$,

$$(14.2.6)\quad |\partial_x^\alpha \partial_\xi^\beta a(\varrho)| \leq C_{\alpha,\beta} m(\varrho) \lambda_{\mathsf{T},\tau}(\varrho')^{-|\beta|},\qquad \varrho = (x,\xi,\tau) \in \mathbb{R}^{2d} \times [1,+\infty),\quad \varrho' = (x,\xi',\tau).$$

Consider $m(\varrho,\gamma)$ a positive g^γ-admissible order function. Let $a(\varrho,\gamma) \in \mathscr{C}^\infty(\mathbb{R}^{2d})$ with $\tau, \gamma \in [1,+\infty)$, acting as parameters. One says that $a(\varrho,\gamma) \in S(m,g^\gamma)$ if one has, for $\alpha, \beta \in \mathbb{N}^d$,

$$(14.2.7)\quad |\partial_x^\alpha \partial_\xi^\beta a(\varrho,\gamma)| \leq C_{\alpha,\beta} m(\varrho,\gamma) (\tau\gamma^{-1})^{(|\alpha|+|\beta|)/2} \lambda_{\mathsf{T},\tau}(\varrho')^{-|\beta|},\qquad \varrho \in \mathbb{R}^{2d} \times [1,+\infty),\ \gamma \in [1,+\infty),\ \tau \geq \gamma.$$

Here, we are interested in the counterpart tangential symbols for the metrics g and g^γ. Consider $m_{\mathsf{T}}(\varrho')$ a positive g-admissible order function. Recall that $\varrho' = (x,\xi',\tau)$. Let $a(\varrho') \in S(m_{\mathsf{T}},g)$. Observe that Definition 14.11 reads in this case

$$(14.2.8)\quad |\partial_x^\alpha \partial_{\xi'}^{\beta'} a(\varrho')| \leq C_{\alpha,\beta'} m_{\mathsf{T}}(\varrho') \lambda_{\mathsf{T},\tau}(\varrho')^{-|\beta'|},\qquad \varrho' \in \mathbb{R}^{2d-1} \times [1,+\infty),$$

for $\alpha \in \mathbb{N}^d$ and $\beta' \in \mathbb{N}^{d-1}$.

Finally, for $m_{\mathsf{T}}(\varrho',\gamma)$ a positive g^γ-admissible order function, if $a(\varrho',\gamma) \in S_{\mathsf{T}}(m_{\mathsf{T}},g^\gamma)$, then

$$(14.2.9)\quad |\partial_x^\alpha \partial_{\xi'}^{\beta'} a(\varrho',\gamma)| \leq C_{\alpha,\beta'} m_{\mathsf{T}}(\varrho',\gamma) (\tau\gamma^{-1})^{(|\alpha|+|\beta'|)/2} \lambda_{\mathsf{T},\tau}(\varrho')^{-|\beta'|},\qquad \varrho' \in \mathbb{R}^{2d-1} \times [1,+\infty),\ \gamma \in [1,+\infty),\ \tau \geq \gamma \geq 1,$$

for $\alpha \in \mathbb{N}^d$ and $\beta' \in \mathbb{N}^{d-1}$.

REMARK 14.12. In the case $m_{\mathsf{T}}(\varrho') = \lambda_{\mathsf{T},\tau}^m$, (14.2.8) means that $S(\lambda_{\mathsf{T},\tau}^m, g) = S_{\mathsf{T},\tau}^m$, with the notation of Definition 2.39 of Volume 1.

Note that in the case $m(\varrho) = \lambda_\tau^m$, one does not have $S(\lambda_\tau^m, g) \neq S_\tau^m$, with S_τ^m as in Definition 2.1 (also in Volume 1).

REMARK 14.13. Observe that if $m(\varrho)$ (respectively, $m_{\mathsf{T}}(\varrho')$) is both g-admissible and g^γ-admissible, then one has $S(m,g) \subset S(m,g^\gamma)$ (respectively, $S(m_{\mathsf{T}},g) \subset S(m_{\mathsf{T}},g^\gamma)$). An example is given by $m(\varrho) = \tau^a\gamma^b\lambda_\tau(\varrho)$ (respectively, $m_{\mathsf{T}}(\varrho') = \tau^a\gamma^b\lambda_{\mathsf{T},\tau}(\varrho')$).

For $a(\varrho) \in S(m,g)$ (respectively, $a(\varrho,\gamma) \in S(m,g^\gamma)$), one sets the pseudo-differential operator

$$a(x,D,\tau)u(x) = \mathrm{Op}(a)u(x) := (2\pi)^{-d} \int_{\mathbb{R}^d} e^{ix\cdot\xi} a(\varrho)\ \hat{u}(\xi)\ d\xi$$

$$(\text{resp. } a(x,D,\tau,\gamma)u(x) = \mathrm{Op}(a)u(x) := (2\pi)^{-d} \int_{\mathbb{R}^d} e^{ix\cdot\xi} a(\varrho,\gamma)\ \hat{u}(\xi)\ d\xi).$$

We denote by $\Psi(m, g)$ (respectively, $\Psi(m, g^\gamma)$) the space of these operators.

For $m_\mathsf{T}(\varrho')$ a tangential g-admissible order function and $a(\varrho') \in S(m_\mathsf{T}, g)$, one sets the tangential pseudo-differential operator

$$\begin{aligned} a(x, D', \tau)u(x) &= \operatorname{Op}_\mathsf{T}(a)u(x) \\ &:= (2\pi)^{1-d} \int_{\mathbb{R}^{d-1}} e^{ix'\cdot\xi'} a(\varrho')\, \hat{u}(\xi', x_d)\, d\xi' \\ &= (2\pi)^{1-d} \iint_{\mathbb{R}^{d-1}\times\mathbb{R}^{d-1}} e^{i(x'-y')\cdot\xi'} a(\varrho')\, u(y', x_d)\, dy'\, d\xi'. \end{aligned}$$

We denote by $\Psi(m_\mathsf{T}, g)$ the space of these operators.

For $m_\mathsf{T}(\varrho', \gamma)$ a tangential g^γ-admissible order function and $a(\varrho, \gamma) \in S(m_\mathsf{T}, g^\gamma)$, one sets

$$\begin{aligned} a(x, D', \tau, \gamma)u(x) = \operatorname{Op}_\mathsf{T}(a)u(x) &:= (2\pi)^{1-d} \int_{\mathbb{R}^{d-1}} e^{ix'\cdot\xi'} a(\varrho', \gamma)\, \hat{u}(\xi', x_d)\, d\xi' \\ &= (2\pi)^{1-d} \iint_{\mathbb{R}^{d-1}\times\mathbb{R}^{d-1}} e^{i(x'-y')\cdot\xi'} a(\varrho', \gamma)\, u(y', x_d)\, dy'\, d\xi'. \end{aligned}$$

We denote by $\Psi(m_\mathsf{T}, g^\gamma)$ the space of these operators.

In what follows, we shall only be interested in tangential symbols and operators.

14.2.3. Symbol Calculus. For the results of this section, we refer to Sections 18.4 and 18.5 in [175].

PROPOSITION 14.14 (Formal Adjoint). *Let $m_\mathsf{T}(\varrho', \gamma)$ be a g^γ-admissible order function and $a(\varrho', \gamma) \in S(m_\mathsf{T}, g^\gamma)$, with $\varrho' = (x, \xi', \tau)$. Then $\operatorname{Op}_\mathsf{T}(a)^* = \operatorname{Op}_\mathsf{T}(a^*)$ for a certain $a^* \in S(m_\mathsf{T}, g^\gamma)$ and*

$$a^* \sim \sum_\alpha \frac{1}{i^{|\alpha|}\alpha!} \partial_{x'}^\alpha \partial_{\xi'}^\alpha \bar{a},$$

meaning, for any $N \in \mathbb{N}$,

$$a^* = \sum_{|\alpha|\leq N} \frac{1}{i^{|\alpha|}\alpha!} \partial_{x'}^\alpha \partial_{\xi'}^\alpha \bar{a} + R_N,$$

with $R_N(\varrho', \gamma) \in S(m_\mathsf{T}(h^\gamma)^{N+1}, g^\gamma) = S(m_\mathsf{T}(\tau/\gamma)^{N+1}\lambda_{\mathsf{T},\tau}^{-N-1}, g^\gamma)$.

In particular, $a^* - \overline{a} \in S(m_\mathsf{T}\tau(\gamma\lambda_{\mathsf{T},\tau})^{-1}, g^\gamma)$.

PROPOSITION 14.15. *Let $m_{\mathsf{T},1}(\varrho', \gamma)$ and $m_{\mathsf{T},2}(\varrho', \gamma)$ be two g^γ-admissible order functions and $a_1(\varrho', \gamma) \in S(m_{\mathsf{T},1}, g^\gamma)$ and $a_2(\varrho', \gamma) \in S(m_{\mathsf{T},2}, g^\gamma)$. Then, $\operatorname{Op}_\mathsf{T}(a_1) \circ \operatorname{Op}_\mathsf{T}(a_2) = \operatorname{Op}_\mathsf{T}(a)$ with $a \in S(m_{\mathsf{T},1}m_{\mathsf{T},2}, g^\gamma)$ and*

$$a = a_1 \circ a_2 \sim \sum_{\alpha'\in\mathbb{N}^{d-1}} \frac{1}{i^{|\alpha'|}(\alpha')!}\, \partial_{\xi'}^{\alpha'} a_1\, \partial_{x'}^{\alpha'} a_2, \tag{14.2.10}$$

meaning, for any $N \in \mathbb{N}$,

$$a(\varrho', \gamma) = \sum_{|\alpha'| \leq N} \frac{1}{i^{|\alpha'|}(\alpha')!} \partial_{\xi'}^{\alpha'} a_1 \, \partial_x^{\alpha'} a_2 + R_N,$$

with the remainder $R_N \in S(m_{\mathsf{T},1} m_{\mathsf{T},2} (h^\gamma)^{N+1}, g^\gamma) = S(m_{\mathsf{T},1} m_{\mathsf{T},2} (\tau/\gamma)^{N+1} \lambda_{\mathsf{T},\tau}^{-N-1}, g^\gamma)$.

We use the notation "$\circ$" as introduced in the statement of Theorem 2.22 of Volume 1.

REMARK 14.16. Observe that $h^\gamma = \gamma^{-1} \tau \lambda_{\mathsf{T},\tau}^{-1} \eqsim \gamma^{-1}$ if $|\xi'| \leq \tau$. In such a region, one has R_N behaving as a symbol in $S(m_{\mathsf{T},1} m_{\mathsf{T},2} \gamma^{-N-1}, g^\gamma)$. Thus, for example, for $m_{\mathsf{T},1} = \lambda_{\mathsf{T},\tau}^a$ and $m_{\mathsf{T},1} = \lambda_{\mathsf{T},\tau}^b$, one finds the behavior of a symbol in $S(\lambda_{\mathsf{T},\tau}^{a+b} \gamma^{-N-1}, g^\gamma)$. This is in contrast with the symbol calculus in $S_{\mathsf{T},\tau}^m = S(\lambda_{\mathsf{T},\tau}^m, g)$ for which the counterpart remainder is in $S(\lambda_{\mathsf{T},\tau}^{a+b-N-1}, g) = S_{\mathsf{T},\tau}^{a+b-N-1}$. Consequently, the symbol calculus associated with the metric g^γ yields a weaker gain in the symbol decay in the microlocal region $|\xi'| \leq \tau$ since there $\gamma^{-1} \geq \tau^{-1} \geq \lambda_{\mathsf{T},\tau}^{-1}$.

In sections below, we shall need to compose operators associated with different calculi, say for an operator in $\Psi(m_{\mathsf{T},1}, g)$ and an operator in $\Psi(m_{\mathsf{T},2}, g^\gamma)$. This is provided by the following theorem.

THEOREM 14.17. *Let $m_{\mathsf{T},1}(\varrho')$ be a g-admissible order function, and let $m_{\mathsf{T},2}(\varrho', \gamma)$ be a g^γ-admissible order function. If $a_1(\varrho') \in S(m_{\mathsf{T},1}, g)$ and $a_2(\varrho', \gamma) \in S(m_{\mathsf{T},2}, g^\gamma)$, then $\operatorname{Op}_{\mathsf{T}}(a_1) \circ \operatorname{Op}_{\mathsf{T}}(a_2) = \operatorname{Op}_{\mathsf{T}}(a)$ and $\operatorname{Op}_{\mathsf{T}}(a_2) \circ \operatorname{Op}_{\mathsf{T}}(a_1) = \operatorname{Op}_{\mathsf{T}}(\tilde{a})$ with both $a = a_1 \circ a_2$ and $\tilde{a} = a_2 \circ a_1$ in $S(m_{\mathsf{T},1} m_{\mathsf{T},2}, g^\gamma)$, with an asymptotic expansion as in (14.2.10) with, in each case, a remainder term in $S(m_{\mathsf{T},1} m_{\mathsf{T},2} (\tau/\gamma)^{(N+1)/2} \lambda_{\mathsf{T},\tau}^{-N-1}, g^\gamma)$.*
In particular, one has

(1) $a_1 \circ a_2 = a_1 a_2 \mod S(m_{\mathsf{T},1} m_{\mathsf{T},2} (\tau/\gamma)^{1/2} \lambda_{\mathsf{T},\tau}^{-1}, g^\gamma)$.
(2) $a_2 \circ a_1 = a_1 a_2 \mod S(m_{\mathsf{T},1} m_{\mathsf{T},2} (\tau/\gamma)^{1/2} \lambda_{\mathsf{T},\tau}^{-1}, g^\gamma)$.
(3) $a_1 \circ a_2 = a_1 a_2 - i \nabla_{\xi'} a_1 \cdot \nabla_{x'} a_2 \mod S(m_{\mathsf{T},1} m_{\mathsf{T},2} \tau \gamma^{-1} \lambda_{\mathsf{T},\tau}^{-2}, g^\gamma)$.
(4) $a_2 \circ a_1 = a_1 a_2 - i \nabla_{\xi'} a_2 \cdot \nabla_{x'} a_1 \mod S(m_{\mathsf{T},1} m_{\mathsf{T},2} \tau \gamma^{-1} \lambda_{\mathsf{T},\tau}^{-2}, g^\gamma)$.
(5) $a_1 \circ a_2 - a_2 \circ a_1 = -i \{a_1, a_2\} \mod S(m_{\mathsf{T},1} m_{\mathsf{T},2} \tau \gamma^{-1} \lambda_{\mathsf{T},\tau}^{-2}, g^\gamma)$.

The notation $\{.,.\}$ refers to the Poisson bracket (see Sect. 15.7.2). For a proof, we refer to Appendix 14.B.3.

REMARK 14.18. If $m_{\mathsf{T},1}$ is both g-admissible and g^γ-admissible and $a_1 \in S(m_{\mathsf{T},1}, g)$, then one also has $a_1 \in S(m_{\mathsf{T},1}, g^\gamma)$ by Remark 14.13. Then, the composition symbol $a_1 \circ a_2$ can also be analyzed by means of Proposition 14.15. However, the results therein are weaker than those provided by Theorem 14.17 that exploits the differences in the metrics.

14.2.4. Sobolev Bounds and Positivity Inequalities. Concerning Sobolev bounds, the basic result is the following one.

THEOREM 14.19. *Let $a(\varrho',\gamma) \in S(1, g^\gamma)$, then $\mathrm{Op}_{\mathsf{T}}(a)$ is bounded on $L^2(\mathbb{R}^d_+)$.*

For a proof in the case of a general slowly varying σ-temperate metric, we refer to Theorem 18.6.3 in [175].

COROLLARY 14.20. *Let $s, r \in \mathbb{R}$ and $a(\varrho',\gamma) \in S(\lambda^r_{\mathsf{T},\tau}, g^\gamma)$. Then, there exists $C > 0$ such that*

$$\|\mathrm{Op}_{\mathsf{T}}(a)u\|_{\tau,0,s} \leq C\|u\|_{\tau,0,s+r}, \qquad u \in \overline{\mathscr{S}}(\mathbb{R}^d_+), \quad \tau \geq \gamma \geq 1.$$

The tangential norm $\|.\|_{\tau,0,s}$ is defined in Sect. 6.2.

For the operator classes we have introduced, the Gårding inequality takes the following form.

THEOREM 14.21 (Gårding Inequality). *Let $m_{\mathsf{T}}(\varrho',\gamma)$ be a g^γ-admissible order function, and let $a(\varrho',\gamma) \in S(m_{\mathsf{T}}, g^\gamma)$ be such that $\mathrm{Re}\, a(\varrho',\gamma) \geq C_0 m_{\mathsf{T}}(\varrho',\gamma)$, for $\varrho' = (x,\xi',\tau)$, $\tau \geq \gamma \geq 1$, with $|\xi'| \geq R$ for some $C_0 > 0$ and $R > 0$. Then, for any $0 < C_1 < C_0$, there exist $\tau_* \geq 1$ and $\gamma_* \geq 1$ such that*

$$\mathrm{Re}(\mathrm{Op}_{\mathsf{T}}(a)u, u)_+ \geq C_1\|\mathrm{Op}_{\mathsf{T}}(m_{\mathsf{T}}^{1/2})u\|_+, \quad u \in \overline{\mathscr{S}}(\mathbb{R}^d_+), \quad \gamma \geq \gamma_*,$$
$$\tau \geq \max(\tau_*, \gamma).$$

Such an inequality is however not sufficient to achieve the Carleman estimates in the present chapter. Instead, we shall rely on much stronger inequalities, namely, a sharp Fefferman–Phong inequality, as proven by J.-M. Bony [77, Theorem 3.2].

THEOREM 14.22 (Fefferman–Phong–Bony Inequality). *Let $a(\varrho',\gamma) \in \mathscr{C}^\infty(\mathbb{R}^d \times \mathbb{R}^{d-1})$ with τ and γ as parameters in $[1,+\infty)$. We assume that there exists $C > 0$ such that $a(\varrho',\gamma)$ satisfies*

1. $a \geq 0$.
2. $|a| \leq C\lambda^2_{\mathsf{T},\tau}$.
3. $|\partial_x^\alpha \partial_{\xi'}^\beta a| \leq C\lambda_{\mathsf{T},\tau}^{2-|\beta|}$ *if* $|\alpha| + |\beta| = 2$.
4. $|\partial_x^\alpha \partial_{\xi'}^\beta a| \leq C_{\alpha,\beta}\lambda_{\mathsf{T},\tau}^{(|\alpha|-|\beta|)/2}$ *if* $|\alpha| + |\beta| \geq 4$ *where* $C_{\alpha,\beta} > 0$.

Then, there exists $C_0 > 0$ such that

$$\mathrm{Re}(\mathrm{Op}_{\mathsf{T}}(a)u, u)_+ + C_0\|u\|^2_+ \geq 0, u \in \overline{\mathscr{S}}(\mathbb{R}^d_+). \tag{14.2.11}$$

As the statement of this inequality we give here is different from that of J.-M. Bony, in particular because the pseudo-differential calculus and operators are not written in the same quantification, we show in Appendix 14.B.4 how this theorem follows from J.-M. Bony's results.

REMARK 14.23. Observe that a as given in the statement of Theorem 14.22 does not lie in the symbol classes introduced above. In fact, $a(\varrho', \gamma) \in S(\lambda^2_{\mathsf{T},\tau}, \tilde{g})$ with $\tilde{g} = \lambda_{\mathsf{T},\tau}|dx|^2 + \lambda^{-1}_{\mathsf{T},\tau}|d\xi'|^2$. The metric $\tilde{g}$ is slowly varying and σ-temperate and $\lambda_{\mathsf{T},\tau}$ is a $\tilde{g}$-admissible order function, thus yielding a proper symbol class and calculus. In particular, $\mathrm{Op}_{\mathsf{T}}(a)$ is well defined. However, the usual Fefferman–Phong inequality (see Theorem 18.6.8 in [175]) in this symbol/operator class does not yield any useful estimate.

14.3. Proof of the First-Order Estimate

Here, we prove Proposition 14.6, relying on the material introduced in Sect. 14.2. The proof we provide is based on a multiplier method and uses both the g and g^γ-calculus and the Fefferman–Phong–Bony inequality of Theorem 14.22.

PROOF OF PROPOSITION 14.6. Since $\mathbb{S}_{\overline{\mathscr{U}}}$ is compact, we deduce from (14.1.7) that there exist $C_0 > 0$ and $C_1 > 0$ such that (see Lemma 3.8 of Volume 1 and its proof)

$$C_0 \lambda^{-1}_{\mathsf{T},\tau} \ell_1(\varrho')^2 + \{\ell_2, \ell_1\}(\varrho') \geq C_1 \lambda_{\mathsf{T},\tau}, \quad \varrho' \in \overline{\mathscr{U}}. \tag{14.3.1}$$

We introduce a $\mathscr{C}^\infty$-function σ_0 that is a regularized version of the sign function, viz.,

$$\sigma_0' \geq 0, \quad \sigma_0 \text{ is odd}, \quad \sigma_0(t) = 1 \text{ if } t \geq 1, \quad \sigma(t) = 2t, \text{ if } |t| \leq \frac{1}{4},$$

and we set

$$k = \sigma_0\big((\tau/\gamma)^{1/2} \lambda^{-1}_{\mathsf{T},\tau} \operatorname{Im} f\big) \ \text{ and } \ \tilde{k} = \sigma_0'\big((\tau/\gamma)^{1/2} \lambda^{-1}_{\mathsf{T},\tau} \operatorname{Im} f\big). \tag{14.3.2}$$

The function σ_0 is illustrated in Fig. 14.1.

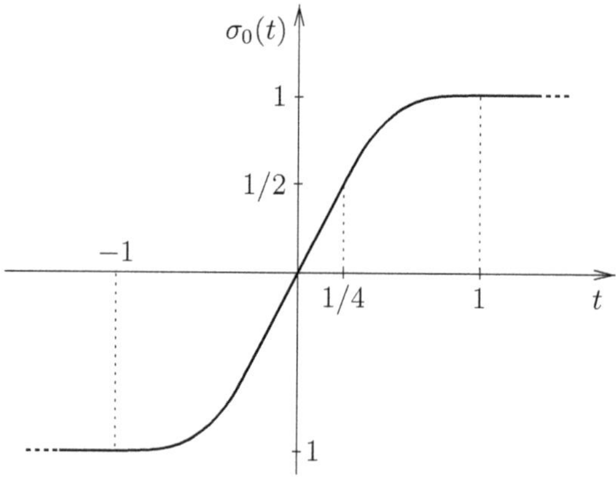

FIGURE 14.1. Regularized version of the sign function used to form symbol in the Hörmander calculus associated with the metric g^γ

An important point in what follows lies in the following observation.

LEMMA 14.24. *One has* $k, \tilde{k} \in S(1, g^\gamma)$. *Moreover,*

$$\tilde{k} \operatorname{Im} f \in S\big((\tau/\gamma)^{-1/2} \lambda_{\mathsf{T},\tau}, g^\gamma\big).$$

For $\alpha \in \mathbb{N}^d$ *and* $\beta \in \mathbb{N}^{d-1}$, *with* $|\alpha| = |\beta| = 1$, *one also has*

$$\partial_x^\alpha (k \operatorname{Im} f) \in S(\lambda_{\mathsf{T},\tau}, g^\gamma) \quad \text{and} \quad \partial_{\xi'}^\beta (k \operatorname{Im} f) \in S(1, g^\gamma).$$

A proof is given in Appendix 14.B.1.

REMARK 14.25. Note that the last two properties are equivalent to having

$$|\partial_x^\alpha \partial_{\xi'}^\beta (k \operatorname{Im} f)| \leq C_{\alpha,\beta} (\tau/\gamma)^{(|\alpha|+|\beta|-1)/2} \lambda_{\mathsf{T},\tau}^{1-|\beta|}, \tag{14.3.3}$$

for $\alpha \in \mathbb{N}^d$ and $\beta \in \mathbb{N}^{d-1}$ such that $|\alpha| + |\beta| \geq 1$.

With $K = (\operatorname{Op}_{\mathsf{T}}(k) + \operatorname{Op}_{\mathsf{T}}(k)^*)/2$ (formally) selfadjoint, we use $-iK$ as a multiplier. For $v = \operatorname{Op}_{\mathsf{T}}(\chi) u$, we write

$$\begin{aligned} 2\operatorname{Re}(Lv, -iKv)_+ &= (i \operatorname{Im} F v, iKv)_+ + (iKv, i \operatorname{Im} F v)_+ \\ &\quad - 2\operatorname{Re}((D_d - \operatorname{Re} F)v, iKv)_+ \\ &= \big((K \operatorname{Im} F + \operatorname{Im} F K - i[D_d - \operatorname{Re} F, K])v, v\big)_+ \\ &\quad - (Kv_{|x_d=0^+}, v_{|x_d=0^+})_\partial. \end{aligned} \tag{14.3.4}$$

Observe that the two terms on the right-hand side are real.

We have, writing formally $D_d^* = D_d$,

$$\begin{aligned} 2[D_d, K] &= [D_d, \operatorname{Op}_{\mathsf{T}}(k)] + [D_d, \operatorname{Op}_{\mathsf{T}}(k)^*] = \operatorname{Op}_{\mathsf{T}}(D_d k) - [D_d, \operatorname{Op}_{\mathsf{T}}(k)]^* \\ &= \operatorname{Op}_{\mathsf{T}}(D_d k) - \operatorname{Op}_{\mathsf{T}}(D_d k)^*, \end{aligned}$$

yielding

$$\begin{aligned} 2\big(i[D_d, K]v, v\big)_+ &= \big(i\operatorname{Op}_{\mathsf{T}}(D_d k)v, v\big)_+ - \big(i\operatorname{Op}_{\mathsf{T}}(D_d k)^* v, v\big)_+ \\ &= \big(i\operatorname{Op}_{\mathsf{T}}(D_d k)v, v\big)_+ + \big(v, i\operatorname{Op}_{\mathsf{T}}(D_d k)v\big)_+ \\ &= 2\operatorname{Re}\big(i\operatorname{Op}_{\mathsf{T}}(D_d k)v, v\big)_+. \end{aligned}$$

One has $i\operatorname{Op}_{\mathsf{T}}(D_d k) = \operatorname{Op}_{\mathsf{T}}(\partial_d k) = \operatorname{Op}_{\mathsf{T}}(\{\xi_d, k\})$. By symbol calculus, we have $\operatorname{Re} F = \operatorname{Op}_{\mathsf{T}}(\operatorname{Re} f) \mod \Psi(1, g)$, and from Theorem 14.17, we have $i[\operatorname{Re} F, K] = \operatorname{Op}_{\mathsf{T}}(\{\operatorname{Re} f, k\}) \mod \Psi(\tau\gamma^{-1}\lambda_\tau^{-1}, g^\gamma)$. We thus have

$$(i[D_d - \operatorname{Re} F, K]v, v)_+ = \operatorname{Re}(\operatorname{Op}_{\mathsf{T}}(\{\xi_d - \operatorname{Re} f, k\})v, v)_+ + (\operatorname{Op}_{\mathsf{T}}(r_0)v, v)_+ \tag{14.3.5}$$

with $r_0 \in S(\tau\gamma^{-1}\lambda_\tau^{-1}, g^\gamma)$. Observing that

$$\begin{aligned} &\operatorname{Re}\big((\operatorname{Op}_{\mathsf{T}}(k)^* \operatorname{Im} F + \operatorname{Im} F \operatorname{Op}_{\mathsf{T}}(k)^*)v, v\big)_+ \\ &= \operatorname{Re}\big((\operatorname{Op}_{\mathsf{T}}(k) \operatorname{Im} F + \operatorname{Im} F \operatorname{Op}_{\mathsf{T}}(k))v, v\big)_+, \end{aligned}$$

we have

$$\big((K \operatorname{Im} F + \operatorname{Im} F K)v, v\big)_+ = \operatorname{Re}\big((\operatorname{Op}_{\mathsf{T}}(k) \operatorname{Im} F + \operatorname{Im} F \operatorname{Op}_{\mathsf{T}}(k))v, v\big)_+. \tag{14.3.6}$$

From usual symbol calculus, Lemma 14.24, and Theorem 14.17, we have

$$\begin{aligned}
&\operatorname{Im} F = \operatorname{Op}_{\mathsf{T}}(\operatorname{Im} f) \mod \Psi(1,g),\\
&k \circ \operatorname{Im} f = k \operatorname{Im} f - i\nabla_{\xi'} k \cdot \nabla_{x'} \operatorname{Im} f \mod S(\tau\gamma^{-1}\lambda_{\mathsf{T},\tau}^{-1}, g^\gamma),\\
&\operatorname{Im} f \circ k = k \operatorname{Im} f - i\nabla_{\xi'} \operatorname{Im} f \cdot \nabla_{x'} k \mod S(\tau\gamma^{-1}\lambda_{\mathsf{T},\tau}^{-1}, g^\gamma).
\end{aligned}$$

Then, we deduce that

$$\operatorname{Op}_{\mathsf{T}}(k) \operatorname{Im} F + \operatorname{Im} F \operatorname{Op}_{\mathsf{T}}(k) = \operatorname{Op}_{\mathsf{T}}(2k \operatorname{Im} f + i(\tau/\gamma)^{1/2} r_1 + r_1'), \tag{14.3.7}$$

where $r_1, r_1' \in S(1, g^\gamma)$ with r_1 *real* valued. From calculus in the symbol classes associated with the metric g^γ, we have

$$2\operatorname{Re}(i\operatorname{Op}_{\mathsf{T}}(r_1)v, v)_+ = i((\operatorname{Op}_{\mathsf{T}}(r_1) - \operatorname{Op}_{\mathsf{T}}(r_1)^*)v, v)_+ = (\operatorname{Op}_{\mathsf{T}}(r_2)v, v)_+, \tag{14.3.8}$$

where $r_2 \in S(\tau/(\gamma\lambda_{\mathsf{T},\tau}), g^\gamma)$. From (14.3.5)–(14.3.8), we obtain

$$\begin{aligned}
&\big((K \operatorname{Im} F + \operatorname{Im} F K - i[D_d - \operatorname{Re} F, K])v, v\big)_+ \\
&\qquad \geq \operatorname{Re}\big(\operatorname{Op}_{\mathsf{T}}(2k \operatorname{Im} f - \{\xi_d - \operatorname{Re} f, k\})v, v\big)_+ - C(\tau^{1/2}\gamma^{-3/2} + 1)\|v\|_+^2.
\end{aligned} \tag{14.3.9}$$

We compute, with the form of k and $\tilde{k}$ given in (14.3.2),

$$\{\operatorname{Re} f, k\} = (\tau/\gamma)^{1/2}\Big(\lambda_{\mathsf{T},\tau}^{-1}\{\operatorname{Re} f, \operatorname{Im} f\} - \operatorname{Im} f\{\operatorname{Re} f, \lambda_{\mathsf{T},\tau}^{-1}\}\Big)\tilde{k}.$$

From Lemma 14.24, the term $(\tau/\gamma)^{1/2}\tilde{k} \operatorname{Im} f \nabla_{x'} \operatorname{Re} f \cdot \nabla_{\xi'}(\lambda_{\mathsf{T},\tau}^{-1})$ is in $S(1, g^\gamma)$. We also have $\{\xi_d, k\} = (\tau/\gamma)^{1/2}\lambda_{\mathsf{T},\tau}^{-1}\tilde{k}\{\xi_d, \operatorname{Im} f\}$. From (14.3.9), we thus have

$$\begin{aligned}
&\big((K \operatorname{Im} F + \operatorname{Im} F K - i[D_d - \operatorname{Re} F, K])v, v\big)_+ \\
&\qquad\qquad \geq \operatorname{Re}\big(\operatorname{Op}_{\mathsf{T}}(q_0)v, v\big)_+ - C(\tau^{1/2}\gamma^{-3/2} + 1)\|v\|_+^2,
\end{aligned} \tag{14.3.10}$$

where

$$q_0 = 2k \operatorname{Im} f - (\tau/\gamma)^{1/2}\lambda_{\mathsf{T},\tau}^{-1}\{\xi_d - \operatorname{Re} f, \operatorname{Im} f\}\tilde{k} \in S(\lambda_{\mathsf{T},\tau}, g^\gamma). \tag{14.3.11}$$

We now study the positivity of the symbol q_0 in $\overline{\mathscr{U}}$, and we split the analysis into two cases.

Case $\varrho \in \overline{\mathscr{U}}$ and $|\operatorname{Im} f| \leq (\tau/\gamma)^{-1/2}\lambda_{\mathsf{T},\tau}/4$. We then have $\tilde{k} = 2$ and $k = 2(\tau/\gamma)^{1/2}\lambda_{\mathsf{T},\tau}^{-1} \operatorname{Im} f$. We choose $\tau_0 > 0$ such that $2 \geq C_0\lambda_{\mathsf{T},\tau}^{-1}$ for $\tau \geq \tau_0$ with C_0 as in (14.3.1). Since $\ell_2 = \xi_d - \operatorname{Re} f$ and $\ell_1 = -\operatorname{Im} f$ in (14.3.1), one finds

$$\begin{aligned}
q_0 &\geq 4(\tau/\gamma)^{1/2}\lambda_{\mathsf{T},\tau}^{-1}(\operatorname{Im} f)^2 - 2(\tau/\gamma)^{1/2}\lambda_{\mathsf{T},\tau}^{-1}\{\xi_d - \operatorname{Re} f, \operatorname{Im} f\}\\
&= 2(\tau/\gamma)^{1/2}\lambda_{\mathsf{T},\tau}^{-1}\Big(2(\operatorname{Im} f)^2 + \{\xi_d - \operatorname{Re} f, -\operatorname{Im} f\}\Big)\\
&\geq 2(\tau/\gamma)^{1/2}\lambda_{\mathsf{T},\tau}^{-1}\Big(C_0\lambda_{\mathsf{T},\tau}^{-1}(\operatorname{Im} f)^2 + \{\xi_d - \operatorname{Re} f, -\operatorname{Im} f\}\Big)\\
&\geq 2C_1(\tau/\gamma)^{1/2}.
\end{aligned}$$

Case $\varrho \in \overline{\mathscr{U}}$ and $|\operatorname{Im} f| \geq (\tau/\gamma)^{-1/2}\lambda_{\mathsf{T},\tau}/4$. We then have

$$-k \operatorname{Im} f \geq |\operatorname{Im} f|/2 \geq (\tau/\gamma)^{-1/2}\lambda_{\mathsf{T},\tau}/8.$$

Observe that $(\tau/\gamma)^{-1/2}\lambda_{\mathsf{T},\tau}/8 \geq \tau^{1/2}\gamma^{1/2}/8$ and $|(\tau/\gamma)^{1/2}\lambda_{\mathsf{T},\tau}^{-1}\tilde{k}\{\xi_d - \operatorname{Re} f, \operatorname{Im} f\}| \lesssim (\tau/\gamma)^{1/2}$. Thus for $\gamma \geq 1$ chosen sufficiently large, we obtain $q_0 \geq C_2\tau^{1/2}\gamma^{1/2}$, for $C_2 > 0$.

Both cases yield $q_0 \geq C_3(\tau/\gamma)^{1/2}$ in $\overline{\mathscr{U}}$ for some $C_3 > 0$, if $\gamma \geq \gamma_0$ for $\gamma_0 \geq 1$ chosen sufficiently large.

Let $\tilde{\chi} \in S^0_{\mathsf{T},\tau}$, homogeneous of degree 0, be such that $\operatorname{supp}(\tilde{\chi}) \subset \mathscr{U}$ and $\tilde{\chi} \equiv 1$ on $\operatorname{supp}(\chi)$. We then set

$$a = \gamma^{3/2}\tau^{-1/2}\big(q_0 - C_3(\tau/\gamma)^{1/2}\big) \quad \text{and} \quad \tilde{a} = \tilde{\chi} a.$$

One has $a, \tilde{a} \in S(\gamma^{3/2}\tau^{-1/2}\lambda_{\mathsf{T},\tau}, g^\gamma)$. Yet, the most important is the following proposition.

LEMMA 14.26. *The function $\tilde{a}$ fulfills the properties of Theorem 14.22 if $\tau \geq \gamma \geq \gamma_0 \geq 1$ with γ_0 as chosen above.*

A proof is given in Appendix 14.B.2.

With the Fefferman–Phong–Bony inequality of Theorem 14.22, we thus obtain

$$\operatorname{Re}(\operatorname{Op}_{\mathsf{T}}(\tilde{a})v, v)_+ \geq -C_0\|v\|_+^2.$$

By Theorem 14.17, for any $N \in \mathbb{N}$, one has

$$\tilde{a} \circ \chi - a \circ \chi \in S((\tau/\gamma)^{(N+1)/2}\lambda_{\mathsf{T},\tau}^{-N}, g^\gamma) \subset S(\lambda_{\mathsf{T},\tau}^{1-N/2}, g^\gamma).$$

Thus, since $v = \operatorname{Op}_{\mathsf{T}}(\chi)u$, for any $N \in \mathbb{N}$, there exists $C_N > 0$ such that

$$\operatorname{Re}(\operatorname{Op}_{\mathsf{T}}(a)v, v)_+ \geq -C_0\|v\|_+^2 - C_N\|u\|_{\tau,0,-N}^2.$$

Multiplying by $\tau^{1/2}\gamma^{-3/2}$, we deduce

(14.3.12)
$$\begin{aligned}\operatorname{Re}(\operatorname{Op}_{\mathsf{T}}(q_0)v, v)_+ &\geq (C_3\tau^{1/2}\gamma^{-1/2} - C_0\tau^{1/2}\gamma^{-3/2})\|v\|_+^2 - C_N'\|u\|_{\tau,0,-N}^2 \\ &\gtrsim \tau^{1/2}\gamma^{-1/2}\|v\|_+^2 - C_N'\|u\|_{\tau,0,-N}^2,\end{aligned}$$

for some $C_N' > 0$, if $\gamma \geq \gamma_0$ is chosen sufficiently large.

From (14.3.10) and (14.3.12) for $\gamma \geq \gamma_0$ chosen sufficiently large and to be kept fixed in what follows, we obtain

$$\begin{aligned}&\big((K \operatorname{Im} F + \operatorname{Im} F K - i[D_d - \operatorname{Re} F, K])v, v\big)_+ \\ &\qquad\qquad \geq C\tau^{1/2}\|v\|_+^2 - C'\|v\|_+^2 - C_N\|u\|_{\tau,0,-N}^2.\end{aligned}$$

With τ chosen sufficiently large, we conclude that

$$\big((K \operatorname{Im} F + \operatorname{Im} F K - i[D_d - \operatorname{Re} F, K])v, v\big)_+ \gtrsim \tau^{1/2}\|v\|_+^2 - C_N'\|u\|_{\tau,0,-N}^2.$$

With (14.3.4), we find

$$\operatorname{Re}(Lv, -iKv)_+ + |v_{|x_d=0^+}|^2 \gtrsim \tau^{1/2}\|v\|_+^2 - C'_N\|u\|_{\tau,0,-N}^2,$$

using that K is bounded on $L^2(\mathbb{R}^{d-1})$. With the Young inequality, for $\varepsilon > 0$, we write

$$|(Lv, -iKv)_+| \lesssim \varepsilon^{-1}\tau^{-1/2}\|Lv\|_+^2 + \varepsilon\tau^{1/2}\|v\|_+^2.$$

Thus, for $\varepsilon > 0$ chosen sufficiently small, we obtain

$$\tau^{1/2}\|v\|_+ \leq C\Big(\|Lv\|_+ + \tau^{1/4}|v_{|x_d=0^+}|_\partial\Big) + C_N\|u\|_{\tau,0,-N}.$$

Replacing u by $\tau^s u$ and using Proposition 6.11 as $\tau \gtrsim \lambda_{\mathsf{T},\tau}$ in $\overline{\mathscr{U}}$ by (14.1.5), one obtains, for any $N \in \mathbb{N}$,

$$\begin{aligned}\tau^{-1/2}\|v\|_{\tau,0,s+1} - C_N\|u\|_{\tau,0,-N} &\lesssim \tau^s\|Lv\|_+ + \tau^{s+1/4}|v_{|x_d=0^+}|_\partial \\ &\lesssim \|Lv\|_{\tau,0,s} + \tau^{-1/4}|v_{|x_d=0^+}|_{\tau,s+1/2}.\end{aligned}$$

Observing that

$$\|D_d v\|_{\tau,0,s} \lesssim \|Lv\|_{\tau,0,s} + \|v\|_{\tau,0,s+1},$$

one concludes the proof of Proposition 14.6. ∎

14.4. Proof of the Microlocal Estimate

In what follows, to ease the notation, we shall write φ, p, b, and b_φ in place of $\varphi^{\mathcal{C}}$, $p^{\mathcal{C}}$, $b^{\mathcal{C}}$, and $b_\varphi^{\mathcal{C}} = b_{0,\varphi}$, respectively. We recall some of the setting of Sect. 8.3 that is used for the microlocal estimate of Proposition 14.5.

In the chosen coordinates, the principal symbol of P takes the form

$$p(x,\xi) = \xi_d^2 + r(x,\xi'), \qquad r(x,\xi') = \sum_{1\leq i,j\leq d-1} g^{ij}(x)\xi_i\xi_j \gtrsim |\xi'|^2.$$

The symbol $p_{0,\varphi}$ of the conjugated operator $P_{0,\varphi}$ can be written as

$$p_{0,\varphi}(x,\xi',z,\tau) = \big(z - \gamma_1(\varrho')\big)\big(z - \gamma_2(\varrho')\big), \qquad \varrho' = (x,\xi',\tau),$$

with

$$\gamma_j(\varrho') = -i\tau\partial_d\varphi(x) + i(-1)^j\alpha(y'),$$

where

$$\begin{aligned}\alpha(\varrho')^2 &= r(x,\xi' + i\tau d_{x'}\varphi(x)) \\ &= r(x,\xi') - \tau^2 r(x, d_{x'}\varphi(x)) + 2i\tau\tilde{r}(x,\xi',d_{x'}\varphi(x)),\end{aligned}$$

with $\tilde{r}(x,.,.)$ the bilinear form associated with the quadratic form $r(x,.)$. As in Sect. 8.2, α is chosen such that $\operatorname{Re}\alpha \geq 0$.

By the first result of Theorem 8.7, if the Lopatinskiĭ–Šapiro condition holds for (P,B,φ) at m^0, then $\partial_d\varphi(x^0) > 0$, and locally we have $\partial_d\varphi(x) > 0$, implying that $\operatorname{Im}\gamma_1(\varrho') < 0$.

With the definition of $\check{p}_{\varphi}^{+}$ in Sect. 8.2, one has

$$\check{p}_{\varphi}^{+}(m,\omega',z,\tau) = \begin{cases} 1 & \text{if } \operatorname{Im}\gamma_2(\varrho') < 0, \\ z - \gamma_2(\varrho') & \text{if } \operatorname{Im}\gamma_2(\varrho') \geq 0, \end{cases} \qquad \varrho' = (x,\xi',\tau),$$

with x and ξ' the local representatives of $m \in \partial\mathcal{M}$ and $\omega' \in T_m^*\partial\mathcal{M}$. One has $\operatorname{Im}\gamma_2(x,\xi',\tau) \geq 0$ equivalent to having $\operatorname{Re}\alpha(x,\xi',\tau) \geq \tau\partial_d\varphi(x)$.

We prove the estimate of Proposition 14.5 in the following three (exhaustive) cases:

(1) $\operatorname{Im}\gamma_2(\varrho^{0\prime}) > 0$.
(2) or $\operatorname{Im}\gamma_2(\varrho^{0\prime}) = 0$.
(3) or $\operatorname{Im}\gamma_2(\varrho^{0\prime}) < 0$.

14.4.1. Case 1: One Root in the Open Upper Complex Half-Plane.

Here, we have $\operatorname{Im}\gamma_1(x^0,\xi^{0\prime},\tau^0) < 0$, $\operatorname{Im}\gamma_2(x^0,\xi^{0\prime},\tau^0) > 0$, and

$$\check{p}_{\varphi}^{+}(m^0,\omega^{0\prime},z,\tau^0) = z - \gamma_2(x^0,\xi^{0\prime},\tau^0).$$

With the Lopatinskiĭ–Šapiro condition holding at the considered point, by (8.3.11), we have moreover

$$b_\varphi(x^0,\xi^{0\prime},\xi_d=\gamma_2,\tau^0) = b(x^0,\xi^{0\prime}+i\tau^0 d_{x'}\varphi(x^0),\gamma_2+i\tau^0\partial_{x_d}\varphi(x^0)) \neq 0.$$

As the roots γ_1 and γ_2 are locally smooth with respect to (x,ξ',τ) and homogeneous of degree one in (ξ',τ) by Proposition 6.28, there exist $\mathscr{U}$ a conic open neighborhood of $(x^0,\xi^{0\prime},\tau^0)$ in $U^0\times\mathbb{R}^{d-1}\times\mathbb{R}_+$ and $C, C' > 0$ such that $\mathbb{S}_{\overline{\mathscr{U}}}$ is compact and

$$\operatorname{Im}\gamma_2(\varrho') \geq C\lambda_{\mathsf{T},\tau}, \text{ and } \operatorname{Im}\gamma_1(\varrho') \leq -C'\lambda_{\mathsf{T},\tau},$$

and

$$b_\varphi(x,\xi',\xi_d=\gamma_2(\varrho'),\tau) \neq 0, \tag{14.4.1}$$

if $\varrho' = (x,\xi',\tau) \in \overline{\mathscr{U}}$.

We let $\chi \in S^0_{\mathsf{T},\tau}$ be as in the statement and $\tilde{\chi} \in S^0_{\mathsf{T},\tau}$ be homogeneous of degree zero and be such that $\operatorname{supp}(\tilde{\chi}) \subset \mathscr{U}$ and $\tilde{\chi} \equiv 1$ on $\operatorname{supp}(\chi)$. From the smoothness and the homogeneity of the roots, we have $\tilde{\chi}\gamma_j \in S^1_{\mathsf{T},\tau}$, $j=1,2$. We set

$$P^+ = D_d - \operatorname{Op}_{\mathsf{T}}(\tilde{\chi}\gamma_2) \text{ and } P^- = D_d - \operatorname{Op}_{\mathsf{T}}(\tilde{\chi}\gamma_1).$$

From Lemma 6.21, we have, for some $\tau_* > 0$,

$$\begin{aligned} &\|\operatorname{Op}_{\mathsf{T}}(\chi)u\|_{\tau,1,s} \\ &\quad \leq C\Big(\|P^+\operatorname{Op}_{\mathsf{T}}(\chi)u\|_{\tau,0,s} + |\operatorname{Op}_{\mathsf{T}}(\chi)u_{|x_d=0^+}|_{\tau,s+1/2}\Big) + C_N\|u\|_{\tau,0,-N}, \end{aligned} \tag{14.4.2}$$

for $\tau \geq \tau_*$, $u \in \overline{\mathscr{S}}(\mathbb{R}^d_+)$ and $s \in \mathbb{R}$. From Lemma 6.20 (with $s = 0$), we have, for τ chosen sufficiently large,

(14.4.3)
$$\|\mathrm{Op}_\mathsf{T}(\chi)u\|_{\tau,1} + |\mathrm{Op}_\mathsf{T}(\chi)u_{|x_d=0^+}|_{\tau,1/2} \lesssim \|P^-\mathrm{Op}_\mathsf{T}(\chi)u\|_+ + \|u\|_{\tau,0,-N},$$

for $u \in \overline{\mathscr{S}}(\mathbb{R}^d_+)$.

Let now $v \in \overline{\mathscr{S}}(\mathbb{R}^d_+)$. We apply estimate (14.4.3) for $u = P^+v$, yielding

$$\begin{aligned}\|\mathrm{Op}_\mathsf{T}(\chi)P^+v\|_{\tau,1} &+ |\mathrm{Op}_\mathsf{T}(\chi)P^+v_{|x_d=0^+}|_{\tau,1/2}\\ &\lesssim \|P^-\mathrm{Op}_\mathsf{T}(\chi)P^+v\|_+ + \|v\|_{\tau,1,-N}\\ &\|P_{0,\varphi}v\|_+ + \|v\|_{\tau,1},\end{aligned}$$

using that $P^-\mathrm{Op}_\mathsf{T}(\chi)P^+ = \mathrm{Op}_\mathsf{T}(\chi)P^-P^+ \mod \Psi^{1,0}_{\tau,\mathrm{ph}} = \mathrm{Op}_\mathsf{T}(\chi)P_{0,\varphi} \mod \Psi^{1,0}_{\tau,\mathrm{ph}}$. We set $w = \mathrm{Op}_\mathsf{T}(\chi)v$. We observe that we have

$$\begin{aligned}|P^+w_{|x_d=0^+}|_{\tau,1/2} &\lesssim |\mathrm{Op}_\mathsf{T}(\chi)P^+v_{|x_d=0^+}|_{\tau,1/2} + |v_{|x_d=0^+}|_{\tau,1/2}\\ &\lesssim |\mathrm{Op}_\mathsf{T}(\chi)P^+v_{|x_d=0^+}|_{\tau,1/2} + \|v\|_{\tau,1},\end{aligned}$$

using the trace inequality of Proposition 6.9. We also have

$$\|P^+w\|_{\tau,1} \lesssim \|\mathrm{Op}_\mathsf{T}(\chi)P^+v\|_{\tau,1} + \|v\|_{\tau,1}.$$

We thus obtain

$$\|P^+w\|_{\tau,1} + |P^+w_{|x_d=0^+}|_{\tau,1/2} \lesssim \|P_{0,\varphi}v\|_+ + \|v\|_{\tau,1}.$$

Here, since (14.4.1) holds in $\overline{\mathscr{U}}$, Lemma 8.19 applies and yields

$$|\operatorname{tr}(w)|_{\tau,1,1/2} \lesssim |B_{0,\varphi}w_{|x_d=0^+}|_{\tau,3/2-k} + |P^+w_{|x_d=0^+}|_{\tau,1/2} + \|v\|_{\tau,1}.$$

We thus obtain

$$\|P^+w\|_{\tau,1} + |\operatorname{tr}(w)|_{\tau,1,1/2} \lesssim \|P_{0,\varphi}v\|_+ + |B_{0,\varphi}w_{|x_d=0^+}|_{\tau,3/2-k} + \|v\|_{\tau,1}.$$

With estimate (14.4.2) (for $s = 1$), we find

$$\|w\|_{\tau,1,1} + |\operatorname{tr}(w)|_{\tau,1,1/2} \lesssim \|P_{0,\varphi}v\|_+ + |B_{0,\varphi}w_{|x_d=0^+}|_{\tau,3/2-k} + \|v\|_{\tau,1}.$$

With the form of $P_{0,\varphi}$, that is, $P_{0,\varphi} = (D_d + i\tau\partial_d\varphi)^2 \mod \mathscr{D}^2_{\mathsf{T},\tau}$, we can estimate $\|D_d^2u\|_+$ from $\|P_{0,\varphi}v\|_+$ and $\|u\|_{1,1}$, and we obtain

$$\|w\|_{\tau,2} + |\operatorname{tr}(w)|_{\tau,1,1/2} \lesssim \|P_{0,\varphi}v\|_+ + |B_{0,\varphi}w_{|x_d=0^+}|_{\tau,3/2-k} + \|v\|_{\tau,1}.$$

As $w = \mathrm{Op}_\mathsf{T}(\chi)v$, with a commutator argument, we have

(14.4.4)
$$\begin{aligned}|B_{0,\varphi}w_{|x_d=0^+}|_{\tau,3/2-k} &\lesssim |B_{0,\varphi}v_{|x_d=0^+}|_{\tau,3/2-k} + |\operatorname{tr}(v)|_{\tau,1,-1/2}\\ &\lesssim |B_{0,\varphi}v_{|x_d=0^+}|_{\tau,3/2-k} + \|v\|_{\tau,2,-1},\end{aligned}$$

using the trace inequality of Corollary 6.10. We then obtain

$$\|w\|_{\tau,2} + |\operatorname{tr}(w)|_{\tau,1,1/2} \lesssim \|P_{0,\varphi}v\|_+ + |B_{0,\varphi}v_{|x_d=0^+}|_{\tau,3/2-k} + \|v\|_{\tau,2,-1},$$

which is stronger than (14.1.4). This concludes the proof for Case 1.

14.4.2. Case 2: One Root on the Real Axis. Here, $\operatorname{Im}\gamma_1(x^0,\xi^{0\prime},\tau^0) < 0$ and $\operatorname{Im}\gamma_2(x^0,\xi^{0\prime},\tau^0) = 0$. In particular, we have

$$\check{p}_{\varphi}^{+}(m^0,\omega^{0\prime},z,\tau^0) = z - \gamma_2(x^0,\xi^{0\prime},\tau^0).$$

Note that this implies $\tau^0 \neq 0$ (since in the case $\tau = 0$, one has $\operatorname{Im}\gamma_2 > 0$). Thus there exists C_0 such that $\lambda_{\mathsf{T},\tau}(\xi^0,\tau^0) \le C_0\tau^0$.

With the Lopatinskiĭ–Šapiro condition holding at the considered point, by (8.3.11), we have moreover

$$b_{\varphi}(x^0,\xi^{0\prime},\xi_d = \gamma_2,\tau^0) = b(x^0,\xi^{0\prime} + i\tau^0 d_{x'}\varphi(x^0), \gamma_2 + i\tau^0\partial_{x_d}\varphi(x^0)) \neq 0.$$

As the roots γ_1 and γ_2 are locally smooth with respect to (x,ξ',τ) and homogeneous of degree one in (ξ',τ) by Proposition 6.28, there exist $\mathscr{U}$ a conic open neighborhood of $(x^0,\xi^{0\prime},\tau^0)$ in $U^0 \times \mathbb{R}^{d-1} \times \mathbb{R}_+$ and $C, C' > 0$ such that $\mathbb{S}_{\overline{\mathscr{U}}}$ is compact and

$$\gamma_1(\varrho') \neq \gamma_2(\varrho'),\ \operatorname{Im}\gamma_2(\varrho') \ge -C\lambda_{\mathsf{T},\tau},\ \text{and}\ \operatorname{Im}\gamma_1(\varrho') \le -C'\lambda_{\mathsf{T},\tau},$$

and

$$\lambda_{\mathsf{T},\tau}(\varrho') \le 2C_0\tau,$$

and

$$b_{\varphi}(x,\xi',\xi_d = \gamma_2(\varrho'),\tau) \neq 0, \tag{14.4.5}$$

if $\varrho' = (x,\xi',\tau) \in \overline{\mathscr{U}}$.

We let $\chi \in S^0_{\mathsf{T},\tau}$ be as in the statement and $\tilde{\chi} \in S^0_{\mathsf{T},\tau}$ be homogeneous of degree zero and be such that $\operatorname{supp}(\tilde{\chi}) \subset \mathscr{U}$ and $\tilde{\chi} \equiv 1$ on $\operatorname{supp}(\chi)$. From the smoothness and the homogeneity of the roots, we have $\tilde{\chi}\gamma_j \in S^1_{\mathsf{T},\tau}$, $j = 1,2$. We set

$$P^+ = D_d - \operatorname{Op}_{\mathsf{T}}(\tilde{\chi}\gamma_2)\ \text{ and }\ P^- = D_d - \operatorname{Op}_{\mathsf{T}}(\tilde{\chi}\gamma_1).$$

From Lemma 6.20 (with $s = 0$), we have, for τ chosen sufficiently large,

(14.4.6)
$$\|\operatorname{Op}_{\mathsf{T}}(\chi)u\|_{\tau,1} + |\operatorname{Op}_{\mathsf{T}}(\chi)u_{|x_d=0^+}|_{\tau,1/2} \lesssim \|P^-\operatorname{Op}_{\mathsf{T}}(\chi)u\|_+ + \|u\|_{\tau,0,-N},$$

for $u \in \overline{\mathscr{S}}(\mathbb{R}^d_+)$.

We write $P^+ = P_2^+ + iP_1^+$ with

$$P_2^+ = \frac{1}{2}(P^+ + (P^+)^*) = D_d - \frac{1}{2}(\operatorname{Op}_{\mathsf{T}}(\tilde{\gamma}_2) + \operatorname{Op}_{\mathsf{T}}(\tilde{\gamma}_2)^*),$$
$$P_1^+ = \frac{1}{2i}(P^+ - (P^+)^*) = -\frac{1}{2i}(\operatorname{Op}_{\mathsf{T}}(\tilde{\gamma}_2) - \operatorname{Op}_{\mathsf{T}}(\tilde{\gamma}_2)^*).$$

Observe that $P_2^+ \in \Psi_{\tau}^{1,0}$ and $P_1^+ \in \Psi^1_{\mathsf{T},\tau}$ are both formally selfadjoint. Their respective principal symbols are

$$p_2^+(\varrho) = \xi_d - \operatorname{Re}\tilde{\gamma}_2(\varrho') \in S_{\tau}^{1,0}, \qquad p_1^+(\varrho') = -\operatorname{Im}\tilde{\gamma}_2(\varrho') \in S^1_{\mathsf{T},\tau}.$$

The principal symbol of P^+ is $p^+(\varrho) = p_2^+(\varrho) + ip_1^+(\varrho')$. By Lemma 13.12, the operator P^+ is sub-elliptic: there exists a conic open neighborhood $\mathscr{W}$ of $\operatorname{supp}(\chi)$ in $\mathbb{R}^d_+ \times \mathbb{R}^{d-1} \times \mathbb{R}_+$ such that $\tilde{\chi} \equiv 1$ in $\mathscr{W}$ and

$$p^+(\varrho) = 0 \quad \Rightarrow \quad \frac{1}{2i}\{\overline{p^+}, p^+\}(\varrho') = \{p_2^+, p_1^+\}(\varrho') > 0,$$
$$\varrho' \in \overline{\mathscr{W}}, \ \ \varrho = (\varrho', \xi_d).$$

Since we also have $\lambda_{\mathsf{T},\tau} \lesssim \tau$ in $\overline{\mathscr{U}}$, from Lemma 14.6, we have, for some $\tau_* > 0$,

$$\begin{aligned} &\tau^{-1/2}\|\mathrm{Op}_{\mathsf{T}}(\chi)u\|_{\tau,1,s} \\ &\quad\lesssim \|P^+\mathrm{Op}_{\mathsf{T}}(\chi)u\|_{\tau,0,s} + \tau^{-1/4}|\mathrm{Op}_{\mathsf{T}}(\chi)u_{|x_d=0^+}|_{\tau,s+1/2} + \|u\|_{\tau,0,-N}, \end{aligned} \tag{14.4.7}$$

for τ sufficiently large, $u \in \overline{\mathscr{S}}(\mathbb{R}^d_+)$ and $s \in \mathbb{R}$.

Let now $v \in \overline{\mathscr{S}}(\mathbb{R}^d_+)$. We apply estimate (14.4.6) for $u = P^+v$, yielding

$$\begin{aligned} \|\mathrm{Op}_{\mathsf{T}}(\chi)P^+v\|_{\tau,1} &+ |\mathrm{Op}_{\mathsf{T}}(\chi)P^+v_{|x_d=0^+}|_{\tau,1/2} \\ &\lesssim \|P^-\mathrm{Op}_{\mathsf{T}}(\chi)P^+v\|_+ + \|v\|_{\tau,1,-N} \\ &\lesssim \|P_{0,\varphi}v\|_+ + \|v\|_{\tau,1}, \end{aligned}$$

using that $P^-\mathrm{Op}_{\mathsf{T}}(\chi)P^+ = \mathrm{Op}_{\mathsf{T}}(\chi)P^-P^+ \mod \Psi^{1,0}_{\tau,\mathrm{ph}} = \mathrm{Op}_{\mathsf{T}}(\chi)P_{0,\varphi} \mod \Psi^{1,0}_{\tau,\mathrm{ph}}$. We set $w = \mathrm{Op}_{\mathsf{T}}(\chi)v$. We observe that we have

$$\begin{aligned} |P^+w_{|x_d=0^+}|_{\tau,1/2} &\lesssim |\mathrm{Op}_{\mathsf{T}}(\chi)P^+v_{|x_d=0^+}|_{\tau,1/2} + |v_{|x_d=0^+}|_{\tau,1/2} \\ &\lesssim |\mathrm{Op}_{\mathsf{T}}(\chi)P^+v_{|x_d=0^+}|_{\tau,1/2} + \|v\|_{\tau,1}, \end{aligned}$$

using the trace inequality of Proposition 6.9. We also have

$$\|P^+w\|_{\tau,1} \lesssim \|\mathrm{Op}_{\mathsf{T}}(\chi)P^+v\|_{\tau,1} + \|v\|_{\tau,1}.$$

We thus obtain

$$\|P^+w\|_{\tau,1} + |P^+w_{|x_d=0^+}|_{\tau,1/2} \lesssim \|P_{0,\varphi}v\|_+ + \|v\|_{\tau,1}.$$

Here, since (14.4.5) holds in $\overline{\mathscr{U}}$, Lemma 8.19 applies and yields

$$|\operatorname{tr}(w)|_{\tau,1,1/2} \lesssim |B_{0,\varphi}w_{|x_d=0^+}|_{\tau,3/2-k} + |P^+w_{|x_d=0^+}|_{\tau,1/2} + \|v\|_{\tau,1}.$$

We thus obtain

$$\begin{aligned} &\|P^+w\|_{\tau,1} + \tau^{-1/4}|\operatorname{tr}(w)|_{\tau,1,1/2} \\ &\quad\lesssim \|P_{0,\varphi}v\|_+ + \tau^{-1/4}|B_{0,\varphi}w_{|x_d=0^+}|_{\tau,3/2-k} + \|v\|_{\tau,1}. \end{aligned}$$

With estimate (14.4.7) (for $s = 1$), we find

$$\begin{aligned} &\tau^{-1/2}\|w\|_{\tau,1,1} + \tau^{-1/4}|\operatorname{tr}(w)|_{\tau,1,1/2} \\ &\quad\lesssim \|P_{0,\varphi}v\|_+ + \tau^{-1/4}|B_{0,\varphi}w_{|x_d=0^+}|_{\tau,3/2-k} + \|v\|_{\tau,1}. \end{aligned}$$

With the form of $P_{0,\varphi}$ and (14.4.4), we finally have

$$\tau^{-1/2}\|w\|_{\tau,2} + \tau^{-1/4}|\operatorname{tr}(w)|_{\tau,1,1/2} \lesssim \|P_{0,\varphi}v\|_{+} + \tau^{-1/4}|B_{0,\varphi}v_{|x_d=0^+}|_{\tau,3/2-k} + \|v\|_{\tau,2,-1}.$$

This is the sought estimate (14.1.4). This concludes the proof for Case 2.

14.4.3. Case 3: Both Roots in the Open Lower Complex Half-Plane. Here, one applies without any change the argument of Sect. 8.3.4.2 that gives

$$\|\operatorname{Op}_{\mathsf{T}}(\chi)v\|_{\tau,2} + |\operatorname{tr}(\operatorname{Op}_{\mathsf{T}}(\chi)v)|_{\tau,1,1/2} \lesssim \|P_{0,\varphi}v\|_{+} + \|v\|_{\tau,2,-1},$$

for $\tau > 0$ chosen sufficiently large as in (8.3.31). This is stronger than (14.1.4). This concludes the proof for Case 3.

14.5. Optimality Aspect

As already pointed out above, in Theorem 14.1, the trace term in the r.h.s. has an additional factor $\tau^{-1/4}$ if compared to the result of Theorem 8.14. A natural question is then the optimality of such a result. With the following theorem, we answer this question positively.

THEOREM 14.27. *Let $(\mathcal{M}, g)$ be a compact Riemannian manifold with boundary, and let $P = -\Delta_g + R_1$ with R_1 a first-order differential operator with smooth coefficients on $\mathcal{M}$. Let V be an open set of $\mathcal{M}$ with $V_\partial = V \cap \partial\mathcal{M} \neq \emptyset$ and φ be a smooth real function defined on V. Let $k \in \mathbb{N}$ and $B = B^{k-1}\partial_\nu + B^k$ with B^{k-1} and B^k differential operators on $\partial\mathcal{M}$ of order $k-1$ and k, respectively.*

Assume that there exist $C > 0$, $\tau_* > 0$, and $\alpha > 0$ such that

$$\tau^{3/2}\|e^{\tau\varphi}u\|_{L^2(\mathcal{M})} \leq C\Big(\|e^{\tau\varphi}Pu\|_{L^2(\mathcal{M})} + \tau^{-\alpha}|e^{\tau\varphi}Bu_{|\partial\mathcal{M}}|_{\tau,3/2-k}\Big), \tag{14.5.1}$$

for all $u \in \mathscr{C}^\infty(\mathcal{M})$, with $\operatorname{supp}(u) \subset V$, and $\tau \geq \tau_*$. Then, $\alpha \leq 1/4$.

REMARK 14.28. Because of the proof scheme of Theorem 14.1, the optimality of the estimate therein obtained in Theorem 14.27 implies the optimality of estimation for a first-order operator $D_d - \operatorname{Op}_{\mathsf{T}}(f)$ given in Proposition 14.6 with respect to the boundary term.

PROOF. We consider a local chart $\mathcal{C} = (\kappa, O)$ at the boundary with $O \subset V$, that is, $O \cap \partial\mathcal{M} \neq \emptyset$, $\kappa(O) \subset \{x_d \geq 0\}$ and $\kappa(O \cap \partial\mathcal{M}) \subset \{x_d = 0\}$ (see Sect. 15.1). By abuse of notation, we still denote by P and φ the local representatives of the elliptic operator and weight function under consideration. Let U_+ be a bounded open set in $\overline{\mathbb{R}^d_+}$ that intersects $\{x_d = 0\}$ and such that $U_+ \subset \kappa(O)$.

The assumed estimate (14.5.1) reads

$$\tau^{3/2}\|e^{\tau\varphi}u\|_{+} \lesssim \|e^{\tau\varphi}Pu\|_{+} + \tau^{-\alpha}|e^{\tau\varphi}Bu_{|x_d=0^+}|_{\tau,3/2-k}, \tag{14.5.2}$$

$u \in \overline{\mathscr{C}}_c^\infty(U_+)$ and $\tau \geq \tau_*$.

We first proceed as in the proof of Theorem 4.4 of Volume 1 in the case $d \geq 2$ therein. Let $x^0 \in U^+ \cap \{x_d = 0\}$ and $\xi^0 \in \mathbb{R}^d$ be such that $p(x^0, \xi^0 + i d\varphi(x^0)) = 0$, that is,

$$p(x^0, \xi^0) = p(x^0, d\varphi(x^0)) \text{ and } \tilde{p}(x^0, \xi^0, d\varphi(x^0)) = 0. \tag{14.5.3}$$

Without any loss of generality, we can take $x^0 = 0$ and $\varphi(x^0) = 0$. We set $\zeta^0 = \xi^0 + i d\varphi(0)$, $w(x) = x \cdot \zeta^0$ and

$$G(x) = \frac{1}{2} \sum_{j,k} \partial^2_{x_j x_k} \varphi(0) x_j x_k.$$

We pick $f \in \mathscr{C}_c^\infty(\mathbb{R}^d)$, $f \not\equiv 0$, and set $u_\tau(x) = e^{i\tau w(x)} f(\tau^{1/2} x)$. Arguing as in the proof of Theorem 4.4, one finds (see (4.1.21) and (4.1.25))

$$\tau^{3/2} \|e^{\tau\varphi} u_\tau\|_+ \sim \tau^{3/2 - d/4} \|e^G f\|_+ \text{ and } \|e^{\tau\varphi} P u_\tau\|_+ \sim \tau^{3/2-d/4} \|e^G M f\|_+,$$

with

$$M = y \cdot p'_x(0, \zeta^0) + p'_\xi(0, \zeta^0) \cdot D_x.$$

Moreover, one has $Mf \neq 0$ by (4.1.24).

We now give an asymptotic estimation of the boundary term $|e^{\tau\varphi} B u_{\tau|x_d=0^+}|_{\tau, 3/2-k}$.

Lemma 14.29. *There exists $C > 0$ such that*

$$|e^{\tau\varphi} B u_{\tau|x_d=0^+}|_{\tau,3/2} \leq C \tau^{3/2-(d-1)/4} |e^G f_{|x_d=0^+}|_\partial \qquad \text{as } \tau \to +\infty.$$

A proof is given below.

From (14.5.2) and the asymptotic estimate obtained for each term, if $\alpha > 1/4$, letting $\tau \to +\infty$, one obtains

$$\|e^G f\|_+ \lesssim \|e^G M f\|_+,$$

for any $f \in \mathscr{C}_c^\infty(\mathbb{R}^d)$, even in the case $\operatorname{supp}(f)$ intersects $\{x_d = 0\}$. Changing f into $e^{-G} f$ yields

$$\|f\|_+ \lesssim \|Lf\|_+, \tag{14.5.4}$$

where $L = x \cdot p'_x(0, \zeta^0) + p'_\xi(0, \zeta^0) \cdot \big(D_x - (D_x G)\big)$ is a first-order differential operator with a principal part with *constant* coefficients. We shall now prove that the estimate in (14.5.4) cannot hold. This implies the conclusion, that is, $\alpha \leq 1/4$.

Above we considered any local chart $\mathcal{C} = (\kappa, O)$. Yet as in Sect. 8.3.2, we can choose a local chart associated with normal geodesic coordinates.

In such a chart, one has $p(x, \xi) = \xi_d^2 + r(x, \xi')$, with $r(x, \xi')$ a positive definite quadratic form in ξ' associated with the bilinear form $(\xi', \eta') \mapsto \tilde{r}(x, \xi', \eta')$. One thus finds that the principal part of L is given by $2\zeta_d^0 D_d + 2\tilde{r}(0, \zeta^{0,\prime}, D_{x'})$. Recall that $\zeta^0 = \xi^0 + i d\varphi(0)$.

LEMMA 14.30. *One can choose ξ^0 in (14.5.3) so that the principal symbol of L reads $\ell(\xi) = z^0\big(\xi_d - \tilde{r}(0, \eta^{0\prime}, \xi')\big)$, with $z^0 \in \mathbb{C}\setminus\{0\}$ and $\eta^{0\prime} \in \mathbb{C}^{d-1}\setminus\mathbb{R}^{d-1}$.*

A proof is given below.

The principal symbol of L thus reads $\ell(\xi) = z^0\big(\xi_d - b\cdot\xi')\big)$, with $b \in \mathbb{C}^{d-1}\setminus\{0\}$, and since $\eta^{0\prime} \notin \mathbb{R}^{d-1}$ and $\tilde{r}$ has real coefficients, one sees that for a well-chosen $\xi^{1\prime} \in \mathbb{R}^{d-1}\setminus\{0\}$, one has $\operatorname{Im} b\cdot\xi^{1\prime} = \operatorname{Im}(b)\cdot\xi^{1\prime} > 0$. One sets $\rho = b\cdot\xi^{1\prime}$. Observe that L takes the form

$$z_0(D_d - b\cdot D_{x'}) + x\cdot A,$$

where $A \in \mathbb{C}^d$ is constant.

Let $h \in \mathscr{C}_c^\infty(\mathbb{R}^d)$ and $\chi \in \mathscr{C}_c^\infty(-1,1)$ be such that $\chi = 1$ in a neighborhood of 0. For $\lambda \geq 1$, one sets

$$f(x', x_d) = e^{i\lambda(\xi^{1\prime}\cdot x' + \rho x_d)} h(x)\chi(x_d).$$

One has

$$Lf(x) = e^{i\lambda(\xi^{1\prime}\cdot x' + \rho x_d)} g(x), \quad g(x) = \big((Lh)(x)\chi(x_d) + z_0 h(x) D_d\chi(x_d)\big).$$

Estimate (14.5.4) reads

$$\|e^{-\lambda x_d \operatorname{Im}\rho}\chi h\|_+ \lesssim \|e^{-\lambda x_d \operatorname{Im}\rho} g\|_+.$$

For a smooth function w with compact support, one has

$$\|e^{-\lambda x_d \operatorname{Im}\rho} w\|_+^2 = \int_{\mathbb{R}_+} e^{-2\lambda x_d \operatorname{Im}\rho} |w(.,x_d)|^2_{L^2(\mathbb{R}^{d-1})}\, dx_d.$$

With the Taylor formula for $a(x_d) = |w(.,x_d)|^2_{L^2(\mathbb{R}^{d-1})}$, one has

$$\begin{aligned}
\int_{\mathbb{R}_+} e^{-2\lambda x_d \operatorname{Im}\rho} a(x_d) dx_d &= \int_{\mathbb{R}_+} e^{-2\lambda x_d \operatorname{Im}\rho}\big(a(0) + \mathcal{O}(x_d)\big) dx_d \\
&= \frac{1}{2\lambda \operatorname{Im}\rho} a(0) + \frac{1}{\lambda^2}\mathcal{O}(1) \\
&= \frac{1}{2\lambda \operatorname{Im}\rho} |w(.,0)|^2_{L^2(\mathbb{R}^{d-1})} + \frac{1}{\lambda^2}\mathcal{O}(1),
\end{aligned}$$

as $\lambda \to +\infty$. Applied for $w(x) = \chi(x_d)h(x)$ and $w = g$, one obtains

$$|h_{|x_d=0^+}|_{L^2(\mathbb{R}^{d-1})} \lesssim |g_{|x_d=0^+}|_{L^2(\mathbb{R}^{d-1})} = |(Lh)_{|x_d=0^+}|_{L^2(\mathbb{R}^{d-1})}. \tag{14.5.5}$$

With $\chi_1, \chi_2 \in \mathscr{C}_c^\infty(\mathbb{R}^{d-1})$, we set $h(x) = \big(\chi_1(x') + x_d\chi_2(x')\big)\chi(x_d)$, and we compute

$$(Lh)_{|x_d=0^+} = -z_0 b\cdot D_{x'}\chi_1(x') + (x'\cdot A')\chi_1(x') - iz_0\chi_2(x'),$$

where $A' = (A_1, \dots, A_{d-1})$. Observe that if we choose

$$\chi_2 = ib\cdot D_{x'}\chi_1(x') - i(x'\cdot A')\chi_1(x')/z_0,$$

we find $(Lh)_{|x_d=0^+} = 0$. However one has $h_{|x_d=0^+} = \chi_1$. Thus for $\chi_1 \neq 0$, estimate (14.5.5) cannot hold. We have reached a contradiction. ∎

PROOF OF LEMMA 14.29. We consider three cases regarding the order of the boundary operator B: (1) $k = 0$, that is, a Dirichlet-type boundary operator, meaning $B^{k-1} = 0$, (2) $k = 1$, and (3) $k \geq 2$. In all the three cases, the estimation is rather similar, yet it is better to separate them for the sake of exposition.

Case 1: $k = 0$. One has a Dirichlet-type boundary condition, that is, $Bu_{|\partial\mathcal{M}} = B^0 u_{|\partial\mathcal{M}}$, with $B^0 = b^0$ a smooth function, and one writes

$$
\begin{aligned}
|e^{\tau\varphi} B^0 u_{\tau|x_d=0^+}|_{\tau,3/2} &\lesssim \tau^{-1/2} |e^{\tau\varphi} b^0 u_{\tau|x_d=0^+}|_{\tau,2} \\
&\lesssim \tau^{-1/2} \sum_{\substack{\beta'\in\mathbb{N}^{d-1}\\ |\beta'|\leq 2}} \tau^{2-|\beta'|} |D_{x'}^{\beta'} (e^{\tau\varphi} b^0 u_\tau)_{|x_d=0^+}|_\partial \\
&\lesssim \tau^{-1/2} \sum_{\substack{\beta'\in\mathbb{N}^{d-1}\\ |\beta'|\leq 2}} \tau^{2-|\beta'|} |e^{\tau(\varphi+iw)} D_{x'}^{\beta'} \big(f(\tau^{1/2}.)\big)_{|x_d=0^+}|_\partial \\
&\lesssim \tau^{-1/2} \sum_{\substack{\beta'\in\mathbb{N}^{d-1}\\ |\beta'|\leq 2}} \tau^{2-|\beta'|/2} |e^{\tau(\varphi+iw)} (D_{x'}^{\beta'} f)(\tau^{1/2}.)_{|x_d=0^+}|_\partial.
\end{aligned}
$$

Then, arguing as in (4.1.21) in the proof of Theorem 4.4 of Volume 1, yet in dimension $d-1$, one obtains the result.

Case 2: $k = 1$. One has $B = B^1 - iB^0 D_d$, with B^1 and B^0 differential operators of order 1 and 0 on $\partial\mathcal{M}$. One writes

$$
\begin{aligned}
&|e^{\tau\varphi} B u_{\tau|x_d=0^+}|_{\tau,1/2} \\
&\quad \lesssim \tau^{-1/2} |e^{\tau\varphi} B u_{\tau|x_d=0^+}|_{\tau,1} \\
&\quad \lesssim \tau^{-1/2} \sum_{\substack{\beta'\in\mathbb{N}^{d-1}\\ |\beta'|\leq 1}} \tau^{1-|\beta'|} |D_{x'}^{\beta'} (e^{\tau\varphi} B u_\tau)_{|x_d=0^+}|_\partial \\
&\quad \lesssim \tau^{-1/2} \sum_{\substack{\beta'\in\mathbb{N}^{d-1},\ \ell=0,1\\ |\beta'|+\ell\leq 2}} \tau^{2-|\beta'|-\ell} |e^{\tau(\varphi+iw)} D_{x'}^{\beta'} D_d^\ell \big(f(\tau^{1/2}.)\big)_{|x_d=0^+}|_\partial,
\end{aligned}
$$

using that the coefficients of B and their derivatives are bounded. This yields

$$
\begin{aligned}
&|e^{\tau\varphi} B u_{\tau|x_d=0^+}|_{\tau,1/2} \\
&\quad \lesssim \tau^{3/2} \sum_{\substack{\beta'\in\mathbb{N}^{d-1},\ \ell=0,1\\ |\beta'|+\ell\leq 2}} \tau^{-(|\beta'|+\ell)/2} |e^{\tau(\varphi+iw)} (D_{x'}^{\beta'} D_d^\ell f)(\tau^{1/2}.)_{|x_d=0^+}|_\partial.
\end{aligned}
$$

As above, arguing as in (4.1.21), one obtains the result.

Case 3: $k \geq 2$. One has $B = B^k - iB^{k-1}D_d$, with B^k and B^{k-1} differential operators of order k and $k-1$ on $\partial\mathcal{M}$. As $3/2 - k < 0$, one writes

$$\begin{aligned}
&|e^{\tau\varphi} B u_{\tau|x_d=0^+}|_{\tau,3/2-k} \\
&\quad \lesssim \tau^{3/2-k} |e^{\tau\varphi} B u_{\tau|x_d=0^+}|_\partial \\
&\quad \lesssim \tau^{3/2-k} \sum_{\substack{\beta'\in\mathbb{N}^{d-1},\ \ell=0,1 \\ |\beta'|+\ell\leq k}} \tau^{k-|\beta'|-\ell} |e^{\tau(\varphi+iw)} D_{x'}^{\beta'} D_d^\ell \big(f(\tau^{1/2}.)\big)_{|x_d=0^+}|_\partial,
\end{aligned}$$

using that the coefficients of B are bounded. This yields

$$\begin{aligned}
&|e^{\tau\varphi} B u_{\tau|x_d=0^+}|_{\tau,3/2-k} \\
&\quad \lesssim \tau^{3/2} \sum_{\substack{\beta'\in\mathbb{N}^{d-1},\ \ell=0,1 \\ |\beta'|+\ell\leq k}} \tau^{-(|\beta'|+\ell)/2} |e^{\tau(\varphi+iw)} (D_{x'}^{\beta'} D_d^\ell f)(\tau^{1/2}.)_{|x_d=0^+}|_\partial.
\end{aligned}$$

As above, arguing as in (4.1.21), one obtains the result. ■

Proof of Lemma 14.30. In the chosen normal geodesic coordinates, (14.5.3) reads

$$p(0,\xi^0) = p(0, d\varphi(0)), \qquad \xi_d^0 \partial_d\varphi(0) + \tilde{r}(0, \xi^{0\prime}, d_{x'}\varphi(0)) = 0. \tag{14.5.6}$$

If $\partial_d\varphi(0) = 0$, one can choose $\xi^{0\prime} = 0$ and $\xi_d^0 \neq 0$ in (14.5.6). Consequently, one can choose $\zeta_d^0 = \xi_d^0 \neq 0$ and $\zeta^{0\prime} = d_{x'}\varphi(0) \neq 0$ since $d\varphi(0) \neq 0$ by Theorem 4.5 of Volume 1. Thus, with $z^0 = 2\xi_d^0 \neq 0$ and $\eta^{0\prime} = i d_{x'}\varphi(0)/\xi_d^0 \notin \mathbb{R}^{d-1}$, one obtains the sought form for the principal symbol $\ell(\xi)$.

If $\partial_d\varphi(0) \neq 0$ and $d \geq 3$, we choose ξ^0 such that $\xi_d^0 = 0$ and $\xi^{0\prime} \neq 0$ with $\tilde{r}\big(0, \xi^{0\prime}, d_{x'}\varphi(0)\big) = 0$ in (14.5.6). Then with $z^0 = 2i\partial_d\varphi(0) \neq 0$ and $\eta^{0\prime} = \big(\xi^{0\prime} + i d_{x'}\varphi(0)\big)/\big(i\partial_d\varphi(0)\big) \notin \mathbb{R}^{d-1}$, one obtains the sought form for the principal symbol $\ell(\xi)$.

If $d = 2$, then $p(x,\xi)$ takes locally the form $p(x,\xi) = \xi_2^2 + r(x)\xi_1^2$, where $r(x) \geq C$ with $C > 0$. One has $\tilde{p}\big(0, \xi^0, d_{x'}\varphi(0)\big) = \xi_2^0\partial_2\varphi(0) + r(0)\xi_1^0\partial_1\varphi(0)$. The second condition in (14.5.6) reads

$$\xi_2^0\partial_2\varphi(0) + r(0)\xi_1^0\partial_1\varphi(0) = 0.$$

If $\partial_d\varphi(0) = \partial_2\varphi(0) \neq 0$, one finds $\xi_2^0 = -r(0)\xi_1^0\partial_1\varphi(0)/\partial_2\varphi(0)$, with $\xi_1^0 \neq 0$ chosen such that $p(x,\xi^0) = p(x, d\varphi(0))$. Then, with $z^0 = 2\xi_2^0 + 2i\partial_2\varphi(0) \neq 0$ and $\eta^{0\prime} = \big(\xi_1^0 + i\partial_1\varphi(0)\big)/\big(\xi_2^0 + i\partial_2\varphi(0)\big)$, one finds that $\operatorname{Im}\eta^{0\prime} \neq 0$ if and only if

$$\xi_2^0\partial_1\varphi(0) - \xi_1^0\partial_2\varphi(0) \neq 0.$$

With the form of ξ_2^0 given above, one has

$$\xi_2^0\partial_1\varphi(0) - \xi_1^0\partial_2\varphi(0) = -\frac{\xi_1^0}{\partial_2\varphi(0)}\big(r(0)(\partial_1\varphi(0))^2 + (\partial_2\varphi(0))^2\big) \neq 0.$$

Thus, one obtains the sought form for the principal symbol $\ell(\xi)$. ■

14.6. A Refined Estimate Without Any Prescribed Boundary Operator

Here, we give the counterpart of the results obtained in Sect. 8.4.2. One aims to obtain an estimation without prescribing any particular boundary condition. We only wish to obtain bounds by means of norms on the Dirichlet trace $u_{|\partial\mathcal{M}}$ and a first-order trace $Bu_{|\partial\mathcal{M}}$, that is, where $B = \partial_\nu + B'$ is a first-order boundary operator. For this to make sense, one assumes that B fulfills the Lopatinskiĭ–Šapiro condition of Definition 2.2 along with P. Such boundary operators are characterized in Proposition 2.8. Here, the only assumption on the weight function made at the boundary is $\partial_\nu \varphi \neq 0$. Yet, one does not assume that (P, B, φ) fulfills the Lopatinskiĭ–Šapiro condition of Definition 8.1.

THEOREM 14.31. *Let $(\mathcal{M}, g)$ be a compact Riemannian manifold with boundary, and let $P = -\Delta_g + R_1$ with R_1 a first-order differential operator with bounded coefficients on $\mathcal{M}$. Let V be an open set of $\mathcal{M}$ such that $V_\partial = V \cap \partial\mathcal{M} \neq \emptyset$. Let B' be a differential operator of order one on $\partial\mathcal{M}$ such that (P, B) fulfills the Lopatinskiĭ–Šapiro condition of Definition 2.2 on $\overline{V_\partial}$ for $B = \partial_\nu + B'$. Let $\varphi \in \mathscr{C}^\infty(\mathcal{M})$ be such that the pair (P, φ) has the sub-ellipticity property of Definition 5.1 in $\overline{V}$ and $\partial_\nu \varphi \neq 0$ in $\overline{V_\partial}$. Then, there exist C and $\tau_* > 0$ such that*

$$\tau^{-1/2}\|e^{\tau\varphi}u\|_{\tau,2} + \tau^{-1/4}|e^{\tau\varphi}\operatorname{tr}(u)|_{\tau,1,1/2} \leq C\Big(\|e^{\tau\varphi}Pu\|_{L^2(\mathcal{M})} + \tau^{5/4}|e^{\tau\varphi}u_{|\partial\mathcal{M}}|_{L^2(\partial\mathcal{M})} + \tau^{-1/4}|e^{\tau\varphi}Bu_{|\partial\mathcal{M}}|_{\tau,1/2}\Big), \tag{14.6.1}$$

for all $u \in \mathscr{C}^\infty(\mathcal{M})$, with $\operatorname{supp}(u) \subset V$, and $\tau \geq \tau_*$.

In the framework of the normal geodesic coordinates introduced through the local chart $\mathcal{C} = (\kappa, O)$ in Sect. 8.3.2 and briefly recalled in Sect. 14.1.1, we prove the following proposition. Then, the proof of Theorem 14.31 is obtained through a patching procedure[1] of such local estimations as performed in Section 3.5.

PROPOSITION 14.32. *Let $P_0 = \operatorname{Op}(p^{\mathcal{C}})$ and $B_0 = \operatorname{Op}(b^{\mathcal{C}})$. Let $m^0 \in \partial\mathcal{M}$, V^0 be an open neighborhood of m^0 in $\mathcal{M}$. Set $x^0 = \kappa(m^0)$. Assume that (P_0, B_0) satisfies the Lopatinskiĭ–Šapiro condition of Definition 2.2 at x^0. Assume that $(P_0, \varphi^{\mathcal{C}})$ has the sub-ellipticity property of Definition 3.2 of Volume 1 in $\overline{U^0}$ with $U^0 = \kappa(V^0 \cap O)$ and that $\partial_d \varphi^{\mathcal{C}}_{|x_d=0} \neq 0$ in $\overline{U^0} \cap \{x_d = 0\}$. There exist a bounded neighborhood U_+ of x^0 in $\mathbb{R}^d_+$ such that $U_+ \subset U^0$ and*

[1] This patching procedure is also invoked for the proofs of Theorems 5.5, 5.6, and 8.14 for instance.

two constants C and $\tau_* > 0$ such that

$$\tau^{-1/2}\|v\|_{\tau,2,0} + \tau^{-1/4}|\operatorname{tr}(v)|_{\tau,1,1/2} \leq C\Big(\|P_{0,\varphi}v\|_+ + \tau^{5/4}|v_{|x_d=0^+}|_\partial + \tau^{-1/4}|B_{0,\varphi}v_{|x_d=0^+}|_{\tau,1/2}\Big),$$

for all $v \in \overline{\mathscr{C}}_c^\infty(U_+)$ and $\tau \geq \tau_*$.

The proof of Proposition 14.32 relies on a microlocal version of the estimate that we state below and it follows the proof scheme of Proposition 8.16 by patching together such microlocal estimates as in Sect. 8.3.5.

PROPOSITION 14.33. *Let x^0 be as in the statement of Proposition 14.32. Assume that (P_0, B_0) satisfies the Lopatinskiĭ–Šapiro condition of Definition 2.2 at x^0. Assume that $(P_0, \varphi^\mathcal{C})$ has the sub-ellipticity property of Definition 3.2 of Volume 1 in $\overline{U^0}$ with $U^0 = \kappa(V^0 \cap O)$ and that $\partial_d\varphi^\mathcal{C}_{|x_d=0} \neq 0$ in $\overline{U^0} \cap \{x_d = 0\}$. Let $\omega^{0\prime} \in T^*_{m^0}\partial\mathcal{M}$, with local representative $(\omega^{0\prime})^\mathcal{C} = \xi^{0\prime} \in \mathbb{R}^{d-1}$, and $\tau^0 \geq 0$ such that $(\xi^{0\prime}, \tau^0) \neq 0$. Then, there exists $\mathscr{U}$ a conic open neighborhood of $(x^0, \xi^{0\prime}, \tau^0)$ in $U^0 \times \mathbb{R}^{d-1} \times \mathbb{R}_+$ such that for $\chi \in S^0_{\mathsf{T},\tau}$, homogeneous of degree 0, with $\operatorname{supp}(\chi) \subset \mathscr{U}$, there exist $C > 0$ and $\tau_* > 0$ such that*

$$\tau^{-1/2}\|\operatorname{Op}_\mathsf{T}(\chi)v\|_{\tau,2} + \tau^{-1/4}|\operatorname{tr}(\operatorname{Op}_\mathsf{T}(\chi)v)|_{\tau,1,1/2} \leq C\Big(\|P_{0,\varphi}v\|_+ + \tau^{5/4}|v_{|x_d=0^+}|_\partial + \tau^{-1/4}|B_{0,\varphi}v_{|x_d=0^+}|_{\tau,1/2} + \|v\|_{\tau,2,-1}\Big), \tag{14.6.2}$$

for $\tau \geq \tau_*$, $v \in \overline{\mathscr{S}}(\mathbb{R}^d_+)$.

We follow the setting and notation of Sect. 14.4. In particular, to ease the notation, we write φ, p, b, and b_φ in place of $\varphi^\mathcal{C}$, $p^\mathcal{C}$, $b^\mathcal{C}$, and $b^\mathcal{C}_\varphi = b_{0,\varphi}$, respectively.

We write

$$p_{0,\varphi}(x, \xi', z, \tau) = \big(z - \gamma_1(\varrho')\big)\big(z - \gamma_2(\varrho')\big), \qquad \varrho' = (x, \xi', \tau),$$

with

$$\gamma_j(\varrho') = -i\tau\partial_d\varphi(x) + i(-1)^j\alpha(y'),$$

where

$$\begin{aligned}\alpha(\varrho')^2 &= r(x, \xi' + i\tau d_{x'}\varphi(x)) \\ &= r(x, \xi') - \tau^2 r(x, d_{x'}\varphi(x)) + 2i\tau\tilde{r}(x, \xi', d_{x'}\varphi(x)).\end{aligned}$$

For the proof of the microlocal estimate of Proposition 14.33, we shall consider three (exhaustive) cases for $\varrho^{0\prime} = (x^0, \xi^{0,\prime}, \tau^0)$ in the statement:

(1) $\tau^0 = 0$.
(2) $\tau^0 > 0$ and $\gamma_1(\varrho'^0) \neq \gamma_2(\varrho'^0)$.
(3) $\tau^0 > 0$ and $\gamma_1(\varrho'^0) = \gamma_2(\varrho'^0)$.

REMARK 14.34. Here, we assume that $\partial_d\varphi(x^0) \neq 0$. As a result, the roots $\gamma_1(\varrho'^0)$ and $\gamma_2(\varrho'^0)$ cannot be both real; otherwise, one has

$$\operatorname{Im}\gamma_1(\varrho'^0) = -i\tau\partial_d\varphi(x^0) - i\operatorname{Re}\alpha(\varrho'^0) = 0$$
$$\operatorname{Im}\gamma_2(\varrho'^0) = -i\tau\partial_d\varphi(x^0) + i\operatorname{Re}\alpha(\varrho'^0) = 0,$$

implying $\partial_d\varphi(x^0) = 0$ which is excluded.

14.6.1. Case 1: $\tau^0 = 0$. Here we consider $\varrho^{0\prime} = (x^0, \xi^{0,\prime}, \tau^0)$ with $\tau^0 = 0$ and $\xi^{0\prime} \in \mathbb{R}^{d-1} \setminus \{0\}$. Observe that the Lopatinskiĭ–Šapiro condition for (P_0, B_0, φ) (Definition 8.1) holds at $\varrho^{0\prime}$ for any weight function φ, since the Lopatinskiĭ–Šapiro condition of Definition 8.1 coincides with the Lopatinskiĭ–Šapiro condition of Definition 2.2 at such a point. By Proposition 14.5, there exists $\mathscr{U}$ a conic open neighborhood of $\varrho^{0\prime}$ in $U^0 \times \mathbb{R}^{d-1} \times \mathbb{R}_+$ such that for $\chi \in S^0_\tau$, homogeneous of degree 0 with $\operatorname{supp}(\chi) \subset \mathscr{U}$, one has

$$\tau^{-1/2}\|\operatorname{Op}_{\mathsf{T}}(\chi)v\|_{\tau,2} + \tau^{-1/4}|\operatorname{tr}(\operatorname{Op}_{\mathsf{T}}(\chi)v)|_{\tau,1,1/2} \lesssim \|P_{0,\varphi}v\|_+ + \tau^{-1/4}\,|B_{0,\varphi}v_{|x_d=0^+}|_{\tau,1/2} + \|v\|_{\tau,2,-1}, \tag{14.6.3}$$

for $\tau > 0$ chosen sufficiently large. This is stronger than (14.6.2).

14.6.2. Case 2: $\tau^0 > 0$ and $\gamma_1(\varrho'^0) \neq \gamma_2(\varrho'^0)$. By remark 14.34, one of the two roots is not real. Assume here that $\gamma_2(\varrho'^0) \notin \mathbb{R}$ (the other case is treated by exchanging the roles of the two roots).

Since the roots γ_1 and γ_2 are locally smooth with respect to (x, ξ', τ) and homogeneous of degree one in (ξ', τ) by Proposition 6.28, there exist $\mathscr{U}$ a conic open neighborhood of $\varrho^{0\prime} = (x^0, \xi^{0\prime}, \tau^0)$ in $U^0 \times \mathbb{R}^{d-1} \times \mathbb{R}_+$ and $C, C' > 0$ such that $\mathbb{S}_{\overline{\mathscr{U}}}$ is compact and

$$|\gamma_1(\varrho') - \gamma_2(\varrho')| \geq C\lambda_{\mathsf{T},\tau}, \quad |\operatorname{Im}\gamma_2(\varrho')| \geq C\lambda_{\mathsf{T},\tau}, \quad \text{and}$$
$$\lambda_{\mathsf{T},\tau}(\varrho') \leq C'\tau, \qquad \varrho' \in \mathscr{U}.$$

We let $\chi \in S^0_{\mathsf{T},\tau}$ be as in the statement and $\tilde{\chi} \in S^0_{\mathsf{T},\tau}$ be homogeneous of degree zero and be such that $\operatorname{supp}(\tilde{\chi}) \subset \mathscr{U}$ and $\tilde{\chi} \equiv 1$ on $\operatorname{supp}(\chi)$. From the smoothness and the homogeneity of the roots, one has $\tilde{\chi}\gamma_j \in S^1_{\mathsf{T},\tau}$, $j = 1, 2$. We set

$$L_2 = D_d - \operatorname{Op}_{\mathsf{T}}(\tilde{\chi}\gamma_2) \quad \text{and} \quad L_1 = D_d - \operatorname{Op}_{\mathsf{T}}(\tilde{\chi}\gamma_1).$$

If $\operatorname{Im}\gamma_2(\varrho') \gtrsim \lambda_{\mathsf{T},\tau}$, one may apply Lemma 6.21 (with $s = 1$ therein). If $\operatorname{Im}\gamma_2(\varrho') \lesssim -\lambda_{\mathsf{T},\tau}$, one may apply (the stronger) Lemma 6.20 (with $s = 1$ therein). Either way, one has

$$\|\operatorname{Op}_{\mathsf{T}}(\chi)v\|_{\tau,1,1} \lesssim \|\operatorname{Op}(L_2)\operatorname{Op}_{\mathsf{T}}(\chi)v\|_{\tau,0,1} + |\operatorname{Op}_{\mathsf{T}}(\chi)v_{|x_d=0^+}|_{\tau,3/2} + \|v\|_{\tau,0,-N}, \tag{14.6.4}$$

for $\tau \geq \tau_0$, for some $\tau_0 > 0$.

By Lemma 13.25, one sees that the operator L_1 fulfills the required properties of Proposition 14.6. With $s = 0$ therein, one finds

(14.6.5)
$$\tau^{-1/2}\|\mathrm{Op}_\mathsf{T}(\chi)w\|_{\tau,1,0} \lesssim \|\mathrm{Op}(L_1)\mathrm{Op}_\mathsf{T}(\chi)w\|_+ + \tau^{-1/4}|\mathrm{Op}_\mathsf{T}(\chi)w_{|x_d=0^+}|_{\tau,1/2} + \|w\|_{\tau,0,-N},$$

for $\tau \geq \tau_1$ and $w \in \overline{\mathscr{S}}(\mathbb{R}^d_+)$ for some $\tau_1 > 0$.

In (14.6.5), we take $w = \mathrm{Op}(L_2)v$:

$$\tau^{-1/2}\|\mathrm{Op}_\mathsf{T}(\chi)\mathrm{Op}(L_2)v\|_{\tau,1,0} \lesssim \|\mathrm{Op}(L_1)\mathrm{Op}_\mathsf{T}(\chi)\mathrm{Op}(L_2)v\|_+ + \tau^{-1/4}|\mathrm{Op}_\mathsf{T}(\chi)\mathrm{Op}(L_2)v_{|x_d=0^+}|_{\tau,1/2} + \|\mathrm{Op}(L_2)v\|_{\tau,0,-N}.$$

As $[\mathrm{Op}_\mathsf{T}(\chi), \mathrm{Op}(L_2)] \in \Psi^0_{\mathsf{T},\tau}$ and $\mathrm{Op}_\mathsf{T}(\chi)P_{0,\varphi} = \mathrm{Op}(L_1)\mathrm{Op}_\mathsf{T}(\chi)\mathrm{Op}(L_2) \mod \Psi^{1,0}_{\mathsf{T},\tau}$, one obtains

$$\begin{aligned}\tau^{-1/2}&\|\mathrm{Op}(L_2)\mathrm{Op}_\mathsf{T}(\chi)v\|_{\tau,1,0}\\ &\lesssim \|\mathrm{Op}_\mathsf{T}(\chi)P_{0,\varphi}v\|_+ + \tau^{-1/4}|\mathrm{Op}_\mathsf{T}(\chi)\mathrm{Op}(L_2)v_{|x_d=0^+}|_{\tau,1/2} + \|v\|_{\tau,1,0}\\ &\lesssim \|P_{0,\varphi}v\|_+ + \tau^{-1/4}|\mathrm{Op}(L_2)\mathrm{Op}_\mathsf{T}(\chi)v_{|x_d=0^+}|_{\tau,1/2} + \|v\|_{\tau,1,0},\end{aligned}$$

by using the trace inequality of Corollary 6.10. With estimation (14.6.4), one writes

$$\begin{aligned}\tau^{-1/2}\|\mathrm{Op}_\mathsf{T}(\chi)v\|_{\tau,1,1} &+ \tau^{-1/2}\|D_d\mathrm{Op}(L_2)\mathrm{Op}_\mathsf{T}(\chi)v\|_+\\ &\lesssim \|P_{0,\varphi}v\|_+ + \tau^{-1/4}|\mathrm{Op}(L_2)\mathrm{Op}_\mathsf{T}(\chi)v_{|x_d=0^+}|_{\tau,1/2}\\ &\quad + \tau^{-1/2}|\mathrm{Op}_\mathsf{T}(\chi)v_{|x_d=0^+}|_{\tau,3/2} + \|v\|_{\tau,1,0}.\end{aligned}$$

From the forms of L_2 and $B_{0,\varphi}$, one then obtains

$$\begin{aligned}\tau^{-1/2}&\|\mathrm{Op}_\mathsf{T}(\chi)v\|_{\tau,2,0}\\ &\lesssim \|P_{0,\varphi}v\|_+ + \tau^{-1/4}|B_{0,\varphi}\mathrm{Op}_\mathsf{T}(\chi)v_{|x_d=0^+}|_{\tau,1/2}\\ &\quad + \tau^{-1/4}|\mathrm{Op}_\mathsf{T}(\chi)v_{|x_d=0^+}|_{\tau,3/2} + \|v\|_{\tau,1,0}.\end{aligned}$$

Since $B_{0,\varphi} - \partial_\nu \in \Psi^1_{\mathsf{T},\tau}$ and $\partial_\nu = -\partial_d = -iD_d$, one also has

(14.6.6)
$$\begin{aligned}\tau^{-1/2}\|\mathrm{Op}_\mathsf{T}(\chi)v\|_{\tau,2,0} &+ \tau^{-1/4}|\mathrm{Op}_\mathsf{T}(\chi)v_{|x_d=0^+}|_{\tau,1,1/2}\\ &\lesssim \|P_{0,\varphi}v\|_+ + \tau^{-1/4}|B_{0,\varphi}\mathrm{Op}_\mathsf{T}(\chi)v_{|x_d=0^+}|_{\tau,1/2}\\ &\quad + \tau^{-1/4}|\mathrm{Op}_\mathsf{T}(\chi)v_{|x_d=0^+}|_{\tau,3/2} + \|v\|_{\tau,1,0}.\end{aligned}$$

Using that $[B_{0,\varphi}, \mathrm{Op}_{\mathsf{T}}(\chi)] \in \Psi^0_{\mathsf{T},\tau}$ and the trace inequality of Corollary 6.10, one finds

$$\begin{aligned}\tau^{-1/2}\|\mathrm{Op}_{\mathsf{T}}(\chi)v\|_{\tau,2,0} &+ \tau^{-1/4}|\mathrm{Op}_{\mathsf{T}}(\chi)v_{|x_d=0^+}|_{\tau,1,1/2}\\ &\lesssim \|P_{0,\varphi}v\|_+ + \tau^{-1/4}|B_{0,\varphi}v_{|x_d=0^+}|_{\tau,1/2}\\ &\quad + \tau^{-1/4}|\mathrm{Op}_{\mathsf{T}}(\chi)v_{|x_d=0^+}|_{\tau,3/2} + \|v\|_{\tau,1,0}.\end{aligned}$$

Finally, as $\Lambda^{3/2}_{\mathsf{T},\tau}\mathrm{Op}_{\mathsf{T}}(\chi) \in \tau^{3/2}\Psi^0_{\mathsf{T},\tau}$ since $\lambda_{\mathsf{T},\tau} \lesssim \tau$ in $\mathrm{supp}(\chi)$, we achieve the estimate

$$\begin{aligned}\tau^{-1/2}&\|\mathrm{Op}_{\mathsf{T}}(\chi)v\|_{\tau,2,0} + \tau^{-1/4}|\mathrm{Op}_{\mathsf{T}}(\chi)v_{|x_d=0^+}|_{\tau,1,1/2}\\ &\lesssim \|P_{0,\varphi}v\|_+ + \tau^{-1/4}|B_{0,\varphi}v_{|x_d=0^+}|_{\tau,1/2} + \tau^{5/4}|v_{|x_d=0^+}|_{\partial} + \|v\|_{\tau,1,0},\end{aligned}$$

which is only stronger than (14.6.2) with respect to the remainder term.

14.6.3. Case 3: $\tau^0 > 0$ and $\gamma_1(\varrho'^0) = \gamma_2(\varrho'^0)$. By Remark 14.34, $\gamma_1(\varrho'^0) = \gamma_2(\varrho'^0) \notin \mathbb{R}$ and there exist $\mathscr{U}$ a conic open neighborhood of $\varrho^{0\prime} = (x^0, \xi^{0\prime}, \tau^0)$ in $U^0 \times \mathbb{R}^{d-1} \times \mathbb{R}_+$ and $C, C' > 0$ such that $\mathbb{S}_{\overline{\mathscr{U}}}$ is compact and

$$\begin{aligned}|\operatorname{Im}\gamma_1(\varrho')| \geq C\lambda_{\mathsf{T},\tau}, \quad |\operatorname{Im}\gamma_2(\varrho')| \geq C\lambda_{\mathsf{T},\tau}, \quad \text{and} \quad \lambda_{\mathsf{T},\tau}(\varrho') \leq C'\tau,\\ \varrho' \in \mathscr{U}.\end{aligned}$$

We let $\chi \in S^0_{\mathsf{T},\tau}$ be as in the statement. Then, Lemma 8.25 applies to $\mathrm{Op}_{\mathsf{T}}(\chi)v$. One has

$$\begin{aligned}\|\mathrm{Op}_{\mathsf{T}}(\chi)v\|_{\tau,2} + |\operatorname{tr}(\mathrm{Op}_{\mathsf{T}}(\chi)v)|_{\tau,1,1/2} \lesssim \|P_{0,\varphi}v\|_+ + |\operatorname{tr}(\mathrm{Op}_{\mathsf{T}}(\chi)v)|_{\tau,1,1/2}\\ + \|v\|_{\tau,2,-1}.\end{aligned}$$

Since $B_{0,\varphi} - \partial_\nu \in \Psi^1_{\mathsf{T},\tau}$ and $\partial_\nu = -\partial_d = -iD_d$, one also has

$$\begin{aligned}\|\mathrm{Op}_{\mathsf{T}}(\chi)v\|_{\tau,2} &+ |\operatorname{tr}(\mathrm{Op}_{\mathsf{T}}(\chi)v)|_{\tau,1,1/2}\\ &\lesssim \|P_{0,\varphi}v\|_+ + |B_{0,\varphi}\mathrm{Op}_{\mathsf{T}}(\chi)v_{|x_d=0^+}|_{\tau,1/2}\\ &\quad + |\mathrm{Op}_{\mathsf{T}}(\chi)v_{|x_d=0^+}|_{\tau,3/2} + \|v\|^2_{\tau,2,-1}.\end{aligned} \tag{14.6.7}$$

Arguing as we did from (14.6.6) in Case 2, from (14.6.7), we obtain estimate (14.6.2) in Case 3.

14.7. Optimal Estimate with Source Terms of Weaker Regularity

Here, we prove an improved version of Theorem 13.4 that allows one to provide an estimate for H^1-functions solutions to $Pu = F_0 + \operatorname{div}_g F$ with $F_0 \in L^2(\mathcal{M})$, $F \in L^2V(\mathcal{M})$. As above, the (optimal) improvement lies in the presence of a factor $\tau^{-1/4}$ for the trace terms. We let $\mathsf{B}(.,.)$ be as given in (13.1.3), and more generally we use the notation introduced in Chap. 13.

THEOREM 14.35. *Let $(\mathcal{M}, g)$ be a compact Riemannian manifold with boundary, and let $P = -\Delta_g + R_1$ with R_1 a first-order differential operator on $\mathcal{M}$. Let V be an open set of $\mathcal{M}$ and set $V_\partial = V \cap \partial\mathcal{M}$. Let $\varphi \in \mathscr{C}^\infty(\mathcal{M})$ be such that the pair (P, φ) has the sub-ellipticity property of Definition 5.1 in $\overline{V}$. If $V_\partial \neq \emptyset$, consider B a differential operator of order $k \in \mathbb{N}$ in V, and assume moreover that (P, B, φ) satisfies the Lopatinskiĭ–Šapiro condition at all points $m \in \overline{V_\partial}$ (Definition 8.1). Then, there exist C and $\tau_* > 0$ such that*

$$(14.7.1)\quad \tau^{1/2}\|e^{\tau\varphi}u\|_{\tau,1} + \tau^{3/4}|e^{\tau\varphi_{|\partial\mathcal{M}}}u_{|\partial\mathcal{M}}|_{\tau,1/2} + \tau^{3/4}|e^{\tau\varphi_{|\partial\mathcal{M}}}(\partial_\nu u + g(F,\nu))_{|\partial\mathcal{M}}|_{\tau,-1/2} \leq C\big(\|e^{\tau\varphi}F_0\|_{L^2(\mathcal{M})} + \tau\|e^{\tau\varphi}F\|_{L^2V(\mathcal{M})} + \tau^{3/4}|e^{\tau\varphi_{|\partial\mathcal{M}}}\mathsf{B}(u, g(F,\nu))|_{\tau,1/2-k}\big),$$

for $\tau \geq \tau_*$ and $u \in H^1(\mathcal{M})$, $F_0 \in L^2(\mathcal{M})$, $F \in L^2V(\mathcal{M})$ such that

$$Pu = F_0 + \operatorname{div}_g F, \quad \textit{and} \quad \operatorname{supp}(u) \cup \operatorname{supp}(F_0) \cup \operatorname{supp}(F) \subset \overline{V}.$$

Under the setting and notation of Sect. 13.2, we prove microlocal versions of the result in the next section, and the proof of Theorem 14.35 then follows from a patching argument as done in Sect. 13.4.

14.7.1. Microlocal Estimate. With the notation of Sects. 13.2 and 14.1.1, we set $P_0 = \operatorname{Op}(p^{\mathcal{C}})$, $B_0 = \operatorname{Op}(b^{\mathcal{C}})$.

The following proposition is an improved version of Proposition 13.10 with respect to the boundary terms.

PROPOSITION 14.36. *Let $m^0 \in \partial\mathcal{M}$ and V^0 be an open neighborhood of m^0 in $\mathcal{M}$ that meets one connected component of $\partial\mathcal{M}$. Set $x^0 = \kappa(m^0)$. Assume that (P_0, φ) has the sub-ellipticity property of Definition 3.2 of Volume 1 in $\overline{U^0}$ with $U^0 = \kappa(V^0 \cap O)$. Let $\omega^{0\prime} \in T^*_{m^0}\partial\mathcal{M}$, with local representative $(\omega^{0\prime})^{\mathcal{C}} = \xi^{0\prime} \in \mathbb{R}^{d-1}$, and $\tau^0 \geq 0$ such that $(\xi^{0\prime}, \tau^0) \neq 0$, and assume that the Lopatinskiĭ–Šapiro condition of Definition 8.1 holds at $(m^0, \omega^{0\prime}, \tau^0)$. Then, there exists $\mathscr{U}$ a conic open neighborhood of $\varrho^{0\prime} = (x^0, \xi^{0\prime}, \tau^0)$ in $U^0 \times \mathbb{R}^{d-1} \times \mathbb{R}_+$ such that for $\chi \in S^0_{\mathsf{T},\tau}$, homogeneous of degree zero, with $\operatorname{supp}(\chi) \subset \mathscr{U}$, there exist $C > 0$ and $\tau_* > 0$ such that*

$$(14.7.2)\quad \tau^{1/2}\|\operatorname{Op}_{\mathsf{T}}(\chi)v\|_{\tau,1} + \tau^{3/4}|(D_d\operatorname{Op}_{\mathsf{T}}(\chi)v - i\operatorname{Op}_{\mathsf{T}}(\chi)G^d)_{|x_d=0^+}|_{\tau,-1/2} + \tau^{3/4}|\operatorname{Op}_{\mathsf{T}}(\chi)v_{|x_d=0^+}|_{\tau,1/2} \leq C\Big(\|G_0\|_+ + \tau\|G\|_+ + \tau^{3/4}|\mathsf{B}_{0,\varphi}(v, -G^d)_{|x_d=0^+}|_{\tau,1/2-k} + \|v\|_{\tau,1} + \tau^{3/4}|(D_d v - iG^d)_{|x_d=0^+}|_{\tau,-3/2}\Big),$$

for $\tau \geq \tau_*$ and $v \in H^1(\mathbb{R}^d_+)$, $G_0 \in L^2(\mathbb{R}^d_+)$, $G \in L^2V(\mathbb{R}^d_+)$ such that

$$P_{0,\varphi}v = G_0 + \sum_{1\leq i\leq d} \partial_j G^j.$$

We prove the estimate of Proposition 14.36 in the following four (exhaustive) cases:

(1) $\operatorname{Im}\gamma_2(\varrho^{0\prime}) > 0$.
(2) or $\operatorname{Im}\gamma_2(\varrho^{0\prime}) = 0$.
(3) or $\operatorname{Im}\gamma_2(\varrho^{0\prime}) < 0$ and $\gamma_2(\varrho^{0\prime}) \neq \gamma_1(\varrho^{0\prime})$.
(4) or $\operatorname{Im}\gamma_2(\varrho^{0\prime}) < 0$ and $\gamma_2(\varrho^{0\prime}) = \gamma_1(\varrho^{0\prime})$.

14.7.1.1. *Case 1: One Root in the Open Upper Complex Half-Plane.* Here, we have $\operatorname{Im}\gamma_1(x^0, \xi^{0\prime}, \tau^0) < 0$, $\operatorname{Im}\gamma_2(x^0, \xi^{0\prime}, \tau^0) > 0$, and

$$\check{p}_\varphi^+(m^0, \omega^{0\prime}, z, \tau^0) = z - \gamma_2(x^0, \xi^{0\prime}, \tau^0).$$

With the Lopatinskiĭ–Šapiro condition of Definition 8.1 holding at the considered point, by (8.3.11), we have moreover

$$b_\varphi(x^0, \xi^{0\prime}, \xi_d = \gamma_2, \tau^0) = b(x^0, \xi^{0\prime} + i\tau^0 d_{x'}\varphi(x^0), \gamma_2 + i\tau^0\partial_{x_d}\varphi(x^0)) \neq 0.$$

As the roots γ_1 and γ_2 are locally smooth with respect to (x, ξ', τ) and homogeneous of degree one in (ξ', τ) by Proposition 6.28, there exist $\mathscr{U}$ a conic open neighborhood of $(x^0, \xi^{0\prime}, \tau^0)$ in $U^0 \times \mathbb{R}^{d-1} \times \mathbb{R}_+$ and $C, C' > 0$ such that $\mathbb{S}_{\overline{\mathscr{U}}}$ is compact and

$$\operatorname{Im}\gamma_2(\varrho') \geq C\lambda_{\mathsf{T},\tau}, \quad \text{and} \quad \operatorname{Im}\gamma_1(\varrho') \leq -C'\lambda_{\mathsf{T},\tau},$$

and

$$b_\varphi(x, \xi', \xi_d = \gamma_2(\varrho'), \tau) \neq 0,$$

if $\varrho' = (x, \xi', \tau) \in \overline{\mathscr{U}}$.

We let $\chi \in S^0_{\mathsf{T},\tau}$ be as in the statement and $\tilde{\chi} \in S^0_{\mathsf{T},\tau}$ be homogeneous of degree zero and be such that $\operatorname{supp}(\tilde{\chi}) \subset \mathscr{U}$ and $\tilde{\chi} \equiv 1$ on $\operatorname{supp}(\chi)$. From the smoothness and the homogeneity of the roots, we have $\tilde{\chi}\gamma_j \in S^1_{\mathsf{T},\tau}$, $j = 1, 2$. We set

$$P^+ = D_d - \operatorname{Op}_{\mathsf{T}}(\tilde{\chi}\gamma_2) \quad \text{and} \quad P^- = D_d - \operatorname{Op}_{\mathsf{T}}(\tilde{\chi}\gamma_1).$$

From Lemma 6.21, we have, for some $\tau_* > 0$,

$$\begin{aligned} &\|\operatorname{Op}_{\mathsf{T}}(\chi)u\|_{\tau,1,s} \\ &\quad \leq C\Big(\|P^+\operatorname{Op}_{\mathsf{T}}(\chi)u\|_{\tau,0,s} + |\operatorname{Op}_{\mathsf{T}}(\chi)u_{|x_d=0^+}|_{\tau,s+1/2}\Big) + C_N\|u\|_{\tau,0,-N}, \end{aligned} \tag{14.7.3}$$

for $\tau \geq \tau_*$, $u \in \overline{\mathscr{S}}(\mathbb{R}^d_+)$ and $s \in \mathbb{R}$.

Let now $v \in \overline{\mathscr{S}}(\mathbb{R}^d_+)$. We set $w = \operatorname{Op}_{\mathsf{T}}(\chi)v$. Estimate (13.3.7) that reads

$$\begin{aligned} &\|P^+w\|_+ + |(P^+w - i\operatorname{Op}_{\mathsf{T}}(\chi)G^d)_{|x_d=0^+}|_{\tau,-1/2} \\ &\qquad \lesssim \tau^{-1}\|G_0\|_+ + \|G\|_+ + \|v\|_{\tau,1,-1} \end{aligned} \tag{14.7.4}$$

applies here since the present setting is a subcase of Case 1 for the proof of Proposition 13.10. For the same reason, estimate (13.3.9) also holds in the present case:

(14.7.5)
$$|w_{|x_d=0^+}|_{\tau,1/2} + |(D_d w - i\mathrm{Op_T}(\chi)G^d)_{|x_d=0^+}|_{\tau,-1/2} \lesssim |\mathsf{B}_{0,\varphi}(w, -\mathrm{Op_T}(\chi)G^d)|_{\tau,1/2-k} + |(P^+ w - i\mathrm{Op_T}(\chi)G^d)_{|x_d=0^+}|_{\tau,-1/2} + \tau^{-1}\|v\|_{\tau,1} + |(D_d v - iG^d)_{|x_d=0^+}|_{\tau,-3/2}.$$

With (14.7.4) and (14.7.5), we obtain

(14.7.6)
$$\|P^+ w\|_+ + \tau^{-1/4}|w_{|x_d=0^+}|_{\tau,1/2} + \tau^{-1/4}|(D_d w - i\mathrm{Op_T}(\chi)G^d)_{|x_d=0^+}|_{\tau,-1/2} \lesssim \tau^{-1/4}|\mathsf{B}_{0,\varphi}(w, -\mathrm{Op_T}(\chi)G^d)|_{\tau,1/2-k} + \tau^{-1}\|G_0\|_+ + \|G\|_+ + \tau^{-1}\|v\|_{\tau,1} + \tau^{-1/4}|(D_d v - iG^d)_{|x_d=0^+}|_{\tau,-3/2}.$$

With (14.7.3) in the case $s = 0$, we obtain

$$\tau^{-1/4}\|w\|_{\tau,1} + \tau^{-1/4}|w_{|x_d=0^+}|_{\tau,1/2} + \tau^{-1/4}|(D_d w - i\mathrm{Op_T}(\chi)G^d)_{|x_d=0^+}|_{\tau,-1/2} \lesssim \tau^{-1/4}|\mathsf{B}_{0,\varphi}(w, -\mathrm{Op_T}(\chi)G^d)|_{\tau,1/2-k} + \tau^{-1}\|G_0\|_+ + \|G\|_+ + \tau^{-1}\|v\|_{\tau,1} + \tau^{-1/4}|(D_d v - iG^d)_{|x_d=0^+}|_{\tau,-3/2},$$

yielding

$$\tau^{3/4}\|w\|_{\tau,1} + \tau^{3/4}|w_{|x_d=0^+}|_{\tau,1/2} + \tau^{3/4}|(D_d w - i\mathrm{Op_T}(\chi)G^d)_{|x_d=0^+}|_{\tau,-1/2} \lesssim \tau^{3/4}|\mathsf{B}_{0,\varphi}(w, -\mathrm{Op_T}(\chi)G^d)|_{\tau,1/2-k} + \|G_0\|_+ + \tau\|G\|_+ + \|v\|_{\tau,1} + \tau^{3/4}|(D_d v - iG^d)_{|x_d=0^+}|_{\tau,-3/2},$$

which is stronger than (14.7.2). This concludes the proof for Case 1.

14.7.1.2. *Case 2: One Root on the Real Axis.* Here, $\operatorname{Im}\gamma_1(x^0,\xi^{0\prime},\tau^0) < 0$ and $\operatorname{Im}\gamma_2(x^0,\xi^{0\prime},\tau^0) = 0$. In particular, we have

$$\check{p}_\varphi^+(m^0,\omega^{0\prime},z,\tau^0) = z - \gamma_2(x^0,\xi^{0\prime},\tau^0).$$

Note that this implies $\tau^0 \neq 0$ (since in the case $\tau = 0$, one has $\operatorname{Im}\gamma_2 > 0$). Thus, there exists C_0 such that $\lambda_{\mathsf{T},\tau}(\xi^0,\tau^0) \leq C_0\tau^0$.

With the Lopatinskiĭ–Šapiro condition of Definition 8.1 holding at the considered point, by (8.3.11), we have moreover

$$b_\varphi(x^0,\xi^{0\prime},\xi_d = \gamma_2,\tau^0) = b(x^0,\xi^{0\prime} + i\tau^0 d_{x'}\varphi(x^0), \gamma_2 + i\tau^0\partial_{x_d}\varphi(x^0)) \neq 0.$$

As the roots γ_1 and γ_2 are locally smooth with respect to (x,ξ',τ) and homogeneous of degree one in (ξ',τ) by Proposition 6.28, there exist $\mathscr{U}$ a conic open neighborhood of $(x^0, \xi^{0\prime}, \tau^0)$ in $U^0 \times \mathbb{R}^{d-1} \times \mathbb{R}_+$ and $C, C' > 0$ such that $\mathbb{S}_{\overline{\mathscr{U}}}$ is compact and

$$\gamma_1(\varrho') \neq \gamma_2(\varrho'), \ \operatorname{Im} \gamma_2(\varrho') \geq -C\lambda_{\mathsf{T},\tau}, \ \text{and} \ \ \operatorname{Im} \gamma_1(\varrho') \leq -C'\lambda_{\mathsf{T},\tau},$$

and

$$\lambda_{\mathsf{T},\tau}(\varrho') \leq 2C_0\tau,$$

and

$$b_\varphi(x, \xi', \xi_d = \gamma_2(\varrho'), \tau) \neq 0,$$

if $\varrho' = (x, \xi', \tau) \in \overline{\mathscr{U}}$.

We let $\chi \in S^0_{\mathsf{T},\tau}$ be as in the statement and $\tilde{\chi} \in S^0_{\mathsf{T},\tau}$ be homogeneous of degree zero and be such that $\operatorname{supp}(\tilde{\chi}) \subset \mathscr{U}$ and $\tilde{\chi} \equiv 1$ on $\operatorname{supp}(\chi)$. From the smoothness and the homogeneity of the roots, we have $\tilde{\chi}\gamma_j \in S^1_{\mathsf{T},\tau}$, $j = 1, 2$. We set

$$P^+ = D_d - \operatorname{Op}_{\mathsf{T}}(\tilde{\chi}\gamma_2) \ \text{ and } \ P^- = D_d - \operatorname{Op}_{\mathsf{T}}(\tilde{\chi}\gamma_1).$$

As in (14.4.7), we have

$$\begin{aligned}(14.7.7) \quad & \tau^{-1/2}\|\operatorname{Op}_{\mathsf{T}}(\chi)u\|_{\tau,1,s} \\ & \lesssim \|P^+\operatorname{Op}_{\mathsf{T}}(\chi)u\|_{\tau,0,s} + \tau^{-1/4}|\operatorname{Op}_{\mathsf{T}}(\chi)u_{|x_d=0^+}|_{\tau,s+1/2} + \|u\|_{\tau,0,-N},\end{aligned}$$

for τ sufficiently large, $u \in \overline{\mathscr{S}}(\mathbb{R}^d_+)$ and $s \in \mathbb{R}$.

Let now $v \in \overline{\mathscr{S}}(\mathbb{R}^d_+)$. We set $w = \operatorname{Op}_{\mathsf{T}}(\chi)v$. As in Case 1, we have estimates (14.7.4) and (14.7.5) since the present setting is a subcase of Case 1 for the proof of Proposition 13.10. We thus obtain (14.7.6), that is,

$$\begin{aligned}\|P^+w\|_+ &+ \tau^{-1/4}|w_{|x_d=0^+}|_{\tau,1/2} + \tau^{-1/4}|(D_d w - i\operatorname{Op}_{\mathsf{T}}(\chi)G^d)_{|x_d=0^+}|_{\tau,-1/2} \\ &\lesssim \tau^{-1/4}|\mathsf{B}_{0,\varphi}(w, -\operatorname{Op}_{\mathsf{T}}(\chi)G^d)|_{\tau,1/2-k} + \tau^{-1}\|G_0\|_+ + \|G\|_+ \\ &\quad + \tau^{-1}\|v\|_{\tau,1} + \tau^{-1/4}|(D_d v - iG^d)_{|x_d=0^+}|_{\tau,-3/2}.\end{aligned}$$

With (14.7.7) in the case $s = 0$, we obtain

$$\begin{aligned}\tau^{-1/2}&\|\operatorname{Op}_{\mathsf{T}}(\chi)u\|_{\tau,1} + \tau^{-1/4}|w_{|x_d=0^+}|_{\tau,1/2} \\ &+ \tau^{-1/4}|(D_d w - i\operatorname{Op}_{\mathsf{T}}(\chi)G^d)_{|x_d=0^+}|_{\tau,-1/2} \\ &\lesssim \tau^{-1/4}|\mathsf{B}_{0,\varphi}(w, -\operatorname{Op}_{\mathsf{T}}(\chi)G^d)|_{\tau,1/2-k} + \tau^{-1}\|G_0\|_+ + \|G\|_+ \\ &\quad + \tau^{-1}\|v\|_{\tau,1} + \tau^{-1/4}|(D_d v - iG^d)_{|x_d=0^+}|_{\tau,-3/2}.\end{aligned}$$

Upon multiplication by τ, this is the sought estimate (14.7.2). This concludes the proof for Case 2.

14.7.1.3. *Case 3: Roots Are Different and Both in the Open Lower Complex Half-Plane.* Here, one applies without any change the argument of Sect. 13.3.2 that gives a conic neighborhood $\mathscr{U}$ as in the statement and

$$\tau\Big(\|\mathrm{Op_T}(\chi)v\|_{\tau,1} + |\mathrm{Op_T}(\chi)v_{|x_d=0^+}|_{\tau,1/2} + |(D_d\mathrm{Op_T}(\chi)v - i\mathrm{Op_T}(\chi)G^d)_{|x_d=0^+}|_{\tau,-1/2}\Big) \lesssim \|G_0\|_+ + \tau\|G\|_+ + \tau\|v\|_{\tau,1,-1},$$

for $\tau > 0$ chosen sufficiently large and for χ as in the statement. This estimate is stronger than (14.7.2). This concludes the proof for Case 3.

14.7.1.4. *Case 4: Roots Coincide in the Open Lower Complex Half-Plane.* Here, one applies without any change the argument of Sect. 13.3.3. One obtains a conic neighborhood $\mathscr{U}$ as in the statement and estimate (13.3.16) that we reproduce here:

$$\tau\Big(\|\mathrm{Op_T}(\chi)v\|_{\tau,1} + |\mathrm{Op_T}(\chi)v_{|x_d=0^+}|_{\tau,1/2} + |(D_d\mathrm{Op_T}(\chi)v - i\mathrm{Op_T}(\chi)G^d)_{|x_d=0^+}|_{\tau,-1/2}\Big) \lesssim \|G_0\|_+ + \tau\|G\|_+ + \|v\|_{\tau,1} + |(D_dv - iG^d)_{|x_d=0^+}|_{\tau,-N},$$

for $\tau > 0$ chosen sufficiently large and for χ as in the statement. This estimate is stronger than (14.7.2). This concludes the proof for Case 4.

14.7.2. Shifted Estimate. Arguing as for the proof of Theorem 8.24, we deduce from Theorem 14.31 the following shifted estimate.

THEOREM 14.37. *Let $(\mathcal{M}, g)$ be a compact Riemannian manifold with boundary, and let $P = -\Delta_g + R_1$ with R_1 a first-order differential operator on $\mathcal{M}$. Let V be an open set of $\mathcal{M}$ and set $V_\partial = V \cap \partial\mathcal{M}$. Let $\varphi \in \mathscr{C}^\infty(\mathcal{M})$ be such that the pair (P, φ) has the sub-ellipticity property of Definition 5.1 in $\overline{V}$. If $V_\partial \neq \emptyset$, consider B a differential operator of order $k \in \mathbb{N}$ in V, and assume moreover that (P, B, φ) satisfies the Lopatinskiĭ–Šapiro condition at all points $m \in \overline{V_\partial}$ (Definition 8.1). Then, there exist C and $\tau_* > 0$ such that*

(14.7.8)

$$\tau^{1/2}\|e^{\tau\varphi}u\|_{\tau,1} + \tau^{5/4}|e^{\tau\varphi_{|\partial\mathcal{M}}}u_{|\partial\mathcal{M}}|_{\tau,0} + \tau^{5/4}|e^{\tau\varphi_{|\partial\mathcal{M}}}(\partial_\nu u + g(F,\nu))_{|\partial\mathcal{M}}|_{\tau,-1} \leq C\big(\|e^{\tau\varphi}F_0\|_{L^2(\mathcal{M})} + \tau\|e^{\tau\varphi}F\|_{L^2V(\mathcal{M})} + \tau^{5/4}|e^{\tau\varphi_{|\partial\mathcal{M}}}\mathsf{B}(u, g(F,\nu))|_{\tau,-k}\big),$$

for $\tau \geq \tau_*$ and $u \in H^1(\mathcal{M})$, $F_0 \in L^2(\mathcal{M})$, $F \in L^2V(\mathcal{M})$ such that

$$Pu = F_0 + \operatorname{div}_g F, \quad \textit{and} \quad \operatorname{supp}(u) \cup \operatorname{supp}(F_0) \cup \operatorname{supp}(F) \subset \overline{V}.$$

14.8. Notes

In the case of Dirichlet boundary conditions, the results of Theorems 14.1 and 14.35 were obtained in the work of O. Yu. Imanuvilov and J.-P. Puel [179]. As mentioned above, if compared to the results in Chaps. 8 and 13, such estimates yield an improvement with respect to the boundary terms. They have been used to obtain controllability results for the Navier–Stokes equations; see, for instance, the work of E. Fernández-Cara et al. [145].

Together with M. Yamamoto, O. Yu. Imanuvilov and J.-P. Puel further extended the result in the case of the associated parabolic operator [180]. This result is, for instance, used in the work by M. Badra [45] to obtain an estimation of a solution of the Stokes system.

The proof scheme we use here has some similarity with that of [179], for instance, by writing the conjugated operator as a product of two first-order terms. The proof of the critical estimate, that is, the estimate of Proposition 14.6 here relies however on different arguments. Here, we have chosen to rely on the strength of a Fefferman–Phong type inequality [142] as proven by J.-M. Bony [77]. For details on the proofs of such inequalities, we refer to the book by L. Hörmander [175, Sections 18.4-6] and the book by N. Lerner [228, Chapter 2].

Here, the Carleman estimate obtained by O. Yu. Imanuvilov and J.-P. Puel [179] is generalized to any boundary condition such that the Lopatinskiĭ–Šapiro condition holds. Moreover, the optimality of the estimation is proven in Sect. 14.5.

In Sect. 14.6, an estimation is also derived without prescribing any particular boundary condition. We have to assume that $\partial_\nu \varphi \neq 0$. To us, obtaining the optimal estimate if $\partial_\nu \varphi$ is allowed to vanish is an open question, whose study is useful if one wishes to derive global estimates with boundary observation terms; see Sect. 8.5.

In Chapter 3 of Volume 1 and Chaps. 5, 8, and 13 in the present volume, global estimates were derived. This could also be done in the framework of the optimal local estimations we have obtained here.

Appendix

14.A. Hörmander Calculus Properties

14.A.1. Slowly Varying and Temperate Metrics. Here, we prove Proposition 14.7 for the metric g^γ. The proof for g follows by taking $\gamma = \tau$.

Slow Variation. Let $\varepsilon > 0$ whose value will be set below, and let (x,ξ), $(y,\eta) \in \mathbb{R}^{2d}$, and $\tau, \gamma \in [1,+\infty)$ such that $\tau \geq \gamma$ and $g^\gamma_\varrho(y,\eta) < \varepsilon$, for $\varrho = (x,\xi,\tau)$. This reads

(14.A.1)
$$|y|^2 + \frac{|\eta|^2}{\tau^2 + |\xi'|^2} = |y|^2 + \lambda_{\mathsf{T},\tau}(\varrho')^{-2}|\eta|^2 \leq \gamma\varepsilon/\tau \leq \varepsilon, \quad \text{with } \varrho' = (x,\xi',\tau).$$

Set $\tilde{\varrho} = (x+y, \xi+\eta, \tau)$, and observe that it suffices to have $\lambda_{\mathsf{T},\tau}(\varrho) \lesssim \lambda_{\mathsf{T},\tau}(\tilde{\varrho})$ to obtain $g^{\gamma}_{\tilde{\varrho}} \lesssim g^{\gamma}_{\varrho}$. This reads

$$\tau + |\xi'| \lesssim \tau + |\xi' + \eta'|. \tag{14.A.2}$$

By (14.A.1), one has

$$\begin{aligned}\tau + |\xi'| &\lesssim \tau + |\xi' + \eta'| + |\eta'| \lesssim \tau + |\xi' + \eta'| + |\eta| \\ &\lesssim \tau + |\xi' + \eta'| + \varepsilon^{1/2}(\tau + |\xi'|).\end{aligned}$$

Thus, for $\varepsilon > 0$ chosen sufficiently small, one obtains (14.A.2).

Temperate Property. Let $(z, \zeta) \in \mathbb{R}^{2d}$. With

$$\begin{aligned}g^{\gamma}_{\varrho}(z, \zeta) &= \tau\gamma^{-1}|z|^2 + \tau\gamma^{-1}\lambda_{\mathsf{T},\tau}^{-2}(\varrho')|\zeta|^2, \\ g^{\gamma}_{\tilde{\varrho}}(z, \zeta) &= \tau\gamma^{-1}|z|^2 + \tau\gamma^{-1}\lambda_{\mathsf{T},\tau}^{-2}(\tilde{\varrho}')|\zeta|^2,\end{aligned}$$

one sees that the property described in (14.2.3) holds if one has

$$\lambda_{\mathsf{T},\tau}^{-2}(\varrho') \lesssim \big(1 + (g^{\gamma})^{\sigma}_{\varrho}(x - \tilde{x}, \xi - \tilde{\xi})\big)^N \lambda_{\mathsf{T},\tau}^{-2}(\tilde{\varrho}'),$$

for some N with $(g^{\gamma})^{\sigma}_{\varrho}$ given in (14.2.1). With $\lambda_{\mathsf{T},\tau}(\varrho')^2 = \tau^2 + |\xi'|^2$ and $\lambda_{\mathsf{T},\tau}(\tilde{\varrho}')^2 = \tau^2 + |\tilde{\xi}'|^2$, it is then sufficient to prove that

$$|\tilde{\xi}'|^2 \lesssim \big(1 + (g^{\gamma})^{\sigma}_{\varrho}(x - \tilde{x}, \xi - \tilde{\xi})\big)^N (\tau^2 + |\xi'|^2). \tag{14.A.3}$$

In fact, with $\tau \geq 1$ and $\gamma \geq 1$ here, one has

$$\begin{aligned}|\tilde{\xi}'|^2 &\lesssim |\xi' - \tilde{\xi}'|^2 + |\xi'|^2 \lesssim (1 + \tau^{-1}\gamma|\xi' - \tilde{\xi}'|^2)(\tau^2 + |\xi'|^2) \\ &\lesssim (1 + \tau^{-1}\gamma|\xi - \tilde{\xi}|^2)(\tau^2 + |\xi'|^2).\end{aligned}$$

With $(g^{\gamma})^{\sigma}_{\varrho}(x - \tilde{x}, \xi - \tilde{\xi}) = \lambda_{\mathsf{T},\tau}^2(\varrho')\tau^{-1}\gamma|x - \tilde{x}|^2 + \tau^{-1}\gamma|\xi - \tilde{\xi}|^2$, one thus finds

$$|\tilde{\xi}'|^2 \lesssim \big(1 + (g^{\gamma})^{\sigma}_{\varrho}(x - \tilde{x}, \xi - \tilde{\xi})\big) (\tau^2 + |\xi'|^2),$$

that is, (14.A.3) with $N = 1$. ■

14.A.2. Order Functions. Here we prove Proposition 14.10. By Remarks 14.8 and 14.9, it suffices to prove g-continuity and σ, g^{γ}-temperateness.

Note that if $m_1(y)$ and $m_2(y)$ are both g-continuous, then their product has also this property. Similarly, if they are both σ, g^{γ}-temperate, then their product has also this property. We thus concentrate on the properties of λ_{τ} and $\lambda_{\mathsf{T},\tau}$.

g-Continuity. For the g-continuity of λ_τ and $\lambda_{\mathsf{T},\tau}$, it suffices to consider the case $r=1$. See the definition of g-continuity in (14.2.4).

We write the proof for λ_τ. The proof for $\lambda_{\mathsf{T},\tau}$ can be adapted in a straightforward manner.

Let $\varepsilon > 0$ whose value will be set below, and let (x,ξ), $(y,\eta) \in \mathbb{R}^{2d}$ and $\tau \in [1,+\infty)$ such that $g_\varrho(y,\eta) < \varepsilon$, for $\varrho = (x,\xi,\tau)$. This reads

(14.A.4)
$$|y|^2 + \frac{|\eta|^2}{\tau^2 + |\xi'|^2} = |y|^2 + \lambda_{\mathsf{T},\tau}(\varrho')^{-2}|\eta|^2 \leq \varepsilon, \qquad \text{with } \varrho' = (x,\xi',\tau).$$

With $\lambda_\tau(\varrho)^2 = \tau^2 + |\xi|^2$ and $\lambda_\tau(\tilde{\varrho})^2 = \tau^2 + |\xi+\eta|^2$, where $\tilde{\varrho} = (x+y, \xi+\eta, \tau)$, we wish to prove

$$\tau + |\xi| \lesssim \tau + |\xi + \eta| \lesssim \tau + |\xi|, \tag{14.A.5}$$

under condition (14.A.4).

On the one hand, we have $\tau + |\xi+\eta| \leq \tau + |\xi| + |\eta| \leq \tau + |\xi| + \varepsilon^{1/2}(\tau + |\xi'|) \lesssim \tau + |\xi|$, that is, one part of (14.A.5). On the other hand, by (14.A.4), one has

$$\tau + |\xi| \lesssim \tau + |\xi+\eta| + |\eta| \lesssim \tau + |\xi+\eta| + \varepsilon^{1/2}(\tau + |\xi'|).$$

Thus, for $\varepsilon > 0$ chosen sufficiently small, one obtains the second part of (14.A.5).

σ, g^γ-Temperate Property. For the g^γ-temperate property of λ_τ and $\lambda_{\mathsf{T},\tau}$, it suffices to consider the cases $r=1$ and $r=-1$. See the definition of σ, g^γ-temperateness in (14.2.5).

We write the proof for λ_τ. The proof for $\lambda_{\mathsf{T},\tau}$ can be adapted in a straightforward manner.

We write

$$|\xi|^2 \lesssim |\xi - \tilde{\xi}|^2 + |\tilde{\xi}|^2 \lesssim (1 + \tau^{-1}\gamma|\xi - \tilde{\xi}|^2)(\tau^2 + |\tilde{\xi}|^2).$$

With $(g^\gamma)^\sigma_\varrho(x-\tilde{x}, \xi-\tilde{\xi}) = \lambda^2_{\mathsf{T},\tau}(\varrho')\tau^{-1}\gamma|x-\tilde{x}|^2 + \tau^{-1}\gamma|\xi-\tilde{\xi}|^2$, one thus finds

$$\lambda_\tau(\varrho)^2 \lesssim \big(1 + (g^\gamma)^\sigma_\varrho(x-\tilde{x}, \xi-\tilde{\xi})\big)\lambda_\tau(\tilde{\varrho})^2,$$

that is, the result for $r=1$. By symmetry, one also has

$$|\tilde{\xi}|^2 \lesssim (1 + \tau^{-1}\gamma|\xi - \tilde{\xi}|^2)(\tau^2 + |\xi|^2).$$

This yields

$$\lambda_\tau(\tilde{\varrho})^2 \lesssim \big(1 + (g^\gamma)^\sigma_\varrho(x-\tilde{x}, \xi-\tilde{\xi})\big)\lambda_\tau(\varrho)^2,$$

that is, the result for $r=-1$. ■

14.B. Symbol Properties

14.B.1. Properties of k, $\tilde{k}$, $k\operatorname{Im} f$, $\tilde{k}\operatorname{Im} f$. Here we prove Lemma 14.24.

Note that $\partial_x^\alpha \partial_{\xi'}^\beta k$ is a linear combination of terms of the form

$$(\tau/\gamma)^{p/2}\big(\partial_x^{\alpha_1}\partial_{\xi'}^{\beta_1}(\lambda_{\mathsf{T},\tau}^{-1}\operatorname{Im} f)\big)\cdots\big(\partial_x^{\alpha_p}\partial_{\xi'}^{\beta_p}(\lambda_{\mathsf{T},\tau}^{-1}\operatorname{Im} f)\big) \\ \times \sigma_0^{(p)}\big((\tau/\gamma)^{1/2}\lambda_{\mathsf{T},\tau}^{-1}\operatorname{Im} f\big),$$

with $\alpha_1+\cdots+\alpha_p=\alpha$, $\beta_1+\cdots+\beta_p=\beta$, and $p\le|\alpha|+|\beta|$. This gives

$$|\partial_x^\alpha\partial_{\xi'}^\beta k| \lesssim (\tau/\gamma)^{p/2}\lambda_{\mathsf{T},\tau}^{-|\beta_1|}\cdots\lambda_{\mathsf{T},\tau}^{-|\beta_p|} \lesssim (\tau/\gamma)^{p/2}\lambda_{\mathsf{T},\tau}^{-|\beta|} \lesssim (\tau/\gamma)^{(|\alpha|+|\beta|)/2}\lambda_{\mathsf{T},\tau}^{-|\beta|},$$

yielding $k\in S(1,g^\gamma)$. Similarly, $\tilde{k}\in S(1,g^\gamma)$.

Let $\alpha\in\mathbb{N}^d$ and $\beta\in\mathbb{N}^{d-1}$. Observe now that $\partial_x^\alpha\partial_{\xi'}^\beta(\tilde{k}\operatorname{Im} f)$ is a linear combination of terms of the form $\partial_x^{\alpha'}\partial_{\xi'}^{\beta'}(\tilde{k})\partial_x^{\alpha''}\partial_{\xi'}^{\beta''}(\operatorname{Im} f)$, where $\alpha'+\alpha''=\alpha$ and $\beta'+\beta''=\beta$. On the one hand, if $|\alpha'|+|\beta'|\le -1+|\alpha|+|\beta|$, that is, $|\alpha''|+|\beta''|\ge 1$, we have

$$|\partial_x^{\alpha'}\partial_{\xi'}^{\beta'}(\tilde{k})\partial_x^{\alpha''}\partial_{\xi'}^{\beta''}(\operatorname{Im} f)| \lesssim (\tau/\gamma)^{(|\alpha'|+|\beta'|)/2}\lambda_{\mathsf{T},\tau}^{-|\beta'|}\lambda_{\mathsf{T},\tau}^{1-|\beta''|} \\ \lesssim (\tau/\gamma)^{(-1+|\alpha|+|\beta|)/2}\lambda_{\mathsf{T},\tau}^{1-|\beta|}.$$

On the other hand, if $|\alpha''|+|\beta''|=0$, we have

$$|\operatorname{Im} f\,\partial_x^\alpha\partial_{\xi'}^\beta(\tilde{k})| \lesssim (\tau/\gamma)^{(|\alpha|+|\beta|-1)/2}\lambda_{\mathsf{T},\tau}^{1-|\beta|},$$

using that $|\operatorname{Im} f|\le(\tau/\gamma)^{-1/2}\lambda_{\mathsf{T},\tau}$ on the support of $\tilde{k}$. This allows one to conclude that $\tilde{k}\operatorname{Im} f\in S\big((\tau/\gamma)^{-1/2}\lambda_{\mathsf{T},\tau},g^\gamma\big)$.

Let $\alpha\in\mathbb{N}^d$ be such that $|\alpha|=1$. We then write

$$\begin{aligned}\partial_x^\alpha(k\operatorname{Im} f) &= k(\partial_x^\alpha\operatorname{Im} f)+(\partial_x^\alpha k)\operatorname{Im} f\\ &\in S(1,g^\gamma)S(\lambda_{\mathsf{T},\tau},g)+(\tau/\gamma)^{1/2}\lambda_{\mathsf{T},\tau}^{-1}(\partial_x^\alpha\operatorname{Im} f)\tilde{k}\operatorname{Im} f\\ &\subset S(\lambda_{\mathsf{T},\tau},g^\gamma)+(\partial_x^\alpha\operatorname{Im} f)S(1,g^\gamma)\subset S(\lambda_{\mathsf{T},\tau},g^\gamma),\end{aligned}$$

as $S(\lambda_{\mathsf{T},\tau},g)\subset S(\lambda_{\mathsf{T},\tau},g^\gamma)$ and using that $\tilde{k}\operatorname{Im} f\in S\big((\tau/\gamma)^{-1/2}\lambda_{\mathsf{T},\tau},g^\gamma\big)$ as proven above.

Let $\beta\in\mathbb{N}^{d-1}$ be such that $|\beta|=1$. Similarly, we find

$$\begin{aligned}\partial_{\xi'}^\beta(k\operatorname{Im} f) &= k(\partial_{\xi'}^\beta\operatorname{Im} f)+(\partial_{\xi'}^\beta k)\operatorname{Im} f\\ &\in S(1,g^\gamma)S(1,g)+(\tau/\gamma)^{1/2}\partial_{\xi'}^\beta(\lambda_{\mathsf{T},\tau}^{-1}\operatorname{Im} f)\tilde{k}\operatorname{Im} f\\ &\subset S(1,g^\gamma)+(\tau/\gamma)^{1/2}S(\lambda_{\mathsf{T},\tau}^{-1},g)\,\tilde{k}\operatorname{Im} f\\ &\subset S(1,g^\gamma),\end{aligned}$$

as $S(1,g)\subset S(1,g^\gamma)$ and $S(\lambda_{\mathsf{T},\tau}^{-1},g)\subset S(\lambda_{\mathsf{T},\tau}^{-1},g^\gamma)$. ■

14.B.2. Applicability of the Fefferman–Phong–Bony Inequality. Here, we prove Lemma 14.26.

We have $\tilde{a} \geq 0$ if $\gamma \geq \gamma_0$. By Lemma 14.24, we have $k \operatorname{Im} f \in S(\lambda_{\mathsf{T},\tau}, g^\gamma)$ and $(\tau/\gamma)^{1/2} \lambda_{\mathsf{T},\tau}^{-1} \{\xi_d - \operatorname{Re} f, \operatorname{Im} f\} \tilde{k} \in S((\tau/\gamma)^{1/2}, g^\gamma)$ implying $|q_0| \lesssim \lambda_{\mathsf{T},\tau} + (\tau/\gamma)^{1/2}$ yielding

$$\begin{aligned}\tilde{a} &= \gamma^{3/2}\tau^{-1/2}\tilde{\chi}\big(q_0 - C_3(\tau/\gamma)^{1/2}\big) \lesssim \gamma^{3/2}\tau^{-1/2}\lambda_{\mathsf{T},\tau} + \gamma \\ &\lesssim \gamma^{3/2}\tau^{-3/2}\lambda_{\mathsf{T},\tau}^2 + \tau \lesssim \lambda_{\mathsf{T},\tau}^2,\end{aligned}$$

for $\tau \geq \gamma \geq 1$.

We shall now estimate $\partial_x^\alpha \partial_{\xi'}^\beta \tilde{a}$ for $|\alpha| + |\beta| = 2$ and $|\alpha| + |\beta| \geq 4$.

Note that $\partial_x^\alpha \partial_{\xi'}^\beta (\tilde{\chi} k \operatorname{Im} f)$ is a linear combination of terms of the form

$$(\partial_x^{\alpha'} \partial_{\xi'}^{\beta'} \tilde{\chi})\, \partial_x^{\alpha''} \partial_{\xi'}^{\beta''} (k \operatorname{Im} f),$$

with $\alpha' + \alpha'' = \alpha$ and $\beta' + \beta'' = \beta$. We recall that

$$|\partial_x^{\alpha''} \partial_{\xi'}^{\beta''} (k \operatorname{Im} f)| \lesssim (\tau/\gamma)^{(|\alpha''| + |\beta''| - 1)/2} \lambda_{\mathsf{T},\tau}^{1 - |\beta''|},$$

for $|\alpha''| + |\beta''| \geq 1$ as stated in (14.3.3). In such a case, since $|\partial_x^{\alpha'} \partial_{\xi'}^{\beta'} \tilde{\chi}| \lesssim \lambda_{\mathsf{T},\tau}^{-|\beta'|}$, we find

$$|(\partial_x^{\alpha'} \partial_{\xi'}^{\beta'} \tilde{\chi})\, \partial_x^{\alpha''} \partial_{\xi'}^{\beta''} (k \operatorname{Im} f)| \lesssim (\tau/\gamma)^{(|\alpha| + |\beta| - 1)/2} \lambda_{\mathsf{T},\tau}^{1 - |\beta|},$$

as $\tau \geq \gamma$. If $|\alpha''| + |\beta''| = 0$, since $k \operatorname{Im} f \lesssim \lambda_{\mathsf{T},\tau}$, we find

$$|(\partial_x^{\alpha'} \partial_{\xi'}^{\beta'} \tilde{\chi})\, \partial_x^{\alpha''} \partial_{\xi'}^{\beta''} (k \operatorname{Im} f)| = |(\partial_x^{\alpha} \partial_{\xi'}^{\beta} \tilde{\chi})\, k \operatorname{Im} f| \lesssim \lambda_{\mathsf{T},\tau}^{1 - |\beta|}.$$

In both cases, we thus have

$$|\partial_x^{\alpha} \partial_{\xi'}^{\beta} (\tilde{\chi} k \operatorname{Im} f)| \lesssim (\tau/\gamma)^{(|\alpha| + |\beta| - 1)/2} \lambda_{\mathsf{T},\tau}^{1 - |\beta|}.$$

If $|\alpha| + |\beta| = 2$, one thus finds

$$\gamma^{3/2}\tau^{-1/2} |\partial_x^{\alpha} \partial_{\xi'}^{\beta} (\tilde{\chi} k \operatorname{Im} f)| \lesssim \gamma \lambda_{\mathsf{T},\tau}^{1 - |\beta|} \lesssim \lambda_{\mathsf{T},\tau}^{2 - |\beta|}$$

using $\gamma \leq \tau \leq \lambda_{\mathsf{T},\tau}$. If $|\alpha| + |\beta| \geq 4$ one has

$$\gamma^{3/2}\tau^{-1/2} |\partial_x^{\alpha} \partial_{\xi'}^{\beta} (\tilde{\chi} k \operatorname{Im} f)| \lesssim \gamma^{2 - (|\alpha| + |\beta|)/2} \tau^{(|\alpha| + |\beta|)/2 - 1} \lambda_{\mathsf{T},\tau}^{1 - |\beta|} \lesssim \lambda_{\mathsf{T},\tau}^{(|\alpha| - |\beta|)/2},$$

using $\gamma \geq 1$ and $0 < \tau \leq \lambda_{\mathsf{T},\tau}$.

We have $a_1 = \gamma \lambda_{\mathsf{T},\tau}^{-1} \tilde{\chi} \tilde{k} \{\xi_d - \operatorname{Re} f, \operatorname{Im} f\} \in S(\gamma, g^\gamma)$. We thus have from the properties of such symbol classes recalled in Sect. 14.2

$$|\partial_x^{\alpha} \partial_{\xi'}^{\beta} a_1| \lesssim \gamma (\tau/\gamma)^{(|\alpha| + |\beta|)/2} \lambda_{\mathsf{T},\tau}^{-|\beta|} \lesssim \tau^{(|\alpha| + |\beta|)/2} \lambda_{\mathsf{T},\tau}^{-|\beta|} \lesssim \lambda_{\mathsf{T},\tau}^{(|\alpha| - |\beta|)/2},$$

if $|\alpha| + |\beta| \geq 2$ using $\gamma \geq 1$ and $0 < \tau \leq \lambda_{\mathsf{T},\tau}$. Note that if $|\alpha| + |\beta| = 2$, we have $(|\alpha| - |\beta|)/2 = 1 - |\beta| \leq 2 - |\beta|$.

All the estimations written above give the conclusion of the lemma. ∎

14.B.3. Proof of Theorem 14.17. The proof follows that of [175, Theorem 18.5.5] that is carried out within the framework of the Weyl quantification. In this book, we chose to preferably use the so-called classical quantification, as presented in Chapter 2 of Volume 1. Here, we explain how one can deduce Theorem 14.17 from results proven in [175]. In what follows, instead of writing [175, (18.∗.∗)] or [175, Theorem 18.∗.∗], we shall instead write (H18.∗.∗) or Theorem H18.∗.∗ for concision.

In this section, it is sometimes important to put forward in which space the Euclidean inner product is carried out. As opposed to the rest of this book, we shall thus write $\langle x, \xi\rangle_{\mathbb{R}^d}$ instead of $x\cdot\xi$. If there is no ambiguity, we shall simply write $\langle x, \xi\rangle$. Moreover, with this notation, the presentation is consistent with that of Chapter 18 in [175] on which all the argumentation is founded.

For a_1, a_2, and a_3 tangential symbols, if one writes $\operatorname{Op}_{\mathsf{T}}(a_3) = \operatorname{Op}_{\mathsf{T}}(a_2)\operatorname{Op}_{\mathsf{T}}(a_1)$, from (H18.5.21), one finds

$$a_3(\varrho') = e^{i\langle D_\xi, D_{\hat{x}}\rangle} a_1(\varrho') a_2(\hat{\varrho}')_{|\varrho'=\hat{\varrho}'} = e^{i\langle D_{\xi'}, D_{\hat{x}'}\rangle} a_1(\varrho') a_2(\hat{\varrho}')_{|\varrho'=\hat{\varrho}'}, \tag{14.B.1}$$

where $\varrho' = (x, \xi', \tau)$, $x = (x', x_d)$, $\hat{\varrho}' = (\hat{x}, \hat{\xi}', \tau)$, and $\hat{x} = (\hat{x}', \hat{x}_d)$, using that the symbols are independent of ξ_d.

To estimate a_3 and its derivatives, one wishes to apply Theorems H18.4.10' and H18.4.11. This requires however the introduction of a metric G, a quadratic form A, and the computation of the dual metric G^A following the notation of Section H18.4. Observe that even if we are only interested in tangential symbols, that is, not depending on ξ_d, to obtain estimates on the symbols, including the behavior with respect to x_d, it is convenient to consider metrics acting on the whole $\mathbb{R}^{2d}$. For $w = (w, \hat{w})$ with $w = (w^x, w^\xi) \in \mathbb{R}^{2d}$ and $\hat{w} = (w^{\hat{x}}, w^{\hat{\xi}}) \in \mathbb{R}^{2d}$, we set

$$\begin{aligned} G_{\varrho,\hat{\varrho}}(w) = g_\varrho(w) + g^\gamma_{\hat{\varrho}}(\hat{w}) &= |w^x|^2 + \tau\gamma^{-1}|w^{\hat{x}}|^2 + \lambda^{-2}_{\mathsf{T},\tau,\xi'}|w^\xi|^2 \\ &\quad + \tau\gamma^{-1}\lambda^{-2}_{\mathsf{T},\tau,\hat{\xi}'}|w^{\hat{\xi}}|^2, \end{aligned}$$

where $\varrho = (\varrho', \xi_d) = (x, \xi, \tau)$ and $\hat{\varrho} = (\hat{\varrho}', \hat{\xi}_d) = (\hat{x}, \hat{\xi}, \tau)$ and $\lambda^2_{\mathsf{T},\tau,\xi'} = \lambda_{\mathsf{T},\tau}(\varrho')^2 = \tau^2 + |\xi'|^2$ and $\lambda^2_{\mathsf{T},\tau,\hat{\xi}'} = \lambda_{\mathsf{T},\tau}(\hat{\varrho}')^2 = \tau^2 + |\hat{\xi}'|^2$, and we also set

$$A(w) = \langle w^\xi, w^{\hat{x}}\rangle_{\mathbb{R}^d} = \langle Aw, w\rangle_{\mathbb{R}^{4d}},$$

with the $4d \times 4d$ symmetric matrix A given by

$$A = \frac{1}{2}\begin{pmatrix} 0 & 0 & 0 & 0 \\ 0 & 0 & \operatorname{Id}_{\mathbb{R}^d} & 0 \\ 0 & \operatorname{Id}_{\mathbb{R}^d} & 0 & 0 \\ 0 & 0 & 0 & 0 \end{pmatrix}.$$

The matrix A appears naturally from (14.B.1) since one has

$$\langle D_\xi, D_{\hat{x}}\rangle_{\mathbb{R}^d} = \langle A(D_x, D_\xi, D_{\hat{x}}, D_{\hat{\xi}}), (D_x, D_\xi, D_{\hat{x}}, D_{\hat{\xi}})\rangle_{\mathbb{R}^{4d}}.$$

Note that $G_{\varrho,\hat{\varrho}}$ with $\varrho = (\varrho', \xi_d)$ and $\hat{\varrho} = (\hat{\varrho}', \hat{\xi}_d)$ is in fact independent of the values of ξ_d and $\hat{\xi}_d$.

The dual metric of G with respect to A is given by

$$G^A_{\varrho,\hat{\varrho}}(w) = \sup_{\substack{\tilde{w}\in\mathbb{R}^{4d} \\ G_{\varrho,\hat{\varrho}}(A\tilde{w})<1}} \langle w, \tilde{w}\rangle^2_{\mathbb{R}^{4d}}.$$

On the one hand, one has $G^A_{\varrho,\hat{\varrho}}(w) = +\infty$ if $w \notin \ker(A)^\perp$, that is, $w^x \neq 0$ or $w^{\hat{\xi}} \neq 0$. On the other hand, if $w \in \ker(A)^\perp$, that is, $w^x = 0$ and $w^{\hat{\xi}} = 0$, one obtains

$$G^A_{\varrho,\hat{\varrho}}(0, w^\xi, w^{\hat{x}}, 0) = 4\tau^{-1}\gamma|w^\xi|^2 + 4\lambda^2_{\mathsf{T},\tau,\xi'}|w^{\hat{x}}|^2.$$

With (H18.4.19), we thus find

$$h^2_{G,A} = \sup_w G_{\varrho,\hat{\varrho}}(w)/G^A_{\varrho,\hat{\varrho}}(w) = \tau\lambda^{-2}_{\mathsf{T},\tau,\xi'}\gamma^{-1}/4.$$

LEMMA 14.38. *The metric G is slowly varying and A-temperate. The functions $\lambda^2_{\mathsf{T},\tau,\xi'} = \tau^2 + |\xi'|^2$ and $\lambda^2_{\mathsf{T},\tau,\hat{\xi}'} = \tau^2 + |\hat{\xi}'|^2$ are both G-continuous and A, G-temperate.*

A proof is given below.

With Lemma 14.38, we can now apply Theorems H18.4.10' and H18.4.11. This gives the result of Theorem 14.17. ■

PROOF OF LEMMA 14.38.

Slow Variation of G. By slowly varying, one means the existence of $\varepsilon > 0$ and $C > 0$ such that

$$G_{\varrho,\hat{\varrho}}(w) \leq \varepsilon \;\Rightarrow\; G_{\mu,\hat{\mu}} \leq CG_{\varrho,\hat{\varrho}}, \tag{14.B.2}$$

for $\varrho = (x, \xi, \tau)$, $\hat{\varrho} = (\hat{x}, \hat{\xi}, \tau)$, $\mu = (x + w^x, \xi + w^\xi, \tau)$, $\hat{\mu} = (\hat{x} + w^{\hat{x}}, \hat{\xi} + w^{\hat{\xi}}, \tau)$, $w = (w^x, w^\xi, w^{\hat{x}}, w^{\hat{\xi}})$, with $(x, \xi), (\hat{x}, \hat{\xi}), (w^x, w^\xi), (w^{\hat{x}}, w^{\hat{\xi}}) \in \mathbb{R}^{2d}$, and $\tau \geq 1$.

Let $\varepsilon > 0$ and assume $G_{\varrho,\hat{\varrho}}(w) \leq \varepsilon$, that is,

$$g_\varrho(w) + g^\gamma_{\hat{\varrho}}(\hat{w}) \leq \varepsilon. \tag{14.B.3}$$

Yet, by Lemma 14.7, the metrics g and g^γ are both slowly varying. Thus, for ε chosen sufficiently small, one finds

$$g_\mu \lesssim g_\varrho, \qquad g^\gamma_{\hat{\mu}} \lesssim g^\gamma_{\hat{\varrho}},$$

implying (14.B.2).

***A*-Temperate Property of *G*.** By A-temperate, one means that for some $N \in \mathbb{N}$,

$$G_{\varrho,\hat{\varrho}} \leq C\big(1 + G^A_{\varrho,\hat{\varrho}}(w)\big)^N G_{\mu,\hat{\mu}}, \tag{14.B.4}$$

for all ϱ, $\hat{\varrho}$, $\mu, \hat{\mu}$, and w as above and $\tau \geq 1$. Observe that we may assume $w^x = 0$ and $w^{\hat{\xi}} = 0$, since otherwise $G^A_{\varrho,\hat{\varrho}}(w) = +\infty$. Thus, we may simply consider

$$\mu = (y, \eta', \eta_d, \tau) = (x, \xi + w^\xi, \tau) \quad \text{and} \quad \hat{\mu} = (\hat{y}, \hat{\eta}', \hat{\eta}_d, \tau) = (\hat{x} + w^{\hat{x}}, \hat{\xi}, \tau).$$

Recall now that $G_{\varrho,\hat{\varrho}}$ and $G^A_{\varrho,\hat{\varrho}}$ are independent of ξ_d and $G_{\mu,\hat{\mu}}$ is independent of η_d. It is thus sufficient to obtain (14.B.4) in the case

$$w^x = 0, \quad w^\xi = (w^{\xi'}, 0), \quad \text{and} \quad w^{\hat{\xi}} = 0.$$

For $v = (v, \hat{v}) = (v^x, v^\xi, v^{\hat{x}}, v^{\hat{\xi}}) \in \mathbb{R}^{4d}$, we thus need to prove that, for some $N \in \mathbb{N}$,

$$\begin{aligned} &|v^x|^2 + \tau\gamma^{-1}|v^{\hat{x}}|^2 + \lambda^{-2}_{\mathsf{T},\tau,\xi'}|v^\xi|^2 + \tau\gamma^{-1}\lambda^{-2}_{\mathsf{T},\tau,\hat{\xi}'}|v^{\hat{\xi}}|^2 \\ &\quad\lesssim \big(1 + \tau^{-1}\gamma|w^{\xi'}|^2 + \lambda^2_{\mathsf{T},\tau,\xi'}|w^{\hat{x}}|^2\big)^N \\ &\qquad \times \big(|v^x|^2 + \tau\gamma^{-1}|v^{\hat{x}}|^2 + \lambda^{-2}_{\mathsf{T},\tau,\xi'+w^{\xi'}}|v^\xi|^2 + \tau\gamma^{-1}\lambda^{-2}_{\mathsf{T},\tau,\hat{\xi}'}|v^{\hat{\xi}}|^2\big). \end{aligned}$$

For the first, second, and fourth terms on the l.h.s., this estimation clearly holds with $N = 0$. It thus suffices to find some $N \in \mathbb{N}$ such that

$$\lambda^{-2}_{\mathsf{T},\tau,\xi'}|v^\xi|^2 \lesssim \big(1 + \tau^{-1}\gamma|w^{\xi'}|^2\big)^N \lambda^{-2}_{\mathsf{T},\tau,\xi'+w^{\xi'}}|v^\xi|^2,$$

for all $v^\xi \in \mathbb{R}^d$, that is,

$$\tau^2 + |\xi' + w^{\xi'}|^2 \lesssim (\tau^2 + |\xi'|^2)\big(1 + \tau^{-1}\gamma|w^{\xi'}|^2\big)^N. \tag{14.B.5}$$

As here $\tau, \gamma \in [1, +\infty)$, one simply writes

$$\tau^2 + |\xi' + w^{\xi'}|^2 \lesssim \tau^2 + |\xi'|^2 + |w^{\xi'}|^2 \lesssim \tau^2 + |\xi'|^2 + \tau\gamma|w^{\xi'}|^2,$$

which is less than or equal to the r.h.s. of (14.B.5), in the case $N = 1$. This concludes that proof of the A-temperate property of G.

***G*-Continuity of $\boldsymbol{\lambda_{\mathsf{T},\tau,\xi'}}$ and $\boldsymbol{\lambda_{\mathsf{T},\tau,\hat{\xi}'}}$.** For a function $m(\varrho, \hat{\varrho})$, the G-continuity property means the existence of $\varepsilon > 0$ and $C > 0$ such that

$$G_{\varrho,\hat{\varrho}}(w) \leq \varepsilon \quad \Rightarrow \quad m(\varrho, \hat{\varrho})/C \leq m(\mu, \hat{\mu}) \leq Cm(\varrho, \hat{\varrho}), \tag{14.B.6}$$

for $\varrho = (x, \xi, \tau)$, $\hat{\varrho} = (\hat{x}, \hat{\xi}, \tau)$, $\mu = (x + w^x, \xi + w^\xi, \tau)$, $\hat{\mu} = (\hat{x} + w^{\hat{x}}, \hat{\xi} + w^{\hat{\xi}}, \tau)$, $w = (w^x, w^\xi, w^{\hat{x}}, w^{\hat{\xi}})$, with $(x, \xi), (\hat{x}, \hat{\xi}), (w^x, w^\xi), (w^{\hat{x}}, w^{\hat{\xi}}) \in \mathbb{R}^{2d}$, and $\tau \geq 1$.

As seen above, having $G_{\varrho,\hat{\varrho}}(w) \leq \varepsilon$ reads $g_\varrho(w) + g^\gamma_{\hat{\varrho}}(\hat{w}) \leq \varepsilon$. Since $\lambda_{\mathsf{T},\tau,\xi'}$ is g-continuous and $\lambda_{\mathsf{T},\tau,\hat{\xi}'}$ is g^γ-continuous by Proposition 14.10, we find that

$$\lambda_{\mathsf{T},\tau,\xi'} \eqsim \lambda_{\mathsf{T},\tau,\mu'} \quad \text{and} \quad \lambda_{\mathsf{T},\tau,\hat{\xi}'} \eqsim \lambda_{\mathsf{T},\tau,\hat{\mu}'}.$$

A,G-Temperate Property of $\lambda_{\mathsf{T},\tau,\xi'}$ and $\lambda_{\mathsf{T},\tau,\hat{\xi}'}$. For a function $m(\varrho,\hat{\varrho})$, the A,G-temperate property means that, for some $N\in\mathbb{N}$ and $C>0$,

$$m(\varrho,\hat{\varrho})\leq C\big(1+G^A_{\varrho,\hat{\varrho}}(w)\big)^N m(\mu,\hat{\mu}),$$

for all ϱ, $\hat{\varrho}$, μ, $\hat{\mu}$, and w as above and $\tau\geq 1$. As for the A-temperate property of G, it suffices to consider the case

$$w^x=0,\quad w^\xi=(w^{\xi'},0),\quad\text{and}\quad w^{\hat{\xi}}=0.$$

As here $\lambda_{\mathsf{T},\tau,\hat{\xi}'}=\lambda_{\mathsf{T},\tau,\hat{\xi}'+w^{\hat{\xi}'}}$, the property is obvious for $\lambda_{\mathsf{T},\tau,\hat{\xi}'}$. For $\lambda_{\mathsf{T},\tau,\xi'}$, we need to prove

$$\lambda_{\mathsf{T},\tau,\xi'}\lesssim\big(1+\tau^{-1}\gamma|w^{\xi'}|^2+\lambda^2_{\mathsf{T},\tau,\xi'}|w^{\hat{x}}|^2\big)^N\lambda_{\mathsf{T},\tau,\xi'+w^{\xi'}}. \tag{14.B.7}$$

Since $\gamma,\tau\in[1,+\infty)$, we write

$$\begin{aligned}\lambda^2_{\mathsf{T},\tau,\xi'}=\tau^2+|\xi'|^2&\lesssim\tau^2+|\xi'+w^{\xi'}|^2+|w^{\xi'}|^2\\&\lesssim\big(1+\gamma\tau^{-1}|w^{\xi'}|^2\big)\big(\tau^2+|\xi'+w^{\xi'}|^2\big)\\&\lesssim\big(1+\tau^{-1}\gamma|w^{\xi'}|^2+\lambda^2_{\mathsf{T},\tau,\xi'}|w^{\hat{x}}|^2\big)\lambda^2_{\mathsf{T},\tau,\xi'+w^{\xi'}},\end{aligned}$$

and (14.B.7) follows for $N=1$. ∎

14.B.4. Proof of Fefferman–Phong–Bony Inequality. Here, we prove Theorem 14.22 from the version given in [77].

The result given by J.-M. Bony is stated in the context of the Weyl calculus. We denote by $p^w(x,D',\tau)$ or simply by p^w the tangential operator associated with the symbol $p(x,\xi',\tau)$ in the Weyl quantification, that is,

$$\begin{aligned}&p^w(x,D',\tau)u\\&\quad=(2\pi)^{1-d}\iint_{\mathbb{R}^{2d-2}}e^{i(x'-y')\cdot\xi'}p((x'+y')/2,x_d,\xi',\tau)u(y',x_d)\,d\xi'dy',\end{aligned}\tag{14.B.8}$$

in the sense of oscillatory integrals (see Section 2.4 of Volume 1). Here, the tangential symbol $p(x,\xi',\tau)$ used may, for instance, lie in the symbol classes introduced in Section 2.10 of Volume 1 and Sect. 14.2.

THEOREM 14.39 (J.-M. Bony's Theorem 3.2 in [77]). *Let $p(x,\xi',\tau)\in\mathscr{C}^\infty(\mathbb{R}^d\times\mathbb{R}^{d-1})$ be nonnegative and such that for any $\alpha',\beta'\in\mathbb{N}^{d-1}$, with $|\alpha'|+|\beta'|\geq 4$, there exists $C_{\alpha',\beta'}>0$ such that*

$$|\partial^{\alpha'}_{x'}\partial^{\beta'}_{\xi'}p(x,\xi',\tau)|\leq C_{\alpha',\beta'},\qquad(x,\xi')\in\mathbb{R}^{2d-1},\ \tau\geq 1.\tag{14.B.9}$$

Then, for some $C>0$, one has

$$(p^w u,u)_+ + C\|u\|^2_+\geq 0,\qquad u\in\overline{\mathscr{S}}(\mathbb{R}^d_+).$$

This result is in fact adapted to the tangential case; here, x_d is considered a parameter in the symbol property of $p(x,\xi',\tau)$.

We begin by proving that if the result holds true in the Weyl quantification, it holds also true in the classical quantification.

Following [175, Theorem 18.5.10], one can find $b(x, \xi', \tau)$ such that $\operatorname{Op}_{\mathsf{T}}(a) = b^w$. More precisely, one has

$$b(x, \xi', \tau) = e^{-i\langle D_{x'}, D_{\xi'}\rangle/2} a(x, \xi', \tau),$$

or equivalently $a(x, \xi', \tau) = e^{i\langle D_{x'}, D_{\xi'}\rangle/2} b(x, \xi', \tau)$. One important feature of the Weyl quantification lies in having $(p^w)^* = \bar{p}^w$ as can be readily seen formally from the form of the kernel of $p^w(x, D', \tau)$. Then, one has

$$\operatorname{Op}_{\mathsf{T}}(a) + \operatorname{Op}_{\mathsf{T}}(a)^* = 2(\operatorname{Re} b)^w.$$

From the Taylor formula $e^{ix} = 1 + ix - \int_0^1 (1-t) e^{itx} x^2 dt$, for a real symbol $a(x, \xi', \tau)$, one finds $2 \operatorname{Re} b(x, \xi', \tau) = e^{-i\langle D_{x'}, D_{\xi'}\rangle/2} a(x, \xi', \tau) + e^{i\langle D_{x'}, D_{\xi'}\rangle/2}$ $a(x, \xi', \tau)$ yielding

$$\operatorname{Re} b(x, \xi', \tau) = a(x, \xi', \tau) - \int_0^1 \frac{(1-t)}{8} \Big(e^{i\frac{t}{2}\langle D_{x'}, D_{\xi'}\rangle} + e^{-i\frac{t}{2}\langle D_{x'}, D_{\xi'}\rangle} \Big) \\ \times \langle D_{x'}, D_{\xi'}\rangle^2 a(x, \xi', \tau) dt.$$

We now consider the symbol $a(x, \xi', \tau)$ as given in the statement of Theorem 14.22. One has $\langle D_{x'}, D_{\xi'}\rangle^2 a(x, \xi', \tau) \in S(1, \lambda_{\mathsf{T},\tau}|dx|^2 + \lambda_{\mathsf{T},\tau}^{-1}|d\xi'|^2)$.

LEMMA 14.40. *The operators $B = e^{\pm it\langle D_{x'}, D_{\xi'}\rangle/2}$ map $S(1, \lambda_{\mathsf{T},\tau}|dx|^2 + \lambda_{\mathsf{T},\tau}^{-1}|d\xi'|^2)$ onto itself continuously and uniformly with respect to $t \in [0,1]$.*

A proof is given in Sect. 14.B.4.1 below.

With Lemma 14.40, one has

$$e^{\pm it\langle D_{x'}, D_{\xi'}\rangle/2} \langle D_{x'}, D_{\xi'}\rangle^2 a(x, \xi', \tau) \in S(1, \lambda_{\mathsf{T},\tau}|dx|^2 + \lambda_{\mathsf{T},\tau}^{-1}|d\xi'|^2),$$

uniformly with respect to $t \in [0,1]$.

We deduce that $\operatorname{Re} b = a + c$, where $c \in S(1, \lambda_{\mathsf{T},\tau}|dx|^2 + \lambda_{\mathsf{T},\tau}^{-1}|d\xi'|^2)$, yielding $(\operatorname{Op}(a) + \operatorname{Op}(a)^*)/2 = a^w + c^w$, where c^w is bounded on $L^2(\mathbb{R}^d_+)$ from [175, Theorem 18.6.3]. Consequently, in the proof of Theorem 14.22, we may thus replace the operator $\operatorname{Op}_{\mathsf{T}}(a)$ by its Weyl quantification a^w.

Observe that the result of Theorem 14.39 does not apply directly to the symbol $a(x, \xi', \tau)$ given in the statement of Theorem 14.22. It however applies to localized versions of this symbol after the action of a symplectomorphism. Localization is performed in the (ξ', τ) variables.

Let us use a Littlewood–Paley decomposition as introduced in the proof of Proposition 7.3. With φ_k, $k \in \mathbb{N}$ as defined in (7.4.2), one has $\sum_{k=0}^{\infty} \varphi_k = 1$. We recall that φ_0 is supported on $[-1, 1]$, and $\varphi_k(t) = \varphi(2^{1-k}t)$, where φ is supported on $1 \le |t| \le 3$. We set $a_k(x, \xi', \tau) = a(x, \xi', \tau)\varphi_k(\lambda_{\mathsf{T},\tau})$ and

$$q_k = a_k \circ \Sigma_k,$$

where Σ_k is the following symplectomorphism in the (x', ξ') variables:

$$\Sigma_k : (x, \xi', \tau) \mapsto (2^{-k/2}x', x_d, 2^{k/2}\xi', \tau).$$

We have the following result.

PROPOSITION 14.41. *The symbol q_k fulfills the property* (14.B.9) *with constants $C_{\alpha,\beta} > 0$ that are uniform with respect to $k \in \mathbb{N}$.*

A proof is given in Sect. 14.B.4.2 below.

We are now in a position to apply the result of J.-M. Bony given in Theorem 14.22 to the symbol $q_k(x, \xi', \tau)$ in the x', ξ' variables with an additional integration with respect to x_d on $(0, +\infty)$, yielding

$$(q_k^w u, u)_+ + C\|u\|_+^2 \geq 0, \qquad u \in \overline{\mathscr{S}}(\mathbb{R}_+^d),$$

uniformly with respect to $k \in \mathbb{N}$.

For p a symbol in one of the classes considered here, $u \in \overline{\mathscr{S}}(\mathbb{R}_+^d)$, and $\lambda \neq 0$ from (14.B.8), one finds $(p^w u)(\lambda x', x_d) = (p_\lambda^w u_\lambda)(x)$ with

$$p_\lambda(x, \xi', \tau) = p(\lambda x', x_d, \xi'/\lambda, \tau) \quad \text{and} \quad u_\lambda(x) = u(\lambda x', x_d),$$

yielding

$$\frac{(p^w u, u)_+}{\|u\|_+^2} = \frac{(p_\lambda^w u_\lambda, u_\lambda)_+}{\|u_\lambda\|_+^2}.$$

One thus finds

$$(q_{k,\lambda}^w u, u)_+ + C\|u\|_+^2 \geq 0, \qquad u \in \overline{\mathscr{S}}(\mathbb{R}_+^d),$$

for any $\lambda > 0$ with $q_{k,\lambda}(x, \xi', \tau) = q_k(\lambda x', x_d, \xi'/\lambda, \tau)$. For $\lambda = 2^{k/2}$, one has $q_{k,\lambda}(x, \xi', \tau) = a_k(x, \xi', \tau)$. Thus,

$$(a_k^w u, u)_+ + C\|u\|_+^2 \geq 0, \qquad u \in \overline{\mathscr{S}}(\mathbb{R}_+^d), \tag{14.B.10}$$

for $\tau \geq 1$. As above, this inequality is uniform with respect to $k \in \mathbb{N}$.

Let $\psi \in \mathscr{C}_c^\infty(\mathbb{R})$ be such that $\operatorname{supp}(\psi) \subset (1/2, 4)$ $\psi \equiv 1$ in a neighborhood of $\operatorname{supp}(\varphi)$. We set $\psi_k(t) = \psi(2^{1-k}t)$ such that $\psi_k \varphi_k = \varphi_k$ on $\mathbb{R}^+$ for $k \geq 1$. For $k = 0$, let $\psi_0 \in \mathscr{C}_c^\infty(\mathbb{R})$ be such that $\operatorname{supp}(\psi) \subset (-2, 2)$ and $\psi_0 \equiv 1$ in a neighborhood of $\operatorname{supp}(\varphi_0)$. By abuse of notation, we write ψ_k^w instead of $(\psi_k(\lambda_{\mathsf{T},\tau}))^w$. Observe that $\psi_k(\lambda_{\mathsf{T},\tau}) a_k(x, \xi', \tau) = a_k(x, \xi', \tau)$.

Applying inequality (14.B.10) to $\psi_k^w u$, we find

$$(a_k^w \psi_k^w u, \psi_k^w u)_+ + C\|\psi_k^w u\|_+^2 \geq 0.$$

As ψ is real valued, we have $(\psi_k^w)^* = \psi_k^w$ yielding

$$(\psi_k^w a_k^w \psi_k^w u, u)_+ + C\|\psi_k^w u\|_+^2 \geq 0.$$

LEMMA 14.42. *One has $\varphi_k(\lambda_{\mathsf{T},\tau}), \psi_k(\lambda_{\mathsf{T},\tau}) \in S(1, |dx|^2 + 2^{-2k}|d\xi'|^2)$ and $a_k \in S(2^{2k}, 2^k|dx|^2 + 2^{-k}|d\xi'|^2)$.*

A proof is given in Sect. 14.B.4.3 below.

Applying [175, Theorem 18.5.5], we obtain that $a_k^w \psi_k^w = (a_k)^w + c_k^w$, where $c_k \in S(2^{-kN}, 2^k|dx|^2 + 2^{-k}|d\xi'|^2)$, uniformly with Respect to $k \geq 0$ and for all $N > 0$ since derivatives of ψ_k vanish on support of φ_k. Applying the same argument a second time, one obtains $\psi_k^w a_k^w \psi_k^w = a_k^w + d_k^w$, where $d_k \in S(2^{-kN}, 2^k|dx|^2 + 2^{-k}|d\xi'|^2)$, uniformly with respect to $k \geq 0$ and for all $N > 0$. As d_k^w is L^2 continuous from [175, Theorem 18.6.3] with a norm estimated by $C2^{-Nk}$, we deduce that

$$(a_k^w u, u)_+ + C\|\psi_k^w u\|_+^2 + C' 2^{-Nk}\|u\|_+^2 \geq 0. \tag{14.B.11}$$

A summation with respect to $k \in \mathbb{N}$ can be carried out by the following lemma.

LEMMA 14.43. *There exists $C > 0$ such that $\sum_{k\in\mathbb{N}} \|\psi_k^w u\|_+^2 \leq C\|u\|_+^2$, for $u \in \overline{\mathscr{S}}(\mathbb{R}_+^d)$. One also has the convergence of $\sum_k (a_k^w u, u)_+$ with*

$$\sum_{k\in\mathbb{N}} (a_k^w u, u)_+ = (a^w u, u)_+, \qquad u \in \overline{\mathscr{C}}_c^\infty(\mathbb{R}_+^d).$$

A proof is given in Sect. 14.B.4.4 below. Here we recall the definition of the space of function

$$\overline{\mathscr{C}}_c^\infty(\mathbb{R}_+^d) = \{u = v_{|\mathbb{R}_+^d};\ v \in \mathscr{C}_c^\infty(\mathbb{R}^d)\}.$$

For some $C'' > 0$, one thus obtains

$$(a^w u, u)_+ + C''\|u\|_+^2 \geq 0, \qquad u \in \overline{\mathscr{C}}_c^\infty(\mathbb{R}_+^d).$$

From the density of $\overline{\mathscr{C}}_c^\infty(\mathbb{R}_+^d)$ in $\overline{\mathscr{S}}(\mathbb{R}_+^d)$ and the continuity of a^w from $\overline{\mathscr{S}}(\mathbb{R}_+^d)$ into itself, we conclude the proof of Theorem 14.22. ■

14.B.4.1. *Proof of Lemma 14.40.* It suffices to apply [175, Theorem 18.5.10] to prove that B is continuous from $S(1, \lambda_{\mathsf{T},\tau}|dx|^2 + \lambda_{\mathsf{T},\tau}^{-1}|d\xi'|^2)$ onto itself. To prove the uniformity, we have to apply [175, Theorem 18.4.10]. The continuity constants depend on the constants in the estimates (18.4.2)', (18.4.5), 18.4.13), and (18.4.14) of [175]. The estimates (18.4.2)' and (18.4.5) involve the constants related with the metric and the order function that is the constants ε and C in the following estimates:

$$\mathsf{g}_\varrho(y,\eta) < \varepsilon \ \Rightarrow\ C^{-1}\mathsf{g}_\varrho \leq \mathsf{g}_{\tilde\varrho} \leq C\mathsf{g}_\varrho, \qquad \text{for } \tilde\varrho = (x+y, \xi+\eta, \tau)$$

and

$$\mathsf{g}_\varrho(y,\eta) < \varepsilon \ \Rightarrow\ C^{-1}\mathsf{m}(\varrho) \leq \mathsf{m}(\tilde\varrho) \leq C\mathsf{m}(\varrho), \quad \text{for } \tilde\varrho = (x+y, \xi+\eta, \tau).$$

In our case, the metric and the order function do not depend on t. The estimates (18.4.13) and (18.4.14) involve the constants related with dual metric that is, using the previous notation,

$$\mathsf{g}_\varrho \leq C\big(1 + \mathsf{g}_\varrho^A(x-\tilde x, \xi-\tilde\xi)\big)^N \mathsf{g}_{\tilde\varrho}, \tag{14.B.12}$$

and

$$\mathsf{m}(\varrho) \leq C\mathsf{m}(\tilde\varrho)\big(1 + \mathsf{g}_\varrho^A(x-\tilde x, \xi-\tilde\xi)\big)^N, \tag{14.B.13}$$

where A is the quadratic form $A(x',\xi') = t\langle x',\xi'\rangle/2$ associated with the matrix also denoted $A = (t/4)\begin{pmatrix} 0 & Id \\ Id & 0 \end{pmatrix}$. We have to prove that C can be chosen independently of t. In the lemma, the metric is $\tilde{g} = \lambda_{\mathsf{T},\tau}|dx|^2 + \lambda_{\mathsf{T},\tau}^{-1}|d\xi'|^2$ and dual metric

$$\tilde{g}_\varrho^A(w) = \sup_{\substack{w' \in \mathbb{R}^{2(d-1)} \\ \tilde{g}_\varrho(Aw')<1}} (w \cdot w')^2.$$

We obtain $g_\varrho^A = 16t^{-2}\tilde{g}_\varrho$. Thus if (14.B.12) and (14.B.13) are true with a constant C for $t = 1$, they are also true for the same constant C for all $t \in [0,1]$. Of course, (14.B.12) and (14.B.13) are true for $t = 1$ as it is the way to obtain [175, Theorem 18.5.10]. ■

14.B.4.2. *Proof of Proposition 14.41.* One has

$$\begin{aligned} q_k(x,\xi',\tau) &= a_k(2^{-k/2}x', x_d, 2^{k/2}\xi', \tau) \\ &= \varphi_k(|(2^{k/2}\xi',\tau)|)a(2^{-k/2}x', x_d, 2^{k/2}\xi', \tau). \end{aligned}$$

For $\alpha', \beta' \in \mathbb{N}^{d-1}$, computing $\partial_{x'}^{\alpha'}\partial_{\xi'}^{\beta'} q_k(x,\xi',\tau)$ yields a linear combination of terms of the form

$$I_{\alpha',\delta',\hat{\delta}'} = \partial_{\xi'}^{\delta'}\big(\varphi_k(|(2^{k/2}\xi',\tau)|)\big)\partial_{x'}^{\alpha'}\partial_{\xi'}^{\hat{\delta}'}\big(a(2^{-k/2}x', x_d, 2^{k/2}\xi', \tau)\big),$$

with $\delta' + \hat{\delta}' = \beta'$. The conclusion thus holds if one proves that $|I_{\alpha',\delta',\hat{\delta}'}| \lesssim 1$ if $|\alpha'| + |\beta'| \geq 4$.

We use the following lemma whose proof is given below.

Lemma 14.44. *Let $\delta' \in \mathbb{N}^{d-1}$. There exists $C_{\delta'}$ such that*

$$\big|\partial_{\xi'}^{\delta'}\big(\varphi_k(|(2^{k/2}\xi',\tau)|)\big)\big| \leq C_\beta 2^{-k|\delta'|/2}. \tag{14.B.14}$$

We observe that $(|\alpha'| - |\beta'|)/2 = 2 - |\beta'|$ if $|\alpha'| + |\beta'| = 4$. Recalling the assumption made on $a(x,\xi',\tau)$ in the statement of Theorem 14.22, one thus has

$$|\partial_{x'}^{\alpha'}\partial_{\xi'}^{\beta'} a(x,\xi',\tau)| \lesssim \lambda_{\mathsf{T},\tau}^{2-|\beta'|} \quad \text{if } |\alpha'| + |\beta'| \in \{0,2,4\}. \tag{14.B.15}$$

Lemma 14.45. *For $\alpha \in \mathbb{N}^d$ and $\beta' \in \mathbb{N}^{d-1}$ with $|\alpha| + |\beta'| \leq 4$, there exists $C_{\alpha,\beta'} > 0$ such that*

$$|\partial_x^\alpha\partial_{\xi'}^{\beta'} a(x,\xi',\tau)| \leq C_{\alpha,\beta'}\lambda_{\mathsf{T},\tau}^{2-|\beta'|}, \qquad (x,\xi') \in \mathbb{R}^{2d-1},\ \tau \geq 1.$$

A proof is given below.

Choosing $\alpha = (\alpha', 0)$ in Lemma 14.45, one obtains an interpolation of property (14.B.15).

We write

$$\begin{aligned} &\partial_{x'}^{\alpha'}\partial_{\xi'}^{\hat{\delta}'}\big(a(2^{-k/2}x', x_d, 2^{k/2}\xi', \tau)\big) \\ &\quad = 2^{(|\hat{\delta}'|-|\alpha'|)k/2}(\partial_{x'}^{\alpha'}\partial_{\xi'}^{\hat{\delta}'} a)(2^{-k/2}x', x_d, 2^{k/2}\xi', \tau). \end{aligned}$$

If $|\alpha'| + |\hat{\delta}'| \leq 4$, we have

$$|\partial_{x'}^{\alpha'}\partial_{\xi'}^{\hat{\delta}'}\big(a(2^{-k/2}x', x_d, 2^{k/2}\xi', \tau)\big)| \lesssim 2^{(|\hat{\delta}'|-|\alpha'|)k/2}\nu_k^{2-|\hat{\delta}'|}$$

with $\nu_k^2 = \tau^2 + |2^{k/2}\xi'|^2$. On supp $\big(\varphi_k(|(2^{k/2}\xi', \tau)|)\big)$, one has $\nu_k \eqsim 2^k$. Thus, one finds

$$|I_{\alpha',\delta',\hat{\delta}'}| \lesssim 2^{(|\hat{\delta}'|-|\alpha'|)k/2} 2^{2k-|\hat{\delta}'|k} 2^{-k|\delta'|/2} = 2^{k(4-|\alpha'|-|\beta'|)/2}.$$

If $|\alpha'| + |\hat{\delta}'| > 4$, we have

$$|\partial_{x'}^{\alpha'} \partial_{\xi'}^{\hat{\delta}'} \big(a(2^{-k/2}x', x_d, 2^{k/2}\xi', \tau)\big)| \lesssim 2^{(|\hat{\delta}'|-|\alpha'|)k/2} \nu_k^{(|\alpha'|-|\hat{\delta}'|)/2} \lesssim 1,$$

as $\nu_k \eqsim 2^k$. Hence, if $|\alpha'| + |\beta'| \geq 4$, one obtains $|I_{\alpha',\delta',\hat{\delta}'}| \lesssim 1$. ■

Proof of Lemma 14.44. The function $\lambda_{\mathsf{T},\tau}$ is homogeneous of degree one with respect to (ξ', τ), with $\xi' \in \mathbb{R}^{d-1}$ and $\tau > 0$. Thus, for $\alpha' \in \mathbb{N}^{d-1}$, the function $\partial_{\xi'}^{\alpha'} \lambda_{\mathsf{T},\tau}$ is homogeneous of degree $1 - |\alpha'|$. In particular, there exists $C_{\alpha'} > 0$ such that

$$|\partial_{\xi'}^{\alpha'} \lambda_{\mathsf{T},\tau}| \leq C_{\alpha'} \lambda_{\mathsf{T},\tau}^{1-|\alpha'|}, \qquad \xi' \in \mathbb{R}^{d-1}, \ \tau > 0. \tag{14.B.16}$$

The result of the lemma is clear for $k = 0$. For $k \geq 1$, we now write

$$\varphi_k(|(2^{k/2}\xi', \tau)|) = \varphi(2^{1-k}|(2^{k/2}\xi', \tau)|) = \varphi(2^{1-k/2}\lambda_{\mathsf{T},\tau_k}),$$

where $\lambda_{\mathsf{T},\tau_k}^2 = |\xi'|^2 + \tau_k^2$ with $\tau_k = 2^{-k/2}\tau$.

For $\delta' \in \mathbb{N}^{d-1}$, computing $\partial_{\xi'}^{\delta'} \varphi_k(|(2^{k/2}\xi', \tau)|)$ yields a linear combination of terms of the form

$$J_{m_1,\dots,m_j} = 2^{-jk/2} (\partial_{\xi'}^{m_1} \lambda_{\mathsf{T},\tau_k}) \cdots (\partial_{\xi'}^{m_j} \lambda_{\mathsf{T},\tau_k}) \varphi^{(j)}(2^{1-k/2}\lambda_{\mathsf{T},\tau_k}),$$

for $m_1 + \cdots + m_j = \delta'$. With (14.B.16), one finds

$$\begin{aligned} |J_{m_1,\dots,m_j}| &\lesssim 2^{-jk/2} \lambda_{\mathsf{T},\tau_k}^{1-|m_1|} \cdots \lambda_{\mathsf{T},\tau_k}^{1-|m_j|} |\varphi^{(j)}(2^{1-k/2}\lambda_{\mathsf{T},\tau_k})| \\ &\lesssim 2^{-jk/2} \lambda_{\mathsf{T},\tau_k}^{j-|\delta'|} |\varphi^{(j)}(2^{1-k/2}\lambda_{\mathsf{T},\tau_k})|. \end{aligned}$$

In $\varphi^{(j)}(2^{1-k/2}\lambda_{\mathsf{T},\tau_k})$, one has $\lambda_{\mathsf{T},\tau_k} \eqsim 2^{k/2}$ yielding

$$|J_{m_1,\dots,m_j}| \lesssim 2^{-jk/2} 2^{k(j-|\delta'|)/2} \lesssim 2^{-k|\delta'|/2},$$

which yields estimation (14.B.14). ■

Proof of Lemma 14.45. From the assumption on $a(x, \xi', \tau)$ in the statement of Theorem 14.22, one has

$$|\partial_x^\alpha \partial_{\xi'}^{\beta'} a(x, \xi', \tau)| \lesssim \lambda_{\mathsf{T},\tau}^{2-|\beta'|} \quad \text{if } |\alpha| + |\beta'| \in \{0, 2, 4\}.$$

With this property, one only needs to consider the cases $|\alpha| + |\beta'| = 1$ and $|\alpha| + |\beta'| = 3$. First, we consider the case $|\alpha| + |\beta'| = 1$.

Case $|\alpha| = 1$ and $|\beta'| = 0$. Here, one estimates $\nabla_x a(x, \xi', \tau)$. To that purpose, one writes

$$\begin{aligned} a(x + y, \xi', \tau) &= a(x, \xi', \tau) + \nabla_x a(x, \xi', \tau) \cdot y \\ &\quad + \int_0^1 (1 - t) d_x^2 a(x + ty, \xi', \tau)(y, y) \, dt. \end{aligned}$$

One may assume $\nabla_x a(x,\xi',\tau) \neq 0$, since otherwise the estimation is obvious. If one chooses $y = \nabla_x a(x,\xi',\tau)/|\nabla_x a(x,\xi',\tau)|$, one finds

$$|\nabla_x a(x,\xi',\tau)| \lesssim \lambda_{\mathsf{T},\tau}^2.$$

Case $|\alpha| = 0$ and $|\beta'| = 1$. Similarly, one writes

$$a(x,\xi'+\eta',\tau) = a(x,\xi',\tau) + \nabla_{\xi'} a(x,\xi',\tau)\cdot\eta' + \int_0^1 (1-t) d_{\xi'}^2 a(x,\xi'+t\eta',\tau)(\eta',\eta')\, dt.$$

We choose $\eta' = \delta\lambda_{\mathsf{T},\tau,\xi'}\nabla_{\xi'} a(x,\xi',\tau)/|\nabla_{\xi'} a(x,\xi',\tau)|$ with $\delta > 0$. If δ is chosen sufficiently small, one has $\lambda_{\mathsf{T},\tau,\xi'} \approx \lambda_{\mathsf{T},\tau,\xi'+t\eta'}$ uniformly for all $t\in[0,1]$. We then obtain

$$\lambda_{\mathsf{T},\tau,\xi'}\, |\nabla_{\xi'} a(x,\xi',\tau)| \lesssim \lambda_{\mathsf{T},\tau,\xi'}^2,$$

yielding the estimate in this case.

Second, we consider the case $|\alpha| + |\beta'| = 3$. If $|\alpha| \geq 1$, then arguing as in the first case treated above the result follows. The only remaining case is then $|\beta'| = 3$; one argues in the second case treated above. The details are left to the reader. ■

14.B.4.3. *Proof of Lemma 14.42.* For $k \geq 1$, we set $f_k = \varphi_k(\lambda_{\mathsf{T},\tau}) = \varphi(2^{1-k}\lambda_{\mathsf{T},\tau})$. Recall that φ is supported in the interval $[1,3]$. For $\beta' \in \mathbb{N}^{d-1}$, observe that $\partial_{\xi'}^{\beta'} f_k$ is a linear combination of terms of the form

$$f_k^{m_1,\dots,m_j} = 2^{-jk}(\partial_{\xi'}^{m_1}\lambda_{\mathsf{T},\tau})\cdots(\partial_{\xi'}^{m_j}\lambda_{\mathsf{T},\tau})\varphi^{(j)}(2^{1-k}\lambda_{\mathsf{T},\tau}),$$

with $m_1+\cdots+m_j = \beta'$, and one has

$$\begin{aligned}|f_k^{m_1,\dots,m_j}| &\lesssim 2^{-jk}\lambda_{\mathsf{T},\tau}^{1-|m_1|}\cdots\lambda_{\mathsf{T},\tau}^{1-|m_j|}|\varphi^{(j)}(2^{1-k}\lambda_{\mathsf{T},\tau})|\\ &\lesssim 2^{-jk}\lambda_{\mathsf{T},\tau}^{j-|\beta'|}|\varphi^{(j)}(2^{1-k}\lambda_{\mathsf{T},\tau})| \lesssim 2^{-k|\beta'|},\end{aligned}$$

using that $\lambda_{\mathsf{T},\tau} \approx 2^k$ in $\operatorname{supp}\big(\varphi^{(j)}(2^{1-k}\lambda_{\mathsf{T},\tau})\big)$. One thus finds

$$|\partial_{\xi'}^{\beta'}(\varphi_k(\lambda_{\mathsf{T},\tau}))| \lesssim 2^{-k|\beta'|}, \tag{14.B.17}$$

implying the result for $\varphi_k(\lambda_{\mathsf{T},\tau})$. The same proof gives the counterpart result for $\psi_k(\lambda_{\mathsf{T},\tau})$.

Since $a_k = \varphi_k(\lambda_{\mathsf{T},\tau})a$, for $\alpha\in\mathbb{N}^d$, $\beta'\in\mathbb{N}^{d-1}$, observe that $\partial_x^\alpha\partial_{\xi'}^{\beta'} a_k$ is a linear combination of terms of the form

$$b_{\alpha,\delta',\hat{\delta}'} = (\partial_x^\alpha\partial_{\xi'}^{\delta'} a)\partial_{\xi'}^{\hat{\delta}'}(\varphi_k(\lambda_{\mathsf{T},\tau})),$$

with $\delta' + \hat{\delta}' = \beta'$. We consider two cases.

Case $|\alpha| + |\delta'| \leq 4$. In such a case, by Lemma 14.45, one has

$$|\partial_x^\alpha\partial_{\xi'}^{\delta'} a(x,\xi',\tau)| \lesssim \lambda_{\mathsf{T},\tau}^{2-|\delta'|}.$$

In the support of $\varphi_k(\lambda_{\mathsf{T},\tau})$, where $\lambda_{\mathsf{T},\tau} \eqsim 2^k$, one thus finds

$$|\partial_x^\alpha \partial_{\xi'}^{\delta'} a(x,\xi',\tau)| \lesssim 2^{2k-k|\delta'|}.$$

With (14.B.17), one obtains

$$|b_{\alpha,\delta',\hat{\delta}'}| \lesssim 2^{2k-k|\delta'|} 2^{-k|\hat{\delta}'|} \lesssim 2^{2k-k|\beta'|} \lesssim 2^{2k+k(|\alpha|-|\beta'|)/2}. \tag{14.B.18}$$

Case $|\alpha|+|\delta'|>4$. In such a case, by the assumption of Theorem 14.22, one has

$$|\partial_x^\alpha \partial_{\xi'}^{\delta'} a(x,\xi',\tau)| \lesssim \lambda_{\mathsf{T},\tau}^{(|\alpha|-|\delta'|)/2}. \tag{14.B.19}$$

Hence one finds

$$|\partial_x^\alpha \partial_{\xi'}^{\delta'} a(x,\xi',\tau)| \lesssim 2^{k(|\alpha|-|\delta'|)/2},$$

in the support of $\varphi_k(\lambda_{\mathsf{T},\tau})$. With (14.B.17), one obtains

$$|b_{\alpha,\delta',\hat{\delta}'}| \lesssim 2^{k(|\alpha|-|\delta'|)/2} 2^{-k|\hat{\delta}'|} \lesssim 2^{k(|\alpha|-|\beta'|)/2} \lesssim 2^{2k+k(|\alpha|-|\beta'|)/2}. \tag{14.B.20}$$

Together, (14.B.18) and (14.B.20) yield

$$|\partial_x^\alpha \partial_{\xi'}^{\beta'} a_k| \lesssim 2^{2k+k(|\alpha|-|\beta'|)/2},$$

implying the second result of the lemma. ■

14.B.4.4. *Proof of Lemma 14.43.* One has

$$\begin{aligned}\|\psi_k^w u\|_+^2 &= (\psi_k^w \psi_k^w u, u)_+ = \big((\psi_k^2)^w u, u\big)_+ \\ &= (2\pi)^{-(d-1)} \int_{\mathbb{R}_+} \int_{\mathbb{R}^{d-1}} \psi_k^2(\lambda_{\mathsf{T},\tau}) |\hat{u}(\xi', x_d)|^2 \, d\xi' dx_d,\end{aligned}$$

where $\hat{u}$ denotes the partial Fourier transform of u with respect to x', using that ψ_k^w acts as a Fourier multiplier. With the monotone convergence theorem, one finds

$$\sum_{k\in\mathbb{N}} \|\psi_k^w u\|_+^2 \lesssim \int_{\mathbb{R}_+} \int_{\mathbb{R}^{d-1}} \sum_{k\in\mathbb{N}} \psi_k^2(\lambda_{\mathsf{T},\tau}) |\hat{u}(\xi', x_d)|^2 \, d\xi' dx_d \lesssim \|u\|_+^2.$$

One has $a = \sum_{k\in\mathbb{N}} a_k$ pointwise. Denote by $K_k(x,y)$ the kernel of the operator $a_k^w(x,D')$. In the sense of oscillatory integrals, one has

$$K_k(x',y') = (2\pi)^{-(d-1)} \int_{\mathbb{R}^{d-1}} e^{(x'-y')\cdot\xi'} a_k\big((x'+y')/2, x_d, \xi'\big) \, d\xi',$$

with $x_d \geq 0$ acting as a parameter, implying that

$$K_k(x',y') = \check{a}_k\big((x'+y')/2, x_d, x'-y'\big),$$

where $\check{a}_k$ denotes the partial inverse Fourier transform of a_k with respect to ξ'.

Up to change of variables, one sees that $\sum_k K_k(x',y')$ converges in $\mathscr{S}'(\mathbb{R}^{d-1}\times\mathbb{R}^{d-1})$ if and only if $\sum_k \check{a}_k(x',x_d,y')$ converges in $\mathscr{S}'(\mathbb{R}^{d-1}\times\mathbb{R}^{d-1})$, with x_d acting as a parameter. In turn, the latter property holds if and only if $\sum_k a_k(x',x_d,\xi')$ converges in $\mathscr{S}'(\mathbb{R}^{d-1}\times\mathbb{R}^{d-1})$. Since $a_k = \varphi_k(\lambda_{\mathsf{T},\tau})a$, this last convergence holds because of the temperate growth of

$a(x,\xi)$ at infinity and the uniform convergence of $\sum_k a_k(x,\xi)$ to $a(x,\xi)$ on any compact set of $\mathbb{R}^{2(d-1)}$. Thus, $\sum_k K_k(x',y')$ converges to the kernel $K(x',y')$ of $a^w(x,D')$ in $\mathscr{S}'(\mathbb{R}^{d-1}\times\mathbb{R}^{d-1})$. Note that this convergence is uniform for x_d in a compact set.

For $u\in\overline{\mathscr{C}}_c^\infty(\mathbb{R}^d_+)$, one obtains

$$\sum_{k\leq n}\int_{\mathbb{R}_+}\langle K_k,u\otimes\overline{u}\rangle_{\mathscr{S}'(\mathbb{R}^{2(d-1)}),\mathscr{S}(\mathbb{R}^{2(d-1)})}dx_d$$

$$\underset{n\to+\infty}{\to}\int_{\mathbb{R}_+}\langle K,u\otimes\overline{u}\rangle_{\mathscr{S}'(\mathbb{R}^{2(d-1)}),\mathscr{S}(\mathbb{R}^{2(d-1)})}dx_d,$$

yielding the second result of Lemma 14.43. ∎

Part 5

Background Material: Geometry

CHAPTER 15

Elements of Differential Geometry

Contents

This chapter is devoted to the exposition of facts of differential in particular to prepare for a proper presentation of integration and differential operators in the next chapter. In Chapter 9 of Volume 1, we reviewed the action of change of variables on some of the notions that are important

J. Le Rousseau et al., *Elliptic Carleman Estimates and Applications to Stabilization and Controllability, Volume II*, PNLDE Subseries in Control 98, https://doi.org/10.1007/978-3-030-88670-7_15

for the understanding of Carleman estimates. On manifolds, the notion of change of (local) variables, through local charts, is essential for the understanding of geometrical objects such as tangent and cotangent vectors, vector fields, covariant tensors, and p-forms. We only review the essential features. An interested reader should consult a more thorough reference, such as [1, 222, 309], for analysis on manifolds

15.1. Manifolds

Manifolds in the sense used here are topological spaces that have smooth properties. Local representations allow one to characterize this smoothness. We start by providing the precise definitions of manifold without and with boundary that we shall use.

DEFINITION 15.1 (Manifold Without Boundary). Let $d, k \in \mathbb{N}$. A d-dimensional $\mathscr{C}^k$-manifold without boundary is a topological space $\mathcal{M}$ equipped with a set $(O^i, \kappa^i)_{i\in\mathcal{I}}$, with, for each $i \in \mathcal{I}$, O^i is an open subset of $\mathcal{M}$ and κ^i is a map from O^i into $\mathbb{R}^d$ such that

(1) The open sets O^i, $i \in \mathcal{I}$, cover $\mathcal{M}$, that is, $\cup_{i\in\mathcal{I}} O^i = \mathcal{M}$.
(2) $\kappa^i(O^i)$ is an open subset of $\mathbb{R}^d$.
(3) For each $i \in \mathcal{I}$, the map $\kappa^i : O^i \to \kappa^i(O^i)$ is a homeomorphism.
(4) If $O^{ij} = O^i \cap O^j \neq \emptyset$, then for $W^\ell = \kappa^\ell(O^{ij}) \subset \kappa^\ell(O^\ell)$, $\ell = i$ or j, the map $\kappa^{ij} : W^i \to W^j$ given by $\kappa^j \circ (\kappa^i)^{-1}$ is a $\mathscr{C}^k$-diffeomorphism.

A couple (O^i, κ^i) is called a local chart. A set of local charts as above is called an atlas. If the definition holds for any $k \in \mathbb{N}$, one says that $\mathcal{M}$ is a smooth or $\mathscr{C}^\infty$-manifold.

When in a local chart (O, κ), we shall denote by κ_i the map from O to $\mathbb{R}$ such that $\kappa_i(m)$ is the ith coordinate of $\kappa(m)$ in $\mathbb{R}^d$, that is, $\kappa(m) = (\kappa_1(m), \dots, \kappa_d(m))$.

DEFINITION 15.2 (Manifold with Boundary). A topological space $\mathcal{M}$ is said to be a d-dimensional $\mathscr{C}^k$-manifold with boundary if it is equipped with a set $(O^i, \kappa^i)_{i\in\mathcal{I}}$, where O^i is an open subset of $\mathcal{M}$ and κ^i is a map from O^i into $\mathbb{R}^d$ for each $i \in \mathcal{I}$, and such that

(1) The open sets O^i, $i \in \mathcal{I}$, cover $\mathcal{M}$, that is $\cup_{i\in\mathcal{I}} O^i = \mathcal{M}$.
(2) $\kappa^i(O^i)$ is either an open subset of $\mathbb{R}^d$ or an open subset of $\overline{\mathbb{R}^d_+} = \{x \in \mathbb{R}^d; x_d \geq 0\}$.
(3) For each $i \in \mathcal{I}$, the map $\kappa^i : O^i \to \kappa^i(O^i)$ is a homeomorphism.
(4) If $O^{ij} = O^i \cap O^j \neq \emptyset$, then for $W^\ell = \kappa^\ell(O^{ij}) \subset \kappa^\ell(O^\ell)$, $\ell = i$ or j, the map $\kappa^{ij} : W^i \to W^j$ given by $\kappa^j \circ (\kappa^i)^{-1}$ is a $\mathscr{C}^k$-diffeomorphism. In particular, in the case W^i is an open subset of $\overline{\mathbb{R}^d_+}$ that contains part of $\{x_d = 0\}$, then W^j satisfies this property as well, and the map $\kappa^{ij} : W^i \to W^j$ is the restriction of a $\mathscr{C}^k$-map $\underline{\kappa}^{ij} : \underline{W}^i \to \underline{W}^j$, for $\underline{W}^i$ and $\underline{W}^j$ open subsets of $\mathbb{R}^d$ such that $W^i \subset \underline{W}^i$ and $W^j \subset \underline{W}^j$.

A couple (O^i, κ^i) as above is called a local chart.

A point $m \in \mathcal{M}$ is said to be a boundary point of $\mathcal{M}$ if $m \in O$ for a local chart (O, κ) with $\kappa(O)$ an open set of $\mathbb{R}^d_+$ that meets $\{x_d = 0\}$ and $m = \kappa^{-1}(x)$ with $x_d = 0$. We denote by $\partial\mathcal{M}$ the boundary of $\mathcal{M}$, that is, the set of all such boundary points. Observe that $\partial\mathcal{M} \subset \mathcal{M}$ in the definition we have given. The boundary $\partial\mathcal{M}$ can be equipped with the structure of a manifold. It is then a manifold without boundary; we refer to Sect. 15.5.

If Ω is a smooth open set of $\mathbb{R}^d$, then $\Omega \cup \partial\Omega$ is a manifold with boundary in the sense given above. We recall that an open set Ω with $\partial\Omega \neq \emptyset$ is said to be smooth (or regular) if $\partial\Omega$ is smooth and if Ω is only located on one side of $\partial\Omega$. Note however that Ω is not to be considered as a manifold with boundary, since $\partial\Omega \not\subset \Omega$.

A manifold may not be connected. Each of its connected components can however be considered as an individual manifold. If a manifold is compact, the number of its connected components is finite.

For $\mathcal{M}$ and $\tilde{\mathcal{M}}$ two manifolds possibly with boundaries, of respective class $\mathscr{C}^k$ and $\mathscr{C}^{\tilde{k}}$, consider a continuous map $\phi : \mathcal{M} \to \tilde{\mathcal{M}}$. It is said to be of class $\mathscr{C}^\ell$, for $\ell \leq \min(k, \tilde{k})$, if for every $m \in \mathcal{M}$ and every coordinate patch (O, κ) on $\mathcal{M}$ and $(\tilde{O}, \tilde{\kappa})$ on $\tilde{\mathcal{M}}$, such that $m \in O$ and $\phi(m) \in \tilde{O}$, the map $g : \tilde{\kappa} \circ \phi \circ \kappa^{-1}$, which is well defined in a neighborhood of $\kappa(m)$ in $\kappa(O)$, is of class $\mathscr{C}^\ell$. In the case $\mathcal{M}$ is a d-dimensional manifold with a boundary and m is a boundary point, then $\kappa(O)$ is an open subset of $\overline{\mathbb{R}^d_+}$ and $\mathscr{C}^\ell$-regularity for g means such regularity up to the boundary. In particular, g is the restriction to $\overline{\mathbb{R}^d_+}$ of a $\mathscr{C}^\ell$-map defined in a neighborhood of $\kappa(m)$ in $\mathbb{R}^d$ into $\mathbb{R}^{\tilde{d}}$, where $\tilde{d}$ is the dimension of $\tilde{\mathcal{M}}$.

For a function f defined on $\tilde{\mathcal{M}}$, we define its pullback by ϕ, denoted by $\phi^* f$ to be the function on $\mathcal{M}$ given by $\phi^* f = f \circ \phi$.

If f is a function defined on $\mathcal{M}$ with values in some set E, then for every local chart $\mathcal{C} = (O, \kappa)$, $f^{\mathcal{C}} = (\kappa^{-1})^* f : \kappa(O) \to E$ is the local representative of f in this chart. If we now consider two local charts $\mathcal{C}^\ell = (O^\ell, \kappa^\ell)$, $\ell = 1, 2$, such that $O = O^1 \cap O^2 \neq \emptyset$, then we have $f^{\mathcal{C}^1}(x) = (\kappa^{12})^* f^{\mathcal{C}^2}(x)$ for $x \in \kappa^1(O)$ and $\kappa^{12} = \kappa^2 \circ (\kappa^1)^{-1}$ that is a diffeomorphism from $\kappa^1(O)$ onto $\kappa^2(O)$. Conversely, we have the following proposition.

PROPOSITION 15.3. *Let $\mathcal{A} = \{(O^i, \kappa^i);\ i \in \mathcal{I}\}$ be an atlas of a manifold $\mathcal{M}$. Let $(f^i)_{i \in \mathcal{I}}$ be such that*

(1) *For each $i \in \mathcal{I}$, $f^i : \kappa^i(O^i) \to E$, for some set E.*
(2) *For each $(i, j) \in \mathcal{I}^2$ such that $O^{ij} = O^i \cap O^j \neq \emptyset$, we have $f^i(x) = (\kappa^{ij})^* f^j(x)$ for $x \in \kappa^i(O^{ij})$ and $\kappa^{ij} = \kappa^j \circ (\kappa^i)^{-1}$.*

Then, there exists $f : \mathcal{M} \to E$ such that $f^i = f^{\mathcal{C}^i}$ for all $i \in \mathcal{I}$.

15.2. Open Coverings and Partitions of Unity

Let $\mathcal{M}$ be a $\mathscr{C}^k$-manifold ($0 \le k \le \infty$) with or without boundary. Let F be a closed subset of $\mathcal{M}$, and let $(V^j)_{j\in\mathcal{J}}$ be an open covering of F in $\mathcal{M}$, that is, for all $j \in \mathcal{J}$, V^j is an open set of $\mathcal{M}$ and $F \subset \cup_{j\in\mathcal{J}} V^j$.

Definition 15.4. A family of $\mathscr{C}^k$-functions $(\chi^j)_{j\in\mathcal{J}}$ is called a $\mathscr{C}^k$-partition of unity of F subordinated to the open covering $(V^j)_{j\in\mathcal{J}}$ if we have:

(1) For all $j \in \mathcal{J}$, we have $\operatorname{supp}(\chi^j) \subset V^j$.
(2) For all $j \in \mathcal{J}$, we have $0 \le \chi^j \le 1$.
(3) The sum $\sum_{j\in\mathcal{J}} \chi^j$ is locally finite and $\sum_{j\in\mathcal{J}} \chi^j(m) = 1$ for m in a neighborhood of F.

By locally finite, one means that for all $m \in \mathcal{M}$, there exists a neighborhood V_m of m in $\mathcal{M}$ such that $\#\{j \in \mathcal{J};\ \chi^j \not\equiv 0 \text{ in } V_m\} < \infty$. An open covering $\mathcal{V} = (V^j)_{j\in\mathcal{J}}$ is also said to be locally finite if for all $m \in \mathcal{M}$, there exists a neighborhood V_m of m in $\mathcal{M}$ such that $\#\{j \in \mathcal{J};\ V^j \cap V_m \neq \emptyset\} < \infty$.

Let $\mathcal{V} = (V^j)_{j\in\mathcal{J}}$ and $\mathcal{U} = (U^k)_{k\in\mathcal{K}}$ be two open coverings of a closed set F in $\mathcal{M}$. One says that $\mathcal{U}$ is finer than $\mathcal{V}$ if for all $k \in \mathcal{K}$ there exists $j \in \mathcal{J}$ such that $U^k \subset V^j$. We then observe that a partition of unity of F subordinated to $\mathcal{U}$ yields a partition of unity of F subordinated to $\mathcal{V}$ upon summing together some of the functions. Below, we shall use this argument implicitly, as in the process of constructing a partition of unity subordinated to a given open covering, we shall make use of finer open coverings.

15.2.1. Some Topological Results. A manifold $\mathcal{M}$ with or without boundary is σ-compact if it admits an exhaustion by compact sets: there exists a sequence of compact sets $K^0 \subset K^1 \subset \cdots \subset K^n \subset \cdots$ such that

$$K^n \subset \operatorname{int} K^{n+1},\ n \in \mathbb{N}, \quad \text{and} \quad \bigcup_{n\in\mathbb{N}} K^n = \mathcal{M}.$$

One also says that $\mathcal{M}$ is countable at infinity. In such case, for any compact set K of $\mathcal{M}$, there exists $n \in \mathbb{N}$ such that $K \subset \operatorname{int} K^n$. Examples are $\mathbb{R}^d$ and (obviously) compact manifolds. We shall make the assumption that the considered manifold is σ-compact to build a partition of unity. The importance of this property lies in the following lemma.

Lemma 15.5. *Let $\mathcal{M}$ be a σ-compact manifold. Then, for any closed set F and for any open covering $\mathcal{V} = (V^j)_{j\in\mathcal{J}}$ of F in $\mathcal{M}$, there exists a second open covering $\mathcal{U} = (U^k)_{k\in\mathcal{K}}$ of F in $\mathcal{M}$, with $\mathcal{K}$ countable, that is finer than $\mathcal{V}$ and locally finite. In particular, this holds for $F = \mathcal{M}$, precisely meaning that $\mathcal{M}$ is paracompact.*

Proof. Let $(K^n)_{n\in\mathbb{N}}$ be an exhaustive sequence of compact sets for $\mathcal{M}$. For each $n \in \mathbb{N}$, there exists a finite sub-family of $\mathcal{V}$, $(V_n^k)_{0\le k\le k_n}$, such that $F \cap K^n \subset \cup_{j=0}^{k_n} V_n^k$. Note that the family $(V_n^k)_{n,k}$, is countable and finer than

$\mathcal{V}$. We now construct a locally finite refinement of this family. Define the open sets $V^{(0,k)} = V_0^k \cap \operatorname{int} K^1$ for $k = 0, \dots, k_0$, and $V^{(n,k)} = V_n^k \setminus K^{n-1}$, for $n \geq 1$ and $k = 0, \dots, k_n$. We have

$$F \cap K^0 \subset \bigcup_{k=0,\dots,k_0} V^{(0,k)}, \qquad F \cap K^n \setminus K^{n-1} \subset \bigcup_{k=0,\dots,k_n} V^{(n,k)}.$$

Consequently, the family $\mathcal{U} = \{V^{(n,k)};\ n \in \mathbb{N}, k = 0, \dots, k_n\}$ is an open covering of F. We see that $V^{(\ell,k)} \cap K^{n-1} = \emptyset$ if $\ell \geq n$, implying that the open covering $\mathcal{U}$ is locally finite. ■

The following lemma is also useful.

LEMMA 15.6. *Let $\mathcal{M}$ be a σ-compact manifold, F be a closed subset, and $\mathcal{V} = (V^j)_{j\in\mathcal{J}}$ be an open covering of F. There exists an open covering $\mathcal{U}$ of F finer than $\mathcal{V}$ formed by relatively compact open sets. Moreover if $\mathcal{V}$ is locally finite, then $\mathcal{U}$ is also locally finite, and if $\mathcal{V}$ is countable, then $\mathcal{U}$ is also countable.*

PROOF. We introduce the sequence of open sets $(\Omega^n)_{n\in\mathbb{N}}$ given by

$$\Omega^0 = \operatorname{int} K^1, \quad \Omega^n = \left(\operatorname{int} K^{n+1}\right) \setminus K^{n-1}, \text{ for } n \geq 1.$$

We then set $U^{j,n} = V^j \cap \Omega^n$ for $j \in \mathcal{J}$ and $n \in \mathbb{N}$. The family $\mathcal{U} = (U^{j,n})_{j\in\mathcal{J},n\in\mathbb{N}}$ is a covering of F as $\cup_{n\in\mathbb{N}}\Omega^n = \mathcal{M}$ and one sees that it has the required property. ■

In the next two sections, we shall use the previous two lemmata to construct partitions of unity, first, in $\mathbb{R}^d$ and, second, in a manifold.

REMARK 15.7. In many texts, the topological assumption made on the manifold is that of second countability. A manifold $\mathcal{M}$ is second-countable, if there exists a countable basis of open sets defining its topology. In fact, this implies that the manifold is σ-compact by the following lemma.

LEMMA 15.8. *A second-countable manifold $\mathcal{M}$ with or without boundary is σ-compact.*

PROOF. As $\mathcal{M}$ is locally homeomorphic to $\mathbb{R}^d$, we find that there exists a basis of its topology made of open sets with compact closure. As $\mathcal{M}$ is second-countable, we may in fact choose such a basis to be countable, which we denote by $(U^n)_{n\in\mathbb{N}}$. We claim that we can construct a sequence of compact sets satisfying

$$U^n \subset K^n \quad \text{and } K^n \subset \operatorname{int} K^{n+1}, \ n \in \mathbb{N}.$$

We set the compact set $K^0 = \overline{U^0}$. Now assume that the compact sets $K^0, \dots, K^n$ are chosen such that $U^j \subset K^j$, for $j = 0, \dots, n$, and $K^{j-1} \subset \operatorname{int} K^j$, for $j = 1, \dots, n$. As K^n is compact, there exists $N \geq n+1$ such that $K^n \subset \cup_{0\leq j\leq N} U^j$. We define the compact set $K^{n+1} = \cup_{0\leq j\leq N}\overline{U^j}$. By the definition of N and K^{n+1}, we have $U^{n+1} \subset K^{n+1}$. As $\cup_{0\leq j\leq N} U^j \subset K^{n+1}$, we have $\cup_{0\leq j\leq N} U^j \subset \operatorname{int} K^{n+1}$ yielding $K^n \subset \operatorname{int} K^{n+1}$. Finally, as

$\mathcal{M} = \cup_{n\in\mathbb{N}} U^n \subset \cup_{n\in\mathbb{N}} K^n$, we see that we have constructed an exhaustion by compact sets. ■

15.2.2. Partitions of Unity on $\mathbb{R}^d$. An important lemma toward the construction of partitions of unity in $\mathbb{R}^d$ is the following one.

LEMMA 15.9. *Let F be a closed subset of $\mathbb{R}^d$. Let $\mathcal{V} = (V^j)_{j\in\mathcal{J}}$ be a locally finite open covering of F in $\mathbb{R}^d$. For all $j \in \mathcal{J}$, there exists W^j an open subset of $\mathbb{R}^d$ such that $\overline{W^j} \subset V^j$ and $\mathcal{W} = (W^j)_{j\in\mathcal{J}}$ is a locally finite open covering of F in $\mathbb{R}^d$.*

PROOF. Let $(K^n)_{n\in\mathbb{N}}$ be an exhaustive sequence of compact sets for $\mathbb{R}^d$. Let $n \in \mathbb{N}$, and let $\mathcal{J}_n$ be the finite set of indices $j \in \mathcal{J}$ such that $V^j \cap K^n \neq \emptyset$. Let us define the continuous function $d_n(x)$ given by

$$d_n(x) = \max_{j\in\mathcal{J}_n} \operatorname{dist}(x, \mathbb{R}^d \setminus V^j).$$

As a side remark, observe that $d_n(x) \leq d_{n+1}(x)$ as $\mathcal{J}_n \subset \mathcal{J}_{n+1}$. If $x \in K^n$ and if $j \in \mathcal{J} \setminus \mathcal{J}_n$, then $x \notin V^j$ and thus $\operatorname{dist}(x, \mathbb{R}^d \setminus V^j) = 0$. It follows that $d_{n+\ell}(x) = d_n(x)$ for all $\ell \geq 0$ and $x \in K^n$. For all $x \in \mathbb{R}^d$, the sequence $d_n(x)$ is thus nondecreasing and stationary for n sufficiently large.

If $x \in K^n \cap F$, then $x \in V^j$ for some $j \in \mathcal{J}_n$ and thus $\operatorname{dist}(x, \mathbb{R}^d \setminus V^j) > 0$. We thus have $d_n(x) > 0$ on $K^n \cap F$. As $K^n \cap F$ is compact, we have $C_n = \inf_{x\in K^n\cap F} d_n(x) > 0$. For $j \in \mathcal{J}_n$, we now define an open set W_n^j of $\mathbb{R}^d$ such that $W_n^j \subset V^j$ and given by

$$x \in W_n^j \quad \Leftrightarrow \quad x \in V^j \text{ and } \operatorname{dist}(x, \mathbb{R}^d \setminus V^j) > C_n/2. \tag{15.2.1}$$

Note that W_n^j may very well be empty. We moreover set $W_n^j = \emptyset$ for $j \in \mathcal{J} \setminus \mathcal{J}_n$. If $x \in K^n \cap F$, there exists $j \in \mathcal{J}_n$ such that $\operatorname{dist}(x, \mathbb{R}^d \setminus V^j) = d_n(x) \geq C_n$. By the definition of W_n^j, we have $x \in W_n^j$. Hence, $(W_n^j)_{j\in\mathcal{J}}$ is an open covering of $K^n \cap F$.

We introduce the sequence of open sets $(\mathcal{O}^n)_{n\in\mathbb{N}}$ given by $\mathcal{O}^n = \operatorname{int} K_n$ and we have $\cup_{n\in\mathbb{N}} \mathcal{O}^n = \mathbb{R}^d$. The sequence of compact sets $(\overline{\mathcal{O}^n})_{n\in\mathbb{N}}$ is also an exhaustion of $\mathbb{R}^d$. We then introduce the sequence of open sets $(\Omega^n)_{n\in\mathbb{N}}$ given by

$$\Omega^0 = \mathcal{O}^0, \ \Omega^1 = \mathcal{O}^1, \quad \Omega^n = \mathcal{O}^n \setminus \overline{\mathcal{O}^{n-2}}, \text{ for } n \geq 2.$$

We have $\cup_{n\in\mathbb{N}} \Omega_n = \mathbb{R}^d$. We set

$$\tilde{W}_n^j = W_n^j \cap \Omega^n, \quad j \in \mathcal{J}, \ n \in \mathbb{N}.$$

We define the open sets $W^j = \cup_{n\in\mathbb{N}} \tilde{W}_n^j \subset V^j$ for $j \in \mathcal{J}$.

Let $x \in F$. Then, for some $n \in \mathbb{N}$, $x \in \Omega^n \subset \mathcal{O}^n \subset K^n$. As $(W_n^j)_{j\in\mathcal{J}}$ is an open covering of $K^n \cap F$, there exists $j \in \mathcal{J}$ such that $x \in W_n^j$ and thus $x \in \Omega^n \cap W_n^j = \tilde{W}_n^j \subset W^j$. The family of open sets $(W^j)_{j\in\mathcal{J}}$ thus forms an open covering of F in $\mathbb{R}^d$.

If $j \in \mathcal{J}$ and $x \in \overline{W^j}$, then there exists a sequence $(x_p)_{p\in\mathbb{N}} \subset W^j$ such that $x_p \to x$ in $\mathbb{R}^d$. As this sequence is bounded, there exists $n \in \mathbb{N}$ such that $\{x\} \cup (x_p)_{p\in\mathbb{N}} \subset \overline{\mathcal{O}^n} \subset K^n$, and we thus have $(x_p)_{p\in\mathbb{N}} \subset \cup_{\ell\leq n+1}\tilde{W}^j_\ell$. This yields $(x_p)_{p\in\mathbb{N}} \subset \cup_{\ell\leq n+1}W^j_\ell$. As a result, we have $x \in \overline{\cup_{\ell\leq n+1}W^j_\ell} \subset V^j$ by (15.2.1). We have thus obtained that $\overline{W^j} \subset V^j$, which concludes the proof. ∎

LEMMA 15.10. *Let V be an open set of $\mathbb{R}^d$ and $K \subset V$ a compact set. There exists $\varphi \in \mathscr{C}_c^\infty(V)$ such that $0 \leq \varphi \leq 1$ and $\varphi \equiv 1$ in a neighborhood of K.*

PROOF. Set $\alpha = \operatorname{dist}(K, \mathbb{R}^d \setminus V)$. For $0 < \varepsilon < \alpha$, we define the following neighborhood of K:

$$K_\varepsilon = \{x \in \mathbb{R}^d; \operatorname{dist}(x, K) \leq \varepsilon\} \subset V.$$

We let ψ be the characteristic function of $K_{\alpha/2}$. We also pick $\chi \in \mathscr{C}_c^\infty(\mathbb{R}^d)$ supported in the unit ball $B(0,1)$ such that $\chi \geq 0$ and $\int \chi = 1$. For $\varepsilon > 0$, we set $\chi_\varepsilon(x) = \varepsilon^{-d}\chi(x/\varepsilon)$. We have

$$\operatorname{supp}\chi_\varepsilon \subset B(0,\varepsilon), \qquad \int \chi_\varepsilon = 1.$$

We set $\varphi_\varepsilon = \chi_\varepsilon * \psi \in \mathscr{C}_c^\infty(\mathbb{R}^d)$. We have $\varphi_\varepsilon \geq 0$. For $\varepsilon = \alpha/4$, we find $\varphi_\varepsilon \equiv 1$ in $K_{\alpha/4}$ from the definition of the convolution and $\operatorname{supp}(\varphi_\varepsilon) \subset K_{3\alpha/4} \subset V$ by the support theorem for the convolution of functions. ∎

We can now prove the following theorem that states the existence of partitions of unity in $\mathbb{R}^d$.

THEOREM 15.11. *Let F be a closed subset of $\mathbb{R}^d$ and let $(V^j)_{j\in\mathcal{J}}$ be an open covering of F in $\mathbb{R}^d$. There exists a $\mathscr{C}^\infty$-partition of unity of F subordinated to this open covering.*

PROOF. By Lemmata 15.5 and 15.6, we first obtain a finer open covering $(\mathcal{O}^n)_{n\in\mathbb{N}}$ of F in $\mathbb{R}^d$ that is countable, locally finite, and formed by relatively compact open sets. By Lemma 15.9, there exists a second locally finite open covering $\mathcal{W} = (W^n)_{n\in\mathbb{N}}$ of F in $\mathbb{R}^d$ such that $\overline{W^n} \subset \mathcal{O}^n$ for all $n \in \mathbb{N}$. The open set $W = \cup_{n\in\mathbb{N}}W^n$ is a neighborhood of F.

By Lemma 15.10 for all $n \in \mathbb{N}$, there exists $\phi^n \in \mathscr{C}_c^\infty(\mathcal{O}^n)$ such that $0 \leq \phi^n \leq 1$ and $\phi^n \equiv 1$ in a neighborhood of W^n. We then define the following sequence of functions:

$$\varphi^0 = \phi^0, \qquad \varphi^n = \phi^n \prod_{j=0}^{n-1}(1-\phi^j), \text{ for } n \geq 1.$$

We have $\varphi^n \geq 0$, $\operatorname{supp}(\varphi^n) \subset \mathcal{O}^n$, and, as the open covering $(\mathcal{O}^n)_{n\in\mathbb{N}}$ of F is locally finite, we see that $\varphi = \sum_{n\in\mathbb{N}}\varphi^n$ is a well defined $\mathscr{C}^\infty$ function. Let $x_0 \in W$. With the same argument, there exists $k \in \mathbb{N}$ such that a neighborhood $U_0 \subset W$ of x_0 does not intersect $\mathcal{O}^n$ for $n > k$. We thus have

$$\varphi = \sum_{0 \le n \le k} \varphi^n \text{ in } U_0.$$

We claim that $\varphi \equiv 1$ in U_0, which concludes the proof.

In fact, define the functions

$$\rho^k = \phi^k, \qquad \rho^\ell = \phi^\ell + \sum_{n=\ell+1}^{k} \phi^n \prod_{j=\ell}^{n-1} (1 - \phi^j), \text{ for } 0 \le \ell < k.$$

In U_0, we have $\varphi = \rho^0$. We then write, for $\ell = 0, \dots, k-1$,

$$\begin{aligned}\rho^\ell - 1 &= \phi^\ell - 1 + \phi^{\ell+1}(1 - \phi^\ell) + \sum_{n=\ell+2}^{k} \phi^n \prod_{j=\ell}^{n-1} (1 - \phi^j) \\ &= (1 - \phi^\ell)\Big(-1 + \phi^{\ell+1} + \sum_{n=\ell+2}^{k} \phi^n \prod_{j=\ell+1}^{n-1} (1 - \phi^j)\Big) \\ &= (1 - \phi^\ell)(\rho^{\ell+1} - 1).\end{aligned}$$

By induction, we find, for $\ell = 0, \dots, k-1$,

$$\rho^0 - 1 = \prod_{0 \le n \le \ell} (1 - \phi^n)(\rho^{\ell+1} - 1) = - \prod_{0 \le n \le k} (1 - \phi^n). \tag{15.2.2}$$

If $x \in U_0 \subset W$, then there exists $n \in \{1, \dots, k\}$ such that $x \in W^n$ implying $\phi^n(x) = 1$. We thus have $\varphi(x) - 1 = \rho^0(x) - 1 = 0$ by (15.2.2). ■

15.2.3. Partitions of Unity on a σ-Compact Manifold. Let $\mathcal{M}$ be a σ-compact $\mathscr{C}^k$-manifold. Let $\mathcal{A}$ be an atlas of $\mathcal{M}$ with $\mathcal{A} = (O^i, \kappa^i)_{i \in \mathcal{I}}$. Let F be a closed subset of $\mathcal{M}$, and let $\mathcal{V} = (V^j)_{j \in \mathcal{J}}$ be an open covering of F in $\mathcal{M}$. Upon refining both open coverings, according to Lemmata 15.5 and 15.6, we may assume that both the coverings $(O^i)_{i \in \mathcal{I}}$ and $\mathcal{V}$ are locally finite and (at most) countable and formed by relatively compact open sets. We can thus use $\mathcal{I} = \mathcal{J} = \mathbb{N}$ to index the two families of open sets. Setting now $V^{i,j} = O^i \cap V^j$, $i, j \in \mathbb{N}$, we see that we can furthermore assume that the covering $\mathcal{V}$ of F is finer than the covering of $\mathcal{M}$ associated with the atlas.

REMARK 15.12. Observe that if F is compact, then the manifold $\mathcal{M}$ need not be assumed σ-compact as the results of Lemmata 15.5 and 15.6 are clear in this case.

The following lemma is the counterpart to Lemma 15.9 for manifolds.

LEMMA 15.13. *Let $\mathcal{V} = (V^j)_{j \in \mathbb{N}}$ be a countable locally finite open covering of F in $\mathcal{M}$ made of relatively compact open sets. For all $j \in \mathbb{N}$, there exists a relatively compact open subset W^j of $\mathcal{M}$ such that $W^j \Subset V^j$ and $\mathcal{W} = (W^j)_{j \in \mathbb{N}}$ is a locally finite open covering of F in $\mathcal{M}$.*

PROOF. We first consider the open set V^0. It is contained in a relatively compact open set O^{i_0} associated with a local chart (O^{i_0}, κ^{i_0}) of the atlas for some $i_0 \in \mathbb{N}$. Note that different choices of i_0 may be possible. We set

$\check{V}^0 = \cup_{j\geq 1}V^j$. As $F \subset V^0 \cup \check{V}^0$, the closed set $L^0 = F \setminus \check{V}^0$ satisfies $L^0 \subset V^0 \subset \overline{O^{i_0}}$. As $\overline{O^{i_0}}$ is compact, L^0 is itself compact.

We set $\tilde{L}^0 = \kappa^{i_0}(L^0)$ and $\tilde{V}^0 = \kappa^{i_0}(V^0)$ both subsets of $\tilde{O} = \kappa^{i_0}(O^{i_0})$ an open subset of $\mathbb{R}^d$. As κ is a homeomorphism, $\tilde{L}^0$ is a compact set of $\mathbb{R}^d$. Thus there exists an open set $\tilde{W}^0$ that is relatively compact and such that $\tilde{L}^0 \subset \tilde{W}^0 \Subset \tilde{V}^0$. We set $W^0 = (\kappa^{i_0})^{-1}(\tilde{W}^0)$, and we have

$$L^0 \subset W^0 \Subset V^0,$$

implying that

$$\mathcal{W}^0 = \{W^0\} \cup \bigcup_{j\geq 1}\{V^j\} \text{ is an open covering of } F \text{ in } \mathcal{M}.$$

Next starting from $\mathcal{W}^0$, setting $L^1 = F \setminus (W^0 \cup \cup_{j\geq 2}V^j)$, we construct W^1 such that $L^1 \subset W^1 \Subset V^1$ and such that

$$\mathcal{W}^1 = \{W^0, W^1\} \cup \bigcup_{j\geq 2}\{V^j\} \text{ is an open covering of } F \text{ in } \mathcal{M}.$$

By induction, we construct the relatively compact open sets W^j, $j \in \mathbb{N}$, such that $W^j \Subset V^j$ and

$$\forall k \in \mathbb{N},\ \mathcal{W}^k = \bigcup_{0\leq j\leq k}\{W^j\} \cup \bigcup_{j\geq k+1}\{V^j\} \text{ is an open covering of } F \text{ in } \mathcal{M}.$$

Let $x \in F$, then there exists $\mathcal{J}_x = \{j_1, \ldots, j_n\} \subset \mathbb{N}$ such that $x \in V^j$ *if and only if* $j \in \mathcal{J}_x$ as the covering is locally finite. We thus see that there exists $j \in \mathcal{J}_x$ such that $x \in W^j$ using that $\mathcal{W}^k$ is an open covering of F in $\mathcal{M}$ for $k > \max J_x$. Hence $\mathcal{W} = (W^j)_{j\in\mathbb{N}}$ is an open covering of F in $\mathcal{M}$. As $W^j \subset V^j$, this covering is necessarily locally finite. ■

With this lemma, we may then conclude to the existence of a partition of unity subordinated to a given open covering.

THEOREM 15.14. *Let $\mathcal{M}$ be a σ-compact $\mathscr{C}^k$-manifold. Let F be a closed subset of $\mathcal{M}$, and let $\mathcal{V} = (V^j)_{j\in\mathcal{J}}$ be an open covering of F in $\mathcal{M}$. There exists a $\mathscr{C}^k$-partition of unity of F in $\mathcal{M}$ subordinated to this covering.*

PROOF. Upon choosing refinements of the atlas and the covering as in the beginning of this section, we may assume that both are at most countable (and indexed by $\mathbb{N}$), locally finite, and made of relatively compact open sets. We may also assume that the covering is finer than the covering given by the atlas.

We can then apply Lemma 15.13 and have a second locally finite open covering $\mathcal{W} = (W^n)_{n\in\mathbb{N}}$ of F in $\mathcal{M}$ such that $W^n \Subset V^n$, for all $n \in \mathbb{N}$. For each $n \in \mathbb{N}$, there exists $\phi^n \in \mathscr{C}_c^k(V^n)$ such that $\phi^n \equiv 1$ on a neighborhood of W^n. To see this, we consider a chart $\mathcal{C} = (O, \kappa)$ such that $V^n \subset O$, and we set $\tilde{V}^n = \kappa(V^n)$ and $\tilde{W}^n = \kappa(W^n)$. Then, by Lemma 15.10, there exists $\tilde{\phi}^n \in \mathscr{C}_c^\infty(\tilde{V}^n)$ such that $\tilde{\phi}^n \equiv 1$ on $\tilde{W}^n$. Then $\phi^n = \kappa^*\tilde{\phi}^n$ has the required properties.

We then define the following sequence of $\mathscr{C}^k$-functions on $\mathcal{M}$:

$$\varphi^0 = \phi^0, \qquad \varphi^n = \phi^n \prod_{j=0}^{n-1} (1 - \phi^j), \text{ for } n \geq 1.$$

Then, the sum $\sum_{n\in\mathbb{N}} \varphi^n$ is locally finite and forms a partition of unity of F in $\mathcal{M}$ subordinated to the covering $(V^n)_{n\in\mathbb{N}}$, by the same argument as in the end of proof of Theorem 15.11. ■

We shall see below that partitions of unity can be very important to define proper geometrical objects on a manifold.

REMARK 15.15. Proposition 15.3 characterizes functions on a manifold. In fact, if $f : \mathcal{M} \to E$ is a $\mathscr{C}^k$-function and if $\mathcal{A} = (O^n, \kappa^n)_{n\in\mathbb{N}}$ is an atlas chosen countable and the covering $(O^n)_{n\in\mathbb{N}}$ of $\mathcal{M}$ locally finite if $\mathcal{M}$ is assumed σ-compact. The local representative of f in a local chart $\mathcal{C} = (O, \kappa)$ is $(\kappa^{-1})^* f$. This function is $\mathscr{C}^k$ in $\kappa(O)$ but not compactly supported. In some arguments, it can be useful to have some compactly supported representative of f. A partition of unity comes in handy. Assume indeed that $\sum_{n\in\mathbb{N}} \varphi^n$ is a partition of unity subordinated to the covering $(O^n)_{n\in\mathbb{N}}$. We then write $f = \sum_{n\in\mathbb{N}} f^n$ with $f^n = f\varphi^n$ compactly supported in O^n. Upon application of the pullback by $(k^n)^{-1}$, we then obtain some sort of compactly supported representative of f in $\kappa^n(O^n)$. Naturally, this representation is only exact in $\kappa^n\big((\varphi^n)^{-1}(\{1\})\big)$.

We conclude this section by the following useful lemma that is the counterpart on manifolds of Lemma 15.10.

LEMMA 15.16. *Let $\mathcal{M}$ be a d-dimensional manifold. Let V be an open set of $\mathcal{M}$ and $K \subset V$ be a compact set. There exists $\varphi \in \mathscr{C}_c^\infty(V)$ such that $0 \leq \varphi \leq 1$ and $\varphi \equiv 1$ in a neighborhood of K.*

PROOF. By Lemma 15.6, there exists an open set $V' \subset V$ such that V' is relatively compact in $\mathcal{M}$ and $K \subset V'$. Then, the partition of unity provided by Theorem 15.14 yields a single function in $\mathscr{C}_c^\infty(V)$ that fulfills the sought properties. ■

15.3. Tangent Space and Vector Fields

Let $\mathcal{M}$ be a $\mathscr{C}^1$ d-dimensional manifold and $m_0 \in \mathcal{M}$. We shall use equivalent classes of $\mathscr{C}^1$-curves to define tangent vectors at m_0.

15.3.1. Tangent Vectors.

We call a curve γ passing through m_0 on $\mathcal{M}$ any map $\mathbb{R} \supset I \to \mathcal{M}$ of class $\mathscr{C}^1$ for I an open interval such that $\gamma(t_0) = m_0$ for some $t_0 \in I$. Without any loss of generality, we may choose $I = (-\varepsilon, \varepsilon)$, with $\varepsilon > 0$, and $t_0 = 0$. If m_0 is a boundary point, we choose $I = (-\varepsilon, 0]$ or $[0, \varepsilon)$ and derivatives below need to be replaced by one-sided derivatives.

Let $\mathcal{C}^1 = (O^1, \kappa^1)$ be a local chart with $m_0 \in O^1$. We set

$$v_{(1)} = F_{m_0,\mathcal{C}^1}(\gamma) := \frac{d}{dt}\kappa^1(\gamma(t))_{|t=0}.$$

Note that this derivative exists because of the $\mathscr{C}^1$-regularity of γ. This vector in $\mathbb{R}^d$ is the tangent vector of the curve $t \mapsto \kappa^1(\gamma(t))$ at $\kappa^1(m_0)$ (for $t=0$) in $\mathbb{R}^d$. Observe that for any $v \in \mathbb{R}^d$, the local curve $\gamma(t) = (\kappa^1)^{-1}(\kappa^1(m_0)+tv)$ passes through m_0 and is such that $F_{m_0,\mathcal{C}^1}(\gamma) = v$. The map $F_{m_0,\mathcal{C}^1}$ is thus onto $\mathbb{R}^d$.

If $\mathcal{C}^2 = (O^2, \kappa^2)$ is a second local chart with $m_0 \in O^2$, then we may also set $v_{(2)} = F_{m_0,\mathcal{C}^2}(\gamma) := \frac{d}{dt}\kappa^2(\gamma(t))_{|t=0}$. Setting $\kappa^{12} = \kappa^2 \circ (\kappa^1)^{-1}$, which is well defined near $\kappa^1(m_0)$ in $\kappa^1(O^1)$, we write, for t near 0,

$$\kappa^2(\gamma(t)) = \kappa^{12}\big(\kappa^1(\gamma(t))\big).$$

As κ^{12} is a $\mathscr{C}^1$-diffeomorphism, this yields the following relation between $v_{(2)}$ and $v_{(1)}$:

$$v_{(2)} = d\kappa^{12}(\kappa^1(m_0))\big(v_{(1)}\big). \tag{15.3.1}$$

We then observe that if two curves, γ and $\tilde{\gamma}$, both passing through m_0, are such that $F_{m_0,\mathcal{C}^1}(\gamma) = F_{m_0,\mathcal{C}^1}(\tilde{\gamma})$, that is, they share the same tangent vector $v_{(1)}$ in the local chart $\mathcal{C}^1$, then we have $F_{m_0,\mathcal{C}^2}(\gamma) = F_{m_0,\mathcal{C}^2}(\tilde{\gamma})$ in the second local chart $\mathcal{C}^2$. We may thus define the following equivalence relation on the set of local curves passing through m_0, independently of the picked local chart. One says that $\gamma \sim \tilde{\gamma}$ if, in any chart $\mathcal{C} = (O, \kappa)$, they share the same tangent vector v. We denote by $[\gamma]$ the equivalence class. We call $[\gamma]$ a tangent vector. The vectorial property of such an object is the subject of the next paragraph.

We define the tangent set of $\mathcal{M}$ at m_0 as the set of those equivalence classes, and we denote it by $T_{m_0}\mathcal{M}$. For any chart $\mathcal{C} = (O, \kappa)$, we may define the bijective map $\tilde{F}_{m_0,\mathcal{C}} : T_{m_0}\mathcal{M} \to \mathbb{R}^d$, by $\tilde{F}_{m_0,\mathcal{C}}([\gamma]) = F_{m_0,\mathcal{C}}(\gamma)$. This induces a linear structure on $T_{m_0}\mathcal{M}$, which is independent of the considered local chart, as we have the *linear* correspondence (15.3.1) for the local representative of the tangent vectors. The tangent set $T_{m_0}\mathcal{M}$ is then a vector space of dimension d. It is called the tangent vector space of $\mathcal{M}$ at m_0. Naturally, this vectorial structure on $T_{m_0}\mathcal{M}$ makes the bijection $\tilde{F}_{m_0,\mathcal{C}}$ an isomorphism.

Observe that this definition of the tangent space only requires the manifold $\mathcal{M}$ to be $\mathscr{C}^1$.

Let f be a $\mathscr{C}^1$ real valued function defined in a neighborhood of m_0 in $\mathcal{M}$. For $v \in T_{m_0}\mathcal{M}$, we consider γ a representative of v, that is, $v = [\gamma]$. For a local chart $\mathcal{C} = (O, \kappa)$, we define its local representative, that is, the $\mathscr{C}^1$-function $f^{\mathcal{C}} : \kappa(O) \to \mathbb{R}$ by $f^{\mathcal{C}} = (\kappa^{-1})^* f$. The map $t \mapsto \kappa(\gamma(t))$ is a $\mathscr{C}^1$-curve in $\mathbb{R}^d$ that goes through $x_0 = \kappa(m_0)$ at $t = 0$. As we have

$f(\gamma(t)) = f^{\mathcal{C}}(\kappa(\gamma(t)))$ and $\frac{d}{dt}\kappa(\gamma(t))_{|t=0} = F_{m_0,\mathcal{C}}(\gamma) = \tilde{F}_{m_0,\mathcal{C}}(v)$, we obtain

$$\lim_{t\to 0} \frac{d}{dt} f(\gamma(t))_{|t=0} = df^{\mathcal{C}}(x_0)(\tilde{F}_{m_0,\mathcal{C}}(v)). \tag{15.3.2}$$

If we consider a second representative $\tilde{\gamma}$ of v, that is, $\gamma \sim \tilde{\gamma}$, then we also have $\frac{d}{dt}\kappa(\tilde{\gamma}(t))_{|t=0} = \tilde{F}_{m_0,\mathcal{C}}(v)$. This implies that $\frac{d}{dt}f(\gamma(t))_{|t=0} = \frac{d}{dt}f(\tilde{\gamma}(t))_{|t=0}$. We may thus define a linear form $f \mapsto v(f)$ by this common value:

$$v(f) := \frac{d}{dt} f(\gamma(t))_{|t=0} = \frac{d}{dt} f(\tilde{\gamma}(t))_{|t=0}. \tag{15.3.3}$$

Observe that this map is linear. Moreover, from (15.3.2), we find that

$$v(fh) = f(m_0)v(h) + v(f)h(m_0), \qquad f, h \in \mathscr{C}^1.$$

The map $f \mapsto v(f)$ is thus a *derivation* on the $\mathscr{C}^1$-functions defined locally near m_0. Note in particular that v vanishes on constant functions.

REMARK 15.17. In the case of a $\mathscr{C}^\infty$-manifold, the tangent vector space at m_0 can be alternatively defined as the vector space of derivations on $\mathscr{C}^\infty$-functions defined in a neighborhood of m_0, as is often done in the literature. Note, however, that the vector space of derivations on $\mathscr{C}^k$-functions defined in a neighborhood of m_0, with $k \in \mathbb{N}$, does not coincide with $T_{m_0}\mathcal{M}$ as the former is in fact infinite dimensional [263, 316]. In particular, this approach cannot be used for $\mathscr{C}^k$-manifolds. This can however be repaired by additional requirements on the derivations [132].

Let us further express the local representative of a tangent vector v at m_0. Consider a local chart $\mathcal{C} = (O, \kappa)$. Let $v^{\mathcal{C}} = (v^1, \dots, v^d)$ be the coordinates of the local representative of v in $\mathcal{C}$, that is, $v^{\mathcal{C}} = \tilde{F}_{m_0,\mathcal{C}}(v)$ in the notation used above. We denote by X_i the map from $\mathbb{R}^d$ to $\mathbb{R}$ such that $X_i(x) = x_i$. We then set $\kappa_i = \kappa^* X_i$. We have, as seen above, for γ a representative of v, $v^{\mathcal{C}} = \frac{d}{dt}\kappa(\gamma(t))_{|t=0}$, meaning that

$$v^i = \frac{d}{dt}\kappa_i(\gamma(t))_{|t=0} = v(\kappa_i). \tag{15.3.4}$$

This allows one to obtain the action of v on $\mathscr{C}^1$-functions as follows. As above, for a $\mathscr{C}^1$ real valued function f given in a neighborhood of m_0, we set $f^{\mathcal{C}} = (\kappa^{-1})^* f$ from a neighborhood in $\mathbb{R}^d$ of $x_0 = \kappa(m_0)$ into $\mathbb{R}$. Setting $x = \kappa(m) \in \mathbb{R}^d$, yielding $x_i = \kappa_i(m)$, we then write

$$f(m) = f^{\mathcal{C}}(x) = f^{\mathcal{C}}(x_0) + \sum_{1\le j\le d} \partial_{x_j} f^{\mathcal{C}}(x_0)(x_i - x_0) + \varepsilon(x),$$

with $\varepsilon(x) = w(x)|x - x_0|_{\mathbb{R}^d}$ of class $\mathscr{C}^1$, with $\lim w(x) = 0$ as $x \to x_0$. Observe that we have

$$t^{-1}\varepsilon(\kappa(\gamma(t))) = t^{-1}|\kappa(\gamma(t)) - x_0|_{\mathbb{R}^d} w(\kappa(\gamma(t))) \sim |v|\, w(\kappa(\gamma(t))) \sim 0,$$

as $t \to 0$. This implies that $v(\varepsilon \circ \kappa) = \lim_{t\to 0} t^{-1}\varepsilon(\kappa(\gamma(t))) = 0$. By (15.3.4), we then obtain

$$v(f) = \sum_{1\le i\le d} \partial_{x_i} f^{\mathcal{C}}(x_0) v(\kappa_i) = \sum_{1\le i\le d} v^i \partial_{x_i} f^{\mathcal{C}}(x_0). \tag{15.3.5}$$

Let us consider as above two local charts $\mathcal{C}^\ell = (O^\ell, \kappa^\ell)$, $\ell = 1, 2$, with $m_0 \in O = O^1 \cap O^2$ and $\kappa^{12} = \kappa^2 \circ (\kappa^1)^{-1}$. For $v \in T_{m_0}\mathcal{M}$, we let $v^{\mathcal{C}^1}, v^{\mathcal{C}^2} \in \mathbb{R}^d$ be its local representative in the two local charts. Formula (15.3.1) can also be written in the form, with $x_0^{(1)} = \kappa^1(m_0)$,

(15.3.6)
$$V^{\mathcal{C}^2} = (\kappa^{12})'(x_0^{(1)})\, V^{\mathcal{C}^1}, \qquad \text{with } V^{\mathcal{C}^\ell} = {}^t(v^{\mathcal{C}^\ell,1}, \dots, v^{\mathcal{C}^\ell,d}),\ \ell = 1, 2,$$

where $(\kappa^{12})'(x_0^{(1)})$ is the Jacobian matrix of the map $\kappa^{12} : \kappa^1(O) \to \kappa^2(O)$ at $x_0^{(1)}$, that is,

$$v^{\mathcal{C}^2,i} = \sum_{1\le j\le d} \partial_{x_j} \left(\kappa^{12}\right)_i (x_0^{(1)})\, v^{\mathcal{C}^1,j}, \tag{15.3.7}$$

where $\left(\kappa^{12}\right)_i = (\kappa^{12})^* X_i$, with the function X_i defined as above.

15.3.2. Differential of a Map and Action on Tangent Vectors. We consider two $\mathscr{C}^1$-manifolds $\mathcal{M}$ and $\tilde{\mathcal{M}}$ of respective dimensions d and $\tilde{d}$ and a $\mathscr{C}^1$-map ϕ from $\mathcal{M}$ into $\tilde{\mathcal{M}}$ defined in a neighborhood of m_0. We let $v \in T_{m_0}\mathcal{M}$ and γ be such that $v = [\gamma]$. Then, the function $t \mapsto \phi(\gamma(t))$ is a curve that goes through $\tilde{m}_0 = \phi(m_0)$. We set $\tilde{v} = [\phi \circ \gamma] \in T_{\tilde{m}_0}\tilde{\mathcal{M}}$. We denote this tangent vector by $\tilde{v} = T\phi(m_0)(v)$, and the map $T\phi(m_0)$ is called the differential of ϕ at m_0 or sometimes the tangent of ϕ at m_0.

As both v and $\tilde{v}$ are defined independently of local charts, it is sufficient to consider the action of $T\phi(m_0)$ in local charts. Let thus $\mathcal{C} = (O, \kappa)$ be a chart on $\mathcal{M}$ with $m_0 \in O$ and $\tilde{\mathcal{C}} = (\tilde{O}, \tilde{\kappa})$ be a chart on $\tilde{\mathcal{M}}$ with $\tilde{m}_0 \in \tilde{O}$. We then set $\psi = \tilde{\kappa} \circ \phi \circ \kappa^{-1}$ that maps a neighborhood in $\mathbb{R}^d$ of $x_0 = \kappa(m_0)$ into a neighborhood in $\mathbb{R}^{\tilde{d}}$ of $\tilde{x}_0 = \tilde{\kappa}(\tilde{m}_0)$. This function ψ is the local representative of the function ϕ with respect to the two chosen local charts.

Let $v^{\mathcal{C}} = (v^1, \dots, v^d)$ be the coordinates of the local representative of v in $\mathcal{C}$, that is, $v^i = v(\kappa_i)$ in the notation used above. Similarly we let $\tilde{v}^{(\tilde{\mathcal{C}})} = (\tilde{v}^1, \dots, \tilde{v}^{\tilde{d}})$ be the coordinates of the local representative of $\tilde{v}$ in $\tilde{\mathcal{C}}$, that is, $\tilde{v}^i = \tilde{v}(\tilde{\kappa}_i)$. Then, from the standard differential calculus, we have

$$\tilde{v}^{\tilde{\mathcal{C}}} = d\psi(x_0) v^{\mathcal{C}}. \tag{15.3.8}$$

or ${}^t\tilde{v}^{(\tilde{\mathcal{C}})} = \psi'(x_0)\ {}^t v^{\mathcal{C}}$, where $\psi'(x_0)$ is the $\tilde{d} \times d$ Jacobian matrix of ψ at x_0. From this local representative, we see that $T\phi(m_0)$ is a linear map from $T_{m_0}\mathcal{M}$ into $T_{\tilde{m}_0}(\tilde{M})$.

Observe that in the case $\phi = \mathrm{Id}_{\mathcal{M}}$, we recover the change of variables formula (15.3.1). Note also that we have for the chart $\mathcal{C}$:

$$T\kappa(m_0)(v) = \tilde{F}_{m_0,\mathcal{C}}(v) = v^{\mathcal{C}} \in T_{x_0}\mathbb{R}^d \cong \mathbb{R}^d, \quad v \in T_{m_0}\mathcal{M}. \tag{15.3.9}$$

We recall that $T\kappa(m_0)$ is an isomorphism. This gives

$$T\kappa(m_0)(v) = (v(\kappa_1), \dots, v(\kappa_d)).$$

If we have two maps $\phi : \mathcal{M} \to \tilde{\mathcal{M}}$ defined in a neighborhood of $m_0 \in \mathcal{M}$ and $\varphi : \tilde{\mathcal{M}} \to \hat{\mathcal{M}}$ defined in a neighborhood of $\phi(m_0) \in \tilde{\mathcal{M}}$, then $\Phi = \varphi \circ \phi : \mathcal{M} \to \hat{\mathcal{M}}$ is well defined in a neighborhood of m_0 and, from (15.3.8), using local charts, we find

$$T\Phi(m_0) = T\varphi(\phi(m_0)) \circ T\phi(m_0). \tag{15.3.10}$$

15.3.3. The Tangent Bundle, Vector Fields. With the tangent vector space defined at any point m of $\mathcal{M}$, we then set

$$T\mathcal{M} = \bigcup_{m \in \mathcal{M}} \{m\} \times T_m\mathcal{M}$$

to be the tangent bundle associated with $\mathcal{M}$. We denote by π the natural projection $\pi : T\mathcal{M} \to \mathcal{M}$ given by $\pi(m, v) = m$. We have $\pi^{-1}(m) = \{m\} \times T_m\mathcal{M}$, and we refer to $T_m\mathcal{M}$ or to $\pi^{-1}(m)$ as the fiber above m in the tangent bundle.

If $\mathcal{C} = (O, \kappa)$ is a local chart on $\mathcal{M}$ with $m_0 \in O$, we set $TO = \cup_{m\in O}\{m\} \times T_m\mathcal{M}$ and set the map, using (15.3.9),

$$T\kappa : TO \ni (m, v) \mapsto (\kappa(m), T\kappa(m)(v)) \in \mathbb{R}^d \times \mathbb{R}^d.$$

We have

$$T\kappa(m, v) = (\kappa_1(m), \dots, \kappa_d(m), v(\kappa_1), \dots, v(\kappa_d)).$$

We see that $T\kappa$ is a bijection onto $\kappa(O) \times \mathbb{R}^d$. If $\mathcal{C}^j = (O^j, \kappa^j)$, $j = 1, 2$, are two local charts with $O = O^1 \cap O^2 \neq \emptyset$, then $T\kappa^2 \circ (T\kappa^1)^{-1}$ is a diffeomorphism from $\kappa^1(O) \times \mathbb{R}^d$ onto $\kappa^2(O)\mathbb{R}^d$ of class $\mathscr{C}^{k-1}$ if $\mathcal{M}$ is a $\mathscr{C}^k$-manifold $(k \geq 2)$. In fact, with $\kappa^{12} = \kappa^2 \circ (\kappa^1)^{-1}$, we have

$$T\kappa^2 \circ (T\kappa^1)^{-1}(x, w) = \big(\kappa^{12}(x), d\kappa^{12}(x)(w)\big), \quad (x, w) \in \kappa^1(O) \times \mathbb{R}^d.$$

If $(O^i, \kappa^i)_{i\in\mathcal{I}}$ is an atlas for $\mathcal{M}$ and if we equip $T\mathcal{M}$ with a topology that makes the maps $T\kappa^i$ as defined above continuous on TO^i into $\mathbb{R}^{2d}$, we thus obtain that $T\mathcal{M}$ is a $\mathscr{C}^{k-1}$-manifold of dimension $2d$. An atlas is then given by the set of charts $(TO^i, T\kappa^i)_{i\in\mathcal{I}}$.

Let $\mathcal{M}$ be a $\mathscr{C}^k$-manifold and $\tilde{\mathcal{M}}$ a $\mathscr{C}^{\tilde{k}}$-manifold. Let ϕ be a $\mathscr{C}^\ell$-map, $\ell \leq \min(k, \tilde{k})$, from $\mathcal{M}$ into $\tilde{\mathcal{M}}$ defined on an open set U of $\mathcal{M}$. If $\ell \geq 1$, we can define the map $T\phi : TU \to T\tilde{\mathcal{M}}$ by

$$T\phi(m, v) = \big(\phi(m), T\phi(m)(v)\big),$$

which is called the differential or the tangent map of ϕ on TU. As TU is an open set of $T\mathcal{M}$, $T\phi$ is a $\mathscr{C}^{\ell-1}$-map from TU into $T\tilde{\mathcal{M}}$.

Let U be an open set of $\mathcal{M}$, a $\mathscr{C}^k$-manifold. For $\ell \leq k-1$, a $\mathscr{C}^\ell$-vector field v on U is a $\mathscr{C}^\ell$-map from U into TU that satisfies the property $\pi v(m) = m$, that is, $v(m)$ is in the fiber above m in the tangent bundle. A vector field is also called a section of the tangent vector bundle. We denote by $\mathscr{C}^\ell V(U)$ the space of $\mathscr{C}^\ell$-vector fields on U. On a $\mathscr{C}^\infty$-manifold, we set $\mathscr{C}^\infty V(U) = \cap_{\ell\geq 0}\mathscr{C}^\ell V(U)$. For $v \in \mathscr{C}^\ell V(\mathcal{M})$, its action on a $\mathscr{C}^{\ell+1}$-function f on $\mathcal{M}$ yields a $\mathscr{C}^\ell$-function defined by

$$v(f)(m) = v_m(f), \qquad \text{with } v(m) = (m, v_m).$$

In a local chart $\mathcal{C} = (O, \kappa)$, where v is defined, the local representative of v is $(x, v_x^{\mathcal{C}})$ with $v_x^{\mathcal{C}} = T\kappa(m)(v_m)$ if $v(m) = (m, v_m)$ and $x = \kappa(m)$. As $v_x^{\mathcal{C}}$ is a function of x, we have $v_x^{\mathcal{C}} = (v^1(x), \dots, v^d(x))$, with $v^i(x) = v_m(\kappa_i)$ from (15.3.4), and considering the action given in (15.3.5), we write

$$v_x^{\mathcal{C}} = \sum_{1\leq i\leq d} v^i(x)\partial_{x_i}.$$

If we denote by $f^{\mathcal{C}}$ the representative of f in the local chart, that is, $f^{\mathcal{C}}(x) = f(\kappa^{-1}(x))$, we have

$$v(f)(m) = v_m(f) = v_x^{\mathcal{C}}(f^{\mathcal{C}}) = \sum_{1\leq i\leq d} v^i(x)\partial_{x_i} f^{\mathcal{C}}(x), \qquad x = \kappa(m).$$

Let $\mathcal{M}$ be a $\mathscr{C}^k$-manifold and $\tilde{\mathcal{M}}$ a $\mathscr{C}^{\tilde{k}}$-manifold, and let the two manifolds be of the same dimension. If a map ϕ is a $\mathscr{C}^1$-diffeomorphism from U onto $\phi(U)$ with U an open set of $\mathcal{M}$ and if v is a vector field on U, we can define the image of v by ϕ, called the push-forward of v by ϕ, denoted by $\phi_*(v)$,

$$\phi_*(v)(\tilde{m}) = T\phi(v(m)), \quad \text{for } \tilde{m} \in \phi(U), \text{ with } m = \phi^{-1}(\tilde{m}).$$

This means that $\phi_*(v)(\tilde{m}) = T\phi(m, v_m) = (\tilde{m}, T\phi(m)(v_m))$ with $(m, v_m) = v(m) \in T\mathcal{M}$.

We finish with the following consequence of Proposition 15.3.

COROLLARY 15.18. *Let $\mathcal{A} = (TO^i, T\kappa^i)_{i\in\mathcal{I}}$ be an atlas for $T\mathcal{M}$ as defined above, and let $(f^{\mathcal{C}^i})_{i\in\mathcal{I}}$ be such that:*

1. *For each $i \in \mathcal{I}$, $f^{\mathcal{C}^i} : T\kappa^i(TO^i) \to E$, for some set E.*
2. *For each $(i,j) \in \mathcal{I}^2$ such that $O^{ij} = O^i \cap O^j \neq \emptyset$, we have $f^{\mathcal{C}^i}(x,v) = (T\kappa^{ij})^* f^{\mathcal{C}^j}(x,v) = f^{\mathcal{C}^j}(T\kappa^{ij}(x,v))$ for $x \in \kappa^i(O^{ij})$, $v \in \mathbb{R}^d$, and $T\kappa^{ij} = T\kappa^j \circ (T\kappa^i)^{-1}$.*

Then, there exists $f : T\mathcal{M} \to E$ *such that for all* $i \in I$, $f^{\mathcal{C}^i}$ *is the local representative of* f *in the chart* $\mathcal{C}^i = (TO^i, T\kappa^i)$, *that is,* $f^{\mathcal{C}^i} = ((T\kappa^i)^{-1})^* f$.

Recall that $T\kappa^{ij}(x, v) = (\kappa^{ij}(x), d\kappa^{ij}(x)v)$ for $(x, v) \in \kappa^i(O^{ij}) \times \mathbb{R}^d$.

15.4. Cotangent Vectors and Forms

For $m \in \mathcal{M}$, a $\mathscr{C}^k$ d-dimensional manifold, we define the cotangent vector space at m as the dual space of $T_m\mathcal{M}$. It is denoted by $T_m^*\mathcal{M}$.

15.4.1. Cotangent Vectors, Local Representatives.

Let $m_0 \in \mathcal{M}$ and $\mathcal{C} = (O, \kappa)$ be a local chart with $m_0 \in O$ and $x_0 = \kappa(m_0)$. If $\omega \in T_{m_0}^*\mathcal{M}$ and $v \in T_{m_0}\mathcal{M}$, and if $v^{\mathcal{C}} = (v^1, \dots, v^d) = (v(\kappa_1), \dots, v(\kappa_d))$ is the local representative of v, that is, $v^{\mathcal{C}} = T\kappa(m_0)v$, we write

$$\omega(v) = \langle \omega, v \rangle = \langle \omega, \big(T\kappa(m_0)\big)^{-1} v^{\mathcal{C}} \rangle = \langle {}^t\big(T\kappa(m_0)\big)^{-1} \omega, v^{\mathcal{C}} \rangle,$$

which yields the representative $\xi = \omega^{\mathcal{C}} = {}^t\big(T\kappa(m_0)\big)^{-1}\omega$ of ω in the local chart. If we denote by $(dx^1, \dots, dx^d)$ the dual base of $(\partial_{x_1}, \dots, \partial_{x_d})$ of $T_{\kappa(m_0)}\mathbb{R}^d \cong \mathbb{R}^d$, then we have $\xi = \sum_{1\le i\le d} \xi_i dx^i$. Since $\langle \xi, \partial_{x_i} \rangle = \xi_i$, we find with the above duality identity

$$\xi_i = \langle \omega, (\partial_{x_i}^O)_{m_0} \rangle, \quad \text{with } (\partial_{x_i}^O)_{m_0} = \big(T\kappa(m_0)\big)^{-1} (\partial_{x_i})_{x_0}.$$

Let us now consider two local charts $\mathcal{C}^\ell = (O^\ell, \kappa^\ell)$, $\ell = 1, 2$, with $m_0 \in O = O^1 \cap O^2$ and $\kappa^{12} = \kappa^2 \circ (\kappa^1)^{-1}$. If $\omega \in T_{m_0}^*\mathcal{M}$, let $\xi^{(\ell)} = \omega^{\mathcal{C}^\ell} = \sum_{1\le i\le d} \xi_i^{(\ell)} dx^i$ be its local representative in (O^ℓ, κ^ℓ). Let now $v \in T_{m_0}\mathcal{M}$, with $v^{\mathcal{C}^\ell}$ its local representative, $\ell = 1, 2$. We have

$$\langle \omega, v \rangle = \langle \xi^{(1)}, v^{\mathcal{C}^1} \rangle = \langle \xi^{(2)}, v^{\mathcal{C}^2} \rangle.$$

From (15.3.1), this yields, with $x_0^{(1)} = \kappa^1(m_0)$,

$$\xi^{(1)} = {}^t d\kappa^{12}(x_0^{(1)}) \xi^{(2)}. \tag{15.4.1}$$

In terms of coordinates, this reads

$$\Xi^{(1)} = {}^t(\kappa^{12})'(x_0^{(1)})\, \Xi^{(2)}, \qquad \text{with } \Xi^{(\ell)} = {}^t(\xi_1^{(\ell)}, \dots, \xi_d^{(\ell)}),\ \ell = 1, 2, \tag{15.4.2}$$

that is,

$$\xi_i^{(1)} = \sum_{1\le j\le d} \partial_{x_i} \big(\kappa^{12}\big)_j (x_0^{(1)})\, \xi_j^{(2)}, \tag{15.4.3}$$

which is the counterpart of (15.3.6) and (15.3.7) for cotangent vectors.

15.4.2. Action of a Smooth Map and Differential of a Function. We consider now two $\mathscr{C}^1$-manifolds $\mathcal{M}$ and $\tilde{\mathcal{M}}$ of respective dimensions d and $\tilde{d}$ and a $\mathscr{C}^1$-map ϕ from $\mathcal{M}$ into $\tilde{\mathcal{M}}$ defined in a neighborhood of m_0. As $T\phi(m_0)$ maps $T_{m_0}\mathcal{M}$ into $T_{\phi(m_0)}\tilde{\mathcal{M}}$, the transpose map ${}^tT\phi(m_0)$: $T^*_{\phi(m_0)}\tilde{\mathcal{M}} \to T^*_{m_0}\mathcal{M}$ is given by

$$\langle {}^tT\phi(m_0)\tilde{\omega}, v\rangle = \langle \tilde{\omega}, T\phi(m_0)v\rangle, \qquad \tilde{\omega} \in T^*_{\phi(m_0)}\tilde{\mathcal{M}},\ v \in T_{m_0}\mathcal{M}.$$

In particular, observe that if $\mathcal{C} = (O, \kappa)$ is a local chart with $m_0 \in O$, then

$$\xi = \omega^{\mathcal{C}} = {}^t\big(T\kappa(m_0)\big)^{-1}\omega = {}^tT\kappa^{-1}(x_0)\omega, \qquad \text{with } x_0 = \kappa(m_0). \tag{15.4.4}$$

We recall that if f is a $\mathscr{C}^1$-function on $\mathcal{M}$, $m_0 \in \mathcal{M}$, and $v \in T_{m_0}\mathcal{M}$, then, from (15.3.5), the map $v \mapsto v(f)$ is well defined and linear. It thus defines a cotangent vector at m_0 which we denote by $df(m_0)$. In fact, by (15.3.5), we have

$$\langle df(m_0), v\rangle = v(f) = \sum_{1\le i\le d} v^i \partial_{x_i} g(x_0),$$

with $g = (\kappa^{-1})^* f$ and $x_0 = \kappa(m_0)$. Observe that if $\mathcal{M} = \mathbb{R}^d$, then $f = g$, and we recover the usual definition of the differential of a function at one point. In fact, in the local chart (O, κ), the above identity precisely means that the local representative of the cotangent vector $df(m_0)$ is $\sum_{1\le i\le d} \partial_{x_i} g(x_0) dx^i$. By abuse of notation, one often writes $df(m_0) = \sum_{1\le i\le d} \partial_{x_i} f(m_0) dx^i$.

15.4.3. The Cotangent Bundle. With the cotangent vector space defined at any point m of $\mathcal{M}$, we set

$$T^*\mathcal{M} = \bigcup_{m\in\mathcal{M}} \{m\} \times T^*_m\mathcal{M} \tag{15.4.5}$$

to be the cotangent bundle associated with $\mathcal{M}$. We denote by $\tilde{\pi}$ the natural projection $\tilde{\pi} : T^*\mathcal{M} \to \mathcal{M}$ given by $\tilde{\pi}(m, \omega) = m$. We have $(\tilde{\pi})^{-1}(m) = \{m\} \times T^*_m\mathcal{M}$, and we refer to $T^*_m\mathcal{M}$ or to $(\tilde{\pi})^{-1}(m)$ as the fiber above m in the cotangent bundle.

If $\mathcal{C} = (O, \kappa)$ is a local chart on $\mathcal{M}$, we set $T^*O = \cup_{m\in O}\{m\} \times T^*_m\mathcal{M}$, and we define the map, using (15.4.4),

$$\begin{aligned} T^*\kappa : T^*O \ni (m, \omega) &\mapsto (\kappa(m), {}^t(T\kappa(m))^{-1}(\omega)) \\ &\in \mathbb{R}^d \times T_{\kappa(m)}\mathbb{R}^d \cong \mathbb{R}^d \times \mathbb{R}^d. \end{aligned}$$

We have

$$T^*\kappa(m, \omega) = (\kappa_1(m), \dots, \kappa_d(m), \omega((e_1^O)_m), \dots, \omega((e_d^O)_m)),$$

for $(e_j^O)_m = \big(T\kappa(m)\big)^{-1}(\partial_{x_j})_x$ if $x = \kappa(m)$. We see that $T^*\kappa$ is a bijection onto $\kappa(O) \times \mathbb{R}^d$. If $\mathcal{C}^j = (O^j, \kappa^j)$, $j = 1, 2$, are two local charts with $O = O^1 \cap O^2 \neq \emptyset$, then $T^*\kappa^2 \circ (T^*\kappa^1)^{-1}$ is a diffeomorphism from $\kappa^1(O) \times \mathbb{R}^b$

onto $\kappa^2(O) \times \mathbb{R}^b$ of class $\mathscr{C}^{k-1}$ if $\mathcal{M}$ is a $\mathscr{C}^k$-manifold ($k \geq 2$). In fact, with $\kappa^{12} = \kappa^2 \circ (\kappa^1)^{-1}$, we have

$$T^*\kappa^2 \circ (T^*\kappa^1)^{-1}(x,\xi) = \big(\kappa^{12}(x), {}^t d(\kappa^{12})^{-1}(x)\xi\big), \quad (x,\xi) \in \kappa^1(O) \times \mathbb{R}^d.$$

If $(O^i, \kappa^i)_{i\in\mathcal{I}}$ is an atlas for $\mathcal{M}$ and if we equip $T^*\mathcal{M}$ with a topology that makes the maps $T^*\kappa^i$ as defined above continuous on T^*O^i into $\mathbb{R}^{2d}$, we thus obtain that $T^*\mathcal{M}$ is a $\mathscr{C}^{k-1}$-manifold of dimension $2d$. An atlas is then given by the set of charts $(T^*O^i, T^*\kappa^i)_{i\in\mathcal{I}}$.

15.4.4. One-Forms. Let U be an open set of $\mathcal{M}$, a $\mathscr{C}^k$-manifold. For $\ell \leq k-1$, a $\mathscr{C}^\ell$ one-form ω on U is a $\mathscr{C}^\ell$-map from U into T^*U that satisfies the property $\tilde{\pi}\omega(m) = m$, that is, $\omega(m)$ is in the fiber above m in the cotangent bundle. A one-form is also called a section of the cotangent vector bundle. A one-form on U is also called a cotangent vector field on U. A one-form ω acts on a vector field v as follows:

$$\langle \omega, v\rangle(m) = \langle \omega_m, v_m\rangle_{T_m^*\mathcal{M}, T_m\mathcal{M}}, \quad m \in \mathcal{M},$$

for $\omega(m) = (m, \omega_m)$, and $v(m) = (m, v_m)$. One then obtains a function on $\mathcal{M}$. If both ω and v are $\mathscr{C}^\ell$, then $f = \langle \omega, v\rangle$ is $\mathscr{C}^\ell$.

We denote by $\mathscr{C}^\ell\Lambda(U)$ the space of $\mathscr{C}^\ell$-one-forms on U. On a $\mathscr{C}^\infty$-manifold, we set $\mathscr{C}^\infty\Lambda(U) = \bigcap_{\ell\geq 0}\mathscr{C}^\ell\Lambda(U)$.

Recalling that if f is a $\mathscr{C}^{\ell+1}$-function on $\mathcal{M}$, $\ell \leq k-1$, then $df(m) \in T_m^*\mathcal{M}$, we see that the map $\omega : m \mapsto (m, df(m))$ is in fact a $\mathscr{C}^\ell$ one-form:

$$\langle \omega, v\rangle(m) = \langle df(m), v_m\rangle_{T_m^*\mathcal{M}, T_m\mathcal{M}} = v_m(f),$$

for a vector field v. One often identifies $\omega(m)$ and $df(m)$ and uses the notation df for ω given above. We call df the differential of f on $\mathcal{M}$.

Let $\mathcal{M}$ be a $\mathscr{C}^k$-manifold and $\tilde{\mathcal{M}}$ a $\mathscr{C}^{\tilde{k}}$-manifold, and let d and $\tilde{d}$ be their respective dimensions. If ϕ is $\mathscr{C}^1$-map from U into $\tilde{\mathcal{M}}$ with U an open set of $\mathcal{M}$ and if ω is a one-form on $\tilde{\mathcal{M}}$, we can define the pullback of ω by ϕ, denoted by $\phi^*(\omega)$, with $\phi^*(\omega)(m) = (m, \phi^*(\omega)_m)$, by

$$\langle \phi^*(\omega)_m, v\rangle_{T_m^*\mathcal{M}, T_m\mathcal{M}} = \langle \omega_{\phi(m)}, T\phi(m)v\rangle_{T^*_{\phi(m)}\tilde{\mathcal{M}}, T_{\phi(m)}\tilde{\mathcal{M}}},$$

for $v \in T_m\mathcal{M}$, $m \in \mathcal{M}$. Note that as opposed to the definition of push-forwards of vector fields at the end of Sect. 15.3, we need not require the map ϕ to be a diffeomorphism to define pullback of one-forms. If ϕ is a $\mathscr{C}^1$-diffeomorphism, we however have

$$\begin{aligned}\langle \phi^*(\omega), v\rangle(m) = \langle \phi^*(\omega)_m, v_m\rangle_{T_m^*\mathcal{M}, T_m\mathcal{M}} &= \langle \omega_{\phi(m)}, T\phi(m)v_m\rangle_{T^*_{\phi(m)}\tilde{\mathcal{M}}, T_{\phi(m)}\tilde{\mathcal{M}}} \\ &= \langle \omega_{\phi(m)}, \phi_* v_{\phi(m)}\rangle_{T^*_{\phi(m)}\tilde{\mathcal{M}}, T_{\phi(m)}\tilde{\mathcal{M}}} = \langle \omega, \phi_* v\rangle(\phi(m)),\end{aligned}$$

for ω a one-form on $\tilde{\mathcal{M}}$ and v a vector field on $\mathcal{M}$.

We finish with the following consequence of Proposition 15.3.

COROLLARY 15.19. *Let* $\mathcal{A} = (T^*O^i, T^*\kappa^i)_{i\in\mathcal{I}}$ *be an atlas for* $T^*\mathcal{M}$ *as defined above, and let* $(f^{\mathcal{C}^i})_{i\in\mathcal{I}}$ *be such that*

(1) *For each* $i \in \mathcal{I}$, $f^{\mathcal{C}^i} : T^*\kappa^i(T^*O^i) \to E$, *for some set* E.
(2) *For each* $(i,j) \in \mathcal{I}^2$ *such that* $O^{ij} = O^i \cap O^j \neq \emptyset$, *we have* $f^{\mathcal{C}^i}(x,\xi) = (T^*\kappa^{ij})^* f^{\mathcal{C}^j}(x,\xi)$ *for* $x \in \kappa^i(O^{ij})$, $\xi \in \mathbb{R}^d$, *and* $T^*\kappa^{ij} = T^*\kappa^j \circ (T^*\kappa^i)^{-1}$.

Then, there exists $f : T^*\mathcal{M} \to E$ *such that for all* $i \in I$, $f^{\mathcal{C}^i}$ *is the local representative of* f *in the chart* $\mathcal{C}^i = (T^*O^i, T^*\kappa^i)$, *that is,* $f^{\mathcal{C}^i} = ((T^*\kappa^i)^{-1})^* f$.

REMARK 15.20. If we have

$$(y,\eta) = T^*\kappa^{ij}(x,\xi) = T^*\kappa^j \circ (T^*\kappa^i)^{-1}(x,\xi) = (\kappa^{ij}(x), {}^t d(\kappa^{ij})^{-1}(x)\xi)$$

for $(x,\xi) \in \kappa^i(O^{ij}) \times \mathbb{R}^d$, then $\xi = {}^t d\kappa^{ij}(x)\eta$. Thus, property (2) in Corollary 15.19 also reads

$$f^{\mathcal{C}^i}(x, {}^t d\kappa^{ij}(x)\eta) = f^{\mathcal{C}^j}(\kappa^{ij}(x), \eta), \qquad x \in \kappa^i(O^{ij}), \quad \eta \in \mathbb{R}^d.$$

15.5. Submanifold

Let $\mathcal{M}$ be a d-dimensional smooth manifold with or without boundary.

DEFINITION 15.21 (Submanifold). A subset $\mathcal{N}$ such that $\mathcal{N} \cap \partial\mathcal{M} = \emptyset$ is called a smooth submanifold (without boundary) of codimension $n \in \mathbb{N}$ if for every $m_0 \in \mathcal{N}$, there exists a neighborhood V_0 of m_0 in $\mathcal{M}$ and n real valued smooth functions $f_1, \dots, f_n$ such that the rank of $(df_1(m_0), \dots, df_n(m_0))$ in $T^*_{m_0}\mathcal{M}$ is n, and we have $\mathcal{N} \cap V_0 = \{m \in \mathcal{M} \cap V_0;\ f_1(m) = \cdots = f_n(m) = 0\}$.

Such a submanifold is also called an embedded submanifold. Upon choosing an open set $O \subset V_0$, we can furthermore enforce the rank of $(df_1(m), \dots, df_n(m))$ to be constant and equal to n in O.

We provide $\mathcal{N}$ with the induced subset topology. For a point m_0 of $\mathcal{N}$ and in a neighborhood O as above, the n functions $f_1, \dots, f_n$ can be used to generate a diffeomorphism $\kappa : O \to \kappa(O) \subset \mathbb{R}^d$, with a possibly reduced open set O, such that $\kappa(m) = (\tilde{\kappa}(m), f_1(m), \dots, f_n(m))$ with $\tilde{\kappa} : O \to \mathbb{R}^{d-n}$ smooth. The couple (O, κ) can then be used as a local chart if added to any atlas.

In such a local chart, the submanifold is given by $x_d = \cdots = x_{d-n+1} = 0$. Let $(O^i, \kappa^i)_{i\in\mathcal{I}}$ be a family of such local charts that covers $\mathcal{N}$, that is, $\mathcal{N} \subset \cup_{i\in\mathcal{I}} O^i$. For $m \in O^i$, we then have $\kappa^i(m) = (\tilde{\kappa}^i(m), 0_{\mathbb{R}^n})$ if and only if $m \in \mathcal{N}$. If we set $\tilde{O}^i = O^i \cap \mathcal{N}$, then $\tilde{\mathcal{C}}^i = (\tilde{O}^i, \tilde{\kappa}^i)$ is a local chart for $\mathcal{N}$ and $(\tilde{\mathcal{C}}^i)_{i\in\mathcal{I}}$ forms an atlas that yields a smooth $(d-n)$-dimensional manifold structure for $\mathcal{N}$.

If we consider the inclusion map $\phi : \mathcal{N} \to \mathcal{M}$, we see that for any $m \in \mathcal{N}$, the map $T\phi(m) : T_m\mathcal{N} \to T_m\mathcal{M}$ is an injection. Then, any tangent vector $v \in T_m\mathcal{N}$, $m \in \mathcal{N}$, is identified with its image, $T\phi(m)v$. This allows one to consider $T_m\mathcal{N}$ as a linear subspace of $T_m\mathcal{M}$ and, in turn, $T\mathcal{N}$ is itself a submanifold of $T\mathcal{M}$. If $v \in T_m\mathcal{M}$, then $v \in T_m\mathcal{N}$ if and only if, in a local chart $\mathcal{C} = (O, \kappa)$, where $\mathcal{N}$ is given by $x_d = \cdots = x_{d-n+1} = 0$, its local representative $v^{\mathcal{C}}$ is of the form $v^{\mathcal{C}} = (v^1, \dots, v^{d-n}, 0, \dots, 0) = \sum_{1 \leq i \leq d-n} v^i \partial_{x_i}$.

This above identification of $T_m\mathcal{N}$ as a subspace of $T_m\mathcal{M}$ allows cotangent vectors in $T_m^*\mathcal{M}$ to act on tangent vectors in $T_m\mathcal{N}$. This can also be done by pulling back any cotangent vector $\omega \in T_m^*\mathcal{M}$ to $T_m^*\mathcal{N}$ by the inclusion map ϕ. Observe that the pullback ϕ^* is not injective. In fact, we denote by $N_m^*\mathcal{N}$ the cotangent vectors in $T_m^*\mathcal{M}$ that vanish on $T_m\mathcal{N}$. Those vectors are called conormal. With the identification made above, we have $T_m^*\mathcal{M} = T_m^*\mathcal{N} \oplus N_m^*\mathcal{N}$, with $\dim T_m^*\mathcal{N} = d-n$ and $\dim N_m^*\mathcal{N} = n$. We define $N^*\mathcal{N} = \cup_{m \in \mathcal{N}} \{m\} \times N_m^*\mathcal{N}$, called the conormal bundle of $\mathcal{N}$. It is a submanifold of $T^*\mathcal{M}$ of codimension d (and thus of dimension d). In the particular case of a submanifold $\mathcal{N}$ of codimension one, $N_m^*\mathcal{N}$ is of dimension one.

If $\mathcal{M}$ is a manifold with boundary, then in every chart $\mathcal{C} = (O, \kappa)$ such that $O \cap \partial\mathcal{M} \neq \emptyset$, we have $\kappa(m) = (\kappa_1(m), \dots, \kappa_d(m))$ with $\kappa_d(m) \geq 0$, and the boundary $\partial\mathcal{M}$ is locally given by $\kappa_d(m) = 0$. Setting $\tilde{\kappa}(m) = (\kappa_1(m), \dots, \kappa_{d-1}(m))$, we see that the above analysis applies and allows one to give a manifold structure to $\partial\mathcal{M}$. We also say that $\partial\mathcal{M}$ is a submanifold of $\mathcal{M}$ of codimension 1. Then, if Ω is a smooth open subset of $\mathcal{M}$, its boundary $\partial\Omega$ can be seen as a submanifold of codimension one of either $\mathcal{M}$ or $\overline{\Omega}$. Note in particular that $\partial\mathcal{M}$ does not have a boundary.

Let $\mathcal{N}$ be a submanifold of $\mathcal{M}$, and let U be an open subset of $\mathcal{N}$. We define a $\mathscr{C}^\ell$-vector field on $\mathcal{M}$ along the open set U of the submanifold $\mathcal{N}$ to be a $\mathscr{C}^\ell$-map $v : U \to T\mathcal{M}$ that satisfies the property $\pi v(m) = m$. This is not a vector field on $\mathcal{N}$ as $v(m) = (m, v_m)$ and v_m may not be in $T_m\mathcal{N}$, but rather $v_m \in T_m\mathcal{M}$. Similarly, we define that a $\mathscr{C}^\ell$-one-form on $\mathcal{M}$ along the open set U is $\mathscr{C}^\ell$-map $\omega : U \to T^*\mathcal{M}$ that satisfies the property $\tilde{\pi}\omega(m) = m$. In particular, if U is an open set of $\partial\mathcal{M}$, a $\mathscr{C}^\ell$-one-form ω on $\mathcal{M}$ along U such that $\omega_m \in N_m^*\partial\mathcal{M}$ is called a conormal vector field on U.

15.6. Tensors and p-Forms

15.6.1. Covariant Tensors. Let $\mathcal{M}$ be a smooth d-dimensional manifold and $m \in \mathcal{M}$. For $r \in \mathbb{N}$, the vector space of multilinear forms from $(T_m\mathcal{M})^r$ into $\mathbb{C}$ is denoted by $\otimes^r T_m^*\mathcal{M}$. We have $\dim \otimes^r T_m^*\mathcal{M} = d^r$. Such forms are called r-covariant tensors at m. If ω and $\tilde{\omega}$ are r- and s-covariant tensor, respectively, we define their tensor product $\omega \otimes \tilde{\omega}$ as the following p-covariant tensor, with $p = r + s$,

$$\omega \otimes \tilde{\omega}(v, \tilde{v}) = \omega(v)\tilde{\omega}(\tilde{v}), \qquad v \in (T_m\mathcal{M})^r, \tilde{v} \in (T_m\mathcal{M})^s.$$

If $\phi : \mathcal{M} \to \tilde{\mathcal{M}}$ is a smooth map and $\omega \in \otimes^r T^*_{\phi(m)}\tilde{\mathcal{M}}$, for $m \in \mathcal{M}$, we define its pullback $\phi^*\omega \in \otimes^r T^*_m\mathcal{M}$ as

$$\phi^*\omega(v_1, \dots, v_r) = \omega(T\phi(m)(v_1), \dots, T\phi(m)(v_r)), \quad v_1, \dots, v_r \in T_m\mathcal{M}. \tag{15.6.1}$$

Observe then that we have $\phi^*(\omega_1 \otimes \omega_2) = \phi^*\omega_1 \otimes \phi^*\omega_2$ for $\omega_1 \in \otimes^r T^*_{\phi(m)}\tilde{\mathcal{M}}$ and $\omega_2 \in \otimes^s T^*_{\phi(m)}\tilde{\mathcal{M}}$. This allows us to analyze how local representatives change from one chart to the other. If $\omega \in \otimes^r T^*_m\mathcal{M}$, in local charts $\mathcal{C}^j = (O^j, \kappa^j)$, $j = 1, 2$, its representatives are of the form

$$\omega^{\mathcal{C}^j} = \sum_{1 \le i_1, \dots, i_r \le d} \omega^j_{i_1 \dots i_r} dx^{i_1} \otimes \dots \otimes dx^{i_r}.$$

If we set $\kappa = \kappa^{12} = \kappa^2 \circ (\kappa^1)^{-1}$ on $\kappa^1(O)$ with $m \in O = O^1 \cap O^2$, we then have $\omega^{\mathcal{C}^1} = \kappa^*\omega^{\mathcal{C}^2}$, which reads

$$\omega^1_{i_1 \dots i_r} = \sum_{1 \le j_1, \dots, j_r \le d} \partial_{i_1}(\kappa)_{j_1}(x^{(1)}) \cdots \partial_{i_r}(\kappa)_{j_r}(x^{(1)}) \omega^2_{j_1 \dots j_r}, \tag{15.6.2}$$

for $x^{(1)} = \kappa^1(m)$.

We define $\otimes^r T^*\mathcal{M} = \cup_{m \in \mathcal{M}} \{m\} \times \otimes^r T^*_m\mathcal{M}$, which can be given a manifold structure similar to that of T^*M. It is called the r-covariant tensor bundle. If U is an open set of $\mathcal{M}$, a $\mathscr{C}^\ell$-r-covariant tensor field on U is a $\mathscr{C}^\ell$-section of $\otimes^r T^*\mathcal{M}$, that is, a $\mathscr{C}^\ell$-map $\omega : U \to \otimes^r T^*\mathcal{M}$ such that $\omega(m) = (m, \omega_m)$ with $\omega_m \in \otimes^r T^*_m\mathcal{M}$. We denote by $\mathscr{C}^\ell\Lambda^r(U)$ the space of $\mathscr{C}^\ell$-r-covariant tensor fields on U. On a $\mathscr{C}^\infty$-manifold, we set $\mathscr{C}^\infty\Lambda^r(U) = \cap_{\ell \ge 0}\mathscr{C}^\ell\Lambda^r(U)$. For $r = 1$, we recover one-forms as defined in Sect. 15.4, that is, $\mathscr{C}^\ell\Lambda^1(U) = \mathscr{C}^\ell\Lambda(U)$.

If $\phi : \mathcal{M} \to \tilde{\mathcal{M}}$ is a smooth map and $\tilde{\omega}$ is an r-covariant tensor field on $\tilde{\mathcal{M}}$, we define its pullback as $(\phi^*\tilde{\omega})_m = \phi^*(\tilde{\omega}_{\phi(m)})$, which by (15.6.1) gives, for $v_1, \dots, v_r \in T_m\mathcal{M}$,

$$(\phi^*\tilde{\omega})_m(v_1, \dots, v_r) = \tilde{\omega}_{\phi(m)}(T\phi(m)(v_1), \dots, T\phi(m)(v_r)). \tag{15.6.3}$$

15.6.2. p-Forms. Above we defined one-forms as sections of the cotangent bundle.

We now consider $p \ge 1$, and we denote by $\mathscr{A}^p_m\mathcal{M}$ the vector space of p-covariant tensors at m that are alternating, that is, if $\omega \in \mathscr{A}^p_m\mathcal{M}$, then for all $(v_1, \dots, v_p) \in (T_m\mathcal{M})^p$, we have

$$\omega(v_1, \dots, v_p) = 0 \text{ if } v_i = v_j \text{ for some } i, j \in \{1, \dots, p\},\ i \ne j.$$

Consequently, we have

$$\omega(v_{\sigma(1)}, \dots, v_{\sigma(p)}) = \epsilon(\sigma)\omega(v_1, \dots, v_p),$$

with $\sigma \in \mathfrak{S}_p$, that is, a permutation of $\{1, \dots, p\}$ and $\epsilon(\sigma)$ its signature. In particular, we have

$$\sum_{\sigma \in \mathfrak{S}_r} \epsilon(\sigma)\,\omega(v_{\sigma(1)}, \dots, v_{\sigma(p)}) = p!\,\omega(v_1, \dots, v_p). \tag{15.6.4}$$

Observe that we have $\mathscr{A}^1_m\mathcal{M} = T^*_m\mathcal{M}$ and $\mathscr{A}^p_m\mathcal{M} = \{0\}$ if $p > d$. By convention, we also set $\mathscr{A}^0_m\mathcal{M} = \mathbb{C}$, that is, forms that take no argument. For $0 \le p \le d$, the dimension of $\mathscr{A}^p_m\mathcal{M}$ is given by $\binom{d}{p} = \frac{d!}{(d-p)!p!}$.

If $\omega, \tilde{\omega} \in \mathscr{A}^1_m\mathcal{M}$, we see that $\omega \otimes \tilde{\omega} - \tilde{\omega} \otimes \omega \in \mathscr{A}^2_m\mathcal{M}$. More generally, if $\omega \in \mathscr{A}^p_m\mathcal{M}$ and $\tilde{\omega} \in \mathscr{A}^{\tilde{p}}_m\mathcal{M}$, with $p, \tilde{p} \in \mathbb{N}$, we define $\omega \wedge \tilde{\omega} \in \mathscr{A}^r_m\mathcal{M}$, with $r = p + \tilde{p}$, by

(15.6.5)

$$\omega \wedge \tilde{\omega}(v_1, \dots, v_r) = \frac{1}{p!\tilde{p}!} \sum_{\sigma \in \mathfrak{S}_r} \epsilon(\sigma)\, \omega(v_{\sigma(1)}, \dots, v_{\sigma(p)})\, \tilde{\omega}(v_{\sigma(p+1)}, \dots, v_{\sigma(r)}),$$

which we call the wedge or exterior product of ω and $\tilde{\omega}$.

Proposition 15.22. *Let $\omega \in \mathscr{A}^p_m\mathcal{M}$ and $\tilde{\omega} \in \mathscr{A}^{\tilde{p}}_m\mathcal{M}$, with $p, \tilde{p} \in \mathbb{N}$. The wedge product has the following properties:*

(1) *The wedge product is bilinear and associative.*
(2) *If $p = 0$, that is, $\omega = \lambda \in \mathbb{C}$, one has $\omega \wedge \tilde{\omega} = \lambda\tilde{\omega}$.*

For $\omega_j \in \mathscr{A}^{p_j}_m\mathcal{M}$, $j = 1, \dots, N$, one finds by induction, with $r = p_1 + \cdots + p_N$,

$$\begin{aligned} &\omega_1 \wedge \cdots \wedge \omega_N(v_1, \dots, v_r) \\ &\quad = \frac{1}{p_1! \cdots p_N!} \sum_{\sigma \in \mathfrak{S}_r} \epsilon(\sigma) \prod_{j=1}^{N} \omega_j(v_{\sigma(P_{j-1}+1)}, \dots, v_{\sigma(P_{j-1}+p_j)}), \end{aligned}$$

with $P_0 = 0$ and $P_{j+1} = P_j + p_{j+1}$. For $p_1 = \cdots = p_N = 1$, this gives in particular

$$\omega_1 \wedge \cdots \wedge \omega_N(v_1, \dots, v_N) = \sum_{\sigma \in \mathfrak{S}_N} \epsilon(\sigma)\, \omega_1(v_{\sigma(1)}) \cdots \omega_N(v_{\sigma(N)}).$$

This allows us to write, for $s \in \mathfrak{S}_N$,

(15.6.6)

$$\begin{aligned} \omega_{s(1)} \wedge \cdots \wedge \omega_{s(N)}(v_1, \dots, v_N) &= \sum_{\sigma \in \mathfrak{S}_N} \epsilon(\sigma)\, \omega_{s(1)}(v_{\sigma(1)}) \cdots \omega_{s(N)}(v_{\sigma(N)}) \\ &= \sum_{\sigma \in \mathfrak{S}_N} \epsilon(\sigma)\, \omega_1(v_{\sigma \circ s^{-1}(1)}) \cdots \omega_N(v_{\sigma \circ s^{-1}(N)}) \\ &= \epsilon(s) \omega_1 \wedge \cdots \wedge \omega_N(v_1, \dots, v_N). \end{aligned}$$

If $\phi : \mathcal{M} \to \tilde{\mathcal{M}}$ is a smooth map and $\omega \in \mathscr{A}^p_{\phi(m)}\tilde{\mathcal{M}}$, for $m \in \mathcal{M}$, we define its pullback $\phi^*\omega \in \mathscr{A}^p_m\mathcal{M}$ as in (15.6.1). Observe then that we have $\phi^*(\omega_1 \wedge \omega_2) = \phi^*\omega_1 \wedge \phi^*\omega_2$ for $\omega_1 \in \mathscr{A}^p_{\phi(m)}\tilde{\mathcal{M}}$ and $\omega_2 \in \mathscr{A}^q_{\phi(m)}\tilde{\mathcal{M}}$. This allows us to analyze how local representatives change from one chart to the other.

In a local chart, a basis of $\mathscr{A}^p_m\mathcal{M}$ is given by

$$dx^{i_1} \wedge \cdots \wedge dx^{i_p}, \quad \text{with } 1 \le i_1 < \cdots < i_p \le d.$$

With the local representation this basis provides, one deduces the anticommutativity property of the wedge product, which could also be obtained from the computations above.

PROPOSITION 15.23. *The wedge product is anticommutative in the sense that*

$$\omega \wedge \tilde{\omega} = (-1)^{p\tilde{p}}\, \tilde{\omega} \wedge \omega,$$

for $\omega \in \mathscr{A}^p_m\mathcal{M}$ *and* $\tilde{\omega} \in \mathscr{A}^{\tilde{p}}_m\mathcal{M}$.

Note that for $p = d$, one has $\mathscr{A}^d_m\mathcal{M} = \operatorname{span}(dx^1 \wedge \cdots \wedge dx^d)$ and moreover $dx^1 \wedge \cdots \wedge dx^d(v_1, \ldots, v_d) = \det(v_1, \ldots, v_d)$.

If $\omega \in \mathscr{A}^p\mathcal{M}$, in local charts $\mathcal{C}^j = (O^j, \kappa^j)$, $j = 1, 2$, its representatives are of the form

$$\omega^{\mathcal{C}^j} = \sum_{1 \le i_1 < \cdots < i_p \le d} \omega^j_{i_1 \ldots i_p} dx^{i_1} \wedge \cdots \wedge dx^{i_p}.$$

We set $\kappa = \kappa^{12} = \kappa^2 \circ (\kappa^1)^{-1}$ on $\kappa^1(O)$ with $m \in O = O^1 \cap O^2$. We start by computing $\alpha = \kappa^*(dx^{i_1} \wedge \cdots \wedge dx^{i_p})$ for $1 \le i_1 < \cdots < i_p \le d$. We have, with $x_0 = \kappa(m)$,

$$\begin{aligned}
\alpha &= \kappa^* dx^{i_1} \wedge \cdots \wedge \kappa^* dx^{i_p} \\
&= \sum_{1 \le j_1, \ldots, j_p \le d} \partial_{j_1}(\kappa)_{i_1}(x_0) \cdots \partial_{j_p}(\kappa)_{i_p}(x_0) dx^{j_1} \wedge \cdots \wedge dx^{j_p} \\
&= \sum_{1 \le j_1, \ldots, j_p \le d} \epsilon(\sigma_{j_1, \ldots, j_p})\, \partial_{\sigma(k_1)}(\kappa)_{i_1}(x_0) \cdots \partial_{\sigma(k_p)}(\kappa)_{i_p}(x_0) dx^{k_1} \wedge \cdots \wedge dx^{k_p}
\end{aligned}$$

with $k_1 < k_2 < \cdots < k_p$ and $\{k_1, \ldots, k_p\} = \{j_1, \ldots, j_p\}$ and $\sigma_{j_1, \ldots, j_p}$ is the permutation on this set such that $\sigma(k_\ell) = j_\ell$. We then have

$$\begin{aligned}
\alpha = &\sum_{1 \le k_1 < \cdots < k_p \le d} \Big(\sum_{\sigma \in \mathfrak{S}\{k_1, \ldots, k_p\}} \epsilon(\sigma)\, \partial_{\sigma(k_1)}(\kappa)_{i_1}(x_0) \\
&\qquad \cdots \partial_{\sigma(k_p)}(\kappa)_{i_p}(x_0) \Big) dx^{k_1} \wedge \cdots \wedge dx^{k_p} \\
= &\sum_{1 \le k_1 < \cdots < k_p \le d} D^{i_1, \ldots, i_p}_{k_1, \ldots, k_p} dx^{k_1} \wedge \cdots \wedge dx^{k_p},
\end{aligned}$$

with $D^{i_1, \ldots, i_p}_{k_1, \ldots, k_p} = \det(\partial_{k_q} \kappa_{i_\ell}(x_0))_{1 \le \ell, q \le p}$. As $\omega^{\mathcal{C}^1} = \kappa^* \omega^{\mathcal{C}^2}$, we then find

$$\begin{aligned}
\omega^{\mathcal{C}^1} &= \sum_{1 \le i_1 < \cdots < i_p \le d} \omega^2_{i_1 \ldots i_p} \kappa^* dx^{i_1} \wedge \cdots \wedge dx^{i_p} \\
&= \sum_{1 \le k_1 < \cdots < k_p \le d} \Big(\sum_{1 \le i_1 < \cdots < i_p \le d} D^{i_1, \ldots, i_p}_{k_1, \ldots, k_p} \omega^2_{i_1 \ldots i_p} \Big) dx^{k_1} \wedge \cdots \wedge dx^{k_p},
\end{aligned}$$

yielding

$$\omega^1_{k_1 \ldots k_p} = \sum_{1 \le i_1 < \cdots < i_p \le d} D^{i_1, \ldots, i_p}_{k_1, \ldots, k_p} \omega^2_{i_1 \ldots i_p}.$$

An important case is $p = d$. For d-forms, the local representatives take the simple form $\omega^{\mathcal{C}^j} = \omega^j dx^1 \wedge \cdots \wedge dx^d$, and we find

$$\omega^1 = \det(\kappa'(x_0))\omega^2, \qquad \kappa = \kappa^2 \circ (\kappa^1)^{-1}, \tag{15.6.7}$$

where $\kappa'(x_0)$ is the Jacobian matrix of κ at x_0.

We then set $\mathscr{A}^p\mathcal{M} = \cup_{m\in\mathcal{M}}\{m\} \times \mathscr{A}^p_m\mathcal{M}$, and we see that $\mathscr{A}^p\mathcal{M}$ is a submanifold of $\otimes^p T^*\mathcal{M}$ of dimension $d + \binom{d}{p}$. Let U be an open set of $\mathcal{M}$. A $\mathscr{C}^\ell$ p-form on U is then a section of $\mathscr{A}^p\mathcal{M}$, that is, a $\mathscr{C}^\ell$-map $\omega : U \to \mathscr{A}^p\mathcal{M}$ such that $\omega(m) = (m, \omega_m)$ with $\omega_m \in \mathscr{A}^p_m\mathcal{M}$. The vector space of such p-forms on U is denoted by $\mathscr{C}^\ell\Omega^p(U)$. On a $\mathscr{C}^\infty$-manifold, we set $\mathscr{C}^\infty\Omega^p(U) = \cap_{\ell\geq 0}\mathscr{C}^\ell\Omega^p(U)$. Note that 0-forms on U are functions on U. With $p = 1$, we also recover the one-forms defined in Sect. 15.4, that is, $\mathscr{C}^\ell\Omega^1(U) = \mathscr{C}^\ell\Lambda(U)$.

If ω and $\tilde{\omega}$ are p and $\tilde{p}$-forms, the map $m \mapsto (m, \omega_m \wedge \tilde{\omega}_m)$ defines an r-form with $r = p+\tilde{p}$ denoted by $\omega\wedge\tilde{\omega}$. We set $\mathscr{C}^\infty\Omega(M) = \oplus_{0\leq p\leq d}\mathscr{C}^\infty\Omega^p(\mathcal{M})$. This is an algebra.

The exterior derivative d is the *unique* linear map $\mathrm{d} : \mathscr{C}^\infty\Omega(M) \to \mathscr{C}^\infty\Omega(M)$ that satisfies the following properties:

(1) If $\omega \in \mathscr{C}^\infty\Omega^p(M)$, then $\mathrm{d}\omega \in \mathscr{C}^\infty\Omega^{p+1}(M)$.
(2) If $f \in \mathscr{C}^\infty\Omega^0(M)$, that is, f is a smooth function, the one-form $\mathrm{d}f$ is given by the differential of f.
(3) $\mathrm{d} \circ \mathrm{d} = 0$.
(4) If $\omega \in \mathscr{C}^\infty\Omega^p(M)$ and $\tilde{\omega} \in \mathscr{C}^\infty\Omega(M)$, we have $\mathrm{d}(\omega \wedge \tilde{\omega}) = \mathrm{d}\omega \wedge \tilde{\omega} + (-1)^p\omega \wedge \mathrm{d}\tilde{\omega}$.

In a local chart, the exterior derivative of $f dx^{i_1} \wedge \cdots \wedge dx^{i_p}$ is given by

$$df \wedge dx^{i_1} \wedge \cdots \wedge dx^{i_p} = \sum_{1\leq j\leq d} \partial_{x_j} f\, dx^j \wedge dx^{i_1} \wedge \cdots \wedge dx^{i_p}. \tag{15.6.8}$$

If $\phi : \mathcal{M} \to \tilde{\mathcal{M}}$ is a smooth map and $\tilde{\omega} \in \mathscr{C}^\infty\Omega(\tilde{\mathcal{M}})$, we have

$$\phi^*\mathrm{d}\tilde{\omega} = \mathrm{d}(\phi^*\tilde{\omega}),$$

with the pullback as defined in (15.6.3).

15.6.3. General Tensors. For $r, s \in \mathbb{N}$, the vector space of multilinear forms from $(T_m\mathcal{M})^r \times (T^*_m\mathcal{M})^s$ into $\mathbb{C}$ is denoted by $(\otimes^r T^*_m\mathcal{M}) \otimes (\otimes^s T_m\mathcal{M})$. It is of dimension d^{r+s}. Such forms are called r-covariant,s-contravariant tensors at m. Covariant tensors correspond to the case $s = 0$. Tangent vectors at m correspond to the case $r = 0$, $s = 1$; they are 1-contravariant tensors.

If θ and $\tilde{\theta}$ are, respectively, r-covariant,s-contravariant and $\tilde{r}$-covariant,$\tilde{s}$-contravariant tensors at m, then we define the $(r + \tilde{r})$-covariant,$(s + \tilde{s})$-contravariant tensor $\theta \otimes \tilde{\theta}$ by

$$\theta \otimes \tilde{\theta}(v, \tilde{v}, \omega, \tilde{\omega}) = \theta(v, \omega)\tilde{\theta}(\tilde{v}, \tilde{\omega}),$$

for $v \in (T_m\mathcal{M})^r$, $\tilde{v} \in (T_m\mathcal{M})^{\tilde{r}}$, $\omega \in (T^*_m\mathcal{M})^s$, and $\tilde{\omega} \in (T^*_m\mathcal{M})^{\tilde{s}}$.

If $\phi : \mathcal{M} \to \tilde{\mathcal{M}}$ is a smooth diffeomorphism and $\theta \in (\otimes^r T^*_{\phi(m)}\tilde{\mathcal{M}}) \otimes (\otimes^s T_{\phi(m)}\tilde{\mathcal{M}})$, for $m \in \mathcal{M}$, we define its pullback $\phi^*\theta \in (\otimes^r T^*_m\mathcal{M})\otimes(\otimes^s T_m\mathcal{M})$ as

$$\phi^*\theta(v_1, \ldots, v_r, \omega_1, \ldots, \omega_s) = \theta(\tilde{v}_1, \ldots, \tilde{v}_r, \tilde{\omega}_1, \ldots, \tilde{\omega}_s), \tag{15.6.9}$$

for $v_1, \dots, v_r \in T_m\mathcal{M}$ and $\omega_1, \dots, \omega_s \in T_m^*\mathcal{M}$, with $\tilde{v}_i = T\phi(m)(v_i)$, $1 \le i \le r$, and $\tilde{\omega}_j = {}^tT\phi^{-1}(\phi(m))(\omega_j)$, $1 \le j \le s$.

Considering two local charts $\mathcal{C}^\ell = (O^\ell, \kappa^\ell)$, $\ell = 1, 2$, such that $O^1 \cap O^2 \neq \emptyset$, a tensor $\theta \in (\otimes^r T_m^*\mathcal{M}) \otimes (\otimes^s T_m\mathcal{M})$ has the local representatives

$$\theta^{\mathcal{C}^\ell} = \sum_{\substack{1 \le i_1, \dots, i_r \le d \\ 1 \le j_1, \dots, j_s \le d}} {}^\ell\theta_{i_1, \dots, i_r}^{j_1, \dots, j_s} dx^{i_1} \otimes \cdots \otimes dx^{i_r} \otimes \partial_{j_1} \otimes \cdots \otimes \partial_{j_s}.$$

If we set $\kappa = \kappa^{12} = \kappa^2 \circ (\kappa^1)^{-1}$ on $\kappa^1(O)$ with $m \in O = O^1 \cap O^2$, we then have $\theta^{\mathcal{C}^1} = \kappa^* \theta^{\mathcal{C}^2}$, which reads

$$\begin{aligned} (15.6.10) \quad {}^1\theta_{i_1, \dots, i_r}^{j_1, \dots, j_s} = & \sum_{1 \le j_1, \dots, j_r \le d} \partial_{i_1}(\kappa)_{\tilde{i}_1}(x_1) \cdots \partial_{i_r}(\kappa)_{\tilde{i}_r}(x_1) \\ & \times \partial_{\tilde{j}_1}(\kappa^{-1})_{j_1}(x_2) \cdots \partial_{\tilde{j}_s}(\kappa^{-1})_{j_s}(x_2) \, {}^2\theta_{\tilde{i}_1, \dots, \tilde{i}_r}^{\tilde{j}_1, \dots, \tilde{j}_s}, \end{aligned}$$

for $x_1 = \kappa^1(m)$ and $x_2 = \kappa^2(m)$.

We define $(\otimes^r T^*\mathcal{M}) \otimes (\otimes^s T\mathcal{M}) = \cup_{m \in \mathcal{M}} \{m\} \times (\otimes^r T_m^*\mathcal{M}) \otimes (\otimes^s T_m\mathcal{M})$, which can be given a manifold structure. It is called the r-covariant,s-contravariant tensor bundle.

If U is an open set of $\mathcal{M}$, a $\mathscr{C}^\ell$-r-covariant,s-contravariant tensor field is a $\mathscr{C}^\ell$-section of $(\otimes^r T^*\mathcal{M}) \otimes (\otimes^s T\mathcal{M})$, that is, a $\mathscr{C}^\ell$-map $\theta : U \to (\otimes^r T^*\mathcal{M}) \otimes (\otimes^s T\mathcal{M})$ such that $\theta(m) = (m, \theta_m)$ with $\theta_m \in (\otimes^r T_m^*\mathcal{M}) \otimes (\otimes^s T_m\mathcal{M})$. We denote by $\mathscr{C}^\ell \mathscr{T}_r^s(U)$ the space of such $\mathscr{C}^\ell$-r-covariant,s-contravariant tensor fields on U. On a $\mathscr{C}^\infty$-manifold, we set $\mathscr{C}^\infty \mathscr{T}_r^s(U) = \cap_{\ell \ge 0} \mathscr{C}^\ell \mathscr{T}_r^s(U)$. Note that we have $\mathscr{C}^\ell \mathscr{T}_r^0(U) = \mathscr{C}^\ell \Lambda^r(U)$ and $\mathscr{C}^\ell \mathscr{T}_0^1(U) = \mathscr{C}^\ell V(U)$.

If $\phi : \mathcal{M} \to \tilde{\mathcal{M}}$ is a smooth diffeomorphism and θ is a r-covariant,s-contravariant tensor field on $\tilde{\mathcal{M}}$, we define its pullback as $(\phi^*\theta)_m = \phi^*(\theta_{\phi(m)})$ according to (15.6.9).

15.7. Symplectic Structure of the Cotangent Bundle

This section simply aims to generalize to manifolds some of the content of Section 9.2 of Volume 1.

15.7.1. The Symplectic Two-Form. On $T^*\mathcal{M}$, we recall that $\tilde{\pi}$ is the natural projection onto M, that is, $\tilde{\pi}(\gamma) = m$ for $\gamma = (m, \omega) \in T^*\mathcal{M}$. At such a point γ, we can consider the transpose of the tangent map of $\tilde{\pi}$, as introduced in Sect. 15.4.2, that is, ${}^tT\tilde{\pi}(\gamma) : T_m^*\mathcal{M} \to T_\gamma^*(T^*\mathcal{M})$. Setting

$$(15.7.1) \qquad \theta_\gamma = {}^tT\tilde{\pi}(\gamma)(\omega) \in T_\gamma^*(T^*\mathcal{M}),$$

that is, $\langle \theta_\gamma, t \rangle = \langle \omega, T\tilde{\pi}(t) \rangle$ for $t \in T_\gamma(T^*\mathcal{M})$, we define the following one-form $\gamma \mapsto (\gamma, \theta_\gamma)$ on $T^*\mathcal{M}$. One calls θ the canonical one-form on $T^*\mathcal{M}$.

In a local chart $\mathcal{C} = (O, \kappa)$ of $T^*\mathcal{M}$ inherited from one of $\mathcal{M}$ (see Sect. 15.4.3), we let $x = (x_1, \dots, x_d)$ denote the coordinates associated with $m \in \mathcal{M}$ and $\xi = (\xi_1, \dots, \xi_d)$ be the coordinates in the cotangent fiber. Then, for $\gamma = (m, \omega)$, having $\kappa(\gamma) = (x, \xi)$ in the chart means that the representative of ω in $\mathcal{C}$ reads $\omega^{\mathcal{C}} = \sum_{1 \le i \le d} \xi_i dx^i \in T_m^* \mathcal{M}$. If $t \in T_\gamma(T^*\mathcal{M})$ has representative $t^{\mathcal{C}} = \sum_{1\le i\le d} t_x^i \partial_{x_i} + \sum_{1\le i\le d} t_\xi^i \partial_{\xi_i}$, then $T\tilde{\pi}(\gamma)(t) = \sum_{1\le i\le d} t_x^i \partial_{x_i}$. As we have $\langle \omega, T\tilde{\pi}(\gamma)(t)\rangle = \sum_{1\le i\le d} t_x^i \xi_i$, we find that $\theta_\gamma \in T^*(T^*\mathcal{M})$ has representative

$$\theta_\gamma^{\mathcal{C}} = \sum_{1\le i\le d} \xi_i dx^i. \tag{15.7.2}$$

In the chart $\mathcal{C}$, the representative of the one-form θ is thus given by $\theta^{\mathcal{C}} = \sum_{1\le i\le d} \xi_i dx^i$.

The symplectic form on $T^*\mathcal{M}$ is a two-form given by $\sigma = \mathrm{d}\theta$, with the exterior derivative d as introduced in Sect. 15.6.2. In a local chart $\mathcal{C}$ as above, we have

$$\sigma_\gamma^{\mathcal{C}} = \sum_{1\le i\le d} d\xi_i \wedge dx^i, \quad \gamma \in T^*\mathcal{M}. \tag{15.7.3}$$

For $t, s \in T_\gamma(T^*\mathcal{M})$ with (t_x, t_ξ) and (s_x, s_ξ) as local representatives, we have

$$\sigma_\gamma(t, s) = t_\xi \cdot s_x - t_x \cdot s_\xi.$$

One sees that the symplectic two-form σ is anti-symmetric and nondegenerate.

REMARK 15.24. In fact, more generally, a manifold equipped with a closed, anti-symmetric, and nondegenerate two-form is called a symplectic manifold. It is necessarily of even dimension $2d$, and its structure is locally that of $T^*\mathbb{R}^d$.

15.7.2. Hamiltonian Vector Field and Poisson Bracket. If f is a function on $T^*\mathcal{M}$, then for any $\gamma = (m, \omega) \in T^*\mathcal{M}$, as σ is nondegenerate, there exists a unique vector $(H_f)_\gamma \in T_\gamma(T^*\mathcal{M})$ such that

$$\sigma_\gamma(u, (H_f)_\gamma) = df(\gamma)(u), \qquad u \in T_\gamma(T^*\mathcal{M}).$$

In a local chart $\mathcal{C} = (O, \kappa)$ as above, we have

$$(H_f)_\gamma^{\mathcal{C}} = \sum_{1\le i\le d} \big((\partial_{\xi_i} f^{\mathcal{C}})\partial_{x_i} - (\partial_{x_i} f^{\mathcal{C}})\partial_{\xi_i}\big)(x, \xi), \qquad (x, \xi) = \kappa(\gamma).$$

One then defines the vector field on $T^*\mathcal{M}$ given by $\gamma \mapsto (\gamma, (H_f)_\gamma)$ that is called the Hamiltonian vector field of f.

For two functions f and g on $T^*\mathcal{M}$, we define the following function, called the Poisson bracket of f and g,

$$\{f, g\}(\gamma) = \sigma_\gamma((H_f)_\gamma, (H_g)_\gamma), \qquad \gamma \in T^*\mathcal{M}.$$

In a local chart $\mathcal{C}$ as above, we have

$$\{f, g\}^{\mathcal{C}}(x, \xi) = \sum_{1\le i\le d} (\partial_{\xi_i} f^{\mathcal{C}} \partial_{x_i} g^{\mathcal{C}} - \partial_{x_i} f^{\mathcal{C}} \partial_{\xi_i} g^{\mathcal{C}})(x, \xi).$$

Furthermore, we have

$$\{f, g\} = H_f\, g = -H_g\, f,$$

and

$$H_{\{f,g\}} = [H_f, H_g],$$

where $[.,.]$ denotes the Lie bracket of two vector fields, that is, $[u, v](h) = u(v(h)) - v(u(h))$ for a function h; observe that $[u, v]$ is itself a vector field.

CHAPTER 16

Integration and Differential Operators on Manifolds

Contents

This chapter is devoted to the exposition of integration on manifolds. Integration applies either to d-forms or to density functions. We recover results such as the divergence formula, a particular case of the Stokes formula, that are useful in the understanding of weak formulation of elliptic problems for instance. Radon measure densities and distribution densities are also reviewed. We also introduce differential operators on manifolds and describe their action on distributions.

J. Le Rousseau et al., *Elliptic Carleman Estimates and Applications to Stabilization and Controllability, Volume II*, PNLDE Subseries in Control 98, https://doi.org/10.1007/978-3-030-88670-7_16

16.1. Oriented Manifolds, Integration of d-Forms

As there is no canonical measure on a manifold in general, integration of functions cannot be done intrinsically. The case of orientable d-dimensional manifolds yields a natural framework for the integration of d-forms, with the celebrated Stokes formula as a consequence. Functions can then be integrated if identified with d-forms by a choice of volume form. The Riemannian case yields a canonical choice for this volume form once an orientation is chosen, see Chap. 17.

16.1.1. Orientable and Oriented Manifolds. A smooth d-dimensional manifold $\mathcal{M}$ with or without boundary is said to be orientable if there exists an atlas on $\mathcal{M}$ such that for any two charts $\mathcal{C}^j = (O^j, \kappa^j)$, $j = 1, 2$, with $O = O^1 \cap O^2 \neq \emptyset$, the diffeomorphism $\kappa^{12} = \kappa^2 \circ (\kappa^1)^{-1}$ defined from $\kappa^1(O) \subset \mathbb{R}^d$ onto $\kappa^2(O) \subset \mathbb{R}^d$ is such that $\det(\kappa^{12})' > 0$. An oriented manifold is a manifold equipped with such an atlas.

Let now $\mathcal{M}$ be an oriented manifold together with the atlas $\mathcal{A} = (O^i, \kappa^i)_{i\in\mathcal{I}}$. We moreover assume that $\mathcal{M}$ is σ-compact. We then have a partition of unity $(\chi^i)_{i\in\mathcal{I}}$ subordinated to the open covering $(O^i)_{i\in\mathcal{I}}$ by Theorem 15.14.

In $\kappa^i(O^i)$, we consider the d-form $dw^i = \tilde{\chi}^i dx^1 \wedge \cdots \wedge dx^d$, with $\tilde{\chi}^i = ((\kappa^i)^{-1})^* \chi^i \geq 0$. If we pull it back to $\mathcal{M}$ and if we sum over $i \in \mathcal{I}$, we obtain

$$dv = \sum_{i\in\mathcal{I}} (\kappa^i)^* dw^i,$$

yielding a smooth d-form on $\mathcal{M}$ (the sum is locally finite by the properties of the partition of unity). Let $\mathcal{C} = (O, \kappa)$ be a local chart of the atlas, and set $\psi^i = \kappa^i \circ \kappa^{-1}$. The local representative of dv in $\mathcal{C}$ is

$$dv^{\mathcal{C}} = \sum_{i\in\mathcal{I}} (\kappa^{-1})^* (\kappa^i)^* dw^i = \sum_{i\in\mathcal{I}} (\psi^i)^* dw^i.$$

For $i \in \mathcal{I}$, we have, by (15.6.7),

$$(\psi^i)^* dw^i = \big(\det(\psi^i)'\big)\big((\kappa^{-1})^* \chi^i\big) dx^1 \wedge \cdots \wedge dx^d.$$

As the manifold is oriented with the atlas, the local representative of dv in any chart is of the form $\alpha(x) dx^1 \wedge \cdots \wedge dx^d$ with $\alpha > 0$, using that $\sum_i \chi^i = 1$. The d-form dv is called a volume form. In fact, the existence of such a d-form that has a positive representative in any chart of an atlas is equivalent to having the manifold oriented by this choice of atlas. For any local chart (O^i, κ^i) of the considered atlas, we denote by $dv^{\mathcal{C}^i}$ the local representative of dv:

$$dv^{\mathcal{C}^i} = ((\kappa^i)^{-1})^* dv = \alpha^i dx^1 \wedge \cdots \wedge dx^d,$$

with α^i a function on $\kappa^i(O^i)$. Once a volume form dv is given, a volume form $d\tilde{v}$ associated with the opposite orientation is given by

$$d\tilde{v}_m(u_1, \ldots, u_d) = -f(m) dv_m(u_1, \ldots, u_d) = f(m) dv_m(-u_1, \ldots, u_d),$$

with f a positive function on $\mathcal{M}$, for any $(u_1, \ldots, u_d) \in (T_m\mathcal{M})^d$ and $m \in \mathcal{M}$.

In fact, once a volume form dv is given, we say that an ordered set of tangent vectors $\{u_1, \ldots, u_d\}$ of $T_m\mathcal{M}$ is a direct frame at m for the orientation given by dv if $dv_m(u_1, \ldots, u_d) > 0$. Similarly, if $u_1, \ldots, u_d$ are vector fields, we say that the ordered set $\{u_1, \ldots, u_d\}$ is a direct moving frame if $dv_m((u_1)_m, \ldots, (u_d)_m) > 0$ for any $m \in \mathcal{M}$.

Observe that if there exists a set of d vector fields $u_1, \ldots, u_d$ such that at each $m \in \mathcal{M}$,

$$\operatorname{rank}\big((u_1)_m, \ldots, (u_d)_m\big) = d,$$

then this frame allows one to provide an orientation on $\mathcal{M}$. Note, however, that this condition is not necessary as many orientable manifolds do not have such global frames. In fact, many manifolds do not exhibit nonvanishing vector fields. For example, for closed surfaces, that is, compact manifolds without boundary of dimension $d = 2$, the existence of a nonvanishing vector field implies that the Euler characteristic of the surface is zero. This rules out many surfaces, including the sphere, tori with more than one hole, etc. They are yet orientable.

16.1.2. Integration of d-Forms on an Oriented Manifold. Let $\mathcal{M}$ be oriented and dv be a volume form associated with this orientation, that is, dv has a positive representative in each local chart $\mathcal{C}^i$, that is, $dv^{\mathcal{C}^i} = \alpha^i dx^1 \wedge \cdots \wedge dx^d$, with $\alpha^i > 0$. If $m \in \mathcal{M}$, we have $\mathscr{A}^d_m\mathcal{M} = \operatorname{span}(dv_m)$. Consequently, if ω is a smooth d-form on $\mathcal{M}$, we have

$$\omega = f dv, \tag{16.1.1}$$

with f a smooth function on $\mathcal{M}$. Once a volume form is chosen, this allows in particular one to identify d-forms and functions. We may also define $|\omega| = |f| dv$.

We now define the integral of a nonnegative d-form, that is, $\omega = f dv$ with $f : \mathcal{M} \to \mathbb{R}^+$. The local representative of ω and f in $\mathcal{C}^i$ are $\omega^{\mathcal{C}^i} = \omega^i dx^1 \wedge \cdots \wedge dx^d$ and $((\kappa^i)^{-1})^* f$, respectively. We have $\omega^i = f^{\mathcal{C}^i} \alpha^i \geq 0$. If we set $\tilde{\chi}^i = ((\kappa^i)^{-1})^* \chi^i$, we define

$$\int_{\mathcal{M}} \omega = \sum_{i \in \mathcal{I}} \int_{\mathbb{R}^d} \tilde{\chi}^i(x) \omega^i(x) dx \in \mathbb{R}_+ \cup \{+\infty\}. \tag{16.1.2}$$

It also reads

$$\int_{\mathcal{M}} \omega = \int_{\mathcal{M}} f dv = \sum_{i \in \mathcal{I}} \int_{\mathbb{R}^d} \tilde{\chi}^i(x) f^{\mathcal{C}^i}(x) \alpha^i(x) dx. \tag{16.1.3}$$

Using formula (15.6.7) expressing the change of local coordinates for d-forms, one can prove that this definition of the integral of ω is only orientation dependent. Once an orientation is fixed, it is independent of the choice of the oriented atlas and the used partition of unity. Independence of $\int_{\mathcal{M}} \omega$ with respect to the choice of volume form dv is clear in (16.1.2). A nonnegative form ω is called integrable if $\int_{\mathcal{M}} \omega < \infty$.

Next, we observe that if ω is a d-form, such that $|\omega|$ is integrable, then the series in Formulae (16.1.2) and (16.1.3) converge, yielding a proper definition of the integral of ω over $\mathcal{M}$. Note also that we have, from the definition of the orientation,

$$\int_{\mathcal{M}} dv > 0.$$

So far, we only considered smooth d-forms. Yet, one can easily extend the notion of measurable d-forms (for instance, using (16.1.1) and a volume form dv as above and yet noticing that this measurability is independent of the choice made). In particular, zero almost everywhere d-forms can be defined. We then say that a measurable d-form ω is integrable if $\int_{\mathcal{M}} |\omega| < \infty$.

Observe that if ω is integrable and supported in a chart O^i, then using (15.6.7), we have

$$\int_{\mathcal{M}} \omega = \int_{\kappa^i(O^i)} \omega^i(x)dx = \int_{\kappa^i(O^i)} \omega^{\mathcal{C}^i},$$

where $\omega^{\mathcal{C}^i}$ is the local representative of ω in the chart $\mathcal{C}^i = (O^i, \kappa^i)$, that is, $\omega^{\mathcal{C}^i} = \omega^i dx^1 \wedge \cdots dx^d$. In particular, for any integrable d-form ω, this yields

$$\int_{\mathcal{M}} \chi^i \omega = \int_{\kappa^i(O^i)} \tilde{\chi}^i \omega^i(x)dx.$$

For the definition of the integral of an integrable d-form ω, we thus have

$$\int_{\mathcal{M}} \omega = \int_{\mathcal{M}} \sum_{i \in \mathcal{I}} \chi^i \omega = \sum_{i \in \mathcal{I}} \int_{\mathcal{M}} \chi^i \omega. \tag{16.1.4}$$

16.1.3. Orientation of the Boundary of an Oriented Manifold. Let $\mathcal{M}$ be a smooth oriented d-dimensional manifold with boundary and dv a volume form on $\mathcal{M}$. We can provide the boundary $\partial\mathcal{M}$ with an orientation inherited from that of $\mathcal{M}$ as follows. We recall that in a local chart at the boundary, of the form $\mathcal{C} = (O, \kappa)$ with O an open subset of $\mathcal{M}$ and $\kappa : O \to \mathbb{R}^d_+$, we have $m \in O$ equivalent to $\kappa_d(m) \geq 0$.

If $m \in \partial\mathcal{M}$ and $v \in T_m\mathcal{M}$, we say that v is outward-pointing if there exists a path $\gamma : [t^0, t^1] \to \mathcal{M}$ such that $\gamma(t^1) = m$, $\gamma'(t^1) = v$ and $v \notin T_m\partial\mathcal{M}$. Note that this implies that $\gamma(t) \notin \partial\mathcal{M}$ for $t \in [t^1 - \varepsilon, t^1)$, for $\varepsilon > 0$ chosen sufficiently small. In fact, this is equivalent to having $v^d = v(\kappa_d) < 0$, independently of the chosen chart.

We choose a vector field on $\mathcal{M}$ along $\partial\mathcal{M}$, that is, a map $\nu : \partial\mathcal{M} \to T\mathcal{M}$ with $\nu(m) = (m, \nu_m)$ such that $\nu_m \in T_m\mathcal{M}$ is an outward-pointing vector. Such a vector field is called an outward-pointing vector field at the boundary $\partial\mathcal{M}$. This can be first done locally in charts and then globally by using a partition of unity subordinated to local charts that cover the boundary $\partial\mathcal{M}$. This outward-pointing vector field at the boundary ν allows one to define a $(d-1)$-form dv_∂ as follows:

$$(dv_\partial)_m(u_1, \ldots, u_{d-1}) = (dv)_m(\nu_m, u_1, \ldots, u_{d-1}), \; u_1, \ldots, u_{d-1} \in T_m\partial\mathcal{M}.$$

We see that having $(dv_\partial)_m = 0$ implies that $\text{rank}(\nu_m, u_1, \ldots, u_{d-1}) < d$ for any $u_1, \ldots, u_{d-1} \in T_m\partial\mathcal{M}$, which cannot occur as $\nu_m \in T_m\mathcal{M}\backslash T_m\partial\mathcal{M}$. The $(d-1)$-form dv_∂ is thus nonvanishing on $\partial\mathcal{M}$, which provides an orientation on $\partial\mathcal{M}$, and allows one to integrate $(d-1)$-forms on $\partial\mathcal{M}$.

Observe that the orientation given on $\partial\mathcal{M}$ by dv_∂ is such that a positive frame $\{u_1, \ldots, u_{d-1}\}$ at a point $m \in \partial\mathcal{M}$ yields the following positive frame $\{\nu_m, u_1, \ldots, u_{d-1}\}$ for dv. Observe also that we did not rely on any particular atlas compatible with the orientation on $\mathcal{M}$ to define the orientation on $\partial\mathcal{M}$.

In a local chart $\mathcal{C} = (O, \kappa)$ that need not be in an atlas compatible with the orientation on $\mathcal{M}$ and where $\mathcal{M}$ is given by $\kappa_d(m) \geq 0$, considering the $(d-1)$-form $\omega = f dv_\partial$ supported in O, we then have

$$\int_{\partial\mathcal{M}} \omega = \int_{\kappa(O)\cap\{x_d=0\}} f^{\mathcal{C}}(x', 0)|\alpha(x')|dx', \tag{16.1.5}$$

with $f^{\mathcal{C}} = (\kappa^{-1})^* f$ if dv_∂ has $\alpha(x')dx^1 \wedge \cdots \wedge dx^{d-1}$ for local representative in the chart. If $\alpha(x') > 0$, this means that the local chart $(\tilde{O}, \tilde{\kappa})$, with $\tilde{O} = O \cap \partial\mathcal{M}$, and $\tilde{\kappa} = (\kappa_1, \ldots, \kappa_{d-1})$, is in an atlas with an orientation compatible with dv_∂. If however $\alpha(x') < 0$, the orientation of the chart needs to be changed, for instance, by simply changing $\tilde{\kappa}$ into $(\kappa_2, \kappa_1, \kappa_3, \ldots, \kappa_{d-1})$. This is consistent with having $\int_{\partial\mathcal{M}} dv_\partial > 0$.

REMARK 16.1. Here, we have provided $\partial\mathcal{M}$ with an orientation inherited from that of $\mathcal{M}$. Hence, for example, the boundary of a smooth open set of $\mathbb{R}^d$ is orientable. However, note that the boundary of nonorientable manifold may also be nonorientable. An example is given by the Klein bottle that is the boundary of a manifold and is yet nonorientable.

In fact, the solid Klein bottle is homeomorphic to the quotient space obtained by gluing the top disk of a cylinder $D^2 \times I$ to the bottom disk by a reflection across a diameter of the disk. Its boundary is the Klein Bottle. This is illustrated in Fig. 16.1. If ν is an outward-pointing tangent vector at the boundary, in the right-hand side figure, when moving along the blue path on the boundary, from point A to point B, the direct frame (e_1, e_2, ν) at A is changed into a frame with the opposite orientation at B, illustrating the impossibility to provide the Klein bottle with a proper orientation.

16.1.4. The Stokes Formula.

With the orientation of $\partial\mathcal{M}$ provided above, one can obtain the Stokes formula.

THEOREM 16.2 (Stokes' Formula). *Let $\mathcal{M}$ be an oriented smooth d-manifold with boundary. Let ω be a $(d-1)$-form on $\mathcal{M}$ of class $\mathscr{C}^1$ be compactly supported. We have*

$$\int_{\partial\mathcal{M}} \omega = \int_{\mathcal{M}} d\omega.$$

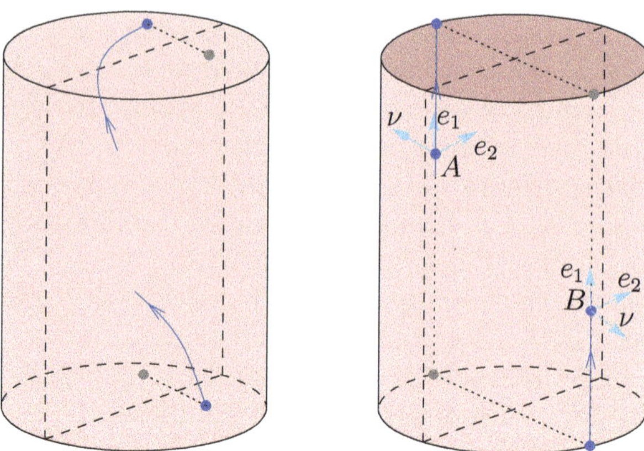

FIGURE 16.1. The solid Klein Bottle on the left-hand side and its boundary, the Klein bottle, on the right-hand side

There is a slight abuse of notation in the Stokes formula as we write it above. With ω a $(d-1)$-form on $\mathcal{M}$, it is understood as a $(d-1)$-form on $\partial\mathcal{M}$. In fact, if one denotes by i the natural injection of $\partial\mathcal{M}$ into $\mathcal{M}$, the formula that is proven below is actually

$$\int_{\partial\mathcal{M}} i^*\omega = \int_{\mathcal{M}} \mathrm{d}\omega,$$

with $i^*\omega$ the pullback of ω by i as defined in (15.6.3).

PROOF. Let $\mathcal{A} = (O^i, \kappa^i)_{i\in\mathcal{I}}$ be an atlas compatible with the orientation. We start by showing that it suffices to consider ω supported in a single chart. Choose by Theorem 15.14 a partition of unity $(\chi^i)_{i\in\mathcal{I}}$ of $\mathcal{M}$ subordinated to the open covering (O^i). On the one hand, by (16.1.4), we have $\int_{\partial\mathcal{M}}\omega = \sum_{i\in\mathcal{I}}\int_{\partial\mathcal{M}}\chi^i\omega$. On the other hand, we have $\mathrm{d}\omega = \sum_{i\in\mathcal{I}}\mathrm{d}(\chi^i\omega)$. We also have

$$\int_{\mathcal{M}} \mathrm{d}\omega = \int_{\mathcal{M}}\sum_{i\in\mathcal{I}} \mathrm{d}(\chi^i\omega) = \sum_{i\in\mathcal{I}}\int_{\mathcal{M}} \mathrm{d}(\chi^i\omega),$$

as the sum is finite because of the compact support of ω. It then suffices to prove the Stokes formula for $\chi^i\omega$ which is localized in the chart O^i.

If ω is now supported in a single chart $\mathcal{C} = (O,\kappa)$, it is only relevant to consider a chart that contains part of the boundary as shown by the computation below. We thus consider the case where $U_+ = \kappa(O)$ is an open set of $\overline{\mathbb{R}^d_+}$. We have $U_+ = \overline{\mathbb{R}^d_+}\cap U$ with U an open set of $\mathbb{R}^d$. As we have $\int_{\partial\mathcal{M}}\omega = \int_{\partial U_+}\omega^{\mathcal{C}}$ and $\int_{\mathcal{M}}\mathrm{d}\omega = \int_{U_+}\mathrm{d}\omega^{\mathcal{C}}$, we shall simply work in U_+. To lighten the notation, we write ω in place of $\omega^{\mathcal{C}}$.

Any smooth $(d-1)$-form on U_+, compactly supported in U_+, is given by

$$\omega = \sum_{1\le j\le d} f_j dx^1\wedge\cdots\wedge\widehat{dx^j}\wedge\cdots\wedge dx^d$$

with $f_j \in \mathscr{C}_c^\infty(U_+)$, $j = 1, \dots, d$, where the hat sign means the omission of the associated term in the wedge product. We then have by (15.6.8)

$$\begin{aligned} d\omega &= \sum_{1 \le j,k \le d} \partial_{x_k} f_j \, dx^k \wedge dx^1 \wedge \cdots \wedge \widehat{dx^j} \wedge \cdots \wedge dx^d \\ &= \sum_{1 \le j \le d} \partial_{x_j} f_j \, dx^j \wedge dx^1 \wedge \cdots \wedge \widehat{dx^j} \wedge \cdots \wedge dx^d \\ &= \Big(\sum_{1 \le j \le d} (-1)^{j-1} \partial_{x_j} f_j \Big) dx^1 \wedge \cdots \wedge dx^d. \end{aligned}$$

As the chart has the proper orientation, on U_+, we can use the volume form $dv = dx^1 \wedge \cdots \wedge dx^d$ yielding

$$\int_{U_+} d\omega = \int_{\mathbb{R}^d_+} g(x) dx, \quad g(x) = \Big(\sum_{1 \le j \le d} (-1)^{j-1} \partial_{x_j} f_j \Big).$$

An integration by parts gives, with $x' = (x_1, \dots, x_{d-1})$,

$$\int_{U_+} d\omega = (-1)^d \int_{\mathbb{R}^{d-1}} f_d(x', 0) dx'.$$

We now compute the volume form dv_∂ on $\partial U_+ = \{x_d = 0\} \cap U_+$. We set $\nu = (0, \dots, 0, -1)$ as the outward-pointing vector on ∂U_+ and we have, for $u_1, \dots, u_{d-1} \in T_{(x',0)} \partial U_+$,

$$\begin{aligned} dv_\partial(u_1, \dots, u_{d-1}) &= dv(\nu, u_1, \dots, u_{d-1}) = \det(\nu, u_1, \dots, u_{d-1}) \\ &= (-1)^d \det(u_1, \dots, u_{d-1}). \end{aligned}$$

Hence, $dv_\partial = (-1)^d dx^1 \wedge \cdots \wedge dx^{d-1}$. We now compute the integral of ω on ∂U_+. As $dx^d{}_{|T\partial U_+} = 0$ we have $i^*\omega = f_d(x', 0) dx^1 \wedge \cdots \wedge dx^{d-1}$, with i the natural injection of ∂U_+ into U_+ with $i^*\omega$ as given by (15.6.3). This yields

$$\begin{aligned} \int_{\partial U_+} i^*\omega &= \int_{\partial U_+} f_d(x', 0) dx^1 \wedge \cdots \wedge dx^{d-1} = \int_{\partial U_+} (-1)^d f_d(x', 0) dv_\partial \\ &= \int_{\mathbb{R}^{d-1}} (-1)^d f_d(x', 0) dx', \end{aligned}$$

invoking (16.1.5) to have the proper orientation on ∂U_+. We have thus found that $\int_{\partial U_+} i^*\omega = \int_U d\omega$.

Note that in the case d is even, we see that the local chart $x' \mapsto (x_1, \dots, x_{d-1})$ on ∂U_+ is correctly oriented, that is, compatible with the orientation given by dv_∂. However in the case d is odd, this chart is not correctly oriented. ■

As an application, we show that the Stokes formula allows one to recover the usual divergence theorem (also known as the Gauss or Ostrogradsky theorem). Consider Ω a bounded smooth open set of $\mathbb{R}^d$. The Lebesgue measure dx can be identified with the volume form $dv = dx_1 \wedge \cdots \wedge dx_d$, meaning that

$$\int_\Omega f(x) dx = \int_\Omega f dv,$$

with the latter integral understood in the sense of the integral on a d-form. Note that

$$dv(w_1, \ldots, w_d) = \det(w_1, \ldots, w_d).$$

This may seem quite artificial as there is only one chart here. However, if ν is the unit normal outward-pointing vector of $\partial\Omega$ (in the Euclidean geometry), then we can define a volume form dv_∂ on $\partial\Omega$ from what precedes:

$$(dv_\partial)_x(w_1, \ldots, w_{d-1}) = dv_x(\nu_x, w_1, \ldots, w_{d-1}) = \det(\nu_x, w_1, \ldots, w_{d-1}),$$

for $w_1, \ldots, w_{d-1} \in T_x\partial\Omega$. Let u be a $\mathscr{C}^1$ compactly supported vector field in $\mathbb{R}^d$. The divergence formula states that

$$\int_\Omega \operatorname{div} u(x)dx = \int_{\partial\Omega} (\nu \cdot u)dv_\partial. \tag{16.1.6}$$

The $(d-1)$-form dv_∂ is then called the surface measure of $\partial\Omega$ and one uses often the notation $d\sigma(x)$ instead. A proof of (16.1.6) goes as follows.

Define the $(d-1)$-form $\omega_x = \sum_{1\leq j\leq d}(-1)^{j-1}u_x^j dx^1 \wedge \cdots \wedge \widehat{dx^j} \wedge \cdots \wedge dx^d$. First, observe that $\mathrm{d}\omega = \operatorname{div} u\, dv$. With the Stokes formula, one then has

$$\int_\Omega \operatorname{div} u(x)dx = \int_\Omega \operatorname{div} u dv = \int_\Omega \mathrm{d}\omega = \int_{\partial\Omega} \omega.$$

Second, observe that for $w_1, \ldots, w_{d-1} \in \mathbb{R}^d$, we have

$$\begin{aligned}
&\omega_x(w_1, \ldots, w_{d-1}) \\
&\quad = \sum_{1\leq j\leq d} (-1)^{j-1}u_x^j \big(dx^1 \wedge \cdots \wedge \widehat{dx^j} \wedge \cdots \wedge dx^d\big)(w_1, \ldots, w_{d-1}) \\
&\quad = \det(u, w_1, \ldots, w_{d-1}),
\end{aligned}$$

since this corresponds to developing the determinant with respect to the first column, identifying the vectors with column vectors. For $x \in \partial\Omega$, we may write $u_x = u_x^\nu \nu_x + u_x'$ where $u_x' \in T_x\partial\Omega$ and $u_x^\nu = u_x \cdot \nu_x$ since ν_x is unitary. If now $w_1, \ldots, w_{d-1} \in T_x\partial\Omega$, we find

$$\omega_x(w_1, \ldots, w_{d-1}) = \det(u, w_1, \ldots, w_{d-1}) = u_x^\nu \det(\nu_x, w_1, \ldots, w_{d-1}),$$

since $\det(u_x', w_1, \ldots, w_{d-1})$ vanishes. We thus conclude that $\omega_x = (\nu_x \cdot u_x)\,(dv_\partial)_x$ and this gives (16.1.6).

16.2. Densities on a Nonoriented Manifold

Above, we saw that d-forms are well suited for integration since their change-of-coordinate formula makes the determinant of the Jacobian matrix appear. The integration of d-forms however requires the atlas to be oriented. Then, the determinant of the Jacobian matrix is positive, which is precisely needed in the change of variable formula for integration in $\mathbb{R}^d$.

On nonoriented manifolds, a proper integration can yet be defined; it coincides with that of d-forms on oriented manifolds. For that we need to introduce the notion of densities, for which the change-of-coordinate formula makes naturally the absolute value of the determinant of the Jacobian matrix appear regardless of any orientation.

Densities can be introduced through their local representative in a chosen atlas. However, the dual space of compactly supported functions on the manifold allows one to introduce them as natural geometrical objects.

16.2.1. Density Radon Measures on a Manifold. Let $\mathcal{M}$ be a smooth manifold with or without boundary. If U is an open set of $\mathcal{M}$, possibly equal to $\mathcal{M}$, we define $\mathscr{D}_c^0(U)$ as the space of continuous compactly supported functions ϕ in U such that moreover $\text{supp}(\phi) \cap \partial\mathcal{M} = \emptyset$. In fact, recall that $\partial\mathcal{M} \subset \mathcal{M}$ and thus ∂U may contain part of $\partial\mathcal{M}$. Note that if U does not meet $\partial\mathcal{M}$, then $\mathscr{C}_c^0(U) = \mathscr{D}_c^0(U)$. However, if U meets $\partial\mathcal{M}$, then $\mathscr{D}_c^0(U) \subsetneq \mathscr{C}_c^0(U)$.

If (O, κ) is a local chart that meets the boundary, then $\kappa(O)$ is an open set of $\overline{\mathbb{R}^d_+}$ but not of $\mathbb{R}^d$. In this case, we shall denote by $\mathscr{D}_c^0(\kappa(O))$ the space of continuous compactly supported functions in $\mathbb{R}^d_+ \cap \kappa(O)$.

Let u be a continuous form on $\mathscr{D}_c^0(U)$, that is, for all compact set $K \subset U$ that does not meet the boundary $\partial\mathcal{M}$, there exists $C > 0$ such that

$$|\langle u, \phi\rangle| \leq C \sup_{m\in U} |\phi(m)|, \qquad \phi \in \mathscr{D}_c^0(U),\ \text{supp}(\phi) \subset K.$$

Such a form is called a density Radon measure on U. The space of density Radon measure on U is denoted[1] by ${}^1\mathscr{D}'^0(U)$.

If $u \in {}^1\mathscr{D}'^0(U)$ and $\mathcal{C} = (O, \kappa)$ is a local chart with $O \cap U \neq \emptyset$, the restriction $u_{|O\cap U}$ of u to $O \cap U$ is given by

$$u_{|O\cap U} : \phi \mapsto \langle u, \phi\rangle, \quad \phi \in \mathscr{D}_c^0(O \cap U).$$

With $\tilde{O} = \kappa(O \cap U)$, we define its local representative $u^{\mathcal{C}}$ as the form on $\mathscr{D}_c^0(\tilde{O})$ given by

$$u^{\mathcal{C}} : \tilde{\phi} \mapsto \langle u_{|O\cap U}, \kappa^*\tilde{\phi}\rangle, \quad \tilde{\phi} \in \mathscr{D}_c^0(\tilde{O}).$$

It is a Radon measure on $\text{int}\,\tilde{O} \subset \mathbb{R}^d$, that is, a zero-order distribution: $u^{\mathcal{C}} \in \mathscr{D}'^0(\text{int}\,\tilde{O})$. We insist here on the fact that the interior is considered for a set in $\mathbb{R}^d$. We recall that the manifold $\mathcal{M}$ contains its boundary $\partial\mathcal{M}$; thus, for a set L of $\mathcal{M}$ contained in a local chart $\mathcal{C} = (O, \kappa)$, we may have $\text{int}\,\kappa(L) \neq \kappa(\text{int}\,L)$ if L meets the boundary.

Note that on $\mathbb{R}^d$, the notions of Radon measures and density Radon measures coincide.[2] In general, this is not the case on a manifold as we shall see below.

We analyze the effect of a change of charts. Let thus $\mathcal{C}^i = (O^i, \kappa^i)$, $i = 1, 2$, be two local charts such that $O = O^1 \cap O^2 \neq \emptyset$ and $O \cap U \neq \emptyset$, and let $u^{\mathcal{C}^1}$ and $u^{\mathcal{C}^2}$ be the local representatives of u in these charts. Let

---

[1] The notation will become clearer in Sect. 16.2.5.

[2] To be more precise, the two notions coincide as long as the trivial chart is used. If change of variables are applied the same issues as those one faces on manifold appear.

$\phi \in \mathscr{D}_c^0(O \cap U)$ and $\phi^i = ((\kappa^i)^{-1})^* \phi \in \mathscr{D}_c^0(\kappa^i(O \cap U))$. We set $\kappa^{ij} = \kappa^j \circ (\kappa^i)^{-1} : \kappa^i(O) \to \kappa^j(O)$. We have $(\kappa^{ij})^* \phi^j = \phi^i$, yielding

$$\langle u, \phi \rangle = \langle u^{\mathcal{C}^1}, \phi^1 \rangle = \langle u^{\mathcal{C}2}, (\kappa^{21})^* \phi^1 \rangle. \tag{16.2.1}$$

We recall the notation $J_{\kappa^{ij}}(x) = |\det(\kappa^{ij})'(x)|$ for $x \in \kappa^i(O)$, where $(\kappa^{ij})'(x)$ is the Jacobian matrix of κ^{ij} at x. Here, $J_{\kappa^{ij}}(x)$ is a smooth function. Letting $\varphi \in \mathscr{D}_c^0(\kappa^1(O \cap U))$, by (16.2.1), we have

$$\langle ((\kappa^{12})^* J_{\kappa^{21}}) u^{\mathcal{C}^1}, \varphi \rangle = \langle u^{\mathcal{C}^1}, ((\kappa^{12})^* J_{\kappa^{21}}) \varphi \rangle = \langle u^{\mathcal{C}^2}, J_{\kappa^{21}} (\kappa^{21})^* \varphi \rangle.$$

Using the formula for pullbacks of distributions by diffeomorphisms on $\mathbb{R}^d$ given in Definition 8.26 of Volume 1, we find

$$\langle ((\kappa^{12})^* J_{\kappa^{21}}) u^{\mathcal{C}^1}, \varphi \rangle = \langle (\kappa^{12})^* u^{\mathcal{C}^2}, \varphi \rangle,$$

for all $\varphi \in \mathscr{D}_c^0(\kappa^1(O \cap U))$, implying that we have $((\kappa^{12})^* J_{\kappa^{21}}) u^{\mathcal{C}^1} = (\kappa^{12})^* u^{\mathcal{C}^2}$, which we write

$$u^{\mathcal{C}^1} = J_{\kappa^{12}} (\kappa^{12})^* u^{\mathcal{C}^2} \text{ in } \kappa^1(\mathcal{O}). \tag{16.2.2}$$

Observe that if $u \in {}^1\mathscr{D}'^0(U)$ and f is a $\mathscr{C}^0$-function on U, then fu defined by

$$\langle fu, \phi \rangle = \langle u, f\phi \rangle, \quad \phi \in \mathscr{D}_c^0(U),$$

is also a density Radon measure on U. Its local representative is simply $fu^{\mathcal{C}^i}$ in each chart $\mathcal{C}^i = (O^i, \kappa^i)$.

The support $\operatorname{supp}(u)$ of a density Radon measure u is defined as $U \setminus \mathcal{O}$, where $\mathcal{O}$ is the largest open set of U such that $\langle u, \phi \rangle = 0$ for $\phi \in \mathscr{D}_c^0(\mathcal{O})$. One says that u is compactly supported if $\operatorname{supp}(u)$ is a compact set of $\mathcal{M}$.

If $\mathcal{A} = (O^i, \kappa^i)_{i \in \mathcal{I}}$ is an atlas and if $(\chi^i)_{i \in \mathcal{I}}$ is a partition of unity subordinated to the open covering $(O^i)_{i \in \mathcal{I}}$ as given by Theorem 15.14, for each function $\phi \in \mathscr{D}_c^0(U)$, we have $\phi = \sum_{i \in \mathcal{I}} \chi_i \phi$, where the sum is actually finite. This gives, for $u \in {}^1\mathscr{D}'^0(U)$,

$$\langle u, \phi \rangle = \sum_{i \in \mathcal{I}} \langle \chi^i u, \phi \rangle,$$

which we write $u = \sum_{i \in \mathcal{I}} \chi^i u$ with each $\chi^i u$ supported in O^i. Moreover, we have $\langle \chi^i u, \phi \rangle = \langle u, \chi^i \phi \rangle = \langle u^{\mathcal{C}^i}, ((\kappa^i)^{-1})^* (\chi^i \phi) \rangle$ yielding a formula providing u from its local representatives

$$\langle u, \phi \rangle = \sum_{i \in \mathcal{I}} \langle u^{\mathcal{C}^i}, ((\kappa^i)^{-1})^* (\chi^i \phi) \rangle, \quad \phi \in \mathscr{D}_c^0(U).$$

This formula shows that an alternative definition of density Radon measures on U can be done as a collection of Radon measures u^i on $\operatorname{int} \kappa^i(O^i \cap U)$ such that

$$u^i = J_{\kappa^{ij}} (\kappa^{ij})^* (u^j) \text{ in } \operatorname{int} \kappa^i(O \cap U), \qquad i, j \in \mathcal{I} \text{ with } O \cap U \neq \emptyset,\ O = O^i \cap O^j. \tag{16.2.3}$$

Then, the map $\phi \mapsto \sum_{i\in\mathcal{I}}\langle u^i, \big((\kappa^i)^{-1}\big)^*(\chi^i\phi)\rangle$ is a density Radon measure, with local representatives given precisely by $u^{\mathcal{C}^i} = u^i$. This density Radon measure is independent of the choice of the partition of unity. This definition is less intrinsic than that we gave above, yet they are equivalent. In fact, if one chooses in each open set $\kappa^i(O^i \cap U) \subset \mathbb{R}^d$ a Radon measure u^i without requiring (16.2.3) and one forms $u : \phi \mapsto \sum_{i\in\mathcal{I}}\langle u^i, \tilde{\chi}^i\big((\kappa^i)^{-1}\big)^*\phi\rangle$, with $\tilde{\chi}^i = \big((\kappa^i)^{-1}\big)^*\chi^i$, then $u \in {}^1\mathscr{D}'^0(U)$ and its local representatives are

$$u^{\mathcal{C}^i} = \tilde{\chi}^i u^i + \sum_{j\in\mathcal{J}_i} J_{\kappa^{ij}}\,(\kappa^{ij})^*(\tilde{\chi}^j u^j),$$

where $\mathcal{J}_i = \{j \in \mathcal{I}; j \neq i \text{ and } O^i \cap O^j \cap U \neq \emptyset\}$. Observe that the sum is well defined since locally finite. Here, the definition of u depends on the partition of unity and $u^{\mathcal{C}^i} \neq u^i$ in general.

16.2.2. Density Functions. Consider $u \in {}^1\mathscr{D}'^0(U)$ such that, given any chart $\mathcal{C}^i = (O^i, \kappa^i)$, its representative $u^{\mathcal{C}^i}$ in this chart can be identified with an L^1_{loc}-function $f^{\mathcal{C}^i}$, that is, L^1 in any compact set away from the boundary. Then, if $\varphi \in \mathscr{D}^0_c(\kappa^i(O^i \cap U))$, as defined in the beginning of Sect. 16.2.1, this means that we have

$$\langle u^{\mathcal{C}^i}, \varphi\rangle = \int_{\kappa^i(O^i\cap U)} f^{\mathcal{C}^i}(x)\varphi(x)\,dx,$$

or $u^{\mathcal{C}^i} = T_{f^{\mathcal{C}^i}}$ with the notation of Chapter 8 of Volume 1 for L^1_{loc}-functions identified with distributions.

The change of variable formula (16.2.2) amounts then to having

$$f^{\mathcal{C}^1} = J_{\kappa^{12}}\,(\kappa^{12})^* f^{\mathcal{C}^2} \text{ in } \kappa^1(O \cap U),$$

as $T_{(\kappa^{12})^* f^2} = (\kappa^{12})^* T_{f^2}$ by (8.1.1). This simply reads

$$f^{\mathcal{C}^1}(x) = J_{\kappa^{12}}(x)\, f^{\mathcal{C}^2}(y), \qquad x \in \kappa^1(O \cap U),\ y = \kappa^{12}(x). \tag{16.2.4}$$

We are thus led to define L^1_{loc}-density functions on U (respectively, $\mathscr{C}^k$) as follows.

DEFINITION 16.3. One says that u is an L^1_{loc}-density function on $U\backslash\partial\mathcal{M}$ if $u \in {}^1\mathscr{D}'^0(U)$ and if for every chart $\mathcal{C}^i = (O^i, \kappa^i)$ with $O^i \cap U \neq \emptyset$, the local representative of u is such that $u^{\mathcal{C}^i} = T_{f^{\mathcal{C}^i}}$ with $f^{\mathcal{C}^i}$ in $L^1_{\text{loc}}(\operatorname{int}\kappa^i(O^i \cap U))$. The local representatives then transform according to (16.2.4). If in addition, for some $k \in \mathbb{N}$, each representative is in $\mathscr{C}^k(\operatorname{int}\kappa^i(O^i \cap U))$, we say that u is $\mathscr{C}^k$.

Alternatively, if given an atlas $\mathcal{A} = (O^i, \kappa^i)_{i\in\mathcal{I}}$, a partition of unity $(\chi^i)_{i\in\mathcal{I}}$ subordinated to the open covering $(O^i)_{i\in\mathcal{I}}$ by Theorem 15.14, and a family of functions $f^i \in L^1_{\text{loc}}(\operatorname{int}\kappa^i(O^i \cap U))$ moreover satisfying

$$f^i(x) = J_{\kappa^{ij}}(x)\, f^j(y), \qquad x \in \kappa^i(O^{ij} \cap U),\ y = \kappa^{ij}(x), \tag{16.2.5}$$

if $O^{ij} \cap U = O^i \cap O^j \cap U \neq \emptyset$, then

$$u : \phi \mapsto \sum_{i \in \mathcal{I}} \int_{\kappa^i(O^i \cap U)} f^i(x)(\chi^i \phi)\big((\kappa^i)^{-1}(x)\big)\, dx, \quad \phi \in \mathscr{D}_c^0(U), \tag{16.2.6}$$

defines an L^1_{loc}-density function u. One can check using (16.2.5) that the measure T_{f^i} is the local representative $u^{\mathcal{C}^i}$ of u in $\mathcal{C}^i = (O^i, \kappa^i)$. A similar argument applies for $\mathscr{C}^k$-density functions.

As is done above for density Radon measures, if given a collection of functions $f^i \in L^1_{\text{loc}}(\operatorname{int} \kappa^i(O^i \cap U))$ (respectively, $\mathscr{C}^k$), possibly not satisfying (16.2.5), then the map

$$\phi \mapsto \sum_{i \in \mathcal{I}} \int_{\kappa^i(O^i \cap U)} f^i(x)(\chi^i \phi)\big((\kappa^i)^{-1}(x)\big)\, dx, \quad \phi \in \mathscr{D}_c^0(U),$$

defines a density function u. Its local representatives are in $L^1_{\text{loc}}(\operatorname{int} \kappa^i(O^i \cap U))$ (respectively, $\mathscr{C}^k$) and given by

$$f^{\mathcal{C}^i}(x) = \tilde{\chi}^i(x) f^i(x) + \sum_{j \in \mathcal{J}_i} J_{\kappa^{ij}}(x)\, (\tilde{\chi}^j f^j)(\kappa^{ij}(x)),$$

where $\mathcal{J}_i = \{j \in \mathcal{I}; j \neq i \text{ and } O^i \cap O^j \cap U \neq \emptyset\}$ and $\tilde{\chi}^i = \big((\kappa^i)^{-1}\big)^* \chi^i$. In general, $f^{\mathcal{C}^i} \neq f^i$. In particular, if we choose $f^i \equiv 1$ for all $i \in \mathcal{I}$, we find $f^{\mathcal{C}^i} > 0$, thus exhibiting a positive density function on $\mathcal{M}$. This density function is not canonical as it depends both on the chosen atlas and on the partition of unity. Once a particular positive density function μ_0 is chosen, we see that we can identify functions and density functions on $\mathcal{M}$ (respectively, U) through the bijective map $f \mapsto f\mu_0$. Note that a Riemannian metric on a manifold allows one to pick a canonical positive density function (see Sect. 17.3).

If the atlas $\mathcal{A} = (O^i, \kappa^i)_{i \in \mathcal{I}}$ is locally finite, and K is a compact set of $\mathcal{M}$, then K meets a finite number of open sets O^i. For a $\mathscr{C}^0$-density function u supported in K, we can define the norm

$$\|u\|_{\mathcal{A},K} = \sup_{i \in \mathcal{I}} \|f^{\mathcal{C}^i}\|_{L^\infty(\kappa^i(O^i))}. \tag{16.2.7}$$

This norm depends on the choice of the atlas; yet, two such atlases yield two equivalent norms.

16.2.3. Radon Measures on a Manifold. If U is an open set of $\mathcal{M}$, we denote by ${}^1\mathscr{D}_c^0(U)$ the space of $\mathscr{C}^0$-density functions u that are compactly supported with $\operatorname{supp}(u) \subset U$ and $\operatorname{supp}(u) \cap \partial U = \emptyset$ (recall that $\partial\mathcal{M} \subset \mathcal{M}$ and that U, and thus ∂U, may contain part of $\partial\mathcal{M}$).

A Radon measure w on U is a continuous form on the space ${}^1\mathscr{D}_c^0(U)$, with respect to the topology given by the norms (16.2.7): given an atlas,

$$\forall K \text{ compact set of } U,\ K \cap \partial\mathcal{M} = \emptyset,\ \exists C_{\mathcal{A},K} > 0,$$
$$|\langle w, u\rangle| \leq C_{\mathcal{A},K} \|u\|_{\mathcal{A},K}, \quad u \in {}^1\mathscr{D}_c^0(U),\ \operatorname{supp}(u) \subset K.$$

We denote[3] by ${}^0\mathscr{D}'^0(U)$ the space of Radon measures on U.

For each chart $\mathcal{C}^i = (O^i, \kappa^i)$, there exists a Radon measure $w^{\mathcal{C}^i}$ on $\operatorname{int}\kappa^i(O^i \cap U) \subset \mathbb{R}^d$ such that

$$\langle w, u\rangle = \langle w^{\mathcal{C}^i}, u^{\mathcal{C}^i}\rangle, \quad u \in {}^1\mathscr{D}^0_c(O^i \cap U)$$

with $u^{\mathcal{C}^i}$ local representative of u in the chart. The second bracket is the duality bracket for Radon measures and compactly supported $\mathscr{C}^0$-functions in an open set of $\mathbb{R}^d$. We say that $w^{\mathcal{C}^i}$ is the local representative of w in the chart. Let then $\mathcal{C}^i = (O^i, \kappa^i)$, $i = 1, 2$, be two local charts such that $O \cap U \neq \emptyset$, $O = O^1 \cap O^2$, and let $w^{\mathcal{C}^1}$ and $w^{\mathcal{C}^2}$ be the local representatives of w in these charts. We find

$$w^{\mathcal{C}^1} = (\kappa^{12})^* w^{\mathcal{C}^2}, \text{ in } \kappa^1(O \cap U), \tag{16.2.8}$$

where the pullback is that for distributions (and thus for measures) introduced in Section 8.1.3 of Volume 1. Comparing with (16.2.2), one sees that the action of a change of chart differs by the presence of the absolute value of the associated Jacobian determinant. If provided with a family of Radon measures in each local chart that satisfies (16.2.8), we see that it determines a Radon measure on the manifold, arguing as we did above for density Radon measures and density functions.

A Radon measure w can be multiplied by functions. It can also be multiplied by a density function u, yielding a density Radon measure: for all $\phi \in \mathscr{D}^0_c(U)$, we have

$$\langle uw, \phi\rangle = \langle w, \phi u\rangle,$$

since ϕu is a density function. Note that given a positive density function μ_0 as exhibited in the previous section, the bijective map $w \mapsto \mu_0 w$ allows one to identify Radon measures and density Radon measures.

16.2.4. Integration of Densities. Let $\mathcal{M}$ be a smooth d-dimensional manifold, $\mathcal{A} = (O^i, \kappa^i)_{i\in\mathcal{I}}$ be an atlas, and $(\chi^i)_{i\in\mathcal{I}}$ be a partition of unity subordinated to the open covering $(O^i)_{i\in\mathcal{I}}$ by Theorem 15.14. We set $\tilde{\chi}^i = \big((\kappa^i)^{-1}\big)^* \chi^i$.

If u is an L^1_{loc}-density function on $\mathcal{M}$, we say that u is nonnegative if for each $i \in \mathcal{I}$ the function $f^{\mathcal{C}^i}$, representative of u in the chart (O^i, κ^i), that is, $T_{f^{\mathcal{C}^i}} = u^{\mathcal{C}^i}$, is nonnegative. For such a nonnegative L^1_{loc}-density function, we define its integral in $\mathcal{M}$ as

$$\int_{\mathcal{M}} u = \sum_{i\in\mathcal{I}} \int_{\kappa^i(O^i)} f^{\mathcal{C}^i}(x)\tilde{\chi}^i(x)\, dx \in \mathbb{R}_+ \cup \{+\infty\}. \tag{16.2.9}$$

One can prove that this definition of the integral of u is independent of the choice of the atlas and the used partition of unity. Such a nonnegative density function is called integrable if $\int_{\mathcal{M}} u < \infty$. This is equivalent to having all the function $f^{\mathcal{C}^i} \in L^1(\kappa^i(O^i))$, including those in charts that

[3]The notation will become clearer in Sect. 16.2.5.

meet the boundary (compare with the definition of L^1_{loc}-density functions in Definition 16.3 above). We say that u is a nonnegative L^1-density function on $\mathcal{M}$.

If u is an L^1_{loc}-density function on $\mathcal{M}$, one says that u is L^1 on $\mathcal{M}$ if the nonnegative L^1_{loc}-density function $|u|$, given in each chart by $|f^{\mathcal{C}^i}|$, is integrable. Formula (16.2.9) then also makes sense in this case and defines the integral of u in $\mathcal{M}$. We say that u is an L^1-density function on $\mathcal{M}$.

Observe that if a density function is supported in a local chart O^i, then using (16.2.4), we find

$$\int_{\mathcal{M}} u = \int_{\kappa^i(O^i)} f^{\mathcal{C}^i}(x)dx.$$

One can introduce the notion of measurable density functions, with each representative measurable. In particular, zero almost everywhere density functions can be defined. We then say that a measurable density function u is integrable if $\int_{\mathcal{M}} |u| < \infty$ and we recover the L^1-density function on $\mathcal{M}$ given above. Moreover, one says that a measurable density function is locally integrable if the density function $\mathbf{1}_K u$ is integrable for all compact sets K of $\mathcal{M}$ that do not meet the boundary $\partial\mathcal{M}$, where $\mathbf{1}_K$ is the characteristic function of K. We then recover the L^1_{loc}-density function on $\mathcal{M}$ given above.

Consider now an orientable manifold $\mathcal{M}$ oriented by the atlas $\mathcal{A} = (O^i, \kappa^i)_{i\in\mathcal{I}}$. Let ω be an integrable d-form. Its integral is well defined; see Sect. 16.1.2. In each chart (O^i, κ^i), the representative of ω is of the form $\omega^{\mathcal{C}^i}(x) = \omega^i(x)dx^1 \wedge \cdots \wedge dx^d$. Because of the orientation given by $\mathcal{A}$, which states that $\det(\kappa^{ij})' > 0$ for $\kappa^{ij} = \kappa^j \circ (\kappa^i)^{-1}$, $i, j \in \mathcal{I}$ such that $O^i \cap O^j \neq \emptyset$, we see that the transformation rule (15.6.7) for ω^i coincides with that of density function given in (16.2.4). With (16.2.6), we then see that

$$u : \phi \mapsto \sum_{i\in\mathcal{I}} \int_{\kappa^i(O^i)} \omega^i(x)(\chi^i\phi)\big((\kappa^i)^{-1}(x)\big)\, dx, \quad \phi \in \mathscr{D}^0_c(\mathcal{M}),$$

defines an L^1-density function u with $u^{\mathcal{C}^i} = T_{\omega^i}$ and

$$\int_{\mathcal{M}} u = \sum_{i\in\mathcal{I}} \int_{\kappa^i(O^i)} \omega^i(x)\tilde{\chi}^i(x)\, dx = \int_{\mathcal{M}} \omega.$$

The integration of density functions thus coincides with that of d-forms on oriented manifolds. In fact, consider $\mathcal{B} = (U^j, \psi^j)_{j\in\mathcal{J}}$ a second atlas in which some charts possibly do not have the same orientation as that of $\mathcal{A}$. In a chart $\mathcal{C}^j = (U^j, \psi^j)$, where $\omega^{\mathcal{C}^j}(x) = \omega^j(x)dx^1 \wedge \cdots \wedge dx^d$, we have just seen that $u^{\mathcal{C}^j} = T_{\omega^j}$ if $\mathcal{C}^j$ has the same orientation as $\mathcal{A}$. However, if $\mathcal{C}^j$ has the opposite orientation, we have $u^{\mathcal{C}^j} = -T_{\omega^j}$. The integral of u on $\mathcal{M}$ is yet independent of the orientation of $\mathcal{M}$.

16.2.5. Density Distributions and a-Densities. For U an open set of $\mathcal{M}$, we introduce the space $\mathscr{D}^k_c(U)$, $0 \leq k \leq \infty$, as the space of compactly supported $\mathscr{C}^k$ functions on $\mathcal{M}$ such that $\operatorname{supp}(u) \subset U$ and $\operatorname{supp}(u) \cap \partial U = \emptyset$.

Note that if U does not meet $\partial\mathcal{M}$, then $\mathscr{C}_c^k(U) = \mathscr{D}_c^k(U)$. However, if U meets $\partial\mathcal{M}$, then $\mathscr{D}_c^k(U) \subsetneq \mathscr{C}_c^k(U)$. Below, we shall see that we shall also denote this space ${}^0\mathscr{D}_c^k(U)$.

Above we saw that ${}^1\mathscr{D}'^0(U)$ the space of density Radon measures can either be defined as the space continuous forms on the space $\mathscr{D}_c^0(U)$ or by a collection of Radon measures in $\kappa^i(O^i \cap U)$ that satisfy the proper change of variable transformation given by (16.2.2).

We give the following definition for density distributions.

Definition 16.4. Let $\mathcal{M}$ be a smooth d-dimensional manifold and U be an open set of $\mathcal{M}$. Let $\mathcal{A} = (O^i, \kappa^i)_{i\in\mathcal{I}}$ be an atlas and $(\chi^i)_{i\in\mathcal{I}}$ be a partition of unity subordinated to the open covering $(O^i)_{i\in\mathcal{I}}$. We set $\tilde{\chi}^i = \big((\kappa^i)^{-1}\big)^* \chi^i$. Let $(u^{\mathcal{C}^i})_{i\in\mathcal{I}}$ be such that, for each $i \in \mathcal{I}$, $u^{\mathcal{C}^i} \in \mathscr{D}'(\operatorname{int}\kappa^i(O^i \cap U))$ and

$$\begin{gathered} u^{\mathcal{C}^i} = J_{\kappa^{ij}}\,(\kappa^{ij})^*(u^{\mathcal{C}^j}) \text{ in } \operatorname{int}\kappa^i(O \cap U), \\ i, j \in \mathcal{I} \text{ with } O \cap U \neq \emptyset,\ O = O^i \cap O^j, \end{gathered}$$

with $\kappa^{ij} = \kappa^j \circ (\kappa^i)^{-1}$, for the pullback of distributions by a diffeomorphism given in Definition 8.26 of Volume 1. We say that the map

$$u : \phi \mapsto \sum_{i\in\mathcal{I}} \langle u^{\mathcal{C}^i}, \tilde{\chi}^i \big((\kappa^i)^{-1}\big)^* \phi \rangle_{\mathscr{D}'(\operatorname{int}\kappa^i(O^i\cap U)), \mathscr{C}_c^\infty(\operatorname{int}\kappa^i(O^i\cap U))},$$

for $\phi \in \mathscr{D}_c^\infty(U)$, is a density distribution on U. One writes $u(\phi) = \langle u, \phi\rangle$. We denote by ${}^1\mathscr{D}'(U)$ the set of density distributions on U.

The map $\phi \mapsto \langle u, \phi\rangle$ is in fact independent of the choice of atlas and partition of unity. The density distribution is said to be of finite order if the supremum of the orders of the distributions $u^{\mathcal{C}^i}$ is finite. This supremum is called the order of u. Otherwise, the density distribution is said to be of infinite order. A zero-order density distribution is a density Radon measure.

The definition we just wrote is not as intrinsic as that given for density Radon measures. In fact, one can define density distributions on U as the dual space of $\mathscr{D}_c^\infty(U)$. This however requires the introduction of a proper topology on this space that involves derivatives of such functions on U. Local charts or vector fields can be used for that purpose. We choose not to do it here.

Note that if $\phi \in \mathscr{D}_c^\infty(U)$ is supported in a single chart (O^i, κ^i), we then find that

$$\langle u, \phi\rangle = \langle u^{\mathcal{C}^i}, \phi^{\mathcal{C}^i} \rangle_{\mathscr{D}'(\operatorname{int}\kappa^i(O^i\cap U)), \mathscr{C}_c^\infty(\operatorname{int}\kappa^i(O^i\cap U))},$$

with $\phi^{\mathcal{C}^i} = \big((\kappa^i)^{-1}\big)^*\phi$. We thus say that $u^{\mathcal{C}^i}$ is the local representative of u in this chart.

Similarly to what is done above for ${}^0\mathscr{D}'^0(U)$ the space of Radon measures on U, we can also define distributions on U that act on ${}^1\mathscr{D}_c^\infty(U)$ the space of smooth density functions that have a compact support away from $\partial\mathcal{M}$. More generally, we define the sets of a-density functions and a-density distributions as follows.

DEFINITION 16.5. Let $a \in [0,1]$. With the notation of Definition 16.4, a family of functions $f^{\mathcal{C}^i} \in L^1_{\mathrm{loc}}(\operatorname{int} \kappa^i(O^i \cap U))$ satisfying

$$f^{\mathcal{C}^i}(x) = (J_{\kappa^{ij}}(x))^a f^{\mathcal{C}^j}(y), \qquad x \in \kappa^i(O \cap U),\ y = \kappa^{ij}(x), \tag{16.2.10}$$

for $i, j \in \mathcal{I}$ with $O \cap U \neq \emptyset$, $O = O^i \cap O^j$, is called a a-density function. We denote the space of $\mathscr{C}^k$ a-density functions compactly supported in U away from $\partial\mathcal{M}$ by ${}^a\mathscr{D}^k_c(U)$ for $0 \le k \le \infty$.

If f is such a a-density function, we say that the functions $f^{\mathcal{C}^i}$ are its local representative in the charts $\mathcal{C}^i = (O^i, \kappa^i)$. If each local representative $f^{\mathcal{C}^i}$ is $\mathscr{C}^k$, $0 \le k \le \infty$, one says that the a-density function is $\mathscr{C}^k$.

DEFINITION 16.6. Let $a \in [0,1]$. With the notation of Definition 16.4, a family $(u^{\mathcal{C}^i})_{i\in\mathcal{I}}$ such that, for each $i \in \mathcal{I}$, $u^{\mathcal{C}^i}$ is Radon measure in $\operatorname{int} \kappa^i(O^i \cap U)$ (respectively, $u^{\mathcal{C}^i} \in \mathscr{D}'(\operatorname{int} \kappa^i(O^i \cap U))$) and

$$u^{\mathcal{C}^i} = (J_{\kappa^{ij}})^a\, (\kappa^{ij})^*(u^{\mathcal{C}^j}) \text{ in } \operatorname{int} \kappa^i(O \cap U),$$
$$i, j \in \mathcal{I} \text{ with } O \cap U \neq \emptyset,\ O = O^i \cap O^j,$$

is called a a-density Radon measure (respectively, distribution) on U. Its action on ${}^{1-a}\mathscr{D}^0_c(U)$ (respectively, ${}^{1-a}\mathscr{D}^\infty_c(U)$) is given by

$$\langle u, \phi\rangle = \sum_{i\in\mathcal{I}} \langle u^{\mathcal{C}^i}, \tilde{\chi}^i \phi^{\mathcal{C}^i}\rangle_{\mathscr{D}'^0(\operatorname{int}\kappa^i(O^i\cap U)),\mathscr{C}^0_c(\operatorname{int}\kappa^i(O^i\cap U))}, \qquad \phi \in {}^{1-a}\mathscr{D}^0_c(U),$$

$$\Big(\text{respectively,} \langle u, \phi\rangle = \sum_{i\in\mathcal{I}} \langle u^{\mathcal{C}^i}, \tilde{\chi}^i \phi^{\mathcal{C}^i}\rangle_{\mathscr{D}'(\operatorname{int}\kappa^i(O^i\cap U)),\mathscr{C}^\infty_c(\operatorname{int}\kappa^i(O^i\cap U))},$$
$$\phi \in {}^{1-a}\mathscr{D}^\infty_c(U)\Big).$$

We denote by ${}^a\mathscr{D}'^0(U)$ and ${}^a\mathscr{D}'(U)$ the space of a-density Radon measure and a-density distributions, respectively, on U.

The definition of a-density functions (respectively, Radon measure and distributions) is independent of the chosen atlas and partition of unity. Density functions (respectively, Radon measure and distributions) on U are the 1-density functions (respectively, Radon measure and distributions). Functions (respectively, Radon measure and distributions) on U are the 0-density functions (respectively, Radon measure and distributions). This justifies some of the notation introduced in the previous sections.

The absolute value of the determinant of the Jacobian matrix J_κ is "carried" by 1-density functions when changing local chart making them suitable for integration; 1-density distribution does the same making them suitable for acting on 0-functions. In the duality between a-density distributions and $(1-a)$-density functions, each term brings its part of J_κ, making this action consistent with integration, for instance, in the case the a-density distributions u identify with a-density functions, say L^1_{loc} in $U \setminus \partial\mathcal{M}$, that is, $u^{\mathcal{C}^i} = T_{f^{\mathcal{C}^i}}$ in each chart. In this latter case, we have

$$\langle u,\phi\rangle = \sum_{i\in\mathcal{I}} \int_{\kappa^i(O^i\cap U)} \tilde{\chi}^i(x) f^{\mathcal{C}^i}(x)\phi^{\mathcal{C}^i}(x)dx, \quad \phi\in {}^{1-a}\mathscr{D}^0_c(U).$$

An important case is $a = 1/2$. One then refers to half-density functions and distributions.

16.3. Differential Operators on Manifolds

16.3.1. Differential Operators. For $\mathcal{M}$ a smooth manifold and $\ell_1, \ell_2 \in \mathbb{R}$. An operator $P : \mathscr{C}^{\ell_1}(\mathcal{M}) \to \mathscr{C}^{\ell_2}(\mathcal{M})$ is said to be local if $\operatorname{supp}(Pf) \subset \operatorname{supp}(f)$ for all $f \in \mathscr{C}^{\ell_1}(\mathcal{M})$.

Let $k, \ell \in \mathbb{N}$. An operator $P : \mathscr{C}^{k+\ell}(\mathcal{M}) \to \mathscr{C}^{\ell}(\mathcal{M})$ is said to be differential of order k and of class $\mathscr{C}^\ell$ if

(1) P is local.
(2) In every chart $\mathcal{C} = (O, \kappa)$, the representative of the operator takes the form $P^{\mathcal{C}}(x, D) = \sum_{|\alpha|\leq k} a_\alpha(x) D^\alpha$ with $a_\alpha \in \mathscr{C}^\ell(\kappa(O))$, with $D = -i\partial$ as in the main text.

By representative, one means that $(\kappa^{-1})^* P \kappa^* g = P^{\mathcal{C}}(x, D)g$ for every $g \in \mathscr{C}^{k+\ell}(\kappa(O))$. Simple examples are $\mathscr{C}^\ell$-vector fields that are first-order differential operators and multiplications by $\mathscr{C}^\ell$-functions that are zero-order differential operators. We shall denote by $\mathscr{D}^k(\mathcal{M})$ the space of differential operators of order k with smooth coefficients.

REMARK 16.7. In fact, if P is a $\mathscr{C}^k$ first-order operator, it is given by the sum of a $\mathscr{C}^k$ vector field and a $\mathscr{C}^k$ function in each chart. With a partition of unity, we see that first-order differential operators on $\mathcal{M}$ are precisely the sums of vector fields and functions.

Local representatives fully characterize the differential operator as shown by the following proposition.

PROPOSITION 16.8. *Let $\mathcal{A} = (\mathcal{C}^i)_{i\in\mathcal{I}}$ be an atlas with $\mathcal{C}^i = (O^i, \kappa^i)$ and $k \in \mathbb{N}$. For each $i \in \mathcal{I}$, let $P^i(x, D)$ be a differential operator of order k with $\mathscr{C}^\ell$ coefficients such that*

$$P^i(x, D) = \kappa^* P^j(x, D)\big(\kappa^{-1}\big)^*,$$

on $\kappa^i(O^i \cap O^j)$, for $i, j \in \mathcal{I}$ such that $O^i \cap O^j \neq \emptyset$ and with $\kappa = \kappa^j \circ (\kappa^i)^{-1}$. Then, there exists a differential operator P of order k and of class $\mathscr{C}^\ell$ on $\mathcal{M}$ such that its representative in every chart $\mathcal{C}^i$ is precisely $P^i(x, D)$.

PROOF. Let $\varphi \in \mathscr{C}^{k+\ell}(\mathcal{M})$. For $i \in \mathcal{I}$, the representative of φ in $\mathcal{C}^i$ is $\varphi^{\mathcal{C}^i} = \big((\kappa^i)^{-1}\big)^*\varphi$. From the assumption made on the operators $P^i(x, D)$, $i \in \mathcal{I}$, one has

$$(\kappa^i)^* P^i(x, D)\varphi^{\mathcal{C}^i}(m) = (\kappa^j)^* P^j(x, D)\varphi^{\mathcal{C}^j}(m),$$

if $m \in O^i \cap O^j \neq \emptyset$. We may thus define unambitiously a function $P\varphi$ by

$$P\varphi(m) = (\kappa^i)^* P^i(x, D)\varphi^{\mathcal{C}^i}(m),$$

where $m \in O^i$ as the right-hand side is independent of the chosen chart that contains the point m. One sees that $P\varphi \in \mathscr{C}^\ell$ and that the operator $\varphi \mapsto P\varphi$ is local, and its representative in the each chart $\mathcal{C}^i$ is precisely P^i. ■

REMARK 16.9. Above, we have only considered differential operators with some smoothness. We may also consider differential operators with bounded coefficients on $\mathcal{M}$. They are local operators that take the form of a differential operator with bounded coefficients in every chart. In particular, with a partition of unity, we find that such a differential operator takes the form $\sum_{j\in\mathcal{J}} a_j v_j + b$, where the v_j are smooth vector fields and the a_j and b are bounded functions on $\mathcal{M}$ and the sum is locally finite.

16.3.2. Principal Symbol. In a local chart $\mathcal{C} = (O, \kappa)$, the symbol of $P^{\mathcal{C}}$ is given by $\sum_{|\alpha|\leq k} a_\alpha(x)\xi^\alpha$ and principal symbol by

$$p^{\mathcal{C}}(x,\xi) = \sigma(P^{\mathcal{C}})(x,\xi) = \sum_{|\alpha|=k} a_\alpha(x)\xi^\alpha.$$

Let us now consider two local charts $\mathcal{C}^\ell = (O^\ell, \kappa^\ell)$, $\ell = 1, 2$, and let $P^{\mathcal{C}^\ell}(x, D)$ be the representative of the operator P in $\mathcal{C}^\ell$. If $f \in \mathscr{C}^\infty(\mathcal{M})$ has support in $O = O^1 \cap O^2$, we have, by the local property of P,

$$(\kappa^1)^* P^{\mathcal{C}^1}(x,D)((\kappa^1)^{-1})^* f = (\kappa^2)^* P^{\mathcal{C}^2}(x,D)((\kappa^2)^{-1})^* f,$$

that is, for any function $g \in \mathscr{C}^\infty(\kappa^2(O))$,

$$P^{\mathcal{C}^1}(x,D)((\kappa^{12})^* g) = (\kappa^{12})^* P^{\mathcal{C}^2}(x,D) g, \qquad \kappa^{12} = \kappa^2 \circ (\kappa^1)^{-1}.$$

Understanding the local representation of differential operator on $\mathcal{M}$ thus amounts to understanding the action of a change of variables for a differential operator on an open set of $\mathbb{R}^d$. This is done in Section 9.1.2 of Volume 1. In particular, we find that the principal symbols of $P^{\mathcal{C}^1}(x, D)$ and $P^{\mathcal{C}^2}(x, D)$ satisfy the identity

$$p^{\mathcal{C}^1}(x, {}^t d\kappa^{12}(x)\eta) = p^{\mathcal{C}^2}(\kappa^{12}(x), \eta), \quad x \in O^1, \ \eta \in \mathbb{R}^d.$$

Then, by Corollary 15.19, we see that the principal symbol of a differential operator is a geometrical object, namely a function on the cotangent bundle $T^*\mathcal{M}$.

PROPOSITION 16.10. *There exists a map $\sigma(P) : T^*\mathcal{M} \to \mathbb{C}$, called the principal symbol of P, that yields, in any local chart $\mathcal{C}$, the principal symbol $p^{\mathcal{C}} = \sigma(P^{\mathcal{C}})$ of the representative $P^{\mathcal{C}}$ of the operator P.*

The following proposition yields a coordinate-free characterization of the principal symbol of P.

PROPOSITION 16.11. *Let P be a differential operator of order k on $\mathcal{M}$. Let $(m_0, \omega_0) \in T^*\mathcal{M}$. If $f \in \mathscr{C}^\infty(\mathcal{M}; \mathbb{R})$ is such that $df(m_0) = \omega_0$, we then have*

$$\sigma(P)(m_0, \omega_0) = \lim_{\lambda\to\infty} \lambda^{-k} e^{-i\lambda f(m_0)} \big(P e^{i\lambda f(m)}\big)\big|_{m=m_0}.$$

PROOF. Let us consider a local chart $\mathcal{C} = (O, \kappa)$ with $m_0 \in O$ and $P^{\mathcal{C}}(x, D) = \sum_{|\alpha| \le k} a_\alpha(x) D^\alpha$ the representative of P in this chart and $p^{\mathcal{C}}(x, \xi) = \sum_{|\alpha|=k} a_\alpha(x)\xi^\alpha$ its principal symbol. If $(x_0, \xi_0) = T^*\kappa(m_0, \omega_0)$, that is, $x_0 = \kappa(m_0)$ and $\omega_0 = {}^tT\kappa(m_0)(\xi_0)$, then $p^{\mathcal{C}}(x_0, \xi_0) = \sigma(P)(m_0, \omega_0)$ by the very definition of $\sigma(P)$. Let $g = (\kappa^{-1})^* f$ implying $e^{i\lambda g(x)} = (\kappa^{-1})^* e^{i\lambda f}(x)$. We have $e^{-i\lambda f(m_0)}\big(Pe^{i\lambda f(m)}\big) = e^{-i\lambda g(x_0)}\big(P^{\mathcal{C}}(x, D)e^{i\lambda g(x)}\big)$ for $x = \kappa(m)$. As $dg(x_0) = \xi_0$, it remains to prove that

$$p^{\mathcal{C}}(x_0, \xi_0) = \lim_{\lambda \to \infty} \lambda^{-k} e^{-i\lambda g(x_0)}\big(P^{\mathcal{C}}(x, D)e^{i\lambda g(x)}\big)\big|_{x=x_0}.$$

Having $\xi_0 = dg(x_0)$ reads $(\xi_0)_i = \partial_{x_i} g(x_0)$. We find that we have

$$\begin{aligned}\big(P^{\mathcal{C}}(x, D)e^{i\lambda g(x)}\big)\big|_{x=x_0} &= \lambda^k e^{i\lambda g(x_0)} \sum_{|\alpha|=k} a_\alpha(x_0)(\partial_{x_1} g(x_0))^{\alpha_1} \cdots (\partial_{x_d} g(x_0))^{\alpha_d} \\ &\quad + O(\lambda^{k-1}) \\ &= \lambda^k e^{i\lambda g(x_0)} \sum_{|\alpha|=k} a_\alpha(x_0)(\xi_0)^\alpha + O(\lambda^{k-1}),\end{aligned}$$

which yields the result. ∎

REMARK 16.12. In the case $\mathcal{M} = \mathbb{R}^d$, with Proposition 16.11, we recover the usual characterization of the principal symbol of a differential operator of order k:

$$\sigma(P)(x_0, \xi_0) = \lim_{\lambda \to \infty} \lambda^{-k} e^{-i\lambda x_0 \cdot \xi_0}\big(P(x, D)e^{i\lambda x \cdot \xi_0}\big)\big|_{x=x_0}.$$

PROPOSITION 16.13. *Let P be a differential operator of order k on $\mathcal{M}$. Let $m \in \mathcal{M}$ and $\omega, \tilde{\omega} \in T^*_m\mathcal{M}$. The map $t \mapsto \sigma(P)(m, t\omega + \tilde{\omega})$ is a polynomial function of degree k in t.*

PROOF. In fact, it suffices to consider a local chart $\mathcal{C} = (O, \kappa)$ with $m \in O$ and we have

$$\sigma(P)(m, t\omega + \tilde{\omega}) = \sigma(P^{\mathcal{C}})(x, t\xi + \tilde{\xi}) = \sum_{|\alpha|=k} a_\alpha(x)(t\xi + \tilde{\xi})^\alpha,$$

where $x = \kappa(m)$, $\xi_1 dx^1 + \cdots + \xi_d dx^d$ and $\tilde{\xi}_1 dx^1 + \cdots + \tilde{\xi}_d dx^d$ are the respective representatives of ω and $\tilde{\omega}$. It is then obvious that $t \mapsto \sigma(P)(m, t\omega + \tilde{\omega})$ is a polynomial function of degree k in t. ∎

If we have $\sigma(P)(m, t\omega + \tilde{\omega}) = \sum_{0 \le j \le k} p_j(m, \omega, \tilde{\omega})t^j$, then the Taylor formula at order k, with respect to t, holds exactly meaning

$$p_j(m, \omega, \tilde{\omega}) = \frac{1}{j!}\frac{d}{dt}^j \sigma(P)(m, t\omega + \tilde{\omega})_{|t=0}.$$

In particular, $p_0(m, \omega, \tilde{\omega}) = \sigma(P)(m, \tilde{\omega})$, and with the proof of the above proposition, we have

$$p_k(m, \omega, \tilde{\omega}) = \sigma(P)(m, \omega). \tag{16.3.1}$$

16.3.3. Transposition, Action on Distributions. In $\mathbb{R}^d$, the action of a differential operator on a distribution is done by duality. The same can be done on a manifold. This requires however to define the transpose of a differential operator.

PROPOSITION 16.14. *Let μ_0 be a smooth positive density function on $\mathcal{M}$. Let $P \in \mathscr{D}^k(\mathcal{M})$, that is, a differential operator of order k on $\mathcal{M}$ with smooth coefficients. There exists $Q \in \mathscr{D}^k(\mathcal{M})$ such that*

$$\int_{\mathcal{M}} \varphi(P\psi)\mu_0 = \int_{\mathcal{M}} (Q\varphi)\psi\mu_0$$

for all $\varphi, \psi \in {}^0\mathscr{D}_c^\infty(\mathcal{M})$.

One denotes ${}^tP = Q$ and says that Q is the transpose operator of P. Note in particular that ${}^t({}^tP) = P$. Note also that ${}^tP$ depends on the choice of the positive density function μ_0. In fact, μ_0 defines the duality bracket. In what follows, for simplicity, we shall always assume that μ_0 is smooth.

DEFINITION 16.15. Let $u \in {}^0\mathscr{D}'(\mathcal{M})$, that is, a 0-density distribution and $P \in \mathscr{D}^k(\mathcal{M})$. We denote by Pu the 0-density distribution given by

$$\langle Pu, \varphi\mu_0\rangle_{{}^0\mathscr{D}'(\mathcal{M}),{}^1\mathscr{D}_c^\infty(\mathcal{M})} = \langle u, ({}^tP\varphi)\mu_0\rangle_{{}^0\mathscr{D}'(\mathcal{M}),{}^1\mathscr{D}_c^\infty(\mathcal{M})}, \quad \varphi \in {}^0\mathscr{D}_c^\infty(\mathcal{M}).$$

PROOF OF PROPOSITION 16.14. Let $(\mathcal{C}^i)_{i\in\mathcal{I}}$ be an atlas with $\mathcal{C}^i = (O^i, \kappa^i)$. Consider first a local chart $\mathcal{C} = (O, \kappa)$ of the atlas and $\varphi \in \mathscr{C}^\infty(\mathcal{M})$. If $\psi \in {}^0\mathscr{D}_c^\infty(\mathcal{M})$ is supported in O, we write, for $f_0^{\mathcal{C}}(x)dx = \mu_0^{\mathcal{C}}$,

$$\int_{\mathcal{M}} \varphi(P\psi)\mu_0 = \int_{\kappa(O)} \varphi^{\mathcal{C}} P^{\mathcal{C}}(x,D)\psi^{\mathcal{C}} f_0^{\mathcal{C}}\, dx = \int_{\kappa(O)} (Q^{\mathcal{C}}(x,D)\varphi^{\mathcal{C}})\psi^{\mathcal{C}} f_0^{\mathcal{C}}\, dx,$$

with $Q^{\mathcal{C}}(x,D) = (f_0^{\mathcal{C}})^{-1}\, {}^tP^{\mathcal{C}}(x,D) f_0^{\mathcal{C}} \in \mathscr{D}^k(\kappa(O))$, where transposition is here to be understood with respect to the Lebesgue measure in $\mathbb{R}^d$.

If now ψ is supported in $O^i \cap O^j$, we have

$$\begin{aligned}\int_{\mathcal{M}} \varphi(P\psi)\mu_0 &= \int_{\kappa^i(O^i)} (Q^{\mathcal{C}^i}(x,D)\varphi^{\mathcal{C}^i})\psi^{\mathcal{C}^i} f_0^{\mathcal{C}^i}\, dx\\ &= \int_{\kappa^j(O^j)} (Q^{\mathcal{C}^j}(x,D)\varphi^{\mathcal{C}^j})\psi^{\mathcal{C}^j} f_0^{\mathcal{C}^j}\, dx,\end{aligned}$$

meaning that $Q^{\mathcal{C}^i}(x,D)\varphi^{\mathcal{C}^i} = \kappa^*\big(Q^{\mathcal{C}^j}(x,D)\varphi^{\mathcal{C}^j}\big)$ with $\kappa = \kappa^j \circ (\kappa^i)^{-1}$, meaning in turn

$$Q^{\mathcal{C}^i}(x,D) = \kappa^* Q^{\mathcal{C}^j}(x,D)\big(\kappa^{-1}\big)^*.$$

By Proposition 16.8, there exists $Q \in \mathscr{D}^k(\mathcal{M})$ with $Q^{\mathcal{C}^i}$ for representative in the chart $\mathcal{C}^i$.

For $\varphi \in \mathscr{C}^\infty(\mathcal{M})$ and $\psi \in {}^0\mathscr{D}_c^\infty(\mathcal{M})$, we obtain that

$$\int_{\mathcal{M}} \varphi(P\psi)\mu_0 = \int_{\mathcal{M}} (Q\varphi)\psi\mu_0, \tag{16.3.2}$$

if ψ is supported in a single chart. If ψ is now any function in ${}^0\mathscr{D}_c^\infty(\mathcal{M})$, a partition of unity subordinated to the covering of its support by the atlas shows that (16.3.2) also holds. ■

16.3.4. Differential Operators with a Large Parameter. Let $k \in \mathbb{N}$. An operator $P_\tau : \mathscr{C}^\infty(\mathcal{M}) \to \mathscr{C}^\infty(\mathcal{M})$, with $\tau > 0$, is said to be differential with a large parameter, with smooth coefficients and of order k, if

(1) P_τ is local.
(2) In every chart $\mathcal{C} = (O, \kappa)$, the representative of the operator is in $\mathscr{D}_\tau^k$, as introduced in Section 2.3 of Volume 1.

We denote by $\mathscr{D}_\tau^k(\mathcal{M})$ the set of all such operators.

We then have

$$P_\tau = \sum_{0 \leq j \leq k} \tau^{k-j} P_j,$$

where P_j, $j = 0, \dots, k$, is a differential operator of j on $\mathcal{M}$. The principal symbol of P_τ is defined as the function $T^*\mathcal{M}$ with τ as a parameter

$$\sigma(P_\tau)(m, \omega) = \sum_{0 \leq j \leq k} \tau^{k-j} \sigma(P_j)(m, \omega), \tag{16.3.3}$$

where $\sigma(P_j)$ denotes the principal symbol of P_j in the sense of the differential operators of order j on $\mathcal{M}$.[4]

The following proposition follows from Proposition 16.11 and yields a coordinate-free characterization of the principal symbol of P_τ.

PROPOSITION 16.16. *Let P_τ be a differential operator with a large parameter of order k on $\mathcal{M}$. Let $(m_0, \omega_0) \in T^*\mathcal{M}$. If $f \in \mathscr{C}^\infty(\mathcal{M}; \mathbb{R})$ is such that $df(m_0) = \omega_0$, we then have*

$$\sigma(P_\tau)(m_0, \omega_0) = \lim_{\lambda \to \infty} \lambda^{-k} e^{-i\lambda f(m_0)} \big(P_{\lambda\tau} e^{i\lambda f(m)}\big)\big|_{m=m_0}.$$

In each local chart, the representative of $\sigma(P_\tau)$ coincides with the principal symbol of the representative of P_τ as defined in Sections 2.2–2.3, in the case of differential operators, and as given in (2.2.3).

Consider two local charts $\mathcal{C}^\ell = (O^\ell, \kappa^\ell)$, $\ell = 1, 2$, and let $P^{\mathcal{C}^1}(x, D, \tau)$ and $P^{\mathcal{C}^2}(x, D, \tau)$ be the representatives of the operator P_τ in these two charts, using the notation of Section 2.3 of Volume 1. The principal symbols of $P^{\mathcal{C}^1}(x, D, \tau)$ and $P^{\mathcal{C}^2}(x, D, \tau)$ are then the representative of $\sigma(P_\tau)$. They thus satisfy the identity

$$\sigma(P^{\mathcal{C}^1})(x, {}^t d\kappa^{12}(x)\eta, \tau) = \sigma(P^{\mathcal{C}^2})(\kappa^{12}(x), \eta, \tau), \quad x \in O^1, \ \eta \in \mathbb{R}^d, \tag{16.3.4}$$

with $\kappa^{12} = \kappa^2 \circ (\kappa^1)^{-1}$.

[4] If P_j is in fact of order less than or equal to $j - 1$, then $\sigma(P_j) = 0$.

An important example for us here is the case of a conjugated differential operator on $\mathcal{M}$.

PROPOSITION 16.17. *Let P be a differential operator of order k on $\mathcal{M}$ and φ be a smooth function on $\mathcal{M}$. Let $P_\varphi = e^{\tau\varphi} P e^{-\tau\varphi}$. It is a differential operator with a large parameter of order k on $\mathcal{M}$, and its principal symbol is given by*

$$\sigma(P_\varphi)(m, \omega, \tau) = \sigma(P)(m, \omega + i\tau d\varphi(m)), \tag{16.3.5}$$

where $\sigma(P)$ is the principal symbol of P as defined in Section 16.3.1.

Note that (16.3.5) makes sense as $t \mapsto \sigma(P)(m, \omega + t d\varphi(m))$ is polynomial by Proposition 16.13.

PROOF. In a local chart $\mathcal{C} = (O, \kappa)$, if $P^{\mathcal{C}}(x, D)$, $P^{\mathcal{C}}_\varphi(x, D, \tau)$, and $\varphi^{\mathcal{C}}$ are the representatives of P, P_φ, and φ, we have $P^{\mathcal{C}}_\varphi = e^{\tau\varphi^{\mathcal{C}}} P^{\mathcal{C}} e^{-\tau\varphi^{\mathcal{C}}}$. The discussion in the beginning of Section 9.3.1 of Volume 1 shows that $P^{\mathcal{C}}_\varphi$ is a differential operator with a large parameter of order k on $\kappa(O)$. Its principal symbol is given by $\sigma(P^{\mathcal{C}}_\varphi)(x, \xi, \tau) = \sigma(P^{\mathcal{C}})(x, \xi + i\tau d\varphi^{\mathcal{C}}(x))$. As P_φ is local, this precisely means that it is a differential operator with a large parameter of order k on $\mathcal{M}$ and that we have (16.3.5). ■

Note that we could have also used either the formula of change of variables for conjugated operators as provided in Section 9.3.1 or Proposition 16.16.

CHAPTER 17

Elements of Riemannian Geometry

Contents

This chapter is devoted to the exposition of some facts of Riemannian geometry. Basic aspects of differential geometry are exposed in Chap. 15. One goal is to properly define the Laplace–Beltrami operator. As second goal is to give the necessary tools to prepare for Chap. 18 where Sobolev spaces on a Riemannian manifold are presented and where the elliptic properties of the Laplace–Beltrami operator are studied.

J. Le Rousseau et al., *Elliptic Carleman Estimates and Applications to Stabilization and Controllability, Volume II*, PNLDE Subseries in Control 98, https://doi.org/10.1007/978-3-030-88670-7_17

An interested reader should consult a more thorough reference, such as [120, 189, 221] for Riemannian geometry.

17.1. Riemannian Structure on a Manifold

We consider a smooth manifold $\mathcal{M}$ of dimension d. Having a Riemannian structure on $\mathcal{M}$ means having a way to measure distances on the manifold.

17.1.1. Metric and Distance. On $\mathbb{R}^d$, the Euclidean distance or anisotropic versions of it can be expressed by a quadratic form. For manifolds, this is generalized by providing a quadratic form at every point $m \in \mathcal{M}$, with some smoothness requirement. A quadratic form in $\mathbb{R}^d$ can be represented by its associated bilinear form that acts on vectors of $\mathbb{R}^d$. The same is done on $\mathcal{M}$ with an action on tangent vectors at m.

DEFINITION 17.1. A Riemannian metric is a smooth 2-covariant tensor field g on $\mathcal{M}$ that satisfies the following properties, for any $m \in \mathcal{M}$ and $v, w \in T_m\mathcal{M}$,

(1) symmetry: we have $g_m(v, w) = g_m(w, v)$;
(2) positivity: we have $g_m(v, v) > 0$, for $v \neq 0$.

A couple $(\mathcal{M}, g)$ is called a Riemannian manifold. The metric g thus yields for each $m \in \mathcal{M}$ an inner product on $T_m\mathcal{M}$ and the following associated norm

$$|v|_{g_m} = g_m(v, v)^{1/2}, \quad v \in T_m\mathcal{M}.$$

Having such a metric allows one to define the length of a $\mathscr{C}^1$-path $\gamma : [a, b] \to \mathcal{M}$:

$$\ell(\gamma) = \int_a^b g_{\gamma(t)}(\dot\gamma(t), \dot\gamma(t))^{1/2} dt, \qquad \text{where } \dot\gamma = \frac{d}{dt}\gamma, \tag{17.1.1}$$

that is independent of the chosen (injective) parametrization γ. For $\mathcal{M}$ connected, which we shall always assume here, we define the distance

$$\operatorname{dist}_g(x, y) = \inf_{\gamma \in \mathcal{P}(x,y)} \ell(\gamma), \quad x, y \in \mathcal{M}, \tag{17.1.2}$$

where $\mathcal{P}(x, y)$ is the set of all the $\mathscr{C}^1$-paths $\gamma : [0, 1] \to \mathcal{M}$ such that $\gamma(0) = x$ and $\gamma(1) = y$. This distance yields a metric space structure for $(\mathcal{M}, g)$ and the induced topology is equivalent to the manifold topology. Connected to the notion of distance is that of geodesics. They are covered in Sect. 17.5.

17.1.2. Representation in Local Charts. In a local chart $\mathcal{C} = (O, \kappa)$, for every $m \in \mathcal{M}$ and $x = \kappa(m)$, the local representative of g_m, that is, $g_x^{\mathcal{C}} = (\kappa^{-1})^* g_m$ identifies with a symmetric positive definite matrix $\big(g_{ij}^{\mathcal{C}}(x)\big)_{1 \leq i,j \leq d}$ and

$$g_m(v, w) = \sum_{1 \leq i,j \leq d} g_{ij}^{\mathcal{C}}(x) v^i w^j,$$

where $(v^i)_{1\le i\le d}$ and $(w^i)_{1\le i\le d}$ are the coordinates of the representatives of v and w in the local chart, that is,

$$g^{\mathcal{C}}(x) = g_x^{\mathcal{C}} = \sum_{1\le i,j\le d} g_{ij}^{\mathcal{C}}(x) dx^i \otimes dx^j .$$

We shall most often use the notation $g_x^{\mathcal{C}}$ close to the notation we use for covariant tensors but at places we shall use $g^{\mathcal{C}}(x)$ if the symbol of the metric has a subscript.

In the case of the Euclidean space $\mathbb{R}^d$, the metric is simply

$$\sum_{1\le i\le d} dx^i \otimes dx^i,$$

often called the *flat* metric.

Let $\mathcal{C}^k = (O^k, \kappa^k)$, $k = 1, 2$, be two local charts such that $m \in O^1 \cap O^2$. For $x^{(k)} = \kappa^k(m)$ let $g_{x^{(k)}}^{\mathcal{C}^k} = \big(g_{ij}^{\mathcal{C}^k}(x^{(k)})\big)_{1\le i,j\le d}$ be the local matrix representatives of g in $\mathcal{C}^k$. The formula expressing the action of a change of coordinates on 2-covariant tensor fields (15.6.2) gives

$$g_{ij}^{\mathcal{C}^1}(x^{(1)}) = \sum_{1\le \ell,q\le d} \big(\partial_{x_i}\kappa_\ell \partial_{x_j}\kappa_q\big)(x^{(1)})\, g_{\ell q}^{\mathcal{C}^2}(x^{(2)}),$$

with $\kappa = \kappa^2 \circ (\kappa^1)^{-1}$. In matrix form this reads

$$g_{x^{(1)}}^{\mathcal{C}^1} = {}^t\kappa'(x^{(1)})\, g_{x^{(2)}}^{\mathcal{C}^2}\, \kappa'(x^{(1)}).$$

In particular, this gives

$$\big(\det g_{x^{(1)}}^{\mathcal{C}^1}\big)^{1/2} = |\det \kappa'(x^{(1)})| \big(\det g_{x^{(2)}}^{\mathcal{C}^2}\big)^{1/2}. \tag{17.1.3}$$

17.1.3. Musical Isomorphisms. For $\omega \in T_m^*\mathcal{M}$ we define $\omega^\sharp \in T_m\mathcal{M}$ by

$$\langle \omega, v\rangle = g_m(\omega^\sharp, v), \quad v \in T_m\mathcal{M}.$$

The map $\sharp : T_m^*\mathcal{M} \to T_m\mathcal{M}$ and its inverse $\flat : T_m\mathcal{M} \to T_m^*\mathcal{M}$ are called musical isomorphisms. In a local chart $\mathcal{C} = (O, \kappa)$ with $m \in O$ and $x = \kappa(m)$, this reads

$$(\omega^\sharp)^i = \sum_{1\le j\le d} g^{\mathcal{C},ij}(x)\omega_j,$$

where $(g^{\mathcal{C},ij}(x))_{1\le i,j\le d}$ is the inverse of the matrix $(g_{ij}^{\mathcal{C}}(x))_{1\le i,j\le d}$. Similarly, for $v \in T_m\mathcal{M}$ we have

$$(v^\flat)_i = \sum_{1\le j\le d} g_{ij}^{\mathcal{C}}(x)v^j.$$

We then define the following bilinear form and associated norm on $T_m^*\mathcal{M}$ (and more generally on $T^*\mathcal{M}$)

$$(\omega, \omega')_{g_m} = g_m(\omega^\sharp, \omega'^\sharp) = \langle \omega, \omega'^\sharp\rangle, \qquad |\omega|_{g_m}^2 = (\omega, \omega)_{g_m}, \tag{17.1.4}$$

for $\omega, \omega' \in T^*_m\mathcal{M}$. In local coordinates, we have

$$(\omega, \omega')_{g_m} = \sum_{1 \le i,j \le d} g^{\mathcal{C},ij}(x)\omega_i\omega'_j.$$

The musical isomorphism $\sharp$ and $\flat$ are introduced here for cotangent and tangent vectors, respectively; their action is naturally extended to one-forms and vector fields.

17.1.4. Volume Form and Integration. The Riemannian structure on $\mathcal{M}$ allows one to choose a particular volume form if $\mathcal{M}$ is orientable. Let us consider an atlas $\mathcal{A} = (O^i, \kappa^i)_{i \in \mathcal{I}}$ for which $\mathcal{M}$ is oriented. In any local chart $\mathcal{C}^i = (O^i, \kappa^i)$ of this atlas, we consider

$$dv^i(x) = (\det g_x^{\mathcal{C}^i})^{1/2} dx^1 \wedge \cdots \wedge dx^d. \tag{17.1.5}$$

By (15.6.7) and (17.1.3) we find that

$$(\kappa^{ij})^* dv^j = dv^i, \qquad \kappa^{ij} = \kappa^j \circ (\kappa^i)^{-1}, \tag{17.1.6}$$

as $\det(d\kappa^{ij}(x)) > 0$ here. For $m \in \mathcal{M}$, we can thus set

$$(dv_g)_m = (\kappa^i)^* dv^i(x) \in \mathscr{A}^d_m\mathcal{M}, \quad x = \kappa^i(m), \tag{17.1.7}$$

if $m \in O^i$, as this d-form at m is independent of the choice of $i \in \mathcal{I}$ by (17.1.6). The d-form $dv_g : m \mapsto (m, (dv_g)_m)$ has the properties of a volume form (see Sect. 16.1). We call it the canonical Riemannian volume form.

If $\mathcal{N}$ is a submanifold of $\mathcal{M}$, the metric g induces a smooth 2-covariant tensor field $g_{\mathcal{N}}$ on $\mathcal{N}$ since $T_m\mathcal{N}$ naturally injects in $T_m\mathcal{M}$ (see Sect. 15.5). One can readily check that the properties of Definition 17.1 are fulfilled, meaning that $g_{\mathcal{N}}$ is a Riemannian metric on $\mathcal{N}$. If $\mathcal{N}$ is orientable this metric can yield a canonical volume form once an orientation is chosen.

The particular case of the boundary $\partial\mathcal{M}$ of $\mathcal{M}$ in the case $\mathcal{M}$ is oriented is of interest. In fact, once a smooth outward pointing vector field ν is chosen along $\partial\mathcal{M}$, the canonical volume form dv_g on $\mathcal{M}$ (associated with the orientation of $\mathcal{M}$, see above) yields a volume form $(dv_g)_\partial$ on $\partial\mathcal{M}$ and thus an orientation (see Sect. 16.1). However, the induced metric g_∂ along with this orientation yields a canonical volume form dv_{g_∂} on $\partial\mathcal{M}$. These two volume forms coincide for the following natural choice of the vector field ν along $\partial\mathcal{M}$.

PROPOSITION 17.2. *Let ν be the unique outward pointing vector field along $\partial\mathcal{M}$ such that, for all $m \in \partial\mathcal{M}$, $g_m(\nu_m, \nu_m) = 1$ and $g_m(\nu_m, u) = 0$ for all $u \in T_m\partial\mathcal{M}$, that is, ν is unitary and orthogonal to $T_m\partial\mathcal{M}$ in the sense of g. Let the volume form $(dv_g)_\partial$ on $\partial\mathcal{M}$ be given at m by*

$$(dv_g)_\partial(u_1, \dots, u_{d-1}) = dv_g(\nu, u_1, \dots, u_{d-1}), \qquad u_1, \dots, u_{d-1} \in T_m\partial\mathcal{M}.$$

Denote by dv_{g_∂} the canonical Riemannian volume form associated with the induced metric g_∂ on $\partial\mathcal{M}$ and with the orientation of $\partial\mathcal{M}$ given by the volume form $(dv_g)_\partial$.

(1) *We have* $(dv_g)_\partial = dv_{g_\partial}$.
Moreover, if $\mathcal{C} = (O, \kappa)$ *is a local chart such that* $O \cap \partial\mathcal{M} \neq \emptyset$ *and* $U_+ = \kappa(O)$ *is an open set of* $\overline{\mathbb{R}^d_+}$, *we have, writing* $\kappa(m) = x = (x', x_d)$,

(2) $\big(\det g^{\mathcal{C}}_{(x',0)}\big)^{1/2} = \big(\det g^{\mathcal{C}}_{\partial}(x')\big)^{1/2} / |\nu^{\mathcal{C},d}(x',0)|$;

(3) $g_m(u,\nu) = -u^{\mathcal{C},d}/|\nu^{\mathcal{C},d}(x',0)|$ *if* $u \in T_m\mathcal{M}$ *and* $m \in \partial\mathcal{M}$.

Proof. To prove the first item, it suffices to prove that $|(dv_g)_\partial| = |dv_{g_\partial}|$ since both volume forms share the same orientation. We choose a local chart $\mathcal{C} = (O, \kappa)$ near $m \in \partial\mathcal{M}$ where $\mathcal{M}$ is given by $\{x_d \geq 0\}$, that is $U_+ = \kappa(O)$ is an open set of $\overline{\mathbb{R}^d_+}$. We set $(x',0) = \kappa(m)$. We set $\partial U_+ = \{x_d = 0\} \cap U_+$. We have $T_{(x',0)}U_+ \cong \mathbb{R}^d$ and let $(e_j)_{1 \leq j \leq d}$ be the canonical basis of $\mathbb{R}^d$. As $T_{(x',0)}\partial U_+ \cong \mathbb{R}^{d-1}$ (canonically) injects in $T_{(x',0)}U_+ \cong \mathbb{R}^d$, a vector $u' \in T_{(x',0)}\partial U_+$ viewed as a vector u of $\mathbb{R}^d$, that is $u = (u^1, \dots, u^d)$ is characterized by $u^d = 0$. Then, $(e_j)_{1 \leq j \leq d-1}$ is a basis of $T_{(x',0)}\partial U_+$. For such a vector $u = (u^1, \dots, u^{d-1}, 0) \in T_{(x',0)}\partial U_+$ we shall write $u' = (u^1, \dots, u^{d-1})$ the associated vector in $\mathbb{R}^{d-1}$.

In what follows we simply write $\nu^{\mathcal{C}}$ in place of $\nu^{\mathcal{C}}(x',0)$. The tangent vector $\nu^{\mathcal{C}} = (\nu^1, \dots, \nu^d)$ is such that $g^{\mathcal{C}}_{(x',0)}(\nu^{\mathcal{C}}, \nu^{\mathcal{C}}) = 1$ and

$$g^{\mathcal{C}}_{(x',0)}(\nu^{\mathcal{C}}, u) = 0, \quad u = (u',0).$$

The matrix $(G_{i,j})_{1 \leq i,j \leq d}$ of the bilinear form $g^{\mathcal{C}}_{(x',0)}(.,.)$ in this basis given by $G_{i,j} = g^{\mathcal{C}}_{ij}(x',0) = g^{\mathcal{C}}_{(x',0)}(e_i, e_j)$. We denote by $(\tilde{G}_{i,j})_{1 \leq i,j \leq d-1}$ the submatrix given by $\tilde{G}_{i,j} = G_{i,j}$, $1 \leq i,j \leq d-1$, that is,

$$G = \left(\begin{array}{ccc|c} & & & G_{1,d} \\ & \tilde{G} & & \vdots \\ & & & G_{d-1,d} \\ \hline G_{1,d} & \cdots & G_{d-1,d} & G_{d,d} \end{array}\right). \tag{17.1.8}$$

At $(x',0)$, we have $dv^{\mathcal{C}}_g = \sqrt{\det(G)} dx^1 \wedge \cdots \wedge dx^d$. Then, on the one hand, at $(x',0)$ we have, for $u'_1, \dots, u'_{d-1} \in T_{(x',0)}\partial U_+$, from Sect. 16.1,

$$\begin{aligned}(dv_g)^{\mathcal{C}}_\partial(u'_1, \dots, u'_{d-1}) &= dv^{\mathcal{C}}_g(\nu^{\mathcal{C}}, u_1, \dots, u_{d-1}) = \sqrt{\det(G)} \det(\nu^{\mathcal{C}}, u_1, \dots, u_{d-1}) \\ &= (-1)^{d+1}\nu^d \sqrt{\det(G)} \det(u'_1, \dots, u'_{d-1}),\end{aligned}$$

since $u^d_1 = \cdots = u^d_{d-1} = 0$. On the other hand, the matrix $\tilde{G}$ is that of the induced metric g_∂ on $\partial\mathcal{M}$ at m. We thus find that the associated volume form dv_{g_∂} on $\partial\mathcal{M}$ is such that $|dv^{\mathcal{C}}_{g_\partial}| = \sqrt{\det(\tilde{G})} |dx^1 \wedge \cdots \wedge dx^{d-1}|$ at $(x',0)$, implying, for $u'_1, \dots, u'_{d-1} \in T_{(x',0)}\partial U_+$,

$$|dv^{\mathcal{C}}_{g_\partial}|(u'_1, \dots, u'_{d-1}) = \sqrt{\det(\tilde{G})}\, |\det(u'_1, \dots, u'_{d-1})|.$$

It thus remains to prove that

(17.1.9) $$\det(\tilde{G}) = \left(\nu^d\right)^2 \det(G),$$

which is precisely the second item of the proposition.

To that purpose, we consider the basis $(e_1, \dots, e_{d-1}, \nu)$ of $\mathbb{R}^d$. The matrix $(H_{i,j})_{1\le i,j\le d}$ of the bilinear form $g^{\mathcal{C}}_{(x',0)}(.,.)$ in this second basis and the invertible matrix P representing the change of basis are

$$H = \begin{pmatrix} \tilde{G} & \begin{matrix} 0 \\ \vdots \\ 0 \end{matrix} \\ \begin{matrix} 0 & \cdots & 0 \end{matrix} & 1 \end{pmatrix}, \qquad P = \begin{pmatrix} \mathrm{Id}_{d-1} & \begin{matrix} \nu^1 \\ \vdots \\ \nu^{d-1} \end{matrix} \\ \begin{matrix} 0 & \cdots & 0 \end{matrix} & \nu^d \end{pmatrix}.$$

(17.1.10)

We have $H = {}^tP\, G\, P$ yielding $\det(\tilde{G}) = \left(\nu^d\right)^2 \det(G)$.

We now prove the last item of the proposition. We note that $|\nu^d| = -\nu^d$ as the vector field ν is outward pointing (see Sect. 16.1.3). We thus need to prove that $u^d/\nu^d = g^{\mathcal{C}}(\nu^{\mathcal{C}}, u)$ for any $u \in \mathbb{R}^d$. Let $\underline{u}$ be the column vector formed by the coordinates of u in the second basis $(e^1, \dots, e^{d-1}, \nu^{\mathcal{C}})$. We have ${}^tu = P\underline{u}$ yielding $u^d = \nu^d \underline{u}^d$. Next, using the second basis we find $g^{\mathcal{C}}(\nu^{\mathcal{C}}, u) = (0, \dots, 0, 1) H \underline{u} = \underline{u}^d$. ∎

REMARK 17.3. Note that this proof can be simplified by using normal geodesic coordinates as introduced in Section 9.4 of Volume 1. However, the proof of the existence of such coordinates is more involved than the above argument that is only based on linear algebra.

17.2. Gradient, Divergence, and Laplace–Beltrami Operators

In the Euclidean space we have $\nabla f(m) \cdot v = df(m)(v) = v(f)$, for any vector v at m, yielding a definition of the gradient vector. On $(\mathcal{M}, g)$, this definition remains unchanged, yet replacing the inner product by the bilinear form g. Hence the gradient at m of a $\mathscr{C}^k$-function $f : \mathcal{M} \to \mathbb{C}$ at m is defined as the unique $\nabla_g f_m \in T_m\mathcal{M}$ such that

(17.2.1) $$g_m(\nabla_g f_m, v) = v(f) = df(m)(v), \quad \forall v \in T_m\mathcal{M}.$$

Letting m vary we obtain a $\mathscr{C}^{k-1}$-vector field $\nabla_g f : m \mapsto \nabla_g f_m$ on $\mathcal{M}$. In a local chart $\mathcal{C} = (O, \kappa)$, the representative of $\nabla_g f_m$ is given by, with $x = \kappa(m)$,

(17.2.2) $$(\nabla_g f_m)^{\mathcal{C},i} = \sum_{1\le j\le d} g^{\mathcal{C},ij}(x) \partial_{x_j}((\kappa^{-1})^* f)(x),$$

where $(g^{\mathcal{C},ij}(x))_{1\le i,j\le d}$ is the inverse of the matrix $(g^{\mathcal{C}}_{ij}(x))_{1\le i,j\le d}$ of the local representative of the metric. With the musical notation (see Sect. 17.1.3) we have $\nabla_g f_m = df(m)^\sharp$.

For two smooth functions f_1 and f_2 one has

(17.2.3) $$\nabla_g(f_1 f_2) = f_1 \nabla_g f_2 + f_2 \nabla_g f_1.$$

For a smooth vector field u on an oriented Riemannian manifold $\mathcal{M}$ we define its divergence $\operatorname{div}_g u$ as the unique function that satisfies

$$\int_{\mathcal{M}} \varphi \operatorname{div}_g u \, dv_g = - \int_{\mathcal{M}} g(\nabla_g \varphi, u) dv_g, \tag{17.2.4}$$

for any $\varphi \in \mathscr{D}^\infty(\mathcal{M})$, with the volume form dv_g defined above. In a local chart $\mathcal{C} = (O, \kappa)$, if $u^i(x)$ are the coordinates of the local representative of u, $u^{\mathcal{C}} = \kappa^* u$, we have, by (17.1.5) and (17.1.7),

$$\kappa^* \operatorname{div}_g u(x) = (\det g_x^{\mathcal{C}})^{-1/2} \sum_{1 \le i \le d} \partial_{x_i} \big((\det g_x^{\mathcal{C}})^{1/2} u^i \big)(x). \tag{17.2.5}$$

From (17.2.3) and (17.2.4) we find, for a smooth function f and a smooth vector field u,

$$\operatorname{div}_g (fu) = f \operatorname{div}_g u + g(\nabla_g f, u) = f \operatorname{div}_g u + u(f). \tag{17.2.6}$$

The Laplace–Beltrami operator acting on $\mathscr{C}^2$-functions is then defined as the composition of the gradient and the divergence operators:

$$\Delta_g f = \operatorname{div}_g \nabla_g f,$$

yielding a function on $\mathcal{M}$. In particular, if f_1 and f_2 are two $\mathscr{C}^2$-functions supported away from the boundary of $\mathcal{M}$, from (17.2.4) one finds

$$\int_{\mathcal{M}} f_1 \Delta_g f_2 dv_g = - \int_{M} g(\nabla_g f_1, \nabla_g f_2) dv_g = \int_{\mathcal{M}} (\Delta_g f_1) f_2 dv_g. \tag{17.2.7}$$

In a local chart $\mathcal{C} = (O, \kappa)$, from (17.2.2) and (17.2.5), we have

$$\begin{aligned} &(\kappa^{-1})^* (\Delta_g f)(x) \\ &\quad = (\det g_x^{\mathcal{C}})^{-1/2} \sum_{1 \le i,j \le d} \partial_{x_i} \big((\det g_x^{\mathcal{C}})^{1/2} g^{\mathcal{C},ij}(x) \partial_{x_j} ((\kappa^{-1})^* f) \big)(x). \end{aligned} \tag{17.2.8}$$

One sees that the Laplace–Beltrami operator fulfills the definition of differential operator on $\mathcal{M}$ in the sense of Sect. 16.3. The principal symbol of Δ_g is a function on $T^*\mathcal{M}$ by Proposition 16.10. It is given by $\sigma(\Delta_g)(m, \omega) = -|\omega|^2_{g_m}$, with $|.|_{g_m}$ as defined in (17.1.4). In the local chart $\mathcal{C}$, it is given by

$$- \sum_{1 \le i,j \le d} g^{\mathcal{C},ij}(x) \xi_i \xi_j.$$

Note that the local representative of the Laplace–Beltrami operator is often written

$$\Delta_g f = (\det g)^{-1/2} \sum_{1 \le i,j \le d} \partial_{x_i} \big((\det g)^{1/2} g^{ij} \partial_{x_j} f \big),$$

by abuse of notation (compare with (17.2.8)). With a similar abuse of notation for the gradient operator one writes

$$(\nabla_g f)^i = \sum_{1 \le j \le d} g^{ij} \partial_{x_j} f, \quad i = 1, \dots, d.$$

(See, for instance, Chap 5).

When considering smooth vector fields, we obtain the classical divergence theorem.

PROPOSITION 17.4. *Let* $(\mathcal{M}, g)$ *be an oriented Riemannian manifold with boundary and let* u *be a smooth compactly supported vector field on* $\mathcal{M}$*. Let* ν *be the unique outward pointing vector field along* $\partial\mathcal{M}$ *such that, for all* $m \in \partial\mathcal{M}$, $g_m(\nu_m, \nu_m) = 1$ *and* $g_m(\nu_m, u) = 0$ *for all* $u \in T_m\partial\mathcal{M}$*, that is,* ν *is unitary and orthogonal to* $T_m\partial\mathcal{M}$ *in the sense of* g*. We have*

$$\int_{\mathcal{M}} \operatorname{div}_g(u) dv_g = \int_{\partial\mathcal{M}} g(u, \nu)(dv_g)_\partial.$$

Compare with (16.1.6) that states this result in the flat case.

PROOF. Arguing as in the beginning of the proof of the Stokes formula (Theorem 16.2), it suffices the consider the vector field u supported in a single local chart $\mathcal{C} = (O, \kappa)$ that meets the boundary in an atlas compatible with the chosen orientation of $\mathcal{M}$. We set $U_+ = \kappa(O)$. It is an open set of $\overline{\mathbb{R}^d_+}$, that is, we have $U_+ = \overline{\mathbb{R}^d_+} \cap U$ with U an open set of $\mathbb{R}^d$.

By abuse of notation, we write u, g, and dv_g to denote the local representatives of the vector field, the metric and the volume form.

We then have

$$\int_{\mathcal{M}} \operatorname{div}_g(u) dv_g = \int_{U_+} \operatorname{div}_g(u)(\det g)^{1/2} dx = \int_{U_+} \operatorname{div}((\det g)^{1/2} u^j) dx.$$

Applying the divergence formula in $\mathbb{R}^d$, see (16.1.6), we obtain

$$\begin{aligned}\int_{\mathcal{M}} \operatorname{div}_g(u) dv_g &= \int_{\partial U_+} (-1)^{d-1} (\det g)^{1/2} u^d dx^1 \wedge \cdots \wedge dx^{d-1} \\ &= -\int_{\partial U_+} u^d (\det g)^{1/2} dx'.\end{aligned}$$

We have $(\det g)^{1/2}(x', 0) = (\det g_\partial(x'))^{1/2} / |\nu^d(x', 0)|$ where g_∂ is the restriction of the metric to ∂U_+ by the second item of Proposition 17.2, yielding

$$\int_{\mathcal{M}} \operatorname{div}_g(u) dv_g = -\int_{\partial U_+} \frac{u^d}{|\nu^d|} (\det g_\partial)^{1/2} dx'.$$

The third item of Proposition 17.2 states $-u^d/|\nu^d| = g(u, \nu)$ implying

$$\begin{aligned}\int_{\mathcal{M}} \operatorname{div}_g(u) dv_g &= \int_{\partial U_+} g(u, \nu)(\det g_\partial)^{1/2} dx' = \int_{\partial\mathcal{M}} g(u, \nu)(dv_{g_\partial}) \\ &= \int_{\partial\mathcal{M}} g(u, \nu)(dv_g)_\partial,\end{aligned}$$

using the first item of Proposition 17.2 for the two equivalent formulations of the volume form $(dv_g)_\partial$, which is the sought result. ∎

We finish this section by the observation that the gradient, the divergence, and the Laplace–Beltrami operators are truly of geometrical nature observing how they transform through the action of a diffeomorphism.

PROPOSITION 17.5. *Let* $(\mathcal{M}_1, g_1)$ *and* $(\mathcal{M}_2, g_2)$ *be two oriented Riemannian manifolds and* $\phi : \mathcal{M}_1 \to \mathcal{M}_2$ *a diffeomorphism such that* $\phi^* g_2 = g_1$. *Then, we have*

$$\kappa_* \nabla_{g_1} \kappa^* = \nabla_{g_2}, \qquad \operatorname{div}_{g_1} = \kappa^* \operatorname{div}_{g_2} \kappa_*, \qquad \Delta_{g_1} \kappa^* = \kappa^* \Delta_{g_2}.$$

17.3. Canonical Positive Density Function and Divergence Formula

Consider an atlas $\mathcal{A} = (O^i, \kappa^i)_{i \in \mathcal{I}}$. In any local chart $\mathcal{C}^i = (O^i, \kappa^i)$, we define the function

$$f^{\mathcal{C}^i}(x) = (\det g^{\mathcal{C}^i})^{1/2} \text{ on } \kappa^i(O^i).$$

By (17.1.3), we see that the change of variable formula (16.2.4) for density functions is satisfied, thus yielding by (16.2.6) a smooth positive density function μ_g such that its local representative is precisely the functions $f^{\mathcal{C}^i}$, that is, $\mu_g^{\mathcal{C}^i} = T_{f^{\mathcal{C}^i}}$ (using the notation of Chapter 8 of Volume 1 for L^1_{loc}-functions identified with distributions). Such a positive density function allows one to identify functions and density functions on $\mathcal{M}$ through the bijective map $\phi \mapsto \phi \mu_g$ in a canonical way.

The canonical density function allows to define L^1-functions (resp. L^1_{loc}) on $\mathcal{M}$ as the functions f such that $f \mu_g$ is an L^1-density functions (resp. L^1_{loc}). We denote by $L^1(\mathcal{M})$ (resp. $L^1_{\text{loc}}(\mathcal{M})$) the space of such functions.

Above, the divergence of a vector field was defined through the integration of d-forms on an oriented Riemannian manifold. On a nonoriented Riemannian manifold, the integration of densities allows one to have a similar definition that yields the same form in local coordinates. Indeed, for u a $\mathscr{C}^1$-vector field, the map

$$S_u : \phi \mapsto -\int_{\mathcal{M}} u(\phi) \mu_g = -\int_{\mathcal{M}} g(\nabla_g \phi, u) \mu_g, \quad \phi \in {}^0\mathscr{D}_c^\infty(\mathcal{M}),$$

is a density distribution. A computation shows that S_u is in fact a density function with $S_u = \operatorname{div}_g u \, \mu_g$ with $\operatorname{div}_g u \in \mathscr{C}^0(\mathcal{M})$ given in local charts by (17.2.5). This yields the following identity defining $\operatorname{div}_g u$ on a nonoriented manifold

$$\int_{\mathcal{M}} \phi \operatorname{div}_g u \, \mu_g = -\int_{\mathcal{M}} u(\phi) \mu_g, \qquad \phi \in {}^0\mathscr{D}_c^\infty(\mathcal{M}).$$

With the canonical positive density μ_g and this definition of the divergence we can state the following form of the divergence theorem in the context of integration of densities on nonoriented manifolds.

PROPOSITION 17.6 (Divergence Formula). *Let* $(\mathcal{M}, g)$ *be a Riemannian manifold with boundary and let* u *be a smooth compactly supported vector field on* $\mathcal{M}$. *Let* ν *be the unique outward pointing vector field along* $\partial\mathcal{M}$ *such that, for all* $m \in \partial\mathcal{M}$, $g_m(\nu_m, \nu_m) = 1$ *and* $g_m(\nu_m, v) = 0$ *for all* $v \in T_m\partial\mathcal{M}$, *that is,* ν *is unitary and orthogonal to* $T_m\partial\mathcal{M}$ *in the sense of* g. *We have*

$$\int_{\mathcal{M}} \operatorname{div}_g(u)\mu_g = \int_{\partial\mathcal{M}} g(u,\nu)\mu_{g_\partial},$$

where g_∂ is the induced Riemannian metric on $\partial\mathcal{M}$ and μ_{g_∂} the resulting canonical positive density function on $\partial\mathcal{M}$.

We insist on the fact that here, as opposed to Proposition 17.4, $\mathcal{M}$ and $\partial\mathcal{M}$ need not be oriented.

PROOF. Arguing as in the beginning of the proof of Theorem 16.2, it suffices the consider the vector field u supported in a single local chart $\mathcal{C} = (O, \kappa)$ that meets the boundary. In this single chart, an orientation can be chosen, allowing to identify density functions and d-forms (see the last paragraph of Sect. 16.2.4). Then the result of Proposition 17.4 applies. ∎

If $X \in \mathscr{C}^\infty V(\mathcal{M})$, that is, X is a smooth vector field it can be viewed as a first-order differential operator on $\mathcal{M}$ and one can compute ${}^tX$, the associated transpose operator as defined in Sect. 16.3.3.

PROPOSITION 17.7 (Transposition of a Vector Field). *Let $X \in \mathscr{C}^\infty V(\mathcal{M})$. Then ${}^tX = -X - \operatorname{div}_g X$. Moreover, we have*

$$\int_{\mathcal{M}} (Xf_1)f_2\mu_g = -\int_{\mathcal{M}} f_1(Xf_2 + f_2 \operatorname{div}_g X)\mu_g + \int_{\partial\mathcal{M}} f_1 f_2\, g(X,\nu)\mu_{g_\partial},$$

for f_1 and f_2 smooth compactly supported functions on $\mathcal{M}$.

PROOF. With (17.2.6) we write

$$\begin{aligned}\operatorname{div}_g(f_1 f_2 X) &= f_1 \operatorname{div}_g(f_2 X) + f_2(Xf_1)\\ &= f_1\big(f_2 \operatorname{div}_g X + X(f_2)\big) + f_2(Xf_1).\end{aligned}$$

With Proposition 17.6 the result follows. ∎

PROPOSITION 17.8 (Green Formula). *We have ${}^t\Delta_g = \Delta_g$. Moreover, we have the Green formula*

$$\int_{\mathcal{M}} f_1\Delta_g f_2\mu_g + \int_{\partial\mathcal{M}} g(\nabla_g f_1,\nu)f_2\mu_{g_\partial} = \int_{\mathcal{M}} (\Delta_g f_1)f_2\mu_g + \int_{\partial\mathcal{M}} f_1 g(\nabla_g f_2,\nu)\mu_{g_\partial},$$

for f_1 and f_2 smooth compactly supported functions on $\mathcal{M}$.

PROOF. We have $\operatorname{div}_g(f_1\nabla_g f_2) = f_1\Delta_g f_2 + g(\nabla_g f_1, \nabla_g f_2)$ by (17.2.6). From Proposition 17.6 we deduce

$$\int_{\mathcal{M}} f_1\Delta_g f_2\mu_g = -\int_{\mathcal{M}} g(\nabla_g f_1, \nabla_g f_2)\mu_g + \int_{\partial\mathcal{M}} f_1 g(\nabla_g f_2,\nu)\mu_{g_\partial}. \tag{17.3.1}$$

Writing the same identity after interchanging the *rôles* of f_1 and f_2 and subtracting the two equations yields the Green formula. For two functions $f_1, f_2 \in {}^0\mathscr{D}_c^\infty(\mathcal{M})$ we naturally recover (17.2.7),

$$\int_{\mathcal{M}} f_1\Delta_g f_2\mu_g = \int_{\mathcal{M}} (\Delta_g f_1)f_2\mu_g. \tag{17.3.2}$$

Yet, here integration concerns densities instead of d-forms. From the definition of the transpose of an operator in Sect. 16.3.3 we have ${}^t\Delta_g = \Delta_g$. ∎

Anticipating upon the definition of the L^2-inner product recalled in Sect. 18.1, changing f_2 into $\overline{f_2}$ the Green formula reads

$$(f_1, \Delta_g f_2)_{L^2(\mathcal{M})} + (\partial_\nu f_{1|\partial\mathcal{M}}, f_{2|\partial\mathcal{M}})_{L^2(\partial\mathcal{M})} \\ = (\Delta_g f_1, f_2)_{L^2(\mathcal{M})} + (f_{1|\partial\mathcal{M}}, \partial_\nu f_{2|\partial\mathcal{M}})_{L^2(\partial\mathcal{M})},$$

with $\partial_\nu w_{|\partial\mathcal{M}} = g(\nabla_g w, \nu)_{|\partial\mathcal{M}}$.

A formula expressing the transposition of the Laplace–Beltrami operator in the sense of distribution is given in Sect. 18.3. An extension of the Green formula for functions in $H^2(\mathcal{M})$ is given in Proposition 18.30 at the end of Sect. 18.5 after traces of Sobolev functions are properly defined. A Green formula for a function f that is only L^2 is given in Lemma 18.34 under the requirement that $\Delta_g f \in L^2(\mathcal{M})$.

We say that a vector field u is L^1 (resp. L^1_{loc}) if $g(u,u)^{1/2}$ is a L^1 (resp. L^1_{loc}) function on $\mathcal{M}$.

For such nonsmooth functions with support away from the boundary we have the following result that will be of use in what follows.

PROPOSITION 17.9. *Let $(\mathcal{M}, g)$ be a Riemannian manifold with boundary and let u be a compactly supported L^1-vector field on $\mathcal{M}$ such that* $\operatorname{supp} u \cap \partial\mathcal{M} = \emptyset$ *and* $\operatorname{div}_g u \in L^1(\mathcal{M})$. *We have* $\int_\mathcal{M} \operatorname{div}_g(u)\mu_g = 0$.

PROOF. Arguing as in the beginning of the proof of Theorem 16.2, it suffices to consider the vector field u supported in a single local chart $\mathcal{C} = (O, \kappa)$. In this chart, using the local form of the divergence given by (17.2.5), we thus have to prove that

$$J = \int_{\kappa(O)} \sum_{1 \leq i \leq d} \partial_{x_i}\big((\det g^\mathcal{C})^{1/2} u^i\big)(x)\, dx$$

vanishes if for each $i = 1, \dots, d$, $\operatorname{supp}(u^i) \subset \operatorname{int} \kappa(O)$, $u^i \in L^1(\operatorname{int} \kappa(O))$, and moreover $\sum_{1 \leq i \leq d} \partial_{x_i} w^i \in L^1(\operatorname{int} \kappa(O))$, with $w^i = (\det g^\mathcal{C})^{1/2} u^i$. For $\chi \in \mathscr{C}^\infty_c(\operatorname{int} \kappa(O))$ such that $\chi \equiv 1$ in a neighborhood of $\operatorname{supp}(u^i)$, $i = 1, \dots, d$ we write

$$J = \sum_{1 \leq i \leq d} \langle \partial_{x_i} w^i, \chi \rangle_{\mathscr{D}'(\mathbb{R}^d), \mathscr{C}^\infty_c(\mathbb{R}^d)} = - \sum_{1 \leq i \leq d} \langle w^i, \partial_{x_i} \chi \rangle_{\mathscr{D}'(\mathbb{R}^d), \mathscr{C}^\infty_c(\mathbb{R}^d)} = 0,$$

as $\partial_{x_i}\chi = 0$ in a neighborhood of $\operatorname{supp}(w^i)$. ■

17.4. Linear Connection and Covariant Derivatives

A linear connection is a natural way to formalize the directional differentiation of a vector field. In particular, it allows one to formulate the differentiation of a vector field along a curve. Recall that $\mathscr{C}^\infty V(\mathcal{M})$ is the space of smooth vector fields over $\mathcal{M}$.

17.4.1. Linear Connection. A linear connection is a map $\mathsf{D} : \mathscr{C}^\infty V(\mathcal{M}) \times \mathscr{C}^\infty V(\mathcal{M}) \to \mathscr{C}^\infty V(\mathcal{M})$, $(u, v) \mapsto \mathsf{D}(u, v) = \mathsf{D}_u\, v$ that satisfies the following properties:

(1) D is linear over $\mathscr{C}^\infty(\mathcal{M})$ w.r.t. its first argument, that is,

$$\mathsf{D}_{fu+g\tilde{u}}\, v = f\, \mathsf{D}_u\, v + g\, \mathsf{D}_{\tilde{u}}\, v,$$

for $f, g \in \mathscr{C}^\infty(\mathcal{M})$ and $u, \tilde{u}, v \in \mathscr{C}^\infty V(\mathcal{M})$.

(2) D is linear over $\mathbb{R}$ w.r.t. its second argument, that is,

$$\mathsf{D}_u(\lambda v + \mu \tilde{v}) = \lambda\, \mathsf{D}_u\, v + \mu\, \mathsf{D}_u\, \tilde{v},$$

for $\lambda, \mu \in \mathbb{R}$ and $u, v, \tilde{v} \in \mathscr{C}^\infty V(\mathcal{M})$.

(3) D acts as a derivation w.r.t. its second argument, that is,

$$\mathsf{D}_u(fv) = f\, \mathsf{D}_u\, v + u(f)\, v,$$

for $f \in \mathscr{C}^\infty(\mathcal{M})$ and $u, v \in \mathscr{C}^\infty V(\mathcal{M})$.

For such a connection, in a local chart $\mathcal{C} = (O, \kappa)$, setting in O the vector fields e_i, $i = 1, \ldots, d$, such that their local representatives are ∂_{x_i}, for $i, j \in \{1, \ldots, d\}$, there exists Γ_{ij}^k, $k = 1, \ldots, d$, such that

$$\mathsf{D}_{e_i}(e_j)(m) = \sum_{1 \leq k \leq d} \Gamma_{ij}^k(m) e_k(m). \tag{17.4.1}$$

The real functions Γ_{ij}^k, for $i, j, k \in \{1, \ldots, d\}$, defined on O are the so-called Christoffel symbols associated with the connection D. The Christoffel symbols fully characterize the connection D. For two vector fields u and v, the coordinates of the representative of $\mathsf{D}_u\, v$ are given by

$$(\mathsf{D}_u\, v)^{\mathcal{C},k} = \sum_{1 \leq i \leq d} u^{\mathcal{C},i} \partial_{x_i} v^{\mathcal{C},k} + \sum_{1 \leq i,j \leq d} u^{\mathcal{C},i} v^{\mathcal{C},j} \Gamma_{ij}^k, \tag{17.4.2}$$

at $x = \kappa(m)$ with $\Gamma_{ij}^k = \Gamma_{ij}^k(m)$.

We observe that for $m \in \mathcal{M}$, the tangent vector $(\mathsf{D}_u\, v)_m$ only depends on u_m. This appears quite clearly in (17.4.2). We may in fact define a connection as a map $T_m\mathcal{M} \times \mathscr{C}^\infty V(\mathcal{M}) \to T_m\mathcal{M}$ for all $m \in \mathcal{M}$.

For $u \in T_m\mathcal{M}$, one calls $\mathsf{D}_u\, v$ the covariant derivative of v with respect u (at m).

17.4.2. Differentiation Along a Curve. As mentioned above a primary use of a connection is to differentiate vector fields along curves.

Let $\gamma : I \to \mathcal{M}$ is a smooth curve, I an interval of $\mathbb{R}$. Consider $v : I \to T\mathcal{M}$ such that $v(t) \in T_{\gamma(t)}\mathcal{M}$, that is, a vector field along γ. For $t^0 \in I$, there exist $\varepsilon > 0$, a neighborhood V of $\gamma(t^0)$ in $\mathcal{M}$ and a vector field $\tilde{v}$ on V such that $v(t) = \tilde{v}_{\gamma(t)}$ for $t \in (t^0 - \varepsilon, t^0 + \varepsilon)$. If given a connection one introduces the following differentiation of v along γ:

$$\frac{\mathsf{D}\, v}{dt} = \mathsf{D}_{\frac{d\gamma}{dt}(t^0)}\, \tilde{v} \quad \text{at the point } \gamma(t^0). \tag{17.4.3}$$

With (17.4.2) one observes that this definition of $\frac{\mathsf{D}\, v}{dt}$ is independent of the choice of $\tilde{v}$.

17.4.3. Extension to Tensors of All Orders. The action of D_u on $\mathscr{C}^\infty V(\mathcal{M})$ can be extended to tensors of all orders. For a tensor of order zero, that is a smooth function on $\mathcal{M}$, for $u \in T_m\mathcal{M}$, we set

$$\mathsf{D}_u\, f(m) = u_m(f) = df(m)(u). \tag{17.4.4}$$

For a one-form ω we set

$$(\mathsf{D}_u\, \omega)(v) = u(\omega(v)) - \omega(\mathsf{D}_u\, v), \qquad v \in \mathscr{C}^\infty V(\mathcal{M}),$$

meaning that we have the product law $u(\omega(v)) = (\mathsf{D}_u\, \omega)(v) + \omega(\mathsf{D}_u\, v)$. We observe that $\mathsf{D}_u \omega$ is a one-form, meaning that its action on v at m only depends on v_m. This can be seen by computing the $(\mathsf{D}_u\, \omega)(v)$ in a local chart $\mathcal{C}$ for instance, with $u^{\mathcal{C}} = u^i \partial_{x_i}$, $v^{\mathcal{C}} = v^j \partial_{x_j}$ and $\omega^{\mathcal{C}} = \omega_k dx^k$,

$$\begin{aligned}(\mathsf{D}_u\, \omega)(v) &= \sum_{1\le i,j\le d} u^j \partial_{x_j}(\omega_i v^i) - \sum_{1\le i,j\le d} \omega_i (u^j \partial_{x_j} v^i + \sum_{1\le k\le d} u^j v^k \Gamma^i_{jk}) \\ &= \sum_{1\le i,k\le d} v^i u^j \partial_{x_j} \omega_i - \sum_{1\le i,j,k\le d} \omega_i v^k u^j \Gamma^i_{jk}.\end{aligned} \tag{17.4.5}$$

In particular, we may define $(\mathsf{D}_u\, \omega)(v)$ at m for $u, v \in T_m\mathcal{M}$ without requiring u and v to be vector fields.

As a generalization, for θ a r-covariant,s-contravariant tensor field we set

$$\begin{aligned}(\mathsf{D}_u\, \theta)(v_1, \dots, v_r, \omega_1, \dots, \omega_s) &= u\big(\theta(v_1, \dots, v_r, \omega_1, \dots, \omega_s)\big) \\ &\quad - \sum_{1\le i\le r} \theta(v_1, \dots, \mathsf{D}_u\, v_i, \dots, v_r, \omega_1, \dots, \omega_s) \\ &\quad - \sum_{1\le j\le s} \theta(v_1, \dots, v_r, \omega_1, \dots, \mathsf{D}_u\, \omega_i, \dots, \omega_s),\end{aligned} \tag{17.4.6}$$

for $v_1, \dots, v_r \in \mathscr{C}^\infty V(\mathcal{M})$ and $\omega_1, \dots, \omega_s$ smooth one-forms. With a computation similar to (17.4.5), we see that $(\mathsf{D}_u\, \theta)(v_1, \dots, v_r, \omega_1, \dots, \omega_s)$ evaluated at $m \in \mathcal{M}$ is only a multilinear function of u_m, $(v_1)_m, \dots, (v_r)_m$ and $(\omega_1)_m, \dots, (\omega_s)_m$. This shows that $\mathsf{D}_u\, \theta$ is a r-covariant,s-contravariant tensor at m and $(u, v_1, \dots, v_r, \omega_1, \dots, \omega_s) \mapsto (\mathsf{D}_u\, \theta)(v_1, \dots, v_r, \omega_1, \dots, \omega_s)$ is a $(r+1)$-covariant,s-contravariant tensor at m.

For a tangent vector field v, we set $\mathsf{D}\, v$ to be the 1-covariant,1-contravariant tensor field given by, at a point m,

$$\mathsf{D}\, v(u, \omega) = (\mathsf{D}_u\, v)(\omega) = \langle \omega, \mathsf{D}_u\, v \rangle,$$

for $u \in T_m\mathcal{M}$ and $\omega \in T^*_m\mathcal{M}$. In a local chart, this yields

$$(\mathsf{D}\, v)^k_i = (\mathsf{D}_{e_i}\, v)^k = \partial_{x_i} v^k + \sum_{1\le j\le d} v^j \Gamma^k_{ij}. \tag{17.4.7}$$

Similarly, for a one-form ω, we set $\mathsf{D}\,\omega$ to be the 2-covariant tensor field by, at a point m,

$$\mathsf{D}\,\omega(v,u) = (\mathsf{D}_u\,\omega)(v) = \langle \mathsf{D}_u\,\omega, v\rangle,$$

for $u, v \in T_m\mathcal{M}$. In a local chart, this yields

$$(\mathsf{D}\,\omega)_{ij} = (\mathsf{D}_{e_j}\,\omega)_i = \partial_{x_j}\omega_i - \sum_{1\le k\le d} \omega_k \Gamma^k_{ji}. \tag{17.4.8}$$

More, generally for θ a r-covariant,s-contravariant tensor field, we set $\mathsf{D}\,\theta$ to be the $(r+1)$-covariant,s-contravariant tensor field given by,

$$\mathsf{D}\,\theta(v_1,\dots,v_r,u,\omega_1,\dots,\omega_s) = (\mathsf{D}_u\,\theta)(v_1,\dots,v_r,\omega_1,\dots,\omega_s),$$

for $v_1,\dots,v_r \in \mathscr{C}^\infty V(\mathcal{M})$ and $\omega_1,\dots,\omega_s$ smooth one-forms. In a local chart, this yields

$$\begin{aligned}(\mathsf{D}\,\theta)^{j_1\dots j_s}_{i_1\dots i_r \ell} &= (\mathsf{D}_{e_\ell}\,\theta)^{j_1\dots j_s}_{i_1\dots i_r}\\ &= \partial_{x_\ell}(\theta^{j_1\dots j_s}_{i_1\dots i_r}) - \sum_{\substack{1\le n\le r\\ 1\le k\le d}} \Gamma^k_{\ell i_n}\theta^{j_1\dots j_s}_{i_1\dots i_{n-1}\,k\,i_{n+1}\dots i_r}\\ &\quad + \sum_{\substack{1\le n\le s\\ 1\le k\le d}} \Gamma^{j_n}_{\ell k}\theta^{j_1\dots j_{n-1}\,k\,j_{n+1}\dots j_s}_{i_1\dots i_r},\end{aligned} \tag{17.4.9}$$

which is a generalization of (17.4.7) and (17.4.8). For a function f we find $\mathsf{D}\,f = df$.

17.4.4. The Levi-Civita Connection. For a connection D on $\mathcal{M}$ its torsion is defined as the map $T : \mathscr{C}^\infty V(\mathcal{M}) \times \mathscr{C}^\infty V(\mathcal{M}) \to \mathscr{C}^\infty V(\mathcal{M})$ given by

$$T(u,v) = \mathsf{D}_u\,v - \mathsf{D}_v\,u - [u,v],$$

where $[.,.]$ denotes the Lie bracket of two vector fields. A connection is said to be torsion free or symmetric if T vanishes identically.

In the case of a Riemannian manifold, one says that the connection D is compatible with the metric g if one has

$$u\big(g(v,w)\big) = \mathsf{D}_u\,\big(g(v,w)\big) = g(\mathsf{D}_u\,v, w) + g(v, \mathsf{D}_u\,w), \tag{17.4.10}$$

for $u, v, w \in \mathscr{C}^\infty V(\mathcal{M})$. With (17.4.6) this means that $\mathsf{D}_u\,g$ vanishes for any vector field u.

The following result is fundamental in Riemannian geometry.

THEOREM 17.10 (Levi-Civita Connection). *If $(\mathcal{M}, g)$ is a smooth Riemannian manifold, there exists a unique torsion free linear connection D that is compatible with the metric g. This connection is called the Levi-Civita connection.*

We refer to [221] for a proof. In particular, one can find therein a clear explanation of the reason for this choice of connection. In each chart $\mathcal{C} = (O, \kappa)$ the Christoffel symbols associated with the Levi-Civita connection are given by

$$\Gamma_{ij}^{k} = \frac{1}{2} \sum_{1 \le \ell \le d} g^{\mathcal{C},k\ell} (\partial_{x_i} g_{j\ell}^{\mathcal{C}} + \partial_{x_j} g_{i\ell}^{\mathcal{C}} - \partial_{x_\ell} g_{ij}^{\mathcal{C}}). \tag{17.4.11}$$

On a Riemannian manifold $(\mathcal{M}, g)$, if D is the Levi-Civita connection, the compatibility between D and g simply reads $\mathsf{D}\, g \equiv 0$. We also have $\nabla_g f = (df)^\sharp = (\mathsf{D}\, f)^\sharp$.

Let ν be the unique outward pointing vector field along $\partial\mathcal{M}$ such that, for all $m \in \partial\mathcal{M}$, $g_m(\nu_m, \nu_m) = 1$ and $g_m(\nu_m, u) = 0$ for all $u \in T_m\partial\mathcal{M}$. In (5.3.1) in Sect. 5.3, for $m \in \partial\mathcal{M}$ we have defined $\partial_\nu f(m)$ to be $\nu_m(f) = df(m)(\nu_m)$. Note that we have $\partial_\nu f(m) = g_m(\nabla_g(f)_m, \nu_m)$. We see that we may also write the normal derivative by means of the connection:

$$\partial_\nu f(m) = \mathsf{D}\, f_m(\nu_m) = \mathsf{D}_{\nu_m} f(m). \tag{17.4.12}$$

17.5. Geodesics and Geodesic Flows

Geodesics can be defined once provided a connection. Here, we consider the Levi-Civita connection given by Theorem 17.10. Some results are stated without proofs and we refer the interested reader to [120, 189, 221].

For a curve $t \mapsto x(t)$, as is done classically, we denote

$$v(t) = \dot{x}(t) = \frac{dx}{dt} \quad \text{and} \quad \dot{v}(t) = \ddot{x}(t) = \frac{d^2x}{dt^2}.$$

Recall that $T\mathcal{M}$ is a smooth manifold; see Sect. 15.3.3. In a local chart $\mathcal{C} = (O, \kappa)$, $T\mathcal{M}$ is locally diffeomorphic to $\kappa(O) \times \mathbb{R}^d$. Then $t \mapsto \big(x(t), v(t)\big)$ is a curve on $T\mathcal{M}$. One deduces that

$$\big(v(t), \dot{v}(t)\big) \in T_{(x(t),v(t))}T\mathcal{M}.$$

The definition of a geodesic is based on the differentiation of vector fields along curves given in (17.4.3). In a local chart, from (17.4.2) and (17.4.3), we have

$$\frac{\mathsf{D}\,\dot{x}}{dt} = \ddot{x} + \sum_{1 \le j,k,\ell \le d} \Gamma_{k\ell}^{j}(x) \dot{x}^k \dot{x}^\ell \partial_j. \tag{17.5.1}$$

DEFINITION 17.11. A curve $\gamma : I \to \mathcal{M}$, with I an interval of $\mathbb{R}$, is called a geodesic if

$$\frac{\mathsf{D}\,\dot{\gamma}}{dt} = 0 \qquad t \in I,$$

for $\dot{\gamma}(t) = \frac{d}{dt}\gamma(t) \in T_{\gamma(t)}\mathcal{M}$.

In other words geodesics are curves with vanishing 'acceleration'.

If $t \mapsto \gamma(t)$ is a geodesic, then $t \mapsto |\dot\gamma(t)|_g$ is constant since, by (17.4.4) and (17.4.10), we have

$$\begin{aligned}\frac{d}{dt}|\dot\gamma(t)|_g^2 &= \frac{d}{dt} g_{\gamma(t)}\big(\dot\gamma(t),\dot\gamma(t)\big) = \mathsf{D}_{\dot\gamma(t)}\, g_{\gamma(t)}\big(\dot\gamma(t),\dot\gamma(t)\big)\\ &= 2g_{\gamma(t)}\big(\mathsf{D}_{\dot\gamma(t)}\,\dot\gamma(t),\dot\gamma(t)\big)\\ &= 2g_{\gamma(t)}\big(\frac{D}{dt}\dot\gamma(t),\dot\gamma(t)\big) = 0.\end{aligned}$$

Setting $c = |\dot\gamma(t)|_g$, with the length of a curve given as in (17.1.1), the length of the geodesic joining $\gamma(t^0)$ and $\gamma(t^1)$ is then simply $c|t^1 - t^0|$.

REMARK 17.12. In Definition 17.11 where a geodesic is characterized as a curve with vanishing 'acceleration' one sees that the parametrization of the curve is essential. Hence a geodesic is not defined here as the set of points visited by γ,

$$G_\gamma = \{\gamma(t);\ t \in I\},$$

but the map itself. If fact, a change of parametrization will not affect G_γ yet can yield $\frac{\mathsf{D}\dot\gamma}{dt} \neq 0$.

Having $\frac{\mathsf{D}\dot\gamma}{dt} = 0$ gives a second-order system; cast into a first-order system, it takes the form

$$\frac{d}{dt}\gamma = v, \qquad \frac{d}{dt}v = -\sum_{1\le j,k,\ell\le d} \Gamma^j_{k\ell}\dot\gamma^k\dot\gamma^\ell\partial_j. \tag{17.5.2}$$

From the Cauchy–Lipschitz theorem, given $(m^0,v^0)\in T\mathcal{M}$ we see that there exists a unique maximal geodesic that goes through (m^0,v^0).

If $t \mapsto \gamma(t)$ is a geodesic we saw above that

$$\big(v(t),\dot v(t)\big) = \big(\dot\gamma(t),\ddot\gamma(t)\big) \in T_{(\gamma(t),v(t))}T\mathcal{M},$$

with $\dot v(t) = -\sum_{1\le j,k,\ell\le d}\Gamma^j_{k\ell}(\gamma(t))v^k(t)v^\ell(t)\partial_j$. Conversely, if $m\in\mathcal{M}$ and $v\in T_m\mathcal{M}$, then

$$\big(v,w\big)\in T_{(m,v)}T\mathcal{M}, \quad \text{with } w = -\sum_{1\le j,k,\ell\le d}\Gamma^j_{k\ell}(m)v^kv^\ell\partial_j.$$

In fact, consider $\gamma(t)$ the unique (maximal) geodesic such that $\gamma(0) = m$ and $\dot\gamma(0) = v$. Then $\big(v,w\big) = \big(\dot\gamma(0),\ddot\gamma(0)\big)\in T_{(m,v)}T\mathcal{M}$.

17.5.1. Tangent Geodesic Flow. For $(m,v)\in T\mathcal{M}$, we saw above for w_v with $-\sum_{1\le j,k,\ell\le d}\Gamma^j_{k\ell}(x)v^kv^\ell\partial_j$ as local representative, then $(v,w_v)\in T_{(m,v)}T\mathcal{M}$. Setting

$$G_{(m,v)} = (v,w_v),$$

one thus defines a vector field on $T\mathcal{M}$. It is called the *(tangent) geodesic vector field.* From (17.5.2) we find that a curve $t\mapsto x(t)$ is a geodesic if and only if $t\mapsto \big(x(t),\dot x(t)\big)$ is integral curve of G on $T\mathcal{M}$. The flow χ_t associated with G is called the *(tangent) geodesic flow.* A geodesic $t \mapsto \gamma(t)$ is thus solution to

$$\frac{d}{dt}\big(\gamma(t), \dot{\gamma}(t)\big) = G_{(\gamma(t),\dot{\gamma}(t))}.$$

For $(x, v) \in T\mathcal{M}$, $t \mapsto \gamma(t) = \chi_t(m, v)$ yields the maximal unique geodesic such that $\gamma(0) = m$ and $\dot{\gamma}(0) = v$.

PROPOSITION 17.13 (Local Geodesic Flow Box). *Let $m^0 \in \mathcal{M}$. There exist an open neighborhood V of m^0, $C > 0$, and $\delta > 0$ such that $\chi_t(m, v)$ is well defined for $t \in (-\delta, \delta)$, $m \in V$ and $v \in T_m\mathcal{M}$ such that $|v|_g \leq C$.*

Observing that the following holds

$$\chi_t(m, \lambda v) = \chi_{\lambda t}(m, v), \quad \lambda > 0,$$

if one side makes sense, Proposition 17.13 thus yields the following result.

PROPOSITION 17.14. *Let $m^0 \in \mathcal{M}$ and $T > 0$. There exist an open neighborhood V_T of m^0, $C_T > 0$ such that $\chi_t(m, v)$ is well defined for $t \in (-T, T)$, $m \in V_T$ and $v \in T_m\mathcal{M}$ such that $|v|_g \leq C_T$.*

For $m^0 \in \mathcal{M}$ and $T > 1$ one can then define the so-called *exponential* map by

$$\exp(m, v) = \chi_1(m, v) = \chi_{|v|_g}(m, v/|v|_g),$$

for $(m, v) \in T_m\mathcal{M}$ with $m \in V$ and $v \in T_m\mathcal{M}$ such that $|v|_g \leq C_T$, where V and $C_T > 0$ are given by Proposition 17.14.

For $m \in V$ one sets $\exp_m(v) = \exp(m, v)$.

PROPOSITION 17.15. *Let $m \in \mathcal{M}$. Define the open ball*

$$B_g(0, \varepsilon) = \{v \in T_m\mathcal{M};\ |v|_g < \varepsilon\}.$$

There exists $\varepsilon > 0$ such that $\exp_m$ *maps $B_g(0, \varepsilon)$ diffeomorphically onto an open subset of $\mathcal{M}$.*

The proof is in fact that of Lemma 4.36 of Volume 1 once connection is made with the Hamiltonian geodesic flow; see Sect. 17.5.3 and Proposition 17.20.

If the exponential map at m, $\exp_m$, is a diffeomorphism from an open neighborhood V of 0 in $T_m\mathcal{M}$ onto an open subset U of $\mathcal{M}$ one says that U is a *normal open neighborhood* of m. Any neighborhood U' of m such that $U' \subset U$ is also called a normal neighborhood of m.

In the previous proposition m is kept fixed. The following theorem provides a locally uniform version of the previous result.

THEOREM 17.16. *Let $m^0 \in \mathcal{M}$. There exists a neighborhood U of m^0 and $\delta > 0$ such that for all $m \in U$ the geodesic ball* $\exp_m\big(B_g(0, \delta)\big)$ *is well defined and contains U.*

For $\varepsilon > 0$ as in Proposition 17.15, one calls $\exp_m\big(B_g(0,\varepsilon)\big)$ a *geodesic ball* of radius ε centered at m. For $0 \leq \delta < \varepsilon$, if we set

$$S_g(0,\delta) = \{v \in T_m\mathcal{M};\ |v|_g = \delta\},$$

then one calls $\exp_m\big(S(0,\delta)\big)$ the geodesic sphere of radius δ centered at m. It is a submanifold of $\mathcal{M}$ of codimension one.

LEMMA 17.17 (Gauss' Lemma). *Let $m \in \mathcal{M}$ and let $\delta > 0$ be such that the geodesic sphere $\mathcal{S}_{m,\delta} = \exp_m\big(S_g(0,\delta)\big)$ is well defined. If $v \in S_g(0,\delta)$ and if $\gamma(t) = \chi_t(m,v)$, then $\dot\gamma(1)$ is orthogonal to $T_{\gamma(1)}\mathcal{S}_{m,\delta}$ in the sense of g.*

17.5.2. Length Minimization. The following results characterize geodesics locally as the curves that minimize the Riemannian distance.

PROPOSITION 17.18. *Let $m \in \mathcal{M}$ and let $B = \exp_m\big(B_g(0,\varepsilon)\big)$ be a geodesic ball centered at m. Let $\gamma(t) = \chi_t(m,v)$ with $|v|_g < \varepsilon$. Let $\rho : [0,1] \to \mathcal{M}$ be a piecewise differential curve such that $\rho(0) = \gamma(0) = m$ and $\rho(1) = \gamma(1) \in B$. Then $\ell(\gamma) \leq \ell(\rho)$ and if equality holds then $\rho([0,1]) = \gamma([0,1])$.*

PROPOSITION 17.19. *Let $\gamma : \mathbb{R} \supset [a,b] \to \mathcal{M}$ be a piecewise differentiable curve such that, for some $c > 0$, $\ell\big(\gamma([a,t])\big) = c|t-a|$ for any $t \in [a,b]$. If for any other piecewise differentiable curve ρ joining $\gamma(a)$ to $\gamma(b)$ one has $\ell(\gamma) \leq \ell(\rho)$, then γ is a geodesic. In particular γ is a smooth curve.*

The first proposition states that *locally* a geodesic is a length minimizing curve. The second proposition is a converse. Moreover, in this second proposition, the result holds *globally*: a length minimizing curve (with parameter proportional to arc length) is a geodesic. However, considering the sphere $\mathbb{S}^{d-1} \subset \mathbb{R}^d$ one sees that geodesics may not minimize the Riemannian distance *globally*.

17.5.3. The Hamiltonian Geodesic Flow. Recall that $T^*\mathcal{M}$ is a manifold; see Sect. 15.4.3. For a function $f : T^*\mathcal{M} \to \mathbb{R}$ one associates the Hamiltonian vector field H_f on $T^*\mathcal{M}$; see Sect. 15.7.2. Here, we consider $f(m,\omega) = g_m(\omega,\omega)/2$, that is, half of the principal symbol of the Laplace–Beltrami operator. In a local chart one has

$$H_f(x,\xi) = \sum_{1\leq i,j\leq d} g^{ij}(x)\xi_j\partial_{x_i} - \frac{1}{2}\sum_{1\leq i,j,k\leq d} \partial_{x_k} g^{ij}(x)\xi_i\xi_j\partial_{\xi_k}.$$

Maximal integral curves of H_f are unique by the Cauchy–Lipschitz theorem and they provide an alternative way to describe geodesics. We call the flow associated with H_f the *Hamiltonian geodesic* flow.

PROPOSITION 17.20. *Let $(m^0,\xi^0) \in T^*\mathcal{M}$ and $\varphi(t) = (m(t),\xi(t))$ be the unique maximal integral curve of H_f such $\varphi(0) = (m^0,\xi^0)$. Set $v^0 = \xi^{0\sharp}$. If $\gamma(t) = \chi_t(m^0,v^0)$ is the unique maximal integral curve of the geodesic vector field G such that $\gamma(0) = m^0$ and $\dot\gamma(0) = v^0$, then $\gamma(t) = (m(t),\xi^\sharp(t))$.*

REMARK 17.21. Because of the uniqueness of the two families of integral curves considered here, the converse follows, that is, if $(m^0, v^0) \in T\mathcal{M}$ and if $\gamma(t) = (m(t), v(t)) = \chi_t(m^0, v^0)$ is the unique maximal integral curve of the geodesic vector field G such that $\gamma(0) = m^0$ and $\dot{\gamma}(0) = v^0$, then $\varphi(t) = (m(t), v^\flat(t))$ is the unique maximal integral curve of H_f such $\varphi(0) = (m^0, v^{0\flat})$.

PROOF. It is sufficient to consider a local chart. An integral curve $t \mapsto (x(t), \xi(t))$ of H_f on $T^*\mathcal{M}$ is solution to $(\dot{x}(t), \dot{\xi}(t)) = H_f(x(t), \xi(t))$, that is,

$$\dot{x}^i(t) = \sum_{1 \leq \ell \leq d} g^{i\ell}(x(t))\xi_\ell(t), \qquad \dot{\xi}_i(t) = -\frac{1}{2} \sum_{1 \leq k,\ell \leq d} \partial_{x_i} g^{k\ell}(x(t))\xi_k(t)\xi_\ell(t).$$

The first equation reads $\dot{x}(t) = \xi^\sharp(t)$. We thus set $v(t) = \xi^\sharp(t)$. For the sake of concision we omit sums in the proof and use the Einstein summation convention on repeated indices. We also write g^{ij}, v, x and ξ in place of $g^{ij}(x(t))$, $v(t)$, $x(t)$ and $\xi(t)$. We have

$$\begin{aligned}\dot{v}^j &= \frac{d}{dt}(g^{ij}\xi_i) = \partial_{x_k} g^{ij} v^k \xi_i + g^{ij}\dot{\xi}_i = \partial_{x_k} g^{ij} v^k \xi_i - \frac{1}{2} g^{ij} \partial_{x_i} g^{k'\ell'} \xi_{k'}\xi_{\ell'} \\ &= \Big((\partial_{x_k} g^{ij}) g_{i\ell} - \frac{1}{2} g^{ij} (\partial_{x_i} g^{k'\ell'}) g_{kk'} g_{\ell\ell'} \Big) v^k v^\ell.\end{aligned}$$

Since we have $g^{ij} g_{i\ell} = \delta^j_\ell$ we find $(\partial_{x_k} g^{ij}) g_{i\ell} = -g^{ij}\partial_{x_k} g_{i\ell}$. We thus obtain

$$\begin{aligned}(\partial_{x_k} g^{ij}) g_{i\ell} - \frac{1}{2} g^{ij} (\partial_{x_i} g^{k'\ell'}) g_{kk'} g_{\ell\ell'} &= -g^{ij}\partial_{x_k} g_{i\ell} + \frac{1}{2} g^{ij} g^{k'\ell'} g_{kk'} \partial_{x_i} g_{\ell\ell'} \\ &= -g^{ij}\partial_{x_k} g_{i\ell} + \frac{1}{2} g^{ij} \partial_{x_i} g_{\ell k}.\end{aligned}$$

We thus have

$$\dot{v}^j = -\frac{1}{2} g^{ij} \Big(2\partial_{x_k} g_{i\ell} - \partial_{x_i} g_{\ell k} \Big) v^k v^\ell = -\frac{1}{2} g^{ij} \Big(\partial_{x_k} g_{i\ell} + \partial_{x_\ell} g_{ik} - \partial_{x_i} g_{\ell k} \Big) v^k v^\ell.$$

With the formula for the Christoffel symbols given in (17.4.11), we thus find $\dot{v}^j = -\Gamma^j_{kl} v^k v^\ell$. By (17.5.2), the proof is complete. ∎

17.6. Normal Geodesic Coordinates at the Boundary

The Laplace–Beltrami operator Δ_g is canonically defined through the metric g. Let $m \in \partial\mathcal{M}$. By Theorem 9.7 of Volume 1 we can design a local chart $\mathcal{C} = (O, \kappa)$ with $m \in O$, such that the principal symbol of the representative of $P = -\Delta_g$ in $\mathcal{C}$ has the form

$$\xi_d^2 + \sum_{1 \leq i,j \leq d-1} p_{ij}(x)\xi_i\xi_j.$$

In fact, if one considers any local chart $\tilde{\mathcal{C}} = (\tilde{O}, \tilde{\kappa})$ with $m \in \tilde{O}$ the representative of P in this local chart is a second-order elliptic operator with smooth coefficients in $\tilde{\kappa}(\tilde{O})$, an open set of $\overline{\mathbb{R}^d_+}$. One can apply Theorem 9.7 to that operator and this generates new coordinates in a possibly smaller open set. This new open set and the new coordinates yield the local chart $\mathcal{C} = (O, \kappa)$.

Since $P = -\Delta_g$ the principal symbol also reads

$$p^{\mathcal{C}}(x,\xi) = \sum_{1\leq i,j\leq d} g^{\mathcal{C},ij}(x)\xi_i\xi_j,$$

implying that in the local chart $\mathcal{C}$ the representative of the metric is such that

$$g^{\mathcal{C}}_{dj}(x) = \delta_{dj}, j = 1,\dots,d, \ x \in \kappa(O),$$

meaning that

$$g^{\mathcal{C}}_x = g^{\mathcal{C}}_x{}' + d_{x_d} \otimes d_{x_d}, \quad \text{with } g^{\mathcal{C}}_x{}' = \sum_{1\leq i,j\leq d-1} g^{\mathcal{C}}_{ij}{}'(x) dx^i \otimes dx^j, \tag{17.6.1}$$

or rather, in matrix form,

$$g^{\mathcal{C}}_x = \begin{pmatrix} & & & 0 \\ & g^{\mathcal{C}}_x{}' & & \vdots \\ & & & 0 \\ 0 & \cdots & 0 & 1 \end{pmatrix}. \tag{17.6.2}$$

Note that $g^{\mathcal{C}}_{(x',x_d)}{}'$ is a metric on $\kappa(O \cap \partial\mathcal{M})$ with x_d acting as a parameter. For $x_d = 0$, $g^{\mathcal{C}}_{(x',0)}{}'$ is the representative of the induced metric g_∂ on the boundary $\partial\mathcal{M}$.

In local coordinates, the representative of Δ_g is given by (17.2.8). Here, this gives

$$\begin{aligned} P^{\mathcal{C}} &= (\det g^{\mathcal{C}}_x{}')^{-1/2} D_d\big((\det g^{\mathcal{C}}_x{}')^{1/2} D_d\big) \\ &\quad + (\det g^{\mathcal{C}}_x{}')^{-1/2} \sum_{1\leq i,j\leq d-1} D_{x_i}\big((\det g^{\mathcal{C}}_x{}')^{1/2} g^{\mathcal{C}\,\prime,ij}(x) D_{x_j}\big). \end{aligned}$$

If we set

$$R(x, D_{x'}) = (\det g^{\mathcal{C}}_x{}')^{-1/2} \sum_{1\leq i,j\leq d-1} D_{x_i}\big((\det g^{\mathcal{C}}_x{}')^{1/2} g^{\mathcal{C}\,\prime,ij}(x) D_{x_j}\big),$$

we then have

$$P^{\mathcal{C}} = D_d^2 + R(x, D_{x'}) + Q_1(x, D), \tag{17.6.3}$$

with

$$Q_1(x, D) = -\frac{i}{2}(\det g^{\mathcal{C}}_x{}')^{-1} \partial_d\big(\det g^{\mathcal{C}}_x{}'\big) D_d. \tag{17.6.4}$$

The first-order operator $Q_1(x, D)$ thus acts only in the normal direction to $\partial\mathcal{M}$ given by the variable x_d in the local chart.

Normal geodesic coordinates as describes above are defined locally. In a bounded part of a Riemannian manifold they can be used to provide a chart that describes an open set near the boundary in a very natural manner.

THEOREM 17.22. *Let $(\mathcal{M}, g)$ be a Riemannian manifold and let Γ be a bounded open subset of $\partial\mathcal{M}$. There exist O an open set of $\mathcal{M}$, $z_0 > 0$, and a diffeomorphism Φ, such that $\Gamma = O \cap \partial\mathcal{M}$ and*

$$\Phi : \Gamma \times [0, z_0) \to O$$
$$(m', z) \mapsto \Phi(m', z),$$

and the following additional properties hold.

(1) *For any $m \in O$ with $m = \Phi(m', z)$ we have*
$$\text{dist}_g(m, \partial\mathcal{M}) = \text{dist}_g(m, \Gamma) = \text{dist}_g(m, m') = z.$$

(2) *There exists a Riemannian metric $g'(z)$ on Γ that smoothly depends on z such that*
$$(\Phi^* g)_{(m',z)} = g'_{m'}(z) \otimes 1_z + d_z \otimes d_z,$$
and $g'(0) = g_\partial$ on Γ.

A point in the open set O is thus diffeomorphically parameterized by its Riemannian distance z to $\partial\mathcal{M}$ and the point m' of $\partial\mathcal{M}$ where this distance is realized. In particular the map $O \ni m = \Phi(m', z) \mapsto m'$ provides a projection onto Γ naturally associated with the Riemannian structure. As seen in the proof below, for $m = \Phi(m', z) \in O$ the curve $t \mapsto \Phi(m', tz)$, $t \in [0, 1]$, is a geodesic that joins m' and m. This geodesic leaves $\partial\mathcal{M}$ in a normal manner and it minimizes the distance between the two points. This result further justifies the name *normal geodesic coordinates* that one also uses for the "chart" (O, Φ^{-1}). The open set O is illustrated in Fig. 17.1.

In the particular case where $\mathcal{M}$ is a compact manifold, one can choose $\Gamma = \partial\mathcal{M}$ and thus obtain a global parameterization of an open neighborhood of $\partial\mathcal{M}$ by means of normal geodesic coordinates.

PROOF. Let m^0 be an arbitrary point of Γ. From the discussion at the beginning of the section there exists a local chart $\hat{\mathcal{C}} = (\hat{O}, \hat{\kappa})$ that yields

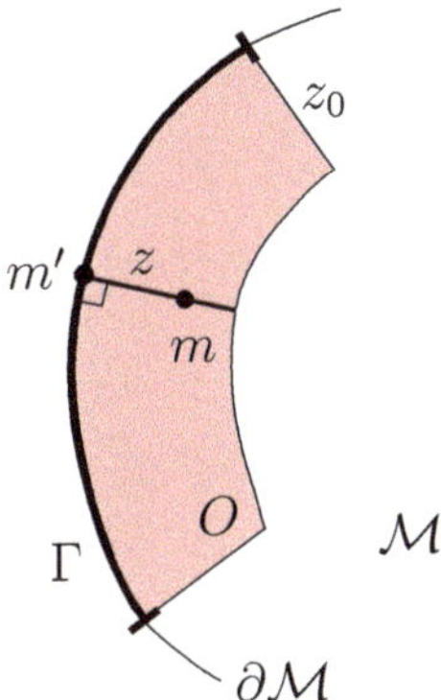

FIGURE 17.1. The open subset Γ of $\partial\mathcal{M}$ and the open subset O of $\mathcal{M}$

normal geodesic coordinates. Without any loss of generality we may assume that $\hat{\kappa}(\hat{O})$ takes the form $\hat{\kappa}(\hat{O}) = \hat{O}' \times [0, \hat{z}_0]$ for $\hat{O}'$ an open set of $\mathbb{R}^{d-1}$ and $\hat{z}_0 > 0$ and $\hat{\kappa}(\hat{O} \cap \partial\mathcal{M}) = \hat{O}'$. We choose $\hat{z}_0$ and $\hat{O}'$ sufficiently small so that $\hat{O}$ lies in a neighborhood U of m^0 given by Theorem 17.16. It then follows that any geodesic that lies in $\hat{O}$ minimizes distances between any pair of its points by Proposition 17.18.

We further reduce the size of $\hat{z}_0$ and $\hat{O}'$ so that for $m \in \hat{O}$ the distance $\operatorname{dist}_g(m, \partial\mathcal{M})$ is only realized by points in $\hat{O} \cap \partial\mathcal{M}$.

Consider $m \in \hat{O}$, $\hat{\kappa}(m) = (x', z)$. The curve $\gamma(t) = (x', tz)$ is a geodesic. In fact, $\dot{\gamma}(t) = (0, z)$, $\ddot{\gamma}(t) = (0, 0)$, and with (17.5.1) we compute

$$\frac{D}{dt}\dot{\gamma}(t) = D_{\dot{\gamma}(t)}\dot{\gamma}(t) = \sum_{1\leq j\leq d} \Gamma^j_{dd}(x) z^2 \partial_j = 0,$$

since $\Gamma^j_{dd} = 0$ by (17.4.11) and the form of the metric in the chosen local chart.

We now claim that for $m \in \hat{O}$, $\hat{\kappa}(m) = (x', z)$ then $\operatorname{dist}_g(m, \partial\mathcal{M}) = z$. In fact, assume that $\tilde{m} \in \partial\mathcal{M}$ is such that $\delta = \operatorname{dist}_g(\tilde{m}, m) = \operatorname{dist}_g(m, \partial\mathcal{M})$ and $\tilde{m} \neq m' = \hat{\kappa}^{-1}(x', 0)$. Since $\hat{O}$ is a normal neighborhood of m there exists a unique geodesic joining m and $\tilde{m}$ by the Gauss lemma (Lemma 17.17), it is orthogonal to the geodesic sphere $\exp_m\big(S_g(m, \delta)\big)$. On the one hand, if this geodesic is orthogonal to $\partial\mathcal{M}$, then the above analysis and the uniqueness of geodesics show that $\tilde{m} = m'$. On the other hand, if this geodesic is not orthogonal to $\partial\mathcal{M}$, then points of $\exp_m\big(S_g(m, \delta)\big)$ lie outside $\mathcal{M}$ and yield points different from $\tilde{m}$ that are closer to m.

As $\overline{\Gamma}$ is compact, there exists a finite number of such charts $(\hat{\mathcal{C}}^i)_{1\leq i\leq n}$, $\hat{\mathcal{C}}^i = (\hat{O}^i, \hat{\kappa}^i)$ such that $\overline{\Gamma} \subset \cup_{1\leq i\leq n} \hat{O}^i$. For each $1 \leq i \leq n$, $\hat{\kappa}^i(\hat{O}^i) = \hat{O}'^i \times [0, \hat{z}^i_0)$ with $\hat{O}'^i$ an open set of $\mathbb{R}^{d-1}$ and $\hat{z}^i_0 > 0$. Set $z_0 = \min_{1\leq i\leq n} \hat{z}^i_0$ and

$$O^i = (\hat{\kappa}^i)^{-1}\big(\hat{O}'^i \times [0, z_0)\big) \subset \hat{O}^i, \quad i = 1, \dots, n, \qquad O = \cup_{1\leq i\leq n} O^i.$$

We have $\overline{\Gamma} \subset O$. We denote by $\mathcal{C}^i$ the modified chart $\mathcal{C}^i = (\kappa^i, O^i)$ with

$$\begin{aligned} \kappa^i : O^i &\to \hat{O}'^i \times [0, z_0) \\ m &\mapsto \hat{\kappa}^i(m). \end{aligned}$$

In the chart $\mathcal{C}^i$, the representative of the metric g has the form

$$g^{\mathcal{C}^i}_x = g^{\mathcal{C}^i\prime}_x + d_{x_d} \otimes d_{x_d} \quad \text{with } g^{\mathcal{C}^i\prime}_x = \sum_{1\leq i,j\leq d-1} g^{\mathcal{C}^i\prime}_{ij}(x) dx^i \otimes dx^j, \tag{17.6.5}$$

where $g^{\mathcal{C}^i\prime}_{(x',x_d)}$ is a metric on $\kappa(O^i \cap \partial\mathcal{M})$ with x_d acting as a parameter. For each chart $\mathcal{C}^i$, we denote by $\kappa^i_d(m)$ the last coordinate associated with the diffeomorphism κ^i, that is, $\kappa^i_d(m) = x_d$ if $x = \kappa^i(m)$ for $m \in O^i$. We also define smooth maps $\psi^i : m \mapsto (\kappa^i)^{-1}(x', 0)$ where $x' \in \mathbb{R}^{d-1}$ is such that $\kappa^i(m) = (x', x_d)$. From the form of the representative of the metric

$g_x^{\mathcal{C}^i}$ given above, the point $m' = \psi^i(m)$ is the unique point in $\partial\mathcal{M}$ such that $\operatorname{dist}_g(m, \partial\mathcal{M}) = \operatorname{dist}_g(m, m')$. Moreover, the path $t \mapsto (x', t)$ for $t \in [0, x_d]$ yields a parameterization of the unique geodesic joining m and m'.

If $m \in O^i \cap O^j$, we have

$$\kappa_d^i(m) = \kappa_d^j(m) = \operatorname{dist}_g(m, \partial\mathcal{M}), \quad \text{and } \psi^i(m) = \psi^j(m), \tag{17.6.6}$$

from the discussion above. We introduce the functions $\psi : O \to O \cap \partial\mathcal{M} = \Gamma$ and $\varphi : O \to [0, z_0)$ by

$$\psi(m) = \psi^i(m), \quad \varphi(m) = \kappa_d^i(m), \qquad \text{if } m \in O^i,$$

and we see that they are both well defined and smooth as the functions ψ^i coincide where the charts overlap.

If we set $\Psi : O \to \Gamma \times [0, z_0)$ with $\Psi(m) = \big(\psi(m), \varphi(m)\big)$ we see that Ψ is a diffeomorphism and the sought diffeomorphism is $\Phi = \Psi^{-1} : \Gamma \times [0, z_0) \to O$.

Finally, for two overlapping charts $\mathcal{C}^i$ and $\mathcal{C}^j$, setting $\kappa = \kappa^j \circ (\kappa^i)^{-1}$, denote by κ_∂ its restriction to $\partial\mathcal{M}$. It is a diffeomorphism from $\kappa^i(O^i \cap \partial\mathcal{M})$ onto $\kappa^j(O^j \cap \partial\mathcal{M})$ and we have

$$\kappa(x', x_d) = (\kappa_\partial(x'), x_d), \qquad (x', x_d) \in \kappa^i(O^i \cap O^j).$$

The formula expressing the action of a change of coordinates on 2-covariant tensor fields (15.6.2) relating $g^{\mathcal{C}^i}$ and $g^{\mathcal{C}^j}$ and (17.6.5) then yields

$$g^{\mathcal{C}^i\prime}(x', x_d) = \kappa_\partial^* \, g^{\mathcal{C}^j\prime}(\kappa_\partial(x'), x_d).$$

Here $z = x_d$ is simply a parameter and this formula shows that there exists a family of metrics $g'(z)$ on Γ, smoothly depending on z, such that its representative in the local chart $\mathcal{C}^i$ is given by $g^{\mathcal{C}^i\prime}(x', z)$. Then, $g'(z) \otimes 1_z$ is 2-covariant tensor field on $\Gamma \times [0, z_0)$. Let $m = \Phi(m', z) \in O$ and $u, v \in T_m\mathcal{M}$. We set $(u', u_z) = \Phi^* u$ and $(v', v_z) = \Phi^* v$ both in $T_{m'}\partial\mathcal{M} \times \mathbb{R}$. From (17.6.5) we conclude that

$$g_m(u, v) = g'_{m'}(z)(u', v') + u^z v^z,$$

meaning that $\Phi^* g_m = g'_{m'}(z) \otimes 1_z + d_z \otimes d_z$. ∎

The following corollary generalizes the local result obtained in (17.6.3)–(17.6.4). It is a consequence of the definition of the Laplace–Beltrami operator.

COROLLARY 17.23. *Let Γ, O and $\Phi : \Gamma \times [0, z_0) \to O$ be as given by Theorem 17.22. Set $\tilde{g} = \Phi^* g$, that is,*

$$\tilde{g}_{(m', z)} = g'_{m'}(z) \otimes 1_z + d_z \otimes d_z.$$

The Laplace–Beltrami operator on $\Gamma \times [0, z_0)$ associated with this metric satisfies $\Delta_{\tilde{g}} = \Phi^* \Delta_g (\Phi^{-1})^*$ and is given by

$$\Delta_{\tilde{g}} = \partial_z^2 + \Delta_{g'(z)} + \frac{1}{2}(\det g'(z))^{-1}\big(\partial_z \det g'(z)\big)\partial_z,$$

where $\Delta_{g'(z)}$ is the Laplace–Beltrami operator on Γ associated with the metric $g'(z)$, with z as a parameter.

Note that the formula $\Delta_{\tilde{g}} = \Phi^* \Delta_g (\Phi^{-1})^*$ follows from Proposition 17.5.

The following result can be easily proven by means of the normal geodesic coordinates obtained in Theorem 17.22.

PROPOSITION 17.24. *Let $(\mathcal{M}, g)$ be a Riemannian manifold. Let Ω be a connected open set of $\mathcal{M}$. Let Γ be a bounded open subset of $\partial\mathcal{M}$ with $\overline{\Gamma} \subset \Omega$. Let also V be an open neighborhood of Γ in $\mathcal{M}$. There exists an open neighborhood W of Γ in $\mathcal{M}$ such that $W \Subset V$ and $\Omega \setminus \overline{W}$ is connected.*

PROOF. Let Γ' be a bounded open neighborhood of $\overline{\Gamma}$ in $\Omega \cap \partial\mathcal{M}$ such that $\overline{\Gamma'} \subset V \cap \partial\mathcal{M}$. By Theorem 17.22 there exist O an open set of $\mathcal{M}$, $z_0 > 0$, and a diffeomorphism Ψ, such that $\Gamma' = O \cap \partial\mathcal{M}$ and

$$\begin{aligned} \Psi : \Gamma' \times [0, z_0) &\to O \\ (m', z) &\mapsto \Psi(m', z). \end{aligned}$$

We claim that there exists $z_1 > 0$ sucht that $\Psi^{-1}\big(\Gamma' \times [0, 2z_1)\big) \subset \Omega$. Otherwise, we can find a sequence $(m_n)_n \subset \mathcal{M} \setminus \Omega$ that converges toward a point of Γ'; a contradiction. We set $O_1 = \Psi^{-1}\big(\Gamma' \times [0, z_1)\big)$.

We now consider a function $h \in \mathscr{C}_c^\infty(\Gamma')$ with $0 \leq h \leq 1$ such that $h \equiv 1$ on a neighborhood of Γ. We then define the following map $H : \Omega \to \Omega$,

$$H(m) = \begin{cases} m & \text{if } m \notin O_1; \\ \Psi(m', (1 - h(m'))z + h(m')z_1) & \text{if } m \in O_1 \text{ and } (m', z) = \Psi^{-1}(m). \end{cases}$$

This map is continuous and thus $H(\Omega)$ is connected. If we set $W = \Omega \setminus H(\Omega)$ we see that W is bounded and fulfills the required properties. ∎

17.7. Higher-Order Covariant Derivatives

For θ a r-covariant,s-contravariant tensor field, setting $\mathsf{D}^0 \theta = \theta$, for $k \in \mathbb{N}^*$, one defines by induction, the $(r+k)$-covariant,s-contravariant tensor field given by,

$$\mathsf{D}^k \theta = \mathsf{D}(\mathsf{D}^{k-1} \theta). \tag{17.7.1}$$

For a function f we saw above that $\mathsf{D} f = df$. The 2-covariant tensor $\mathsf{H} f = \mathsf{D}^2 f$ is called the Hessian of f. In a local chart $\mathcal{C} = (O, \kappa)$ we find the representative of $\mathsf{H} f$ to be $(\mathsf{H} f)^{\mathcal{C}} = \sum_{1 \leq i,j \leq j} (\mathsf{H} f)^{\mathcal{C}}_{ij} dx^i \otimes dx^j$, with

$$(\mathsf{H} f)^{\mathcal{C}}_{ij} = \partial^2_{x_i x_j} f^{\mathcal{C}} - \sum_{1 \leq k \leq d} \Gamma^k_{ij} \partial_{x_k} f^{\mathcal{C}}. \tag{17.7.2}$$

For $\ell \in \mathbb{N}$, following (17.4.12), we define accordingly

$$\partial^\ell_\nu f(m) = \mathsf{D}^\ell f_m(\underbrace{\nu_m, \dots, \nu_m}_{\ell}). \tag{17.7.3}$$

We have the following proposition that shows that this definition is consistent with what one expects from the Euclidean case.

PROPOSITION 17.25. *Let $m^0 \in \partial\mathcal{M}$ and $\mathcal{C} = (O, \kappa)$ be local chart such that $m^0 \in O$ and such that κ provides normal geodesic coordinates in O. If $x = (x_1, \ldots, x_d) = \kappa(m)$ for $m \in O$, we have*

$$\partial_\nu^\ell f(m) = (-\partial_{x_d})^\ell f^{\mathcal{C}}(x), \qquad x = \kappa(m),\ m \in \partial\mathcal{M}.$$

For normal geodesic coordinates we refer to Sections 9.4 and 17.6. We recall that a local chart at a boundary point is such that $\tilde{\kappa}(\tilde{O})$ is an open set of $\overline{\mathbb{R}^d_+}$ and $\tilde{\kappa}(\partial\mathcal{M} \cap \tilde{O}) = \{x_d = 0\} \cap \tilde{\kappa}(\tilde{O})$.

PROOF. In the normal geodesic coordinates provided by κ, the representative of ν in $\mathcal{C}$ is $\nu^{\mathcal{C}} = (0, \ldots, 0, -1)$. We have

$$\partial_\nu^\ell f(m) = \mathsf{D}^\ell f_m(\underbrace{\nu_m, \ldots, \nu_m}_{\ell}) = (-1)^\ell (\mathsf{D}^\ell f_x^{\mathcal{C}})_{\underbrace{d \ldots d}_{\ell}}.$$

For concision we shall write f in place of $f^{\mathcal{C}}$ here. It now suffices to prove that

$$(\mathsf{D}^\ell f_x)_{\underbrace{d \ldots d}_{\ell}} = \partial_{x_d}^\ell f(x) \quad \text{for all } x \in \kappa(O).$$

The proof goes by induction. The result is clear for $\ell = 1$. We assume that the results hold for ℓ. With (17.4.9), we have

$$(\mathsf{D}^{\ell+1} f_x)_{\underbrace{d \ldots d}_{\ell+1}} = \partial_{x_d} (\mathsf{D}^\ell f_x)_{\underbrace{d \ldots d}_{\ell}} - \sum_{\substack{1 \leq n \leq \ell \\ 1 \leq k \leq d}} \Gamma_{dd}^k (\mathsf{D}^\ell f_x)_{\underbrace{d \ldots d}_{n-1} k\, d \ldots d}.$$

For normal geodesic coordinates we have $\Gamma_{dd}^k(x) = 0$ by (17.4.11). The conclusion follows. ∎

CHAPTER 18

Sobolev Spaces and Laplace Problems on a Riemannian Manifold

Contents

J. Le Rousseau et al., *Elliptic Carleman Estimates and Applications to Stabilization and Controllability, Volume II*, PNLDE Subseries in Control 98, https://doi.org/10.1007/978-3-030-88670-7_18

With this chapter, our first goal is to give a geometrical definition of Sobolev spaces on a Riemannian manifold, as the norms associated with those spaces are used in some of the Carleman estimates proven in Chapter 3 of Volume 1 and Chap. 8. A second goal is the study of the elliptic properties of the Laplace–Beltrami operator and diverse boundary value problems associated with that operator.

18.1. L^2 and H^1-Spaces

Let $(\mathcal{M}, g)$ be a smooth d-dimensional Riemannian manifold and let μ_g be the canonical positive density function defined in Sect. 17.3. It allows one to identify functions and density functions, that is, 0-density functions and 1-density functions, 0-density distributions and 1-density distributions, etc.

An L^2-function f on $\mathcal{M}$ (resp. L^2_{loc}) is then a complex valued function f such that $|f|^2\mu_g$ is an L^1-density function (resp. L^1_{loc}).

On $L^2(\mathcal{M})$, the vector space of L^2-functions, we define the inner product and associated norm

$$(f, g)_{L^2(\mathcal{M})} = \int_{\mathcal{M}} f\bar{g}\mu_g, \qquad \|f\|^2_{L^2(\mathcal{M})} = \int_{\mathcal{M}} |f|^2\mu_g,$$

which yields a Hilbert space structure.

REMARK 18.1. If $f \in L^2(\mathcal{M})$, then for every local chart $\mathcal{C} = (O, \kappa)$ its representative is such that $f^{\mathcal{C}} \in L^2(\kappa(O))$. Conversely, if $\mathcal{M}$ is compact, if $f^{\mathcal{C}} \in L^2(\kappa(O))$ for every chart (using charts in a finite atlas suffices), then $f \in L^2(\mathcal{M})$.

PROPOSITION 18.2. *Let $\mathcal{M}$ be σ-compact. The space ${}^0\mathscr{D}_c^\infty(\mathcal{M})$ is dense in $L^2(\mathcal{M})$.*

PROOF. Let $f \in L^2(\mathcal{M})$ and $\varepsilon > 0$. If $(K_n)_n$ is an exhaustive sequence of compact sets for $\mathcal{M}$, we set $f_n = \mathbf{1}_{K_n} f$. The Lebesgue dominated-convergence theorem yield $n_0 \in \mathbb{N}$ such that $\|f_{n_0} - f\|_{L^2(\mathcal{M})} \le \varepsilon$. Then as $\operatorname{supp}(f_{n_0}) \subset K_{n_0}$, working in a finite number of local charts that cover K_{n_0}, there exists a sequence of smooth functions h supported in K_{n_0} such that $\|f_{n_0} - h\|_{L^2(\mathcal{M})} \le \varepsilon$, which yields the result. ■

We recall the notation $\mathscr{C}^\infty V(\mathcal{M})$ for the space of smooth vector fields. We define the following inner product and norm, for $w, v \in \mathscr{C}^\infty V(\mathcal{M})$,

$$(u, v)_{L^2V(\mathcal{M})} = \int_{\mathcal{M}} g(u, \bar{v})\mu_g, \quad \text{and } \|u\|^2_{L^2V(\mathcal{M})} = \int_{\mathcal{M}} g(u, \bar{u})\mu_g.$$

We denote by $L^2V(\mathcal{M})$ the completion of the space

$$\{u \in \mathscr{C}^\infty V(\mathcal{M});\ \|u\|_{L^2V(\mathcal{M})} < \infty\}$$

respect to this norm. If $v \in L^2V(\mathcal{M})$ we say that v is an L^2-vector field.

For a smooth function f, h on $\mathcal{M}$ we define the H^1-inner product and norm

$$(f,h)_{H^1(\mathcal{M})} = (f,h)_{L^2(\mathcal{M})} + (\nabla_g f, \nabla_g h)_{L^2V(\mathcal{M})},$$
$$\|f\|^2_{H^1(\mathcal{M})} = \|f\|^2_{L^2(\mathcal{M})} + \|\nabla_g f\|^2_{L^2V(\mathcal{M})}.$$

We then define the space $H^1(\mathcal{M})$ as the completion of

$$\{f \in \mathscr{C}^\infty(\mathcal{M});\ \|f\|_{H^1(\mathcal{M})} < \infty\}$$

with respect to this norm. We then obtain a Hilbert space. The map $f \to f_{|\partial\mathcal{M}}$ on $\mathscr{C}^\infty(\mathcal{M})$ extends uniquely as a map from $H^1(\mathcal{M})$ into $L^2_{\text{loc}}(\partial\mathcal{M})$ and is referred to as the trace map as in Euclidean spaces. For H^1-functions we use the classical notation $f_{|\partial\mathcal{M}}$ for its trace on $\partial\mathcal{M}$.

REMARK 18.3. Similarly to Remark 18.1 we see that if $f \in H^1(\mathcal{M})$, then for every local chart $\mathcal{C} = (O, \kappa)$ its representative is such that $f^{\mathcal{C}} \in H^1(\kappa(O))$. Conversely, if $\mathcal{M}$ is compact, if $f^{\mathcal{C}} \in H^1(\kappa(O))$ for every chart (using charts in a finite atlas suffices), then $f \in H^1(\mathcal{M})$.

The space $H^1_0(\mathcal{M})$ is defined as the closure for the norm $\|.\|_{H^1(\mathcal{M})}$ of ${}^0\mathscr{D}^\infty_c(\mathcal{M})$.

PROPOSITION 18.4. *Let $(\mathcal{M}, g)$ be a complete Riemannian manifold without boundary. Then $H^1_0(\mathcal{M}) = H^1(\mathcal{M})$.*

A Riemannian manifold is said to be complete if it is complete for its natural distance as given in (17.1.2). This result is based on the exhaustion of $\mathcal{M}$ by compact balls with respect to the Riemannian distance and a Lipschitz cutoff of H^1-functions that reduces the analysis to compactly supported H^1 functions. In the case of a compact manifold without boundary the proof is simply based on the use a finite partition of unity subordinated to an atlas covering, allowing one to use the result known for bounded regular open sets in $\mathbb{R}^d$.

A similar argument, in the case of a compact Riemannian manifold with boundary yields the following density result (anticipating the notion of traces that we present in Sect. 18.5 below).

PROPOSITION 18.5. *Let $(\mathcal{M}, g)$ be a compact Riemannian manifold with boundary. Then $H^1_0(\mathcal{M})$ is the space of H^1-functions f such that $f_{|\partial\mathcal{M}} = 0$.*

REMARK 18.6. Note that the property of Proposition 18.4 is open for Sobolev spaces of higher order if additional assumptions are not made on the manifold [167].

THEOREM 18.7 (Rellich-Kondrachov). *Given χ continuous with compact support on $\mathcal{M}$ the map $u \mapsto \chi u$ is compact from $H^1(\mathcal{M})$ into $L^2(\mathcal{M})$.*

This can be proven using a finite number of local charts that cover the support of χ and by using the counterpart result on $\mathbb{R}^d$ (see, for instance, [90, Theorem 9.16]).

We have the following Poincaré inequalities.

LEMMA 18.8 (Poincaré Inequality). *Let $(\mathcal{M}, g)$ be a connected Riemannian manifold. Let U be an open subset of $\mathcal{M}$ such that $\overline{U}$ is a compact of $\mathcal{M}$ and $\overline{U} \neq \mathcal{M}$. There exists $C > 0$ such that*

$$\|f\|_{L^2(\mathcal{M})} \leq C \|\nabla_g f\|_{L^2 V(\mathcal{M})}, \tag{18.1.1}$$

for $f \in H^1(\mathcal{M})$ supported in $\overline{U}$.

PROPOSITION 18.9 (Poincaré Inequalities). *Let $(\mathcal{M}, g)$ be a connected compact Riemannian manifold. If Γ is an open subset of $\partial\mathcal{M}$, there exists $C > 0$ such that*

$$\|f\|_{L^2(\mathcal{M})} \leq C \|\nabla_g f\|_{L^2 V(\mathcal{M})}, \tag{18.1.2}$$

for $f \in H^1(\mathcal{M})$ such that $f_{|\Gamma} = 0$. A similar inequality holds for H^1-functions with zero mean value, that is, $\int_{\mathcal{M}} f \mu_g = 0$. If $\Gamma = \partial\mathcal{M}$, we call Poincaré constant C_g the largest positive constant such that

$$C_g^{1/2} \|f\|_{L^2(\mathcal{M})} \leq \|\nabla_g f\|_{L^2 V(\mathcal{M})}$$

for $f \in H^1(\mathcal{M})$.

If $A \subset U$ is of positive measure, there exists also $C > 0$ such that

$$\|f\|_{L^2(\mathcal{M})} \leq C \big(\|f\|_{L^2(A)} + \|\nabla_g f\|_{L^2 V(\mathcal{M})} \big), \tag{18.1.3}$$

for $f \in H^1(\mathcal{M})$.

PROOFS OF LEMMA 18.8 AND PROPOSITION 18.9. Assume that inequality (18.1.1) does not hold. Then, there exists a sequence $(f^n)_{n \in \mathbb{N}} \subset H^1(\mathcal{M})$ with $\operatorname{supp}(f^n) \subset \overline{U}$ and such that

$$\|f^n\|_{L^2(\mathcal{M})} = 1, \quad \|\nabla_g f^n\|_{L^2 V(\mathcal{M})} \to 0. \tag{18.1.4}$$

In particular, f^n is bounded in $H^1(\mathcal{M})$ and, thus, there exists $f \in H^1(\mathcal{M})$ such that $f^n \rightharpoonup f$ in $H^1(\mathcal{M})$ up to a subsequence, since the unit ball of a Hilbert space is weakly compact by the Banach Alaoglu theorem [90, Theorem 3.16]. Choose $\chi \in \mathscr{C}_c^\infty(\mathcal{M})$ be such that $\chi = 1$ in a neighborhood of $\overline{U}$ using Lemma 15.16. Then, $f^n = \chi f^n$ converges strongly to $f = \chi f$ in $L^2(\mathcal{M})$ by the Rellich–Kondrachov theorem (see Theorem 18.7). As we have $\nabla_g f^n \rightharpoonup \nabla_g f$ in $L^2 V(\mathcal{M})$, we find that $\nabla_g f = 0$. Because of the form of ∇_g given in (17.2.2) this implies that $f \equiv \text{Cst}$ in every chart and it thus constant on the whole $\mathcal{M}$ as $\mathcal{M}$ is connected. As $\operatorname{supp}(f) \subset \overline{U} \neq \mathcal{M}$ we have $f \equiv 0$. We have thus found that χf^n strongly converges to 0 in $L^2(\mathcal{M})$, which yields a contradiction as $\|\chi f^n\|_{L^2(\mathcal{M})} = \|f^n\|_{L^2(\mathcal{M})} = 1$. This concludes the proof of Lemma 18.8.

For the proof of Proposition 18.9 the manifold $\mathcal{M}$ is compact and we can use the same setting with $U = \mathcal{M}$ and $\chi \equiv 1$. Considering a sequence $(f^n)_{n\in\mathbb{N}} \subset H^1(\mathcal{M})$ that contradicts (18.1.2) by fulfilling (18.1.4) we find as above that, up to a subsequence, $f^n \rightharpoonup f$ in $H^1(\mathcal{M})$ and $f^n \to f$ in $L^2(\mathcal{M})$ with f a constant function.

If $f^n_{|\Gamma} = 0$ since the map $h \mapsto h_{|\Gamma}$ is continuous on $H^1(\mathcal{M})$ into $L^2_{\text{loc}}(\Gamma)$ we find that $f^n_{|\Gamma} \rightharpoonup f_{|\Gamma}$. Since $f^n_{|\Gamma} = 0$ then $f_{|\Gamma} = 0$, which implies that $f \equiv 0$ on $\mathcal{M}$.

If $\int_{\mathcal{M}} f^n \mu_g = 0$ we have

$$0 = \int_{\mathcal{M}} f^n \mu_g = (f^n, 1)_{L^2(\mathcal{M})} \to (f, 1)_{L^2(\mathcal{M})} = \int_{\mathcal{M}} f \mu_g.$$

As $f \equiv \text{Cst}$, we find $f \equiv 0$.

We now consider a sequence $(f^n)_{n\in\mathbb{N}} \subset H^1(\mathcal{M})$ that contradicts (18.1.3), that is,

$$\|f^n\|_{L^2(\mathcal{M})} = 1, \quad \|f^n\|_{L^2(A)} \to 0, \quad \|\nabla_g f^n\|_{L^2V(\mathcal{M})} \to 0.$$

As above we find that, up to a subsequence, $f^n \rightharpoonup f$ in $H^1(\mathcal{M})$ and $f^n \to f$ in $L^2(\mathcal{M})$ with f a constant function. We also find gives $0 = \|f^n\|_{L^2(A)} \to \|f\|_{L^2(A)}$ implying that $f \equiv 0$ since $|A| > 0$.

In all cases the same contradiction as above is reached. ∎

For $s \in \mathbb{R}$, we also define the space $H^s_{\text{loc}}(\mathcal{M})$ as the space of (0-density) distributions u on $\mathcal{M}$ such its local representative is H^s_{loc} in every chart. More precisely, for $u \in {}^0\mathscr{D}'(\mathcal{M})$, we say that $u \in H^s_{\text{loc}}(\mathcal{M})$ if for every chart $\mathcal{C} = (O, \kappa)$ we have $u^{\mathcal{C}} \in H^s_{\text{loc}}(\operatorname{int} \kappa(O))$. Recall that $\operatorname{int} \kappa(O)$ here denotes that interior of $\kappa(O)$ in $\mathbb{R}^d$.

PROPOSITION 18.10. *Let $u \in H^1(\mathcal{M})$ and $\varphi \in {}^0\mathscr{D}^\infty_c(\mathcal{M})$. One has*

$$\int_{\mathcal{M}} g(\nabla_g u, \nabla_g \varphi)\, \mu_g + \int_{\mathcal{M}} u \Delta_g \varphi\, \mu_g = 0.$$

PROOF. One simply applies the divergence formula of Proposition 17.9 to the L^1-vector field $u\nabla_g\varphi$ whose divergence is also L^1. ∎

18.2. Sobolev Spaces

18.2.1. L^2-Based Spaces. Let $s \in \mathbb{R}$. We recall that given a (0-density) distribution u on $\mathcal{M}$, we say that $u \in H^s_{\text{loc}}(\mathcal{M})$, if for each chart $\mathcal{C} = (O, \kappa)$ its representative is in $H^s_{\text{loc}}(\kappa(O))$. We see immediately that if Q is a differential operator of order $m \in \mathbb{N}$ it maps $H^s_{\text{loc}}(\mathcal{M})$ into $H^{s-m}_{\text{loc}}(\mathcal{M})$ for all $s \in \mathbb{R}$.

In contrast, to introduce the Sobolev space $H^s(\mathcal{M})$, a proper norm needs to be defined. On a compact manifold, given a finite atlas and a subordinated partition of unity, one can define Sobolev norms. However, such norms are not intrinsic. On Riemannian manifolds the notion of covariant derivative allows one to define intrinsic Sobolev norms that appear as natural extensions of those used on $\mathbb{R}^d$.

On a Riemannian manifold $(\mathcal{M}, g)$, for $m \in \mathcal{M}$, as seen above in Sect. 17.1, the metric g yields a natural norm on tangent vectors at m, that is, $|v|^2_{g_m} = g_m(v, \bar{v})$ if $v \in T_m\mathcal{M}$ and a natural norm on cotangent vectors at m, that is $|\omega|^2_{g_m} = g_m(\omega^\sharp, \bar{\omega}^\sharp)$ if $\omega \in T^*_m\mathcal{M}$. In a local chart $\mathcal{C} = (O, \kappa)$, for $x = \kappa(m)$, it reads

$$|v|^2_{g_m} = \sum_{1 \le j,j' \le d} g^{\mathcal{C}}_{jj'}(x) v^{\mathcal{C},j} \bar{v}^{\mathcal{C},j'}, \qquad \text{and} \quad |\omega|^2_{g_m} = \sum_{1 \le i,i' \le d} g^{\mathcal{C},ii'}(x) \omega^{\mathcal{C}}_i \bar{\omega}^{\mathcal{C}}_{i'}.$$

A generalization of these norms to r-covariant,s-contravariant tensors θ at m is the following norm

(18.2.1)

$$|\theta|^2_{g_m} = \sum_{\substack{1 \le i_1, i'_1, \dots, i_r, i'_r, \le d \\ 1 \le j_1, j'_1, \dots, j_s, j'_s \le d}} \big(g^{\mathcal{C}, i_1 i'_1} \cdots g^{\mathcal{C}, i_r i'_r} g^{\mathcal{C}}_{j_1 j'_1} \cdots g^{\mathcal{C}}_{j_s j'_s} \, \theta^{j_1 \dots j_s}_{i_1 \dots i_r} \bar{\theta}^{j'_1 \dots j'_s}_{i'_1 \dots i'_r} \big)(x).$$

This norm is independent of the chosen local chart. The associated inner product is

$$(\theta, \vartheta)_{g_m} = \sum_{\substack{1 \le i_1, i'_1, \dots, i_r, i'_r, \le d \\ 1 \le j_1, j'_1, \dots, j_s, j'_s \le d}} \big(g^{\mathcal{C}, i_1 i'_1} \cdots g^{\mathcal{C}, i_r i'_r} g^{\mathcal{C}}_{j_1 j'_1} \cdots g^{\mathcal{C}}_{j_s j'_s} \, \theta^{j_1 \dots j_s}_{i_1 \dots i_r} \bar{\vartheta}^{j'_1 \dots j'_s}_{i'_1 \dots i'_r} \big)(x).$$

A r-covariant, s-contravariant tensor fields θ is said to be L^2 if $|\theta|_{g_m} \in L^2(\mathcal{M})$. We denote by $L^2\mathscr{T}_r^s(\mathcal{M})$ the space of such tensor fields. Equipped with the norm

$$\|\theta\|^2_{L^2\mathscr{T}_r^s(\mathcal{M})} = \int_{\mathcal{M}} |\theta_m|^2_{g_m} \mu_g,$$

and the associated inner product

$$(\theta, \vartheta)_{L^2\mathscr{T}_r^s(\mathcal{M})} = \int_{\mathcal{M}} (\theta_m, \vartheta_m)_{g_m} \mu_g,$$

one obtains a Hilbert space.

We recall the notation $\mathscr{C}^\infty\mathscr{T}_r^s(\mathcal{M})$ introduced in Sect. 15.6.3 for the space of smooth r-covariant, s-contravariant tensor fields on $\mathcal{M}$. The space $L^2\mathscr{T}_r^s(\mathcal{M})$ is in fact the completion of

$$\{\theta \in \mathscr{C}^\infty\mathscr{T}_r^s(\mathcal{M});\ \|\theta\|_{L^2\mathscr{T}_r^s(\mathcal{M})} < \infty\}$$

with respect to $\|.\|_{L^2\mathscr{T}_r^s(\mathcal{M})}$.

If $\theta \in \mathscr{C}^\infty\mathscr{T}_r^s(\mathcal{M})$, and $j \in \mathbb{N}$, one has $\mathsf{D}^j\,\theta \in \mathscr{C}^\infty\mathscr{T}_{r+j}^s(\mathcal{M})$. For $k \in \mathbb{N}$ we define the norm

$$\|\theta\|^2_{H^k\mathscr{T}_r^s(\mathcal{M})} = \sum_{j=0}^{k} \|\,\mathsf{D}^j\,\theta\|^2_{L^2\mathscr{T}_{r+j}^s(\mathcal{M})},$$

and the Sobolev space $H^k \mathscr{T}_r^s$ of H^k r-covariant,s-contravariant tensor fields as the completion of

$$\{\theta \in \mathscr{C}^\infty \mathscr{T}_r^s(\mathcal{M});\ \|\theta\|_{H^k \mathscr{T}_r^s(\mathcal{M})} < \infty\},$$

with respect to this norm. Then, for $\theta \in H^k \mathscr{T}_r^s(\mathcal{M})$ we have $\mathsf{D}^j \theta \in H^{k-j} \mathscr{T}_{r+j}^s(\mathcal{M})$ for $j = 0, \dots, k$. Here, the action of D^j can be understood as a unique extension or in a weak sense (the latter can be defined in local charts). The inner product associated with $\|.\|_{H^k \mathscr{T}_r^s(\mathcal{M})}$ is given by

$$(\theta, \vartheta)_{H^k \mathscr{T}_r^s(\mathcal{M})} = \sum_{j=0}^{k} (\mathsf{D}^j \theta, \mathsf{D}^j \vartheta)_{L^2 \mathscr{T}_{r+j}^s(\mathcal{M})},$$

for $\theta, \vartheta \in H^k \mathscr{T}_r^s(\mathcal{M})$. With this inner product $H^k \mathscr{T}_r^s(\mathcal{M})$ is a Hilbert space.

In the case $r = s = 0$ we simply write $L^2(\mathcal{M})$ and $H^k(\mathcal{M})$, and we recover the L^2- and H^1-functions introduced in Sect. 18.1. Given a function f on $\mathcal{M}$, $\mathsf{D}^k f$ is a k-covariant tensor field. As $(\mathsf{D} f)^\sharp = (df)^\sharp = \nabla_g f$ we see that the H^1-norm defined in Sect. 18.1 coincides with the H^1-norm given here.

Similarly to Remarks 18.1 and 18.3 we see that if $f \in H^k(\mathcal{M})$, then for every local chart $\mathcal{C} = (O, \kappa)$ its representative is such that $f^{\mathcal{C}} \in H^k(\kappa(O))$. Conversely, if $\mathcal{M}$ is compact, if $f^{\mathcal{C}} \in H^k(\kappa(O))$ for every chart, then $f \in H^k(\mathcal{M})$. In fact, we can prove the following result.

PROPOSITION 18.11. *Let $(\mathcal{M}, g)$ be a compact manifold (with or without boundary). Let $\mathcal{A} = (\mathcal{C}^i)_{1 \leq i \leq N}$ be a finite atlas with $\mathcal{C}^i = (O^i, \kappa^i)$. There exists $C > 0$ such that*

$$C^{-1} \|f\|_{H^k(\mathcal{M})} \leq \sum_{1 \leq i \leq N} \|f^{\mathcal{C}^i}\|_{H^k(\kappa^i(O^i))} \leq C \|f\|_{H^k(\mathcal{M})}. \tag{18.2.2}$$

The constant C depends on the choice of the atlas. Moreover, if $i \in \{1, \dots, N\}$, for some $C_i > 0$, we have

$$C_i^{-1} \|f\|_{H^k(\mathcal{M})} \leq \|f^{\mathcal{C}^i}\|_{H^k(\kappa^i(O^i))} \leq C_i \|f\|_{H^k(\mathcal{M})} \tag{18.2.3}$$

if $\operatorname{supp}(f) \subset O^i$.

With the atlas used in this proposition and with the density of smooth functions in each local chart on see that smooth functions are dense in $H^k(\mathcal{M})$ if $\mathcal{M}$ is compact with or without boundary.

In the case of vector fields, we used the notation $L^2 V(\mathcal{M})$ in Sect. 18.1. It coincides with $L^2 \mathscr{T}_0^1(\mathcal{M})$ here and similarly we use $H^k V(\mathcal{M})$ to denote $H^k \mathscr{T}_0^1(\mathcal{M})$.

In the case of one-forms, we use the notation $L^2 \Lambda\ (\mathcal{M})$ instead of $L^2 \mathscr{T}_1^0(\mathcal{M})$ and similarly we use $H^k \Lambda(\mathcal{M})$ to denote $H^k \mathscr{T}_1^0(\mathcal{M})$. In the case of r-covariant tensor fields, we use the notation $L^2 \Lambda^r(\mathcal{M})$ instead of $L^2 \mathscr{T}_r^0(\mathcal{M})$ and similarly we use $H^k \Lambda^r(\mathcal{M})$ to denote $H^k \mathscr{T}_r^0(\mathcal{M})$.

The Sobolev spaces defined above coincide with the usual spaces if $\mathcal{M}\backslash\partial\mathcal{M}$ is an open set[1] of $\mathbb{R}^d$. The action of differential operator on Sobolev space is consistent with the results on $\mathbb{R}^d$. For simplicity, to avoid boundedness assumptions on the coefficients of the operator we consider compact manifold in the following proposition.

PROPOSITION 18.12. *Let $(\mathcal{M}, g)$ be a smooth Riemannian manifold such that $\mathcal{M}$ is compact (with or without boundary). Let Q be a differential operator of order k with smooth coefficients on $\mathcal{M}$. Then, for $\ell \in \mathbb{N}$, the operator Q maps $H^{\ell+k}(\mathcal{M})$ into $H^{\ell}(\mathcal{M})$ continuously.*

Finally, for $k \in \mathbb{N}$, we define $H_0^k(\mathcal{M})$ as the closure for the norm $\|.\|_{H^m(\mathcal{M})}$ of ${}^0\mathscr{D}_c^\infty(\mathcal{M})$. Here, $H_0^k(\mathcal{M})$ is a space 0-densities. Its dual space $\big(H_0^k(\mathcal{M})\big)'$ is a space of 1-density distributions. Using the canonical 1-density μ_g on $(\mathcal{M}, g)$ we can identify $\big(H_0^k(\mathcal{M})\big)'$ with a space of 0-density distributions that we denote by $H^{-k}(\mathcal{M})$.

Using the space $L^2(\mathcal{M})$ as a *pivot* space, we find that

$$\langle u, \bar{v}\rangle_{H^{-k}(\mathcal{M}),H_0^k(\mathcal{M})} = (u, v)_{L^2(\mathcal{M})} = \int_{\mathcal{M}} u\bar{v}\, \mu_g,$$

if $v \in H_0^k(\mathcal{M})$ and $u \in L^2(\mathcal{M})$. Note that using 1/2-density spaces instead is an alternative approach. The norm $\|.\|_{H^{-k}(\mathcal{M})}$ is naturally defined as the norm of a dual space of normed vector space and yields a Hilbert space structure on $H^{-k}(\mathcal{M})$.

18.2.2. L^p-Based Spaces. Above we considered Sobolev spaces that are based on the L^2-structure. In the present book we need very few aspects of Sobolev spaces based on a L^p-structure, with $1 \leq p \leq \infty$.

First, we treat the case $1 \leq p < \infty$. One says that f is a L^p function if $|f|^p$ is a L^1-function and one sets

$$\|f\|^p_{L^p(\mathcal{M})} = \int_{\mathcal{M}} |f|^p \mu_g.$$

With (18.2.1) defining the function $|\theta_m|_{g_m}$ for a r-covariant,s-contravariant tensor field θ, one says that θ is L^p on $\mathcal{M}$ if $|\theta_m|_{g_m}$ is a L^p function on $\mathcal{M}$. One writes $\theta \in L^p\mathscr{T}_r^s(\mathcal{M})$ and one defines the norm

$$\|\theta\|^p_{L^p\mathscr{T}_r^s(\mathcal{M})} = \int_{\mathcal{M}} |\theta_m|^p_{g_m} \mu_g.$$

If $\theta \in \mathscr{C}^\infty\mathscr{T}_r^s(\mathcal{M})$ and $k \in \mathbb{N}$ we define the norm

$$\|\theta\|^p_{W^{k,p}\mathscr{T}_r^s(\mathcal{M})} = \sum_{j=0}^{k} \|\, \mathsf{D}^j\, \theta\|^p_{L^p\mathscr{T}_{r+j}^s(\mathcal{M})},$$

[1]Recall that here manifolds with boundary are defined such that $\partial\mathcal{M} \subset \mathcal{M}$; see Definition 15.2.

and the Sobolev space $W^{k,p}\mathscr{T}_r^s$ of $W^{k,p}$ r-covariant,s-contravariant tensor fields as the completion of

$$\{\theta \in \mathscr{C}^\infty \mathscr{T}_r^s(\mathcal{M});\ \|\theta\|_{W^{k,p}\mathscr{T}_r^s(\mathcal{M})} < \infty\},$$

with respect to this norm. One finds that if $\theta \in W^{k,p}\mathscr{T}_r^s(\mathcal{M})$, then $\mathsf{D}^j\,\theta \in W^{k-j,p}\mathscr{T}_{r+j}^s(\mathcal{M})$ for $1 \leq j \leq k$. For $k \in \mathbb{N}$ and $1 \leq p < \infty$, one can check that $W^{k,p}\mathscr{T}_r^s(\mathcal{M})$ is a Banach space if equipped with the above norm.

Second, we treat the case $p = \infty$. Given an atlas $\mathcal{A} = (\mathcal{C}^i)_{i\in\mathcal{I}}$, with $\mathcal{C}^i = (O^i, \kappa^i)$, for a function f defined on $\mathcal{M}$, a L^∞-norm can be defined as

$$\|f\|_{L^\infty(\mathcal{M})} = \sup_{i\in\mathcal{I}} \|f^{\mathcal{C}^i}\|_{L^\infty(\kappa^i(O^i))},$$

where $f^{\mathcal{C}^i}$ is the representative of f in $\mathcal{C}^i$. One can check that this norm is in fact independent of the chosen atlas.

For θ a r-covariant,s-contravariant tensor field, one says that θ is L^∞ on $\mathcal{M}$ if $|\theta_m|_{g_m} \in L^\infty(\mathcal{M})$. One writes $\theta \in L^\infty\mathscr{T}_r^s(\mathcal{M})$ and one defines the norm

$$\|\theta\|_{L^\infty\mathscr{T}_r^s(\mathcal{M})} = \||\theta_m|_{g_m}\|_{L^\infty(\mathcal{M})}.$$

If $\theta \in L^\infty\mathscr{T}_r^s(\mathcal{M})$ and if $\mathsf{D}^j\,\theta \in L^\infty\mathscr{T}_{r+j}^s(\mathcal{M})$ for $1 \leq j \leq k$ one says that $\theta \in W^{k,\infty}\mathscr{T}_r^s(\mathcal{M})$. Here, the action of D^j is to be understood in a weak sense, which can be defined in local charts. One sets

$$\|\theta\|_{W^{k,\infty}\mathscr{T}_r^s(\mathcal{M})} = \max_{0\leq j\leq k} \|\,\mathsf{D}^j\,\theta\|_{L^\infty\mathscr{T}_{r+j}^s(\mathcal{M})}.$$

For $k \in \mathbb{N}$ one can check that $W^{k,\infty}\mathscr{T}_r^s(\mathcal{M})$ is a Banach space if equipped with the above norm.

Finally, in the case of functions, that is $r = s = 0$, for $1 \leq p \leq \infty$, one write $W^{k,p}(\mathcal{M})$ in place of $W^{k,p}\mathscr{T}_0^0(\mathcal{M})$. Note also that one has $W^{k,2}\mathscr{T}_s^r(\mathcal{M}) = H^k\mathscr{T}_s^r(\mathcal{M})$.

18.3. Transposition of the Laplace–Beltrami Operator and Action on H^1 Functions

In Sect. 17.3 we saw that the transpose of Δ_g (for the density μ_g) is ${}^t\Delta_g = \Delta_g$.

If $u \in {}^0\mathscr{D}'(\mathcal{M})$ and $\varphi \in {}^0\mathscr{D}_c^\infty(\mathcal{M})$ we thus have

$$\langle \Delta_g u, \varphi\mu_g\rangle_{{}^0\mathscr{D}'(\mathcal{M}),{}^1\mathscr{D}_c^\infty(\mathcal{M})} = \langle u, (\Delta_g\varphi)\mu_g\rangle_{{}^0\mathscr{D}'(\mathcal{M}),{}^1\mathscr{D}_c^\infty(\mathcal{M})}. \tag{18.3.1}$$

Let now $v \in L^2V(\mathcal{M})$. For $\varphi \in {}^0\mathscr{D}_c^\infty(\mathcal{M})$, the map $S_v : \varphi \mapsto -\int_\mathcal{M} v(\varphi)\mu_g$ is a 1-density distribution and we define the 0-density distribution $\operatorname{div}_g v$ such that $S_v = \operatorname{div}_g v\,\mu_g$ following Sect. 17.3. If $u \in H^1(\mathcal{M})$, then $\nabla_g u \in L^2V(\mathcal{M})$ and we set $\tilde{\Delta}_g u = \operatorname{div}_g \nabla_g u$. Observe that for $\varphi \in {}^0\mathscr{D}_c^\infty(\mathcal{M})$ we have

$$\langle(-\tilde{\Delta}_g u)\mu_g, \varphi\rangle_{{}^1\mathscr{D}'(\mathcal{M}),{}^0\mathscr{D}_c^\infty(\mathcal{M})} = \int_\mathcal{M} \nabla_g u(\varphi)\mu_g = \int_\mathcal{M} g(\nabla_g u, \nabla_g\varphi)\mu_g.$$

By Proposition 18.10 and (18.3.1) this yields

$$\begin{aligned}\langle(-\tilde{\Delta}_g u)\mu_g, \varphi\rangle_{{}^1\mathscr{D}'(\mathcal{M}),{}^0\mathscr{D}_c^\infty(\mathcal{M})} &= -\int_{\mathcal{M}} u\Delta_g\varphi\,\mu_g\\ &= \langle u\mu_g, -\Delta_g\varphi\rangle_{{}^1\mathscr{D}'(\mathcal{M}),{}^0\mathscr{D}_c^\infty(\mathcal{M})}\\ &= \langle -\Delta_g u\mu_g, \varphi\rangle_{{}^1\mathscr{D}'(\mathcal{M}),{}^0\mathscr{D}_c^\infty(\mathcal{M})},\end{aligned}$$

meaning that $\tilde{\Delta}_g u = \Delta_g u$, for $u \in H^1(\mathcal{M})$ with the latter defined as the action of Δ_g on a distribution. We then find

$$|\langle(\Delta_g u)\mu_g, \varphi\rangle_{{}^1\mathscr{D}'(\mathcal{M}),{}^0\mathscr{D}_c^\infty(\mathcal{M})}| \lesssim \|u\|_{H^1(\mathcal{M})}\|\varphi\|_{H^1(\mathcal{M})},$$

and $\Delta_g u \in H^{-1}(\mathcal{M})$. By density one obtains the following proposition.

PROPOSITION 18.13. *Let $u \in H^1(\mathcal{M})$. Then, $\Delta_g u \in H^{-1}(\mathcal{M})$ and*

$$\langle -\Delta_g u, w\rangle_{H^{-1}(\mathcal{M}),H_0^1(\mathcal{M})} = \int_{\mathcal{M}} g(\nabla_g u, \nabla_g w)\mu_g, \qquad w \in H_0^1(\mathcal{M}).$$

18.4. The Laplace Problem on a Compact Manifold Without Boundary

In the case of a compact Riemannian manifold $(\mathcal{M}, g)$ without boundary, we have $H^k(\mathcal{M}) = H_0^k(\mathcal{M})$. This property is not clear for noncompact manifold; see the discussion below Proposition 18.4. However, in the case of a compact manifold without boundary, as in the case $k = 1$, the proof is simply based on the use a finite partition of unity subordinated to an atlas covering, allowing one to use the result known for bounded regular open sets in $\mathbb{R}^d$.

We consider the Laplace–Beltrami problem

$$-\Delta_g u + qu = f,$$

where $q \in L^\infty(\mathcal{M};\mathbb{R})$ and $f \in H^{-1}(\mathcal{M})$, with the definition of $H^{-1}(\mathcal{M})$ given at the end of Sect. 18.2. If $q = 0$ we have a kernel formed by constant functions on $\mathcal{M}$. As is classical in this case, we thus choose a nonnegative potential function q that exhibits some positivity in a domain of positive measure.

PROPOSITION 18.14 (Laplace Problem on a Manifold without Boundary). *Let $(\mathcal{M}, g)$ be a compact connected Riemannian d-dimensional manifold without boundary. Let $q \in L^\infty(\mathcal{M};\mathbb{R})$ be such that $q \geq 0$ and $q \geq C_0 > 0$ on some subset A of $\mathcal{M}$ with $|A| > 0$. There exists $C > 0$ such that, for any $f \in H^{-1}(\mathcal{M})$, there exists a unique $u \in H^1(\mathcal{M})$ such that*

(18.4.1)

$$\int_{\mathcal{M}} g(\nabla_g u, \nabla_g\varphi)\mu_g + \int_{\mathcal{M}} qu\varphi\mu_g = \langle f, \varphi\rangle_{H^{-1}(\mathcal{M}),H^1(\mathcal{M})}, \qquad \varphi \in H^1(\mathcal{M}),$$

with

$$C^{-1}\|f\|_{H^{-1}(\mathcal{M})} \leq \|u\|_{H^1(\mathcal{M})} \leq C\|f\|_{H^{-1}(\mathcal{M})}.$$

Moreover, the (0-density) distribution $\Delta_g u$ is in $H^{-1}(\mathcal{M})$ and $-\Delta_g u + qu = f$.

Let $k \in \mathbb{N}$ and assume that in addition $q \in W^{k,\infty}(\mathcal{M};\mathbb{R})$. There exists $C_k > 0$, such that if $f \in H^k(\mathcal{M})$, then moreover $u \in H^{k+2}(\mathcal{M})$ and

$$C_k^{-1}\|f\|_{H^k(\mathcal{M})} \leq \|u\|_{H^{k+2}(\mathcal{M})} \leq C_k\|f\|_{H^k(\mathcal{M})}.$$

and the equation $-\Delta_g u + qu = f$ holds in $H^k(\mathcal{M})$.

PROOF. On $H^1(\mathcal{M})$, we consider the continuous bilinear form

$$a(u,v) = \int_{\mathcal{M}} g(\nabla_g u, \nabla_g \bar{v})\mu_g + \int_{\mathcal{M}} qu\bar{v}\mu_g.$$

It is coercive as $a(v,v) \gtrsim (\nabla_g v, \nabla_g v)_{L^2V(\mathcal{M})} + (v,v)_{L^2(A)} \gtrsim \|v\|^2_{H^1(\mathcal{M})}$, by the Poincaré inequality (18.1.3) given in Proposition 18.9. As $v \mapsto \langle f, \bar{v}\rangle_{H^{-1}(\mathcal{M}),H^1(\mathcal{M})}$ is continuous, the Lax–Milgram theorem then gives a unique $u \in H^1(\mathcal{M})$ such that (18.4.1) holds. It also yields the existence of $C > 0$ such that

$$C^{-1}\|f\|_{H^{-1}(\mathcal{M})} \leq \|u\|_{H^1(\mathcal{M})} \leq C\|f\|_{H^{-1}(\mathcal{M})}.$$

For a function in $H^1(\mathcal{M})$ the action of Δ_g yields a 0-distribution in $H^{-1}(\mathcal{M})$ by Proposition 18.13. With this proposition we moreover find that $-\Delta_g u + qu = f$ holds in $H^{-1}(\mathcal{M})$.

If now $f \in H^k(\mathcal{M})$ with $k \in \mathbb{N}$, the regularity result can be obtained by using Proposition 18.11, a finite atlas, and the counterpart of the result in open sets of $\mathbb{R}^d$; see [161, Section 8.4] and [90, Section 9.6]. ∎

We define P_0 as an operator on $L^2(\mathcal{M})$ with domain

$$D(\mathsf{P}_0) = \{u \in H^1(\mathcal{M});\ \Delta_g u \in L^2(\mathcal{M})\},$$

given by $\mathsf{P}_0 u = -\Delta_g u$. As ${}^0\mathscr{D}_c^\infty(\mathcal{M}) \subset D(\mathsf{P}_0)$ we see by Proposition 18.2 that $D(\mathsf{P}_0)$ is dense in $L^2(\mathcal{M})$.

We set $L_q = \mathsf{P}_0 + q$ with $q \geq 0$ and $q \geq C_0 > 0$ on some subset A of $\mathcal{M}$ with $|A| > 0$, in agreement with the properties required in Proposition 18.14. If $u \in D(\mathsf{P}_0)$, then we set $f = L_q u \in L^2(\mathcal{M})$. Then, the function u is precisely the solution of the variational problems of the above propositions since the operator $L_q : D(\mathsf{P}_0) \to L^2(\mathcal{M})$ is injective by the following lemma.

LEMMA 18.15. *Let $u \in D(\mathsf{P}_0)$ be such that $L_q u = 0$. Then $u = 0$.*

PROOF. Consider $\varphi \in {}^0\mathscr{D}_c^\infty(\mathcal{M})$. By the definition of Δ_g for a H^1 function we find that

$$0 = \int_{\mathcal{M}} g(\nabla_g u, \nabla_g \bar{\varphi})\mu_g + \int_{\mathcal{M}} qu\bar{\varphi}\mu_g. \tag{18.4.2}$$

By the density of ${}^0\mathscr{D}_c^\infty(\mathcal{M})$ in $H^1(\mathcal{M})$ by Proposition 18.4, we see that (18.4.2) holds for all $\varphi \in H^1(\mathcal{M})$. The function u is thus the unique solution of the corresponding homogeneous variational problem. This yields $u = 0$. ∎

We thus conclude that $D(\mathsf{P}_0) = H^2(\mathcal{M})$.

The analysis recalled in Section 10.1 of Volume 1 can be adapted to the case of the operator $-\Delta_g$ on $\mathcal{M}$. We denote by S_q the map $L^2(\mathcal{M})$ into $H^2(\mathcal{M})$ such that $S_q(f)$ provides the unique solution $u \in H^2(\mathcal{M})$ of $L_q u = f$. If ι is the natural injection of $H^2(\mathcal{M})$ into $L^2(\mathcal{M})$, we find that the resolvent map $R_q = \iota \circ S_q : L^2(\mathcal{M}) \to L^2(\mathcal{M})$ is compact by the Rellich–Kondrachov theorem (see Theorem 18.7).

Let $f_1, f_2 \in L^2(\mathcal{M})$ and $u_1, u_2 \in D(\mathsf{P}_0)$ the associated solutions, that is, $u_j = S_q(f_j)$, $j = 1, 2$. The variational form of elliptic problem gives in particular

$$(f_1, u_2)_{L^2(\mathcal{M})} = \int_{\mathcal{M}} g(\nabla_g u_1, \nabla_g \overline{u_2})\mu_g + \int_{\mathcal{M}} q u_1 \overline{u_2} \mu_g = (u_1, f_2)_{L^2(\mathcal{M})}, \tag{18.4.3}$$

which reads $(f_1, R_q f_2)_{L^2(\mathcal{M})} = (R_q f_1, f_2)_{L^2(\mathcal{M})}$. The resolvent map is thus selfadjoint. As R_q is closed we immediately have that L_q and thus P_0 are closed operators. Moreover, (18.4.3) also reads $(L_q u_1, u_2)_{L^2(\mathcal{M})} = (u_1, L_q u_2)_{L^2(\mathcal{M})}$, that is,

$$(\mathsf{P}_0 u_1, u_2)_{L^2(\mathcal{M})} = (u_1, \mathsf{P}_0 u_2)_{L^2(\mathcal{M})},$$

using the properties of the solution of the elliptic problem given by Proposition 18.14. The operator $\mathsf{P}_0 = (-\Delta_g, D(\mathsf{P}_0))$ is thus symmetric. Observe also that P_0 is monotone and maximal (see Definitions 12.22 and 12.10 of volume 1) as can be seen by choosing $q = 1$ in the variational problem. By Proposition 12.28 of Volume 1 we obtain that P_0 is selfadjoint.

Let $u \in \ker \mathsf{P}_0$. We then have

$$0 = (-\Delta_g u, u)_{L^2(\mathcal{M})} = \|\nabla_g u\|^2_{L^2 V(\mathcal{M})},$$

meaning that $u \equiv \mathrm{Cst}$ as $\mathcal{M}$ is connected. The kernel of P_0 is thus precisely the one dimensional space of constant functions.

With the features listed above for P_0, following the lines of Section 10.1 of Volume 1 in the case of a bounded open set of $\mathbb{R}^d$, we obtain the following properties:

Hilbert basis of eigenfunctions: There exists a nondecreasing sequence of real eigenvalues $0 = \mu_0 < \mu_1 \leq \mu_2 \leq \cdots \leq \mu_n$ (counted with their multiplicity) and an associated sequence of eigenfunctions of P_0, denoted by $(\phi_j)_{j \in \mathbb{N}}$, that forms a Hilbert basis of $L^2(\mathcal{M})$.

Weyl Law: The eigenvalues satisfy the Weyl law of Theorem 10.3 of Volume 1.

18.5. Continuous Sobolev Scale and Traces

With the Sobolev spaces $H^k(\mathcal{M})$ for $k \in \mathbb{N}$ defined above on a Riemannian manifold the spaces $H^s(\mathcal{M})$ for $s > 0$ can be defined by interpolation [74, 236].

If the manifold $\mathcal{M}$ does not have a boundary we may use the spectral decomposition of the Laplace–Beltrami operator on $\mathcal{M}$ to define $H^s(\mathcal{M})$ for all $s \in \mathbb{R}$.

Let us denote the eigenvalues of $\mathrm{Id} + \mathsf{P}_0$ by $1 = 1+\mu_0 < 1+\mu_1 \leq 1+\mu_2 \leq \cdots \leq 1+\mu_n$ as in Sect. 18.4 with associated eigenfunctions ϕ_j that form a Hilbert basis of $L^2(\mathcal{M})$.

Following Section 10.1.3, for $s \geq 0$ we define

$$K^s(\mathcal{M}) = \{u \in L^2(\mathcal{M});\ (1+\mu_j)^{s/2} u_j \in \ell^2(\mathbb{C})\},$$

where $u_j = (u, \phi_j)_{L^2(\mathcal{M})}$ and we have

$$u = \sum_{j\in\mathbb{N}} u_j \phi_j,$$

with convergence in $L^2(\mathcal{M})$. We set

$$\|u\|^2_{K^s(\mathcal{M})} = \sum_{j\in\mathbb{N}} (1+\mu_j)^s |u_j|^2.$$

With this norm $K^s(\mathcal{M})$ is a Hilbert space.

As ϕ_j is smooth by the elliptic regularity result of Proposition 18.14, we see that smooth functions are dense in $K^s(\mathcal{M})$.

For $k \in \mathbb{N}$, we have $K^{2k}(\mathcal{M}) = D((\mathrm{Id} + \mathsf{P}_0)^k)$ and $\|u\|_{K^{2k}(\mathcal{M})} = \|(\mathrm{Id} + \mathsf{P}_0)^k u\|_{L^2(\mathcal{M})}$.

PROPOSITION 18.16. *Let $\mathcal{M}$ be a compact Riemannian manifold without boundary. For $k \in \mathbb{N}$ we have $H^k(\mathcal{M}) = K^k(\mathcal{M})$. Moreover, the norms $\|.\|_{H^k(\mathcal{M})}$ and $\|.\|_{K^k(\mathcal{M})}$ are equivalent.*

PROOF. The equivalence is obvious for $k = 0$. For $k = 1$, for u smooth, using formula (17.3.1), we write

$$\begin{aligned}\|u\|^2_{K^1(\mathcal{M})} &= (u, (\mathrm{Id} + \mathsf{P}_0)u) = \|u\|^2_{L^2(\mathcal{M})} + \|\nabla_g u\|^2_{L^2V(\mathcal{M})} \\ &= \|u\|^2_{L^2(\mathcal{M})} + \|Du\|^2_{L^2\Lambda^1(\mathcal{M})} = \|u\|^2_{H^1(\mathcal{M})}.\end{aligned}$$

As smooth functions are dense in both $H^1(\mathcal{M})$ and $K^1(\mathcal{M})$ we obtain the equality of the two norms and that $H^1(\mathcal{M}) = K^1(\mathcal{M})$.

We next proceed by induction. Assuming the result holds for $k \in \{0, \ldots, N\}$, $N \geq 1$, we write, for u smooth,

$$\|u\|_{K^{N+1}(\mathcal{M})} = \|(\mathrm{Id} + \mathsf{P}_0)u\|_{K^{N-1}(\mathcal{M})} \eqsim \|(\mathrm{Id} + \mathsf{P}_0)u\|_{H^{N-1}(\mathcal{M})},$$

and we conclude that $\|u\|_{K^{N+1}(\mathcal{M})} \eqsim \|u\|_{H^{N+1}(\mathcal{M})}$ using the elliptic regularity result of Proposition 18.14. We then conclude by density as in the case $k = 1$. ■

Because of the previous proposition, we are led to define the Sobolev as follows.

DEFINITION 18.17. Let $\mathcal{M}$ be a compact Riemannian manifold without boundary. For $s \geq 0$, we define the space $H^s(\mathcal{M})$ to be the Hilbert space $K^s(\mathcal{M})$ equipped with the norm $\|.\|_{K^s(\mathcal{M})}$, that is,

$$\|u\|^2_{H^s(\mathcal{M})} = \sum_{j\in\mathbb{N}} (1+\mu_j)^s |u_j|^2.$$

In particular note that smooth functions are dense in $H^s(\mathcal{M})$.

With this definition we recover the spaces we could also obtain by interpolation starting from the Sobolev spaces defined in Sect. 18.2 as $K^s(\mathcal{M})$ is the natural interpolation of $K^k(\mathcal{M})$ and $K^{k+1}(\mathcal{M})$ if $s \in (k, k+1)$.

Sobolev space of negative orders can be defined too. For $s \geq 0$, the dual space $\big(H^s(\mathcal{M})\big)'$ is a space of 1-density distributions since smooth functions are dense in $H^s(\mathcal{M})$. Using the canonical density function μ_g on $(\mathcal{M}, g)$ we can identify $\big(H^s(\mathcal{M})\big)'$ with a space of 0-density distributions that we denote by $H^{-s}(\mathcal{M})$.

DEFINITION 18.18. Let $\mathcal{M}$ be a compact Riemannian manifold without boundary. For $s \geq 0$, we define the space $H^{-s}(\mathcal{M})$ as

$$H^{-s}(\mathcal{M}) = \{u \in {}^0\mathscr{D}'(\mathcal{M}); \mu_g u \in \big(H^s(\mathcal{M})\big)'\}.$$

We define

$$\|u\|_{H^{-s}(\mathcal{M})} = \|\mu_g u\|_{(H^s(\mathcal{M}))'}.$$

Equipped with this norm $H^{-s}(\mathcal{M})$ is a Banach space.

The space $H^{-s}(\mathcal{M})$ can be characterized in a very natural manner.

PROPOSITION 18.19. *Let $s > 0$ and $(u_j)_j \subset \mathbb{C}$ be such that $\big((1+\mu_j)^{-s/2}u_j\big)_j \in \ell^2(\mathbb{C})$. Then, the series $\sum_j u_j\phi_j$ converges in $H^{-s}(\mathcal{M})$. Conversely, let $u \in H^{-s}(\mathcal{M})$ and $v = \mu_g u \in \big(H^s(\mathcal{M})\big)'$. Set $u_j = \langle v, \overline{\phi_j}\rangle_{(H^s(\mathcal{M}))', H^s(\mathcal{M})}$, $j \in \mathbb{N}$. Then, one has $\big((1+\mu_j)^{-s/2}u_j\big)_j \in \ell^2(\mathbb{C})$ and $u = \sum_j u_j\phi_j$ where convergence is understood in $H^{-s}(\mathcal{M})$, with respect to the norm $\|.\|_{H^{-s}(\mathcal{M})}$. Moreover,*

$$\|u\|^2_{H^{-s}(\mathcal{M})} \eqsim \sum_j (1+\mu_j)^{-s}|u_j|^2.$$

The proof can be adapted from what is done in Section 10.1.3 of Volume 1 in the case of an open set of $\mathbb{R}^d$, the only difficulty being the switching between 1-densities and 0-densities by means of μ_g.

For $u \in H^{-s}(\mathcal{M})$ and $v \in H^s(\mathcal{M})$, we define

$$\langle u, v\rangle_{H^{-s}(\mathcal{M}), H^s(\mathcal{M})} = \langle \mu_g u, v\rangle_{(H^s(\mathcal{M}))', H^s(\mathcal{M})}.$$

Apart from the case $s = 0$, we do not identify $H^{-s}(\mathcal{M})$ and $H^s(\mathcal{M})$ through the inner product of $H^s(\mathcal{M})$ and the Riesz theorem but rather use $L^2(\mathcal{M})$ as a pivot space. We then have, for $s \geq 0$,

$$\langle u, \bar{v}\rangle_{H^{-s}(\mathcal{M}),H^s(\mathcal{M})} = (u, v)_{L^2(\mathcal{M})}, \qquad u \in L^2(\mathcal{M}),\ v \in H^s(\mathcal{M}).$$

REMARK 18.20. Note that in the case of a compact manifold with $\partial\mathcal{M} \neq \emptyset$ along with homogeneous Dirichlet boundary conditions, the iterated domains $K^s(\mathcal{M}) = D(-\Delta_g^{s/2})$ form a scale of Sobolev-like spaces. As in the case of a bounded regular open set in $\mathbb{R}^d$, $K^s(\mathcal{M})$ do not coincide with the natural Sobolev spaces as given in Sect. 18.2 above.

PROPOSITION 18.21. *Let $\mathcal{M}$ be a compact Riemannian manifold without boundary. Let $s \in \mathbb{R}^+$ and let $u \in L^2(\mathcal{M})$. We have $u \in H^s(\mathcal{M})$ if and only if for every local chart $\mathcal{C} = (O, \kappa)$ its representative is such that $u^{\mathcal{C}} \in H^s(\kappa(O))$.*

If $\mathcal{A} = (\mathcal{C}^i)_{1\leq i\leq N}$ is a finite atlas with $\mathcal{C}^i = (O^i, \kappa^i)$, there exists $C = C_{\mathcal{A}} > 0$ such that

$$C^{-1}\|u\|_{H^s(\mathcal{M})} \leq \sum_{1\leq i\leq N} \|u^{\mathcal{C}^i}\|_{H^s(\kappa^i(O^i))} \leq C\|u\|_{H^s(\mathcal{M})},$$

if $u \in H^s(\mathcal{M})$. Moreover, if $i \in \{1, \dots, N\}$, for some $C_i > 0$, we have

$$C_i^{-1}\|u\|_{H^s(\mathcal{M})} \leq \|u^{\mathcal{C}^i}\|_{H^s(\kappa^i(O^i))} \leq C_i\|u\|_{H^s(\mathcal{M})} \tag{18.5.1}$$

if $\operatorname{supp}(u) \subset O^i$.

This result follows from Proposition 18.11 and an interpolation argument. The following proposition treats the case $s < 0$.

PROPOSITION 18.22. *Let $\mathcal{M}$ be a compact Riemannian manifold without boundary, let $\mathcal{A} = (\mathcal{C}^i)_{1\leq i\leq N}$ be a finite atlas with $\mathcal{C}^i = (O^i, \kappa^i)$, and let $(\psi_i)_{1\leq i\leq N}$ be a smooth partition of unity subordinated to the open covering $(O^i)_{1\leq i\leq N}$.*

Let $s < 0$ and let $u \in {}^0\mathscr{D}'(\mathcal{M})$. We have $u \in H^s(\mathcal{M})$ if and only if for all $i \in \{1, \dots, N\}$, $(\psi_i u)^{\mathcal{C}^i} \in H^s(\mathbb{R}^d)$.

Moreover, there exists $C = C_{\mathcal{A}} > 0$ such that

$$C^{-1}\|u\|_{H^s(\mathcal{M})} \leq \sum_{1\leq i\leq N} \|(\psi_i u)^{\mathcal{C}^i}\|_{H^s(\mathbb{R}^d)} \leq C\|u\|_{H^s(\mathcal{M})},$$

if $u \in H^s(\mathcal{M})$. Moreover, if $i \in \{1, \dots, N\}$, for some $C_i > 0$, we have

$$C_i^{-1}\|u\|_{H^s(\mathcal{M})} \leq \|u^{\mathcal{C}^i}\|_{H^s(\mathbb{R}^d)} \leq C_i\|u\|_{H^s(\mathcal{M})} \tag{18.5.2}$$

if $\operatorname{supp}(u) \subset O^i$.

PROOF. One first proves (18.5.2). We thus let $u \in H^s(\mathcal{M})$ with $\operatorname{supp}(u) \subset O^i$. One has $u \in {}^0\mathscr{D}'(O^i)$ and thus $u^{\mathcal{C}^i} \in \mathscr{D}'(\kappa^i(O^i))$. If $v^i \in \mathscr{C}_c^\infty(\kappa^i(O^i))$ we write

$$\langle u^{\mathcal{C}^i}, v^i\rangle_{\mathscr{D}'(\mathbb{R}^d),\mathscr{C}_c^\infty(\mathbb{R}^d)} = \langle \mu_g^{\mathcal{C}^i} u^{\mathcal{C}^i}, (\mu_g^{\mathcal{C}^i})^{-1} v^i\rangle_{\mathscr{D}'(\mathbb{R}^d),\mathscr{C}_c^\infty(\mathbb{R}^d)}.$$

With $v = (\kappa^i)^*\big((\mu_g^{\mathcal{C}i})^{-1} v^i\big) \in \mathscr{C}_c^\infty(O^i)$, understood as a 0-density function on $\mathcal{M}$, one has

$$\begin{aligned}\langle u^{\mathcal{C}i}, v^i\rangle_{\mathscr{D}'(\mathbb{R}^d),\mathscr{C}_c^\infty(\mathbb{R}^d)} &= \langle \mu_g u, v\rangle_{{}^1\mathscr{D}'(O^i),{}^0\mathscr{D}_c^\infty(O^i)} = \langle \mu_g u, v\rangle_{(H^{-s}(\mathcal{M}))',H^{-s}(\mathcal{M})}\\ &= \langle u, v\rangle_{H^s(\mathcal{M}),H^{-s}(\mathcal{M})}.\end{aligned}$$

We thus find

$$\begin{aligned}\big|\langle u^{\mathcal{C}i}, v^i\rangle_{\mathscr{D}'(\mathbb{R}^d),\mathscr{C}_c^\infty(\mathbb{R}^d)}\big| &\leq \|u\|_{H^s(\mathcal{M})}\|v\|_{H^{-s}(\mathcal{M})} \lesssim \|u\|_{H^s(\mathcal{M})}\|v^{\mathcal{C}i}\|_{H^{-s}(\mathbb{R}^d)}\\ &\lesssim \|u\|_{H^s(\mathcal{M})}\|v^i\|_{H^{-s}(\mathbb{R}^d)},\end{aligned}$$

implying that $u^{\mathcal{C}i} \in H^s(\mathbb{R}^d)$ and

$$\|u^{\mathcal{C}i}\|_{H^s(\mathbb{R}^d)} \lesssim \|u\|_{H^s(\mathcal{M})},$$

from the density of $\mathscr{C}_c^\infty(\kappa^i(O^i))$ in the set of function in $H^{-s}(\mathbb{R}^d)$ with compact support in $\kappa^i(O^i)$.

Similarly, for $v \in \mathscr{C}_c^\infty(O^i)$ one has

$$\langle u, v\rangle_{H^s(\mathcal{M}),H^{-s}(\mathcal{M})} = \langle u^{\mathcal{C}i}, v^i\rangle_{\mathscr{D}'(\mathbb{R}^d),\mathscr{C}_c^\infty(\mathbb{R}^d)} = \langle u^{\mathcal{C}i}, v^i\rangle_{H^s(\mathbb{R}^d),H^{-s}(\mathbb{R}^d)}$$

with $v^i = \mu_g^{\mathcal{C}i} v^{\mathcal{C}i}$. One thus finds

$$\|\langle u, v\rangle_{H^s(\mathcal{M}),H^{-s}(\mathcal{M})}\| \leq \|u^{\mathcal{C}i}\|_{H^s(\mathbb{R}^d)}\|v^i\|_{H^{-s}(\mathbb{R}^d)} \lesssim \|u^{\mathcal{C}i}\|_{H^s(\mathbb{R}^d)}\|v\|_{H^{-s}(\mathcal{M})},$$

yielding

$$\|u\|_{H^s(\mathcal{M})} \lesssim \|u^{\mathcal{C}i}\|_{H^s(\mathbb{R}^d)},$$

from the density of $\mathscr{C}_c^\infty(O^i)$ in the set of function in $H^{-s}(\mathcal{M})$ with compact support in O^i.

The first part of the proof follows by using a partition of unity associated with the chosen atlas. ■

This property is used in Chap. 8, allowing us to use H^s-norms in local charts that can be obtained easily by means of Fourier multipliers. There, the manifold without boundary under consideration is the boundary of a compact manifold.

With an interpolation argument we also prove the following continuity result for differential operators.

PROPOSITION 18.23. *Let $(\mathcal{M}, g)$ be a smooth Riemannian manifold such that $\mathcal{M}$ is compact without boundary. Let $s \in \mathbb{R}$ and let Q be a differential operator of order k with smooth coefficients on $\mathcal{M}$. Then, the operator Q maps $H^s(\mathcal{M})$ into $H^{s-k}(\mathcal{M})$ continuously.*

We conclude this section with the following trace inequality.

PROPOSITION 18.24. *Let $(\mathcal{M}, g)$ be a compact Riemannian manifold. The trace $u \mapsto u_{|\partial\mathcal{M}}$ defined for smooth functions can be uniquely extended to functions in $H^1(\mathcal{M})$. Moreover, there exists $C > 0$ such that*

$$|u_{|\partial\mathcal{M}}|_{H^{1/2}(\partial\mathcal{M})} \leq C\|u\|_{H^1(\mathcal{M})}, \quad u \in \mathscr{C}^\infty(\mathcal{M}).$$

PROOF. On $\mathbb{R}^d_+$ arguing as in the proof of Proposition 6.9 (taking $\tau = 1$, for instance) we obtain, for $s > 0$,

$$|u_{|x_d=0^+}|_{H^{1/2}(\mathbb{R}^{d-1})} \lesssim \|u\|_{H^1(\mathbb{R}^d_+)}, \qquad u \in \overline{\mathscr{S}}(\mathbb{R}^d_+). \tag{18.5.3}$$

Let $\mathcal{A} = (\mathcal{C}^i)_{1\leq i\leq N}$, $\mathcal{C}^i = (O^i, \kappa^i)$ be a finite atlas of $\mathcal{M}$. Choose a partition of unity $(\chi^i)_{1\leq i\leq N}$ of $\mathcal{M}$ subordinated to the open covering $(O^i)_{1\leq i\leq N}$ and set $u_i = \chi^i u$. We denote by $u_i^{\mathcal{C}^i}$ the representative of u_i in the chart $\mathcal{C}^i$.

We set $J = \{1 \leq i \leq N;\ O^i \cap \partial\mathcal{M} \neq \emptyset\}$. For $j \in J$, we have by (18.5.3), with a density argument,

$$|u_j^{\mathcal{C}^j}{}_{|x_d=0^+}|_{H^{1/2}(\kappa^j(O^j))} = |u_j^{\mathcal{C}^j}{}_{|x_d=0^+}|_{H^{1/2}(\mathbb{R}^{d-1})} \lesssim \|u_j^{\mathcal{C}^j}\|_{H^1(\mathbb{R}^d_+)} = \|u_j^{\mathcal{C}^j}\|_{H^1(\kappa^j(O^j))}.$$

By (18.2.3) in Proposition 18.11 and (18.5.1) in Proposition 18.21 we thus have

$$|\chi^j u_{|\partial\mathcal{M}}|_{H^{1/2}(\partial\mathcal{M})} \lesssim \|\chi^j u\|_{H^1(\mathcal{M})}.$$

We then write

$$|u_{|\partial\mathcal{M}}|_{H^{1/2}(\partial\mathcal{M})} \lesssim \sum_{j\in J} |\chi^j u_{|\partial\mathcal{M}}|_{H^{1/2}(\partial\mathcal{M})} \lesssim \sum_{j\in J} \|\chi^j u\|_{H^1(\mathcal{M})} \lesssim \|u\|_{H^1(\mathcal{M})},$$

using Proposition 18.23 in the last inequality. ∎

More generally, we have the following result.

THEOREM 18.25 (Trace Theorem). *Let $k \in \mathbb{N}$, $k \geq 1$. The map T defined by*

$$T : H^k(\mathcal{M}) \to \prod_{j=0}^{k-1} H^{k-j-1/2}(\partial\mathcal{M})$$

$$u \mapsto (u_{|\partial\mathcal{M}}, \ldots, \partial_\nu^{k-1} u_{|\partial\mathcal{M}})$$

is well defined, continuous, and surjective. Moreover, T has a linear and continuous right-inverse.

Recall that ∂_ν^ℓ is defined in (17.7.3). We refer to [236, Theorem 8.3 in Chapter 1] and [2, Theorem 7.39 and Lemma 7.41]. In fact, the following lemmata provide explicit forms of right inverses in the case of a half-space. A proof of Theorem 18.25 in the cases $k = 1, 2$ can then be deduced by working locally at the boundary.

LEMMA 18.26. *Let $\chi \in \mathscr{C}_c^\infty(\mathbb{R})$ be even and such that $\int_{\mathbb{R}} \chi = 2\pi$. Let $v \in \mathscr{S}(\mathbb{R}^{d-1})$ and denote by $\hat{v}$ its Fourier transform. The function*

$$\xi = (\xi', \xi_d) \mapsto \langle\xi'\rangle^{-1}\chi(\xi_d/\langle\xi'\rangle)\hat{v}(\xi')$$

is in $\mathscr{S}(\mathbb{R}^d)$ *and if we define* u *as its inverse Fourier transform in* $\mathbb{R}^d$*, then* $u \in \mathscr{S}(\mathbb{R}^d)$ *and we have*

$$u_{|x_d=0} = v, \quad \partial_{x_d} u_{|x_d=0} = 0, \quad \text{and } \|u\|_{H^{s+1/2}(\mathbb{R}^d)} \approx |v|_{H^s(\mathbb{R}^{d-1})},$$

for $s \in \mathbb{R}$.

LEMMA 18.27. *Let* $\chi \in \mathscr{C}_c^\infty(\mathbb{R})$ *be even, such that* $0 \notin \operatorname{supp}(\chi)$ *and* $\int_\mathbb{R} \chi = 2\pi$*. Let* $v \in \mathscr{S}(\mathbb{R}^{d-1})$ *and denote by* $\hat{v}$ *its Fourier transform. The function*

$$\xi = (\xi', \xi_d) \mapsto (i\xi_d\langle\xi'\rangle)^{-1}\chi(\xi_d/\langle\xi'\rangle)\hat{v}(\xi'),$$

is in $\mathscr{S}(\mathbb{R}^d)$ *and if we define* u *as its inverse Fourier transform in* $\mathbb{R}^d$*, then* $u \in \mathscr{S}(\mathbb{R}^d)$ *and we have*

$$u_{|x_d=0} = 0, \quad \partial_{x_d} u_{|x_d=0} = v, \quad \text{and } \|u\|_{H^{s+3/2}(\mathbb{R}^d)} \approx |v|_{H^s(\mathbb{R}^{d-1})},$$

for $s \in \mathbb{R}$.

PROOF OF LEMMA 18.26. Set $w(\xi) = \langle\xi'\rangle^{-1}\chi(\xi_d/\langle\xi'\rangle)\hat{v}(\xi')$. As χ is compactly supported we have $|\xi_d| \lesssim \langle\xi'\rangle$ in $\operatorname{supp}(w)$ and thus

$$\langle\xi\rangle \approx \langle\xi'\rangle, \qquad \xi \in \operatorname{supp}(w). \tag{18.5.4}$$

We then deduce that $w \in \mathscr{S}(\mathbb{R}^d)$ since $v \in \mathscr{S}(\mathbb{R}^{d-1})$. For $s \in \mathbb{R}$, we have,

$$\begin{aligned}
\|\langle\xi\rangle^{s+1/2}w\|^2_{L^2(\mathbb{R}^d)} &= \int_{\mathbb{R}^d} \langle\xi\rangle^{2s+1}\langle\xi'\rangle^{-2}\chi^2(\xi_d/\langle\xi'\rangle)|\hat{v}(\xi')|^2\, d\xi \\
&\lesssim \int_{\mathbb{R}^{d-1}} \langle\xi'\rangle^{2s}|\hat{v}(\xi')|^2\Big(\int_\mathbb{R}\langle\xi'\rangle^{-1}\chi^2(\xi_d/\langle\xi'\rangle)d\xi_d\Big)d\xi' \\
&= \Big(\int_\mathbb{R}\chi^2(t)dt\Big)\int_{\mathbb{R}^{d-1}}\langle\xi'\rangle^{2s}|\hat{v}(\xi')|^2 d\xi' \\
&\approx |\langle\xi'\rangle^s v|^2_{L^2(\mathbb{R}^{d-1})} \approx |v|^2_{H^s(\mathbb{R}^{d-1})}.
\end{aligned}$$

If u is the inverse Fourier transform of w, that is, $u(x) = \frac{1}{(2\pi)^d}\int_{\mathbb{R}^d} e^{ix\cdot\xi}w(\xi)d\xi$, we then have $\|u\|_{H^{s+1/2}(\mathbb{R}^d)} \approx \|\langle\xi\rangle^{s+1/2}w\|_{L^2(\mathbb{R}^d)} \approx |v|_{H^s(\mathbb{R}^{d-1})}$.

Now, since we have

$$u(x', 0) = \frac{1}{(2\pi)^d}\int_{\mathbb{R}^d} e^{ix'\cdot\xi'}w(\xi', \xi_d)\, d\xi'\, d\xi_d,$$

we compute

$$\frac{1}{2\pi}\int_\mathbb{R} w(\xi', \xi_d)d\xi_d = \frac{\hat{v}(\xi')}{2\pi}\int_\mathbb{R}\langle\xi'\rangle^{-1}\chi(\xi_d/\langle\xi'\rangle)d\xi_d = \frac{\hat{v}(\xi')}{2\pi}\int_\mathbb{R}\chi(t)dt = \hat{v}(\xi').$$

This implies that $u(x', 0) = v(x')$. We also have

$$\partial_{x_d}u(x', 0) = \frac{i}{(2\pi)^d}\int_{\mathbb{R}^d} e^{ix'\cdot\xi'}\xi_d w(\xi', \xi_d)\, d\xi'\, d\xi_d.$$

We compute

$$\int_\mathbb{R}\xi_d w(\xi', \xi_d)\, d\xi_d = \hat{v}(\xi')\int_\mathbb{R}(\xi_d/\langle\xi'\rangle)\chi(\xi_d/\langle\xi'\rangle)d\xi_d = \langle\xi'\rangle\hat{v}(\xi')\int_\mathbb{R} t\chi(t)dt = 0,$$

since χ is even. This implies $\partial_{x_d} u_{|x_d=0} = 0$, which concludes the proof. ∎

PROOF OF LEMMA 18.27. Set $w(\xi) = (i\xi_d\langle\xi'\rangle)^{-1}\chi(\xi_d/\langle\xi'\rangle)\hat{v}(\xi')$. Arguing as in the proof of Lemma 18.26 we find that $w \in \mathscr{S}(\mathbb{R}^d)$. For $s \in \mathbb{R}$, we have,

$$\begin{aligned}
\|\langle\xi\rangle^{s+3/2}w\|^2_{L^2(\mathbb{R}^d)} &= \int_{\mathbb{R}^d} \langle\xi\rangle^{2s+3}(|\xi_d|\langle\xi'\rangle)^{-2}\chi^2(\xi_d/\langle\xi'\rangle)|\hat{v}(\xi')|^2\,d\xi \\
&\lesssim \int_{\mathbb{R}^{d-1}} \langle\xi'\rangle^{2s}|\hat{v}(\xi')|^2\Big(\int_{\mathbb{R}}\langle\xi'\rangle^{-1}|\xi_d/\langle\xi'\rangle|^{-2}\chi^2(\xi_d/\langle\xi'\rangle)d\xi_d\Big)d\xi' \\
&= \Big(\int_{\mathbb{R}}|t|^{-2}\chi^2(t)dt\Big)\int_{\mathbb{R}^{d-1}}\langle\xi'\rangle^{2s}|\hat{v}(\xi')|^2d\xi' \\
&\eqsim |\langle\xi'\rangle^s v|^2_{L^2(\mathbb{R}^{d-1})} \eqsim |v|^2_{H^s(\mathbb{R}^{d-1})}.
\end{aligned}$$

If u is the inverse Fourier transform of w, we then have

$$\|u\|_{H^{s+3/2}(\mathbb{R}^d)} \eqsim \|\langle\xi\rangle^{s+3/2}w\|_{L^2(\mathbb{R}^d)} \eqsim |v|_{H^s(\mathbb{R}^{d-1})}.$$

Now, since we have

$$u(x',0) = \frac{1}{(2\pi)^d}\int_{\mathbb{R}^d} e^{ix'\cdot\xi'}w(\xi',\xi_d)\,d\xi'\,d\xi_d,$$

we compute

$$\begin{aligned}
\frac{1}{2\pi}\int_{\mathbb{R}} w(\xi',\xi_d)d\xi_d &= -i\frac{\hat{v}(\xi')}{2\pi\langle\xi'\rangle}\int_{\mathbb{R}}\langle\xi'\rangle^{-1}(\xi_d/\langle\xi'\rangle)^{-1}\chi(\xi_d/\langle\xi'\rangle)d\xi_d \\
&= -i\frac{\hat{v}(\xi')}{2\pi}\int_{\mathbb{R}} t^{-1}\chi(t)dt = 0,
\end{aligned}$$

as χ is even. This implies that $u(x',0) = 0$. We also have

$$\partial_{x_d}u(x',0) = \frac{1}{(2\pi)^d}\int_{\mathbb{R}^d} e^{ix'\cdot\xi'}i\xi_d w(\xi',\xi_d)\,d\xi'\,d\xi_d.$$

We compute

$$\frac{1}{2\pi}\int_{\mathbb{R}} i\xi_d w(\xi',\xi_d)d\xi_d = \frac{\hat{v}(\xi')}{2\pi}\int_{\mathbb{R}}\langle\xi'\rangle^{-1}\chi(\xi_d/\langle\xi'\rangle)d\xi_d = \frac{\hat{v}(\xi')}{2\pi}\int_{\mathbb{R}}\chi(t)dt = \hat{v}(\xi').$$

This implies that $\partial_{x_d}u(x',0) = v(x')$, which concludes the proof. ∎

Observe that the divergence formula stated in Proposition 17.6 for smooth vector fields extends to H^1-vector fields by density.

PROPOSITION 18.28 (Divergence Formula for H^1-Vector Fields). *Let $(\mathcal{M}, g)$ be a Riemannian manifold with boundary and let $u \in H^1V(\mathcal{M})$ be compactly supported on $\mathcal{M}$. Let ν be the unique outward pointing vector field along $\partial\mathcal{M}$ such that, for all $m \in \partial\mathcal{M}$, $g_m(\nu_m,\nu_m) = 1$ and $g_m(\nu_m, v) = 0$ for all $v \in T_m\partial\mathcal{M}$, that is, ν is unitary and orthogonal to $T_m\partial\mathcal{M}$ in the sense of g. We have*

$$\int_{\mathcal{M}} \operatorname{div}_g(u)\mu_g = \int_{\partial\mathcal{M}} g(u_{|\partial\mathcal{M}},\nu)\,\mu_{g_\partial},$$

where g_∂ is the induced Riemannian metric on $\partial\mathcal{M}$ and μ_{g_∂} the resulting canonical positive density function on $\partial\mathcal{M}$.

Since $\operatorname{div}_g(\bar{w}\nabla_g u) = \bar{w}\Delta_g u + g(\nabla_g\bar{w}, \nabla_g u)$, for $u \in H^2(\mathcal{M})$ and $w \in \mathscr{C}^\infty(\mathcal{M})$ we obtain as an application

$$(\Delta_g u, w)_{L^2(\mathcal{M})} = -(\nabla_g u, \nabla_g w)_{L^2V(\mathcal{M})} + (\partial_\nu u_{|\partial\mathcal{M}}, w_{|\partial\mathcal{M}})_{L^2(\partial\mathcal{M})}.$$

By density we find

$$(\Delta_g u, w)_{L^2(\mathcal{M})} = -(\nabla_g u, \nabla_g w)_{L^2V(\mathcal{M})} + (\partial_\nu u_{|\partial\mathcal{M}}, w_{|\partial\mathcal{M}})_{L^2(\partial\mathcal{M})}, \tag{18.5.5}$$

for $u \in H^2(\mathcal{M})$ and $w \in H^1(\mathcal{M})$, both with compact supports.

By density of $\mathscr{C}^\infty(\mathcal{M}) \cap H^1(\mathcal{M})$ in $H^1(\mathcal{M})$ we obtain the following 'integration by parts' formula from Proposition 17.7.

PROPOSITION 18.29. *Let $X \in \mathscr{C}^\infty V(\mathcal{M})$ and let $f_1, f_2 \in H^1(\mathcal{M})$ with compact supports. Then*

$$(Xf_1, f_2)_{L^2(\mathcal{M})} = -(f_1, \bar{X}f_2 + f_2 \operatorname{div}_g \bar{X})_{L^2(\mathcal{M})} + (f_1 g(X,\nu)_{|\partial\mathcal{M}}, f_{2|\partial\mathcal{M}})_{L^2(\partial\mathcal{M})}.$$

From (18.5.5), we can extend the Green formula to compactly supported H^2-functions.

PROPOSITION 18.30. *Let $f_1, f_2 \in H^2(\mathcal{M})$ have compact supports. We have*

$$(f_1, \Delta_g f_2)_{L^2(\mathcal{M})} + (\partial_\nu f_{1|\partial\mathcal{M}}, f_{2|\partial\mathcal{M}})_{L^2(\partial\mathcal{M})} \\ = (\Delta_g f_1, f_2)_{L^2(\mathcal{M})} + (f_{1|\partial\mathcal{M}}, \partial_\nu f_{2|\partial\mathcal{M}})_{L^2(\partial\mathcal{M})}.$$

18.6. The Dirichlet-Laplace Problem

If $(\mathcal{M}, g)$ is a compact Riemannian manifold with boundary we are interested in solving the following problem

$$-\Delta_g u + qu = f \text{ in } \mathcal{M}, \qquad u_{|\partial\mathcal{M}} = h \text{ on } \partial\mathcal{M},$$

that is, an elliptic problem with Dirichlet boundary condition. The potential function q will be assumed bounded in what follows.

18.6.1. Homogeneous Dirichlet Boundary Conditions. We denote by P_0 the differential operator $P_0 = -\Delta_g$. We start by considering the case of homogeneous Dirichlet boundary conditions, that is,

$$P_0 u + qu = f, \qquad u_{|\partial\mathcal{M}} = 0,$$

where $q \in L^\infty(\mathcal{M}; \mathbb{R})$ and $f \in H^{-1}(\mathcal{M})$, with the definition of $H^{-1}(\mathcal{M})$ given at the end of Sect. 18.2.

PROPOSITION 18.31 (Dirichlet-Laplace Problem). *Let $(\mathcal{M}, g)$ be a compact connected Riemannian d-dimensional manifold with boundary. Let $q \in L^\infty(\mathcal{M}; \mathbb{R})$. Let $C_g > 0$ be the Poincaré constant. There exists $C > 0$ such that, if $q > -C_g$, for any $f \in H^{-1}(\mathcal{M})$, there exists a unique $u \in H_0^1(\mathcal{M})$ such that*

(18.6.1)
$$\int_{\mathcal{M}} g(\nabla_g \varphi, \nabla_g \bar{u})\mu_g + \int_{\mathcal{M}} q\varphi\bar{u}\,\mu_g = \langle \bar{f}, \varphi\rangle_{H^{-1}(\mathcal{M}),H_0^1(\mathcal{M})}, \qquad \varphi \in H_0^1(\mathcal{M}),$$

with

$$C^{-1}\|f\|_{H^{-1}(\mathcal{M})} \leq \|u\|_{H^1(\mathcal{M})} \leq C\|f\|_{H^{-1}(\mathcal{M})}.$$

Moreover, the (0-density) distribution $P_0 u$ is in $H^{-1}(\mathcal{M})$ and $P_0 u + qu = f$ holds in $H^{-1}(\mathcal{M})$. Conversely, if $f \in H^{-1}(\mathcal{M})$ and if $u \in H_0^1(\mathcal{M})$ is such that $P_0 u + qu = f$ holds in $H^{-1}(\mathcal{M})$, then u is the unique solution of the variational form of the homogeneous Dirichlet problem (18.6.1).

Let $k \in \mathbb{N}$. For $q \in W^{k,\infty}(\mathcal{M}; \mathbb{R})$, there exists $C_k > 0$, such that if $f \in H^k(\mathcal{M})$, then moreover $u \in H^{k+2}(\mathcal{M})$ and

$$C_k^{-1}\|f\|_{H^k(\mathcal{M})} \leq \|u\|_{H^{k+2}(\mathcal{M})} \leq C_k\|f\|_{H^k(\mathcal{M})}$$

and the equation $P_0 u + qu = f$ holds in $H^k(\mathcal{M})$.

PROOF. On $H_0^1(\mathcal{M})$, we consider the continuous Hermitian form

$$a(v, u) = \int_{\mathcal{M}} g(\nabla_g v, \nabla_g \bar{u})\mu_g + \int_{\mathcal{M}} qv\bar{u}\mu_g = (\nabla_g v, \nabla_g u)_{L^2V(\mathcal{M})} + (qv, u)_{L^2(\mathcal{M})}.$$

If $C_g > 0$ is the Poincaré constant, then

$$C_g^{1/2}\|v\|_{L^2(\mathcal{M})} \leq \|\nabla_g v\|_{L^2V(\mathcal{M})}, \qquad v \in H_0^1(\mathcal{M}).$$

by Proposition 18.9. Thus, if $q > -C_g$ we obtain that $a(v, v) \gtrsim \|v\|^2_{H^1(\mathcal{M})}$ and consequently $a(.,.)$ is coercive on $H_0^1(\mathcal{M})$. As $v \mapsto (v, \bar{f})_{L^2(\mathcal{M})}$ is continuous, the Lax–Milgram theorem then gives a unique $u \in H_0^1(\mathcal{M})$ such that (18.6.1) holds. It also yields the existence of $C > 0$ such that

$$C^{-1}\|f\|_{H^{-1}(\mathcal{M})} \leq \|u\|_{H^1(\mathcal{M})} \leq C\|f\|_{H^{-1}(\mathcal{M})}.$$

We then find that $P_0 u + qu = f$ holds in $H^{-1}(\mathcal{M})$ by Proposition 18.13. Conversely, if $f \in H^{-1}(\mathcal{M})$ and if $u \in H_0^1(\mathcal{M})$ is such that $P_0 u + qu = f$ holds in $H^{-1}(\mathcal{M})$, then by applying (18.5.5), consequence of the divergence formula of Proposition 18.28, one finds that u is the unique solution of (18.6.1).

If now $f \in H^k(\mathcal{M})$ with $k \in \mathbb{N}$, the regularity result can be obtained by using Proposition 18.11, a finite atlas, and the counterpart of the result in open sets of $\mathbb{R}^d$; see [161, Section 8.4] and [90, Section 9.6]. ∎

Following Sects. 10.1 and 18.4, define P_0 as an unbounded operator on $L^2(\mathcal{M})$ with domain

$$D(\mathsf{P}_0) = \{u \in H_0^1(\mathcal{M}); P_0 u \in L^2(\mathcal{M})\}, \tag{18.6.2}$$

given by $\mathsf{P}_0 u = P_0 u$. We have $D(\mathsf{P}_0) = H^2(\mathcal{M}) \cap H_0^1(\mathcal{M})$. We find that P_0 is closed, selfadjoint and maximal monotone. Moreover, $\ker \mathsf{P}_0 = \{0\}$. Proposition 18.31 thus shows that the map, with $k \in \mathbb{N}$,

$$\begin{aligned} L_q : H^{k+2}(\mathcal{M}) \cap H_0^1(\mathcal{M}) &\to H^k(\mathcal{M}) \\ u &\mapsto \mathsf{P}_0 u + qu \end{aligned} \tag{18.6.3}$$

is an isomorphism if $q \in W^{k,\infty}(\mathcal{M};\mathbb{R})$ and $q > -C_g$. We define the resolvent map $\mathsf{R}_0 : L^2(\mathcal{M}) \to H^2(\mathcal{M}) \cap H_0^1(\mathcal{M})$, as the inverse of the above map in the case $q = 0$. It is compact map if considered as a map from $L^2(\mathcal{M})$ into itself.

We then obtain the following properties:

Hilbert basis of eigenfunctions: There exists a nondecreasing sequence of real eigenvalues $0 < \mu_0 < \mu_1 \leq \mu_2 \leq \cdots \leq \mu_n$ (counted with their multiplicity) and an associated sequence of eigenfunctions of P_0, denoted by $(\phi_j)_{j \in \mathbb{N}}$, that forms a Hilbert basis of $L^2(\mathcal{M})$.

Weyl Law: The eigenvalues satisfy the Weyl law of Theorem 10.3 of Volume 1.

Second, we wish to consider the nonhomogeneous Dirichlet problem on a compact Riemannian manifold. To obtain case of Dirichlet data with low regularity we first consider some subspaces of $L^2(\mathcal{M})$.

18.6.2. Spaces with Remarkable Trace Properties. Let $P = P_0 + R_1$ with $P_0 = -\Delta_g$ and R_1 a first-order differential operator with smooth coefficients. We introduce the following subspace of $L^2(\mathcal{M})$ associated with P:

$$\mathcal{W}_P(\mathcal{M}) = \{u \in L^2(\mathcal{M});\ Pu \in L^2(\mathcal{M})\}. \tag{18.6.4}$$

In the case $P = P_0$ we simply write $\mathcal{W}_g(\mathcal{M})$. Here, the action of P on u is to be understood in the sense of distributions, as expressed in (18.3.1), and not as that of the unbounded operator $(\mathsf{P}_0 + R_1, D(\mathsf{P}_0))$ on $L^2(\mathcal{M})$.

One can readily check that $\mathcal{W}_P(\mathcal{M})$ is a Hilbert space if equipped with the inner product

$$(u,v)_{\mathcal{W}_P(\mathcal{M})} = (u,v)_{L^2(\mathcal{M})} + (Pu,Pv)_{L^2(\mathcal{M})}, \quad u,v \in \mathcal{W}_P(\mathcal{M}),$$

and the associated norm

$$\|u\|^2_{\mathcal{W}_P(\mathcal{M})} = \|u\|^2_{L^2(\mathcal{M})} + \|Pu\|^2_{L^2(\mathcal{M})}, \quad u \in \mathcal{W}_P(\mathcal{M}).$$

We find that both the Dirichlet and Neumann trace operators extend to functions in $\mathcal{W}_P(\mathcal{M})$.

LEMMA 18.32. *The Dirichlet trace map $\gamma^{\mathsf{D}} : u \mapsto u_{|\partial\mathcal{M}}$ and the Neumann trace map $\gamma^{\mathsf{N}} : u \mapsto \partial_\nu u_{|\partial\mathcal{M}}$, both well defined on $H^2(\mathcal{M})$, admit a unique* *extension to $\mathcal{W}_P(\mathcal{M})$, that we still denote by γ^{D} and γ^{N}. If $u \in \mathcal{W}_P(\mathcal{M})$ we have*

$$\gamma^{\mathsf{D}}(u) \in H^{-1/2}(\partial\mathcal{M}) \quad \text{and} \quad \gamma^{\mathsf{N}}(u) \in H^{-3/2}(\partial\mathcal{M}),$$

and the linear maps $\gamma^{\mathsf{D}} : \mathcal{W}_P(\mathcal{M}) \to H^{-1/2}(\partial\mathcal{M})$ and $\gamma^{\mathsf{N}} : \mathcal{W}_P(\mathcal{M}) \to H^{-3/2}(\partial\mathcal{M})$ are both bounded.

Finally, the Leibniz rule is compatible with these trace extensions, that is, for $a \in \mathscr{C}^\infty(\mathcal{M})$, we have

$$\gamma^{\mathsf{N}}(au) = \gamma^{\mathsf{D}}(a)\gamma^{\mathsf{N}}(u) + \gamma^{\mathsf{N}}(a)\gamma^{\mathsf{D}}(u).$$

A proof is given in Appendix 18.A.1.

If $R_1 = X + f$ with X a vector field and f a function, the transpose operator of P is given by ${}^tP = -\Delta_g - X - \operatorname{div}_g X + f$; see Sect. 16.3.3, Propositions 17.7 and 17.8. For functions in $\mathcal{W}_P(\mathcal{M})$, we have a Green-like formula.

LEMMA 18.33. *Let $R_1 = X + f$ with X a smooth vector field and f a smooth function and $P = -\Delta_g + R_1$. Let $u \in \mathcal{W}_P(\mathcal{M})$ and $w \in H^2(\mathcal{M})$. We have*

$$\begin{aligned}(Pu, w)_{L^2(\mathcal{M})} &+ \langle \gamma^{\mathsf{N}}(u), \gamma^{\mathsf{D}}(\bar{w})\rangle_{H^{-3/2}(\partial\mathcal{M}),H^{3/2}(\partial\mathcal{M})} \\ &= (u, {}^t\bar{P}w)_{L^2(\mathcal{M})} + \langle \gamma^{\mathsf{D}}(u), \gamma^{\mathsf{N}}(\bar{w})\rangle_{H^{-1/2}(\partial\mathcal{M}),H^{1/2}(\partial\mathcal{M})} \\ &\quad + \langle \gamma^{\mathsf{D}}(g(X,\nu)u), \gamma^{\mathsf{D}}(\bar{w})\rangle_{H^{-1/2}(\partial\mathcal{M}),H^{1/2}(\partial\mathcal{M})}.\end{aligned}$$

A proof is given in Appendix 18.A.2. For functions in $\mathcal{W}_g(\mathcal{M})$, that is, in the case $P = P_0 = -\Delta_g$, this gives the following Green formula.

LEMMA 18.34. *Let $u \in \mathcal{W}_g(\mathcal{M})$ and $w \in H^2(\mathcal{M})$. We have*

$$\begin{aligned}(P_0 u, w)_{L^2(\mathcal{M})} &+ \langle \gamma^{\mathsf{N}}(u), \gamma^{\mathsf{D}}(\bar{w})\rangle_{H^{-3/2}(\partial\mathcal{M}),H^{3/2}(\partial\mathcal{M})} \\ &= (u, P_0 w)_{L^2(\mathcal{M})} + \langle \gamma^{\mathsf{D}}(u), \gamma^{\mathsf{N}}(\bar{w})\rangle_{H^{-1/2}(\partial\mathcal{M}),H^{1/2}(\partial\mathcal{M})}.\end{aligned}$$

We also consider the following subspace of $L^2(\mathcal{M})$:

$$\mathcal{W}_{g,-1}(\mathcal{M}) = \{u \in L^2(\mathcal{M});\ Pu \in H^{-1}(\mathcal{M})\}. \tag{18.6.5}$$

Note that this space is independent of the first-order term R_1 in P. We have $\mathcal{W}_g(\mathcal{M}) \cup H^1(\mathcal{M}) \subset \mathcal{W}_{g,-1}(\mathcal{M})$. The inner product

$$(u, v)_{\mathcal{W}_{g,-1}(\mathcal{M})} = (u, v)_{L^2(\mathcal{M})} + (P_0 u, P_0 v)_{H^{-1}(\mathcal{M})}, \quad u, v \in \mathcal{W}_{g,-1}(\mathcal{M}),$$

and the associated norm,

$$\|u\|^2_{\mathcal{W}_{g,-1}(\mathcal{M})} = \|u\|^2_{L^2(\mathcal{M})} + \|P_0 u\|^2_{H^{-1}(\mathcal{M})}, \quad u \in \mathcal{W}_{g,-1}(\mathcal{M}),$$

yield a Hilbert space structure on $\mathcal{W}_{g,-1}(\mathcal{M})$.

As the Neumann trace of a H^1-function makes no sense in general the same is true for a function in $\mathcal{W}_{g,-1}(\mathcal{M})$. However, the Dirichlet trace makes sense in $H^{-1/2}(\partial\mathcal{M})$ like we found above for functions in $\mathcal{W}_g(\mathcal{M})$.

LEMMA 18.35. *The Dirichlet trace map $\gamma^{\mathsf{D}} : u \mapsto u_{|\partial\mathcal{M}}$ well defined on $H^1(\mathcal{M})$ admits a unique extension to $\mathcal{W}_{g,-1}(\mathcal{M})$, that we still denote by γ^{D}. If $u \in \mathcal{W}_{g,-1}(\mathcal{M})$ we have*

$$\gamma^{\mathsf{D}}(u) \in H^{-1/2}(\partial\mathcal{M})$$

and the linear maps $\gamma^{\mathsf{D}} : \mathcal{W}_{g,-1}(\mathcal{M}) \to H^{-1/2}(\partial\mathcal{M})$ is bounded. For any $w \in H^2(\mathcal{M}) \cap H_0^1(\mathcal{M})$ we have

$$\langle P_0 u, \bar{w}\rangle_{H^{-1}(\mathcal{M}),H_0^1(\mathcal{M})} = (u, P_0 w)_{L^2(\mathcal{M})} + \langle \gamma^{\mathsf{D}}(u), \gamma^{\mathsf{N}}(\bar{w})\rangle_{H^{-1/2}(\partial\mathcal{M}),H^{1/2}(\partial\mathcal{M})}.$$

A proof is given in Appendix 18.A.3.

18.6.3. The Dirichlet Lifting Map. We are interested in defining a map that acts as a solution operator for the following elliptic problem

$$P_0 u = 0 \text{ in } \mathcal{M} \quad \text{and} \quad \gamma^{\mathsf{D}}(u) = h \text{ on } \partial\mathcal{M},$$

for $h \in H^{-1/2}(\partial\mathcal{M})$. The Hilbert space $\mathcal{W}_g(\mathcal{M})$ appears as a natural space for solutions of this problem. In fact, we have the following result.

PROPOSITION 18.36. *Let $h \in H^{-1/2}(\partial\mathcal{M})$. There exists a unique $u \in \mathcal{W}_g(\mathcal{M})$ such that $P_0 u = 0$ and $\gamma^{\mathsf{D}}(u) = h$. Moreover, the map $h \mapsto u$ from $H^{-1/2}(\partial\mathcal{M})$ into $\mathcal{W}_g(\mathcal{M})$ is bounded.*

DEFINITION 18.37 (Dirichlet Lifting Map). The bounded map $\mathsf{D} : h \mapsto u$ from $H^{-1/2}(\partial\mathcal{M})$ into $\mathcal{W}_g(\mathcal{M})$ given by Proposition 18.36 is called the Dirichlet lifting map.

PROOF OF PROPOSITION 18.36. The proof makes use of the resolvent map $\mathsf{R}_0 : L^2(\mathcal{M}) \to D(\mathsf{P}_0)$ introduced below (18.6.3).

First, we address uniqueness. By linearity, we consider $u \in \mathcal{W}_g(\mathcal{M})$ is such that $P_0 u = 0$ and $\gamma^{\mathsf{D}}(u) = 0$. Then, by Lemma 18.34 we have $(u, P_0 v)_{L^2(\mathcal{M})} = 0$ for all $v \in H^2(\mathcal{M}) \cap H_0^1(\mathcal{M})$. Let $f \in L^2(\mathcal{M})$; we choose $v = \mathsf{R}_0 f$. We obtain $(u, f)_{L^2(\mathcal{M})} = 0$. As $f \in L^2(\mathcal{M})$ is arbitrary we conclude that $u = 0$.

Second, we address existence. We consider the following form:

$$\begin{aligned} U : L^2(\mathcal{M}) &\to \mathbb{C} \\ f &\mapsto -\langle \gamma^{\mathsf{N}} \mathsf{R}_0 f, \bar{h}\rangle_{H^{1/2}(\partial\mathcal{M}),H^{-1/2}(\partial\mathcal{M})}. \end{aligned}$$

By the trace formula of Theorem 18.25, this form is bounded and we have

$$|U(f)| \leq |h|_{H^{-1/2}(\partial\mathcal{M})} \|\mathsf{R}_0 f\|_{H^2(\mathcal{M})} \lesssim |h|_{H^{-1/2}(\partial\mathcal{M})} \|f\|_{L^2(\mathcal{M})}.$$

By the Riesz theorem, there exists $u \in L^2(\mathcal{M})$ such that $U(f) = (f, u)_{L^2(\mathcal{M})}$ for all $f \in L^2(\mathcal{M})$. Moreover, $\|u\|_{L^2(\mathcal{M})} \lesssim |h|_{H^{-1/2}(\partial\mathcal{M})}$. If $\varphi \in {}^0\mathscr{D}_c^\infty(\mathcal{M})$ and $f = P_0\varphi$, then $\varphi = \mathsf{R}_0 f$ and we find

$$U(f) = -\langle \gamma^{\mathsf{N}}(\varphi), \bar{h}\rangle_{H^{1/2}(\partial\mathcal{M}),H^{-1/2}(\partial\mathcal{M})} = 0.$$

Consequently

$$\begin{aligned}\langle \overline{P_0 u}, \varphi \mu_g \rangle_{{}^0\mathscr{D}'(\mathcal{M}),{}^1\mathscr{D}^\infty_c(\mathcal{M})} &= \langle \bar{u}, P_0 \varphi \mu_g \rangle_{{}^0\mathscr{D}'(\mathcal{M}),{}^1\mathscr{D}^\infty_c(\mathcal{M})} \\ &= (P_0\varphi, u)_{L^2(\mathcal{M})} = U(f) = 0.\end{aligned}$$

Thus, in the sense of distributions we have $P_0 u = 0$. Thus $u \in \mathcal{W}_g(\mathcal{M})$ and $\|u\|_{\mathcal{W}_g(\mathcal{M})} \lesssim |h|_{H^{-1/2}(\partial\mathcal{M})}$.

LEMMA 18.38. *There exists a bounded map* $\mathsf{N}_0 : H^{1/2}(\partial\mathcal{M}) \to H^2(\mathcal{M}) \cap H^1_0(\mathcal{M})$ *such that* $\gamma^{\mathsf{N}} \circ \mathsf{N}_0 = \mathrm{Id}_{H^{1/2}(\partial\mathcal{M})}$.

This is a particular case of Theorem 18.25. A proof follows from Lemma 18.27 by working locally at the boundary.

Let $\psi \in H^{1/2}(\partial\mathcal{M})$. For $v = \mathsf{N}_0\psi \in H^2(\mathcal{M}) \cap H^1_0(\mathcal{M})$ as given by Lemma 18.38, we have

$$(P_0 v, u)_{L^2(\mathcal{M})} + \langle \psi, \gamma^{\mathsf{D}}(\bar{u}) \rangle_{H^{1/2}(\partial\mathcal{M}),H^{-1/2}(\partial\mathcal{M})} = 0,$$

by Lemma 18.34. We have

$$\begin{aligned}(P_0 v, u)_{L^2(\mathcal{M})} &= (\mathsf{P}_0 v, u)_{L^2(\mathcal{M})} = U(\mathsf{P}_0 v) = -\langle \gamma^{\mathsf{N}} \mathsf{R}_0 \mathsf{P}_0 v, \bar{h} \rangle_{H^{1/2}(\mathcal{M}),H^{-1/2}(\mathcal{M})} \\ &= -\langle \gamma^{\mathsf{N}} v, \bar{h} \rangle_{H^{1/2}(\mathcal{M}),H^{-1/2}(\mathcal{M})} = -\langle \psi, \bar{h} \rangle_{H^{1/2}(\mathcal{M}),H^{-1/2}(\mathcal{M})}.\end{aligned}$$

For all $\psi \in H^{1/2}(\partial\mathcal{M})$ we thus find $\langle \psi, \bar{h} - \gamma^{\mathsf{D}}(\bar{u}) \rangle_{H^{1/2}(\partial\mathcal{M}),H^{-1/2}(\partial\mathcal{M})} = 0$, which concludes the existence part of the proof. ∎

If h is more regular than $H^{-1/2}$, then additional regularity is transferred to $\mathsf{D}(h)$, with consistency with the trace formula of Theorem 18.25.

PROPOSITION 18.39. *Let* $m \in \mathbb{N}$. *If* $h \in H^{m-1/2}(\partial\mathcal{M})$, *then* $\mathsf{D}(h) \in H^m(\mathcal{M})$. *Moreover, for some* $C > 0$, *we have*

$$\|\mathsf{D}(h)\|_{H^m(\mathcal{M})} \leq C |h|_{H^{m-1/2}(\partial\mathcal{M})}, \quad h \in H^{m-1/2}(\partial\mathcal{M}). \tag{18.6.6}$$

PROOF. The case $m = 0$ was treated above. Let $m \geq 1$ and $h \in H^{m-1/2}(\mathcal{M})$. By Theorem 18.25, there exists $v \in H^m(\mathcal{M})$ such that $\gamma^{\mathsf{D}}(v) = h$ and $\|v\|_{H^m(\mathcal{M})} \lesssim |h|_{H^{m-1/2}(\partial\mathcal{M})}$. We have $f = P_0 v \in H^{m-2}(\mathcal{M})$ with

$$\|f\|_{H^{m-2}(\mathcal{M})} \lesssim |h|_{H^{m-1/2}(\partial\mathcal{M})},$$

and there exists a unique $\tilde{v} \in H^m(\mathcal{M}) \cap H^1_0(\mathcal{M})$ such that $P_0\tilde{v} = f$ by Proposition 18.31. We have

$$\|\tilde{v}\|_{H^m(\mathcal{M})} \lesssim |h|_{H^{m-1/2}(\partial\mathcal{M})}.$$

Consider next $w = v - \tilde{v} \in H^m(\mathcal{M})$. We have $P_0 w = 0$ and $\gamma^{\mathsf{D}}(w) = \gamma^{\mathsf{D}}(v) = h$. By the uniqueness part of Proposition 18.36 we have $w = \mathsf{D}(h)$ and (18.6.6) holds. ∎

The above results allow one to fully treat the nonhomogeneous Dirichlet-Laplace problem on a compact manifold with boundary exploiting the linearity of the problem.

THEOREM 18.40. *Let $(\mathcal{M}, g)$ be a compact connected Riemannian d-dimensional manifold with boundary and let $q \in L^\infty(\mathcal{M}; \mathbb{R})$ with $q > -C_g$, where C_g is as given in Proposition 18.31. The map*

$$\begin{aligned} \mathcal{W}_{g,-1}(\mathcal{M}) &\to H^{-1}(\mathcal{M}) \oplus H^{-1/2}(\partial\mathcal{M}) \\ u &\mapsto \big(P_0 u + qu, \gamma^{\mathsf{D}}(u)\big) \end{aligned}$$

is an isomorphism. Moreover, for $k \in \mathbb{N}$, if $q \in W^{k,\infty}(\mathcal{M}; \mathbb{R})$, the map

$$\begin{aligned} {}^{\mathsf{D}}L : H^{k+1}(\mathcal{M}) &\to H^{k-1}(\mathcal{M}) \oplus H^{k+1/2}(\partial\mathcal{M}) \\ u &\mapsto \big(P_0 u + qu, \gamma^{\mathsf{D}}(u)\big) \end{aligned}$$

is an isomorphism.

We also have the following regularity results.

COROLLARY 18.41. *Let $u \in L^2(\mathcal{M})$.*

(1) *Assume that $P_0 u \in H^{-1}(\mathcal{M})$, that is, $u \in \mathcal{W}_{g,-1}(\mathcal{M})$ and assume yet that $\gamma^{\mathsf{D}}(u) \in H^{1/2}(\partial\mathcal{M})$. Then $u \in H^1(\mathcal{M})$.*
(2) *More generally, let $m, m' \in \mathbb{N}$ and assume that $P_0 u \in H^{m-1}(\mathcal{M})$ and $\gamma^{\mathsf{D}}(u) \in H^{m'+1/2}(\partial\mathcal{M})$. Then $u \in H^{\min(m,m')+1}(\mathcal{M})$.*

PROOF. Let $u \in \mathcal{W}_{g,-1}(\mathcal{M})$ with $\gamma^{\mathsf{D}}(u) \in H^{1/2}(\mathcal{M})$. Set $f = P_0 u \in H^{-1}(\mathcal{M})$ and $h = \gamma^{\mathsf{D}}(u) \in H^{1/2}(\mathcal{M})$. By Theorem 18.40 there exists a unique $v \in H^1(\mathcal{M})$ such that $P_0 v = f$ and $\gamma^{\mathsf{D}}(v) = h$. Then $w = u - v \in L^2(\mathcal{M})$ is such that $P_0 w = 0$, that is $w \in \mathcal{W}_g(\mathcal{M})$ and moreover $\gamma^{\mathsf{D}}(w) = 0$. From the uniqueness part of Proposition 18.36 we conclude that $w = 0$.

Let now $u \in L^2(\mathcal{M})$ such that $f = P_0 u \in H^{m-1}(\mathcal{M})$ and $h = \gamma^{\mathsf{D}}(u) \in H^{m'+1/2}(\partial\mathcal{M})$. Set $n = \min(m, m')$. We have $f \in H^{n-1}(\mathcal{M})$ and $h \in H^{n+1/2}(\partial\mathcal{M})$. By Theorem 18.40 there exists a unique $v \in H^{n+1}(\mathcal{M})$ such that $P_0 v = f$ and $\gamma^{\mathsf{D}}(v) = h$. The conclusion then follows as above. ■

In particular, with this last corollary, we see that the domain of the operator P_0 with domain defined in (18.6.2) can equivalently be defined by

$$D(\mathsf{P}_0) = \{u \in L^2(\mathcal{M});\ \Delta_g u \in L^2(\mathcal{M}) \text{ and } \gamma^{\mathsf{D}}(u) = 0\}, \tag{18.6.7}$$

which is a more natural definition as the operator P_0 is an unbounded operator on $L^2(\mathcal{M})$. This, however, requires the knowledge of the existence of a Dirichlet trace for functions in $\mathcal{W}_g(\mathcal{M})$.

18.6.4. More Spaces with Remarkable Trace Properties. By Lemma 18.32 if $u \in \mathcal{W}_g(\mathcal{M})$, then

$$\gamma^{\mathsf{D}}(u) \in H^{-1/2}(\partial\mathcal{M}) \quad \text{and} \quad \gamma^{\mathsf{N}}(u) \in H^{-3/2}(\partial\mathcal{M}).$$

If u is moreover in $H^1(\mathcal{M})$ we naturally have $\gamma^{\mathsf{D}}(u) \in H^{1/2}(\partial\mathcal{M})$ by the trace formula of Theorem 18.25. In fact, there is also an improvement of regularity for the Neumann trace.

LEMMA 18.42. *Let $u \in \mathcal{W}_g(\mathcal{M}) \cap H^1(\mathcal{M})$. Then $\gamma^{\mathsf{N}}(u) \in H^{-1/2}(\partial\mathcal{M})$ and*

$$|\gamma^{\mathsf{N}}(u)|_{H^{-1/2}(\partial\mathcal{M})} \leq C\big(\|u\|_{H^1(\mathcal{M})} + \|P_0 u\|_{L^2(\mathcal{M})}\big),$$

for some $C > 0$ (independent of u).

A proof based on the Dirichlet lifting map introduced in Sect. 18.6.3 is given in Appendix 18.A.4. Lemma 18.42 is also a consequence of a normal trace result on the following Hilbert space of vector fields

$$H(\operatorname{div}_g, \mathcal{M}) = \{U \in L^2V(\mathcal{M});\ \operatorname{div}_g U \in L^2(\mathcal{M})\},$$

equipped with the inner product

$$(U, V)_{H(\operatorname{div}_g,\mathcal{M})} = (U, V)_{L^2V(\mathcal{M})} + (\operatorname{div}_g U, \operatorname{div}_g V)_{L^2(\mathcal{M})}.$$

In the definition of $H(\operatorname{div}_g, \mathcal{M})$ the action div_g is to be understood in the sense of distributions.

LEMMA 18.43. *The trace map $U \mapsto g(U, \nu)_{|\partial\mathcal{M}}$, well defined on the space of H^1-vector fields, admit a unique extension to $H(\operatorname{div}_g, \mathcal{M})$. If $U \in H(\operatorname{div}_g, \mathcal{M})$ one has $g(U, \nu)_{|\partial\mathcal{M}} \in H^{-1/2}(\partial\mathcal{M})$ and for some $C > 0$,*

$$\|g(U, \nu)_{|\partial\mathcal{M}}\|_{H^{-1/2}(\partial\mathcal{M})} \leq C\|U\|_{H(\operatorname{div}_g,\mathcal{M})}, \qquad U \in H(\operatorname{div}_g, \mathcal{M}).$$

A proof is given in Appendix 18.A.5.

18.7. The Neumann-Laplace Problem

Here, for $q \in L^\infty(\mathcal{M}; \mathbb{R})$ such that $q \geq 0$, we consider the elliptic problem

$$(P_0 + q)u = f \text{ in } \mathcal{M}, \qquad \gamma^{\mathsf{N}}(u) = h \text{ on } \partial\mathcal{M}. \tag{18.7.1}$$

If $f \in L^2(\mathcal{M})$ and if we seek $u \in H^2(\mathcal{M})$, then the trace formula of Theorem 18.25 indicates that the proper space for h is $H^{1/2}(\partial\mathcal{M})$. Next, with the divergence formula of Proposition 18.28, we observe that

$$\int_{\mathcal{M}} f \mu_g + \int_{\partial\mathcal{M}} \gamma^{\mathsf{N}}(u)\, \mu_{g_\partial} = \int_{\mathcal{M}} qu\, \mu_g. \tag{18.7.2}$$

We refer to Sects. 16.2.4 and 17.3 for integration on Riemannian manifolds. In particular, μ_g (resp. μ_{g_∂}) is the canonical positive density function on $\mathcal{M}$ (resp. $\partial\mathcal{M}$). If $q \neq 0$ we see that the integral of qu is immediately known. On $H^1(\mathcal{M})$ we define the following Hermitian form:

$$\begin{aligned} a_0(v, u) &= (\mathsf{D}\, v, \mathsf{D}\, u)_{L^2\Lambda^1(\mathcal{M})} = (\nabla_g v, \nabla_g u)_{L^2V(\mathcal{M})} \\ &= \int_{\mathcal{M}} g_m(\nabla_g v(m), \nabla_g \bar{u}(m))\, \mu_g, \end{aligned} \tag{18.7.3}$$

and

$$a_q(v, u) = a_0(v, u) + (v, qu)_{L^2(\mathcal{M})} \tag{18.7.4}$$

PROPOSITION 18.44. *Let $q \in L^\infty(\mathcal{M};\mathbb{R})$ be such that $q \geq 0$ and $q \geq C > 0$ on $A \subset U$ of positive measure. If $f \in L^2(\mathcal{M})$ and $h \in H^{1/2}(\partial\mathcal{M})$, there exists a unique $u \in H^1(\mathcal{M})$ such that*

$$a_q(v,u) = (v,f)_{L^2(\mathcal{M})} + (v_{|\partial\mathcal{M}}, h)_{L^2(\partial\mathcal{M})}, \quad v \in H^1(\mathcal{M}). \tag{18.7.5}$$

Moreover, $u \in H^2(\mathcal{M})$ and we have

$$(P_0 + q)u = f \text{ in } \mathcal{M}, \qquad \gamma^{\mathsf{N}}(u) = h \text{ on } \partial\mathcal{M}.$$

and there exists $C > 0$ such that

$$\|u\|_{H^2(\mathcal{M})} \leq C\big(\|f\|_{L^2(\mathcal{M})} + |h|_{H^{1/2}(\partial\mathcal{M})}\big).$$

If $k \in \mathbb{N}$, if $q \in W^{k,\infty}(\mathcal{M};\mathbb{R})$ in addition to the above requirements, there exists $C > 0$ such that if $f \in H^k(\mathcal{M})$ and $h \in H^{k+1/2}(\mathcal{M})$, then $u \in H^{k+2}(\mathcal{M})$ and

$$\|u\|_{H^{k+2}(\mathcal{M})} \leq C\big(\|f\|_{H^k(\mathcal{M})} + |h|_{H^{k+1/2}(\partial\mathcal{M})}\big).$$

One says that (18.7.5) is the variational form of the Neumann problem (18.7.1).

PROOF. By the trace formula of Theorem 18.25 there exists $\mathsf{h} \in H^2(\mathcal{M})$ such that $\gamma^{\mathsf{N}}(\mathsf{h}) = h$ and $\|\mathsf{h}\|_{H^2(\mathcal{M})} \lesssim |h|_{H^{1/2}(\partial\mathcal{M})}$. With (18.5.5) consequence of the divergence formula of Proposition 18.28 we then have, for all $v \in H^1(\mathcal{M})$

$$a_q(v,\mathsf{h}) = (v,H)_{L^2(\mathcal{M})} + (v_{|\partial\mathcal{M}}, h)_{L^2(\partial\mathcal{M})},$$

where $H = (q + P_0)\mathsf{h} \in L^2(\mathcal{M})$, with $\|H\|_{L^2(\mathcal{M})} \lesssim |h|_{H^{1/2}(\partial\mathcal{M})}$. By linearity, it thus suffices to solve Problem (18.7.5) for $h = 0$ and f replaced by $f - H$.

One then observes that $a_q(.,.)$ is coercive on $H^1(\mathcal{M})$ by the third Poincaré inequality given in Proposition 18.9. As $v \mapsto (v,f)_{L^2(\mathcal{M})}$ is continuous and linear on $H^1(\mathcal{M})$, the Lax–Milgram theorem yields the existence of $u \in H^1(\mathcal{M})$ solution to (18.7.5).

The remainder of the proof is classical. We refer, for instance, to [90, Section 9.5, Example 4] that can be adapted to the manifold case. ∎

COROLLARY 18.45 (Uniqueness of Strong Solutions). *Let $f \in L^2(\mathcal{M})$ and $h \in H^{1/2}(\partial\mathcal{M})$. Let $u \in H^2(\mathcal{M})$ be such that $P_0 u + qu = f$ in $\mathcal{M}$ and $\gamma^{\mathsf{N}}(w) = h$ on $\partial\mathcal{M}$. Then u is the unique solution of the variational form of the Neumann problem* (18.7.5).

PROOF. It suffices to observe that u is itself a solution of (18.7.5) by (18.5.5) consequence of the divergence formula of Proposition 18.28. ∎

We now consider the case $q = 0$ in (18.7.1). Then, we see from (18.7.2) that a necessary condition for the solvability of the elliptic problem is

$$\int_{\mathcal{M}} f \mu_g + \int_{\partial\mathcal{M}} h\, \mu_{g_\partial} = 0. \tag{18.7.6}$$

Moreover, in the case $q = 0$, uniqueness cannot be obtained in the space $H^2(\mathcal{M})$ since $u + \text{Cst}$ is a solution if u solves the problem. We shall recall below that this is the only obstruction to well-posedness under condition (18.7.6).

Let $k \in \mathbb{N}$ and set $\mathsf{C} = \{u \in H^k(\mathcal{M});\ u \equiv \text{Cst}\} \cong \mathbb{C}$. With the above observations, we define the equivalence relation on $H^k(\mathcal{M})$

$$u \sim u' \quad \Leftrightarrow \quad u - u' \in \mathsf{C}, \tag{18.7.7}$$

and we denote by $H^k(\mathcal{M})/\mathsf{C}$ or rather $H^k(\mathcal{M})/\ \mathbb{C}$ the quotient space $H^k(\mathcal{M})/\sim$, and by $\Phi : H^k(\mathcal{M}) \to H^k(\mathcal{M})/\mathbb{C}$ the surjective linear application that maps u to its class $[u]$ in $H^k(\mathcal{M})/\mathbb{C}$. The subspace C is precisely the kernel of Φ. We have a Banach space structure on $H^k(\mathcal{M})/\mathbb{C}$ (see e.g. [320, Chapter 11]) with the norm given by

$$\|\mathsf{u}\|_{H^k(\mathcal{M})/\mathbb{C}} = \inf_{\substack{u \in H^k(\mathcal{M}) \\ [u] = \mathsf{u}}} \|u\|_{H^k(\mathcal{M})},$$

that makes the map Φ continuous. We define the following linear continuous map

$$\begin{aligned} \underline{\Pi} : H^k(\mathcal{M}) &\to H^k(\mathcal{M}) \\ u &\mapsto u - \fint_{\mathcal{M}} u\, \mu_g, \end{aligned}$$

where $\fint_{\mathcal{M}} u\, \mu_g = \frac{1}{|\mathcal{M}|} \int_{\mathcal{M}} u\, \mu_g$ is the average of u on $\mathcal{M}$. We observe that

$$[u] = [\underline{\Pi} u], \qquad \text{for } u \in H^k(\mathcal{M}).$$

We set $\underline{H}^k(\mathcal{M}) = \operatorname{Ran}(\underline{\Pi})$ and note that

$$\underline{H}^k(\mathcal{M}) = \{u \in H^k(\mathcal{M});\ \fint_{\mathcal{M}} u\, \mu_g = 0\}. \tag{18.7.8}$$

In particular, $\underline{H}^k(\mathcal{M})$ is a closed linear subspace of $H^k(\mathcal{M})$. For $k = 0$, we denote $\underline{H}^0(\mathcal{M}) = \underline{L}^2(\mathcal{M})$.

LEMMA 18.46. *We have* $\mathsf{C} \oplus \underline{H}^k(\mathcal{M}) = H^k(\mathcal{M})$ *and* $\underline{\Pi}$ *is the orthogonal projector onto* $\underline{H}^k(\mathcal{M})$.

PROOF. As $\underline{H}^k(\mathcal{M})$ is closed, it suffices to prove that $\mathsf{C} = \underline{H}^k(\mathcal{M})^\perp$ to obtain the first part of the result. The second part follows as $\underline{\Pi}$ vanishes on C and coincides with the identity on $\underline{H}^k(\mathcal{M})$.

Let u be in $\underline{H}^k(\mathcal{M})^\perp$. We then have $0 = (u, \underline{\Pi} u)_{H^k(\mathcal{M})}$ yielding

$$\begin{aligned} \|u\|^2_{L^2(\mathcal{M})} \leq \|u\|^2_{H^k(\mathcal{M})} = (\overline{\fint_{\mathcal{M}} u\, \mu_g})(u, 1)_{H^k(\mathcal{M})} &= \frac{1}{|\mathcal{M}|} \Big| \int_{\mathcal{M}} u\, \mu_g \Big|^2 \\ &\leq \frac{1}{|\mathcal{M}|} \Big(\int_{\mathcal{M}} |u|\, \mu_g \Big)^2. \end{aligned}$$

With the Cauchy–Schwarz inequality, we have

$$\frac{1}{|\mathcal{M}|}\Big(\int_{\mathcal{M}} |u|\,\mu_g\Big)^2 \le \|u\|^2_{L^2(\mathcal{M})}.$$

As a result $\int_{\mathcal{M}} |u|\,\mu_g = |\mathcal{M}|^{1/2}\|u\|_{L^2(\mathcal{M})}$, meaning equality in the Cauchy–Schwarz inequality, which implies that $u \equiv \mathrm{Cst}$. Conversely, if $u \equiv \mathrm{Cst}$, we immediately see that $u \in \underline{H}^k(\mathcal{M})^{\perp}$. ∎

With this lemma, we see that $H^k(\mathcal{M})/\mathbb{C}$ is to be identified with $\underline{H}^k(\mathcal{M})$. In fact, if we denote by $\underline{\Phi}$ the map

$$\begin{aligned} \underline{\Phi} : \underline{H}^k(\mathcal{M}) &\to H^k(\mathcal{M})/\mathbb{C} \\ u &\mapsto \Phi(u) = [u], \end{aligned} \tag{18.7.9}$$

we have the following result formalizing this identification.

LEMMA 18.47. *The map $\underline{\Phi}$ is an isometry of $\underline{H}^k(\mathcal{M})$ onto $H^k(\mathcal{M})/\mathbb{C}$.*

In particular, this yields a Hilbert space structure on $H^k(\mathcal{M})/\mathbb{C}$ with the inner product

$$\big([u],[v]\big)_{H^k(\mathcal{M})/\mathbb{C}} = \big((\underline{\Phi})^{-1}[u], (\underline{\Phi})^{-1}[v]\big)_{H^k(\mathcal{M})}, \quad \text{for } [u],[v] \in H^k(\mathcal{M})/\mathbb{C}. \tag{18.7.10}$$

This is summarized in the following diagram.

$$\begin{array}{ccc} H^k(\mathcal{M}) & & \\ \big\downarrow{\scriptstyle \underline{\Pi}} & \searrow^{\Phi} & \\ \underline{H}^k(\mathcal{M}) & \overset{\underline{\Phi}}{\underset{(\underline{\Phi})^{-1}}{\rightleftarrows}} & H^k(\mathcal{M})/\mathbb{C} \end{array}$$

For $[u] \in H^k(\mathcal{M})/\mathbb{C}$, we shall denote by $\underline{u}$ its unique representative in $\underline{H}^k(\mathcal{M})$, that is, $\underline{u} = (\underline{\Phi})^{-1}[u]$, and we have

$$\|[u]\|_{H^k(\mathcal{M})/\mathbb{C}} = \|\underline{u}\|_{H^k(\mathcal{M})}. \tag{18.7.11}$$

For $u \in H^k(\mathcal{M})$, since $[u] = [\underline{\Pi}u]$, we note that we have

$$\underline{\Pi}u = (\underline{\Phi})^{-1}[u] = \underline{u}.$$

We may now state how the Neumann problem can be solved in the case $q = 0$.

PROPOSITION 18.48. *Let $f \in L^2(\mathcal{M})$ and $h \in H^{1/2}(\partial\mathcal{M})$ be such that*

$$\int_{\mathcal{M}} f\mu_g + \int_{\partial\mathcal{M}} h\,\mu_{g_\partial} = 0. \tag{18.7.12}$$

There exists a unique $[u] \in H^1(\mathcal{M})/\mathbb{C}$ such that,

$$a_0(v,u) = (v,f)_{L^2(\mathcal{M})} + (v_{|\partial\mathcal{M}}, h)_{L^2(\partial\mathcal{M})}, \quad u \in [u],\ v \in H^1(\mathcal{M}), \tag{18.7.13}$$

with the Hermitian form $a_0(.,,.)$ *given in* (18.7.3). *Moreover,* $[u] \in H^2(\mathcal{M})/\mathbb{C}$ *and we have, for* $u \in [u]$,

$$P_0 u = f \text{ in } \mathcal{M}, \qquad \gamma^{\mathsf{N}}(u) = h \text{ on } \partial\mathcal{M}.$$

Finally, there exists $C > 0$ *such that*

$$\|\underline{u}\|_{H^2(\mathcal{M})} = \|[u]\|_{H^2(\mathcal{M})/\mathbb{C}} \leq C\big(\|f\|_{L^2(\mathcal{M})} + |h|_{H^{1/2}(\partial\mathcal{M})}\big). \tag{18.7.14}$$

If $k \in \mathbb{N}$ *there exists* $C_k > 0$ *such that if moreover* $f \in H^k(\mathcal{M})$ *and* $h \in H^{k+1/2}(\partial\mathcal{M})$, *then* $[u] \in H^{k+2}(\mathcal{M})/\mathbb{C}$ *and*

$$\|\underline{u}\|_{H^{k+2}(\mathcal{M})} = \|[u]\|_{H^{k+2}(\mathcal{M})/\mathbb{C}} \leq C\big(\|f\|_{H^k(\mathcal{M})} + |h|_{H^{k+1/2}(\partial\mathcal{M})}\big).$$

PROOF. By the trace formula of Theorem 18.25 there exists $\mathsf{h} \in H^2(\mathcal{M})$ such that $\gamma^{\mathsf{N}}(\mathsf{h}) = h$ and $\|\mathsf{h}\|_{H^2(\mathcal{M})} \lesssim |h|_{H^{1/2}(\partial\mathcal{M})}$. With (18.5.5) consequence of the divergence formula of Proposition 18.28 we then have, for all $v \in H^1(\mathcal{M})$,

$$a_0(v, \mathsf{h}) = (v, H)_{L^2(\mathcal{M})} + (v_{|\partial\mathcal{M}}, h)_{L^2(\partial\mathcal{M})},$$

where $H = P_0\mathsf{h} \in L^2(\mathcal{M})$, with $\|H\|_{L^2(\mathcal{M})} \lesssim |h|_{H^{1/2}(\partial\mathcal{M})}$. Choosing $v = 1$ yields, with (18.7.12),

$$\int_{\mathcal{M}} H\mu_g + \int_{\partial\mathcal{M}} h\,\mu_{g_\partial} = 0, \text{ that is, } \int_{\mathcal{M}} f\mu_g = \int_{\mathcal{M}} H\mu_g. \tag{18.7.15}$$

By linearity, it thus suffices to solve Problem (18.7.13) for $h = 0$ and f replaced by $f - H$. By (18.7.15), we then assume that

$$\int_{\mathcal{M}} f\mu_g = 0, \tag{18.7.16}$$

which is consistent with choosing $h = 0$ in (18.7.12).

The map $\ell : v \mapsto (v, f)_{L^2(\mathcal{M})}$ is continuous on $H^1(\mathcal{M})$ and $\ell(v) = \ell(v')$ if $v \sim v'$ (in the sense of (18.7.7)) as $\int_{\mathcal{M}} f\mu_g = 0$. Thus, there exists a continuous form $\tilde{\ell}$ on $H^1(\mathcal{M})/\mathbb{C}$ such that $\ell = \tilde{\ell} \circ \Phi$. On $H^1(\mathcal{M})/\mathbb{C}$ we also consider the well defined continuous Hermitian form

$$b_0([v],[u]) = a_0(\underline{v}, \underline{u}), \quad [u],[v] \in H^1(\mathcal{M})/\mathbb{C}.$$

By the Poincaré inequality of Proposition 18.9, we have

$$b_0([v],[v]) = \|\,\mathsf{D}\,\underline{v}\|_{L^2\Lambda^1(\mathcal{M})} \gtrsim \|\underline{v}\|_{H^1(\mathcal{M})} = \|[v]\|_{H^1(\mathcal{M})/\mathbb{C}}, \quad [v] \in H^1(\mathcal{M})/\mathbb{C},$$

implying that $b_0(.,.)$ is coercive on $H^1(\mathcal{M})/\mathbb{C}$. As it is Hermitian symmetric, by the Riesz representation theorem, there exists a unique $[u] \in H^1(\mathcal{M})/\mathbb{C}$ such that $b_0([v],[u]) = \tilde{\ell}([v])$ for all $[v] \in H^1(\mathcal{M})/\mathbb{C}$. This reads

$$a_0(v, \underline{u}) = (v, f)_{L^2(\mathcal{M})}, \qquad v \in H^1(\mathcal{M}).$$

We may then prove that $\underline{u} \in H^2(\mathcal{M})$, $\gamma^{\mathsf{N}}(\underline{u}) = 0$, and prove inequality (18.7.14) following, for instance, [90] that can be adapted to the manifold case. ∎

COROLLARY 18.49 (Uniqueness of Strong Solutions Modulo Constants). *Let $f \in L^2(\mathcal{M})$ and $h \in H^{1/2}(\partial\mathcal{M})$ be such that*

$$\int_{\mathcal{M}} f \mu_g + \int_{\partial\mathcal{M}} h \, \mu_{g_\partial} = 0. \tag{18.7.17}$$

Let $w \in H^2(\mathcal{M})$ be such that $P_0 w = f$ in $\mathcal{M}$ and $\gamma^{\mathsf{N}}(w) = h$ on $\partial\mathcal{M}$. If $[u] \in H^2(\mathcal{M})/\mathbb{C}$ is the unique solution of the variational form of the Neumann problem (18.7.13)*, then $w \in [u]$, that is, $w = \underline{u} + \mathrm{Cst}$.*

PROOF. It suffices to observe that w is itself a solution of (18.7.13) by (18.5.5) consequence of the divergence formula of Proposition 18.28. ∎

The case of a homogeneous Neumann boundary condition, $h \equiv 0$, naturally leads to defining the unbounded operator P_{N} on $L^2(\mathcal{M})$, with domain $D(\mathsf{P}_{\mathsf{N}}) = \{u \in H^2(\mathcal{M});\ \gamma^{\mathsf{N}}(u) = 0\}$, by $\mathsf{P}_{\mathsf{N}} u = P_0 = -\Delta_g u$. For $\lambda > 0$, by Proposition 18.48 we have

$$\|u\|_{H^2(\mathcal{M})} \eqsim \|(P_0 + \lambda \, \mathrm{Id}) u\|_{L^2(\mathcal{M})}, \qquad u \in D(\mathsf{P}_{\mathsf{N}}). \tag{18.7.18}$$

Let $[u] \in H^2(\mathcal{M})/\mathbb{C}$ be such that $\gamma^{\mathsf{N}}(\underline{u}) = 0$ (recall that $\underline{u} = (\underline{\Phi})^{-1}[u]$). Then, by (18.7.6) we have $\int_{\mathcal{M}} P_0 \underline{u} \, \mu_g = 0$ meaning that by (18.7.8) $P_0 \underline{u}$ is in the space

$$\underline{L}^2(\mathcal{M}) = \underline{\Pi} L^2(\mathcal{M}) \overset{\Phi}{\cong} L^2(\mathcal{M})/\mathbb{C}.$$

We can then define the unbounded operator $\mathsf{P}_{\mathsf{N},*}$ on $L^2(\mathcal{M})/\mathbb{C}$, with domain

$$D(\mathsf{P}_{\mathsf{N},*}) = \big\{[u] \in H^2(\mathcal{M})/\mathbb{C};\ \gamma^{\mathsf{N}}(\underline{u}) = 0\big\} \quad \text{and} \quad \mathsf{P}_{\mathsf{N},*}[u] = \underline{\Phi}(P_0 \underline{u}). \tag{18.7.19}$$

By Proposition 18.48, if $[u] \in D(\mathsf{P}_{\mathsf{N},*})$, we have

$$\|[u]\|_{H^2(\mathcal{M})/\mathbb{C}} = \|\underline{u}\|_{H^2(\mathcal{M})} \eqsim \|P_0 \underline{u}\|_{L^2(\mathcal{M})} \eqsim \|[P_0 \underline{u}]\|_{L^2(\mathcal{M})/\mathbb{C}}. \tag{18.7.20}$$

Moreover, the same proposition shows that $\mathsf{P}_{\mathsf{N},*} : H^2(\mathcal{M})/\mathbb{C} \to L^2(\mathcal{M})/\mathbb{C}$ is surjective.

With the continuity of the trace $u \mapsto \gamma^{\mathsf{N}}(u)$ as given in Theorem 18.25 for $u \in H^2(\mathcal{M})$ and (18.7.18) (resp. (18.7.20)) we have the following result.

PROPOSITION 18.50. *The unbounded operators $(\mathsf{P}_{\mathsf{N}}, D(\mathsf{P}_{\mathsf{N}}))$ and $(\mathsf{P}_{\mathsf{N},*}, D(\mathsf{P}_{\mathsf{N},*}))$ are closed operators on $L^2(\mathcal{M})$ and $L^2(\mathcal{M})/\mathbb{C}$, respectively.*

We conclude this section with the following lemma that is of use on several occasions in what follows.

LEMMA 18.51. *Let $u \in L^2(\mathcal{M})$ be such that, for all $v \in H^2(\mathcal{M})$ with $\gamma^{\mathsf{N}}(v) = 0$,*

$$(u, P_0 v)_{L^2(\mathcal{M})} = 0.$$

Then, $u \equiv \mathrm{Cst}$.

PROOF. Let $f \in \underline{L}^2(\mathcal{M}) = \Pi L^2(\mathcal{M})$, that is $f \in L^2(\mathcal{M})$ such that $⨍_{\mathcal{M}} f \mu_g = 0$; see (18.7.8). By Proposition 18.48, there exists $[v] \in H^2(\mathcal{M})/\mathbb{C}$ such that $P_0 v = f$ and $\gamma^{\mathsf{N}}(v) = 0$ for $v \in [v]$. We thus obtain, by the assumption made on u,

$$(u, f)_{L^2(\mathcal{M})} = (u, P_0 v)_{L^2(\mathcal{M})} = 0.$$

As f is arbitrary in $\underline{L}^2(\mathcal{M})$, the conclusion follows by Lemma 18.46. ∎

18.7.1. The Neumann Lifting Map. We introduce here a map that is the counterpart of the Dirichlet lifting map for the Neumann problem. We are interested in defining a map that acts as a solution operator for the following elliptic problem

$$P_0 u + u = 0 \text{ in } \mathcal{M} \quad \text{and} \quad \gamma^{\mathsf{N}}(u) = h \text{ on } \partial\mathcal{M},$$

for $h \in H^{-3/2}(\partial\mathcal{M})$. Note that we consider $P_0 + \mathrm{Id}$ instead of P_0 to enforce uniqueness without having to deal with classes of functions modulo constants (see the previous section). Here also the Hilbert space $\mathcal{W}_g(\mathcal{M})$ introduced in Sect. 18.6.2 plays an essential *rôle*.

PROPOSITION 18.52. *Let $h \in H^{-3/2}(\partial\mathcal{M})$. There exists a unique $u \in \mathcal{W}_g(\mathcal{M})$ such that $P_0 u + u = 0$ and $\gamma^{\mathsf{N}}(u) = h$. Moreover, the map $h \mapsto u$ from $H^{-3/2}(\partial\mathcal{M})$ into $\mathcal{W}_g(\mathcal{M})$ is bounded.*

With appropriate changes the proof is an adaptation of that of Proposition 18.36. In fact, it is a particular case of the proof of Proposition 18.59 in Sect. 18.8. We thus omit it.

DEFINITION 18.53 (Neumann Lifting Map). The bounded map $\mathsf{N} : h \mapsto u$ from $H^{-3/2}(\partial\mathcal{M})$ into $\mathcal{W}_g(\mathcal{M})$ given by Proposition 18.52 is called the Neumann lifting map.

If h is more regular than $H^{-3/2}$, then additional regularity is transferred to $\mathsf{N}(h)$, with consistency with the trace formula of Theorem 18.25.

PROPOSITION 18.54. *Let $m \in \mathbb{N}$. If $h \in H^{m-3/2}(\partial\mathcal{M})$, then $\mathsf{N}(h) \in H^m(\mathcal{M})$. Moreover, for some $C > 0$, we have*

$$\|\mathsf{N}(h)\|_{H^m(\mathcal{M})} \leq C|h|_{H^{m-3/2}(\partial\mathcal{M})}, \quad h \in H^{m-3/2}(\partial\mathcal{M}).$$

PROOF. The case $m = 0$ was treated above. The proof can be adapted from that of Proposition 18.39 for $m \geq 2$. The case $m = 1$ follows by an interpolation argument [74, 236]. ∎

The above results allow one to fully treat the nonhomogeneous Neumann-Laplace problem on a compact manifold with boundary, using the linearity of the equation.

THEOREM 18.55. *Let $(\mathcal{M}, g)$ be a compact connected Riemannian d-dimensional manifold with boundary and let $q \in L^\infty(\mathcal{M}; \mathbb{R})$ be such that $q \geq 0$ and $q \geq C > 0$ on $A \subset \mathcal{M}$ of positive measure. Then, the maps*

$$\mathcal{W}_g(\mathcal{M}) \to L^2(\mathcal{M}) \oplus H^{-3/2}(\partial\mathcal{M})$$
$$u \mapsto (P_0 u + qu, \gamma^{\mathsf{N}}(u))$$

and

$$\mathcal{W}_g(\mathcal{M}) \cap H^1(\mathcal{M}) \to L^2(\mathcal{M}) \oplus H^{-1/2}(\partial\mathcal{M})$$
$$u \mapsto (P_0 u + qu, \gamma^{\mathsf{N}}(u))$$

are isomorphisms. Let $k \in \mathbb{N}$, if moreover $q \in W^{k,\infty}(\mathcal{M}; \mathbb{R})$, the map

$$^{\mathsf{N}}L : H^{k+2}(\mathcal{M}) \to H^k(\mathcal{M}) \oplus H^{k+1/2}(\partial\mathcal{M})$$
$$u \mapsto (P_0 u + qu, \gamma^{\mathsf{N}}(u))$$

is also an isomorphism.

Note that for the definition of the second map and its properties we use Lemma 18.42.

In the case $q = 0$ both existence and uniqueness are lost in the Neumann problem. One, however, finds the following result with the above results, again exploiting the linearity of the problem.

THEOREM 18.56. *Let $(\mathcal{M}, g)$ be a compact connected Riemannian d-dimensional manifold with boundary. For $k \in \mathbb{N}$, the map*

$$^{\mathsf{N}}L : H^{k+2}(\mathcal{M}) \to H^k(\mathcal{M}) \oplus H^{k+1/2}(\partial\mathcal{M})$$
$$u \mapsto (P_0 u, \gamma^{\mathsf{N}}(u))$$

has a one dimensional kernel formed by the set of constant functions and its range is given by the set of $(f, h) \in H^k(\mathcal{M}) \times H^{k+1/2}(\partial\mathcal{M})$ such that

$$\int_{\mathcal{M}} f\mu_g + \int_{\partial\mathcal{M}} h\mu_{g_\partial} = 0.$$

The same holds for the maps

$$\mathcal{W}_g(\mathcal{M}) \to L^2(\mathcal{M}) \oplus H^{-3/2}(\partial\mathcal{M})$$
$$u \mapsto (P_0 u, \gamma^{\mathsf{N}}(u))$$

and

$$\mathcal{W}_g(\mathcal{M}) \cap H^1(\mathcal{M}) \to L^2(\mathcal{M}) \oplus H^{-1/2}(\partial\mathcal{M})$$
$$u \mapsto (P_0 u, \gamma^{\mathsf{N}}(u)),$$

with the range condition replaced by

$$\int_{\mathcal{M}} f\mu_g + \langle h, 1\rangle_{H^{-3/2}(\partial\mathcal{M}), H^{3/2}(\partial\mathcal{M})} = 0$$

and

$$\int_{\mathcal{M}} f\mu_g + \langle h, 1\rangle_{H^{-1/2}(\partial\mathcal{M}),H^{1/2}(\partial\mathcal{M})} = 0,$$

respectively.

18.8. The Laplace Problem with Mixed Neumann-Dirichlet Boundary Conditions

If $(\mathcal{M}, g)$ is compact and has a boundary $\partial\mathcal{M}$, the latter may be composed of a finite number of connected components. We are interested in solving an elliptic problem with Dirichlet boundary conditions on some of the connected components of $\partial\mathcal{M}$ and Neumann boundary conditions on others. We exclude the case of pure Dirichlet or Neumann boundary conditions here: there is a least one connected component of $\partial\mathcal{M}$ where we impose Dirichlet boundary conditions and one connected component of $\partial\mathcal{M}$ where we impose Neumann boundary conditions. The treatment of the Dirichlet problem and the Neumann problem can be found in Sects. 18.6 and 18.7, respectively. We denote by ${}^{\mathsf{D}}\partial\mathcal{M}$ the union of the connected components of $\partial\mathcal{M}$ where we impose Dirichlet boundary conditions and by ${}^{\mathsf{N}}\partial\mathcal{M}$ the union of the other connected components of $\partial\mathcal{M}$ where we impose Neumann boundary conditions. As already mentioned we assume ${}^{\mathsf{D}}\partial\mathcal{M} \neq \emptyset$ and ${}^{\mathsf{N}}\partial\mathcal{M} \neq \emptyset$. The problem under consideration then reads, with $P_0 = -\Delta_g$,

$$P_0 u + qu = f, \qquad u_{|{}^{\mathsf{D}}\partial\mathcal{M}} = {}^{\mathsf{D}}h, \quad \partial_\nu u_{|{}^{\mathsf{N}}\partial\mathcal{M}} = {}^{\mathsf{N}}h.$$

The potential function q will be assumed bounded in what follows.

If $f \in L^2(\mathcal{M})$ and if we seek $u \in H^2(\mathcal{M})$, then the trace formula of Theorem 18.25 indicates that the proper space for ${}^{\mathsf{D}}h$ is $H^{3/2}({}^{\mathsf{D}}\partial\mathcal{M})$ and that for ${}^{\mathsf{N}}h$ is $H^{1/2}({}^{\mathsf{N}}\partial\mathcal{M})$.

We first treat the homogeneous case.

18.8.1. Homogeneous Case.

We define the following closed subset of $H^1(\mathcal{M})$

$$H^1_{\mathsf{D}}(\mathcal{M}) = \{u \in H^1(\mathcal{M});\ u_{|{}^{\mathsf{D}}\partial\mathcal{M}} = 0\}.$$

If $q \in L^\infty(\mathcal{M};\mathbb{R})$, on $H^1(\mathcal{M})$ we define the Hermitian form: $a_q(.,.)$ as in (18.7.4).

PROPOSITION 18.57 (Homogeneous Mixed Dirichlet–Neumann Laplace Problem). *Let $q \in L^\infty(\mathcal{M};\mathbb{R})$. There exist $C_0 > 0$ and $C > 0$ such that, if $q > -C_0$, for any $f \in L^2(\mathcal{M})$, there exists a unique $u \in H^1_{\mathsf{D}}(\mathcal{M})$ such that*

$$a_q(\varphi, u) = (\varphi, f)_{L^2(\mathcal{M})}, \qquad \varphi \in H^1_{\mathsf{D}}(\mathcal{M}). \tag{18.8.1}$$

Moreover, $u \in H^2(\mathcal{M})$, $\partial_\nu u_{|{}^{\mathsf{N}}\partial\mathcal{M}} = 0$ and

$$C^{-1}\|f\|_{L^2(\mathcal{M})} \le \|u\|_{H^2(\mathcal{M})} \le C\|f\|_{L^2(\mathcal{M})}.$$

Moreover, the equation $P_0 u + qu = f$ holds in $L^2(\mathcal{M})$.

Let $k \in \mathbb{N}$. For $q \in W^{k,\infty}(\mathcal{M};\mathbb{R})$, there exists $C_k > 0$, such that if $f \in H^k(\mathcal{M})$, then moreover $u \in H^{k+2}(\mathcal{M})$ and

$$C_k^{-1}\|f\|_{H^k(\mathcal{M})} \leq \|u\|_{H^{k+2}(\mathcal{M})} \leq C_k\|f\|_{H^k(\mathcal{M})}$$

and the equation $P_0 u + qu = f$ holds in $H^k(\mathcal{M})$.

Note that the constant C_0 depends on the metric g but also on the choice of connected components of $\partial\mathcal{M}$ where Dirichlet boundary conditions are imposed.

PROOF. Let $C_0 > 0$ be the best possible constant in the first Poincaré inequality of Proposition 18.9, that is,

$$C_0^{1/2}\|v\|_{L^2(\mathcal{M})} \leq \|\nabla_g v\|_{L^2V(\mathcal{M})}, \qquad v \in H^1_{\mathsf{D}}(\mathcal{M}).$$

Thus, if $q > -C_g$ we obtain that $a_q(v,v) \gtrsim \|v\|^2_{H^1(\mathcal{M})}$ and consequently $a_q(.,.)$ is coercive on $H^1_{\mathsf{D}}(\mathcal{M})$. As $v \mapsto (v,\bar{f})_{L^2(\mathcal{M})}$ is continuous on $H^1_{\mathsf{D}}(\mathcal{M})$, the Lax–Milgram theorem gives a unique $u \in H^1_{\mathsf{D}}(\mathcal{M})$ such that (18.8.1) holds. It also yields the existence of $C > 0$ such that

$$C^{-1}\|f\|_{L^2(\mathcal{M})} \leq \|u\|_{H^1(\mathcal{M})} \leq C\|f\|_{L^2(\mathcal{M})}.$$

The remainder of the proof is classical. ∎

COROLLARY 18.58 (Uniqueness of Strong Solutions). *Let $f \in L^2(\mathcal{M})$ and let $u \in H^2(\mathcal{M}) \cap H^1_{\mathsf{D}}(\mathcal{M})$ be such that $P_0 u + qu = f$ in $\mathcal{M}$ and $\partial_\nu u_{|^{\mathsf{N}}\partial\mathcal{M}} = 0$. Then u is the unique solution of the variational form of the homogeneous mixed Dirichlet–Neumann Laplace problem* (18.7.5).

PROOF. It suffices to observe that u is itself a solution of (18.8.1) by (18.5.5) consequence of the divergence formula of Proposition 18.28. ∎

18.8.2. The Mixed Dirichlet–Neumann Lifting Map. We introduce here a map that is the counterpart of the Dirichlet lifting map for the present mixed boundary problem. The Hilbert space $\mathcal{W}_g(\mathcal{M})$ introduced in Sect. 18.6.2 plays once again an essential *rôle*. For $r, s \in \mathbb{R}$ we set

$$H^{r,s}_{\mathsf{DN}}(\partial\mathcal{M}) = H^r({}^{\mathsf{D}}\partial\mathcal{M}) \oplus H^s({}^{\mathsf{N}}\partial\mathcal{M}). \tag{18.8.2}$$

For a function $w \in H^2(\mathcal{M})$, we set

$$\gamma(w) = (w_{|^{\mathsf{D}}\partial\mathcal{M}}, \partial_\nu w_{|^{\mathsf{N}}\partial\mathcal{M}}), \tag{18.8.3}$$

and we have $\gamma(w) \in H^{3/2,1/2}_{\mathsf{DN}}(\partial\mathcal{M})$. By Lemma 18.42 this trace map extends to a function $w \in \mathcal{W}_g(\mathcal{M}) \cap H^1$ and $\gamma(w) \in H^{1/2,-1/2}_{\mathsf{DN}}(\partial\mathcal{M})$; and finally, by Lemma 18.32, for a function $w \in \mathcal{W}_g(\mathcal{M})$ we have $\gamma(w) \in H^{-1/2,-3/2}_{\mathsf{DN}}(\partial\mathcal{M})$.

PROPOSITION 18.59. *Let $h \in H_{\mathsf{DN}}^{-1/2,-3/2}(\partial\mathcal{M})$. There exists a unique $u \in \mathcal{W}_g(\mathcal{M})$ such that $P_0 u = 0$ and $\gamma(u) = h$. Moreover, the map*

$$\begin{aligned} H_{\mathsf{DN}}^{-1/2,-3/2}(\partial\mathcal{M}) &\to \mathcal{W}_g(\mathcal{M}) \\ h &\mapsto u \end{aligned}$$

is bounded.

DEFINITION 18.60 (Mixed Dirichlet–Neumann Lifting Map). We call the bounded map $\mathsf{M} : H_{\mathsf{DN}}^{-1/2,-3/2}(\partial\mathcal{M}) \to \mathcal{W}_g(\mathcal{M})$ given by Proposition 18.59 the mixed Dirichlet–Neumann lifting map.

PROOF OF PROPOSITION 18.59. With appropriate changes the proof is an adaptation of that of Proposition 18.36. Denote by $\mathsf{R} : L^2(\mathcal{M}) \to H^2(\mathcal{M})$ the resolvent map associated with the homogeneous problem of Proposition 18.57.

First, we address uniqueness. By linearity, we consider $u \in \mathcal{W}_g(\mathcal{M})$ such that $P_0 u = 0$ and $\gamma(u) = 0$. Let $f \in L^2(\mathcal{M})$ and choose $w = \mathsf{R}f$. We have $w \in H^2(\mathcal{M})$ and $\gamma(w) = 0$. By Lemma 18.34 we obtain $(u, f)_{L^2(\mathcal{M})} = (u, P_0 w)_{L^2(\mathcal{M})} = 0$. As $f \in L^2(\mathcal{M})$ is arbitrary we conclude that $u = 0$.

Second, we address existence. To ease notation we set $H = H_{\mathsf{DN}}^{-1/2,-3/2}(\partial\mathcal{M})$. For a function $w \in H^2(\mathcal{M})$ set

$$\gamma'(w) = (\partial_\nu w_{|{}^{\mathsf{D}}\partial\mathcal{M}}, w_{|{}^{\mathsf{N}}\partial\mathcal{M}}) \in H' = H_{\mathsf{DN}}^{1/2,3/2}(\partial\mathcal{M}).$$

For $h = ({}^{\mathsf{D}}h, {}^{\mathsf{N}}h) \in H$ and $k = ({}^{\mathsf{D}}k, {}^{\mathsf{N}}k) \in H'$, we set

$$\langle k, h\rangle_{H',H} = -\langle {}^{\mathsf{D}}k, {}^{\mathsf{D}}h\rangle_{H^{1/2}({}^{\mathsf{D}}\partial\mathcal{M}),H^{-1/2}({}^{\mathsf{D}}\partial\mathcal{M})} + \langle {}^{\mathsf{N}}k, {}^{\mathsf{N}}h\rangle_{H^{3/2}({}^{\mathsf{N}}\partial\mathcal{M}),H^{-3/2}({}^{\mathsf{N}}\partial\mathcal{M})}.$$

For $h \in H$, consider then the map

$$\begin{aligned} U : L^2(\mathcal{M}) &\to \mathbb{C} \\ f &\mapsto \langle \gamma' \mathsf{R}f, \bar{h}\rangle_{H',H}. \end{aligned}$$

By the trace formula of Theorem 18.25, this form is bounded and we have

$$|U(f)| \leq |h|_H \|\mathsf{R}f\|_{H^2(\mathcal{M})} \lesssim |h|_H \|f\|_{L^2(\mathcal{M})}.$$

By the Riesz theorem, there exists $u \in L^2(\mathcal{M})$ such that $U(f) = (f, u)_{L^2(\mathcal{M})}$ for all $f \in L^2(\mathcal{M})$. Moreover, $\|u\|_{L^2(\mathcal{M})} \lesssim |h|_H$. If $\varphi \in {}^0\mathscr{D}_c^\infty(\mathcal{M})$ and $f = P_0\varphi$, then $\varphi = \mathsf{R}f$ and we find

$$U(f) = \langle \gamma(\varphi), \bar{h}\rangle_{H',H} = 0.$$

Consequently

$$\begin{aligned} \langle \overline{P_0 u}, \varphi\mu_g\rangle_{{}^0\mathscr{D}'(\mathcal{M}),{}^1\mathscr{D}_c^\infty(\mathcal{M})} &= \langle \bar{u}, P_0\varphi\mu_g\rangle_{{}^0\mathscr{D}'(\mathcal{M}),{}^1\mathscr{D}_c^\infty(\mathcal{M})} \\ &= (P_0\varphi, u)_{L^2(\mathcal{M})} = U(f) = 0. \end{aligned}$$

Thus, in the sense of distributions we have $P_0 u = 0$. Thus $u \in \mathcal{W}_g(\mathcal{M})$ and $\|u\|_{\mathcal{W}_g(\mathcal{M})} \lesssim |h|_H$.

LEMMA 18.61. *There exists a bounded map* $\mathsf{M}_0 : H' \to H^2(\mathcal{M})$ *such that* $\gamma' \circ \mathsf{M}_0 = \mathrm{Id}_{H'}$ *and* $\gamma \circ \mathsf{M}_0 = 0$.

This is a particular case of Theorem 18.25. A proof follows from Lemmata 18.26 and 18.27 by working locally at the boundary.

Let $k \in H'$. For $v = \mathsf{M}_0 k \in H^2(\mathcal{M})$ as given by Lemma 18.61, we have

$$(P_0 v, u)_{L^2(\mathcal{M})} = \langle k, \gamma(\bar{u})\rangle_{H',H},$$

by Lemma 18.34. We also have

$$\begin{aligned}(P_0 v, u)_{L^2(\mathcal{M})} &= U(P_0 v) = \langle \gamma' \mathsf{R}(P_0 v), \bar{h}\rangle_{H',H}\\ &= \langle \gamma'(v), \bar{h}\rangle_{H',H} = \langle k, \bar{h}\rangle_{H',H}.\end{aligned}$$

For all $k \in H'$ we thus find $\langle k, \bar{h} - \gamma(\bar{u})\rangle_{H',H} = 0$, implying $\gamma(u) = h$, which concludes the existence part of the proof. ■

PROPOSITION 18.62. *Let* $k \in \mathbb{N}$. *If* $h \in H_{\mathsf{DN}}^{k-1/2,k-3/2}(\partial\mathcal{M})$, *then* $\mathsf{M}(h) \in H^k(\mathcal{M})$. *Moreover, for some* $C > 0$, *we have*

$$\|\mathsf{M}(h)\|_{H^k(\mathcal{M})} \leq C|h|_{H_{\mathsf{DN}}^{k-1/2,k-3/2}(\partial\mathcal{M})}.$$

PROOF. The case $m = 0$ was treated above. The proof can be adapted from that of Proposition 18.39 for $m \geq 2$. The case $m = 1$ follows by an interpolation argument [74, 236]. ■

With the mixed Dirichlet–Neumann lifting map and Proposition 18.57 we obtain the following theorem.

THEOREM 18.63. *Let* $(\mathcal{M}, g)$ *be a compact connected Riemannian d-dimensional manifold with boundary and let* $q \in L^\infty(\mathcal{M};\mathbb{R})$ *with* $q > -C_0$, *where* C_0 *is as given in Proposition 18.31. Then, the maps*

$$\begin{aligned}\mathcal{W}_g(\mathcal{M}) &\to L^2(\mathcal{M}) \oplus H_{\mathsf{DN}}^{-1/2,-3/2}(\partial\mathcal{M})\\ u &\mapsto \big(P_0 u + qu, \gamma(u)\big)\end{aligned}$$

and

$$\begin{aligned}\mathcal{W}_g(\mathcal{M}) \cap H^1(\mathcal{M}) &\to L^2(\mathcal{M}) \oplus H_{\mathsf{DN}}^{1/2,-1/2}(\partial\mathcal{M})\\ u &\mapsto \big(P_0 u + qu, \gamma(u)\big)\end{aligned}$$

are isomorphisms. Let $k \in \mathbb{N}$, *if moreover* $q \in W^{k,\infty}(\mathcal{M};\mathbb{R})$, *the map*

$$\begin{aligned}{}^{\mathsf{M}}L : H^{k+2}(\mathcal{M}) &\to H^k(\mathcal{M}) \oplus H_{\mathsf{DN}}^{k+3/2,k+1/2}(\partial\mathcal{M})\\ u &\mapsto \big(P_0 u + qu, \gamma(u)\big)\end{aligned}$$

is also an isomorphism.

Recall that the trace map γ is given in (18.8.3). Note that for the definition of the second map and its properties we use Lemma 18.42.

18.9. Second-Order Elliptic Operators in the Euclidean Space

In this section, we explain how the analysis of a second-order elliptic operator in an open set of $\mathbb{R}^d$ can be made with the point of view of Riemannian geometry. As a typical example, we consider below the definition of the Dirichlet map D.

On Ω a regular open set of $\mathbb{R}^d$, consider the second-order elliptic operator

$$Q_0(x,D) = \sum_{1\leq i,j\leq d} D_i(q^{ij}(x)D_j), \qquad \text{with} \quad \sum_{1\leq i,j\leq d} q^{ij}(x)\xi_i\xi_j \geq C|\xi|^2,$$

where $q^{ij} \in \mathscr{C}^\infty(\overline{\Omega};\mathbb{R})$ such that $q^{ij} = q^{ji}$, $1 \leq i,j \leq d$.

At each point $x \in \Omega$, define by $g_x = \big(g_{ij}(x)\big)$ the inverse matrix of $\big(q^{ij}(x)\big)$. If $\mathcal{M} = \overline{\Omega}$ is equipped with the metric g this yields the structure of Riemannian manifold, with boundary if $\overline{\Omega} \neq \mathbb{R}^d$. In what follows we shall use the notation $\mathcal{M}$, $\partial\mathcal{M}$, etc. for the notions associated with the Riemannian structure and Ω, $\partial\Omega$ for those associated with the Euclidean structure and the operator $Q_0(x,D)$.

Note that the operator $Q_0(x,D)$ is not the Laplace–Beltrami operator in general. In fact, we have the following result.

PROPOSITION 18.64. *Define $a(x) = (\det g_x)^{1/4}$. Then,*

$$a^{-1} \circ Q_0(x,D) \circ a = P_0 + \lambda, \qquad P_0 = -\Delta_g,$$

for $\lambda \in \mathscr{C}^\infty(\Omega)$ given by $\lambda(x) = a^{-1}(x)Q_0(x,D)a(x)$.

On $\mathbb{R}^d$ the Lebesgue measure identifies with the volume form $dv = dx_1 \wedge \cdots \wedge dx_d$ or the Riemannian density μ_g. The L^2-inner product then reads

$$(u,v)_{L^2(\Omega)} = \int_\Omega u(x)\bar{v}(x)dx = \int_\Omega u\bar{v}dv. \tag{18.9.1}$$

Note that $Q_0(x,D)$ is symmetric for this inner product.

The Laplace–Beltrami operator Δ_g is, however, symmetric for the L^2-inner product associated with the volume form $dv_g = (\det g_x)^{1/2}dv = a^2 dv$, that is,

$$(u,v)_{L^2(\mathcal{M})} = \int_\Omega u\bar{v}dv_g. \tag{18.9.2}$$

These observations lead to the following proof.

PROOF OF PROPOSITION 18.64. The operators $a^{-1}\circ Q_0(x,D)\circ a$ and P_0 share the same principal symbol. Their difference $a^{-1} \circ Q_0(x,D) \circ a - P_0 = R_1(x,D)$ is thus a first-order operator with real coefficients. For $u,v \in \mathscr{C}_c^\infty(\Omega) = {}^0\mathscr{D}_c^\infty(\mathcal{M})$ we write

$$\begin{aligned}(a^{-1}Q_0(x,D)(au),v)_{L^2(\mathcal{M})} &= \int_\Omega Q_0(x,D)(au)\bar{v}\, a(x)dx \\ &= (Q_0(x,D)(au), av)_{L^2(\Omega)} \\ &= (au, Q_0(x,D)av)_{L^2(\Omega)} \\ &= (u, a^{-1}Q_0(x,D)(av))_{L^2(\mathcal{M})}.\end{aligned}$$

One sees that $a^{-1} \circ Q_0(x, D) \circ a$ is symmetric for $(.,.)_{L^2(\mathcal{M})}$ just as P_0 and thus, so is $R_1(x, D)$. This shows that $R_1(x, D)$ is in fact a function $\lambda(x)$ as its coefficients are real. Then, we simply have $\lambda = a^{-1}Q_0(x, D)a$ since P_0 vanishes on constant functions. ∎

We now wish to understand the connection between the two volume forms dv_∂ and $dv_{g\partial}$ on $\partial\Omega = \partial\mathcal{M}$ that are associated with dv and dv_g, respectively. Let ν be the unique outward pointing vector field along $\partial\Omega$ such that, for all $x \in \partial\Omega$, $\nu_x \cdot \nu_x = 1$ and $\nu_x \cdot t = 0$ for all $t \in T_x\partial\Omega$. As the Euclidean metric is characterized by the identity matrix at every point we may identify the tangent vector $\nu = \sum_{1\leq i\leq d} \nu^i \partial_i$ with its covariant version $n = \sum_{1\leq i\leq d} n_i \partial^i$, that is $n_i = \nu^i$, $i = 1, \ldots, d$. Let also ν_g be the unique outward pointing vector field along $\partial\mathcal{M}$ such that, for all $x \in \partial\mathcal{M}$, $g_x((\nu_g)_x, \nu_x) = 1$ and $g_x((\nu_g)_x, t) = 0$ for all $t \in T_x\partial\Omega$. For $w \in H^2(\Omega)$ we set

$$\gamma_g^{\mathsf{N}}(w) = \partial_{\nu_g} w_{|\partial\mathcal{M}} = g(\nu_g, \nabla_g w)_{|\partial\mathcal{M}},$$

and

$$\gamma^{\mathsf{N}}(w) = \partial_\nu w_{|\partial\Omega} = \sum_{1\leq i,j\leq d} g^{ij} n_i \partial_j w_{|\partial\Omega}.$$

In this section, we use the notation γ_g^{N} to avoid any possible confusion. The maps γ_g^{N} and γ^{N} are related yet different in general.

PROPOSITION 18.65. *The volume forms $dv_{g\partial}$ and $(dv)_\partial$ are related according to*

$$dv_{g\partial} = (dv_g)_\partial = |n|_g (\det g)^{1/2}_{|\partial\mathcal{M}} (dv)_\partial.$$

We also have $\nu_g = |n|_g^{-1} n^\sharp$ and $\gamma_g^{\mathsf{N}} = |n|_g^{-1} \gamma^{\mathsf{N}}$.

In this proposition, the musical isomorphism is related to the Riemannian structure associated with the metric g. Using the coordinates (that are global here) we have

$$(\nu_g)_x^i = \Big(\sum_{1\leq k,l\leq d} g^{kl}(n_x)_k (n_x)_l \Big)^{-1/2} \sum_{1\leq j\leq d} g^{ij}(x)(n_x)_i,$$

for $x \in \partial\mathcal{M}$.

PROOF. If we set $n_g = n/|n|_g$, then as $(n_g)_x$ is conormal to $T_x\partial\mathcal{M}$ and unitary, then $n_g^\sharp$ fulfills the characterizing properties of ν_g. Next, we write

$$|n|_g \gamma_g^{\mathsf{N}}(w) = |n|_g g(\nu_g, \nabla_g w)_{|\partial\mathcal{M}} = g(n^\sharp, \nabla_g w) = \langle n, \nabla_g w\rangle_{|\partial\mathcal{M}} = \partial_\nu w_{|\partial\mathcal{M}}.$$

Since $dv_g = (\det g)^{1/2} dv$, we find

$$(dv)_\partial(w_1, \ldots, w_{d-1}) = dv(\nu, w_1, \ldots, w_{d-1}),$$

and

$$\begin{aligned}(dv_g)_\partial(w_1,\dots,w_{d-1}) &= dv_g(\nu_g, w_1,\dots,w_{d-1})\\ &= (\det g)^{1/2}_{|\partial\mathcal{M}} dv(\nu_g, w_1,\dots,w_{d-1}),\end{aligned}$$

for $w_1,\dots,w_{d-1} \in \mathscr{C}^1V(\partial\mathcal{M})$.

As ν is unitary for the Euclidean metric we have $\nu_g = (\nu_g \cdot \nu_x)\nu + \nu'$ with $\nu'_x \in (\nu_x)^\perp = (n_x)^\perp = T_x\partial\mathcal{M}$. Observing that $dv(\nu', w_1,\dots,w_{d-1}) = 0$, if $w_1,\dots,w_{d-1} \in \mathscr{C}^1V(\partial\mathcal{M})$ we find

$$(dv_g)_\partial(w_1,\dots,w_{d-1}) = (\det g)^{1/2}_{|\partial\mathcal{M}}(\nu_g\cdot\nu)dv(\nu, w_1,\dots,w_{d-1}),$$

that is, $(dv_g)_\partial = (\det g)^{1/2}_{|\partial\mathcal{M}}(\nu_g\cdot\nu)(dv)_\partial$. Finally, we write

$$(\nu_g\cdot\nu)(x) = \langle n_x, (\nu_g)_x\rangle = |n_x|_g^{-1}\langle n_x, n_x^\sharp\rangle = |n_x|_g,$$

which yields the result. ∎

If $u, v \in H^2(\Omega)$, the following Green type formula for the operator $Q_0(x,D)$ follows from two applications of the divergence formula given in (16.1.6) for the Euclidean case:

$$\begin{aligned}(Q_0(x,D)u,v)_{L^2(\Omega)} &+ \int_{\partial\Omega}\gamma^{\mathsf{N}}(u)\bar{v}_{|\partial\Omega}\,(dv)_\partial\\ &= (u, Q_0(x,D)v)_{L^2(\Omega)} + \int_{\partial\Omega} u_{|\partial\Omega}\gamma^{\mathsf{N}}(\bar{v})\,(dv)_\partial.\end{aligned} \tag{18.9.3}$$

In what follows, we shall use this formula to show how the correspondences we put forward above can be used to transfer properties obtained for $P_0 = -\Delta_g$ to $Q_0(x,D)$ and *vice versa*.

Let $u, v \in H^2(\Omega)$ and with $a(x) = (\det g_x)^{1/4}$ as above let $\tilde{u}, \tilde{v}$ be such that $u = a\tilde{u}$ and $v = a\tilde{v}$. With Proposition 18.65 and the Green formula of Proposition 18.30 we have

$$\begin{aligned}(Q_0(x,D)u,v)_{L^2(\Omega)} &= ((P_0+\lambda)\tilde{u},\tilde{v})_{L^2(\mathcal{M})}\\ &= (\tilde{u},(P_0+\lambda)\tilde{v})_{L^2(\mathcal{M})} - (\gamma_g^{\mathsf{N}}(\tilde{u}),\gamma^{\mathsf{D}}(\tilde{v}))_{L^2(\partial\mathcal{M})}\\ &\quad + (\gamma^{\mathsf{D}}(\tilde{u}),\gamma_g^{\mathsf{N}}(\tilde{v}))_{L^2(\partial\mathcal{M})}\\ &= (u, Q_0(x,D)v)_{L^2(\Omega)} - (\gamma_g^{\mathsf{N}}(\tilde{u}),\gamma^{\mathsf{D}}(\tilde{v}))_{L^2(\partial\mathcal{M})}\\ &\quad + (\gamma^{\mathsf{D}}(\tilde{u}),\gamma_g^{\mathsf{N}}(\tilde{v}))_{L^2(\partial\mathcal{M})}.\end{aligned}$$

With the next lemma we see that the Green formula of Proposition 18.30 and (18.9.3) can be deduced from one another.

LEMMA 18.66. *One has*

$$\begin{aligned}(\gamma^{\mathsf{D}}(\tilde{u}),\gamma_g^{\mathsf{N}}(\tilde{v}))_{L^2(\partial\mathcal{M})} &- (\gamma_g^{\mathsf{N}}(\tilde{u}),\gamma^{\mathsf{D}}(\tilde{v}))_{L^2(\partial\mathcal{M})}\\ &= (\gamma^{\mathsf{D}}(u),\gamma^{\mathsf{N}}(v))_{L^2(\partial\Omega)} - (\gamma^{\mathsf{N}}(u),\gamma^{\mathsf{D}}(v))_{L^2(\partial\Omega)},\end{aligned}$$

where $\gamma^{\mathsf{N}}(w) = \partial_\nu w_{|\partial\Omega} = \sum_{1\le i,j\le d} g^{ij} n_i \partial_j w_{|\partial\Omega}$.

PROOF. We write

$$\begin{aligned}I &= (\gamma^{\mathsf{D}}(\tilde{u}), \gamma_g^{\mathsf{N}}(\tilde{v}))_{L^2(\partial\mathcal{M})} - (\gamma_g^{\mathsf{N}}(\tilde{u}), \gamma^{\mathsf{D}}(\tilde{v}))_{L^2(\partial\mathcal{M})} \\ &= \int_{\partial\mathcal{M}} \big(a^{-1}\gamma^{\mathsf{D}}(u)\, g(\nu_g, \nabla_g(a^{-1}\bar{v}))_{|\partial\mathcal{M}} \\ &\qquad - a^{-1}\gamma^{\mathsf{D}}(\bar{v})\, g(\nu_g, \nabla_g(a^{-1}u))_{|\partial\mathcal{M}}\big)\, dv_{g_\partial}.\end{aligned}$$

Due to some cancellation, one obtains

$$u\, g(\nu_g, \nabla_g(a^{-1}\bar{v})) - \bar{v}\, g(\nu_g, \nabla_g(a^{-1}u)) = a^{-1}\big(u\, g(\nu_g, \nabla_g\bar{v}) - \bar{v}\, g(\nu_g, \nabla_g u)\big),$$

yielding

$$I = \int_{\partial\mathcal{M}} \big(\gamma^{\mathsf{D}}(u)\, \gamma_g^{\mathsf{N}}(\bar{v}) - \gamma^{\mathsf{D}}(\bar{v})\, \gamma_g^{\mathsf{N}}(u)\big)\, a_{|\partial\mathcal{M}}^{-2} dv_{g_\partial}.$$

By proposition 18.65, we have $a_{|\partial\mathcal{M}}^{-2} dv_{g_\partial} = |n|_g (dv)_\partial$ and $\gamma_g^{\mathsf{N}} = |n|_g^{-1}\gamma^{\mathsf{N}}$. We thus obtain

$$I = \int_{\partial\mathcal{M}} \big(\gamma^{\mathsf{D}}(u)\, \gamma^{\mathsf{N}}(\bar{v}) - \gamma^{\mathsf{D}}(\bar{v})\, \gamma^{\mathsf{N}}(u)\big)\, (dv)_\partial,$$

which is the result. ■

We now show how the above observation can be used to see how the extended Green formula of Proposition 18.34 can be transposed to the operator $Q_0(x, D)$ on Ω.

First we need to consider the duality bracket for Sobolev spaces on $\partial\mathcal{M}$. Let $s \in \mathbb{R}$. Recall that on $\partial\mathcal{M}$ equipped with the metric g_∂ the duality between $H^s(\partial\mathcal{M})$ and $H^{-s}(\partial\mathcal{M})$ is understood using $L^2(\partial\mathcal{M})$ as a pivot space with the inner product

$$(\varphi, \psi)_{L^2(\partial\mathcal{M})} = \int_{\partial\mathcal{M}} \varphi\bar{\psi}\, (dv_g)_\partial.$$

The same can be done on $\partial\Omega$ for the inner product

$$(\varphi, \psi)_{L^2(\partial\Omega)} = \int_{\partial\Omega} \varphi\bar{\psi}\, (dv)_\partial,$$

and we denote these spaces by $H^s(\partial\Omega)$ and $H^{-s}(\partial\Omega)$. In fact, these spaces can be identified: if $\varphi \in H^s(\partial\Omega)$ and $\psi \in H^{-s}(\partial\Omega)$, then $\tilde{\varphi} = b^{-1}\varphi \in H^s(\partial\mathcal{M})$ and $\tilde{\psi} = b^{-1}\psi \in H^{-s}(\partial\mathcal{M})$, with $b = |n|_g^{1/2}(\det g)_{\partial\mathcal{M}}^{1/4} = |n|_g^{1/2}\gamma^{\mathsf{D}}(a)$, and

$$\langle \varphi, \psi\rangle_{H^s(\partial\Omega), H^{-s}(\partial\Omega)} = \langle \tilde{\varphi}, \tilde{\psi}\rangle_{H^s(\partial\mathcal{M}), H^{-s}(\partial\mathcal{M})}. \tag{18.9.4}$$

Second, we recall the space

$$\mathcal{W}_g(\Omega) = \{u \in L^2(\Omega);\ Q_0(x, D)u \in L^2(\Omega)\}$$

introduced in Section 10.5 of Volume 1. With the counterpart space $\mathcal{W}_g(\mathcal{M})$ introduced in (18.6.4), with Proposition 18.64 we see that

$$u \in \mathcal{W}_g(\Omega) \quad \Leftrightarrow \quad \tilde{u} = a^{-1}u \in \mathcal{W}_g(\mathcal{M}).$$

As a is smooth, from Lemma 18.32 we obtain the result of Lemma 10.54 of Volume 1, that is the existence of both the Dirichlet and Neumann traces of functions in $\mathcal{W}_g(\Omega)$.

With Lemma 18.34 we may then obtain its counterpart on $\mathcal{W}_g(\Omega)$.

LEMMA 18.67. *Let $u \in \mathcal{W}_g(\Omega)$ and $w \in H^2(\Omega)$. We have*

$$(Q_0(x,D)u, w)_{L^2(\Omega)} + \langle \gamma^{\mathsf{N}}(u), \gamma^{\mathsf{D}}(\bar{w})\rangle_{H^{-3/2}(\partial\Omega),H^{3/2}(\partial\Omega)}$$
$$= (u, Q_0(x,D)w)_{L^2(\Omega)} + \langle \gamma^{\mathsf{D}}(u), \gamma^{\mathsf{N}}(\bar{w})\rangle_{H^{-1/2}(\partial\Omega),H^{1/2}(\partial\Omega)}.$$

PROOF. We set $\tilde{u} = a^{-1}u$ and $\tilde{v} = a^{-1}v$ with $a(x) = (\det g_x)^{1/4}$. By Proposition 18.64 we have

$$(Q_0(x,D)u, w)_{L^2(\Omega)} - (u, Q_0(x,D)w)_{L^2(\Omega)}$$
$$= (P_0\tilde{u}, \tilde{v})_{L^2(\mathcal{M})} - (\tilde{u}, P_0\tilde{v})_{L^2(\mathcal{M})}.$$

With Lemma 18.34 it thus suffices to prove that

$$\langle \gamma^{\mathsf{N}}(u), \gamma^{\mathsf{D}}(\bar{w})\rangle_{H^{-3/2}(\partial\Omega),H^{3/2}(\partial\Omega)} - \langle \gamma^{\mathsf{D}}(u), \gamma^{\mathsf{N}}(\bar{w})\rangle_{H^{-1/2}(\partial\Omega),H^{1/2}(\partial\Omega)}$$
$$= \langle \gamma_g^{\mathsf{N}}(\tilde{u}), \gamma^{\mathsf{D}}(\overline{\tilde{w}})\rangle_{H^{-3/2}(\partial\mathcal{M}),H^{3/2}(\partial\mathcal{M})}$$
$$- \langle \gamma^{\mathsf{D}}(\tilde{u}), \gamma_g^{\mathsf{N}}(\overline{\tilde{w}})\rangle_{H^{-1/2}(\partial\mathcal{M}),H^{1/2}(\partial\mathcal{M})}.$$

One cannot apply Lemma 18.66 since there is no density argument to invoke here. One rather reproduces the argument of the proof of Lemma 18.66 yet, here, on the level of the duality bracket and using the trace properties of the functions in $\mathcal{W}_g(\Omega)$ and in $\mathcal{W}_g(\mathcal{M})$. In particular, using Proposition 18.65 one has

$$H^{-3/2}(\partial\Omega) \ni \gamma^{\mathsf{N}}(u) = |n|_g \gamma^{\mathsf{D}}(a)\gamma_g^{\mathsf{N}}(\tilde{u}) + \gamma^{\mathsf{N}}(a)\gamma^{\mathsf{D}}(\tilde{u}),$$

by the Leibniz-like formula for the traces given in Lemma 18.32, where the first term on the r.h.s. is in $H^{-3/2}(\partial\mathcal{M})$ and the second term in $H^{-1/2}(\partial\mathcal{M})$. As we also have

$$H^{1/2}(\partial\Omega) \ni \gamma^{\mathsf{N}}(w) = |n|_g \gamma^{\mathsf{D}}(a)\gamma_g^{\mathsf{N}}(\tilde{w}) + \gamma^{\mathsf{N}}(a)\gamma^{\mathsf{D}}(\tilde{w}),$$

where the first term on the r.h.s. is in $H^{1/2}(\partial\mathcal{M})$ and the second term in $H^{3/2}(\partial\mathcal{M})$, we obtain

$$I = \langle \gamma^{\mathsf{N}}(u), \gamma^{\mathsf{D}}(\bar{w})\rangle_{H^{-3/2}(\partial\Omega),H^{3/2}(\partial\Omega)} - \langle \gamma^{\mathsf{D}}(u), \gamma^{\mathsf{N}}(\bar{w})\rangle_{H^{-1/2}(\partial\Omega),H^{1/2}(\partial\Omega)}$$
$$= \langle |n|_g \gamma^{\mathsf{D}}(a)\gamma_g^{\mathsf{N}}(\tilde{u}), \gamma^{\mathsf{D}}(a\overline{\tilde{w}})\rangle_{H^{-3/2}(\partial\Omega),H^{3/2}(\partial\Omega)}$$
$$- \langle \gamma^{\mathsf{D}}(a\tilde{u}), |n|_g \gamma^{\mathsf{D}}(a)\gamma_g^{\mathsf{N}}(\overline{\tilde{w}})\rangle_{H^{-1/2}(\partial\Omega),H^{1/2}(\partial\Omega)}$$

since

$$\langle \gamma^{\mathsf{N}}(a)\gamma^{\mathsf{D}}(\tilde{u}), \gamma^{\mathsf{D}}(a\overline{\tilde{w}})\rangle_{H^{-3/2}(\partial\Omega),H^{3/2}(\partial\Omega)}$$
$$= \langle \gamma^{\mathsf{D}}(a\tilde{u}), \gamma^{\mathsf{N}}(a)\gamma^{\mathsf{D}}(\overline{\tilde{w}})\rangle_{H^{-1/2}(\partial\Omega),H^{1/2}(\partial\Omega)}.$$

With $b = |n|_g^{1/2}\gamma^{\mathsf{D}}(a)$ as above, we have

$$I = \langle b\gamma_g^{\mathsf{N}}(\tilde{u}), b\gamma^{\mathsf{D}}(\overline{\tilde{w}})\rangle_{H^{-3/2}(\partial\Omega),H^{3/2}(\partial\Omega)} - \langle b\gamma^{\mathsf{D}}(\tilde{u}), b\gamma_g^{\mathsf{N}}(\overline{\tilde{w}})\rangle_{H^{-1/2}(\partial\Omega),H^{1/2}(\partial\Omega)},$$

which gives

$$I = \langle \gamma_g^{\mathsf{N}}(\tilde{u}), \gamma^{\mathsf{D}}(\overline{\tilde{w}})\rangle_{H^{-3/2}(\partial\mathcal{M}),H^{3/2}(\partial\mathcal{M})} - \langle \gamma^{\mathsf{D}}(\tilde{u}), \gamma_g^{\mathsf{N}}(\overline{\tilde{w}})\rangle_{H^{-1/2}(\partial\mathcal{M}),H^{1/2}(\partial\mathcal{M})}$$

by (18.9.4). ■

With these results we can then extend *mutatis mutandis* the results of Section 18.6.3 and prove results of Section 10.5 of Volume 1: Proposition 10.55 that allows one to define the Dirichlet map as in Definition 10.56 therein; similarly, the proof of Proposition 10.57 in Volume 1 is identical to that of Proposition 18.39.

Appendix

18.A. Traces Extension: Technical Aspects

18.A.1. L^2 Functions with L^2 'Laplacian'. Here, we provide a proof of Lemma 18.32.

We use normal geodesic coordinates as given by Theorem 17.22: there exist O an open set of $\mathcal{M}$ neighborhood of $\partial\mathcal{M}$, $z_0 > 0$, and a diffeomorphism Φ, such that

$$\begin{aligned}\Phi : \mathcal{N} = \partial\mathcal{M} \times [0, z_0) &\to O \\ (m', z) &\mapsto \Phi(m', z).\end{aligned} \tag{18.A.1}$$

The pullback of the metric takes the form

$$\Phi^* g_{(m',z)} = g'_{m'}(z) \otimes 1_z + d_z \otimes d_z,$$

where $g'(z)$ is a Riemannian metric on $\partial\mathcal{M}$ that smoothly depends on z and such that $g'(0) = g_\partial$. By Corollary 17.23 the pullback of the Laplace–Beltrami operator on $\mathcal{N}$ is given by

$$P_0 = (\det g'(z))^{-1/2} D_z (\det g'(z))^{1/2} D_z) - \Delta_{g'(z)}. \tag{18.A.2}$$

The operator we consider $P = P_0 + R_1$ with R_1 a first-order differential operator with smooth coefficients.

If $u \in \mathcal{W}_P(\mathcal{M})$, by standard elliptic theory $u \in H^2_{\mathrm{loc}}(\mathcal{M})$, that is, H^2 away from the boundary. Let $\chi \in \mathscr{C}_c^\infty(-z_0, z_0)$ such that $\chi \equiv 1$ in a neighborhood of 0. Observe then that $v = \chi(z)\Phi^* u$ is such that

$$v \in L^2(\mathcal{N}) \text{ and } (P_0 + R_1)v \in L^2(\mathcal{N}),$$

and we have $\|v\|_{\mathcal{W}_P(\mathcal{N})} \lesssim \|u\|_{\mathcal{W}_P(\mathcal{M})}$, where $\|v\|^2_{\mathcal{W}_P(\mathcal{N})} = \|v\|^2_{L^2(\mathcal{N})} + \|Pv\|^2_{L^2(\mathcal{N})}$. It now suffices to prove the result for v supported in $\mathcal{N}$.

LEMMA 18.68. *We have* $v \in H^2(\mathbb{R}_+; H^{-2}(\partial\mathcal{M})) \cap H^1(\mathbb{R}_+; H^{-1}(\partial\mathcal{M}))$. *Moreover, there exists* $C > 0$ *(independent of* v*) such that*

$$\|v\|_{H^2(\mathbb{R}_+;H^{-2}(\partial\mathcal{M}))} + \|v\|_{H^1(\mathbb{R}_+;H^{-1}(\partial\mathcal{M}))} \leq C\|v\|_{\mathcal{W}_P(\mathcal{N})}.$$

A proof is given below.

Consequently, the traces $v_{|z=0^+}$ and $\partial_z v_{|z=0^+}$ are both uniquely defined

$$v_{|z=0^+} = \lim_{z\to 0+} v(z,.) \in H^{-1}(\partial\mathcal{M}), \tag{18.A.3}$$

and

$$\partial_z v_{|z=0^+} = \lim_{z\to 0+} \partial_z v(z,.) \in H^{-2}(\partial\mathcal{M}). \tag{18.A.4}$$

We shall now improve upon these two results.

Let $\psi \in \mathscr{C}^\infty(\partial\mathcal{M})$. According to the trace formula of Theorem 18.25 we can pick $w \in \mathscr{C}^\infty(\mathcal{N})$ be such that $w_{|z=0^+} = \psi$ and $|\psi|_{H^{1/2}(\partial\mathcal{M})} \eqsim \|w\|_{H^1(\mathcal{N})}$. With the regularity of v obtained in Lemma 18.68 we can write

$$\begin{aligned}
&\langle v_{|z=0^+}, \psi\rangle_{H^{-1}(\partial\mathcal{M}),H^1(\partial\mathcal{M})} \\
&\quad = -\int_0^{+\infty} \big(\langle \partial_z v, w\rangle_{H^{-1}(\partial\mathcal{M}),H^1(\partial\mathcal{M})} + \langle v, \partial_z w\rangle_{H^{-1}(\partial\mathcal{M}),H^1(\partial\mathcal{M})}\big) dz \\
&\quad = -\int_0^{+\infty} \langle \partial_z v, w\rangle_{H^{-1}(\partial\mathcal{M}),H^1(\partial\mathcal{M})}\, dz - \langle v, \partial_z w\rangle_{L^2(\mathcal{N}),L^2(\mathcal{N})},
\end{aligned}$$

which gives

$$\begin{aligned}
|\langle v_{|z=0^+}, \psi\rangle_{H^{-1}(\partial\mathcal{M}),H^1(\partial\mathcal{M})}| &\lesssim \big(\|v\|_{H^1(\mathbb{R}_+;H^{-1}(\partial\mathcal{M}))} + \|v\|_{L^2(\mathcal{N})}\big)\|w\|_{H^1(\mathcal{N})} \\
&\lesssim \|v\|_{\mathcal{W}_P(\mathcal{N})}\|\psi\|_{H^{1/2}(\partial\mathcal{M})}.
\end{aligned}$$

As ψ is arbitrary, it follows that $v_{|z=0^+} \in H^{-1/2}(\partial\mathcal{M})$ and moreover we have $|v_{|z=0^+}|_{H^{-1/2}(\partial\mathcal{M})} \lesssim \|v\|_{\mathcal{W}_P(\mathcal{N})}$.

Again with $\psi \in \mathscr{C}^\infty(\partial\mathcal{M})$, according to the trace formula of Theorem 18.25 we can pick $w \in \mathscr{C}^\infty(\mathcal{N})$ be such that $w_{|z=0^+} = \psi$, $\partial_z w_{|z=0^+} = 0$, and $|\psi|_{H^{3/2}(\partial\mathcal{M})} \eqsim \|w\|_{H^2(\mathcal{N})}$. With the regularity of v obtained in Lemma 18.68 we can write

$$\begin{aligned}
&\langle \partial_z v_{|z=0^+}, \psi\rangle_{H^{-2}(\partial\mathcal{M}),H^2(\partial\mathcal{M})} \\
&\quad = -\int_0^{+\infty} \big(\langle \partial_z^2 v, w\rangle_{H^{-2}(\partial\mathcal{M}),H^2(\partial\mathcal{M})} + \langle \partial_z v, \partial_z w\rangle_{H^{-2}(\partial\mathcal{M}),H^2(\partial\mathcal{M})}\big) dz \\
&\quad = -\int_0^{+\infty} \big(\langle \partial_z^2 v, w\rangle_{H^{-2}(\partial\mathcal{M}),H^2(\partial\mathcal{M})} + \langle \partial_z v, \partial_z w\rangle_{H^{-1}(\partial\mathcal{M}),H^1(\partial\mathcal{M})}\big) dz,
\end{aligned}$$

which gives

$$\begin{aligned}
|\langle \partial_z v_{|z=0^+}, \psi\rangle_{H^{-2}(\partial\mathcal{M}),H^2(\partial\mathcal{M})}| &\lesssim \big(\|v\|_{H^2(\mathbb{R}_+;H^{-2}(\partial\mathcal{M}))} + \|v\|_{H^1(\mathbb{R}_+;H^{-1}(\partial\mathcal{M}))}\big)\|w\|_{H^2(\mathcal{N})} \\
&\lesssim \|v\|_{\mathcal{W}_P(\mathcal{N})}\|\psi\|_{H^{3/2}(\partial\mathcal{M})}.
\end{aligned}$$

As ψ is arbitrary, it follows that $v_{|z=0^+} \in H^{-3/2}(\partial\mathcal{M})$ and moreover we have $|v_{|z=0^+}|_{H^{-3/2}(\partial\mathcal{M})} \lesssim \|v\|_{\mathcal{W}_P(\mathcal{N})}$.

Finally, since $v \in H^2(\mathbb{R}_+; H^{-2}(\partial\mathcal{M}))$, if a is a smooth function, then $\partial_z(av) = v\partial_z a + a\partial_z v \in H^1(\mathbb{R}_+; H^{-2}(\partial\mathcal{M}))$ and we find that $\gamma^{\mathsf{N}}(av)$ is also well defined in $H^{-2}(\partial\mathcal{M})$ and moreover $\partial_z(av)_{|z=0^+} = (v\partial_z a)_{|z=0^+} + (a\partial_z v)_{|z=0^+}$. In fact, $(v\partial_z a)_{|z=0^+} \in H^{-1/2}(\partial\mathcal{M})$ and $(a\partial_z v)_{|z=0^+} \in H^{-3/2}(\partial\mathcal{M})$, implying that $\partial_z(av)_{|z=0^+} \in H^{-3/2}(\partial\mathcal{M})$. This concludes the proof of Lemma 18.32. ■

PROOF OF LEMMA 18.68. The operator R_1 reads $R_1 = \alpha(m', z)D_z + R_1'(z)$, where α is a smooth function and $R_1'(z)$ is a family of first-order differential operators on $\partial\mathcal{M}$ smoothly parameterized by z.

With $\beta(m', z) = \int_0^z \alpha(m', \sigma)d\sigma$, we write

$$\begin{aligned} e^{-i\beta}D_z\big(e^{i\beta}(\det g'(z))^{1/2}D_z v\big) &= D_z\big((\det g'(z))^{1/2}D_z v\big) + (\det g'(z))^{1/2}\alpha D_z v \\ &= (\det g'(z))^{1/2}\big(Pv + \Delta_{g'(z)}v - R_1'(z)v\big) \\ &\in L^2(\mathbb{R}_+; H^{-2}(\partial\mathcal{M})), \end{aligned}$$

and $\|D_z\big(e^{i\beta}(\det g'(z))^{1/2}D_z v\big)\|_{L^2(\mathbb{R}_+;H^{-2}(\partial\mathcal{M}))} \lesssim \|v\|_{\mathcal{W}_P(\mathcal{N})}$.

With Lemma 18.69 below, for $r = 0$ and $s = -2$, we find that $D_z v \in H^1(\mathbb{R}_+; H^{-2}(\partial\mathcal{M}))$ and

$$\|D_z v\|_{H^1(\mathbb{R}_+;H^{-2}(\partial\mathcal{M}))} \lesssim \|v\|_{\mathcal{W}_P(\mathcal{N})}.$$

Applying Lemma 18.69 a second time for $r = 1$ and $s = -2$, we find that $v \in H^2(\mathbb{R}_+; H^{-2}(\partial\mathcal{M}))$ and

$$\|v\|_{H^2(\mathbb{R}_+;H^{-2}(\partial\mathcal{M}))} \lesssim \|v\|_{\mathcal{W}_P(\mathcal{N})}.$$

Finally, we conclude that $v \in H^1(\mathbb{R}_+; H^{-1}(\partial\mathcal{M}))$ and

$$\|v\|_{H^1(\mathbb{R}_+;H^{-1}(\partial\mathcal{M}))} \lesssim \|v\|_{\mathcal{W}_P(\mathcal{N})},$$

with an interpolation argument [74, 236]. ■

LEMMA 18.69. *Let $\mathscr{V}$ be $\mathbb{R}^n$ or some compact Riemannian manifold. Let also $r \in \mathbb{N}$ and $s \in \mathbb{R}$ and $w \in {}^0\mathscr{D}'(\mathbb{R}_+ \times \mathscr{V})$ be such that $D_z w \in H^r\big(\mathbb{R}_+; H^s(\mathscr{V})\big)$ and $w_{|(a,+\infty)\times\mathscr{V}} \in H^{r+1}\big((a,+\infty); H^s(\mathscr{V})\big)$ for some $a > 0$. Then, $w \in H^{r+1}\big(\mathbb{R}_+; H^s(\mathscr{V})\big)$ and*

$$\|w\|_{H^{r+1}(\mathbb{R}_+;H^s(\mathscr{V}))} \leq C\big(\|D_z w\|_{H^r(\mathbb{R}_+;H^s(\mathscr{V}))} + \|w\|_{H^{r+1}((a,+\infty);H^s(\mathscr{V}))}\big),$$

for some $C > 0$.

PROOF. We pick $\varphi \in \mathscr{C}_c^\infty(\mathbb{R})$ such that $\varphi \equiv 1$ in a neighborhood of $[0, a]$ with $\operatorname{supp}(\varphi) \subset [-b, b]$ with $b > a$. We set $\theta(x) = \varphi(z)$. We compute

$$D_z(\theta w) = \theta D_z w + (D_z\theta)w.$$

As $D_z\theta$ is supported in $(a, +\infty) \times \mathscr{V}$ we have

$$(D_z\theta)w \in H^{r+1}(\mathbb{R}_+; H^s(\mathscr{V})) \subset H^r(\mathbb{R}_+; H^s(\mathscr{V})).$$

We thus have $D_z(\theta w) \in H^r\big(\mathbb{R}_+; H^s(\mathscr{V})\big)$ and, because of its support, we have $\theta w \in H^{r+1}\big(\mathbb{R}_+; H^s(\mathscr{V})\big)$ and

$$\theta w(.,z) = -\int_z^b \partial_d(\theta w)(.,s)\, ds.$$

Moreover, we have

$$\begin{aligned}\|\theta w\|_{H^{r+1}(\mathbb{R}_+;H^s(\mathscr{V}))} &\lesssim \|\theta D_z w\|_{H^r(\mathbb{R}_+;H^s(\mathscr{V}))} + \|(D_z\theta)w\|_{H^r(\mathbb{R}_+;H^s(\mathscr{V}))} \\ &\lesssim \|D_z w\|_{H^r(\mathbb{R}_+;H^s(\mathscr{V}))} + \|w\|_{H^{r+1}((a,+\infty);H^s(\mathscr{V}))}.\end{aligned}$$

Since $(1-\theta)w \in H^{r+1}(\mathbb{R}_+; H^s(\mathscr{V}))$ the result follows. ■

18.A.2. Green-Like Formula in $\mathcal{W}_P(\mathcal{M})$. Here we prove Lemma 18.33. We use the diffeomorphism Φ defined in (18.A.1). Let $\psi \in \mathscr{C}_c^\infty(-z_0, z_0)$ such that $\psi \equiv 1$ in a neighborhood of 0 and set $\tilde{\psi}(m', z) = \psi(z)$ and $\chi = (\Phi^{-1})^*\tilde{\psi}$. This function $\chi \equiv 1$ in a neighborhood of $\partial\mathcal{M}$.

Let $u \in \mathcal{W}_P(\mathcal{M})$. We have $P(\chi u) = P\big((\chi - 1)u\big) + Pu$. Since $u \in H^2_{\text{loc}}(\mathcal{M})$, that is, H^2 away from the boundary by standard elliptic theory, then $\chi u \in \mathcal{W}_P(\mathcal{M})$. Observe that

$$(P\big((1-\chi)u\big), w)_{L^2(\mathcal{M})} = ((1-\chi)u, {}^tPw)_{L^2(\mathcal{M})}$$

by Propositions 18.29 and 18.30. It thus suffices to prove the result with u replace by χu. In turn, if $v = \Phi^*(\chi u)$ it suffices to prove that

$$\begin{aligned}(Pv, w)_{L^2(\mathcal{N})} &+ \langle \gamma^{\mathsf{N}}(v), \gamma^{\mathsf{D}}(\bar{w})\rangle_{H^{-3/2}(\partial\mathcal{M}),H^{3/2}(\partial\mathcal{M})} \\ &= (v, {}^t\bar{P}w)_{L^2(\mathcal{N})} + \langle \gamma^{\mathsf{D}}(v), \gamma^{\mathsf{N}}(\bar{w})\rangle_{H^{-1/2}(\partial\mathcal{M}),H^{1/2}(\partial\mathcal{M})} \\ &\quad + \langle \gamma^{\mathsf{D}}(g(X,\nu)v), \gamma^{\mathsf{D}}(\bar{w})\rangle_{H^{-1/2}(\partial\mathcal{M}),H^{1/2}(\partial\mathcal{M})},\end{aligned} \tag{18.A.5}$$

for $w \in \mathscr{C}^\infty(\mathcal{N})$. By abuse of notation we also denote by P_0 the Laplace–Beltrami operator of $\mathcal{N}$ associated with the pulled-back metric. Similarly, the pullbacks of the vector field X and the function f are also denoted by X and f, respectively. With the form of the metric we have $\mu_g = \mu_{g'(z)}dz$ and $\mu_{g'(0)} = \mu_{g_\partial}$. Observe that

$$(\det g'(z))^{-1/2}\mu_{g'(z)} = (\det g'(0))^{-1/2}\mu_{g'(0)} = (\det g_\partial)^{-1/2}\mu_{g_\partial}.$$

This can be simply seen in a local chart for $\partial\mathcal{M}$. In particular

$$\rho_z = (\det g_\partial)^{-1/2}(\det g'(z))^{1/2}$$

is to be understood as a function on $\partial\mathcal{M}$ with z acting as a parameter. We write

$$(Pv, w)_{L^2(\mathcal{N})} = \int_{\mathcal{N}} (Pv)\bar{w}\mu_{g'(z)}dz = \int_{\mathbb{R}_+}\int_{\partial\mathcal{M}} \rho_z(Pv)\bar{w}\mu_{g_\partial}dz.$$

As $v \in H^2(\mathbb{R}_+; H^{-2}(\partial\mathcal{M})) \cap H^1(\mathbb{R}_+; H^{-1}(\partial\mathcal{M})) \cap L^2(\mathbb{R}_+ \times \partial\mathcal{M})$ by Lemma 18.68 and $Pv = P_0 v + Xv + fv \in L^2(\mathcal{M})$, we may write Pv as a sum one term in $L^2(\mathbb{R}_+; H^{-2}(\partial\mathcal{M}))$, that is, $P_0 v$, one term in $L^2(\mathbb{R}_+; H^{-1}(\partial\mathcal{M}))$, that is, Xv, and one term in $L^2(\mathbb{R}_+ \times \partial\mathcal{M})$, that is, fv. Recalling that the duality bracket between $H^{-k}(\partial\mathcal{M})$ and $H^k(\partial\mathcal{M})$ is understood using $L^2(\partial\mathcal{M})$ as a pivot space with the inner product on $L^2(\partial\mathcal{M})$ based on the density μ_{g_∂}, we may then write

$$(Pv, w)_{L^2(\mathcal{N})} = I_1 + I_2 + I_3,$$

with

$$I_1 = \int_{\mathbb{R}_+} \langle \rho_z P_0 v, \bar{w}\rangle_{H^{-2}(\partial\mathcal{M}),H^2(\partial\mathcal{M})} dz, \quad I_2 = \int_{\mathbb{R}_+} \langle \rho_z X v, \bar{w}\rangle_{H^{-1}(\partial\mathcal{M}),H^1(\partial\mathcal{M})} dz,$$
$$I_3 = \int_{\mathbb{R}_+} \langle \rho_z f v, \bar{w}\rangle_{L^2(\partial\mathcal{M}),L^2(\partial\mathcal{M})} dz.$$

We write $I_1 = I_1^a - I_1^b$, with

$$I_1^a = \int_{\mathbb{R}_+} \langle (\det g_\partial)^{-1/2} D_z (\det g'(z))^{1/2} D_z v), \bar{w}\rangle_{H^{-2}(\partial\mathcal{M}),H^2(\partial\mathcal{M})} dz,$$
$$I_1^b = \int_{\mathbb{R}_+} \langle \rho_z \Delta_{g'(z)} v, \bar{w}\rangle_{H^{-2}(\partial\mathcal{M}),H^2(\partial\mathcal{M})} dz.$$

Integrations by parts with respect to z yield

$$\begin{aligned}
I_1^a &= \int_{\mathbb{R}_+} \langle (\det g_\partial)^{-1/2} D_z (\det g'(z))^{1/2} D_z v), \bar{w}\rangle_{H^{-2}(\partial\mathcal{M}),H^2(\partial\mathcal{M})} dz \\
&= \int_{\mathbb{R}_+} \langle (\det g_\partial)^{-1/2} D_z v, (\det g'(z))^{1/2} \overline{D_z w}\rangle_{H^{-2}(\partial\mathcal{M}),H^2(\partial\mathcal{M})} dz \\
&\quad + \langle \partial_z v_{|z=0^+}, \bar{w}_{|z=0^+}\rangle_{H^{-2}(\partial\mathcal{M}),H^2(\partial\mathcal{M})} \\
&= \int_{\mathbb{R}_+} \langle (\det g_\partial)^{-1/2} v, \overline{D_z((\det g'(z))^{1/2} D_z w)}\rangle_{H^{-2}(\partial\mathcal{M}),H^2(\partial\mathcal{M})} dz \\
&\quad + \langle \partial_z v_{|z=0^+}, \bar{w}_{|z=0^+}\rangle_{H^{-2}(\partial\mathcal{M}),H^2(\partial\mathcal{M})} \\
&\quad - \langle v_{|z=0^+}, \partial_z \bar{w}_{|z=0^+}\rangle_{H^{-2}(\partial\mathcal{M}),H^2(\partial\mathcal{M})}.
\end{aligned}$$

Using that $v \in L^2(\mathbb{R}_+ \times \partial\mathcal{M})$ and using Lemma 18.32 we may then write

$$\begin{aligned}
&I_1^a + \langle \gamma^{\mathsf{N}}(v), \gamma^{\mathsf{D}}(\bar{w})\rangle_{H^{-3/2}(\partial\mathcal{M}),H^{3/2}(\partial\mathcal{M})} - \langle \gamma^{\mathsf{D}}(v), \gamma^{\mathsf{N}}(\bar{w})\rangle_{H^{-1/2}(\partial\mathcal{M}),H^{1/2}(\partial\mathcal{M})} \\
&\quad = \int_{\mathcal{N}} (\det g_\partial)^{-1/2} v \, \overline{D_z((\det g'(z))^{1/2} D_z w)} \mu_{g_\partial} dz \\
&\quad = (v, (\det g'(z))^{-1/2} D_z((\det g'(z))^{1/2} D_z w))_{L^2(\mathcal{N})}.
\end{aligned}$$

Almost everywhere on $\mathbb{R}_+$, we have, using (18.3.1),

$$\begin{aligned}
\langle \rho_z \Delta_{g'(z)} v, \bar{w}\rangle_{H^{-2}(\partial\mathcal{M}),H^2(\partial\mathcal{M})} &= \langle \rho_z \Delta_{g'(z)} v, \bar{w}\mu_{g_\partial}\rangle_{{}^0\mathscr{D}'(\partial\mathcal{M}),{}^1\mathscr{D}_c^\infty(\partial\mathcal{M})} \\
&= \langle \Delta_{g'(z)} v, \bar{w}\mu_{g'(z)}\rangle_{{}^0\mathscr{D}'(\partial\mathcal{M}),{}^1\mathscr{D}_c^\infty(\partial\mathcal{M})} \\
&= \langle v, (\Delta_{g'(z)} \bar{w})\mu_{g'(z)}\rangle_{{}^0\mathscr{D}'(\partial\mathcal{M}),{}^1\mathscr{D}_c^\infty(\partial\mathcal{M})} \\
&= \int_{\partial\mathcal{M}} v (\Delta_{g'(z)} \bar{w}) \mu_{g'(z)}.
\end{aligned}$$

yielding $I_1^b = (v, \Delta_{g'(z)} w)_{L^2(\mathcal{N})}$. Combining the computations of I_1^a and I_1^b we then obtain

$$
\begin{aligned}
(18.A.6) \qquad I_1 = &-\langle \gamma^{\mathsf{N}}(v), \gamma^{\mathsf{D}}(\bar{w})\rangle_{H^{-3/2}(\partial\mathcal{M}), H^{3/2}(\partial\mathcal{M})} \\
&+ \langle \gamma^{\mathsf{D}}(v), \gamma^{\mathsf{N}}(\bar{w})\rangle_{H^{-1/2}(\partial\mathcal{M}), H^{1/2}(\partial\mathcal{M})} + (v, P_0 w)_{L^2(\mathcal{N})}.
\end{aligned}
$$

Setting $X = X' + X^z\partial_z$ with $X'_{(m',z)} \in T_{m'}\partial\mathcal{M}$, we find $I_2 = I_2^a + I_2^b$ with

$$
\begin{aligned}
I_2^a &= \int_{\mathbb{R}_+} \langle \rho_z X^z \partial_z v, \bar{w}\rangle_{H^{-1}(\partial\mathcal{M}), H^1(\partial\mathcal{M})} dz, \\
I_2^b &= \int_{\mathbb{R}_+} \langle \rho_z X' v, \bar{w}\rangle_{H^{-1}(\partial\mathcal{M}), H^1(\partial\mathcal{M})} dz.
\end{aligned}
$$

We write

$$
\begin{aligned}
I_2^a &= \int_{\mathbb{R}_+} \langle \rho_z X^z \partial_z v, \bar{w}\rangle_{H^{-1}(\partial\mathcal{M}), H^1(\partial\mathcal{M})} dz \\
&= -\langle \gamma^{\mathsf{D}}(v), \gamma^{\mathsf{D}}(X^z \bar{w})\rangle_{H^{-1}(\partial\mathcal{M}), H^1(\partial\mathcal{M})} \\
&\quad - \int_{\mathbb{R}_+} \langle v, \partial_z(\rho_z X^z \bar{w})\rangle_{L^2(\partial\mathcal{M}), L^2(\partial\mathcal{M})} dz. \\
&= \langle \gamma^{\mathsf{D}}(g(X,\nu)v), \gamma^{\mathsf{D}}(\bar{w})\rangle_{H^{-1/2}(\partial\mathcal{M}), H^{1/2}(\partial\mathcal{M})} \\
&\quad - \int_{\mathbb{R}_+} \langle v, \partial_z(\rho_z X^z \bar{w})\rangle_{L^2(\partial\mathcal{M}), L^2(\partial\mathcal{M})} dz,
\end{aligned}
$$

using that $\rho_z = 1$ for $z = 0$. Observe that

$$
\int_{\mathbb{R}_+} \langle v, \partial_z(\rho_z X^z \bar{w})\rangle_{L^2(\partial\mathcal{M}), L^2(\partial\mathcal{M})} = \int_{\mathcal{N}} v \rho_z^{-1} \partial_z(\rho_z X^z \bar{w}) \mu_g
$$

and

$$
\rho_z^{-1}\partial_z(\rho_z X^z \bar{w}) = X^z \partial_z \bar{w} + (\det g'(z))^{-1/2} \partial_z\big((\det g'(z))^{1/2} X^z\big)\bar{w}.
$$

Observe now that

$$
\begin{aligned}
I_2^b &= \int_{\mathbb{R}_+} \langle \rho_z X' v, \bar{w}\rangle_{H^{-1}(\partial\mathcal{M}), H^1(\partial\mathcal{M})} dz \\
&= \int_{\mathbb{R}_+} \langle \rho_z X' v, \bar{w}\mu_{g_\partial}\rangle_{{}^0\mathscr{D}'(\partial\mathcal{M}), {}^1\mathscr{D}_c^\infty(\partial\mathcal{M})} dz \\
&= \int_{\mathbb{R}_+} \langle X' v, \bar{w}\mu_{g'(z)}\rangle_{{}^0\mathscr{D}'(\partial\mathcal{M}), {}^1\mathscr{D}_c^\infty(\partial\mathcal{M})} dz \\
&= \int_{\mathbb{R}_+} \langle v, {}^t X'(\bar{w})\mu_{g'(z)}\rangle_{{}^0\mathscr{D}'(\partial\mathcal{M}), {}^1\mathscr{D}_c^\infty(\partial\mathcal{M})} dz \\
&= (v, {}^t\overline{X'} w)_{L^2(\mathcal{N})}.
\end{aligned}
$$

We have ${}^tX' = -X' - \operatorname{div}_{g'(z)} X'$. As we have

$$
\operatorname{div}_{g'(z)} X' + (\det g'(z))^{-1/2} \partial_z\big((\det g'(z))^{1/2} X^z\big) = \operatorname{div}_g X,
$$

by Proposition 17.7, with (17.2.5) we thus obtain

(18.A.7)
$$I_2 = \langle \gamma^{\mathsf{D}}(g(X,\nu)v), \gamma^{\mathsf{D}}(\bar{w})\rangle_{H^{-1/2}(\partial\mathcal{M}),H^{1/2}(\partial\mathcal{M})} - (v, (\bar{X} + \operatorname{div}_g \bar{X})w)_{L^2(\mathcal{N})}.$$

Finally, we have $I_3 = (v, \bar{f}w)_{L^2(\mathcal{M})}$. Together with (18.A.6) and (18.A.7) this yields the result as ${}^tX = -X - \operatorname{div}_g X$ by Proposition 17.7. ■

18.A.3. L^2 Functions with H^{-1} Laplacian. Here, we provide a proof of Lemma 18.35.

Let $u \in \mathcal{W}_{g,-1}(\mathcal{M})$. Set $f = P_0 u \in H^{-1}(\mathcal{M})$. By Proposition 18.31 there exists $\tilde{u} \in H_0^1(\mathcal{M})$ such that $P_0\tilde{u} = f$ and

$$\|\tilde{u}\|_{H^1(\mathcal{M})} \lesssim \|f\|_{H^{-1}(\mathcal{M})} \lesssim \|u\|_{\mathcal{W}_{g,-1}(\mathcal{M})}.$$

Then $v = u - \tilde{u} \in L^2(\mathcal{M})$ and $P_0 v = 0$ meaning that $v \in \mathcal{W}_g(\mathcal{M})$. As a result v admit a Dirichlet trace $\gamma^{\mathsf{D}}(u)$ in $H^{-1/2}(\partial\mathcal{M})$ and

$$|\gamma^{\mathsf{D}}(v)|_{H^{-1/2}(\partial\mathcal{M})} \lesssim \|v\|_{\mathcal{W}_g(\mathcal{M})} \lesssim \|u\|_{L^2(\mathcal{M})} + \|\tilde{u}\|_{L^2(\mathcal{M})} \lesssim \|u\|_{\mathcal{W}_{g,-1}(\mathcal{M})}.$$

Since $\gamma^{\mathsf{D}}(\tilde{u}) = 0$ we find that the Dirichlet trace of u makes sense and $\gamma^{\mathsf{D}}(u) = \gamma^{\mathsf{D}}(v) \in H^{-1/2}(\partial\mathcal{M})$.

From Lemma 18.34, if $w \in H^2(\mathcal{M}) \cap H_0^1(\mathcal{M})$ we find

$$0 = (v, P_0 w)_{L^2(\mathcal{M})} + \langle \gamma^{\mathsf{D}}(v), \gamma^{\mathsf{N}}(\bar{w})\rangle_{H^{-1/2}(\partial\mathcal{M}),H^{1/2}(\partial\mathcal{M})}.$$

which we write

$$(\tilde{u}, P_0 w)_{L^2(\mathcal{M})} = (u, P_0 w)_{L^2(\mathcal{M})} + \langle \gamma^{\mathsf{D}}(v), \gamma^{\mathsf{N}}(\bar{w})\rangle_{H^{-1/2}(\partial\mathcal{M}),H^{1/2}(\partial\mathcal{M})}.$$

By Proposition 18.13 applied twice we have

$$(\tilde{u}, P_0 w)_{L^2(\mathcal{M})} = \int_{\mathcal{M}} g(\nabla_g \tilde{u}, \nabla_g w)\mu_g = (P_0\tilde{u}, w)_{L^2(\mathcal{M})},$$

which concludes the proof. ■

18.A.4. H^1 Functions with L^2 Laplacian. Here, we prove Lemma 18.42. By Lemma 18.32 we have $\gamma^{\mathsf{N}}(u) \in H^{-3/2}(\partial\mathcal{M})$. Let $w \in H^2(\mathcal{M})$. Lemma 18.34 gives

$$\begin{aligned}(P_0 u, w)_{L^2(\mathcal{M})} &+ \langle \gamma^{\mathsf{N}}(u), \gamma^{\mathsf{D}}(\bar{w})\rangle_{H^{-3/2}(\partial\mathcal{M}),H^{3/2}(\partial\mathcal{M})} \\ &= (u, P_0 w)_{L^2(\mathcal{M})} + \langle \gamma^{\mathsf{D}}(u), \gamma^{\mathsf{N}}(\bar{w})\rangle_{H^{-1/2}(\partial\mathcal{M}),H^{1/2}(\partial\mathcal{M})}.\end{aligned}$$

The formula (18.5.5) consequence of the divergence formula of Proposition 18.28 yields

$$(u, P_0 w)_{L^2(\mathcal{M})} + \langle \gamma^{\mathsf{D}}(u), \gamma^{\mathsf{N}}(\bar{w})\rangle_{H^{-1/2}(\partial\mathcal{M}),H^{1/2}(\partial\mathcal{M})} = (\nabla_g u, \nabla_g w)_{L^2V(\mathcal{M})}.$$

We thus obtain

$$\langle \gamma^{\mathsf{N}}(u), \gamma^{\mathsf{D}}(\bar{w})\rangle_{H^{-3/2}(\partial\mathcal{M}),H^{3/2}(\partial\mathcal{M})} = -(P_0 u, w)_{L^2(\mathcal{M})} + (\nabla_g u, \nabla_g w)_{L^2V(\mathcal{M})}.$$

Let now $\psi \in H^{3/2}(\partial\mathcal{M})$ and set $w = \mathsf{D}(\psi)$. We have $w \in H^2(\mathcal{M})$ by Proposition 18.39 and $\|w\|_{H^2(\mathcal{M})} \lesssim |\psi|_{H^{3/2}(\partial\mathcal{M})}$. Moreover, $\|w\|_{H^1(\mathcal{M})} \lesssim |\psi|_{H^{1/2}(\partial\mathcal{M})}$ according to the same proposition. We thus find

$$\begin{aligned}|\langle\gamma^{\mathsf{N}}(u), \bar{\psi}\rangle_{H^{-3/2}(\partial\mathcal{M}),H^{3/2}(\partial\mathcal{M})}| &\lesssim \big(\|P_0 u\|_{L^2(\mathcal{M})} + \|u\|_{H^1(\mathcal{M})}\big)\|w\|_{H^1(\mathcal{M})}\\ &\lesssim \big(\|P_0 u\|_{L^2(\mathcal{M})} + \|u\|_{H^1(\mathcal{M})}\big)|\psi|_{H^{1/2}(\partial\mathcal{M})}.\end{aligned}$$

This gives the result.

18.A.5. Normal Traces for Vector Fields in $H(\operatorname{div}_g, \mathcal{M})$. Here, we prove Lemma 18.43.

We use normal geodesic coordinates as given by Theorem 17.22: there exist O an open set of $\mathcal{M}$ neighborhood of $\partial\mathcal{M}$, $z_0 > 0$, and a diffeomorphism Φ, such that

$$\begin{aligned}\Phi : \mathcal{N} = \partial\mathcal{M} \times [0, z_0) &\to O\\ (m', z) &\mapsto \Phi(m', z).\end{aligned}$$

The pullback of the metric takes the form

$$\tilde{g} = \Phi^* g_{(m',z)} = g'_{m'}(z) \otimes 1_z + d_z \otimes d_z,$$

where $g'(z)$ is a Riemannian metric on $\partial\mathcal{M}$ that smoothly depends on z and such that $g'(0) = g_\partial$.

If V is a smooth vector field on $\mathcal{N}$ it can be written as $V = (V', V^z)$; the divergence operator takes the form

$$\begin{aligned}\operatorname{div}_{\tilde{g}} V &= (\det g'(z))^{-1/2}\partial_z\big((\det g'(z))^{1/2}V^z\big) + \operatorname{div}_{g'(z)} V'\\ &= \partial_z V^z + \operatorname{div}_{g'(z)} V' + \frac{1}{2}(\det g'(z))^{-1}\big(\partial_z \det g'(z)\big)V^z,\end{aligned}$$

where $\operatorname{div}_{g'(z)}$ is the divergence operator on $\partial\mathcal{M}$ associated with the metric $g'(z)$, with z as a parameter.

For such a vector field V one has

$$\begin{aligned}\|V\|^2_{L^2V(\mathcal{N})} &= \|V^z\|^2_{L^2(\mathcal{N})} + \int_0^{z_0}\int_{\partial\mathcal{M}} g'(z)\big(V'(m',z), V'(m',z)\big)\,\mu_{g'(z)}\,dz\\ &= \|V^z\|^2_{L^2(\mathcal{N})} + \|(0, V')\|^2_{L^2V(\mathcal{N})}.\end{aligned}$$

Let $\chi \in \mathscr{C}_c^\infty(-z_0, z_0)$ such that $\chi \equiv 1$ in a neighborhood of 0. Let U be as in the statement of the lemma. Observe that it suffices to prove the result for $V = \chi(z)\Phi^* U$.

As above the vector field has the form $V = (V', V^z)$. It is in $L^2V(\mathcal{N})$ and $\operatorname{div}_{\tilde{g}} V \in L^2(\mathcal{N})$. One has $\operatorname{div}_{g'(z)} V' \in L^2(\mathbb{R}_+; H^{-1}(\partial\mathcal{M}))$ and

$$\|\operatorname{div}_{g'(z)} V'\|_{L^2(\mathbb{R}_+;H^{-1}(\partial\mathcal{M}))} \lesssim \|(0, V')\|_{L^2V(\mathcal{N})} \lesssim \|V\|_{L^2V(\mathcal{N})}.$$

Since

$$\partial_z V^z = \operatorname{div}_{\tilde{g}} V - \operatorname{div}_{g'(z)} V' - \frac{1}{2}(\det g'(z))^{-1}\big(\partial_z \det g'(z)\big)V^z,$$

One finds $\partial_z V^z \in L^2(\mathbb{R}_+; H^{-1}(\partial\mathcal{M}))$ and

$$\begin{aligned}
&\|\partial_z V^z\|_{L^2(\mathbb{R}_+;H^{-1}(\partial\mathcal{M}))} \\
&\quad \lesssim \|\operatorname{div}_{\tilde{g}} V\|_{L^2(\mathcal{N})} + \|\operatorname{div}_{g'(z)} V'\|_{L^2(\mathbb{R}_+;H^{-1}(\partial\mathcal{M}))} + \|V^z\|_{L^2(\mathcal{N})} \\
&\quad \lesssim \|\operatorname{div}_{\tilde{g}} V\|_{L^2(\mathcal{N})} + \|V\|_{L^2V(\mathcal{N})} \\
&\quad \lesssim \|V\|_{H(\operatorname{div}_{\tilde{g}},\mathcal{N})}.
\end{aligned}$$

Since $V^z \in L^2(\mathcal{N})$ one has $V^z \in H^1(\mathbb{R}_+; H^{-1}(\partial\mathcal{M}))$. In particular, the trace $V^z_{|z=0^+} = -\tilde{g}(\nu, V)$ is uniquely defined and $V^z_{|z=0^+} \in H^{-1}(\partial\mathcal{M})$. We shall now improve upon this result.

Let $\psi \in \mathscr{C}^\infty(\partial\mathcal{M})$. According to the trace formula of Theorem 18.25 we can pick $w \in \mathscr{C}^\infty(\mathcal{N})$ be such that $w_{|z=0^+} = \psi$ and $|\psi|_{H^{1/2}(\partial\mathcal{M})} \eqsim \|w\|_{H^1(\mathcal{N})}$. With the regularity of V^z obtained above we can write

$$\begin{aligned}
&\langle V^z_{|z=0^+}, \psi\rangle_{H^{-1}(\partial\mathcal{M}),H^1(\partial\mathcal{M})} \\
&\quad = -\int_0^{+\infty} \big(\langle \partial_z V^z, w\rangle_{H^{-1}(\partial\mathcal{M}),H^1(\partial\mathcal{M})} + \langle V^z, \partial_z w\rangle_{H^{-1}(\partial\mathcal{M}),H^1(\partial\mathcal{M})}\big) dz \\
&\quad = -\int_0^{+\infty} \langle \partial_z V^z, w\rangle_{H^{-1}(\partial\mathcal{M}),H^1(\partial\mathcal{M})}\, dz - \langle V^z, \partial_z w\rangle_{L^2(\mathcal{N}),L^2(\mathcal{N})},
\end{aligned}$$

which gives

$$\begin{aligned}
|\langle V^z_{|z=0^+}, \psi\rangle_{H^{-1}(\partial\mathcal{M}),H^1(\partial\mathcal{M})}| &\lesssim \big(\|V^z\|_{H^1(\mathbb{R}_+;H^{-1}(\partial\mathcal{M}))} + \|V^z\|_{L^2(\mathcal{N})}\big)\|w\|_{H^1(\mathcal{N})} \\
&\lesssim \|V\|_{H(\operatorname{div}_{\tilde{g}},\mathcal{N})}\|\psi\|_{H^{1/2}(\partial\mathcal{M})}.
\end{aligned}$$

As ψ is arbitrary, it follows that $V^z_{|z=0^+} \in H^{-1/2}(\partial\mathcal{M})$ and moreover we have $|V^z_{|z=0^+}|_{H^{-1/2}(\partial\mathcal{M})} \lesssim \|V\|_{H(\operatorname{div}_{\tilde{g}},\mathcal{N})}$. ■

Bibliography

1. Abraham, R., Marsden, J. E., and Ratiu, T. *Manifolds, tensor analysis, and applications*, second ed., vol. 75 of *Applied Mathematical Sciences*. Springer-Verlag, New York, 1988.
2. Adams, R., and Fournier, J. *Sobolev Spaces*, second ed. Academic Press, New York, 2003.
3. Agmon, S., Douglis, A., and Nirenberg, L. Estimates near the boundary for solutions of elliptic partial differential equations satisfying general boundary conditions. I. *Comm. Pure Appl. Math. 12* (1959), 623–727.
4. Agmon, S., Douglis, A., and Nirenberg, L. Estimates near the boundary for solutions of elliptic partial differential equations satisfying general boundary conditions. II. *Comm. Pure Appl. Math. 17* (1964), 35–92.
5. Agranovič, M. S., and Dynin, A. S. General boundary-value problems for elliptic systems in higher-dimensional regions. *Dokl. Akad. Nauk SSSR 146* (1962), 511–514.
6. Alabau-Boussouira, F. A two-level energy method for indirect boundary observability and controllability of weakly coupled hyperbolic systems. *SIAM J. Control Optim. 42* (2003), 871–906.
7. Alabau-Boussouira, F. On some recent advances on stabilization for hyperbolic equations. In *Control of partial differential equations*, vol. 2048 of *Lecture Notes in Math.* Springer, Heidelberg, 2012, pp. 1–100.
8. Alabau-Boussouira, F., and Ammari, K. Sharp energy estimates for nonlinearly locally damped PDEs via observability for the associated undamped system. *J. Funct. Anal. 260* (2011), 2424–2450.

J. Le Rousseau et al., *Elliptic Carleman Estimates and Applications to Stabilization and Controllability, Volume II*, PNLDE Subseries in Control 98, https://doi.org/10.1007/978-3-030-88670-7

9. Alabau-Boussouira, F., Cannarsa, P., and Fragnelli, G. Carleman estimates for degenerate parabolic operators with applications to null controllability. *J. Evol. Equ. 6* (2006), 161–204.
10. Alabau-Boussouira, F., and Léautaud, M. Indirect stabilization of locally coupled wave-type systems. *ESAIM Control Optim. Calc. Var. 18* (2012), 548–582.
11. Alabau-Boussouira, F., Nicaise, S., and Pignotti, C. Exponential stability of the wave equation with memory and time delay. In *New prospects in direct, inverse and control problems for evolution equations*, vol. 10 of *Springer INdAM Ser.* Springer, Cham, 2014, pp. 1–22.
12. Alabau-Boussouira, F., Privat, Y., and Trélat, E. Nonlinear damped partial differential equations and their uniform discretizations. *J. Funct. Anal. 273* (2017), 352–403.
13. Alessandrini, G. Morassi, A. Strong unique continuation for the Lamé system of elasticity. *Comm. Partial Differential Equations 26*, 9-10 (2001), 1787–1810.
14. Alessandrini, G. Strong unique continuation for general elliptic equations in 2D. *J. Math. Anal. Appl. 386*, 2 (2012), 669–676.
15. Alessandrini, G., and Escauriaza, L. Null-controllability of one-dimensional parabolic equations. *ESAIM Control Optim. Calc. Var. 14* (2008), 284–293.
16. Alessandrini, G., Morassi, A., Rosset, E., and Vessella, S. On doubling inequalities for elliptic systems. *J. Math. Anal. Appl. 357*, 2 (2009), 349–355.
17. Ali Ziane, T., Ouzzane, H., and Zair, O. A Carleman estimate for the two dimensional heat equation with mixed boundary conditions. *C. R. Math. Acad. Sci. Paris 351* (2013), 97–100.
18. Ali Ziane, T., Ouzzane, H., and Zair, O. Controllability results for the two-dimensional heat equation with mixed boundary conditions using Carleman inequalities: a linear and a semilinear case. *Appl. Anal. 97* (2018), 2412–2430.
19. Alinhac, S. Non-unicité pour des opérateurs différentiels à caractéristiques complexes simples. *Ann. Sci. École Norm. Sup. (4) 13*, 3 (1980), 385–393.
20. Alinhac, S. Non-unicité du problème de Cauchy. *Ann. of Math. (2) 117*, 1 (1983), 77–108.
21. Alinhac, S. Unicité du problème de Cauchy pour des opérateurs du second ordre à symboles réels. *Ann. Inst. Fourier (Grenoble) 34*, 2 (1984), 89–109.
22. Alinhac, S., and Baouendi, M. S. Construction de solutions nulles et singulières pour des opérateurs de type principal. In *Séminaire Goulaouic-Schwartz (1978/1979).* École Polytech., Palaiseau, 1979, pp. Exp. No. 22, 6.

23. ALINHAC, S., AND BAOUENDI, M. S. Uniqueness for the characteristic Cauchy problem and strong unique continuation for higher order partial differential inequalities. *Amer. J. Math. 102*, 1 (1980), 179–217.
24. ALINHAC, S., AND BAOUENDI, M. S. A counterexample to strong uniqueness for partial differential equations of Schrödinger's type. *Comm. Partial Differential Equations 19*, 9-10 (1994), 1727–1733.
25. ALINHAC, S., AND BAOUENDI, M. S. A nonuniqueness result for operators of principal type. *Math. Z. 220*, 4 (1995), 561–568.
26. ALINHAC, S., AND GÉRARD, P. *Opérateurs Pseudo-Différentiels et Théorème de Nash-Moser.* Editions du CNRS, 1991.
27. ALINHAC, S., AND LERNER, N. Unicité forte à partir d'une variété de dimension quelconque pour des inégalités différentielles elliptiques. *Duke Math. J. 48*, 1 (1981), 49–68.
28. ALINHAC, S., AND ZUILY, C. Unicité et non unicité du problème de Cauchy pour des opérateurs hyperboliques à charactéristiques doubles. *Comm. Partial Differential Equations 6*, 7 (1981), 799–828.
29. AMMAR KHODJA, F., BENABDALLAH, A., DUPAIX, C., AND GONZÁLEZ-BURGOS, M. A generalization of the Kalman rank condition for time-dependent coupled linear parabolic systems. *Differ. Equ. Appl. 1*, 3 (2009), 427–457.
30. AMMAR-KHODJA, F., BENABDALLAH, A., DUPAIX, C., AND GONZÁLEZ-BURGOS, M. A Kalman rank condition for the localized distributed controllability of a class of linear parbolic systems. *J. Evol. Equ. 9*, 2 (2009), 267–291.
31. AMMAR-KHODJA, F., BENABDALLAH, A., GONZÁLEZ-BURGOS, M., AND DE TERESA, L. Recent results on the controllability of linear coupled parabolic problems: a survey. *Math. Control Relat. Fields 1*, 3 (2011), 267–306.
32. ANANTHARAMAN, N. Spectral deviations for the damped wave equation. *Geom. Funct. Anal. 20*, 3 (2010), 593–626.
33. ANANTHARAMAN, N., AND LÉAUTAUD, M. Sharp polynomial decay rates for the damped wave equation on the torus. *Anal. PDE 7* (2014), 159–214. With an appendix by Stéphane Nonnenmacher.
34. ANG, D. D., IKEHATA, M., TRONG, D. D., AND YAMAMOTO, M. Unique continuation for a stationary isotropic Lamé system with variable coefficients. *Comm. Partial Differential Equations 23*, 1-2 (1998), 371–385.
35. ANG, D. D., TRONG, D. D., AND YAMAMOTO, M. Unique continuation and identification of boundary of an elastic body. *J. Inverse Ill-Posed Probl. 3*, 6 (1996), 417–428.
36. ANGULO-ARDOY, P., FARACO, D., AND GUIJARRO, L. Sufficient conditions for the existence of limiting Carleman weights. *Forum Math. Sigma 5* (2017), e7, 25.
37. ANGULO-ARDOY, P., FARACO, D., GUIJARRO, L., AND RUIZ, A. Obstructions to the existence of limiting Carleman weights. *Anal. PDE 9* (2016), 575–595.

38. Apraiz, J., and Escauriaza, L. Null-control and measurable sets. *ESAIM Control Optim. Calc. Var. 19*, 1 (2013), 239–254.
39. Apraiz, J., Escauriaza, L., Wang, G., and Zhang, C. Observability inequalities and measurable sets. *J. Eur. Math. Soc. (JEMS) 16*, 11 (2014), 2433–2475.
40. Aronszajn, N. Sur l'unicité du prolongement des solutions des équations aux dérivées partielles elliptiques du second ordre. *C. R. Acad. Sci. Paris 242* (1956), 723–725.
41. Aronszajn, N. A unique continuation theorem for solutions of elliptic partial differential equations or inequalities of second order. *J. Math. Pures Appl. (9) 36* (1957), 235–249.
42. Aronszajn, N., Krzywicki, A., and Szarski, J. A unique continuation theorem for exterior differential forms on Riemannian manifolds. *Ark. Mat. 4* (1962), 417–453 (1962).
43. Atiyah, M. F., and Singer, I. M. The index of elliptic operators on compact manifolds. *Bull. Am. Math. Soc. 69* (1963), 422–433.
44. Atiyah, M. F., and Singer, I. M. The index of elliptic operators. III. *Ann. Math. (2) 87* (1968), 546–604.
45. Badra, M. Global Carleman inequalities for Stokes and penalized Stokes equations. *Math. Control Relat. Fields 1* (2011), 149–175.
46. Badra, M., and Takahashi, T. Feedback boundary stabilization of 2D fluid-structure interaction systems. *Discrete Contin. Dyn. Syst. 37* (2017), 2315–2373.
47. Bahouri, H. Non unicité du problème de Cauchy pour des opérateurs à symbole principal réel. *Comm. Partial Differential Equations 8*, 14 (1983), 1521–1547.
48. Bahouri, H. Non prolongement unique des solutions d'opérateurs "somme de carrés". *Ann. Inst. Fourier (Grenoble) 36*, 4 (1986), 137–155.
49. Bahouri, H. Dépendance non linéaire des données de Cauchy pour les solutions des équations aux dérivées partielles. *J. Math. Pures Appl. (9) 66*, 2 (1987), 127–138.
50. Bahouri, H., Chemin, J.-Y., and Danchin, R. *Fourier analysis and nonlinear partial differential equations*, vol. 343 of *Grundlehren der Mathematischen Wissenschaften [Fundamental Principles of Mathematical Sciences]*. Springer, Heidelberg, 2011.
51. Bahouri, H., and Robbiano, L. Unicité de Cauchy pour des opérateurs faiblement hyperboliques. *Hokkaido Math. J. 16*, 3 (1987), 257–275.
52. Barbu, V. Exact controllability of the superlinear heat equation. *Appl. Math. Optim. 42* (2000), 73–89.
53. Barbu, V., Răşcanu, A., and Tessitore, G. Carleman estimates and controllability of linear stochastic heat equations. *Appl. Math. Optim. 47* (2003), 97–120.

54. Barceló, B., Kenig, C. E., Ruiz, A., and Sogge, C. D. Weighted Sobolev inequalities and unique continuation for the Laplacian plus lower order terms. *Illinois J. Math. 32*, 2 (1988), 230–245.
55. Bardos, C., Lebeau, G., and Rauch, J. Un exemple d'utilisation des notions de propagation pour le contrôle et la stabilisation de problèmes hyperboliques. *Rend. Sem. Mat. Univ. Politec. Torino 46*, Special Issue (1988), 11–31 (1989). Nonlinear hyperbolic equations in applied sciences.
56. Bardos, C., Lebeau, G., and Rauch, J. Sharp sufficient conditions for the observation, control, and stabilization of waves from the boundary. *SIAM J. Control Optim. 30* (1992), 1024–1065.
57. Batty, C. J. K. Asymptotic behaviour of semigroups of operators. In *Functional analysis and operator theory (Warsaw, 1992)*, vol. 30 of *Banach Center Publ.* Polish Acad. Sci. Inst. Math., Warsaw, 1994, pp. 35–52.
58. Batty, C. J. K., and Duyckaerts, T. Non-uniform stability for bounded semi-groups on Banach spaces. *J. Evol. Equ.* (2008), 765–780.
59. Baudouin, L., and Ervedoza, S. Convergence of an inverse problem for a 1-D discrete wave equation. *SIAM J. Control Optim. 51* (2013), 556–598.
60. Baudouin, L., Ervedoza, S., and Osses, A. Stability of an inverse problem for the discrete wave equation and convergence results. *J. Math. Pures Appl. (9) 103*, 6 (2015), 1475–1522.
61. Beauchard, K. Null controllability of Kolmogorov-type equations. *Math. Control Signals Systems 26*, 1 (2014), 145–176.
62. Beauchard, K., Cannarsa, P., and Guglielmi, R. Null controllability of Grushin-type operators in dimension two. *J. Eur. Math. Soc. (JEMS) 16*, 1 (2014), 67–101.
63. Beauchard, K., Helffer, B., Henry, R., and Robbiano, L. Degenerate parabolic operators of Kolmogorov type with a geometric control condition. *ESAIM Control Optim. Calc. Var. 21*, 2 (2015), 487–512.
64. Beauchard, K., Miller, L., and Morancey, M. 2D Grushin-type equations: minimal time and null controllable data. *J. Differential Equations 259* (2015), 5813–5845.
65. Beauchard, K., and Pravda-Starov, K. Null-controllability of hypoelliptic quadratic differential equations. *J. Éc. polytech. Math. 5* (2018), 1–43.
66. Beauchard, K., and Zuazua, E. Some controllability results for the 2D Kolmogorov equation. *Ann. Inst. H. Poincaré Anal. Non Linéaire 26*, 5 (2009), 1793–1815.
67. Bellassoued, M. Distribution of resonances and decay rate of the local energy for the elastic wave equation. *Comm. Math. Phys. 215*, 2 (2000), 375–408.

68. BELLASSOUED, M. Carleman estimates and distribution of resonances for the transparent obstacle and application to the stabilization. *Asymptotic Anal. 35* (2003), 257–279.
69. BELLASSOUED, M. Decay of solutions of the elastic wave equation with a localized dissipation. *Ann. Fac. Sci. Toulouse Math. (6) 12*, 3 (2003), 267–301.
70. BELLASSOUED, M., AND LE ROUSSEAU, J. Carleman estimates for elliptic operators with complex coefficients. Part I: Boundary value problems. *J. Math. Pures Appl. (9) 104* (2015), 657–728.
71. BELLASSOUED, M., AND LE ROUSSEAU, J. Carleman estimates for elliptic operators with complex coefficients. Part II: Transmission problems. *J. Math. Pures Appl. (9) 115* (2018), 127–186.
72. BENABDALLAH, A., BOYER, F., AND GONZÁLEZ-BURGOS, M.AND OLIVE, G. Sharp estimates of the one-dimensional boundary control cost for parabolic systems and application to the N-dimensional boundary null controllability in cylindrical domains. *SIAM J. Control Optim. 52*, 5 (2014), 2970–3001.
73. BENABDALLAH, A., DERMENJIAN, Y., AND LE ROUSSEAU, J. Carleman estimates for the one-dimensional heat equation with a discontinuous coefficient and applications to controllability and an inverse problem. *J. Math. Anal. Appl. 336* (2007), 865–887.
74. BERGH, J., AND LÖFSTRÖM, J. *Interpolation spaces. An introduction.* Springer-Verlag, Berlin-New York, 1976. Grundlehren der Mathematischen Wissenschaften, No. 223.
75. BERTHIER, A.-M., AND GEORGESCU, V. Sur la propriété de prolongement unique pour l'opérateur de Dirac. *C. R. Acad. Sci. Paris Sér. A-B 291*, 11 (1980), A603–A606.
76. BONY, J.-M. Calcul symbolique et propagation des singularités pour les équations aux dérivées partielles non linéaires. *Ann. Sci. École Norm. Sup. (4) 14*, 2 (1981), 209–246.
77. BONY, J.-M. Sur l'inégalité de Fefferman-Phong. In *Seminaire: Équations aux Dérivées Partielles, 1998–1999*, Sémin. Équ. Dériv. Partielles. École Polytech., Palaiseau, 1999, pp. Exp. No. III, 16.
78. BORICHEV, A., AND TOMILOV, Y. Optimal polynomial decay of functions and operator semigroups. *Math. Ann. 347* (2010), 455–478.
79. BOSI, R., KURYLEV, Y., AND LASSAS, M. Stability of the unique continuation for the wave operator via Tataru inequality and applications. *J. Differential Equations 260*, 8 (2016), 6451–6492.
80. BOURGEOIS, L. About stability and regularization of ill-posed elliptic Cauchy problems: the case of $C^{1,1}$ domains. *ESAIM: Math. Mod. Numer. Anal. 44* (2010), 715–735.
81. BOURGEOIS, L. Quantification of the unique continuation property for the heat equation. *Math. Control and Relat. Fields 7* (2017), 347–367.
82. BOURGEOIS, L., AND DARDÉ, J. About stability and regularization of ill-posed elliptic Cauchy problems: the case of Lipschitz domains. *Applicable Analysis 89* (2010), 1745–1768.

83. Boutet de Monvel-Berthier, A. An optimal Carleman-type inequality for the Dirac operator. In *Stochastic processes and their applications in mathematics and physics (Bielefeld, 1985)*, vol. 61 of *Math. Appl.* Kluwer Acad. Publ., Dordrecht, 1990, pp. 71–94.
84. Boyer, F. On the penalised HUM approach and its applications to the numerical approximation of null-controls for parabolic problems. *ESAIM, Proc. 41* (2013), 15–58.
85. Boyer, F., Hubert, F., and Le Rousseau, J. Discrete Carleman estimates and uniform controllability of semi-discrete parabolic equations. *J. Math. Pures Appl. 93* (2010), 240–276.
86. Boyer, F., Hubert, F., and Le Rousseau, J. Discrete Carleman estimates for elliptic operators in arbitrary dimension and applications. *SIAM J. Control Optim. 48* (2010), 5357–5397.
87. Boyer, F., Hubert, F., and Le Rousseau, J. Uniform controllability properties for space/time-discretized parabolic equations. *Numer. Math. 118*, 4 (2011), 601–661.
88. Boyer, F., and Le Rousseau, J. Carleman estimates for semi-discrete parabolic operators and application to the controllability of semi-linear semi-discrete parabolic equations. *Ann. Inst. H. Poincaré Anal. Non Linéaire 31*, 5 (2014), 1035–1078.
89. Boyer, F., and Olive, G. Approximate controllability conditions for some linear 1D parabolic systems with space-dependent coefficients. *Math. Control Relat. Fields 4*, 3 (2014), 263–287.
90. Brezis, H. *Functional Analysis, Sobolev Spaces and Partial Differential Equations.* Universitext. Springer, New York, 2011.
91. Buffe, R. *Inégalités de Carleman près du bord, d'une interface et pour des problèmes singuliers.* Theses, Université d'Orléans, 2017.
92. Buffe, R. Stabilization of the wave equation with Ventcel boundary condition. *J. Math. Pures Appl. (9) 108* (2017), 207–259.
93. Burq, N. Mesures semi-classiques et mesures de défaut. *Astérisque 1996/97*, 245 (1997), Exp. No. 826, 4, 167–195. Séminaire Bourbaki.
94. Burq, N. Décroissance de l'énergie locale de l'équation des ondes pour le problème extérieur et absence de résonance au voisinage du réel. *Acta Math.* (1998), 1–29.
95. Burq, N., and Gérard, P. Condition nécessaire et suffisante pour la contrôlabilité exacte des ondes. *C. R. Acad. Sci. Paris Sér. I Math. 325*, 7 (1997), 749–752.
96. Burq, N., and Hitrik, M. Energy decay for damped wave equations on partially rectangular domains. *Math. Res. Lett. 14* (2007), 35–47.
97. Burq, N., and Lebeau, G. Mesures de défaut de compacité, application au système de Lamé. *Ann. Sci. École Norm. Sup. (4) 34*, 6 (2001), 817–870.
98. Burq, N., and Zuily, C. Laplace eigenfunctions and damped wave equation on product manifolds. *Appl. Math. Res. Express. AMRX* (2015), 296–310.

99. Burq, N., and Zuily, C. Concentration of Laplace eigenfunctions and stabilization of weakly damped wave equation. *Comm. Math. Phys. 345* (2016), 1055–1076.
100. Calderón, A.-P. Uniqueness in the Cauchy problem for partial differential equations. *Amer. J. Math. 80* (1958), 16–36.
101. Cannarsa, P., Martinez, P., and Vancostenoble, J. Null controllability of the heat equation in unbounded domains by a finite measure control region. *ESAIM Control Optim. Calc. Var. 10* (2004), 381–408.
102. Cannarsa, P., Martinez, P., and Vancostenoble, J. Null controllability of degenerate heat equations. *Adv. Differential Equations 10*, 2 (2005), 153–190.
103. Cannarsa, P., Martinez, P., and Vancostenoble, J. Carleman estimates for a class of degenerate parabolic operators. *SIAM J. Control Optim. 47* (2008), 1–19.
104. Cannarsa, P., Martinez, P., and Vancostenoble, J. Global Carleman estimates for degenerate parabolic operators with applications. *Mem. Amer. Math. Soc. 239* (2016), ix+209.
105. Carleman, T. Sur une problème d'unicité pour les systèmes d'équations aux dérivées partielles à deux variables indépendantes. *Ark. Mat. Astr. Fys. 26B*, 17 (1939), 1–9.
106. Chanillo, S., and Sawyer, E. Unique continuation for $\Delta + v$ and the C. Fefferman-Phong class. *Trans. Amer. Math. Soc. 318*, 1 (1990), 275–300.
107. Colombini, F., and Grammatico, C. Some remarks on strong unique continuation for the Laplace operator and its powers. *Comm. Partial Differential Equations 24*, 5-6 (1999), 1079–1094.
108. Cornilleau, P., and Robbiano, L. Carleman estimates for the Zaremba boundary condition and stabilization of waves. *Amer. J. Math. 136*, 2 (2014), 393–444.
109. Coron, J.-M. *Control and nonlinearity*, vol. 136 of *Mathematical Surveys and Monographs.* American Mathematical Society, Providence, RI, 2007.
110. Courant, R., and Hilbert, D. *Methods of mathematical physics. Vol. I.* Interscience Publishers, Inc., New York, N.Y., 1953.
111. Dardé, J., and Ervedoza, S. On the reachable set for the one-dimensional heat equation. *SIAM J. Control Optim. 56* (2018), 1692–1715.
112. Davies, E. B. *One-parameter semigroups*, vol. 15 of *London Mathematical Society Monographs.* Academic Press, Inc. [Harcourt Brace Jovanovich, Publishers], London-New York, 1980.
113. De Giorgi, E. Un esempio di non-unicità della soluzione del problema di Cauchy, relativo ad una equazione differenziale lineare a derivate parziali di tipo parabolico. *Rend. Mat. e Appl. (5) 14* (1955), 382–387.

114. DEHMAN, B. Unicité du problème de Cauchy pour une classe d'opérateurs quasi-homogènes. *J. Math. Kyoto Univ. 24* (1984), 453–471.
115. DEHMAN, B. Construction de solutions nulles pour des opérateurs non fuchsiens. *J. Differential Equations 67*, 3 (1987), 330–343.
116. DEHMAN, B., AND ROBBIANO, L. La propriété du prolongement unique pour un système elliptique. Le système de Lamé. *J. Math. Pures Appl. (9) 72*, 5 (1993), 475–492.
117. DENCKER, N. The resolution of the Nirenberg-Treves conjecture. *Ann. of Math. (2) 163* (2006), 405–444.
118. DI CRISTO, M., FRANCINI, E., LIN, C.-L., VESSELLA, S., AND WANG, J.-N. Carleman estimate for second order elliptic equations with Lipschitz leading coefficients and jumps at an interface. *J. Math. Pures Appl. (9) 108*, 2 (2017), 163–206.
119. DIMASSI, M., AND SJÖSTRAND, J. *Spectral Asymptotics in the Semi-classical Limit*, vol. 268 of *London Mathematical Society Lecture Note Series*. Cambridge University Press, Cambridge, 1999.
120. DO CARMO, M. P. *Riemannian geometry*. Mathematics: Theory & Applications. Birkhäuser Boston, Inc., Boston, MA, 1992. Translated from the second Portuguese edition by Francis Flaherty.
121. DOLECKI, S., AND RUSSELL, D. L. A general theory of observation and control. *SIAM J. Control Optimization 15*, 2 (1977), 185–220.
122. DOS SANTOS FERREIRA, D. Sharp L^p Carleman estimates and unique continuation. *Duke Math. J. 129*, 3 (2005), 503–550.
123. DOS SANTOS FERREIRA, D. *Carleman Estimates and Applications to Inverse Problems and Unique Continuation*. PhD thesis, Université Paris 13, 2011.
124. DOS SANTOS FERREIRA, D., KENIG, C. E., SALO, M., AND UHLMANN, G. Limiting Carleman weights and anisotropic inverse problems. *Invent. Math. 178*, 1 (2009), 119–171.
125. DOUBOVA, A., FERNÁNDEZ-CARA, E., AND GONZÁLEZ-BURGOS, M. On the controllability of the heat equation with nonlinear boundary Fourier conditions. *J. Differential Equations 196* (2004), 385–417.
126. DOUBOVA, A., FERNÁNDEZ-CARA, E., GONZÁLEZ-BURGOS, M., AND ZUAZUA, E. On the controllability of parabolic systems with a nonlinear term involving the state and the gradient. *SIAM J. Control Optim. 41* (2002), 798–819.
127. DOUBOVA, A., OSSES, A., AND PUEL, J.-P. Exact controllability to trajectories for semilinear heat equations with discontinuous diffusion coefficients. *ESAIM: Control Optim. Calc. Var. 8* (2002), 621–661.
128. DOUGLAS, R. G. On majorization, factorization, and range inclusion of operators on Hilbert space. *Proc. Amer. Math. Soc. 17* (1966), 413–415.

129. DUISTERMAAT, J. J., AND KOLK, J. A. C. *Distributions.* Cornerstones. Birkhäuser Boston, Inc., Boston, MA, 2010. Theory and applications.
130. DUYCKAERTS, T. Optimal decay rates of the energy of a hyperbolic-parabolic system coupled by an interface. *Asymptot. Anal. 51* (2007), 17–45.
131. ELLER, M. Gårding's inequality on manifolds with boundary. In *Control theory of partial differential equations*, vol. 242 of *Lect. Notes Pure Appl. Math.* Chapman & Hall/CRC, Boca Raton, FL, 2005, pp. 87–99.
132. ELLIS, H. G. Directional differentiation in the plane and tangent vectors on C^r manifolds. *Amer. Math. Monthly 82* (1975), 641–645.
133. ENGEL, K.-J., AND NAGEL, R. *One-parameter semigroups for linear evolution equations*, vol. 194 of *Graduate Texts in Mathematics.* Springer-Verlag, New York, 2000. With contributions by S. Brendle, M. Campiti, T. Hahn, G. Metafune, G. Nickel, D. Pallara, C. Perazzoli, A. Rhandi, S. Romanelli and R. Schnaubelt.
134. ERDÉLYI, A. *Asymptotic expansions.* Dover publications, 1956.
135. ERVEDOZA, S. Control and stabilization properties for a singular heat equation with an inverse-square potential. *Comm. Partial Differential Equations 33*, 10-12 (2008), 1996–2019.
136. ERVEDOZA, S., AND DE GOURNAY, F. Uniform stability estimates for the discrete Calderón problems. *Inverse Problems 27*, 12 (2011), 125012, 37.
137. ERVEDOZA, S., GLASS, O., GUERRERO, S., AND PUEL, J.-P. Local exact controllability for the one-dimensional compressible Navier-Stokes equation. *Arch. Ration. Mech. Anal. 206*, 1 (2012), 189–238.
138. ESCAURIAZA, L. Carleman inequalities and the heat operator. *Duke Math. J. 104*, 1 (2000), 113–127.
139. ESCAURIAZA, L., AND VEGA, L. Carleman inequalities and the heat operator. II. *Indiana Univ. Math. J. 50*, 3 (2001), 1149–1169.
140. FABRE, C., AND LEBEAU, G. Prolongement unique des solutions de l'equation de Stokes. *Comm. Partial Differential Equations 21* (1996), 573–596.
141. FATTORINI, H. O., AND RUSSELL, D. L. Exact controllability theorems for linear parabolic equations in one space dimension. *Arch. Rational Mech. Anal. 43* (1971), 272–292.
142. FEFFERMAN, C., AND PHONG, D. H. On positivity of pseudo-differential operators. *Proc. Nat. Acad. Sci. 75* (1978), 4673–4674.
143. FERNÁNDEZ-CARA, E., GONZÁLEZ-BURGOS, M., GUERRERO, S., AND PUEL, J.-P. Exact controllability to the trajectories of the heat equation with Fourier boundary conditions: the semilinear case. *ESAIM Control Optim. Calc. Var. 12* (2006), 466–483.
144. FERNÁNDEZ-CARA, E., GONZÁLEZ-BURGOS, M., GUERRERO, S., AND PUEL, J.-P. Null controllability of the heat equation with boundary Fourier conditions: the linear case. *ESAIM Control Optim. Calc. Var. 12*, 3 (2006), 442–465.

145. FERNÁNDEZ-CARA, E., GUERRERO, S., IMANUVILOV, O. Y., AND PUEL, J.-P. Local exact controllability of the Navier-Stokes system. *J. Math. Pures Appl. 83* (2004), 1501–1542.
146. FERNÁNDEZ-CARA, E., GUERRERO, S., IMANUVILOV, O. Y., AND PUEL, J.-P. Some controllability results for the N-dimensional Navier-Stokes and Boussinesq systems with $N-1$ scalar controls. *SIAM J. Control Optim. 45* (2006), 146–173.
147. FERNÁNDEZ-CARA, E., AND MÜNCH, A. Numerical null controllability of semi-linear 1D heat equations : Fixed point, least squares and Newton methods. *Mathematical Control and Related Fields 2*, 3 (2012), 217–246.
148. FERNÁNDEZ-CARA, E., AND ZUAZUA, E. The cost of approximate controllability for heat equations: the linear case. *Adv. Differential Equations 5* (2000), 465–514.
149. FERNÁNDEZ-CARA, E., AND ZUAZUA, E. Null and approximate controllability for weakly blowing up semilinear heat equations. *Ann. Inst. H. Poincaré, Analyse non lin. 17* (2000), 583–616.
150. FILONOV, N. Second-order elliptic equation of divergence form having a compactly supported solution. *J. Math. Sci. (New York) 106* (2001), 3078–3086. Function theory and phase transitions.
151. FRANCINI, E., AND VESSELLA, S. Carleman estimates for the parabolic transmission problem and Hölder propagation of smallness across an interface. *J. Differential Equations 265*, 6 (2018), 2375–2430.
152. FRIEDLANDER, F. G. *Introduction to the theory of distributions*, second ed. Cambridge University Press, Cambridge, 1998. With additional material by M. Joshi.
153. FU, X. Null controllability for the parabolic equation with a complex principal part. *J. Funct. Anal. 257*, 5 (2009), 1333–1354.
154. FU, X. Longtime behavior of the hyperbolic equations with an arbitrary internal damping. *Z. Angew. Math. Phys. 62*, 4 (2011), 667–680.
155. FU, X. Stabilization of hyperbolic equations with mixed boundary conditions. *Math. Control Relat. Fields 5*, 4 (2015), 761–780.
156. FURSIKOV, A., AND IMANUVILOV, O. Y. *Controllability of evolution equations*, vol. 34. Seoul National University, Korea, 1996. Lecture notes.
157. FURSIKOV, A. V., AND IMANUVILOV, O. Y. Exact controllability of the Navier-Stokes and Boussinesq equations. *Uspekhi Mat. Nauk 54*, 3(327) (1999), 93–146.
158. GAROFALO, N., AND LIN, F.-H. Monotonicity properties of variational integrals, A_p weights and unique continuation. *Indiana Univ. Math. J. 35* (1986), 245–268.
159. GÉRARD, P. Microlocal defect measures. *Comm. Partial Differential Equations 16*, 11 (1991), 1761–1794.
160. GÉRARD, P., AND LEICHTNAM, E. Ergodic properties of eigenfunctions for the Dirichlet problem. *Duke Math. J. 71*, 2 (1993), 559–607.

161. Gilbarg, D., and Trudinger, N. S. *Elliptic Partial Differential Equations of Second Order.* Springer-Verlag, New York, 1977.
162. Grigis, A., and Sjöstrand, J. *Microlocal Analysis for Differential Operators.* Cambridge University Press, Cambridge, 1994.
163. Grothendieck, A. Produits tensoriels topologiques et espaces nucléaires. *Mem. Amer. Math. Soc. 1955*, 16 (1955), 140.
164. Guglielmi, R. Indirect stabilization of hyperbolic systems through resolvent estimates. *Evol. Equ. Control Theory 6* (2017), 59–75.
165. Haraux, A. Une remarque sur la stabilisation de certains systèmes du deuxième ordre en temps. *Portugal. Math. 46*, 3 (1989), 245–258.
166. Hartmann, A., Kellay, K., and Tucsnak, M. From the reachable space of the heat equation to hilbert spaces of holomorphic functions. *preprint* (2018).
167. Hebey, E. *Nonlinear analysis on manifolds: Sobolev spaces and inequalities*, vol. 5 of *Courant Lecture Notes in Mathematics.* New York University, Courant Institute of Mathematical Sciences, New York; American Mathematical Society, Providence, RI, 1999.
168. Hérau, F. Une inégalité de Gårding à bord. In *Journées "Équations aux Dérivées Partielles" (La Chapelle sur Erdre, 2000).* Univ. Nantes, Nantes, 2000, pp. Exp. No. V, 12.
169. Hille, E., and Phillips, R. S. *Functional analysis and semi-groups.* American Mathematical Society, Providence, R. I., 1974. Third printing of the revised edition of 1957, American Mathematical Society Colloquium Publications, Vol. XXXI.
170. Holmgren, E. Über Systeme von linearen partiellen Differentialgleichungen. *Öfversigt af Kongl. Vetenskaps-Academien Förhandlinger 58* (1901), 91–103.
171. Hörmander, L. On the uniqueness of the Cauchy problem. *Math. Scand. 6* (1958), 213–225.
172. Hörmander, L. *Linear Partial Differential Operators.* Springer-Verlag, Berlin, 1963.
173. Hörmander, L. The Weyl calculus of pseudo-differential operators. *Comm. Pure Appl. Math. 32* (1979), 359–443.
174. Hörmander, L. *The Analysis of Linear Partial Differential Operators*, vol. IV. Springer-Verlag, 1985.
175. Hörmander, L. *The Analysis of Linear Partial Differential Operators*, vol. III. Springer-Verlag, 1985. Second printing 1994.
176. Hörmander, L. *The Analysis of Linear Partial Differential Operators*, second ed., vol. I. Springer-Verlag, 1990.
177. Hörmander, L. A uniqueness theorem for second order hyperbolic differential equations. *Comm. Partial Differential Equations 17*, 5-6 (1992), 699–714.
178. Imanuvilov, O. Controllability of evolution equations of fluid dynamics. In *International Congress of Mathematicians. Vol. III.* Eur. Math. Soc., Zürich, 2006, pp. 1321–1338.

179. IMANUVILOV, O. Y., AND PUEL, J.-P. Global carleman estimates for weak solutions of elliptic nonhomogeneous Dirichlet problems. *Int. Math. Res. Not. 16* (2003), 883–913.
180. IMANUVILOV, O. Y., PUEL, J.-P., AND YAMAMOTO, M. Carleman estimates for parabolic equations with nonhomogeneous boundary conditions. *Chin. Ann. Math. Ser. B 30*, 4 (2009), 333–378.
181. IMANUVILOV, O. Y., AND YAMAMOTO, M. Carleman estimate for a parabolic equation in a Sobolev space of negative order and its applications. In *Lecture Notes in Pure and Applied Mathematics*, vol. 218. Dekker, New York, 2001, pp. 113–137.
182. IMANUVILOV, O. Y., AND YAMAMOTO, M. Carleman estimates for the Lamé system with stress boundary condition. *Publ. Res. Inst. Math. Sci. 43* (2007), 1023–1093.
183. ISAKOV, V. Carleman type estimates in an anisotropic case and applications. *J. Differential Equations 105* (1993), 217–238.
184. ISAKOV, V. M. On the uniqueness of the solution of the Cauchy problem. *Dokl. Akad. Nauk SSSR 255*, 1 (1980), 18–21.
185. JERISON, D., AND KENIG, C. E. Unique continuation and absence of positive eigenvalues for Schrödinger operators. *Ann. of Math. (2) 121*, 3 (1985), 463–494. With an appendix by E. M. Stein.
186. JERISON, D., AND LEBEAU, G. *Harmonic analysis and partial differential equations (Chicago, IL, 1996)*. Chicago Lectures in Mathematics. The University of Chicago Press, Chicago, 1999, ch. Nodal sets of sums of eigenfunctions, pp. 223–239.
187. JOHN, F. On linear partial differential equations with analytic coefficients. Unique continuation of data. *Comm. Pure Appl. Math. 2* (1949), 209–253.
188. JOHN, F. Continuous dependence on data for solutions of partial differential equations with a presribed bound. *Comm. Pure Appl. Math. 13* (1960), 551–585.
189. JOST, J. *Riemannian geometry and geometric analysis*, sixth ed. Universitext. Springer, Heidelberg, 2011.
190. KAMOUN-FATHALLAH, I. Logarithmic decay of the energy for an hyperbolic-parabolic coupled system. *ESAIM Control Optim. Calc. Var. 17*, 3 (2011), 801–835.
191. KANNAI, Y. Off diagonal short time asymptotics for fundamental solutions of diffusion equations. *Commun. Partial Differ. Equations 2* (1977), 781–830.
192. KATO, T. *Perturbation Theory for Linear Operators*. Springer-Verlag, Berlin, 1980.
193. KAVIAN, O., AND DE TERESA, L. Unique continuation principle for systems of parabolic equations. *ESAIM Control Optim. Calc. Var. 16*, 2 (2010), 247–274.
194. KENIG, C. E., RUIZ, A., AND SOGGE, C. D. Uniform Sobolev inequalities and unique continuation for second order constant coefficient differential operators. *Duke Math. J. 55*, 2 (1987), 329–347.

195. Kenig, C. E., Sjöstrand, J., and Uhlmann, G. The Calderón problem with partial data. *Ann. of Math. (2) 165* (2007), 567–591.
196. Khalgui Ounaies, H. Unicité du problème de Cauchy pour les opérateurs quasi-homogènes. *Ann. Scuola Norm. Sup. Pisa Cl. Sci. (4) 15* (1988), 567–582 (1989).
197. Koch, H., and Tataru, D. Carleman estimates and unique continuation for second-order elliptic equations with nonsmooth coefficients. *Comm. Pure Appl. Math. 54* (2001), 339–360.
198. Koch, H., and Tataru, D. Carleman estimates and unique continuation for second order parabolic equations with nonsmooth coefficients. *Comm. Partial Differential Equations 34*, 4-6 (2009), 305–366.
199. Komornik, V. *Exact controllability and stabilization.* RAM: Research in Applied Mathematics. Masson, Paris; John Wiley & Sons, Ltd., Chichester, 1994. The multiplier method.
200. Kumano-go, H. Pseudo-differential operators and the uniqueness of the cauchy problem. *Comm. Pure Appl. Math. 22* (1969), 73–129.
201. Labbé, S., and Trélat, E. Uniform controllability of semidiscrete approximations of parabolic control systems. *Systems Control Lett. 55* (2006), 597–609.
202. Lascar, R., and Zuily, C. Unicité et non unicité du problème de Cauchy pour une classe d'opérateurs différentiels à caractéristiques doubles. *Duke Math. J. 49*, 1 (1982), 137–162.
203. Lasiecka, I., and Tataru, D. Uniform boundary stabilization of semilinear wave equations with nonlinear boundary damping. *Differential Integral Equations 6* (1993), 507–533.
204. Laurent, C., and Léautaud, C. Observability of the heat equation, geometric constants in control theory, and a conjecture of Luc Miller. *preprint: arXiv:1806.00969* (2018).
205. Laurent, C., and Léautaud, M. Quantitative unique continuation for operators with partially analytic coefficients. application to approximate control for waves. *J. Eur. Math. Soc.* 21 (2019), 957–1069.
206. Lax, P. D. Asymptotic solutions of oscillatory initial value problems. *Duke Math. J. 24* (1957), 627–646.
207. Le Borgne, P. Unicité forte pour le produit de deux opérateurs elliptiques d'ordre 2. *Indiana Univ. Math. J. 50*, 1 (2001), 353–381.
208. Le Rousseau, J. Lebeau, G. On carleman estimates for elliptic and parabolic operators. applications to unique continuation and control of parabolic equations. *ESAIM: Control, Optimisation and Calculus of Variations 18* (2012), 712–747.
209. Le Rousseau, J. Carleman estimates and controllability results for the one-dimensional heat equation with *BV* coefficients. *J. Differential Equations 233* (2007), 417–447.
210. Le Rousseau, J., Léautaud, M., and Robbiano, L. Controllability of a parabolic system with a diffusive interface. *J. Eur. Math. Soc. 15* (2013), 1485–1574.

211. Le Rousseau, J., and Lerner, N. Carleman estimates for anisotropic elliptic operators with jumps at an interface. *Anal. PDE 6* (2013), 1601–1648.
212. Le Rousseau, J., and Moyano, I. Null-controllability of the kolmogorov equation in the whole phase space. *J. Differential Equations 260* (2016), 3193–3233.
213. Le Rousseau, J., and Robbiano, L. Carleman estimate for elliptic operators with coefficents with jumps at an interface in arbitrary dimension and application to the null controllability of linear parabolic equations. *Arch. Rational Mech. Anal. 105* (2010), 953–990.
214. Le Rousseau, J., and Robbiano, L. Local and global Carleman estimates for parabolic operators with coefficients with jumps at interfaces. *Invent. Math.* (2011), 245–336.
215. Le Rousseau, J., and Robbiano, L. Spectral inequality and resolvent estimate for the bi-laplace operator. *J. Eur. Math. Soc. 22* (2020), 1003–1094.
216. Léautaud, M., and Lerner, N. Energy decay for a locally undamped wave equation. *Ann. Fac. Sci. Toulouse Math. (6) 26* (2017), 157–205.
217. Lebeau, G. Équation des ondes amorties. In *Algebraic and geometric methods in mathematical physics (Kaciveli, 1993)*, vol. 19 of *Math. Phys. Stud.* Kluwer Acad. Publ., Dordrecht, 1996, pp. 73–109.
218. Lebeau, G., and Robbiano, L. Contrôle exact de l'équation de la chaleur. *Comm. Partial Differential Equations 20* (1995), 335–356.
219. Lebeau, G., and Robbiano, L. Stabilisation de l'équation des ondes par le bord. *Duke Math. J. 86* (1997), 465–491.
220. Lebeau, G., and Zuazua, E. Null-controllability of a system of linear thermoelasticity. *Arch. Rational Mech. Anal. 141* (1998), 297–329.
221. Lee, J. M. *Riemannian manifolds, an introduction to curvature*, vol. 176 of *Graduate Texts in Mathematics.* Springer-Verlag, New York, 1997.
222. Lee, J. M. *Introduction to smooth manifolds*, second ed., vol. 218 of *Graduate Texts in Mathematics.* Springer, New York, 2013.
223. Lerner, N. Unicité de Cauchy pour des opérateurs différentiels faiblement principalement normaux. *J. Math. Pures Appl. (9) 64*, 1 (1985), 1–11.
224. Lerner, N. Carleman's and subelliptic estimates. *Duke Math. J. 56*, 2 (1988), 385–394.
225. Lerner, N. Sufficiency of condition (ψ) for local solvability in two dimensions. *Ann. of Math. (2) 128* (1988), 243–258.
226. Lerner, N. Uniqueness for an ill-posed problem. *J. Differential Equations 71*, 2 (1988), 255–260.
227. Lerner, N. Nonsolvability in L^2 for a first order operator satisfying condition (ψ). *Ann. of Math. (2) 139* (1994), 363–393.
228. Lerner, N. *Carleman Inequalities: An Introduction and More*, vol. 353

of Grundlehren der mathematischen Wissenschaften. Springer, Cham, 2019.

229. Lerner, N. *Carleman Inequalities: An Introduction and More*, vol. 353 of *Grundlehren der mathematischen Wissenschaften.* Springer, Cham, 2019.
230. Lerner, N., and Robbiano, L. Unicité de Cauchy pour des opérateurs de type principal. *J. Analyse Math. 44* (1984/85), 32–66.
231. Lerner, N., and Saint Raymond, X. Une inégalité de Gårding sur une variété à bord. *J. Math. Pures Appl. (9) 77*, 9 (1998), 949–963.
232. Li, H., and Lü, Q. A quantitative boundary unique continuation for stochastic parabolic equations. *J. Math. Anal. Appl. 402*, 2 (2013), 518–526.
233. Liimatainen, T., and Salo, M. Nowhere conformally homogeneous manifolds and limiting Carleman weights. *Inverse Probl. Imaging 6* (2012), 523–530.
234. Lin, C.-L., Nakamura, G., and Sini, M. Unique continuation for the elastic transversely isotropic dynamical systems and its application. *J. Differential Equations 245* (2008), 3008–3024.
235. Lions, J.-L. *Contrôlabilité Exacte, Perturbations et Stabilisation de Systèmes Distribués*, vol. 1. Masson, Paris, 1988.
236. Lions, J.-L., and Magenes, E. *Problèmes aux Limites Non Homogènes*, vol. 1. Dunod, 1968.
237. Liu, Z., and Rao, B. Characterization of polynomial decay rate for the solution of linear evolution equation. *Z. Angew. Math. Phys. 56* (2005), 630–644.
238. Lopatinskiĭ, Y. B. On a method of reducing boundary problems for a system of differential equations of elliptic type to regular integral equations (Russian). *Ukrain. Mat. Ž. 5* (1953), 123–151.
239. Lü, Q. Some results on the controllability of forward stochastic heat equations with control on the drift. *J. Funct. Anal. 260* (2011), 832–851.
240. Lü, Q., and Yin, Z. The L^∞-null controllability of parabolic equation with equivalued surface boundary conditions. *Asymptot. Anal. 83*, 4 (2013), 355–378.
241. Mandache, N. On a counterexample concerning unique continuation for elliptic equations in divergence form. *Math. Phys. Anal. Geom. 1* (1998), 273–292.
242. Maniar, L., Meyries, M., and Schnaubelt, R. Null controllability for parabolic equations with dynamic boundary conditions. *Evol. Equ. Control Theory 6* (2017), 381–407.
243. Martin, P., Rosier, L., and Rouchon, P. On the reachable states for the boundary control of the heat equation. *Appl. Math. Res. Express. AMRX* (2016), 181–216.
244. Martinez, A. *An Introduction to Semiclassical and Microlocal Analysis.* Springer-Verlag, 2002.

245. Massey, W. S. *A basic course in algebraic topology*, vol. 127 of *Graduate Texts in Mathematics*. Springer-Verlag, New York, 1991.
246. Melrose, R. B., and Sjöstrand, J. Singularities of boundary value problems. I. *Comm. Pure Appl. Math. 31* (1978), 593–617.
247. Melrose, R. B., and Sjöstrand, J. Singularities of boundary value problems. II. *Comm. Pure Appl. Math. 35* (1982), 129–168.
248. Meyer, Y. Régularité des solutions des équations aux dérivées partielles non linéaires (d'après J.-M. Bony). In *Bourbaki Seminar, Vol. 1979/80*, vol. 842 of *Lecture Notes in Math.* Springer, Berlin-New York, 1981, pp. 293–302.
249. Meyer, Y. Remarques sur un théorème de J.-M. Bony. *Rend. Circ. Mat. Palermo (2)*, suppl. 1 (1981), 1–20.
250. Meyer, Y., and Coifman, R. R. *Ondelettes et opérateurs. III.* Actualités Mathématiques. [Current Mathematical Topics]. Hermann, Paris, 1991. Opérateurs multilinéaires. [Multilinear operators].
251. Miller, K. Nonunique continuation for uniformly parabolic and elliptic equations in self-adjoint divergence form with Hölder continuous coefficients. *Arch. Rational Mech. Anal. 54* (1974), 105–117.
252. Miller, L. Refraction of high-frequency waves density by sharp interfaces and semiclassical measures at the boundary. *J. Math. Pures Appl. (9) 79* (2000), 227–269.
253. Miller, L. Geometric bounds on the growth rate of null-controllability cost for the heat equation in small time. *J. Differential Equations 204*, 1 (2004), 202–226.
254. Miller, L. Controllability cost of conservative systems: resolvent condition and transmutation. *J. Funct. Anal. 218*, 2 (2005), 425–444.
255. Miller, L. On the null-controllability of the heat equation in unbounded domains. *Bull. Sci. Math. 129* (2005), 175–185.
256. Miller, L. The control transmutation method and the cost of fast controls. *SIAM J. Control Optim. 45*, 2 (2006), 762–772.
257. Miller, L. On exponential observability estimates for the heat semigroup with explicit rates. *Atti Accad. Naz. Lincei Rend. Lincei Mat. Appl. 17* (2006), 351–366.
258. Miller, L. On the controllability of anomalous diffusions generated by the fractional laplacian. *Mathematics of Control, Signals, and Systems 3* (2006), 260–271.
259. Miller, L. A direct Lebeau-Robbiano strategy for the observability of heat-like semigroups. *Discrete Contin. Dyn. Syst. Ser. B 14* (2009), 1465–1485.
260. Milnor, J. *Morse theory.* Annals of Mathematics Studies, No. 51. Princeton University Press, 1963.
261. Mizohata, S. Unicité du prolongement des solutions pour quelques opérateurs différentiels paraboliques. *Mem. Coll. Sci. Univ. Kyoto. Ser. A. Math. 31* (1958), 219–239.

262. Mizohata, S. Some remarks on the Cauchy problem. *J. Math. Kyoto Univ. 1* (1961/1962), 109–127.
263. Newns, W. F., and Walker, A. G. Tangent planes to a differentiable manifold. *J. London Math. Soc. 31* (1956), 400–407.
264. Nicaise, S., and Pignotti, C. Stability and instability results of the wave equation with a delay term in the boundary or internal feedbacks. *SIAM J. Control Optim. 45* (2006), 1561–1585.
265. Nishiyama, H. Polynomial decay for damped wave equations on partially rectangular domains. *Math. Res. Lett. 16* (2009), 881–894.
266. Ouhabaz, E. M. *Analysis of heat equations on domains*, vol. 31 of *London Mathematical Society Monographs Series*. Princeton University Press, Princeton, NJ, 2005.
267. Ouksel, L. Inégalité d'observabilité du type logarithmique et estimation de la fonction de coût des solutions des équations hyperboliques. *ESAIM Control Optim. Calc. Var. 14*, 2 (2008), 318–342.
268. Ouksel, L. Logarithmic stabilisation of a multidimensional structure by the boundary. *Asymptot. Anal. 80*, 3-4 (2012), 347–376.
269. Payne, L. E. *Improperly Posed Problems in Partial Differential Equations*. SIAM, Philadelphia, 1975.
270. Pazy, A. *Semigroups of Linear Operators and Applications to Partial Differential Equations*. Springer-Verlag, New York, 1983.
271. Phung, K.-D. Observability and control of Schrödinger equations. *SIAM J. Control Optim. 40* (2001), 211–230.
272. Phung, K.-D. Remarques sur l'observabilité pour l'équation de Laplace. *ESAIM Control Optim. Calc. Var. 9* (2003), 621–635.
273. Phung, K.-D. Polynomial decay rate for the dissipative wave equation. *J. Differential Equations 240* (2007), 92–124.
274. Phung, K. D., Wang, L., and Zhang, C. Bang-bang property for time optimal control of semilinear heat equation. *Ann. Inst. H. Poincaré Anal. Non Linéaire 31*, 3 (2014), 477–499.
275. Pliś, A. The problem of uniqueness for the solution of a system of partial differential equations. *Bull. Acad. Polon. Sci. Cl. III. 2* (1954), 55–57.
276. Pliś, A. Non-uniqueness in Cauchy's problem for differential equations of elliptic type. *J. Math. Mech. 9* (1960), 557–562.
277. Pliś, A. A smooth linear elliptic differential equation without any solution in a sphere. *Comm. Pure Appl. Math. 14* (1961), 599–617.
278. Pliś, A. On non-uniqueness in Cauchy problem for an elliptic second order differential equation. *Bull. Acad. Polon. Sci. Sér. Sci. Math. Astronom. Phys. 11* (1963), 95–100.
279. Rauch, J., and Taylor, M. Penetrations into shadow regions and unique continuation properties in hyperbolic mixed problems. *Indiana Univ. Math. J. 22* (1972), 277–285.
280. Rauch, J., and Taylor, M. Exponential decay of solutions to hyperbolic equations in bounded domains. *Indiana Univ. Math. J. 24* (1974), 79–86.

281. Rauch, J., Zhang, X., and Zuazua, E. Polynomial decay for a hyperbolic-parabolic coupled system. *J. Math. Pures Appl. (9) 84* (2005), 407–470.
282. Reed, R., and Simon, B. *Methods of Modern Mathematical Physics*, vol. 1. Academic Press, San Diego, 1980.
283. Regbaoui, R. Strong unique continuation results for differential inequalities. *J. Funct. Anal. 148*, 2 (1997), 508–523.
284. Regbaoui, R. Strong uniqueness for second order differential operators. *J. Differential Equations 141*, 2 (1997), 201–217.
285. Regbaoui, R. Unique continuation for differential equations of Schrödinger's type. *Comm. Anal. Geom. 7*, 2 (1999), 303–323.
286. Robbiano, L. Non unicité du problème de Cauchy pour des opérateurs non elliptiques à symboles complexes. *J. Differential Equations 57*, 2 (1985), 200–223.
287. Robbiano, L. Sur les conditions de pseudo-convexité et l'unicité du problème de Cauchy. *Indiana Univ. Math. J. 36*, 2 (1987), 333–347.
288. Robbiano, L. Théorème d'unicité adapté au contrôle des solutions des problèmes hyperboliques. *Comm. Partial Differential Equations 16* (1991), 789–800.
289. Robbiano, L. Fonction de coût et contrôle des solutions des équations hyperboliques. *Asymptotic Anal. 10* (1995), 95–115.
290. Robbiano, L., and Zuily, C. Uniqueness in the Cauchy problem for operators with partially holomorphic coefficients. *Invent. Math. 131*, 3 (1998), 493–539.
291. Robert, D. *Autour de l'Approximation Semi-Classique*, vol. 68 of *Progress in Mathematics*. Birkhuser Boston, Boston, MA, 1987.
292. Roe, J. *Elliptic operators, topology and asymptotic methods*, second ed., vol. 395 of *Pitman Research Notes in Mathematics Series*. Longman, Harlow, 1998.
293. Rozendaal, J., Seifert, D., and R., S. Optimal rates of decay for operator semigroups on hilbert spaces. *Adv. Math. 346* (2019), 359–388.
294. Rudin, W. *Real and complex analysis*, third ed. McGraw-Hill, 1987.
295. Russell, D., and Weiss, G. A general necessary condition for exact observability. *SIAM J. Control Optim. 32* (1994), 1–23.
296. Russell, D. L. A unified boundary controllability theory for hyperbolic and parabolic partial differential equations. *Studies in Appl. Math. 52* (1973), 189–221.
297. Saint Raymond, X. L'unicité pour des problèmes de Cauchy caractéristiques. *Comm. Partial Differential Equations 7*, 5 (1982), 559–579.
298. Saint Raymond, X. Non-unicité de Cauchy pour des opérateurs principalement normaux. *Indiana Univ. Math. J. 33*, 6 (1984), 847–858.

299. Saint Raymond, X. Unicité de Cauchy à partir de surfaces faiblement pseudo-convexes. *Ann. Inst. Fourier (Grenoble) 39*, 1 (1989), 123–147.
300. Saut, J.-C., and Scheurer, B. Unique continuation for some evolution equations. *J. Differential Equations 66* (1987), 118–139.
301. Schulz, F. On the unique continuation property of elliptic divergence form equations in the plane. *Math. Z. 228* (1998), 201–206.
302. Schwartz, L. Sur l'impossibilité de la multiplication des distributions. *C. R. Acad. Sci. Paris 239* (1954), 847–848.
303. Schwartz, L. *Théorie des distributions.* Hermann, Paris, 1966.
304. Shubin, M. A. *Pseudodifferential Operators and Spectral Theory*, second ed. Springer-Verlag, Berlin Heidelberg, 2001.
305. Sjöstrand, J. Asymptotic distribution of eigenfrequencies for damped wave equations. *Publ. Res. Inst. Math. Sci. 36*, 5 (2000), 573–611.
306. Sogge, C. D. Oscillatory integrals and unique continuation for second order elliptic differential equations. *J. Amer. Math. Soc. 2*, 3 (1989), 491–515.
307. Sogge, C. D. Strong uniqueness theorems for second order elliptic differential equations. *Amer. J. Math. 112*, 6 (1990), 943–984.
308. Sogge, C. D. A unique continuation theorem for second order parabolic differential operators. *Ark. Mat. 28*, 1 (1990), 159–182.
309. Spivak, M. *A comprehensive introduction to differential geometry. Vol. I–V*, second ed. Publish or Perish, Inc., Wilmington, Del., 1979.
310. Steenrod, N. *The Topology of Fibre Bundles.* Princeton Mathematical Series, vol. 14. Princeton University Press, Princeton, N. J., 1951.
311. Tang, S., and Zhang, X. Null controllability for forward and backward stochastic parabolic equations. *SIAM J. Control Optim. 48* (2009), 2191–2216.
312. Tartar, L. H-measures, a new approach for studying homogenisation, oscillations and concentration effects in partial differential equations. *Proc. Roy. Soc. Edinburgh Sect. A 115*, 3-4 (1990), 193–230.
313. Tataru, D. Unique continuation for solutions to PDE's; between Hörmander's theorem and Holmgren's theorem. *Comm. Partial Differential Equations 20* (1995), 855–884.
314. Tataru, D. Carleman estimates and unique continuation for solutions to boundary value problems. *J. Math. Pures Appl. 75* (1996), 367–408.
315. Tataru, D. Unique continuation for operators with partially analytic coefficients. *J. Math. Pures Appl. (9) 78*, 5 (1999), 505–521.
316. Taylor, L. E. The tangent space to a C^k manifold. *Bull. Amer. Math. Soc. 79* (1973), 746.
317. Taylor, M. E. *Pseudodifferential Operators.* Princeton University Press, Princeton, New Jersey, 1981.
318. Taylor, M. E. *Tools for PDE*, vol. 81 of *Mathematical Surveys and Monographs.* American Mathematical Society, Providence, RI, 2000. Pseudodifferential operators, paradifferential operators, and layer potentials.

319. TENENBAUM, G., AND TUCSNAK, M. On the null-controllability of diffusion equations. *ESAIM Control Optim. Calc. Var. 17* (2011), 1088–1100.
320. TREVES, F. *Topological Vector Spaces, Distributions and Kernels.* Academic Press, New York, 1967.
321. TUCSNAK, M., AND WEISS, G. *Observation and Control for Operator Semigroups.* Birkhäuser Verlag, Basel, 2009.
322. VANCOSTENOBLE, J., AND ZUAZUA, E. Null controllability for the heat equation with singular inverse-square potentials. *J. Funct. Anal. 254*, 7 (2008), 1864–1902.
323. ŠAPIRO, Z. Y. On general boundary problems for equations of elliptic type (Russian). *Izvestiya Akad. Nauk SSSR. Ser. Mat. 17* (1953), 539–562.
324. WANG, W. Carleman inequalities and unique continuation for higher-order elliptic differential operators. *Duke Math. J. 74*, 1 (1994), 107–128.
325. WOLFF, T. H. Unique continuation for $|\Delta u| \leq V|\nabla u|$ and related problems. *Rev. Mat. Iberoamericana 6*, 3-4 (1990), 155–200.
326. WOLFF, T. H. Note on counterexamples in strong unique continuation problems. *Proc. Amer. Math. Soc. 114* (1992), 351–356.
327. WOLFF, T. H. A property of measures in $\mathbf{R}^N$ and an application to unique continuation. *Geom. Funct. Anal. 2*, 2 (1992), 225–284.
328. ZHANG, X. A unified controllability/observability theory for some stochastic and deterministic partial differential equations. In *Proceedings of the international congress of mathematicians (ICM 2010), Hyderabad, India, August 19–27, 2010. Vol. IV: Invited lectures.* Hackensack, NJ: World Scientific; New Delhi: Hindustan Book Agency, 2011, pp. 3008–3034.
329. ZHANG, X., AND ZUAZUA, E. Long-time behavior of a coupled heat-wave system arising in fluid-structure interaction. *Arch. Ration. Mech. Anal. 184* (2007), 49–120.
330. ZUILY, C. *Uniqueness and Non Uniqueness in the Cauchy Problem.* Birkhauser, Progress in mathematics, 1983.
331. ZUILY, C. *Problems in distributions and partial differential equations*, vol. 143 of *North-Holland Mathematics Studies.* North-Holland Publishing Co., Amsterdam; Hermann, Paris, 1988. Translated from the French.
332. ZWORSKI, M. *Semiclassical analysis*, vol. 138 of *Graduate Studies in Mathematics.* American Mathematical Society, Providence, RI, 2012.

Index

N.B. A page number in normal type refers to Volume 1, in italic type to Volume 2.

J. Le Rousseau et al., *Elliptic Carleman Estimates and Applications to Stabilization and Controllability, Volume II*, PNLDE Subseries in Control 98, https://doi.org/10.1007/978-3-030-88670-7

Index of notation

N.B. A page number in normal type refers to Volume 1, in italic type to Volume 2.

J. Le Rousseau et al., *Elliptic Carleman Estimates and Applications to Stabilization and Controllability, Volume II*, PNLDE Subseries in Control 98, https://doi.org/10.1007/978-3-030-88670-7

GPSR Compliance
The European Union's (EU) General Product Safety Regulation (GPSR) is a set of rules that requires consumer products to be safe and our obligations to ensure this.

If you have any concerns about our products, you can contact us on

ProductSafety@springernature.com

In case Publisher is established outside the EU, the EU authorized representative is:

Springer Nature Customer Service Center GmbH
Europaplatz 3
69115 Heidelberg, Germany

www.ingramcontent.com/pod-product-compliance
Ingram Content Group UK Ltd.
Pitfield, Milton Keynes, MK11 3LW, UK
UKHW050138280726
14058UKWH00006B/707
9783030886721